PLEASE STAMP DATE DUE BOTH BELOW A

The Physics of Star Formation
and Early Stellar Evolution

NATO ASI Series

Advanced Science Institutes Series

A Series presenting the results of activities sponsored by the NATO Science Committee, which aims at the dissemination of advanced scientific and technological knowledge, with a view to strengthening links between scientific communities.

The Series is published by an international board of publishers in conjunction with the NATO Scientific Affairs Division

A	Life Sciences	Plenum Publishing Corporation
B	Physics	London and New York
C	Mathematical and Physical Sciences	Kluwer Academic Publishers Dordrecht, Boston and London
D	Behavioural and Social Sciences	
E	Applied Sciences	
F	Computer and Systems Sciences	Springer-Verlag
G	Ecological Sciences	Berlin, Heidelberg, New York, London,
H	Cell Biology	Paris and Tokyo
I	Global Environmental Change	

NATO-PCO-DATA BASE

The electronic index to the NATO ASI Series provides full bibliographical references (with keywords and/or abstracts) to more than 30000 contributions from international scientists published in all sections of the NATO ASI Series.
Access to the NATO-PCO-DATA BASE is possible in two ways:

– via online FILE 128 (NATO-PCO-DATA BASE) hosted by ESRIN,
Via Galileo Galilei, I-00044 Frascati, Italy.

– via CD-ROM "NATO-PCO-DATA BASE" with user-friendly retrieval software in English, French and German (© WTV GmbH and DATAWARE Technologies Inc. 1989).

The CD-ROM can be ordered through any member of the Board of Publishers or through NATO-PCO, Overijse, Belgium.

Series C: Mathematical and Physical Sciences - Vol. 342

The Physics of Star Formation and Early Stellar Evolution

edited by

Charles J. Lada
Smithsonian Astrophysical Observatory,
Cambridge, MA, U.S.A.

and

Nikolaos D. Kylafis
Department of Physics,
University of Crete,
Heraklion, Crete, Greece

Kluwer Academic Publishers

Dordrecht / Boston / London

Published in cooperation with NATO Scientific Affairs Division

Proceedings of the NATO Advanced Study Institute on
The Physics of Star Formation and Early Stellar Evolution
Agia Pelagia, Crete, Greece
May 27 – June 8, 1990

ISBN 0–7923–1349–6

Published by Kluwer Academic Publishers,
P.O. Box 17, 3300 AA Dordrecht, The Netherlands.

Kluwer Academic Publishers incorporates the publishing programmes of
D. Reidel, Martinus Nijhoff, Dr W. Junk and MTP Press.

Sold and distributed in the U.S.A. and Canada
by Kluwer Academic Publishers,
101 Philip Drive, Norwell, MA 02061, U.S.A.

In all other countries, sold and distributed
by Kluwer Academic Publishers Group,
P.O. Box 322, 3300 AH Dordrecht, The Netherlands.

Printed on acid-free paper

All Rights Reserved
© 1991 Kluwer Academic Publishers
No part of the material protected by this copyright notice may be reproduced or utilized in any form or by any means, electronic or mechanical, including photocopying, recording or by any information storage and retrieval system, without written permission from the copyright owner.

Printed in the Netherlands

Table of Contents

Preface xix
Authors xxiii
Participants xxvi

I. PHYSICS OF GIANT MOLECULAR CLOUDS: ORIGIN, STRUCTURE, AND EVOLUTION

Star Forming Giant Molecular Clouds
Leo Blitz 3

1. Introduction 3
2. Global Properties of Giant Molecular Clouds 4
3. Do All GMCS Form Stars? 10
4. Angular Momentum 12
5. Clouds in Different Evolutionary States 15
6. Relationship to Atomic Hydrogen 19
7. The Clumpy Structure of GMCs 21
8. Boundedness of the Clumps 27
9. Star Formation in the Clumps 29
10. The Evolution of Giant Molecular Clouds 29

The Origin and Evolution of Giant Molecular Clouds
Bruce G. Elmegreen 35

1. Introduction 35
2. Overview of Cloud Formation 36
3. Diffuse and Self-Gravitating Clouds 39
4. Basic Equations for Cloud Formation 40
   - 4.1. Thermal Instability 40
   - 4.2. Gravitational Instability 43
   - 4.3. The Random Collisional Build-Up Model of Cloud Formation 47
   - 4.4. Cloud Formation by Direct Compression of Low Density Gas 48
   - 4.5. Cloud Formation by Turbulent and Magnetic Wave Interactions 49
5. Cloud Formation Models and Cloud Lifetimes 52
6. What Do We Do Next? 55

Cosmic Magnetism and the Basic Physics of the Early Stages of Star Formation
Telemachos Ch. Mouschovias 61

1. Introduction - Overview 61
   1.1. The Approach, and a Viewpoint 62
   1.2. Key Developments: Elements of a Modern Theory of Star Formation 62
   1.3. Outline and Perspective 64
2. Elements of Multi-Fluid Magnetohydrodynamics 66
   2.1. Motivation, and the Key Questions 66
   2.2. Flux Freezing in the Plasma, and Reduction of the Three-Fluid Equations 67
   2.3. Physical Considerations and the Basic Two-Fluid Equations: Ambipolar Diffusion 69
      2.3.1. Quasistatic Contraction: General Treatment 73
      2.3.2. Dynamical Contraction: Its Effect on Ambipolar Diffusion 74
      2.3.3. Self-Initiated Collapse due to Ambipolar Diffusion 76
3. Single-Fluid Magnetohydrodynamics 78
   3.1. Cosmic Rays as a Fluid 79
   3.2. Formation of Giant Clouds and Cloud Complexes 81
      3.2.1. The Jeans Instability 81
      3.2.2. Thermal Instability 82
      3.2.3. Magnetic Rayleigh-Taylor (or Parker) Instability 82
         3.2.3 (a). The Zero-Order State 83
         3.2.3 (b). Stability Analysis 83
   3.3. Redistribution of Angular Momentum by Magnetic Braking 87
      3.3.1. Observational Constraints 87
      3.3.2. Single, Aligned Rotator 89
         3.3.2 (a). Rigid Rotator 89
         3.3.2 (b) Differential Rotator 91
         3.3.2 (c). Results and Applications. 92
         3.3.2. (d) The Effect of Field Lines "Fanning Out" Away from Aligned Rotator 96
      3.3.3. Single, Perpendicular Rotator 96
         3.3.3 (a). Formulation of the Problem 96
         3.3.3 (b). The Dimensionless Equations 98

		3.3.3 (c)	Results and Applications 98

                3.3.3 (d)   Application to a Fragment -
                            Retrograde Spin 100
        3.3.4.   Alternative Forms of the Magnetic Braking
                Time Scales 101
        3.3.5.   Magnetically Linked, Aligned Rotators......................... 102
        3.3.6.   Conclusion 103
4.   The Effect of Ambipolar Diffusion on Magnetic Braking 104
5.   Magnetohydrostatics: Equilibria of Interstellar Clouds................... 106
     5.1.   Large-Scale Condensations or Cloud Complexes..................... 107
        5.1.1.   Motivation 107
        5.1.2.   Self-Consistent Formulation of the
                Problem.. 107
        5.1.3.   Determination of Equilibrium States........................... 109
     5.2.   Self Gravitating Clouds 111
        5.2.1.   Self-Consistent Formulation of the
                Problem.. 111
        5.2.2.   Equilibrium States and Gravitational
                Collapse 113
        5.2.3.   Physical Origin and Validity of the
                Relation $B_c \propto \rho_c^{1/2}$.. 114
6.   The Scalar Virial Theorem and Cloud Collapse 116
7.   Conclusion-- Summary 118

II. THE PHYSICS OF STAR FORMATION

OB Associations and the Fossil Record of Star Formation
Adriaan Blaauw. 125

1.   Introduction, Definition, Nomenclature. 125
2.   The Nearest Associations OSCA and CAS-TAU; ORI-OB1 126
     2.1.   The Ophiuchus-Scorpius-Centaurus Association (OSCA) 126
     2.2.   Ages in OSCA and Progression of Star Formation 128
     2.3.   OSCA and the Model of Sequential Star Formation 128
     2.4.   Stellar Content of OSCA 128
     2.5.   Relation of OSCA to the Interstellar Medium 129
     2.6.   Kinematic Properties of OSCA 132
     2.7.   Double Star Properties and Run-Away Star In OSCA 132
     2.8.   The Cassiopeia-Taurus Association (CAS-TAU) 132
     2.9.   The Association ORI-OB1 135
3.   The Sample Within 1500 PC 138

	3.1.	Global Properties	139
	3.2.	Formation of Massive Stars in Associations as Compared to Clusters	139
	3.3.	Ages and Subgroups; Stellar Content	140
	3.4.	Evaporation Function and Evolved Stars	142
	3.5.	Associated Molecular Clouds	144
	3.6.	Internal Motions; Initial Shapes and Sizes of Subgroups	144
	3.7.	Statistics of Close Binaries	145
	3.8.	Run-Away OB Stars (RAOBS)	148
	3.9.	RAOBS and Stochastic Star Formation	150
4.	The Gould Belt System	150
5.	Promises for Future Research	152

Physical Conditions and Heating/Cooling Processes in High Mass Star Formation Regions
Reinhard Genzel 155

1.	Introduction	155
2.	An Example of a Star Forming Cloud: Orion	156
	2.1.	Global Structure of the Cloud	156
	2.2.	Infrared and Submillimeter Emission in Star Forming Cores	156
	2.3.	Cloud Structure and Clumpiness	158
	2.4.	The Relation Between Dense Gas and Star Formation	161
	2.5.	The Magnetic Field	163
	2.6.	A Close-Up View of the Environment of Massive Young Stars: The BN-KL Area	164
3.	Tools for Studying Dense Interstellar Clouds	164
	3.1.	Excitation and Interpretation of Infrared and Radio Spectral Lines	164
		a. Simple Radiative Transport	165
		b. Optical Depths of Infrared and Radio Lines	167
		c. Rotational and Ro-Vibrational Emission of Simple Linear Molecules	168
		d. Fine Structure Lines of Atoms and Ions	170
		e. Collisional Excitation	171
		i. Two Level System	171
		ii. Multilevel System	173
		f. Spectral Lines as Probes of Physical Conditions in Interstellar Clouds	173
		g. Radiative Excitation	175

		h.	Line Trapping 177

             h.   Line Trapping 177
             i.   Line Profiles 178
      3.2.   Infrared and Submillimeter Continuum Emission 178
      3.3.   Estimates of Cloud Column Densities and Masses 180
             a.   Submillimeter Dust Emission 180
             b.   C O Line Emission 181
                   i.   Tracing Mass With CO Isotopes 181
                   ii.   Tracing Mass with ^{12}CO $1 \to 0$ 181
             c.   Tracing Mass with MeVγ-rays 183
      3.4   Measuring Temperature and Density in Molecular
             Clouds 184
4.   Heating and Cooling Processes in Star Forming Clouds 184
      4.1.   Photodissociation Regions 188
             a.   Theory 188
             b.   Observations of the Orion Photodis-
                   sociation Region 194
      4.2.   Shocks in Dense Interstellar Clouds 194
             a.   Theory 199
                   i.   J-Shocks 199
                   ii.   C-Shocks 202
             b.   Observations of Shocks 202
                   i.   Chemical Probes 204
                   ii.   J-Shocks vs. C-Shocks vs. PDR's 204
      4.3.   Heating by Energetic Particles 206
      4.4.   Heating by Ambipolar Diffusion 208
5.   Two Puzzles 211
      5.1.   How to Find Massive Protostars 211
      5.2.   Warm Quiescent Gas at the Surfaces of Molecular
             Clouds 213

Newly Formed Massive Stars
Ed Churchwell 221

1.   Introduction 221
2.   Evolutionary Perspective 222
3.   Observed Properties 224
      3.1.   The Continuous Spectrum 224
      3.2.   Spectral Types 229
      3.3.   Morphologies 231
      3.4.   Clustering and Location 232
      3.5.   Galactic Population and Distribution 235
4.   Physical Properties 241

	4.1. The Ionized Gas	241
	4.2. The Warm Dust Cocoon	242
	4.3. The Molecular Environment	248
5.	Bow Shocks	251
	5.1. Motivations for Bow Shocks	251
	5.2. Stellar Wind Supported Bow Shocks	254
	5.3. Comparison with Observations	258
	5.4. Comparison with Champagne Flows	258
	5.5. Warm Molecular Cloud Cores and the Lifetime Problem	261
	5.6. Bipolar Molecular Outflows	262
	5.7. Summary	262
6.	Future Directions	263
	6.1. Global Issues	263
	6.2. Local Properties	264

Masers and Star Formation
Nikolaos D. Kylafis 269

1.	Introduction	269
2.	Basic Concepts	270
	2.1. Amplification	271
	2.2. Saturation	271
	2.3. Thermalization	272
	2.4. Beaming	272
	2.5. Geometry and Apparent Size	273
	2.5.1. Elongated Structures	273
	2.5.2. Nearly Spherical Structures	273
	2.6. Variability	275
	2.7. Spectra and Line Widths	275
	2.8. Polarization	276
3.	Laboratory verses Astronomical Masers	276
4.	Usefulness of Masers	277
5.	Maser Models	278
	5.1. Requirements	278
	5.2. Basic Equations	278
	5.3. Collisional and Radiative Pumping	279
6.	Masers in Star-Forming Regions	280
	6.1. H_2O Masers	280
	6.2. OH Masers	282
7.	Conclusions	283

The Physical Conditions of Low Mass Star Forming Regions
J. Cernicharo 287

1. Introduction 287
2. The Determination of the Physical Conditions of Low Mass Star Forming Regions 291
    2.1. Temperature 292
    2.2. Density 295
        2.2.1. Density Determinations 295
        2.2.2. The Density Profile of a Self-Gravitating Cloud 300
    2.3. Molecular Abundances and Exotic Molecules in Dense Cores 302
    2.4. Mass 304
3. The Mass Distribution and th Physical Structure of Low Mass Star Forming Clouds 305
    3.1. Mass Distribution and Fragmentation 305
    3.2. Dense Cores: Low Mass Protostars? 310
4. Low Mass Stars and Low Mass Clouds: Bok Globules 314
5. The Interaction of Newly Born Stars with the Ambient Gas 319
6. Conclusions 323

The Formation of Low Mass Stars: Observations
Charles J. Lada..329

1. Introduction..329
2. The Initial Mass Spectrum 330
3. Clusters vs. Associations: Two Modes of Low Mass Star Formation 337
    3.1. The Nature of Nearby Embedded Clusters 338
4. The Nature of Low Mass Young Stellar Objects 343
    4.1. Infrared Energy Distributions and Spectral Classification 343
    4.2. The Nature of Class III Sources 346
    4.3. The Nature of Class II Sources 347
    4.4. The Nature of Class I Sources 351
        4.4.1. Interpreting Spectral Energy Distributions 351
        4.4.2. Protostellar Models 353
        4.4.3. Extreme Class I Sources..354
5. Molecular Outflows and the Spectral Evolution of YSO's 357

The Formation of Low Mass Stars: Theory
Frank H. Shu 365

1. Overview 365
2. Bimodal Star Formation 368
3. The Bipolar Outflow Phase: Observations 369
4. Rotating, Magnetized, Molecular Cloud Cores 374
5. Protostar Formation from Collapsing Cloud Cores 376
6. Infrared Appearance of Rotating Protostellar Objects 380
7. Protostar Formation By Disk Accretion 383
8. Stellar Winds and Bipolar Flows: Theory 385
9. Revealed T Tauri Stars 392
10. The Disks Inferred for T Tauri Stars 396
11. Binary Stars and Planetary Systems 402

Numerical Studies of Cloud Collapse
W.M. Tscharnuter 411

1. Introduction 411
2. Classical Collapse Models 413
    - 2.1. General Features of Collapse Flows 414
    - 2.2. Timescales 415
3. Basic Physics 417
    - 3.1. Equation of State 417
    - 3.2. Opacity 419
    - 3.3. Turbulent Viscosity 419
4. Structure Equations 421
5. Numerical Tools 423
    - 5.1. Finite Differences 423
    - 5.2. Artificial Viscosity 426
    - 5.3. Adaptive Grid 427
6. Recent Results 428
    - 6.1. Spherically Symmetric (1-D) Models 428
    - 6.2. Axially Symmetric (2-D) Models 431
    - 6.3. 3-D Models 432
7. Conclusions 434

Binary Star Formation
J.E. Pringle 437

1. Introduction 437

2.	Fission	438
3.	Capture	441
4.	Independent Condensations/Separate Nuclei	442
5.	Fragmentation/Continued Fragmentation	444
6.	Concluding Remarks	446

Single-Stage Fragmentation and a Modern Theory of Star Formation
Telemachos Ch. Mouschovias 449

1. Introduction 449
2. Basic Building Blocks of a Theory of Star Formation 450
    2.1. Consequences of Magnetic Support of Clouds 450
    2.2. Critical Length Scales and Protostellar Masses 452
    2.3. The $B_c - \rho_c$ Relation Revisited 455
    2.4. Fragmentation and Its effect on the $B_c - r_c$ Relation: A Simple Theory 457
    2.5. Numerical Modeling of Axisymmetric Collapse due to Ambipolar Diffusion 460
3. A Modern Theory of Star Formation 462
    3.1. Cloud Formation and Early Evolution: Magnetic Breaking and the Onset of Collapse 462
    3.2. Consequences of Ambipolar Diffusion: Fragmentation and Selection of Protostellar Masses - 1 $M_\odot$ 463
        3.2.1. Disk Formation and Bipolar Outflows   464
        3.2.2. Formation of Binary Stars and of Intermediate- and Low Mass Stars in Cloud Cores   464
    3.3. Low-*versus* High-Mass Star Formation   465
4. Summary   466

III. PHYSICS OF EARLY STELLAR EVOLUTION AND STELLAR WINDS

Molecular Outflows: Observed Properties
John Bally and Adair P. Lane 471

1. Introduction 471
2. Molecular Outflow Characteristics 472

3. Recent Developments 480
   3.1. EHV CO Outflows 480
   3.2. Luminosity Dependence of Flow Properties and Statistics 484
   3.3 Optical and Near-IR Observations of Molecular Outflows 488
4. Outflow Models 490

Herbig-Haro Objects
Bo Reipurth 497

1. Introduction 497
2. The Early Years: Observations and Models 498
3. Herbig-Haro Objects at Different Wavelengths 501
   3.1. Optical Spectra 501
   3.2. Infrared Spectra 504
   3.3. Ultraviolet Spectra 505
   3.4. Radio Observations 506
4. A Picture Gallery of Herbig-Haro Objects 508
5. The Highly Collimated Herbig-Haro Jets 512
   5.1. Jets and Bow Shocks 512
       a. Morphological Properties of Jets 512
       b. Kinematics, Physical Conditions and Entrainment of Ambient Material 514
       c. Jets and Magnetic Fields 515
       d. Models of Jet Structure 515
       e. The Working Surface of a Jet 517
       f. Origin and Collimation of Jets 518
   5.2. The HH 46/47 Jet Complex 519
   5.3. The HH 111 Jet Complex 523
   5.4. Herbig-Haro Jets and Molecular Outflows 527
6. Herbig-Haro Energy Sources and Disk Accretion Events 528
   6.1. Surveys and Source Properties 528
   6.2. Three Well Studied Energy Sources 529
   6.3. HH Objects and FU Orionis Eruptions 530

The Physics of Disk Winds
Ralph E. Pudritz, Guy Pelletier and Ana I. Gomez de Castro 539

1. Introduction 539
2. Observational Constraints 539
   2.1. Correlation Between Winds, Stars, And Disks 539

	2.2.	Observed Disk Properties; Mass and Temperature 541
3.	2-D Disk Winds: Basic Physics 542	
	3.1.	Centrifugally Driven Winds...543
	3.2.	Conservations Laws: Flow Along Field Lines 545
	3.3.	1-D Solutions 547
	3.4.	2-D Solutions 548
	3.5.	Blandford and Payne Similarity Solutions 549
	3.6.	Magnetic Collimation of Outflows..................................551
	3.7.	Historical Interlude 553
4.	Angular Momentum Extraction from Disks 555	
5.	Beyond Self Similar Solutions...557	
	5.1.	Control Parameters 557
	5.2.	Alfvén Surface 558
	5.3.	Terminal Speed of the Flow 559
	5.4.	Wind Loss Rate 559
	5.5.	Mechanical Luminosity of the Wind 560
	5.6.	Wind Thrust...560
6.	The Origin and Evolution of Optical Jets 560	
7.	Conclusions 562	

Ionized Winds From Young Stellar Objects
Nino Panagia..565

1.	Introduction 565	
2.	Emission from Extended Outflows: A Theoretical Overview...565	
	2.1.	Assumptions -- Basic Transfer Problem.........................566
	2.2.	The Continuum 569
	2.3.	The Emission Lines..572
	2.4.	Extension to More General Cases 575
		2.4.1. Incomplete Ionization 575
		2.4.2. The Outflow is not Spherically Summetric 576
		2.4.3. Time Variability 576
		2.4.4. Accretion..577
	2.5.	Non-Thermal Emission 578
3.	Radio and Infrared Observations of Young Stars 579	
	3.1.	Ionized Winds from YSO's? 579
	3.2.	The "Well Behaved" Case of S 106-IRS4 581
	3.3.	General Results of Radio Observations.........................582
	3.4.	Radio Surveys..583
	3.5.	Infrared Observations 584
4.	Statistical Properties of Stellar Winds from Young Stars 586	

The Physics of Neutral Winds from Low Mass Young Stellar Objects
A. Natta and C. Giovanardi 595

1. Introduction 595
2. Hydrogen Ionization and Excitation 598
   - 2.1. Ionization State 599
   - 2.2. Dominant Ionization Processes 601
   - 2.3. Excitation State and Line Intensities 605
3. Ionization of Trace Elements: the Case of Sodium 606
4. An Example of Diagnostic of $\dot{M}$ and T_g 609
5. The Gas Temperature 612
   - 5.1. The Heat Equation 612
   - 5.2. Radiative Heating and Cooling 613
     - 5.2.1. H^- Bound-Free Transitions 613
     - 5.2.2. H Bound-Free and Free-Free 614
     - 5.2.3. Ly-α Cooling 614
     - 5.2.4. Ca II H and K Lines 615
   - 5.3. Adiabatic Expansion Cooling 616
   - 5.4. Friction Heating 616
   - 5.5. Heating Due to Dissipation of Alfvén Waves 617
   - 5.6. Heating due to H_2 Formation 618
   - 5.7. Comparison with the Observations 619
6. Conclusions 619

Episodic Phenomena In Early Stellar Evolution
L. Hartmann 623

1. Introduction 623
2. Observed Properties of Fuors 624
3. Steady Disk Models 628
   - 3.1. Steady Disk Spectrum 628
   - 3.2. Differential Rotation 631
   - 3.3. Line Profiles 631
   - 3.4. Inferred Properties of the Central Stars 633
4. Disk Properties 635
5. Boundary Layer 637
6. Outburst Mechanisms 638
7. Mass Loss and Bipolar Flows 642

Properties and Models of T Tauri Stars
Claude Bertout and Gibor Basri 649

1. Historical Background and Key Observations 649
2. Basic Physics of Accretion Disks 656
3. Comparisons with Observations 661
4. The Evolutionary Status of CTTS and WTTS 668
5. Summary..671

The X-Ray and Radio Properties of Low-Mass Pre-Main Sequence Stars
Thierry Montmerle 675

1. Introduction 675
2. Evidence for Magnetic Activity in PMS stars: X-Rays 676
    2.1. X-Ray Emission of PMS Stars: General Properties and Solar-Like Activity 676
    2.2. A New Class of PMS Stars 679
    2.3. Physical Mechanisms of X-Ray Emission 679
    2.4. Derived Magnetic Structures 681
3. Evidence of Magnetic Activity in PMS Stars: Radio (CM Range) ...682
    3.1. Basic Physical Mechanisms 682
        3.1.1. Bremsstrahlung (Free-Free) Emission..........................682
        3.1.2. "Non-Thermal" Mechanisms..683
        3.1.3. Absorption Processes 686
    3.2. Basic Observational Results 686
    3.3. Interpretation 687
    3.4. Derived and Observed Magnetic Structures 689
        3.4.1. Radio Polarization Data 689
        3.4.2. VLBI Measurements 690
4. Evidence for Magnetic Activity in PMS Stars: Optical and UV 691
    4.1. Flares 691
    4.2. Starspots 691
5. Origin of Magnetism 692
    5.1. The Dynamo Effect 692
    5.2. Links with Rotation 693
    5.3. Dependence with Age 694
    5.4. Toward Higher Masses 694
6. Possible Influence of Stellar Magnetic Fields on the Circumstellar Material 696
    6.1. An Evolutionary Problem 696
    6.2. Cold Dust Around Young Stellar Objects 697

		6.3.	Magnetic Fields and Stability of Accretion Disks	701
7.	Conclusions			701

Polarization of Light and Models of the Circumstellar Environment of Young Stellar Objects
P. Bastien 709

1.	Polarization of Light	709
	1.1. What Polarized Light is	710
	1.2. Representation of Polarized Light	711
2.	Polarization Mechanisms	713
	2.1. Dichrotic Extinction	713
	2.2. Scattering of Light	716
	2.2.1. Single Scattering	716
	2.2.2. Multiple Scattering	720
	2.2.3. How Is the Intensity Affected?	722
3.	Models of the Circumstellar Environment	723
	3.1. Observations Relating to the Circumstellar Environment	723
	3.1.1. Observations in the Visible and Near-Infrared Through a Diaphragm	724
	3.1.2. Maps: Intensity, Color, and Polarization Maps	724
	3.2. Theoretical Density Distributions in Disks and Envelopes	726
	3.2.1. Main Stages in the Star Formation Process	726
	3.2.2. Spherical Symmetry	726
	3.2.3. Thin Disks	727
	3.2.4. Thick Disks	728
	3.3. Single Scattering Models	728
	3.4. Polarization and Binary Stars	729
	3.5. Models With Multiple Scattering	729
4.	Conclusions and Prospects	732

Index 739

PREFACE

The origin of stars is one of the principle mysteries of nature. During the last two decades advances in technology have enabled more progress to be made in the quest to understand stellar origins than at any other time in history. The study of star formation has developed into one of the most important branches of modern astrophysical research. A large body of observational data and a considerable literature now exist concerning this topic and a large community of international astronomers and physicists devote their efforts attempting to decipher the secrets of stellar birth. Yet, the young astronomer/physicist or more advanced researcher desiring to obtain a basic background in this area of research must sift through a very diverse and sometimes bewildering literature. A literature which includes research in many disciplines and sub disciplines of classical astrophysics from stellar structure to the interstellar medium and encompasses the entire range of the electromagnetic spectrum from radio to gamma rays. Often, the reward of a successful foray through the current literature is the realization that the results can be obsolete and outdated as soon as the ink is dry in the journal or the conference proceeding in which they are published. The study of star formation is indeed a dynamic one, yet, fueled by the new knowledge provided by technological progress, it has now reached a level of maturity in which the basic empirical and theoretical foundations for an understanding of the physical process of star formation and early stellar evolution are beginning to emerge. For this reason we felt that the time was right to produce a book on this subject which would simultaneously provide a broad and systematic overview of, as well as a rigorous introduction to, the fundamental physics and astronomy at the heart of modern research in star formation and early evolution.

To accomplish this goal we organized a NATO Advanced Study Institute on the topic of the Physics of Star Formation and Early Stellar Evolution. The institute was conducted as an advanced school for graduate students and young researchers in this field. Our institute brought together an international group of 21 distinguished researchers to critically review and update in a systematic fashion the current state of knowledge concerning the entire scope of our present understanding of star formation from the origin of giant molecular clouds to the arrival of young stars on the main sequence. These individuals were given the task of preparing a series of graduate level tutorial review lectures for the school. These lectures formed the basis for the individual chapters of this book which we hope will serve the function of a graduate-level text on the subject of star formation and early stellar evolution. The book contains 22 chapters and is divided into three sections: The Physics of Giant Molecular Clouds: Origin, Structure and Evolution, The Physics of Star Formation, and the Physics of Stellar Winds - Early Stellar Evolution. In addition, we have divided the star formation section into two parts. The first is a discussion of what we know about the formation of high mass stars and the second concentrates on the formation of low mass stars. Although we have somewhat artificially subdivided the chapters into these various sections, there is of course considerable overlap between papers in the various sections of this book. Finally, we have attempted to present a balanced presentation with respect to observational and theoretical review chapters, nearly half of the contributions are theoretical in nature.

A few comments about the NATO Advanced Study Institute which led to this book are in order. The ASI was held at the beautiful Capsis Beach Hotel on the Agia Pelagia Peninsula on the northern coast of Crete. The format of the ASI was that of a summer school for advanced graduate students and post doctoral fellows. The ASI consisted of ten working days spread out over a two week period during the last week of May and first week of June 1990. Each working day was divided into three class sessions of at least 90 minutes duration. Two sessions were held in the morning and one in the evening. All participants shared breakfast before the first morning session and dinner at the end of the evening session. There was a 6 hour free period between 13:00 and 19:00 hours each day for lunch, recreation and special discussion sections organized by the students. This scheduled break turned out to be extremely beneficial for the school because it helped prevent the inevitable mental saturation that would accompany a series of two or more advanced lectures, especially when such intense lectures were given on consecutive days over a two week period! Moreover, this daily break provided informal time for discussion over lunch between the students and the review lecturers while the material taught was still fresh on everyone's mind. Considerable interaction between the participants and a significant amount of scientific discourse occurred during these periods.

On most days, the three classroom sessions each consisted of a 90 minute invited review lecture with additional time for questions and discussion. However, once during each week a mid-morning classroom session was devoted to formal poster presentations by the students and other participants on their own research in star formation and early stellar evolution. The poster sessions were an important and integral part of the ASI. Although, there was only one formal poster session each week, the posters from each session were on display in the main lecture hall for the entire week in which the session was held, allowing ample time for all to view and discuss the posters during the ASI. Moreover, substantial prizes were awarded to the two best posters (based on quality of science and presentation) in each session. The posters were reviewed by two committees of distinguished judges. Professors Blaauw, Shu and Blitz judged the first poster session and Professors Montmerle, Pudritz and Genzel judged the second session. Dr. Alyssa Goodman and Mr. Jochen Eislöfel were awarded the first prizes in their respective sessions and Dr. Scott Kenyon and Mr. Fabien Malbet were awarded the second prizes. The prizes were presented at the closing banquet for the conference.

In the middle of the ASI, a special one day seminar was devoted to a series of shorter invited talks describing the most recent observational results obtained at many of the world's major new observatories. Presentations were made by Drs. John Bally (Bell Labs), Rachel Padman (JCMT), Rolf Güsten (IRAM 30m), S. Guilloteau (IRAM interferometer), Michael Olberg (SEST), Nino Panagia (Space Telescope), and Alex Rudolph (BIMA). Together this seminar and the two poster sessions infused the ASI with some of the most recent and important observational data obtained at a number of the most prominent new astronomical facilities in the world. In addition, two special seminars were organized by the students to discuss in more detail the topics of bipolar outflows and open cluster formation which the students found warranted more discussion than was possible in the review lectures.

The Scientific Organizing Committee for the ASI consisted of Nick Kylafis (Director), Charles Lada (Scientific Director), Claude Bertout, Leo Blitz, Reinhard Genzel, Harm Habing, Lee Hartmann, Nino Panagia, Frank Shu and Hans

Zinnecker.

We are grateful to the NATO Scientific Affairs Division for its generous funding of this Advanced Study Institute. We are also grateful to the University of Crete for providing valuable assistance and funds for the overall organization of the conference. The company MITOS S.A. secured for us low cost accomondation, help with travel, nice excursions and wonderful barbeque and banquet nights. The National Science Foundation provided travel grants to enable 5 students from the U.S. to attend the ASI. We are particularly grateful to Mrs. Lia Papadopoulou whose tireless efforts in handling nearly all the logistical aspects of the local organization from the very beginning to end contributed mightily to the success of the Institute. We also thank Ms. Karen Woodward at the Smithsonian Astrophysical Observatory for able assistance with editing the manuscript for this book.

Of course, the editors owe particular thanks to the authors of this book who clearly took seriously their charge to produce in-depth tutorial level reviews of the basic problems in star formation research. We hope that students and researchers interested in undertaking serious study of stellar formation and early evolution will find this book a useful and stimulating guide to this most interesting field of astronomical inquiry.

Charles J. Lada   Cambridge, Massachusetts

Nikolaos D. Kylafis   Heraklion, Crete

April 11, 1991

Photo By Inge Heyer

AUTHORS

John Bally
L-245 AT&T Bell Labs,
Holmdel, NJ 07733
USA

Gibor Basri
Astronomy Department
University of California
Berkeley, CA 94720
USA

Pierre Bastien
Departement de Physique
Universite de Montreal
B.P. 6128, Succ. A., Montreal
Quebec H3C 3J7
CANADA

Claude Bertout
Institut d'Astrophysique
98 bis, Blvd. Arago
F-75014, Paris
FRANCE

Adriaan Blaauw
Kapteyn Astronomical Lab.
Postbus 800, NL-9700 AV
Groningen
NETHERLANDS

Leo Blitz
Astronomy Program,
University of Maryland
College Park, MD 20742
USA

Jose Cernicharo
IRAM
Av. Divina Pastora N7, NC
18012 Granada,
SPAIN

Ed Churchwell
Astronomy Department
University of Wisconsin
475 N. Charter Street
Madison, WI 53706
USA

Bruce Elmegreen
IBM Watson Research Center
P.O. Box 218
Yorktown Heights, NY 10598
USA

Reinhard Genzel
Max-Planck Institut für
Extraterrestrische Physik
D-8046 Garching
FRG

C. Giovanardi
Osservatorio di Arcetri
Largo Fermi 5
50125 Firenze
ITALY

A. I. Gomez de Castro
Department of Physics
McMaster University
Hamilton
L85 4M1 Ontario
CANADA

Lee Hartmann
Center for Astrophysics
60 Garden Street
Cambridge, MA 02138
USA

Nick Kylafis
University of Crete
Physics Department
P.O. Box 1470
Heraklion, Crete
GREECE

Charles J. Lada
Center for Astrophysics
60 Garden Street
Cambridge, MA 02138
USA

Adair P. Lane
Boston University
Department of Astronomy
725 Commonwealth Avenue
Boston, MA 02215
USA

Thierry Montmerle
Service d'Astrophysique
Centre d'Etudes Nucleaires de
Saclay
91191 Gif-sur-Yvette Cedex
FRANCE

Telemachos Mouschovias
Department of Astronomy
University of Illinois
1002 West Green Street
Urbana, IL 61801
USA

Antonella Natta
Osservatorio di Arcetri
Largo Fermi 5, 50125 Firenze
ITALY

Nino Panagia
Space Telescope Science Inst.
Academic Affairs
3700 San Martin Dr.
Baltimore, MD 21218
USA

Guy Pelletier
Dept. of Physics
McMastor University,
Hamilton, Ontario L8S 4M1
CANADA

J.E. Pringle
Space Telescope Science Inst.
3700 San Martin Drive
Baltimore, MD 21217
USA

Ralph E. Pudritz
Dept. of Physics
McMastor University,
Hamilton, Ontario L8S 4M1
CANADA

Bo Reipurth
ESO/LA SILLA
Karl-Schwarzchild-st. 2
D-8046 Garching bei
Muenchen
FRG

Frank Shu
Astronomy Department
University of California
Berkeley, CA 94720
USA

Werner M. Tscharnuter
Institut für Theoretische
Astrophysik der Universitat
Im Neuenheimer Feld 561
D-6900   Heidelburg 1
FRG

PARTICIPANTS

Nels Anderson
Department of Physics
University of Illinois
1110 W. Green Street
Urbana, IL 61801
USA

Colin Aspin
Joint Astronomy Center
665 Komohana Street
Hilo, HI 96720
USA

Martin Balluch
Inst. fur Theoretische Astrophysik
Univ. Heidelburg,
Im Neuenheimer Feld 561,
D-6900 Heidelburg,
FRG

Dale Barker
Astronomy Centre
University of Sussex, Falmer
Brighton, BN1, East Sussex
ENGLAND

Mary Barsony
Department of Astronomy
University of California, Berkeley
Berkeley, CA 94720
USA

K. Robbins Bell
Lick Observatory
University of California
Santa Cruz, CA 95064
USA

Frank Bertoldi
Princeton University Observatory
Peyton Hall
Princeton, NJ 08544
USA

Sylvie Cabrit
Service d'Astrophysique
CEN Saclay, 91191 Gif-sur-Yvette
FRANCE

Hector Castaneda
Instituto de Astrofisica de Canarias
38200 La Laguna, Tenerife
SPAIN

Alain Castets
Groupe d'Astrophysique,
CERMO
B.P. 53X, 38041 Grenoble Cedex
FRANCE

Cathie Clarke
Institute of Astronomy
Cambridge CB3 OHA
ENGLAND

Vincent Coude du Foresto
Observatoire de Meudon,
DESPA
5 Place Jules Jansen
92195 Meudon, Principal Cedex
FRANCE

Paul M. Cray
Department of Physics
Queen Mary and Westfield College
Mile End Road, London E1 4NS
U.K.

Mike Disney
Physics
Univ. College Cardiff
P.O. Box 913,
Cardiff CF1 3TH
U.K.

M-L Djurhuus
Copenhagen Astronomical Inst.
Oster Voldgade 3,
DK-1350 Copenhagen K
DENMARK

Jochen Eislöffel
Max-Planck Inst. für Astronomie
Königstuhl
D-6900 Heidelburg
FRG

Andrea Ferrara
Dipartimento di Astronomia
Univ. di Firenze
Largo E. Fermi, 5
50125 Firenze
ITALY

Matilde Fernandez
Observatorio Astron. de Madrid
Alfonso XII, 3,
28014 Madrid
SPAIN

Jorge Figueiredo
Astrophysical Institute
Vrije Universiteit Brussel,
Pleinlaan 2, B-1050 Brussel
BELGIUM

Gary Fuller
Center for Astrophysics
60 Garden Street
Cambridge, MA 02138
USA

Daniele Galli
Osservatorio Astrofisico di Arcetri
Largo E. Fermi 5
50125 Firenze
ITALY

V. S. Geroyannis
Astronomy Laboratory,
Department of Physics
University of Patras
GR-26110 Patras
GREECE

Alyssa A. Goodman
Astronomy Department
U. C. Berkeley
Berkeley, CA 94720
USA

Eduardo Martin Guerrero de
  Escalante
Institut d'Astrophysique de Paris
98 bis Bd Arago
75014 Paris
FRANCE

S. Guilloteau
IRAM
300 Rue de la Piscine
38406 Saint Martin d'Heres
FRANCE

Rolf Güsten
MPI Radioastronomie
Auf dem Hugel 69
5300 Bonn 1
FRG

Martin Haas
MPI f. Astronomie
D-6900 Heidelberg
FRG

Ingeborg Heyer
University of Hawaii
Institute for Astronomy
2680 Woodlawn Drive
Honolulu, HI 96822
USA

Isabelle Joncour
CEN Saclay
DPhG SAp Bat. 528
91120 Gif/Yvette Cedex
FRANCE

Ilya Kazes
Observatoire de Meudon, DERAD
5 place Jules Janssen
92195 Meudon Principal
Cedex
FRANCE

Scott Kenyon
Center for Astrophysics
60 Garden Street
Cambridge, MA 02138
USA

Stan Kurtz
University of Wisconsin,
Physics Department
1150 University Ave.
Madison, WI 53706
USA

Elizabeth Lada
Center for Astrophysics
60 Garden Street
Cambridge, MA 02138
USA

Roland Lemke
Max-Planck Institute für Radio-
    astronomie
Auf dem Hugel 69
D-5300 Bonn 1
FRG

Joao Jose Graca Lima
Centro de Astrofisica
Universidade de Porto
Rua do Campo Alegre 823
4100 Porto
PORTUGAL

Rene Liseau
CNR-Istituto di Fisica Dell Spazio
    Interplanetario,
CP 27, I-00044 Frascati (Rome)
ITALY

Susana Lizano
Instituto de Astronomia, Apdo.
70-264
UNAM Cd. Universitaria
04510 Mexico D.F.,
MEXICO

Mark McCaughrean
Steward Observatory
University of Arizona
Tucson, AZ 85721
USA

Peter McCullough
Astronomy Department
UC Berkeley
Berkeley, CA 94720
USA

Mordecai-Mark Mac Low
Mail Stop 245-3
NASA Ames Research Center
Moffett Field, CA 94035
USA

Suzanne Madden
Max-Planck Institut für Physik
    und Astrophysik
Institut fur Extraterrestrische
    Physik
D-8046 Garching bei Muenchen
FRG

Fabien Malbet
Institut d'Astrophysique de Paris
98b, bd. Arago
F-75014 Paris
FRANCE

S.E. Maravelias
Univ. of Athens, Dept. of Physics
Section of Astrophysics,
Astronomy & Mechanics
Panepistimioupolis
GR-15783 Zographos, Athens
GREECE

Francois Menard
Institut d'Astrophysique
98 bis, bd. Arago
75014 Paris
FRANCE

Michael Olberg
Onsala Space Observatory
S-43900 Onsala
SWEDEN

Livia Origlia
ESO- European Southern Observ.
Karl Schwarzschild Str. 2
D-8045 Garching bei Muenchen
FRG

Rachael Padman
Cavendish Laboratory
Madingley Rd., Cambridge, CB3
OHE
ENGLAND

C. Paleologou
University of Crete
Physics Department
P.O. Box 1470
Heraklion, Crete
GREECE

Joseph Pastor
Dept. Fisica de l'Atmosfera,
Astronomia & Astrofisica
Universitat de Barcelona,
Diagonal 647
08028 Barcelona
SPAIN

Randy Phelps
Boston University
Department of Astronomy
725 Commonwealth Avenue
Boston, MA 02215
USA

Thanasis Pitsavas
University of Athens
Dept. of Physics
Section of Astrophysics,
Astronomy, & Mechanics
Panepistimoupolis
Zografos GR-157 83 Athens
GREECE

Timo Prusti
Laboratory for Space Research
P.O. Box 800
9700 AV Groningen
THE NETHERLANDS

Angels Riera
Inst. de Astrofisica de Canarias
38200 La Laguna
Tenerife
SPAIN

Dimitra Rigopoulou
Physics Dept.
Division of Astrogeophysics
University of Ioannina
45110 Ioannina
GREECE

Steven Ruden
Physics Department
University of California Irvine
Irvine, CA 92717
USA

Alexander Rudolph
Astronomy Program
University of Maryland
College Park, MD 20742
USA

Stephen C. Russell
Dublin Inst. for Advanced Studies
5 Merrion Sq
Co. Dublin
IRELAND

Christophe Sauty
University of Crete
Physics Department
P.O. Box 1470
Heraklion, Crete
GREECE

John Scalo
Department of Astronomy
University of Texas
Austin, TX 78712
USA

Flavio Scappini
Ist. do Spettroscopia Molecolare
Via de Castagnoli, 1
40126 Bologna
ITALY

Peter Schlike
Max-Planck Institut für Radio-
    astronomie
Auf dem Hugel 69
D-5300 Bonn 1
FRG

Angie Schultz
Physics Department
Campus Box 1105
Washington University
St. Louis, MO 63130
USA

Nikolaos Solomos
Astronomical Laboratory
Department of Physics
University of Patras
Patras 261 10
GREECE

Luigi Spinoglio
IFSI-CNR, CP 27
00044 Frascati
ITALY

Tomasz Stepinski
University of Arizona
Space Sciences Building
Tucson, AZ 85721
USA

Tom Theuns
University of Antwerp, R.U.C.A.
Astrophysics Research Group
Groenenborgerlaan 171
B-2020 Antwerpen
BELGIUM

Christopher Tout
Lick Observatory
Board of Studies in Astronomy
and Astrophysics
University of California
Santa Cruz, CA 95064
USA

John C. Tsai
6-110
Department of Physics
Massachusetts Inst. of Technology
Cambridge, MA 02139
USA

Kanaris Tsiganos
University of Crete
Physics Department
P.O. Box 1470
Heraklion, Crete
GREECE

Enrique C. Vasquez
Astronomy Department
The University of Texas
Austin, TX 78712
USA

Dolores Walther
Joint Astronomy Centre
665 Komohana Str.
Hilo, HI 96720
USA

Bruce A. Wilking
Physics Department
University of Missouri-St. Louis
8001 Natural Bridge Rd.
St. Louis, MO 63121
USA

Jonathan Williams
Astronomy Department
Univ. of California at Berkeley
Berkeley, CA 94720
USA

David Wilner
Astronomy Department
University of California
Berkeley, CA 94720
USA

Günther Wuchterl
Inst. für Theoretische Astrophysik
Universitat Heidelberg
Im Neuenheimer Feld 561
D-6900 Heidelburg 1
FRG

Werner Verschueren
University of Antwerp, R.U.C.A.
Astrophysics Research Group
Groenenborgerlaan 171
B-2020 Antwerpen
BELGIUM

Evangelia Xiradaki
Section of Astrophysics,
Astronomy and Mechanics,
Department of Physics
University of Athens
Panepistimioupolis
157 71 Athens
GREECE

Joao Lin Yun
Boston University
Department of Astronomy
725 Commonwealth Avenue
Boston, MA 02215
USA

Annie Zavagno
Observatoire de Marseille
2 Place le Verrier
13248 Marseille Cedex 4
FRANCE

Shudong Zhou
2680 Woodlawn Drive
Honolulu, HI 96822
USA

Hans Zinnecker
Inst. f. Astron & Astrophys. Univ.
8700 Würzburg, Am Hubland
Karl-Schwarzschild-Strasse 1
D-8046
FRG

Robert Zylka
Max-Planck Inst. für Radio-
    astronomie
Auf dem Hugel 69
D-5300 Bonn 1, FRG

STAR FORMATION AND EARLY EVOLUTION IN THE
OMEGA NEBULA

Infrared 3-color (J,H,K) mosaic image of M17-- one of the most active sites of massive star formation in the galaxy. (This image was obtained by C. Lada, I. Gatley, D. Depoy and M. Merrill using the 2.1 meter telescope of NOAO.) A cluster of more than 100 massive OB stars provides the energy to excite the luminous HII region. Remarkably and mysteriously, all of these OB stars display excess infrared emission suggestive both of the presence of circumstellar disks, and the extreme youth of the system. The Phaistos Disk (circa 1700 BC), unearthed in Crete more than eighty years ago, has never been deciphered.

I. PHYSICS OF GIANT MOLECULAR CLOUDS: ORIGIN, STRUCTURE, AND EVOLUTION

David Wilner, Elizabeth Lada, Bruce Wilking and John Scalo

Star Forming Giant Molecular Clouds

LEO BLITZ
Astronomy Program
University of Maryland
College Park, MD 20742
USA

ABSTRACT. The properties of galactic giant molecular clouds (GMCs) in the solar vicinity and in the inner Galaxy are reviewed. Special attention is given to the role of the clouds in forming stars. The question of whether all GMCs form stars is raised and it is shown that there is little evidence that GMCs anywhere in the Galaxy are devoid of star formation, even O star formation. The angular momentum of local GMCs is then discussed. It is shown quantitatively that the specific angular momentum of GMCs is within the range expected if the clouds condense out of the diffuse interstellar medium. At least four GMCs in the solar neighborhood, however, have retrograde rotation in an inertial frame of reference, placing significant constraints on how they could have formed. Three GMCs in different evolutionary states are identified, and and some of the differences in their properties are identified. GMCs are shown to have atomic envelopes with masses comparable to their molecular masses. These envelopes are likely to pervade the interclump medium and be responsible for most of its mass.

The structure of GMCs is then discussed in some detail. It is argued that clumps are the fundamental units in which the GMC mass is collected. Clumps are shown to have a small volume filling fraction, and that the clump-interclump density contrast is high. The kinematics of the clump ensemble in one GMC, the Rosette Molecular Cloud, is shown to be consistent with the idea that they have have undergone dynamical evolution; the most massive clumps are concentrated toward the midplane of the complex and also possess a lower clump-to-clump velocity dispersion. The clump mass spectrum for five GMCs is compared and shown to be remarkably similar; it is a power law with an index of -1.6 over more than three orders of magnitude in mass with a spread from cloud to cloud of only 10%. Although the majority of clumps do not appear to be gravitationally bound, these clumps can be confined by the pressure of an atomic intercloud medium. The observations presented in this review are used to suggest a tentative outline for the evolution of GMCs.

1. INTRODUCTION

The fundamental goal of the study of molecular clouds is to understand how they form stars. All present day star formation takes place in molecular clouds, so we may think of them as providing the initial conditions for the process of star formation. Yet, while all stars may form in molecular clouds, there are at least some small molecular clouds that do not form stars. Furthermore, even within those clouds that do form stars, it appears that only a small fraction of the mass of a cloud actively takes part in the star formation process. What controls both the presence and the absence of star formation in molecular clouds? There seems to be an inevitability to the star formation process. Observational evidence from surveys of stars and gas indicates that almost all of the giant molecular clouds (GMCs) in the solar vicinity are currently forming stars. How is this inevitability to be reconciled with the absence of star formation in most of the molecular mass of the Galaxy?

We begin by asking a few broad questions about molecular clouds. 1) How do molecular clouds, especially the GMCs, form? 2) Once a molecular cloud forms, how does it evolve to generate the entities that eventually become stars? 3) What is the basic unit of a molecular cloud that produces stars, and how does it evolve to form a star? None of these questions has been answered as yet, but partial answers are beginning to emerge. This paper will review some of the observational material relevant to these questions and will concentrate on the larger scale aspects of the structure and evolution of giant molecular clouds. It will be argued that an understanding of the evolution of the interstellar medium into stars requires a detailed knowledge of the clumpy structure of molecular clouds. It will be shown that it is possible to identify molecular clouds in different evolutionary states, and that evolutionary effects within a particular molecular cloud, the Rosette Molecular Cloud, are observable in the dispersion of relative velocities of the identifiable clumps. The observations of a number of star-forming molecular clouds are reviewed with the aim of providing generalizations that are useful for future work. Among the topics not covered is the energy balance within molecular clouds. A good review of this subject is given by Genzel (1991). We begin with a review of the large scale properties of GMCs.

2. GLOBAL PROPERTIES OF GIANT MOLECULAR CLOUDS

It is rather amazing that 15 years since the identification of giant molecular clouds, there is no generally accepted definition of what a GMC is. Most observers seem to view GMCs much in the same way that Senator Jesse Helms views pornography, "I may not be able to define it, but I know it when I see it." That is, there seems to be little disagreement about the classification of the largest clouds as GMCs, but an all inclusive definition of what a GMC is has proven elusive. A large part of the problem is that the various studies of the mass spectrum of molecular clouds indicate that the spectrum is well fit by a power law (see below) and there is consequently no natural size or mass scale for molecular clouds. What we call a GMC is therefore largely a question of taste. For the purposes of this paper, I will arbitrarily define any molecular cloud that has a mass $\gtrsim 10^5$ $M_\odot$ definitely to be a GMC, a cloud with a mass 10^5 $M_\odot \gtrsim M(\text{cloud}) \gtrsim 10^4$ $M_\odot$ probably to be a GMC, and a cloud with a mass $\lesssim 10^4$ $M_\odot$ probably not to be a GMC. In most cases, this fuzziness does not cause serious problems.

Giant molecular clouds have been studied as a whole largely through their CO emission in the radio portion of the spectrum. The studies have been mainly of two kinds. 1) Observations of individual objects, where a cloud is identified, often by its association with a visible HII region, and then mapped to its outer boundaries. 2) Surveys of the Galactic plane in CO, where GMCs are identified through some objective criterion. In the second case, GMCs are generally defined down to some contour level because of the possibility of confusion with other gas along the line of sight. Surprisingly, the general properties of GMCs defined in this way are not significantly different from those identified by the first method. Table 1 gives the properties derived for clouds in the solar vicinity (e.g. Blitz 1987b). Inner Galaxy molecular clouds may be somewhat denser and more opaque (see McKee 1989; Solomon, *et al.* 1987), but there is no evidence that they form a separate population from the local clouds. This important conclusion suggests that detailed studies of the molecular clouds near the Sun can tell us about the ensemble properties of GMCs everywhere in the disk. An exception is likely to be in the innermost regions of the Galaxy such as the molecular disk within 400 pc of the center and (possibly) the 3 kpc arm of the Galaxy.

TABLE 1. Global Properties of Solar Neighborhood GMCs

Mass	$1 - 2 \times 10^5$ $M_\odot$
Mean diameter	45 pc
Projected surface area	2.1×10^3 pc^2
Volume	9.6×10^4 pc^3
Volume averaged n(H$_2$)	~ 50 cm^{-3}
Mean N(H$_2$)	$3 - 6 \times 10^{21}$ cm^{-2}
Local surface density	~ 4 kpc^{-2}
Mean separation	~ 500 pc

An example of a local GMC is shown in Figure 1, the L1641 cloud in Orion, in which the Orion Nebula is located. The figure shows the emission from the ^{13}CO J = 1-0 transition in various velocity bins, and in the last panel, the emission integrated over the cloud as a whole is shown. The figure shows several important features that are common to many local GMCs. i) The ridge line of the final panel closely parallels the galactic plane (it is located at $b = -19°4$). Most local GMCs are similarly elongated (Blitz 1978; Stark and Blitz 1978). ii) The ^{13}CO emission shows a strong velocity gradient along the length of the cloud which is generally interpreted as rotation. The subject of angular momentum in GMCs is discussed in Section 4. iii) The ^{13}CO emisison is seen to break up into discrete clumps. In the L1641 cloud, the clumpiness is apparently quite filamentary. The clumpiness of GMCs is discussed in detail in Sections 7, 8, and 9.

From the study of Local GMCs, the following general conclusions may be drawn:

1) GMCs are discrete objects with well defined boundaries (see *e.g.* Blitz and Thaddeus 1980 for a quantitative discussion of this point - see especially Appendix B). The well defined boundaries suggest that there is a phase transition at the edges of a molecular cloud. This point is discussed in greater detail below.

2) GMCs are not uniform entities, but are always composed of numerous dense clumps and have small volume filling fractions (see Section 7 for a quantitative discussion). These clumps appear to have a range of geometries from spherical to highly filamentary (see the maps in the references to Table 4 as well as Bally *et*

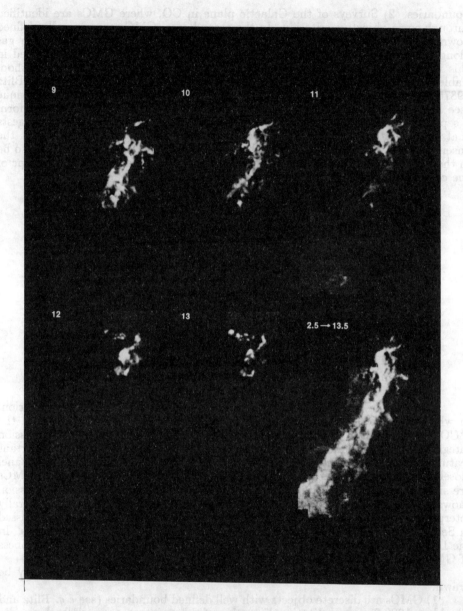

Fig. 1. A series of maps, each showing ^{13}CO emission integrated over a 1 km s^{-1} wide velocity range illustrating the internal structure of the Orion A cloud. The number in each subpanel is the LSR velocity of the center of the velocity range shown. Each panel shows a region roughly 2° × 5° in extent with an angular resolution of about 100″. The galactic plane closely parallels the ridge line in the final panel. From Bally *et al.* (1987b).

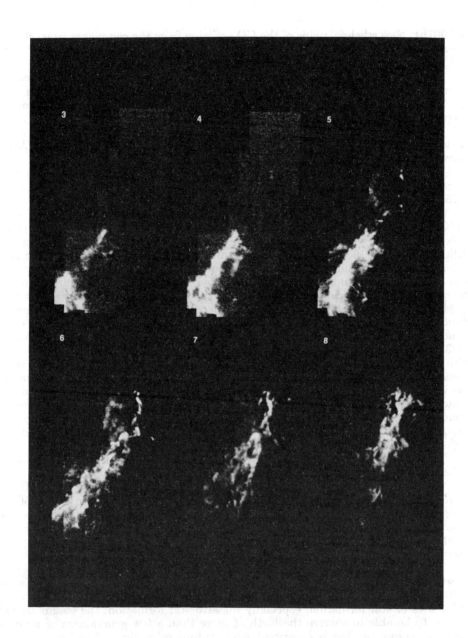

al. 1987b). Nevertheless, maps of the CO emission from the complexes suggest that the surface filling fraction of the clumps is almost always near unity (Blitz 1980 and references therein).

3) The CO and ^{13}CO line widths of the clumps are *always* wider than the thermal widths implied by the excitation temperature. The linewidths are usually interpreted as the result of bulk motions associated with turbulence and/or magnetic fields.

4) GMCs are gravitationally bound (*e.g.* Kutner *et al.* 1977; Elmegreen *et al.* 1979; Blitz 1980). GMC masses are orders of magnitude larger than a Jeans mass.

5) All OB associations form from GMCs; thus GMCs are the nucleation sites for nearly all star formation in the Milky Way (Zuckerman and Palmer 1974; Blitz 1978, 1980 and references therein).

6) Within the uncertainties, the cloud-to-cloud velocity dispersion of local GMCs (with typical masses of $\sim 2 \times 10^5$ M$_\odot$) is the same as that of the small molecular clouds (with typical masses of ~ 50 M$_\odot$) found at high galactic latitude (Blitz 1978, Stark 1984, Magnani, Blitz and Mundy 1985). The velocity dispersion of local molecular clouds therefore appears to be independent of mass over three or four orders of magnitude. If clouds formed by collisional aglomeration of smaller independently moving clouds, then cloud velocity dispersions would be proportional to $M^{-0.5}$, unless there were some way to selectively re-energize the more massive clouds, an unlikely prospect. Furthermore, the velocity dispersion of molecular clouds seems to vary only weakly with galactocentric distance (Liszt and Burton 1983, Stark 1984, Clemens 1985) implying that the constancy of velocity dispersion with mass is likely to hold everywhere in the disk of the Galaxy. For the most massive clouds, Stark (1983) has shown that the scale height, and by implication the vertical velocity dispersion of GMCs, exhibits a decrease in the inner Galaxy: cloud-cloud collisions may therefore be important for the largest, most massive clouds. All of these results taken together suggest that collisional processes are not important in the formation of GMCs except possibly in the inner Galaxy at the highest masses ($\sim 10^6$ M$_\odot$).

It is noteworthy, however, that within a GMC, collisional processes do seem to be important. This point is discussed in more detail in Section 7.

7) The survey of local molecular material within 1 kpc of the Sun (Dame *et al.* 1986) finds no GMCs without star formation. In fact, within 3 kpc of the Sun, only one GMC is found without evidence of star formation (Maddalena and Thaddeus 1986). This cloud is discussed in more detail below. In the solar vicinity at least, molecular clouds without star formation are quite rare. It therefore appears that star formation is quite rapid after the formation of a GMC.

8) The above result, when combined with calculations of the destructive processes associated with the formation of massive stars, implies that the GMCs are quite young, $\sim 3 \times 10^7 y$ (Blitz and Shu 1980). That is, since all GMCs appear to be sites of star formation, especially massive star formation, the clouds do not appear to be able to survive the birth of more than a few generations of massive stars, an argument that is consistent with various other lines of reasoning (Blitz and Shu 1980). Direct observational evidence for lifetimes of this order was first presented by Bash, Green and Peters (1977), and subsequently by Leisewitz, Bash, and Thaddeus (1986) from the association (or lack thereof) of molecular material with young clusters of vaying ages. Cohen (*et al.* 1980), have come to the same conclusion from the confinement of the molecular arms in the outer Galaxy to narrow

kinematic ranges.

9) A large fraction of the stars formed in GMCs do not return their material to the interstellar medium in a Hubble time. Therefore, either the GMCs were more numerous in the past, or the interstellar medium is replenished by infall or inflow from the outer reaches of the Galaxy (e.g. Lacey and Fall 1985).

Studies of GMCs based on inner Galaxy CO surveys run into the problem of finding an objective way to identify the molecular clouds. The degree of blending of the spectral lines, the large amount of foreground and background emission, and the necessity of using kinematic criteria for identifying the clouds add a significant amount of uncertainty to the identification of GMCs (see *e.g.* Adler 1988, Blitz 1987a). Nevertheless, a number of properties of GMCs are inferred from the Galactic CO surveys, and the searches for molecular clouds in the outer Galaxy. Some of the more important of these are the following:

1) There appears to be a linewidth-size relation for GMCs in the inner Galaxy. From various studies, the results may be given as follows:

TABLE 2. Linewidth - Size Relation for GMCs

Dame et al.(1986)	$\Delta_V = 1.20 R^{0.50}$
Solomon et al.(1987)	$\sigma_V = 1.0 S^{0.50}$
Scoville et al.(1987)	$\sigma_V = 0.31 D^{0.55}$
Sanders et al.(1985)	$\Delta_V = 0.88 D^{0.62}$

where the quantity on the left is either the full width at half maximum or the one dimensional velocity dispersion in km s^{-1}, and the quantity on the right is some measure of the mean linear size of a GMC in pc. The scatter in all of the studies is not terribly large, but note that the second and the third determinations are obtained from the same data set. The close agreement between the various observers in the value of the power law exponent in the linewidth-size relation is so striking, that unless there is a selection effect common to all of the studies (see *e.g.* Blitz 1987a), it suggests that the relation underlies some fundamental property of GMCs.

A serious challenge to the reality of the linewidth-size relation comes from an analysis of published survey data by Issa et al. (1990). These authors showed that if the Solomon et al. (1987) method of analysis is used on their own survey data, but at a) positions not centered on their clouds, and b) at random locations in the survey, one obtains a linewidth-size relation similar to that of the original analysis. This result is puzzling unless the linewidth-size relation is an artifact of the data set itself, rather than of the dynamical state of the clouds. In addition, Issa et al. also lowered the threshold of their analysis to a point where one expects clouds to be blended together into apparent clouds that are unrelated to one another. Again, Issa et al. obtained a linewidth-size relation indistinguishable from that of Solomon et al.. Thus, there is evidence that there is no physical significance to the linewidth size realtion, and its application to clouds to obtain, for example, their distances, may be flawed. Analysis of the Bell Labs ^{13}CO Galactic plane survey, should be able to resolve this issue becuase of the higher contrast in the ^{13}CO data compared to that of CO.

2) If the clouds are in virial equilibrium, then

$$(\Delta V)^2 = \alpha GM/R.$$

where α is a constant near unity. If $(\Delta V) \sim R^{0.5}$, then $M \sim R^2$ which in turn implies that the mean H_2 column density of GMCs is constant. This simple conclusion is presumably telling us something important about either how GMCs form or how they regulate themselves. McKee (1989) has assumed the latter in his recent theory of photo-regulated star formation in GMCs.

3) GMCs in the outer Galaxy appear to have lower excitation temperatures than GMCs in the inner Galaxy (Mead and Kutner 1988). It is unclear at present whether this is due to a lower external heating rate, decreased star formation rate or both.

4) The distribution of masses for GMCs in the inner Galaxy is found to be (for masses in excess of 10^5 $M_\odot$), $dN(M)/dM \propto M^{-1.5}$ (Solomon et al. 1987). The power law exponent is the same as that found for the distribution of clump masses in a GMC (see section 7). If one tries to determine at what cloud mass half of the H_2 mass in the Galaxy resides, the various analyses of the inner Galaxy CO give results that fall within a factor of two of the range $1 - 2 \times 10^5 M_\odot$ when account is made of the different assumptions used and assuming that the power law can be extrapolated to very low masses. This is very close to the mean value in the solar vicinity (Stark and Blitz 1978).

One of the glaring deficiencies in Galactic studies of GMCs is a quantitative study of how the properties of GMCs vary with galactic radius. For example, we might expect that to be stable against the larger tidal forces and the larger energy density of dissociating radiation, that inner Galaxy clouds would be denser than the clouds at larger galactocentric distance. Such a conclusion was reached from an indirect analysis of Liszt, Burton, and Xiang (1984), however, no direct confirmation has been made. Such a study would be particularly useful in trying to understand how different Galactic environments affect the formation and evolution of GMCs. It has been known since the Altenhoff *et al.* (1970, 1978) 5 GHz surveys of the Galactic plane, for example, that giant HII regions are much more common in what later became known as the molecular ring than they are at larger Galactic radii. Surely the greater efficiency with which the inner Galaxy GMCs convert molecular gas into O stars must be a reflection of differing molecular cloud properties, but which properties? There have been a few studies of molecular clouds in the outer Galaxy that suggest that there are differences between them and the inner Galaxy clouds (*e.g.* Mead and Kutner 1988; Digel *et al.* 1990), but other than these, systematic studies of the gradients of the large scale properties of GMCs as a function of Galactic radius are almost totally lacking. It would seem imperative that even a rudimentary understanding of extragalactic CO emission would require an understanding of how GMC properties vary within the Milky Way.

3. DO ALL GMCS FORM STARS?

As mentioned above, various surveys of the molecular gas in the vicinity of the Sun have turned up only one GMC within 3 kpc that is devoid of star formation. On the other hand, Mooney and Solomon (1988) argue that at least 25% of the GMCs in the inner Galaxy are devoid of OB star formation. Therefore, on the basis of local observations, the answer to the question posed in the heading is essentially yes, and on the basis of inner Galaxy observations, the answer would be no. The resolution of this apparent paradox appears to reside in the sensitivity and confusion limitations inherent in dealing with inner Galaxy data. To see this, it is useful to refer ahead to Figure 3a which shows the 100 μm surface brightness of the cloud complex associated with the Rosette Nebula as observed by IRAS. The peak brightness of the map, which has had a background removed, is 720 MJy sr^{-1}, and the infrared luminosity of the cloud is $\sim 5 \times 10^5$ L$_\odot$. Surprisingly, the HII region is located at a relative minimum on the map, centered on the the hole at $l = 206.°2$, $b = -2°$.

Mooney and Solomon find that there is a class of IR-quiet molecular clouds, many with masses $> 10^5$ M$_\odot$, which have peak 100 μm surface brightnesses only 1/20 that of the IR-strong clouds. The IR-strong clouds are generally defined by strong localized IR emission which, in most cases, seems to result from the action of an HII region. The mean peak surface brightness of the IR-quiet clouds is $\sim 2 \times 10^2$ MJy sr^{-1}. If it were located in the inner Galaxy, the Rosette Molecular Cloud (RMC) would be classified as infrared quiet because the HII region would not appear in inner Galaxy catalogs (see below), and because the HII region itself is actually a minimum on the infrared maps. Furthermore, its peak 100 μm surface brightness is only 3.5 times the mean of the IR-quiet clouds. Nevertheless, the RMC is particularly rich in O stars even though its size and mass are typical of GMCs in the solar vicinity. The nebula itself is illuminated by 5 O stars and the stellar association has 13 stars of spectral type O9.5 or earlier. The association is about as rich as the Orion OB1 association, the most spectacular star-forming region within 1 kpc of the Sun. Therefore, in spite of the richness of its OB star formation, the RMC would have been classified by Mooney and Solomon as having been devoid of OB star formation. It is quite possible, that all of the clouds they have classified as IR-quiet are Rosette-like objects.

A similar problem exists if one tries to use HII regions to trace the star formation in inner Galaxy GMCs as was done by Myers et al. (1986). They used the Altenhoff et al. (1970, 1978) 5 GHz surveys to find HII regions associated with molecular clouds, and found a few GMCs (a smaller percentage than Mooney and Solomon) with very low star formation efficiencies based on the absence of detectable 5 GHz flux. However, the Rosette Nebula is known to have an emission measure of about 3000 cm^{-6} pc (Bottinelli and Gouguenheim 1964), well below the detection limit at which HII regions are unambiguously identified in the Altenhoff survey. Other solar vicinity OB associations (e.g. the Mon OB1 association which contains the well known young cluster NGC 2264) have HII regions with even lower emission measures. Low level IR or radiocontinuum emission may simply signal that the conditions in inner Galaxy molecular clouds without giant HII regions are like those in the Rosette Nebula, and not due to the absence of O stars, or star formation in general. Such situations can arise when the action of O stars destroys the dust, when the angle subtended by the molecular cloud at the ionizing stars is small, or both.

Lockman (1990) has considered the detectability of the Rosette Nebula in the inner Galaxy in a recent review. He concludes that the Rosette Nebula would not have been detected in *any* discrete source survey of the inner Galaxy to date, including his own sensitive recombination line survey (1989). Consequently, the abundance of low surface brightness HII regions in the inner Galaxy is unknown, and may account for the clouds identified by Mooney and Solomon and by Myers et al. that have little apparent star formation.

There is therefore no apparent discrepancy between the fraction of GMCs that form stars in the solar vicinity and in the inner Galaxy. The Maddalena-Thaddeus cloud remains the only known molecular cloud where the star formation rate within the cloud is known to be a small fraction of the rate observed in local OB associations. It is possible with directed observations to set better limits on the stellar content of individual inner Galaxy molecular clouds than are currently available, and it may be possible to find clouds with star formation rates significantly smaller than those found in local GMCs. On the other hand, if the inner Galaxy clouds form stars with the same efficiency as local GMCs (Mooney and Solomon 1988), we might expect that finding GMCs with significantly depressed star formation efficiencies inside the solar circle will prove difficult.

4. ANGULAR MOMENTUM

That GMCs rotate has long been known from the velocity gradients observed across them (e.g. Kutner et al. 1977). The angular momentum of a GMC should reflect the angular momentum of the interstellar medium from which the cloud formed, thus the specific angular momentum of a GMC (that is, the total angular momentum per unit mass of a cloud complex) is an important parameter for understanding its history. In spite of this, little is known about the angular momentum of GMCs. Mestel (1966) considered the problem of the angular momentum of a planet condensing from a protoplanetary disk; this work is applicable in part to the problem of GMC formation form the disk of the Galaxy. In the case of a GMC that condenses from the general interstellar medium, however, it is possible, perhaps even likely that magnetic fields play a role (e.g. Mouschovias, Shu and Woodward 1974), a possibility not considered by Mestel and which adds some complexity to the problem. Let us, in any event, consider how the specific angular momentum of a GMC compares to that of the material from which a cloud formed.

Consider a typical GMC in the solar vicinity with a mass of 2×10^5 M$_\odot$. Such clouds have a typical maximum dimension of 100 pc and are elongated along the Galactic plane (Stark and Blitz 1978; Blitz 1980). If the true shape of the clouds is that of a cigar, the specific angular momentum is $1/3R^2\Omega$; if it is a disk, then the specific angular momentum is $1/2R^2\Omega$. The Table 3 gives the values of the largest known velocity gradients for entire GMCs

TABLE 3. Measurable Velocity Gradients for GMCs

Cloud	Gradient km s^{-1}pc^{-1}	Reference
Rosette	-0.18	Blitz and Thaddeus (1980)
Mon R1	~ -0.2	Blitz (1978)
W3	-0.13	Thronson et al. (1985)
Orion A	-0.10	Kutner et al. (1977)

The negative sign indicates that the angular velocity is antiparallel to the angular velocity vector of the galactic disk. All other velocity gradients published for GMCs are significantly smaller in absolute value than the values quoted above. Typically, the velocity gradient is less than half that of the L1641 cloud over an entire GMC, and published maps make it difficult to discern gradients smaller than about 0.02 km s^{-1}pc^{-1}. In a survey of the literature, Blitz (1980) found that only half of the GMCs known at that time had a measurable velocity gradient. Thus, if the velocity gradient is entirely due to rotation, Ω has an extreme value for GMCs of 0.2 km s^{-1}pc^{-1} in the solar vicinity, but has a typical value somewhat less than 0.05 km s^{-1}pc^{-1}. The specific angular momentum of a typical GMC is therefore somewhat less than about 60 pc km s^{-1} or 1.22×10^{26} cm^2 s^{-1}. The specific angular momentum of the Rosette GMC is however -225 pc km s^{-1}. The uncertainty in these numbers is relatively small (see below).

For the clouds with measurable gradients, the sense of the rotation is retrograde with respect to the galactic rotation in a rotating coordinate frame of reference centered on the local standard of rest (LSR). That is, all four clouds have higher velocities on the low longitude side of the cloud. To transform to an inertial frame of reference we must add the angular velocity of the LSR, Ω_0, to the observed angular velocities; about 0.025 km s^{-1}pc^{-1}. This value is small compared to the measured angular velocites for the GMCs in the table above. Thus, the sense of the angular momenta of the four clouds above is retrograde even in an inertial frame of reference. The remaining GMCs have gradients consistent with no angular momentum in the inertial reference frame. Any theory of the formation of GMCs must account for this fundamental observational fact. It was noted above that most of the mass in GMCs typically resides in dense clumps which may themselves be spinning. However, becuase of the R^2 dependence of the angular momentum, and because clumps do not, as a general rule show large velocity gradients (although there are many specific counterexamples), the angular momenta of the clumps are not likely to contribute significantly to the total angular momentum of a GMC.

The specific angular momentum of the general ISM with which one wants to compare that of a GMC depends in detail on how a cloud forms and the angular momentum distribution of the disk of the galaxy in which it forms. We might consider, for example, that a cloud forms by having some disklike region become gravitationally unstable and collapse to form the cloud. In this case, we expect a result similar to that considered analytically by Mestel (1966). Mestel showed that for a cloud that has condensed from a disk with a flat or rising rotation curve, the angular momentum is always prograde in an inertial frame of reference. If the rotation curve is falling, then the rotation can be either prograde or retrograde depending on the details of the cloud formation. Furthermore, it is the form of the rotation law *locally* that determines the initial angular momentum of the cloud.

Let us first calculate the specific angular momentum of the general ISM in the rotating frame of reference centered on the LSR, which I will call the LSR frame. This quantity is given by $1/2R^2\Omega$, where Ω depends on the geometry of the collapsing cloud, and can be derived either from the Oort A constant or a flat rotation curve, giving values that differ from one another by 20%. In the case of a needle-like collapse parallel to a galactic radius vector, the value of Ω in the solar neighborhood is -0.025 km s^{-1}pc^{-1}; in the inertial frame, Ω is zero. For a needle-like collapse along a line of constant radius, the value of Ω in the solar neighborhood is zero, and is +0.025 km s^{-1}pc^{-1} in the inertial frame. We now ask from what radius must a cloud contract to obtain a mass of 2×10^5 M$_\odot$? The surface density of atomic gas in the vicinity of the Sun is 5 M$_\odot$ pc^{-2} (Henderson, Jackson and Kerr (1982). We may neglect the contribution from the molecular gas because most of that is already due to GMCs (see e.g. Solomon, *et al.* 1987), and the contribution to the midplane density of small molecular clouds can be ignored (Blitz, Bazell and Desert 1989). Assuming that the local effective scale height of the atomic gas is

200 pc (Falgarone and Lequeux 1973), then the midplane density of atomic gas is 0.5 cm^{-3}.

Although we do not know the initial configuration of the condensing gas, we may assume for the moment that it is cylindrical with a diameter equal to the height. The radius of a cylinder that would contract to form a typical GMC is 140 pc. If the collapse is along a line of Galactic radius, the specific angular momentum of the gas that forms a GMC is -250 km s^{-1}pc in the LSR frame; if the collapse is along a line of Galactic azimuth, the angular momentum is zero. The initial angular momentum of the interstellar medium in the LSR frame is therefore expected to lie between these two values for reasonable collapse geometries, although somewhat higher negative values are allowed. The observations indicate that the angular momenta of the GMCs are all within the allowable range. Because the actual values depend on unknown initial geometries, nothing can be said about whether angular momentum is shed during the formation of a GMC.

It is a puzzle, however, why some of the GMCs should be so strongly counter-rotating. As Mestel (1966) pointed out, retrograde rotation can only occur for a falling rotation curve, and then only if the geometry is favorable. A falling rotation curve can occur locally in the vicinity of a spiral arm. In fact, the locally measured value of the Oort A constant of 15 km $s^{-1}kpc^{-1}$, implies that indeed the rotation curve is locally falling. Thus the retrograde rotation of at least some GMCs suggests that the clouds formed when a local gravitational perturbation such as a spiral arm caused the rotation curve to be locally falling.

It is important to recognize that the observed values of J/M quoted above are reasonably well determined. For a molecular cloud, the velocity differences across a cloud can be measured to an accuracy of about 1 km s^{-1} (measurement errors are considerably smaller). Large velocity gradients are therefore measurable to an accuracy of 10-20%. The major source of uncertainty is probably the distance to a cloud, which, for clouds like Orion and the Rosette are probably as low as about 20%. For the ISM, the largest uncertainties are the value of the Oort A constant, which is probably known to an accuracy of 20%, and the midplane density of the atomic hydrogen gas, which translates into the distance R out to which the ISM must be collected to form a GMC. The uncertainty in the density is probably less than 50%, but in any event, the distance R depends only on the 1/3 power of the midplane density, and thus J/M on the the 2/3 power. The major uncertainty for the specific angular momentum of the ISM is the actual process by which GMCs are formed. For example, if the efficiency of transformation of the ISM into molecular material is low, then R may be bigger, and the resulting J/M may be larger.

Studies of the angular momentum distribution among the GMCs simply do not exist at present in spite of the importance that they have for understanding how the molecular clouds form from the interstellar medium. It is unknown, however, whether there are any examples of clouds with measurable prograde rotation in the LSR frame. Even the results quoted above come partly from unpublished analyses of molecular clouds and from personal communications. There is a wealth of information, however, in the large scale surveys of the CO distribution in the Galaxy. For example, there have been hundreds of GMCs that have been catalogued (Solomon, et al. 1987; Scoville et al. 1987) in the inner Galaxy. It would be an easy task to determine to what degree GMCs which exhibit measurable velocity gradients will have the sense of the angular momentum antiparallel to the angular momentum of the Galaxy, yet this has never been done. Nor has a detailed study of the angular momentum of GMCs as a function of Galactic radius been done.

Many other questions about the distribution of angular momentum are answerable with existing data.

5. CLOUDS IN DIFFERENT EVOLUTIONARY STATES

Within one or two kpc of the Sun, all GMCs have been found to have at least some traces of star formation. However, in 1986, Maddalena and Thaddeus found a cloud in the outer Galaxy with a mass of $\sim 10^6$ $M_\odot$, a longest diameter of about 150 pc (both of these numbers assume a kinematic distance of 3 kpc), without any obvious traces of star formation activity. Moreover, the cloud has relatively weak, broad CO lines which appear to be different from those seen in most of the clouds observed in the local solar neighborhood. Maddalena and Thaddeus speculated that the cloud is so young that it has not yet had time to form stars.

If this hypothesis is correct, there should not be any buried or embedded population of stars within the cloud. It is possible, for example, that there is a large HII region on the far side of the cloud which is obscured by the intervening dust (even this hypothesis is unlikely since HII regions are almost always accompanied by strong CO peaks). In order to look for the effects of heating from an embedded population of stars, maps of the cloud have been made using the IRAS 100 μm data base, and the result is shown in Figure 2a (Puchalsky and Blitz in preparation). Figure 2b shows the CO emission associated with the Maddalena-Thaddeus cloud taken from their paper. For comparison, a similar map of the 100 μm emission from the Rosette Molecular Cloud is shown in Figure 3a, and the molecular cloud is shown in Figure 3b. Both maps have the zodiacal emission removed and have had a background subtracted. Remarkably, the highest contour in Figure 2a is lower than the *lowest* contour in Figure 3a. The average 100 μm emissivity for the Maddalena-Thaddeus cloud is more than *two orders of magnitude* lower than that of the Rosette Molecular Cloud. Since the far infrared emission is generally thought to come from reradiated starlight, by comparison with the Rosette, which has an infrared emissivity typical of GMCs in the solar vicinity (Boulanger and Perault 1988), the Maddalena-Thaddeus cloud is extremely deficient in embedded stars.

This evidence supports the hypothesis that the Maddalena-Thaddeus cloud is so young that it has not yet had time to form stars. It is therefore an excellent candidate to examine the differences in structure between it and a more evolved cloud like the Rosette Cloud, or the Orion Molecular Cloud. But if the Maddalena-Thaddeus cloud is too young to have yet formed stars, and the Rosette and Orion clouds are middle aged clouds still in the throes of star formation, is it possible to find a demonstrably old cloud, one which is now showing only the last vestiges of star formation? The answer appears to be yes. Probably the best candidate for a remnant molecular cloud is the small cloud associated with the Lac OB1 association. Although no systematic study of the molecular gas in this region has yet been undertaken, the overall morphology can be seen in Figure 4 which shows an IRAS map of the region at a wavelength of 100 μm .

The Lac OB1 association is one of the oldest OB associations in the solar vicinity with an age of $\sim 2 \times 10^7$ y (Blaauw 1964). There are two subassociations, the younger of which has an age of $\sim 6 \times 10^6$ y. Very sparsely sampled observations (Blitz, unpublished) indicate that there is almost no molecular gas remaining in the region; what there is appears to be concentrated in the knot centered at $\alpha = 22^\circ 30'$, $\delta = 40^\circ\!.5$. This knot is the location of the reflection nebula DG187. The stars, on the other hand, are spread throughout the area of the map. At the 500

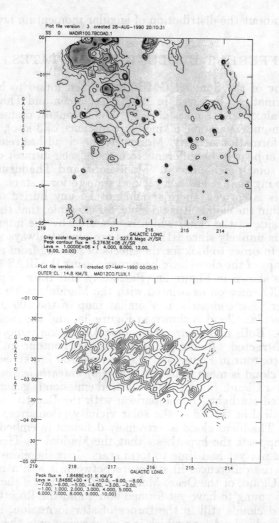

Fig. 2. Figures 2 and 3 show IRAS 100 μm emission and CO emission from two GMCs in apparently different evolutionary states. **2a.** IRAS 100 μm emission from the region of the Maddalena-Thaddeus GMC. The molecular cloud is located in the range $-1°45' < b < -3°30'$ and $214° > l > 219°$. There is no IRAS flux detectable in the figure that can be definitely associated with the molecular cloud. The largest flux detected in the direction of the cloud is about 20 MJy/sr. **2b.** CO emission integrated over all velocities associated with the Maddalena-Thaddeus Cloud. Figures 2a and 2b are on very nearly the same angular scale. Note the almost complete absence of IR emission from the region where the CO emission is detected.

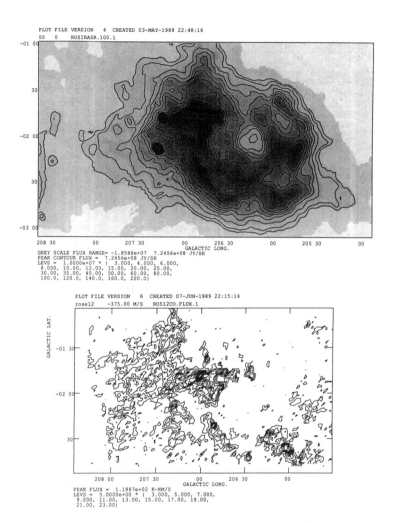

Fig. 3. **3a.** IRAS 100 μm emission associated with the Rosette Molecular Cloud. Note that the *lowest* contour in the emission is 50% higher than the *highest* contour in Figure 2a. The difference in the mean emission from the two clouds is at least two orders of magnitude. The optical nebula is centered on the hole in the infrared emission at $l=206°\!.2$ $b=-2°$. **3b.** Map of the CO emission integrated over all velocities associated with the Rosette Molecular Cloud from data taken with the Bell Labs 7 m antenna. The angular scale is very nearly the same as that shown in Figure 3a (From Blitz, Stark and Long (1991).

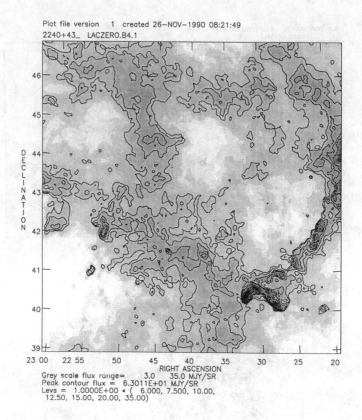

Fig. 4. IRAS 100 μm emission fom the region of Lac OB1. The only location from which CO has been detected to date is the small knot at $\alpha = 22^h 30^m \delta = 40°.5$ which is associated with the reflection nebula DG187. This figure was produced by Eugene de Geus.

pc distance of the OB association, the remnant cloud appears to be no more than 10 pc in extent. The large loop seen in the figure is probably dust associated with a shell of atomic hydrogen, and is evidently the result of the stellar winds, and possibly supernovae from the stellar association, which, together with the ionizing radiation from the O stars, appear to have effectively destroyed the molecular cloud. Very little is known about the overall gas and dust content of the region, but it presents a good opportunity to study what appears to be the last gasp of the star formation process. Studies of such regions promise to explain how the gas and dust in a molecular cloud that has not been converted into stars is ultimately returned to the interstellar medium.

6. RELATIONSHIP TO ATOMIC HYDROGEN

The relatively sharp boundaries of GMCs are inferred from the quantitative analysis of Blitz and Thaddeus (1980), and the appearance of many GMCs on the Palomar Observatory Sky Survey prints. In the latter case, a well defined region of dust obscuration is observed for many GMCs, and these follow the outermost contours of the CO emission quite closely. This is especially true for GMCs within 1 kpc of the Sun where the contrast between the foreground dust obscuration and the background stars is the highest. Good examples are the Orion Molecular Cloud (especially the boundary at the lowest galactic latitudes - see the CO maps of Kutner et al. 1977 and the ^{13}CO maps of Bally et al. 1987b), the Mon OB1 molecular clouds (see Blitz 1980), and the Ophiuchus molecular clouds (see the CO maps made by deGeus 1988 and Loren 1989). There are many other examples that illustrate this point.

These sharp boundaries suggest that there is a phase transition that takes place at the boundaries of the clouds, but what then is the state of the gas at the low density side of the phase transition of the GMC? Wannier, Lichten and Morris (1983) have shown that for a few GMCs, there is a thin layer of atomic hydrogen in a transition zone probably associated with a photodissiation region. In a larger scale unpublished analysis of the atomic clouds associated with local GMCs, Blitz and Terndrup have analyzed the HI emission from the velocity range detected in CO for a number of GMCs in the solar vicinity. One such map was shown in Blitz (1987a). Figure 5 is another such map; it shows the HI column density associated with the molecular cloud accompanying the Per OB2 association (this is the cloud that contains the NGC 1333 star forming region). The map contours are in units of 10^{20} cm^{-2}; the contour marked 10 therefore is associated with an extinction (A_v) of about 0.5 mag. The map itself shows the HI emission integrated over the velocity range 4.2 - 12.7 km s^{-1}, the velocity range associated with the Per OB2 cloud, which is shown as the shaded area in the figure. The association of the atomic gas with the molecular cloud is quite obvious. All of the GMCs surveyed to date show atomic envelopes similar to that shown in Figure 5.

In order to quantify the relationship between the atomic and molecular gas, Blitz and Terndrup estimated the mass of atomic gas associated with the molecular clouds they studied by estimating out to what distance the atomic gas shows an excess over the background in the relevant velocity range. These masses are then plotted as a function of M(H$_2$) derived from the CO maps using the CO/H$_2$ conversion ratio of Bloemen et al. (1986). The results are plotted in Figure 6. The error bars are from estimates of the uncertainty in defining the background level for the atomic clouds, and in determining the CO/H$_2$ conversion ratio for an individual molecular cloud. What the figure clearly shows is that the molecular clouds have atomic envelopes that are as massive as the molecular clouds in most cases. Note, however, that the atomic clouds are much more distended; the HI masses pertain

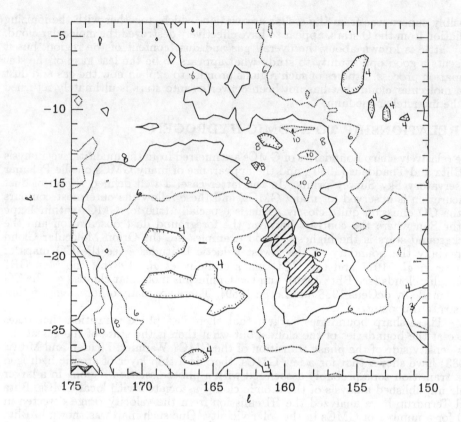

Fig. 5. Map of the atomic hydrogen emission associated with the Per OB2 molecular cloud. The emission is integrated over the velocity range 4.2-12.7 km s^{-1}, and the contours are in units of 10^{20} cm^{-2}. The molecular cloud is shown as the shaded region. There is clearly enhanced emission in the vicinity of the molecular cloud. This map is typical of all of the GMCs in the solar vicinity.

to the entire region in which HI is seen above the background. The individual maps show a very small range in the peak column density of the atomic gas associated with the molecular clouds. That is, only $1 - 2 \times 10^{20}$ cm^{-2}, or an A_v of 0.5 - 1.0 mag. A similar result has been found for diffuse molecular clouds in the Milky Way (Savage, *et al.* 1977), and for a GMC in M31 (Lada *et al.* 1988).

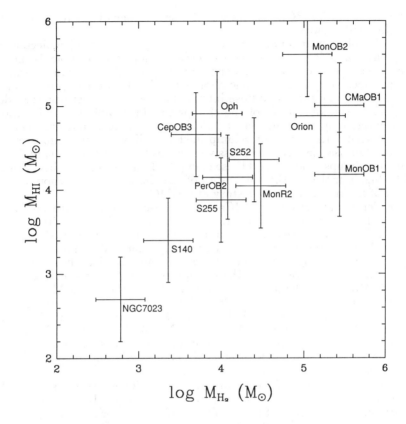

Fig. 6. Plot of the mass of atomic hydrogen emission associated with a number of GMCs in the solar vicinity. The atomic hydrogen mass associated with the GMC is taken over a narrow velocity range centered on the molecular cloud velocity and the molecular mass is taken from the CO emission integated over the area of the molecular cloud.

7. THE CLUMPY STRUCTURE OF GMCs

As early as 1980, Blitz and Shu noted that all GMCs that have been observed up until that time have exhibited clumpy structure. The evidence at the time suggested that the density contrast between the clumps and the interclump medium is large because the densities required to detect CO at the observed antenna temperatures are an order of magnitude greater than the mean densities inferred if the clouds have dimensions along the line of sight comparable to their dimensions in the plane of the sky and the CO is uniformly distributed within it. Thus, if the H_2 is clumped so that the volume filling fraction is ~ 0.1 or less, implying a clump/interclump density contrast $\gtrsim 10$, the discrepancy disappears. This conclusion was qualitatively confirmed from partial ^{13}CO mapping of the dense ridge of the Rosette Molecular Cloud (Blitz and Thaddeus 1980).

In principle, an understanding of the clumpy structure of molecular clouds holds a great deal of information about their evolution. For example, is the clumpy structure a result of the process of star formation, or is the clumpiness primordial? If the clumpiness is primordial, one should be able to see the process of clump formation taking place in the diffuse interstellar medium. In either case, detailed observations of the kinematics of the clumps should be able to resolve the issue. Does the interclump medium play a role in the structure and evolution of the GMCs? In what form is the interclump medium? How do the clumps eventually form stars? OB associations and massive star clusters must eventually form from large massive clumps. Can we find these clumps? Because collisions between clumps should be highly inelastic, the kinematics of the clump ensemble should provide a great deal of information on the history of a GMC.

The density structure of GMCs is best seen in a pervasive optically thin molecule. The best current candidate is ^{13}CO, but receivers sensitive enough to observe ^{13}CO over the large angular extents needed to fully map a GMC have been available only since the early 1980s. Thus, fully quantitative information on the density structure of GMCs has not become available until fairly recently. Large scale ^{13}CO maps which delineate the clumpiness of GMCs are now available for the Rosette Molecular Cloud (Blitz and Stark 1986), the Orion Molecular Cloud (Bally, et al. 1987b), parts of the Cep OB3 molecular cloud (Carr 1987), parts of the Ophiuchus molecular cloud (Loren 1989; Nozawa, et al. 1991), and some anonymous CO clouds toward $l=90°$, $b=3°$ (Perault, Falgarone and Puget 1985). Extensive observations of the clumpiness of molecular clouds also have been made in the Orion B molecular cloud using CS (Lada 1989) and in part of the M17 molecular cloud using $C^{18}O$ (Stutzki and Güsten 1990).

We review here some of the results from the ^{13}CO observations of the Rosette Molecular Cloud to be submitted for publication shortly (Blitz, Stark and Long in preparation) and compare them to other published work on clumpiness. Blitz and Stark (1986) showed that the clumpy structure of the Rosette Molecular Cloud becomes especially clear when analyzed using position-velocity diagrams. Such plots are superior to channel maps (maps made in two spatial dimensions at a particular velocity interval) because small differences in clump velocities are not as apparent in the channel maps. Because the velocity differences between clumps are frequently smaller than the line width of the CO emission from an individual clump, channel maps tend to blend the emission from several adjacent clumps into one large entity. On the other hand, because the velocity resolution of the position-velocity maps is generally high enough that the individual line profiles and the velocity differences between clumps is easily seen, the identification of separate kinematic units is easier to accomplish.

For the Rosette Molecular Cloud, analysis by means of position-velocity maps has made it possible to produce a catalogue of individual clumps because in that cloud the separation between the clumps is large enough to make the identifications of individual objects possible by eye. A similar procedure has been carried out by Loren (1989) for the streamers in Ophiuchus, and by Carr (1987) for a piece of the Cep OB3 molecular cloud. What makes the Rosette study different is that enough of the molecular cloud has been observed to make conclusions about the structure of the cloud as a whole. It is unclear, however, to what degree the procedure of identifying clumps by eye will work for other GMCs. This is simply another way of saying that it is unclear at present to what degree the structure of the Rosette Molecular Cloud is typical of GMCs as a whole. It should be pointed out, however, that in terms of its size and mass, the Rosette Molecular Cloud is quite typical of other GMCs in the solar vicinity.

Blitz, Stark and Long reach the following conclusions from an analysis of their data:

1) Between 60% and 90% of the H_2 mass resides in the clumps they have identified in their catalogue. Therefore, most of the molecular mass is in the clumps and is not in a distributed component.

2) The mean H_2 density for all of the clumps is $\sim 1 \times 10^3$ cm^{-3}. The volume averaged density for the entire complex is ~ 25 cm^{-3} (Blitz and Thaddeus 1980); therefore the volume filling fraction of the clumps is $\sim 2.5\%$. A similarly low volume filling fraction of 3% is derived for the G90+03 molecular cloud complex observed by Perault, Falgarone and Puget (1985). If we assume that 10 - 50% of the H_2 mass is in a diffuse component not related to the clumps, then the interclump density is ~ 2.5 - 12.5 cm^{-3}. The existence of the HI envelopes suggests that the interclump medium in GMCs also contains an atomic component, but because the envelopes are much more extended than the molecular clouds, the average HI density within the volume of the molecular cloud is unlikely to be more than about ~ 15 cm^{-3}. Thus, the clumps are not small density enhancements in a relatively smooth substrate, but are dense blobs moving through a tenuous interclump medium which are held together by their mutual gravitational attraction.

It should be noted that although the ^{13}CO observations provide the best quantitative estimates of the density contrast of the clumps, important confirmation of this contrast comes from the large spatial extent of the CII/CI regions around O stars (Stützki et al. 1988; Genzel 1990). That is, the large penetration of the CII emisison regions into molecular clouds requires a clump/interclump density ratio of 10 to 100, similar to what is found from ^{13}CO mapping of the Rosette and G90+03. Thus, very different kinds of observations of widely separated GMCs reach similar conclusions about the density contrast of their clumps.

An interesting question arises as to whether the clumps themselves have internal structure. Because the clumps have a different mass spectrum than the Salpeter or Miller-Scalo Initial Mass Function (see below), the implication is that the clumps in which the star formation takes place must be fragmented on still smaller scales. Direct evidence for this comes from the $C^{18}O$ mapping of two clumps in G90+03 by Perault, Falgarone and Puget (1985), and from unpublished maps of ^{13}CO, $C^{18}O$, and CS of a large clump in the Rosette Molecular Cloud (Blitz and Puchalsky in preparation). However, to date, this subclumping has not been well characterized quantitatively. Presumably it is the subclumping within a GMC, that is, the fragmentation of the individual clumps, that is the direct precursor of individual stars (see Section 9).

3) The distribution of clump mass follows a power law such that

$$dN(m) = N_0 M^{-1.54} dM$$

where $dN(m)/dM$ is the number of clouds per solar mass interval, and N_0 is 460 $M_\odot^{-1}$.

There are a number of studies that have been done to date which indicate that the clump mass spectrum is a power law and that the power law exponent is similar to that obtained for the Rosette Molecular Cloud. The observations are summarized in Table 4 below.

TABLE 4. Clump Mass Spectrum Power Law

Exponent	Tracer	Source	Reference
-1.5	^{13}CO	Rosette	Blitz et al. (1991)
-1.6	CS	Orion B	Lada (1989)
-1.4	^{13}CO	Cep OB3	Carr (1987)
-1.7	$C^{18}O$	M17SW	Stutzki and Güsten (1990)
-1.7	^{13}CO	Opiuchus	Loren (1989)
-1.7	^{13}CO	Ohiuchus	Nozawa et al. (1991)

The Cep OB3 spectrum was calculated from the original unpublished list kindly provided by John Carr and does not include the lowest mass bin, which is likely to suffer from incompleteness. The Loren value, quoted in the original paper as -1.1, was recomputed from the data presented in his paper.

That all five studies should obtain such a similar power law is even more remarkable than the similarity of the power law exponent in the linewidth-size relation for GMCs given in Section 2. The data for the different clouds are sensitive to different ranges of mass and density, and use different reduction and analysis techniques. Furthermore, unlike the data used to get the molecular cloud distribution, all of the measurements are determined from fundamentally different sources. Given the uncertainties in the determinations of the slopes, it is reasonable to conclude that for the molecular clouds that are well studied to date, there is a universal mass spectrum for the clumps within a GMC, and that the spectrum is a power law with an exponent of -1.6 with an uncertainty of $\sim$ 10% and that the power law seems to hold over three orders of magnitude in mass, *i.e.* from about 1 $M_\odot$ to about 3000 $M_\odot$. A reasonable inference is that the processes that determine the distribution of clump masses are rather similar from cloud to cloud. Clouds that show significantly different values of the mass spectrum are then likely to have had different dynamical histories.

It is noteworthy that none of the power law exponents resemble the initial mass function, however. This suggests that the clumps that are observed in these studies are not the ones that form individual stars. Unless the clumps somehow have a star formation efficiency that varies with the mass of the clump (Zinnecker 1989), the formation of individual stars lies deeper within the clumps that have so far been identified. Some of the larger clumps have sufficient mass to form entire clusters or OB associations, so this result is not entirely surprising. At some level, one expects to find the IMF mirrored in the spectrum of condensations within a GMC, but it appears that these condensations will be identified as the substructure within individual clumps. This concept has meaning only if it is possible to identify a scale or a parameter that can differentiate the clumps within a complex from the structure within the clumps, but this has not yet been accomplished. It is likely that high resolution interferometric observations will be needed to observe the true star forming condensations in most GMCs.

4) Although most of the mass in the Rosette Molecular Cloud (RMC) is in clumps that are at or near the point of being gravitationally bound, most of the clumps are not gravitationally bound. That is, half the mass is in the 10 most massive clumps of the 86 catalogued; however, the 40 or so lowest mass clumps do

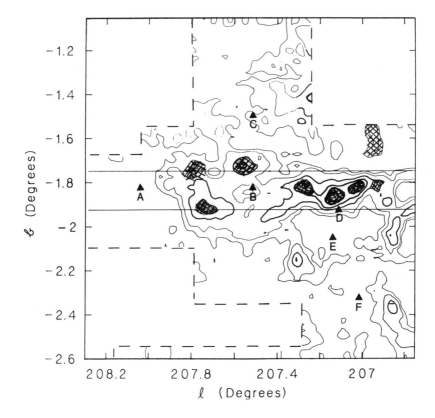

Fig. 7. Map of the ^{13}CO emission from the Rosette Molecular Cloud taken from Blitz and Stark (1986). The shaded areas are the locations of the eight most massive clumps associated with the molecular cloud complex.

not appear to be gravitationally bound. A similar conclusion was reached by Carr (1987) and Loren (1989) who found that virtually none of the clumps in the clouds they observed are gravitationally bound. This point is discussed in greater detail in the following section.

5) The most massive clumps in the RMC lie close to the midplane. This can be seen from Figure 7 which shows the eight most massive clumps shaded on the map of the velocity integrated ^{13}CO emission published by Blitz and Stark. These clumps as a group are the most gravitationally bound of all of the clumps as measured by the ratio of their gravitational to internal kinetic energy. Thus, the next cluster of stars to form in the cloud complex is most likely to form from one of these clumps.

6) The clumps show a strong velocity segregation with mass. Figure 8 is

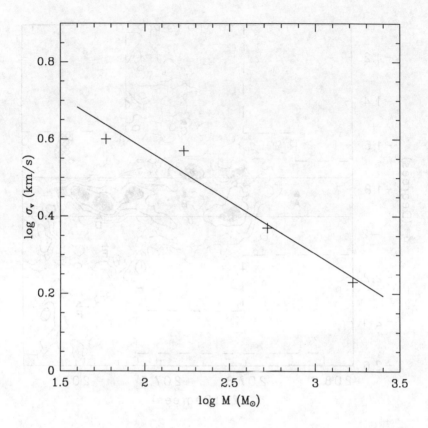

Fig. 8. Plot of the clump-to-clump velocity dispersion as a function of mass in the Rosette Molecular Cloud. The straight line is a least squares fit to the date and has a power law index of -0.27.

a plot of the clump-to-clump velocity dispersion as a function of mass for the 86 catalogued clumps. The most massive clumps, the ones that lie closest to the midplane of the complex also show the smallest velocity dispersion. *These results strongly suggest that the clumps have undergone dynamical evolution in the time since the complex has formed.* That is, the clump maps and the clump kinematics show a clear signature of inelastic collisions that indicate that the cloud has evolved considerably since it first formed. *Because the observations were made in a part of the complex that has not been affected by the Rosette Nebula itself, and because there is little evidence for star formation in the mapped area, it appears that the structure and kinematics that are observed is the result of the dynamics of the clumps themselves without the intervention of energetic phenomena associated with star formation (e.g. HII regions, stellar winds, protostellar outflows, supernova explosions).*

7) There is a strong correlation between the embedded IRAS point sources

and the clumps identified in ^{13}CO (Cox et al. 1991). The brightest infrared sources are associated with the most massive clumps. In the RMC, most of the sources have flux densities that are strongly increasing functions of wavelength suggesting that they are very young. Except for GL-961, none of the sources has been investigated in any detail and it is unclear whether the infrared sources represent embedded objects or heating of the dust by collapse that is taking place in the clumps.

8. BOUNDEDNESS OF THE CLUMPS

In the Blitz, Stark and Long (1991) study of the clumps associated with the RMC, at least half of the clumps are not gravitationally bound. The precise fraction is difficult to assess because of the uncertainties in the luminous masses derived from the ^{13}CO line strengths. Carr (1987) and Loren (1989) find that *none* of the clumps in the clouds they observed are gravitationally bound. The latter two authors conclude that because the clumps are so far from being gravitationally bound that the clumps must be expanding. This conclusion is however difficult to accept. First, the clumps would be expanding on the scale of a crossing time which, for the clumps they measured is less than 10^6 y. The clumps would all be dissolving unless there were also clumps that were forming at the same rate, but such clumps have not been identified. Second, unless all three clouds were observed at some special epoch, then mass conservation would require that

$$\frac{n(clump)}{n(interclump)} = \frac{t(dissolve)}{t(form)}.$$

Direct evidence from the RMC indicates that the right hand side of the equation is of the order of 100 or more, which requires that the formation timescale is no more than about 10^4 y if the structure of the RMC and the Ophiuchus clouds is similar. This improbably short time would seem to require that the interclump medium would be filled with stars that could sweep the interclump gas together quickly to form the clumps as quickly as they dissolve, a requirement which does not seem to be supported by observation.

On the other hand, the clumps could be bound together by the pressure of the interclump medium. Such a medium would be in hydrostatic equilibrium with the gravitational potential of the GMC. Let us calculate what is required and compare it to the observations of the Rosette GMC. The pressure inside a clump is given by:

$$\frac{P(clump)}{k} = \frac{nm(H_2)<v_{1D}^2>}{k}$$

where we obtain from observations of the RMC the following mean values,

$$n = 500 - 1000 \ cm^{-3}$$

$$<v_{1D}^2>^{0.5} = 0.70 \ kms^{-1}$$

$$P/k = 6 - 12 \times 10^4 \ Kcm^{-3}$$

This pressure is a kinetic pressure determined from the linewidth of a typical clump. The thermal pressure, given by nT, is generally less than the value quoted above away from regions of active star formation; typical excitation temperatures are 5-15 K as derived from the optically thick CO line. Therefore, the linewidth and pressures within a clump are generally dominated by non-thermal bulk motions of the gas. The mean pressure within a GMC due to its self-gravity is given by:

$$\frac{P_{grav}}{k} = \frac{2GM<n>m(H_2)}{3Rk}$$

where M and R are the mass and radius of the GMC and $<n>$ is the volume averaged density of the complex. Again for the RMC we have:

$$<n> = 23 \ cm^{-3}$$

$$M = 1.1 \times 10^5 \ M_\odot$$

$$R = 23 \ pc$$

$$\frac{P_{grav}}{k} = 8 \times 10^4 \ Kcm^{-3}$$

where R is found by taking the square root of the projected surface area.

Thus, the pressure that the interclump gas would have if it were in pressure equilibrium with the self-gravity of the complex is equal, within the uncertainties, to the pressure needed to confine the clumps. It was argued in Section 6 above that the atomic hydrogen is very likely to be the interclump gas. Indeed, detailed observations of the atomic hydrogen associated with the Rosette (Bania and Kuchar, in preparation) made with the Arecibo telescope suggest that there is a detailed anticorrelation between CO and HI maxima in the cloud, lending support to the idea that warm HI is likely to be the interclump gas. The pressure of this gas can be obtained from published observations of the HI associated with the RMC at low resolution made by Raimond (1966). These are as follows:

$$\frac{P(HI)}{k} = \frac{nm(HI)<v_{1D}^2>}{k}$$

where n in this case is the mean HI interclump density estimated from both the Raimond (1966) and the Bania and Kuchar observations.

$$<n> = 7 \ cm^{-3}$$

$$<v_{1D}^2>^{0.5} = 10.6 \ kms^{-1}$$

$$\frac{P(HI)}{k} = 10 \times 10^4 \ Kcm^{-3}$$

Thus it seems quite reasonable that the clumps are confined by the pressure of the interstellar gas.

But why is it then, that *all* of the clumps in the Ophiuchus and Cep OB3 observations are found not to be bound by gravity, whereas many of the clumps in the RMC are self-gravitating? The answer is the differing linear resolutions of the observations. In both the Loren and Carr observations, the clouds are considerably nearer than the Rosette; for the Ophiuchus observations, the difference is an order of magnitude. With one exception, the most massive clumps in the Loren and Carr samples are all within the mass range for which only unbound clumps are found in the Rosette. Their observations are therefore quite compatible with the Rosette observations. In fact, the three sets of observations can be combined to conclude that there seems to be a mass below which the clumps are not gravitationally bound, and above which the clumps are bound. This mass appears to be $\sim 300-500$ $M_\odot$. For small masses such as dense cores (*e.g.* Myers and Benson 1983) to be gravitationally bound, apparently clumps must be much denser than the clumps observed in ^{13}CO.

9. STAR FORMATION IN THE CLUMPS

If the clump mass spectrum suggests that the clumps (at least those that have been identified in studies made to date), are not the progenitors of individual stars, then what is the role of the clumps in the star formation process? Since Roberts (1957) first showed that most stars form in clusters and associations, a result that was later confirmed by Miller and Scalo (1978), it seems natural to suppose that the more massive clumps give rise to these stellar groups. The masses of the largest clumps in the RMC, for example, are comparable to the masses of OB associations (assuming that they form with a normal IMF) and open clusters.

A significant advance in understanding the star formation in GMCs was recently made by Lada (1989) who examined the Orion B molecular cloud for CS clumps and also did a near infrared survey to look for embedded stellar objects. She was able to identify 39 dense clumps of molecular gas in her survey; of these nearly all the star formation is limited to the three most massive clumps. These three clumps account for only 30% of the dense clumps and $\leq$ 8% of the total gas mass in the area she surveyed. Yet when account is taken of background sources, she finds that $\sim$ 96% of the of the star formation in the surveyed region is associated with the dense clumps. This work is discussed in more detail elsewhere in this volume.

The conclusion seems to be that the clumps, especially massive, dense clumps are the units of star formation in GMCs. It will be important to extend Lada's work to other clouds to see to what degree her results can be generalized. But the evidence seems to indicate that regions of incipient star formation can be identified from the molecular observations. The key to understanding how molecular clouds form stars now seems to require detailed investigations into the structure of the dense clumps with an eye to understanding how density inhomogeneities develop in these objects and what is the cause of their internal motions.

10. THE EVOLUTION OF GIANT MOLECULAR CLOUDS

The observations cited in this review, taken together, suggest a tentative picture of the evolution of a GMC. We imagine first, that by some process, the interstellar medium collects enough mass together to form a GMC. Two methods by which this can in principle be done is through spiral arm shocks (Roberts 1969; Cowie 1981; Elmegreen 1982), and through "supershells" driven by multiple supernovae (Heiles 1979, McCray and Kafatos 1987). Evidence that some other method is also at work comes from observations of other galaxies which suggest that the star formation rate is not strongly dependent on spiral arm morphology; even galaxies without coherent spiral arms form stars quite efficiently (Stark, Elmegreen and Chance 1987). In any event, the dominant process by which the diffuse interstellar medium is collected into clouds in the solar vicinity has not been identified.

As enough material is collected together to form a cloud, when the column density of atomic hydrogen exceeds a threshold of about 10^{21} cm^{-2}, the associated visual extinction of about 0.5 mag is sufficient to shield the gas from the UV radiation that dissociates the molecules that form, and the cloud begins to turn molecular. It is presumably early in the process of turning molecular that the clumps have begun to form. Evidence for this comes not only from the Rosette Molecular Cloud, where the kinematics suggest that clumpiness precedes the star formation process and then evolves, but from observations of high latitude molecular clouds very close to the Sun. Many of these clouds appear to be very young ($\leq 2 \times 10^6 y$), and all show evidence of clumpiness that must surely be primordial (Magnani 1986). Furthermore, the properties of these clouds are quite similar to the gravitationally unbound clumps in the RMC, Cep OB3, Ophiuchus and M17.

It is not necessary for the interstellar medium to produce proto-GMCs that are gravitationally bound. A gravitationally neutral cloud can become gravitationally bound, because the clumps that have formed will collide, and the kinetic energy of the ensemble of clumps can be radiated away because the clump collisions are so inelastic. The details of such a process are amenable to numerical modeling. If a GMC forms in this way, then it is natural that the atomic gas from which it formed remains as the interclump gas. The HI envelope that remains in this case is largely the remnant of the primordial GMC. All of the gas will remain in pressure equilibrium, and as the cloud becomes gravitationally bound, the interclump medium will respond to the gravitational potential well of the GMC. The pressure of the interclump medium will therefore increase above the nominal interstellar value, putting the clumps within the cloud under higher pressure than they had initially.

The magnetic field can support a clump against gravity up to a certain critical mass (*e.g.* Mouschovias 1987; McKee 1989). Once a clump becomes magnetically supercritical, then even the magnetic field cannot prevent collapse (see *e.g.* McKee 1989). Thus star formation will take place in the clumps that have grown to be the largest and densest though collisions. Some support of the clumps themselves can also come from the energy input from protostellar and neostellar winds (see *e.g.* Margulis, Lada and Snell 1988). It may very well be that a process of self-regulating star formation similar to that described by McKee (1989) may then take place, until the cloud is ultimately consumed from the dissociating effects of the HII regions, stellar winds, and supernova remnants of the stars that formed within it.

Note, however, that the process of molecular cloud evolution described above has no need for sequential star formation. This is not to say that sequential star formation is unimportant, but that some of the difficulties with the original theory of Elmegreen and Lada (1977) can be avoided with the evolutionary scheme outlined above. For example, Elmegreen and Lada noted that there is a need for an initial trigger to generate the first subgroup of O stars in their theory. Furthermore, the theory is based on the assumption that an ionization front propagates into an initially *homogeneous* molecular cloud; as discussed above, such clouds are not observed. Furthermore, in some cases, young OB subgroups have been identified that lie *between* older subgroups; as for example in Cep OB3 (Sargent 1977), and in the Sco-Cen OB association (Blaauw 1964; deGeus 1988). In the outline of GMC evolution described here, star formation proceeds in a quasi-random way within a molecular cloud. That is, the collisions between clumps in a molecular cloud proceed until one of the clumps becomes unstable to the process of formation of an OB subgroup. The stellar activity will ionize the molecular gas in its vicinity, and in a time probably less than a million years (depending on a number of variables), this subgroup and its attendant HII region will appear to be at the edge of the molecular cloud. The next large clump to form an OB association will then be determined by the collisional processes between it and the other clumps in the complex, and will be independent of the previous star formation history of the cloud. In this way, there is no need to postulate a special event to produce the first generation of stars, and the apparently random orientation of the subgroups associated with many GMCs appears quite naturally. Cases like the ordered orientation of the subgroups of the Orion complex therefore would be the result of chance (and are not very unlikely).

The degree to which this evolutionary picture is true will depend on the results obtained from the ^{13}CO mapping of a number of GMC complexes, and the identification of more complexes in different evolutionary states and the identification of embedded cluster of stars within the complexes. Once the process of the formation and evolution of GMCs is better understood in the Milky Way, we can then apply our understanding to other galaxies where the environmental conditions are markedly different.

ACKNOWLEDGEMENTS

This work was partially supported by NSF grant AST-8918912, and funding from the state of Maryland to the Laboratory for Millimeter-wave Astronomy. I wish to thank Charlie Lada, Jim Pringle, Lyman Spitzer, and John Williams for a critical reading of the manuscript, and Charles Gammie for a stimulating conversation.

REFERENCES

Adler, D., 1988, Ph.D. Dissertation, University of Virginia.
Altenhoff, W., Downes, D., Goad, L., Maxwell, A., and Rinehart, R., 1970, *Astr. Ap. Suppl.*, **1**, 319.
Altenhoff, W., Downes, D., Pauls, T., and Schraml, J., 1978, *Astr. Ap. Suppl.*, **35**, 23.
Bally, J., Stark, A.A., Wilson, R.W., and Henkel, C., 1987a, *Ap. J. Suppl.*, **65**, 13.
Bally, J., Langer, W.D., Stark, A.A., and Wilson, R.W., 1987b, *Ap. J. (Letters)*, **312**, L45.
Bash, F.N., Green, E., and Peters, W.L., 1977, *Ap. J.*, **217**, 464.
Blaauw, A., 1964, *Ann. Rev. Astron. Ap.*, **2**, 213.
Blitz, L., 1978, Ph.D. Dissertation, Columbia University.
Blitz, L., 1980, in *Giant Molecular Clouds in the Galaxy*, Solomon and Edmunds, eds., Pergammon:Oxford, p.1.
Blitz, L., 1987a, in *Millimetre and Submillimetre Astronomy*, Wolstencroft and Burton, eds., (Kluwer:Dordrecht), p.269.
Blitz, L., 1987b, in *Physical Processes in Interstellar Clouds*, Morfill and Scholer, eds., (Reidel:Dordrecht), p.35.
Blitz, L. and Shu, F.H., 1980, *Ap. J.*, **238**, 148.
Blitz, L. and Thaddeus, P., 1980, *Ap. J.*, **241**, 676.
Blitz, L. and Stark, A.A., 1986, *Ap. J. (Letters)*, **300**, L89.
Blitz, L, Bazell, D., and Desert, F.X., 1989, *Ap. J. (Letters)*, **352**, L13.

Blitz, L. Stark, A.A., and Long, K., 1991, in preparation.
Bloemen, J.B.G.M., et al., 1986, Astron. Ap., **154**, 25.
Bottinelli, L., and Gouguenheim, L., 1964, Ann. d'Ap., **27**, 685.
Boulanger, F., and Perault, M., 1988, Ap. J., , **330**, 964.
Carr, J.S., 1987, Ap. J., , **323**, 170.
Cohen, R.S., Cong, H-I., Dame, T.M., and Thaddeus, P., 1980, Ap. J. (Letters), **239**, L53.
Clemens, D.P., 1985, Ap. J., **295**, 402.
Cowie, L.L., 1981, Ap. J., **245**, 66.
Cox, P., Deharveng, L., and Leene, A., 1991, Astr. Ap, in press.
Dame, T.M., Elmegreen, B.G., Cohen, R.S., and Thaddeus, P., 1986, Ap. J., **305**, 892.
deGeus, E., 1988, Ph.D. Dissertaiton, Leiden University.
Digel, S., Bally, J., and Thaddeus, P., 1990, Ap. J. (Letters), **357**, L29.
Elmegreen, B.G., and Lada, C.J., 1977, Ap. J., **214**, 725.
Elmegreen, B.G., and Lada, C.J. and Dickinson, D.F., 1979, Ap. J., **230**, 415.
Elmegreen, B.G., 1982, Ap. J., **253**, 655.
Falgarone, E., and Lequeux, J., 1973 Astron. Ap., **25**, 253.
Genzel, R., 1991, in *Molecular Clouds*, R. James, and T. Millar, eds., Cambridge University Press: Cambridge, in press.
Heiles, C. 1979, Ap. J., **229**, 533.
Heiles, C., 1988, Ap. J., **324**, 321.
Henderson, A.P., Jackson, P.D., and Kerr, F.J., 1982, Ap. J., **263**, 116.
Issa, M., MacLaren, I., and Wolfendale. A.W., 1990, Ap. J., **352**, 132.
Kutner, M.L., Tucker, K.D., Chin, G., and Thaddeus, P., 1977, Ap. J., **215**, 521.
Lacey, C.G., and Fall, S.M., 1985, Ap. J., **290**, 154.
Lada, C.J., Margulis, M., Sofue, Y., Nakai, M., and Handa, T., 1988, Ap. J., **328**, 143.
Lada, E.A., 1990, Ph.D. Dissertation, University of Texas.
Leisewitz, D., Bash, F.N., and Thaddeus, P., 1989, Ap. J. Suppl., **70**, 731.
Liszt, H.S., Burton, W.B., and Xiang, D.,1981, Astron. Ap., **140**, 303.
Liszt, H.S., and Burton, W.B., 1983, in *Kinematics, Dynamics, and Structure of the Milky Way*, W.L. Shuter, ed., (Reidel:Dordrecht), p.135.
Lockman, J., 1989, Ap. J. Suppl., **71**, 469.
Lockman, J., 1990 in *Radio Recombination Lines: 25 years of Investigation*, Gordon and Sorochenko, eds., Kluwer: Dordrecht, p.225.
Loren, R.B., 1989, Ap. J., **338**, 902.
Maddalena, R., and Thaddeus, P., 1985, Ap. J., **294**, 231.
Magnani, L., 1986, Ph.D. Dissertaion, University of Maryland.
Magnani, L., Blitz, L., and Mundy, L., 1985, Ap. J., **295**, 402.
Margulis, M., Lada, C.J., and Snell, R.L., 1988, Ap. J., **333**, 316.
McCray, R., and Kafatos, M., 1987, Ap. J., **317**, 190.
McKee, C.F., 1989, Ap. J., **345**, 782.
Mead, K.M., and Kutner, M.L., 1988, Ap. J., **330**, 399.
Mestel, L., 1966, M.N.R.A.S., **131**, 307.
Miller, G.E., and Scalo, J.M., 1978, Pub. A.S.P., **90**, 506.
Mooney, T.J., and Solomon, P.M., 1988, Ap. J. (Letters), **334**, L51.
Mouschovias, T., 1985, Astron. Ap., **142**, 41.
Mouschovias, T., 1987, in *Physical Processes in Interstellar Clouds*, Morfill and Scholer, eds., (Reidel:Dordrecht), p.453.

Mouschovias, T., Shu, F.H., and Woodward, 1974, *Astr. Ap*, , **33**, 73.
Myers, P.C., and Benson, P.J., 1983, *Ap. J.*, **266**, 309.
Myers, P.C., Dame, T.M., Thaddeus, P., Cohen, R.S., Silverberg, R.F., Dwek, E., and Hauser, M.G., 1986, *Ap. J.*, **301**, 398.
Nozawa, S., Mizuno, A., Teshima, Y., Ogawa, A., and Fukui, Y., 1991, *Ap. J. Suppl.*, in press.
Perault, M., Falgarone, E., and Puget, J.L., 1985, *Astron. Ap.*, **152**, 371.
Raimond, E., 1966, *B.A.N.*, **18**, 191.
Roberts, M.S., 1957, *Pub. A.S.P.*, **69**, 59.
Sanders, D.B., Scoville, N.Z., and Solomon, P.M., 1985, *Ap. J.*, **289**, 323.
Sargent, A.I., 1977, *Ap. J.*, **218**, 736.
Savage, B.D., Bohlin, R.C., Drake, J.F., and Budich, W., 1977, *Ap. J.*, , **216** 291.
Scoville, N.Z., Yun, M.S., Clemens, D.P., Sanders, D.B., and Waller, W.H., 1987, *Ap. J. Suppl.*, **63**, 821.
Shu, F.H., Adams, F.C., and Lizano, S., 1987, *Ann. Rev. Astron. Ap.*, **25**, 23.
Solomon, P.M., Rivolo, A.R., Barrett, J., and Yahil, A., 1987, *Ap. J.*, **319**, 730.
Stark, A.A., and Blitz, L., 1978, *Ap. J. (Letters)*, **225**, L15.
Stark, A.A., 1983, in *Kinematics, Dynamics, and Structure of the Milky Way*, W.L. Shuter, ed., (Reidel:Dordrecht), p.127
Stark, A.A., 1984, *Ap. J.*, **281**, 624.
Stark, A.A., Elmegreen, B.G., and Chance, D., 1987, *Ap. J.*, **322**, 64.
Stutzki, J., and Güsten, R., 1990, *Ap. J.*, in press
Thronson, H.A., Lada, C.J., and Hewagama, T, 1985, *Ap. J.*, **297**, 662.
Zinnecker, H., 1989, in *Evolutionary Phenomena in Galaxies*, J. Beckman, ed., Cambridge University Press: Cambridge p.201.
Zuckerman, B. and Palmer P., 1974, *Ann. Rev. Astron. Ap.*, **12**, 279.

Leo Blitz makes his point at special session on clusters.

THE ORIGIN AND EVOLUTION OF GIANT MOLECULAR CLOUDS

Bruce G. Elmegreen
IBM Research Division, Thomas J. Watson Research Center
P.O. Box 218, Yorktown Heights, N.Y. 10598 USA

ABSTRACT - Observations and theories related to cloud formation are discussed. Distinctions between hole and tunnel models for the interstellar gas distribution and cloud and filament models for this distribution are suggested, as are distinctions between diffuse and self-gravitating clouds. The basic equations for five cloud formation models are summarized. These models consider thermal instabilities, gravitational instabilities, random sticking collisions between pre-existing clouds, compression of ambient gas behind shock fronts, and turbulence or magnetic wave interactions. As part of this discussion, the dispersion relation for gravitational instabilities in a shearing disk with an azimuthal magnetic field and time dependent energy equation is derived, and a numerical solution for the interaction between strong magnetic waves is used to illustrate the spontaneous formation and destruction of dense clumps.

The lifetimes of molecular clouds, defined to be the total time from the formation to the destruction of a particular cloud, are probably less than about 40 million years. This result is based on the fraction of clouds with young stars and the duration of star formation inside them. Short lives permit models in which clouds continuously form and disperse in 40 million years, as well as models in which the gas in a cloud remains there indefinitely but the cloud cycles through dense and puffed-up phases with a 40 million year period. In either case, molecular clouds probably contract toward dense cores and star formation on the time scale for energy dissipation rather than the free fall time. A list of seven major unsolved problems, both observational and theoretical, concludes this review.

1. Introduction

The first step in the process of star formation is the creation of a dense cloud. On a large scale, cloud formation involves spiral density waves, the galactic dynamo, multiple supernova explosions in OB associations, photoionization and dissociation by background star light, and so on. Many of these processes are not observed in sufficient detail or understood well enough to model the large scale aspects of cloud formation accurately. The triggering mechanisms for star formation in spiral density waves are not understood, for example, nor is the prevalence of shells, chimneys, and other swept-up structures which can form clouds by direct compression of ambient gas.

On a small scale, cloud formation involves gas dynamics in clumpy, "turbulent," magnetic media, whose components have largely supersonic and super-Alfvénic mutual interactions. The macroscopic equation of state for such media is not understood either. Thus the variation of pressure with density -- even the meaning of pressure -- are not known. It is possible, for example, that a large part of the binding pressure for some diffuse clouds is not just static thermal pressure, as in the two-phase thermal instability model (Field et al. 1969), but a highly variable inertial force from flows around the cloud -- analogous to the $v \cdot \nabla v$ force in a turbulent medium. Such variability can give some clouds a short lifetime, and make others appear unbound when in fact stochastic pressure fluctuations keep them bound on average.

Even with these uncertainties, several qualitative properties of the gas relevant to cloud formation may be inferred from general principles, such as the inevitability of energy dissipation and self-gravity, which combine to make the gas unstable to form dense clouds in almost every environment. Momentum exchange along magnetic field lines also seems important, leading to dense clumps and ropes that are pushed and twisted by distortions in the field. And of course the pressurized accumulation of ambient low-density gas into discrete clouds and shells is well observed. These qualitative properties of the gas have led to various scenarios for cloud formation that are probably on the right track, even though they have little detailed predictive value.

Several other reviews of cloud formation are in the recent literature. The four most comprehensive of my own reviews are in various other conference proceedings. The first, in *Star Forming Regions*, summarizes the observations and formation theories of the largest cloud complexes in galaxies, including density wave triggered star formation and long range propagating star formation (Elmegreen 1987a). The second, in *Interstellar Processes*, contains a more comprehensive review of the observations and theory of compressional (i.e., shock-related) cloud formation, as well as a general discussion of other theories (Elmegreen 1987b). A third summary, in *The Evolution of the Interstellar Medium*, critically reviews the existing theories of spontaneous cloud formation and compares the predictions of these processes to observations (Elmegreen 1990a). The fourth, to be published in *Protostars and Planets III* (Elmegreen 1991a), reviews all aspects of cloud formation and is contemporary with the present review. Reviews of cloud formation by other authors may be found in Larson (1988), Balbus (1990), and Franco (1990), who are concerned with galactic scale processes, by Kwan (1988), who discusses primarily the collisional build-up model, and by Tenorio-Tagle and Bodenheimer (1988), who emphasize large-scale shell formation around pressure sources. These reviews together contain a nearly complete bibliography for twenty years of theoretical and observational work on interstellar gas dynamics and cloud formation.

2. Overview of Cloud Formation

A spiral galaxy like ours contains approximately 5% to 10% of its total mass in the form of a relatively thin disk of gas. This gas is highly structured on a wide range of scales (see reviews in Scalo 1985 and Efremov 1989), with features and patterns that cannot be described easily (Scalo 1990). The earliest classifications considered only roundish, disconnected clouds in a low density intercloud medium (Clark 1965; Spitzer 1968). More detailed studies later revealed shells and filaments too (Heiles 1967; Ames and Heiles 1970; Verschuur 1970).

Today, few observers would claim that the gas has such simple structures. Figure 1 shows an IRAS map of the dust column density in the Taurus clouds, from Scalo (1990). These clouds appear to contain clumps and filaments in a dark background if one looks only at the bright areas, but they can also be imagined to contain holes and tunnels in a bright background if one concentrates on only the dark areas. It is not clear what the structure really is. If it is a collection of clumps and filaments, then coagulation and instability scenarios may apply to its formation, e.g., it could be a collection of clumps that have been brought

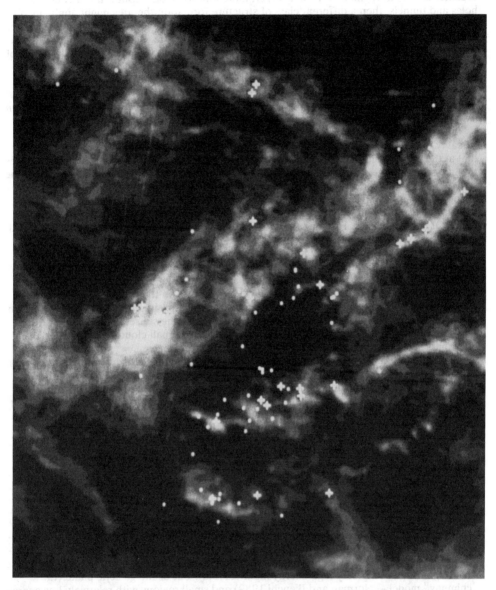

Figure 1. IRAS 100 micron column density map of the L1495 region in Taurus (from Scalo 1990, with permission). The map shows a complex structure, possibly composed of clumps, filaments, holes, and tunnels. Dots are T Tauri stars and pluses are IRAS 60 micron sources.

together or which fragmented from a larger, more uniform cloud. But if it is a collection of holes and tunnels, then a different class of structure formation theories might apply, involving expansion around field and embedded stars, for example. Field stars moving through the gas at higher than turbulent speeds could presumably make the tunnels, and embedded stars that are not moving relative to the gas could presumably make the holes. A similar range of structures is seen in cirrus clouds (Low *et al.* 1984) and in many other maps that have a sufficiently large number of resolution elements ($> 10^5$; see Blitz, this conference). It is likely that interstellar processes make both types of structures, and that *the distribution of gas is a combination of clumps, filaments, holes, tunnels and other shapes which continuously form and reform by coagulation and fragmentation and by expansion around stellar sources.*

This difficulty with describing the structure and topology of the interstellar gas is presumably only temporary -- extensive observations should solve the problem. But it means that at the present time we cannot determine unambiguously the origin of particular features that are seen in large scale sky maps. We can only understand at a general level some of the processes by which the gas *could* change from a more-or-less isotropic distribution of clumpy/filamentary/tunnelled material (the "ambient medium") into centrally condensed and strongly self-gravitating objects ("giant cloud complexes") in which stars eventually form. Of importance are 1. stellar energy sources, whose pressures can move the gas around, directly compressing it into cloud complexes on a wide range of scales, 2. relatively strong self-gravity, which is particularly important on large scales for low density regions but also important on small scales once the density gets large, 3. a strong tendency to dissipate random kinetic energy, especially on small scales, and 4. a tendency for different regions, possibly separated by large distances, to exchange momentum by the propagation of magnetic waves. Also important in some galaxies are 5. large scale density waves that compress parts of the interstellar gas into a shock or cloud collision front; the subsequent collapse of this gas, either on the scale of the Jeans mass or in the form of small clouds which coalesce, presumably leads to the formation of individual cloud complexes.

By combining these basic ingredients of cloud formation, one can understand, in the most general terms: (a) the formation of irregular shells and regions of expansion around supernovae and OB associations, and the consequent regeneration of star formation in this gas but at remote locations following this motion, (b) the formation of the largest clouds in galaxies ($10^7 M_\odot$) by gravitationally-driven condensation and energy loss in Jeans-length kiloparsec-size regions, preferentially occurring in the densest gas, such as spiral arms; (c) the secondary condensation or fragmentation of these large clouds into a hierarchy of substructure following continued energy loss, and (d) the lack of significant spin in most cloud complexes.

The tendency to dissipate kinetic energy by, for example, Alfvén wave propagation or wave steepening and damping, or by gaseous shocks and random clump-clump interactions, implies that *the bulk interstellar gas has to be stirred constantly* to keep the scale height in the galaxy relatively uniform from place to place. Such stirring is presumably accomplished by supernovae and OB association pressures and by cloud-cloud or gas-star gravitational interactions. The uniformity of this stirring is unknown from observations; if supernovae generally occur in or near OB associations, then the *stirring can be very non-uniform,* in which case small regions with relatively high energy input can blow out of the galactic plane (the "chimney" model -- Norman and Ikeuchi 1989) and small regions with relatively low energy input can condense out of the ambient medium and form cloud complexes (Elmegreen 1989c). Without local stirring from stellar energy, some of these condensations can be much smaller than the average Jeans mass (*ibid.*). They form by the combined action of gravitational self-attraction and (ambient) external compression, unresisted by internal pressure because of the lack of heating. This is a cooling instability if the heating rate is far below equilibrium, and it is a thermal instability if heating is still present but too weakly dependent on density to turn around the condensation process. Perhaps the instability is more appropriately named a *macroscopic gravitational-thermal instability,* because thermal pressures and temperatures are not important -- only the macroscopic motions of clouds. Such instabilities may have some relevance to the formation of cloud complexes, or perhaps the cores of cloud complexes, on length scales larger than the clump collisional mean free path or wave damping length and smaller than the Jeans length. More detailed discussions of this process are in Struck-Marcell and Scalo (1984), Tomisaka (1987), and Elmegreen (1989c, 1991b).

An important consideration in the study of molecular cloud formation is the fact that the *molecule formation part of the scenario is independent of the cloud formation part.* The conditions for molecule formation differ from the conditions for the collection of gas into giant cloud complexes. Thus there are nearly identical clouds in terms of mass and structure in the Carina arm of our Galaxy (Grabelsky et al. 1987) and in the spiral arms of M51 (Rand and Kulkarni 1990), but the Carina clouds are mostly atomic while the M51 clouds are mostly molecular. Largely atomic star-forming clouds (with dense molecular cores in which the stars actually form) also occur in Magellanic irregular galaxies (e.g., Skillman 1987). Because cloud formation is concerned primarily with the structural properties of gas, and not with the chemical properties, we shall not distinguish here between molecular and atomic clouds, but only between diffuse and self-gravitating clouds (see the more extensive discussion of this point in Elmegreen 1991a).

3. Diffuse and Self-Gravitating Clouds

The distinction between strongly self-gravitating clouds, such as those in which stars form, and weakly self-gravitating clouds, such as diffuse clouds, filaments, and shells, may be written in terms of the dimensionless parameter $P/G\sigma_c^2$ for boundary pressure P, gravitational constant G, and cloud mass column density σ_c. If this parameter is large, then the external pressure confines the cloud and self-gravity is relatively unimportant. If the quantity is small, then self-gravity binds the cloud, although the external pressure is still important in determining the cloud radius and the average cloud properties. These two limits of large and small $P/G\sigma_c^2$ correspond to diffuse and self-gravitating clouds, respectively. We expect that clouds or cloud parts with large $P/G\sigma_c^2$ will have overall shapes determined by external pressure gradients (e.g., shells, sheets, holes, and tunnels), and that clouds with small $P/G\sigma_c^2$ will have overall shapes determined more by self-attraction to a center of mass (e.g., clumps and filaments).

Typically the total ambient pressure in a galaxy disk is determined by the potential well from the stars and the gas, and may be estimated from the expression

$$P \approx \frac{\pi}{2} G\sigma_g \left(\sigma_g + \sigma_s \frac{c_{g,tot}}{c_s} \right)$$

where σ_g is the total gas mass column density in the disk, σ_s is the total stellar mass column density, $c_{g,tot}$ is the total effective velocity dispersion of the gas, including magnetic and cosmic ray pressures, and c_s is the mean (mass-weighted) stellar velocity dispersion. This expression for P is an approximation to a numerical solution (from Elmegreen 1989a). In the solar neighborhood, $P \sim 10^4 k_B \text{cm}^{-3}\text{K}$ for Boltzmann constant k_B. Note that supernovae and other energetic events affect the pressure by increasing the gas velocity dispersion $c_{g,tot}$. This raises the scale height and increases the mass of stars inside the gas layer. The pressure is essentially G times the product of the gas column density and the total gas+star column density within the gas layer.

This equation implies that the pressure boundary condition for a cloud, and therefore the distinction (using $P/G\sigma_c^2$) between diffuse and self-gravitating clouds for a given cloud mass, should vary with position in a galaxy. For example, the pressure should decrease with increasing galactocentric radius, and from the arm to the interarm regions. Decreasing pressures generally correspond to more atomic and less molecular gas (Elmegreen 1991a).

If we consider the local value of $P = 10^4 k_B$, then a cloud with an atomic hydrogen column density of 6×10^{20}cm^{-2}, corresponding to a color excess of $E(B - V) = 0.1$ mag, has a mass column density of $\sigma_c = 1.4 \times 10^{-3}$gm cm^{-2} and a dimensionless pressure $P/G\sigma_c^2 = 10$. This value places it in the diffuse cloud range, as expected for these typical diffuse cloud parameters (e.g., Spitzer 1978). A giant molecular cloud in the Galaxy, in contrast, has a typical mass column density of $170 M_\odot$pc^{-2} (Solomon *et al.* 1987) and so $\sigma_c = 0.036$gm cm^{-2}, making the dimensionless pressure $P/G\sigma_c^2 = 0.01$. Thus giant molecular clouds in our Galaxy are strongly self-gravitating.

It follows from this simple distinction between diffuse and self-gravitating clouds that *diffuse clouds probably form where pressure gradients are important,* as in thermal instabilities or by the collection of intercloud gas around a high pressure source, and *molecular clouds probably form where self-gravity is important,* as in regions with relatively large gas densities such as spiral arms and the periphery of shells of various sizes. All clouds can presumably grow, once they form, by accreting pieces from other clouds and swept-up regions.

4. Basic Equations for Cloud Formation

A good way to understand the theoretical principles of cloud formation is to look at the dominant terms in the equations. We consider in this section the five most commonly discussed processes of cloud formation: thermal instability (4.1), gravitational instability (4.2), random coagulation of small clouds into large clouds (4.3), pressurized accumulation of shells (4.4), and the accumulation or squeezing of gas by non-linear magnetic waves (4.5). References to more complete discussions and reviews are given.

4.1. THERMAL INSTABILITY

The thermal instability can be reduced to three equations and three variables: velocity v, density ρ, and pressure P. The equation of motion is

$$\rho \frac{Dv}{Dt} = -\nabla P,$$

where $D/Dt = \partial/\partial t + v \cdot \nabla$. The equation of continuity is

$$\frac{D\rho}{Dt} = -\rho \nabla \cdot v,$$

and the equation of energy from the first law of thermodynamics is

$$\frac{DP}{Dt} = \frac{\gamma P}{\rho} \frac{D\rho}{Dt} + (\gamma - 1)(\Gamma - \Lambda),$$

where γ is the ratio of specific heats for the fluid, which depends on the number of degrees of freedom of motion for interactions between fluid elements (e.g., atoms or tiny clouds), and where Γ and Λ are bulk heating and cooling rates, measured in energy per unit volume per unit time. These energy functions depend on the velocity dispersion of the fluid, c, which can sometimes be defined by the equation for a perfect gas

$$P = \rho c^2.$$

Now consider an equilibrium state with $v = 0$, ρ = constant, and P = constant, and perturb the equilibrium, using perturbation variables $\hat{\rho}$, $\hat{P}$, and $\hat{c}$ (v is already a perturbation variable). The equations can then be rewritten in terms of the perturbed variables, by substituting $\rho + \hat{\rho}$ for ρ, and so on, and the first order perturbed quantities (e.g. $\hat{\rho}$) can be separated from the zero order quantities (e.g., ρ) and from the second and higher order quantities (e.g., $\hat{\rho}v$). This leads to equations that are entirely first order:

$$\rho \frac{\partial v}{\partial t} = - \nabla \hat{P}$$

$$\frac{\partial \hat{\rho}}{\partial t} = - \rho \nabla \cdot v$$

$$\frac{\partial \hat{P}}{\partial t} = \gamma c^2 \frac{\partial \hat{\rho}}{\partial t} + (\gamma - 1)(\hat{\Gamma} - \hat{\Lambda})$$

$$\hat{P} = \hat{\rho} c^2 + 2\rho c \hat{c}.$$

These equations for v, $\hat{\rho}$, $\hat{P}$, and $\hat{c}$ have constant coefficients so the solutions are exponential in space and time. Fourier analysis tells us that any shape perturbation can be represented by a sum of pure waves, which in this case would take the form e^{ikx} for position x and wavenumber k. A generalized time dependence of the form $e^{\omega t}$ can be assumed too, for growth rate ω. Then the spatial derivatives can be replaced by ik and the time derivative by ω, so the first two equations give

$$\omega^2 = - k^2 \frac{\hat{P}}{\hat{\rho}},$$

which is the dispersion relation. The ratio $\hat{P}/\hat{\rho}$ comes from the energy equation. Now we see that negative $\hat{P}/\hat{\rho}$ gives instability, i.e., that all perturbed quantities grow as $\exp(k[-\hat{P}/\hat{\rho}]^{1/2}t)$. This implies that when the pressure decreases for increasing density, a slightly overdense region loses pressure and caves in, leading to an even higher density.

In general, $\hat{P}/\hat{\rho}$ depends on ω, which has to be solved as a function of k in the dispersion relation. To determine $\hat{P}/\hat{\rho}$, we need to know the dependence of Γ and Λ on density and velocity dispersion. Suppose that in the equilibrium state, $\Gamma = \Lambda = \Lambda_0$, and that near equilibrium Γ and Λ vary as

$$\Gamma \propto \rho^r c^s \text{ and } \Lambda \propto \rho^l c^m$$

for constants r, s, l, and m. Then

$$\hat{\Gamma} - \hat{\Lambda} = \Lambda_0 \left[(r - l)\left(\frac{\hat{\rho}}{\rho}\right) + (s - m)\left(\frac{\hat{c}}{c}\right) \right].$$

Now substitute $\hat{c}/c = 0.5(\hat{P}/P - \hat{\rho}/\rho)$ from above and solve for $\hat{P}/\hat{\rho}$ to get:

$$\frac{\hat{P}}{\hat{\rho}} \equiv \gamma_{eff} c^2 = \gamma c^2 \left[\frac{\omega - \omega_c(2l + s - m - 2r)}{\omega + \gamma \omega_c(m - s)} \right],$$

where

$$\omega_c = \frac{(\gamma - 1)\Lambda_0}{2\gamma P}$$

is approximately the cooling rate. Here we have defined an effective adiabatic index γ_{eff}. Evidently, $\hat{P}/\hat{\rho} < 0$, or $\gamma_{eff} < 0$, for *some* values of positive ω when either $(2l + s - m - 2r) > 0$ or $(m - s) < 0$, the first giving a negative numerator and the second a negative denominator.

These conditions for instability are the same as in Field (1965), where they are written in terms of a generalized cooling function, $\mathscr{L}$, equal to $\Lambda/\rho - \Gamma/\rho$ in the notation here. Field writes $(\partial\mathscr{L}/\partial T)_P < 0$ for thermal instabilities at constant pressure, and this is the same as $(l - r) > 0.5(m - s)$ if we substitute $T \propto c^{0.5}$ in the definitions for Λ and Γ. Note that the constraint of constant pressure corresponds to $\hat{P} = 0$ in our notation, in which case $\omega = \omega_c(2l + s - m - 2r)$ from the equation $\hat{P}/\hat{\rho} = 0$, and this implies that $(l - r) > 0.5(m - s)$ for positive ω. This is actually a solution to the dispersion relation in the limit of $k \to \infty$, which is one way to get a constant pressure perturbation (i.e., by considering a very small region).

Field also derives $(\partial\mathscr{L}/\partial T)_\rho < 0$ as an instability condition for constant density perturbations, and this gives $m < s$ for $\Lambda \propto c^m$ and $\Gamma \propto c^s$ in our notation. This follows from the dispersion relation too because the constraint on density corresponds to $\hat{\rho} = 0$, which gives $\omega = \gamma\omega_c(s - m) > 0$ as a solution under the same conditions, $m < s$, and in the limit of large sizes, i.e., $k \to 0$. Large sizes are important for the case $\hat{\rho} = 0$ because when $k << \omega_c/c$ the time scale for pressure adjustments is much larger than the cooling time. Then a region that becomes cool and at low pressure initially will not have enough time for the force and continuity equations to drive the convergence of gas that changes the density.

Other perturbations may have constant entropy, which implies that $\hat{P}/\hat{\rho} = \gamma c^2$ in our notation. Then $\omega - \omega_c(2l + s - m - 2r) = \omega + \gamma\omega_c(m - s)$ in the expression for $\hat{P}/\hat{\rho}$ and this implies that $(l - r) = -0.5(m - s)(\gamma - 1)$. Note that Field (1965) writes $(\partial\mathscr{L}/\partial T)_\rho + (1/[\gamma - 1])(\partial\mathscr{L}/\partial\rho)_T < 0$ for instability at constant entropy, and this is the same as $(l - r) < -0.5(m - s)(\gamma - 1)$ for Λ and Γ as defined above.

A criterion for thermal instability in a system out of thermal equilibrium ($\mathscr{L} \neq 0$) was derived by Balbus (1986): $(\partial[\mathscr{L}/T]/\partial s)_A < 0$ for entropy s and thermodynamic variable A, such as pressure. If we write $\mathscr{L} = (\Lambda - \Gamma)/\rho$ as above but take this to be $\Lambda(1 - E)/\rho$ at the time of the perturbation (to denote a lack of equilibrium, where $E = \Gamma/\Lambda$) and if $P = \rho c^2$ as above, then Balbus' criterion becomes $l - Er > 0.5(m - Es)$ for constant pressure, and $0.5(m - Es) < 1 - E$ for constant density. These conditions could also have been derived from $\hat{P}/\hat{\rho}$, as above, if $\Gamma = E\Lambda$ at equilibrium had been used.

These results may be understood more easily using only the ratio Λ/Γ, which is proportional to $\rho^x T^y$ for $x = (l - r)$ and $y = 0.5(m - s)$. In the case of constant pressure, $n \propto 1/T$ so $\Lambda/\Gamma \propto T^{y-x}$. Then $x > y$ (i.e., $[l - r] > 0.5[m - s]$) corresponds to instability. Then regions with low temperatures cool faster than they heat up, so they decrease their temperatures even more (at approximately the cooling rate, modified by a PdV type mechanical energy exchange) and cave in according to the equation of motion and the continuity equation. In the case of constant density, $\Lambda/\Gamma \propto T^y$ and instability requires $y < 0$ (i.e., $m < s$). And for adiabatic perturbations, $\Lambda/\Gamma \propto T^{x/(\gamma - 1) + y}$ because $n \propto T^{1/(\gamma - 1)}$; then instability requires $x/(\gamma - 1) + y < 0$, or $(l - r) < -0.5(m - s)(\gamma - 1)$, as shown above.

In general, an initial perturbation will be neither at constant pressure, constant entropy, nor constant density, but whatever the initial values of $\hat{P}$, $\hat{\rho}$, $\hat{c}$, or v, the dispersion relation

$\omega^2 = -k^2\hat{P}/\hat{\rho}$ still gives the condition for stability ($\omega > 0$) in the linear regime. If, for example, the fluid contains many small clouds or clumps, then the relevant dispersion is not the sound speed but the dispersion between clumps. Density fluctuations (in the number of clumps) could be Poisson from random uncorrelated motions, or fractal from correlated motions as in turbulence, or highly structured as in a shell. The corresponding fluctuations in the dispersion could also be either high or low, depending on the way this region formed.

What do these conditions tell us about the likelihood of thermal instabilities? For a uniform gas in which thermal pressure dominates, the thermal instability is apparently important in the warm HI phase (e.g., $10^3 K$) and in the moderately hot phase ($10^5 K$). Various applications of this instability have been discussed in a review by Begelman (1990). Parravano et al. (1987, 1988, 1989, 1990) and Lioure and Chièze (1990) discuss cloud formation models in which classical thermal instabilities start the condensation process.

Do similar processes operate in a cloudy fluid? For an idealized fluid composed of numerous small clouds and clumps, the dissipation is mostly from cloud collisions and this gives $l = 2$ and $m = 3$. This follows from the idea that collisions remove some fraction of the energy density, which is $1.5\rho c^2$, at the collision rate, which is $\rho c/\sigma_c$ for mean cloud column density σ_c. Here σ_c/ρ is the cloud mean free path. Thus $\Lambda \sim \rho^2 c^3/\sigma_c$. For initial perturbations in an equilibrium system ($E = 1$) at constant pressure, entropy, and density, respectively, the conditions for instability become $r < 0.5(1 + s)$, $r + s/3 > 3$, and $s > 3$. Out of equilibrium, the first and third of these are $r < 0.5(1 + sE)/E$ and $s > (1 + 2E)/E$. When $E < 1$, meaning that heating is initially too low for equilibrium, instability is much more likely for the first case than when $E = 1$. In general, we don't expect s to be large for the various cloud stirring processes, and we don't expect it to be very negative either. Then only the first case of a constant pressure perturbation gives a clear instability, and this happens when r or E are small. This differs from the case of the pure thermal instability, which can have some temperature regions where m is much smaller than 3. All three cases can be thermally unstable if m is less than s. For a cloudy gas, m is probably too large to give instability except in the constant pressure case, and then it does so only if the energy input from bulk stirring processes are relatively insensitive to density (i.e., r small) or if the heating rate is temporarily low ($\Gamma < \Lambda$).

The required lack of sensitivity of the heating rate Γ to the density (i.e., low r) is not unreasonable. Heating for a cloudy fluid is probably dominated by supernovae explosions and other pressure sources which involve shock fronts. Because these shock fronts radiate more at high density, the pressure which drives the motion of clouds decreases faster with time at high density than at low density, and so the velocity dispersion of the clouds can vary somewhat inversely with density, giving $r < 1$.

The interstellar fluid is also likely to be out of equilibrium almost everywhere because cooling by cloud collisions is probably more uniform than heating by clustered supernovae and other large-scale stirring processes, which tend to be very patchy. In that case $E < 1$ in the regions between OB associations, and then a macroscopic thermal instability follows even for large r, as long as $r < 0.5(1 + sE)/E$, as discussed above.

4.2. GRAVITATIONAL INSTABILITY

When self-gravitational forces are added, the equation of motion becomes

$$\rho \frac{Dv}{Dt} = -\nabla P + g\rho,$$

where the gravitational acceleration g satisfies

$$\nabla \cdot g = -4\pi G\rho.$$

The dispersion relation then has an additional term from gravity, which in three dimensions is $4\pi G\rho$:

$$\omega^2 = -k^2\frac{\hat{P}}{\hat{\rho}} + 4\pi G\rho.$$

In a two-dimensional sheet, the gravity term is replaced by $2\pi G\sigma k$ for column density σ. In a cylinder, it is $4\pi G\rho(1 - kRK_1[kR])$ for radius R and Bessel function K_1. For this cylinder, kR is much less than 1 when the wavelength is much longer than the thickness of the cylinder, and then the gravity term is $2\pi G\rho k^2 R^2 \ln(2/kR)$, or $2Gk^2\mu \ln(2/kR)$ for mass per unit length μ. When $kR >> 1$ for a wavelength much shorter than the thickness, the gravity term is $4\pi G\rho$ as for a three dimensional distribution. These gravity terms can be determined by direct integration over the perturbed density.

An important case for the gravitational instability is a shearing disk with an azimuthal magnetic field. This is relevant to the collapse of ambient gas in galaxies. Then the equation of motion is (Elmegreen 1991b)

$$\rho\left(\frac{\partial w}{\partial t} + w\cdot\nabla w + 2\Omega\times w - \Omega^2 R\right) = \rho g_T - \nabla P - \frac{1}{8\pi}\nabla B^2 + \frac{1}{4\pi}B\cdot\nabla B.$$

This is combined, as before, with the continuity and energy equations, but now an additional equation for the conservation of magnetic flux is needed:

$$\frac{\partial B}{\partial t} = \nabla\times(w\times B).$$

In these equations, w is the total velocity including shear, Ω is the rotation rate of the frame of reference, R is the galactocentric distance, and g_T is the total gravitational acceleration.

To simplify the equations, we write $2Ax\hat{e}_y$ for the component of velocity due to shear, where $A = 0.5R\partial\Omega/\partial R$ ($= -0.5\Omega$ for a flat rotation curve) is the Oort rotation constant and (x,y) is a coordinate system with x in the radial direction and y in the direction opposite the rotation (i.e., $\Omega < 0$). This shear component is separated from w, leaving v as a residual perturbation around shear:

$$w = v + 2Ax\hat{e}_y.$$

Now all of the time derivatives in the linearized equations will appear in association with derivatives of the form $2Ax\partial/\partial y$ so we can simplify the equations further by introducing shearing perturbations of the form $\exp(iky - 2Atikx)$ (cf. Goldreich and Lynden-Bell 1965). These perturbation are wave packets that begin as leading waves for negative times and then sweep back with increasing time until they are trailing. At the time $t = 0$, they point in the radial (x) direction in the galaxy.

The total gravitational acceleration is

$$g_T = g - \Omega(R)^2 R\hat{e}_x$$

for self-gravity of the gas g, defined by $\nabla\cdot g = -4\pi G\rho$. For simplicity here, the zero-order magnetic field, B, is taken to lie entirely in the y direction, i.e., azimuthal in the galaxy.

Now the equations are linearized as before. Because of shear, the coefficients of the perturbed variables depend on time so there are no simple exponential solutions $e^{\omega t}$. In general, the equations have to be integrated over t numerically to follow the development of the instability. But the importance of the magnetic field can be seen if we look at the instantaneous growth rate at the time t=0, when the wave crest is pointing in the radial direction. This is approximately when the shearing perturbation has its maximum growth rate. In this case, we can substitute ω for $\partial/\partial t$ and set $t = 0$. Then the equations reduce to a dispersion relation (cf. Elmegreen 1991b)

$$\omega^2 = -k^2 \gamma_{eff} c^2 + 2\pi G \sigma k - \frac{\omega^2 \kappa^2}{\omega^2 + k^2 v_A^2}, \qquad (1)$$

where $\kappa = 2\Omega(1 + A/\Omega)^{1/2}$ is the epicyclic frequency and σ is the mass column density through the disk (assumed to be thin).

The entire effect of the magnetic field in this approximation is contained in the Alfvén velocity $v_A = B/(4\pi\rho)^{1/2}$, which appears in association with κ. The denominator of this last term, if set equal to zero, would be the dispersion relation for an Alfvén wave in the plane. If there were no magnetic field, then the last term in the dispersion relation would be κ^2, which is the usual result for a shearing disk.

The nonmagnetic case with $\gamma_{eff} = 1$ has been extensively studied from the time of Safronov (1960). The peak growth rate is found by differentiating the right hand side of the dispersion relation with respect to k, setting the result to zero, solving for k, and substituting this k back into the equation. The result is

$$\omega_{peak} = \frac{\pi G \sigma}{c}(1 - Q^2)^{1/2} \quad [\ldots \text{ for } \gamma_{eff} = 1, \ v_A = 0]$$

where $Q = c\kappa/(\pi G \sigma)$. Instability ($\omega > 0$) requires $Q < 1$. This condition is sometimes called the Toomre condition, after Toomre (1964) found the analogous condition for stars in a galaxy.

The dispersion relations with γ_{eff} a function of ω (cf. Section 4.1) and $v_A \neq 0$ differ significantly from this classical result. If $\gamma_{eff} c^2 = \hat{P}/\hat{\rho} < 0$ because the thermal instability is present, then all wavelengths are unstable in the gravitational problem (down to the mean free path for cloud or clump collisions -- baring diffusive processes), and not just wavelengths larger than a critical Jeans length. In the short wavelength limit, the growth rate approaches the value for the pure thermal instability, which is the value giving $\gamma_{eff} = 0$ (to cancel the infinite k in the $k^2 \hat{P}/\hat{\rho}$ term in this limit). This limiting growth rate is $\omega = \omega_c(2l + s - m - 2r)$ in the case of a uniform pressure. Thus the growth rate is about equal to the dissipation rate (ω_c) in the short wavelength limit when thermal instabilities are present, and it equals the classical Jeans rate, $(4\pi G\rho)^{1/2}$ or $(2\pi G \sigma k)^{1/2}$, depending on geometry, in the long wavelength limit.

In the solar neighborhood, these two rates are approximately equal if the dominant dissipation mechanism is collisions between diffuse clouds. The corresponding time scales are each several times 10^7 years. In general, the ratio of ω_c to $(4\pi G \rho)^{1/2}$ for cloud collisional cooling, which gives $\Lambda \sim 2\rho^2 c^3/\sigma_c$ as discussed above, is approximately $0.1(P/G\sigma_c^2)^{1/2}$ where σ_c is the column density of a typical cloud in the fluid. Note that $P/G\sigma_c^2$ is approximately the ratio of the pressure external to a cloud to the gravitational binding energy density, as discussed in Section 3. For a fluid composed of diffuse clouds, this ratio is near unity or larger. For a fluid composed of only strongly self-gravitating clouds, this ratio is much less than 1. Thus the ratio of the growth rates in the short and long wavelength limits depends on the type of clouds which dominate the dissipation of energy in the fluid.

Sample solutions to equation (1) are shown in Figure 2. This equation is fifth order in ω and was solved numerically. We consider $Q = 0.5, 2$, and 5 with $v_A/c = 1$ (solid lines) and 0.1 (dashed lines). We also assume that $\omega_c/(\pi G \sigma/c)$, which is the cooling rate divided by the conventional instability rate, equals 1, that $l = 2$ and $m = 3$ for a cloudy fluid, and that $s = 0$ and $r = 0$ and 2 for thermal instability and no instability, respectively. The wavenumber is normalized to the value at the peak growth rate in the classical analysis, $k_J = \pi G \sigma/c^2$, and the growth rate is normalized to this peak, $\omega_J = \pi G \sigma/c$.

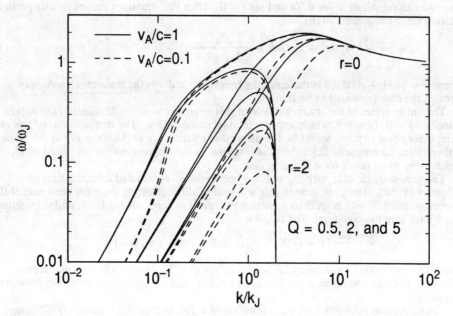

Figure 2. The normalized instability growth rate ω/ω_J, is shown as a function of the normalized wavenumber k/k_J. The solid and dashed lines are for strong and weak magnetic fields, as indicated. Curves that approach $\omega/\omega_J \to 1$ at high k/k_J are for a case with a macroscopic thermal instability ($r = 0$) and curves that fall to zero at $k/k_J = 2$ are for a thermally stable case ($r = 2$). Three dispersion relations $\omega(k)$ are shown for each case, with values of the conventional Q parameter equal to 0.5 (top curves), 2, and 5 (bottom curves).

The solutions show the tendency for ω/ω_J to approach $\omega_c(2l + s - m - 2r)/\omega_J = 1$ in the limit of large wavenumbers when the thermal instability operates, and for ω/ω_J to approach 0 and give stability at $k/k_J = -2(m - s)/(2l + s - m - 2r) = 2$ when the thermal instability does not operate. The solutions also show instability over some k range for $Q >> 1$ even when the gas is thermally stable, because of the destabilizing effect of the magnetic field. Thus $Q < 1$ is not a condition for azimuthal instability when magnetic fields (even weak ones) are present (for more discussion, see Elmegreen 1991b).

The mass spectrum of condensations that form by an instability has been written by DeFazio (1986) as the mass-dependent growth rate multiplied by the phase space density of such clouds, which is the maximum number density that can fit into a region if clouds of this mass are closely packed together. For a three dimensional region, the spectrum becomes $\omega(k)dk^3$; for a two-dimensional region, it is $\omega(k)dk^2$. Considering that mass M scales with wavenumber as k^{-3} and k^{-2} in these two cases, the mass distribution becomes proportional to $\omega(M[k])M^{-2}dM$ in each case. This implies that the slope in the *initial* cloud mass spectrum is $-2 + dlog\omega/dlogM$, which is approximately

$$-2 + \frac{\log[\omega(M_2)/\omega(M_1)]}{\log(M_2/M_1)} \sim -2 + \frac{0.5 \log(25\pi G \sigma_c^2/P)}{\log(M_2/M_1)}$$

for large and small masses M_2 and M_1 in the Jeans and thermal instability limits, respectively (here we assumed $\omega[M_2] \sim \omega_J \sim [4\pi G\rho]^{1/2}$ and $\omega[M_1] \sim \omega_c \sim 0.4\rho c/\sigma_c$). For example, if the full mass range of the instability is 10^3 and $P/G\sigma_c^2 \sim 1$, then the slope of the initial cloud mass spectrum is about -1.7. (For a more extensive discussion of the initial mass distribution function for the magnetic shear instability, see Elmegreen 1991b). Cloud interactions after formation could change this initial slope, but they will not change it much in the random collision model because this also gets about the same value. For reasonable parameters the theoretical slope is close to the observed value.

4.3. THE RANDOM COLLISIONAL BUILD-UP MODEL OF CLOUD FORMATION

The interstellar medium is essentially a cloudy fluid, so most gas dynamics on a large scale involves the motion of gas clouds and clumps. Giant cloud formation is no different. Nearly every process that has been proposed (except for the pure thermal instability) involves the coalescence of small clouds or clumps into larger clouds. In the macroscopic-thermal and gravitational instability models discussed above, this coalescence is driven by self-gravity, pressure, and collisional dissipation. The outcome of an individual collision, i.e., whether it is sticking or disruptive, is not important in these models because all types of collisions dissipate energy, and this dissipation is the single property of the collisions that enables the condensation to proceed. Thus most cloud formation can be viewed as a process in which small pieces come together to make large clouds.

A different use of cloud collisions was assumed for the classical collisional buildup model, discussed by Oort (1954), Field and Saslaw (1965), Kwan (1979), and others. This is a specific model which assumes that the collisions are random (rather than channelled in bulk by large scale gravity or pressure), and that they are sticking (rather than fracturing).

The equation of cloud evolution in this model was derived by Field and Saslaw (1965). If $n(M)dM$ is the number density of clouds of mass between M and $M + dM$, then

$$\frac{dn(M)}{dt} = G(M) - L(M) + F\delta(M, M_L)$$

where the gains, G, and losses, L, are

$$G(M) = \frac{1}{2} \int_{M_L}^{M - M_L} n(M - \mu)n(\mu)\sigma(M - \mu, \mu)v(M - \mu, \mu)d\mu$$

and

$$L(M) = n(M) \int_{M_L}^{M_U} n(\mu)\sigma(M, \mu)v(M, \mu)d\mu,$$

and F is the rate of formation of the lowest mass clouds from the break up of the highest mass clouds. The minimum and maximum masses are denoted by M_L and M_U. The cross section and relative collision velocity for two clouds of mass M and μ are $\sigma(M, \mu)$ and $v(M, \mu)$.

The G and L integrals can be solved exactly for some appropriate σ and v using $n(M) \propto M^{-\alpha}$. When $\sigma v = $ constant, for example, the integrals in G can be solved by substituting $\varepsilon = M/2 - \mu$ and then $\sin\theta = 2\varepsilon/M$; this gives integrals of the form $\int \cos^{2\alpha - 1}\theta d\theta$, which are straightforward when α is a multiple of 0.5. The L integral is also simple in this case. A constant σv corresponds approximately to diffuse cloud collisions, where the clouds all have about the same density ($\sigma \propto M^{2/3}$) and the velocity dispersion decreases with mass in equipartition, $v \propto M^{-1/2}$.

In the case of molecular cloud collisions, we can take σ to be the grazing cross section, $\pi(R_1 + R_2)^2$ and $M \propto R^2$ from the observed correlations (Larson 1981). We can also take

v = constant with mass (Stark and Brand 1989). Then exact solutions to the integrals are again possible, with α = 1.5, 2, 2.5, and so on.

Now, it turns out in the first case, with σv = constant, that the ratio G/L equals 1 for $M \approx M_L$ and α = 2, and that this ratio stays close to 1 for a large range of $M > M_L$ with α = 1.5. This implies that when a Monte Carlo model for the spectrum is evolved in time, the slope will approach an equilibrium value of $\alpha \sim 1.5$ for a large range of $M >> M_L$. This is why the σv = constant case gives an equilibrium cloud spectrum with a slope of around 1.5. In the second case, with $\sigma \propto M$, and v constant, the equilibrium slope is approximately 2 for most masses. Examples of the ratios G/L and the convergence to equilibria are in Elmegreen (1989d). These results imply that diffuse cloud coagulation gives $\alpha \sim 1.5$ and molecular cloud coagulation gives $\alpha \sim 2$, provided the collisions stick.

The hierarchical or fractal structure of interstellar gas implies that the statistics of cloud collisions are not Poisson but are correlated in space, and probably in time also. Nozakura (1990) modelled fracturing collisions and found fractal structure in the resulting distribution. It arises because collisions slow the clouds down and make them smaller. Close neighbors are likely to have just suffered a collision, so they tend to be small and they move at about the same velocity compared to more distant clouds.

4.4. CLOUD FORMATION BY DIRECT COMPRESSION OF LOW DENSITY GAS

Pressure variations in the interstellar medium that are greatly in excess of a factor of 2 can produce shock fronts and dense gas after cooling. Some filamentary dust clouds are apparently formed this way (Schneider and Elmegreen 1979), as are shells, sheets, and other structures, which together resemble a "cosmic bubble bath" (Brand and Zealey 1975). This is one of the most obvious and well-observed mechanisms of cloud formation. It can operate either on small scales in a uniform gas where thermal pressure is important, or on large scales in the general interstellar medium, where the cloudy structure is important.

Much of the small scale structure in molecular clouds could result from driven motions in the presence of embedded stars (Norman and Silk 1980). The same could be true on large scales too, in the general interstellar medium (e.g., Cox and Smith 1974; Chiang and Prendergast 1985). If a substantial amount of structure comes from driven motions, then the gas is likely to have a hole/tunnel structure, rather than a cloud/filament structure, as discussed in section 1. This could be the case also in an abstract sense if the holes and tunnels overlap so much that only the dense islands between them remain, giving what is essentially a cloudy structure. Most likely, the actual structure in the interstellar medium results from a combination of driven motions and cloud interactions. Unfortunately, there are no detailed simulations of such driven, dissipative gases, and the effective equation of state ($\hat{P}/\hat{\rho}$ in Section 4.1) in such a medium is unknown. Indeed, a pervasive hole/tunnel structure created by the constant stirring of numerous small stars would seem to have a different equation of state than a cloudy/filamentary structure, created, for example, by the breakup of isolated self-gravitating clouds. Thus the instability models, which depend strongly on the equation of state, may give different results in the hole/tunnel and cloud/filament cases.

Numerous examples of theoretical solutions to the expansion of giant pressurized regions are in the literature (see the review in Tenorio-Tagle and Bodenheimer 1988). These solutions include the effects of galactic shear (Tenorio-Tagle et al. 1987, 1990; Palous et al. 1990), vertical stratification (Norman and Ikeuchi 1989; Igumentshchev et al. 1990; MacLow et al. 1989), and magnetic fields (Ferriere et al. 1990; Tomisaka 1990). The possibility of blow-out above the plane is an important issue (Cox 1990) because if the high pressure from the explosion is vented into the halo, then the mass of accumulated gas in the shell or chimney can be small.

Sometimes the pressure can persist long enough for the swept-up gas to collapse gravitationally along the dense front, that is, in a direction transverse to the general motion. This would presumably form one or more dense self-gravitating clouds inside the front, and possibly lead to star formation (e.g., McCray and Kafatos 1987). For a large shell, the result could be a giant molecular cloud (e.g., Franco et al. 1988). Such collapse is not well understood because of the large flows parallel to the front that follow from curvature effects and deceleration. A rule of thumb for the collapse to occur is that the age of the front must exceed the inverse of the unstable growth rate inside the front, which is approximately given by a dispersion relation

$$\omega^2 \sim -\alpha k^2 v^2 + 2\pi G\sigma k$$

(cf. Section 4.2), where v is the shock speed and α is slightly smaller than 1. This condition, from Elmegreen (1989b), does not contain the velocity dispersion c as in the ratio $\hat{P}/\hat{\rho}$, because the main resistance to the monotonic growth of self-gravitating condensations is (in this calculation) from the internal flows, which have a speed comparable to the shock speed. For a shock at a constant speed, or a pressurized layer without a shock, v in this equation should be replaced by c, but for a decelerating front, v seems more appropriate from the linear analysis. A result of this disruption from internal flows (which is really an erosion of dense perturbations that emerge from the front of the shock due to its general deceleration) is that the growth of dense, self-gravitating clouds or cloud cores should be delayed significantly beyond the pure Jeans time in the layer, which is $(G\rho_{post})^{-1/2}$ for post-shock density ρ_{post}. A better approximation from the analysis in Elmegreen (1989b) is $0.25(G\rho_{pre})^{-1/2}$ for preshock density ρ_{pre}. This follows from the peak rate in the above equation, $\omega = \pi G\sigma/(\alpha^{1/2}v)$, if we substitute $\sigma = v\rho_{pre}t = v\rho_{pre}\omega^{-1}$.

Spherical expansion of the layer contributes a term comparable to the κ^2 term in the classical dispersion relation for a shearing disk (see Section 4.2), where κ is now approximately v/R for expansion speed v and radius R. Then there is a $Q < 1$ instability condition as for a classical disk, and now $Q \sim \kappa v/\pi G\sigma$, which is $\sim v^2/G\rho_{pre}R^2 \sim (G\rho_{pre}t^2)^{-1}$. Setting this less than 1 gives the same condition as before, namely, $t > (G\rho_{pre})^{-1/2}$ for collapse.

The kinematic instability found by Vishniac (1983) operates early on, but presumably leads only to fragmentation and some turbulence inside the layer. These fragments should still attract each other gravitationally and drive the subsequent collapse to dense clouds (perhaps in analogy to the collapse of a clumpy molecular cloud).

4.5. CLOUD FORMATION BY TURBULENT AND MAGNETIC WAVE INTERACTIONS

Interstellar motions at velocities in excess of the sound speed can produce gas compression and cloudy structure in converging regions, such as the interface between colliding eddies (e.g., Passot et al. 1988). High speed motions can also produce non-linear magnetic waves, which steepen and shock to give moderately dense structures, or which interact and splash to give very dense structures (Elmegreen 1990b). Rotation produces magnetic waves too, and the tension from twisting field lines can squeeze low density regions into dense ropes (Fukui and Mizuno 1990). The primary ingredients for these cloud formation theories are non-linear waves of various types, giving importance to non-linear terms in the equation of motion such as the inertial acceleration $v \cdot \nabla v$ and the perturbed magnetic pressures $\hat{B} \cdot \nabla \hat{B}/4\pi$ and $\nabla \hat{B}^2/8\pi$.

Numerical simulations of the three dimensional structures that are likely to form by these processes are not yet available. The previously unpublished results of the long-time behavior of a 1-dimensional simulation (cf. Elmegreen 1990b) are in Figure 3. This shows the density,

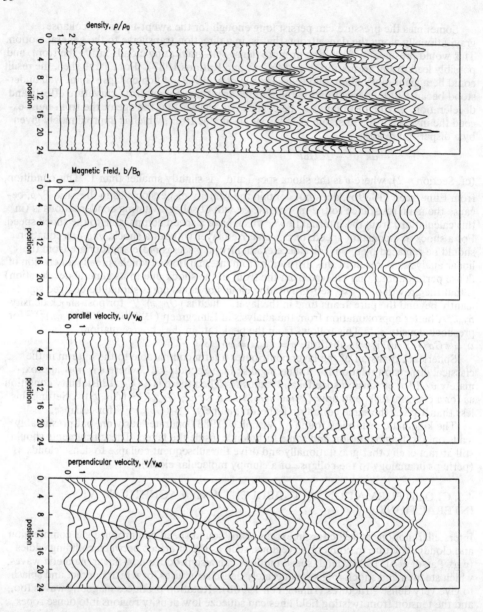

Figure 3. The generation and interaction of plane polarized magnetic waves are shown. Each line is a different time, increasing from the bottom of the figure (i.e., on the left of the page) to the top (page right). The boundary conditions and other parameters are discussed in the text. The density peaks are transient; they may explain some of the clumpy structure in interstellar clouds.

perturbed magnetic field strength, and velocities parallel and perpendicular to the field during the time development of a system of interacting non-linear waves. Time increases from the bottom to the top of the figure (which is from left to right on the page), with a time interval between curves equal to one-half of the time unit, and a time unit equal to one spatial unit on the abscissa divided by the linearized Alfvén speed. There are 2400 grid points in the calculation (100 per spatial unit). The waves are excited by randomly applied accelerations at the endpoints of the grid, with acceleration values equal to 0.5 on the left and 0.3 on the right multiplied by random numbers between -1 and 1 times the Alfvén speed divided by the time step. The accelerations are continuous, with changes to new accelerations at random times in an average interval of 3 time units. The lowest time shown, at the bottom of the figure (page left), equals 10 time units, and the highest time is 32 units, at the top (page right).

The results illustrate the steepening of waves as they move, the interaction between waves when they meet while travelling in different directions, and the generally chaotic behavior of the density and velocity structures at late times. The clump widths are typically one unit in the calculation, which is $\sim 1/k$ or 10%-20% of the (linearized) wavelength, and the clump lifetimes are typically 5 to 10 time units, which equals approximately 3 to 6 Alfvén crossing times inside the clump at the clump density.

This model is presumably scale invariant, and could apply to the formation of cloudy and clumpy structures in both the interiors of molecular clouds and in the general interstellar medium. Because only one polarization is considered, however, the range of possible interactions and wave structures is limited. Compression perpendicular to the mean field, and rotation around this direction are not present, for example. Such rotation could presumably lead to rope-like structures, as envisioned by Fukui and Mizuno (1990).

The equations of motion for a magnetic gas illustrate directly how the non-linear terms channel energy from large scale motions to small scale structures, as shown in the figure. Ignoring gravity, the equations of motion, continuity, pressure, and magnetic convection for a linearly polarized wave in a medium with an unperturbed field B_0 are

$$\left\{\rho \frac{Du}{Dt}\right\} = -\frac{\partial P}{\partial z} - \left\{\frac{1}{8\pi}\frac{\partial b^2}{\partial z}\right\} \quad ; \quad \left\{\rho \frac{Dv}{Dt}\right\} = \frac{B_0}{4\pi}\frac{\partial b}{\partial z} \quad ;$$

$$\frac{D\rho}{Dt} = -\left\{\rho \frac{\partial u}{\partial z}\right\} \quad ; \quad \partial P = \left\{\frac{\gamma P}{\rho}\partial \rho\right\} \quad ; \quad \frac{Db}{Dt} = B_0 \frac{\partial v}{\partial z} - \left\{b \frac{\partial u}{\partial z}\right\}.$$

Here v and u are the velocities perpendicular and parallel to the unperturbed field (which is in the z direction), and b is the field strength in the perpendicular direction. These equations are not linearized but contain all the terms; i.e., v, u and b need not be small. The terms in braces are non-linear.

We can see from these equations that a perturbation of the form $\cos(kz)$ applied to some variable will drive time variations of the form $\cos(kz)^2$, which contains terms in $\cos(2kz)$; i.e., the non-linear terms increase the spatial frequency as time evolves. For example, suppose $v \propto \cos(kz)$ at some initial time, and that $u = 0$, $b = 0$, and $\rho = 1$. The spatial derivative of v drives a time change in the field b, which picks up the same $\cos(kz)$ dependence. But when this is squared in the b^2 term, a $\cos(2kz)$ appears, and this is what determines the time derivative of u. Thus u varies as $\cos(2kz)$ too. It follows that the spatial derivatives of this parallel velocity, $\partial u/\partial z$, also vary as $\cos(2kz)$, and when multiplied by b, which still varies as $\cos(kz)$, they begin to drive the time derivatives of b with terms in $\cos(kz)$ and $\cos(3kz)$. At the same time, $\partial u/\partial z$ also drives the time derivative of ρ, so the density picks up terms in cos($2kz$). The pressure responds with all higher order terms if the gas is adiabatic with γ non-integer. Now there are two cycles of increasing order, one for the field strength and one for the density and pressure. Each feeds back on itself and on the other. These two cycles correspond to steepening sonic waves and steepening Alfvén waves, respectively, and the interaction between them is a steepening magnetosonic wave which eventually forms a shock front.

Evidently, large distortions in any one of the variables ρ, B, v and u can lead to similar distortions in the other variables, and at the same time produce smaller and smaller scale structures until dissipation finally sets in. Such large initial distortions should be present almost everywhere in the interstellar medium because stellar energy sources and self-gravity constantly drive motions at supersonic and super-Alfvénic speeds. The result is probably turbulence of some sort, with a strong connection to the origin of small scale structure (see review in Falgarone and Phillips 1991).

5. Cloud Formation Models and Cloud Lifetimes

There are two very different scenarios for cloud and star formation in the literature. One suggests that because molecular clouds are found in the spiral interarm regions, and because the overall star formation rate in the Galaxy is low, self-gravitating clouds must be relatively stable and long-lived, and star formation in them is triggered by mutual collisions and random supernova explosions. In this model, the self-gravitating clouds live long enough ($10^8 - 10^9$ years) to grow by collisional agglomeration, and the accretion rate may equal approximately the mass conversion rate into stars. Discussions of this "long life" model are in Scoville and Hersh (1979), Scoville, Sanders, and Clemens (1986), and elsewhere.

In a second model, the self-gravitating clouds form continuously from the ambient gas and collapse immediately into stars, on dynamical and dissipation time scales (10^7 years). The stars then disrupt the clouds soon after they form, on similar time scales. Early discussions of this "short life" model are in Bash, Green, and Peters (1977), Elmegreen (1979), and Blitz and Shu (1980).

Other models emphasize the role of pressure sources in forming clouds (Gerola and Seiden 1978), or thermal instabilities in forming diffuse clouds, followed by collisional agglomeration into complexes (Lioure and Chièze 1990; Parravano 1987). Gravitational instabilities have also been considered (see review in Elmegreen 1990a). These other models address only the first step in the formation of the giant complexes, rather than the lifetimes of the complexes, and they can be applied to either the first generation of cloud formation in the long-life scenario, or to the continuous generation of clouds in the short-life scenario. Here we discuss the evidence for long and short cloud lifetimes regardless of the processes that formed the first clouds.

One of the observations cited as evidence for the long life model is the existence of molecular clouds in the regions between the spiral arms of our Galaxy. This observation implies that the clouds survive for the arm transit time if they form or grow mostly in the arms. However, most of the effect of a spiral density wave is probably just to organize the gas and stars into a spiral. The interstellar medium is not dead in the interarm regions; there are usually a few (small) interarm star formation sites along with the (small) interarm clouds. Thus the interarm clouds could have formed recently in shells around ~40 million year (My) old OB associations, for example, in small bursts of secondary star formation following the original burst in the main arms. Moreover, the interarm transit time is often quite short, much shorter than the proposed lifetime of the clouds in the first scenario. In the inner region of the Galaxy, where most of the molecular clouds exist, the arm-to-arm time is less than 100 My if the pattern speed is 13.5 km s^{-1}kpc^{-1} (Yuan 1969), and the *interarm* transit time is likely to be much shorter yet. According to the continuity equation, the relative time spent at a particular phase in a steady spiral is proportional to the density there. If the arm-to-interarm contrast in the gas density is 5, for example, then the gas lingers 5 times longer at high density in the arms than at low density in the interarm regions. Then the interarm transit time is 1/6 of the full arm-to-arm time. It is possible, for example, that the interarm transit time in the 5 kpc region is around 20 My, or at least less than 40 My if the interarm region is taken to be wider than the arms. Such short transit times do not constrain the cloud lifetime enough to distinguish between the two scenarios. In comparison, the arm-to-arm transit time in the solar neighborhood is around 200 My for the same pattern speed, and in fact there are very few interarm clouds at our radius; the Perseus arm gap is an example (Dame et al. 1986). This observation is consistent with short cloud lifetimes.

Another motivation for the long-life scenario is the observation of low star formation rates in galaxies, but this does not prove that clouds are stable and long-lived. Clouds could be short-lived as in the second scenario, but with a low star formation efficiency during this short life because of effective disruption by massive stars. For example, the star formation efficiency in most large clouds is around 5%, although it can be much higher in small cores. Then, if the total formation-to-destruction time of a cloud is around 40 My as suggested below, the overall star formation rate in our Galaxy becomes $0.05 M_{\text{total}}/40$ My for $M_{total} \sim 2 \times 10^9 M_\odot$. This rate equals $2.5 M_\odot/\text{yr}$, which is a reasonable star formation rate. The real physics of the problem in this interpretation comes in the explanation for the 5% efficiency, not in an explanation for cloud stability and long cloud lifetimes, which are not required.

Presumably the low efficiency of star formation is a result of a relatively rapid initial cloud contraction followed by a thorough dispersal driven by a few massive stars. The initial contraction can be rapid because the time scales for energy dissipation and magnetic diffusion are only several times the dynamical time scale in the cloud, which is $(G\rho)^{-1/2}$. And the dispersal can be thorough and efficient because the mass of gas that is pushed around by a massive star is much larger than the star's mass, i.e., it takes only several hundred solar masses of massive stars to disperse a $10^5 M_\odot$ cloud. These two aspects of cloud evolution, the rapid initial contraction and an efficient dispersal, imply that only a small fraction of the gas is able to form stars before the structure of the cloud is changed significantly (see also Larson 1987). If cloud destruction were not so rapid and efficient, because, for example, the most massive stars that form have too low a luminosity or drift away from the cloud too quickly to disperse much gas, then the efficiency of star formation could be higher there and the cloud lifetimes longer (this latter possibility has been considered for the formation of bound clusters in low mass clouds -- Elmegreen 1983).

The lifetimes of clouds can be inferred from the duration of star formation in them and from the fraction of clouds with star formation. Observations suggest that a high fraction, ~0.5, of self-gravitating clouds contain star formation. This is evident from observations of a large number of clouds, both massive and low mass, in the solar neighborhood and throughout the Galaxy (e.g., Beichman et al. 1986; Myers 1987; Mooney and Solomon 1988). The duration of star formation in a cloud comes from observations of the total age span of stars in regions of star formation. This age span ranges between ~20 My for OB associations (Blaauw 1964) and ~ 1 My for smaller clouds containing reflection nebulae and T associations. Such short times are consistent with the rapid destruction of clouds observed in the form of winds and HII regions (e.g., Lada 1985).

These two numbers imply that the total time interval between self-gravitating cloud formation and self-gravitating cloud destruction is

$$\tau = \frac{20 \text{My}}{0.5} \sim 40 \text{ Million years.}$$

We consider this to be the total *lifetime of the star-forming phase of a cloud*.

The lifetimes for all molecular cloud regions that I know of are consistent with the idea that clouds always evolve as rapidly as energy dissipation allows them to evolve, which is several internal crossing times (Blitz and Shu 1980; Scalo and Pumphrey 1982; Elmegreen 1985). The same is apparently true for clumps inside clouds, and perhaps also for subclumps, although the time scale decreases as the density increases. This implies that *star forming clouds may not need internal or external energy sources to prolong their lives. They do not collapse at the free fall rate because they cannot get rid of their energy fast enough* (even with strongly supersonic internal motions). Another way of saying this is that molecular clouds are supported on a large scale by turbulent motions (Bonazzola et al. 1987; Léorat et al. 1990; Pudritz 1990), but they do not have to be supported for very long, and much of this turbulent energy may be gravitational in origin, put in during the formation process. Molecular clouds simply contract to star-forming cores as the initial turbulence dissipates. Short-term turbulent support also implies that shocks promote star formation primarily by increasing the dissipation rate in the gas that is compressed -- not by increasing the free fall collapse rate (Elmegreen 1989b).

This total lifetime of 40 My has important implications for the long and short life scenarios. The long-life scenario as originally stated would be impossible if cloud lifetimes are only 40 My, but the general idea that clouds in one form or another stay together for a long time may still be correct. It is possible, for example, that the gas in a cloud cycles repeatedly through dense star formation phases and low-density, quiescent phases, with an average period of approximately 40 My. The cloud may not be destroyed after the 40 My period, but it could remain loosely bound, puffed-up perhaps, and primarily in the diffuse cloud phase (which may be molecular also, depending on the ambient pressure). Then the collision rate becomes high on average, so collisions could initiate star formation, as could random supernova explosions and spontaneous collapse.

We can obtain a characteristic density for a cloud in the low density phase by setting the collision time equal to the gravitational collapse time. The collision time for molecular clouds, which, in their dense phase have a density n_D, is about 200 My, so if a cloud puffs up to a low density n_P and all puffed clouds have this same density, then the collision time for puffed clouds becomes $\sim 200(n_P/n_D)^{2/3}$ My. The free fall time of a molecular cloud is about 2 My, scaling with $n^{-1/2}$, so the free fall time of the puffed-up cloud is about $2(n_P/n_D)^{-1/2}$ My. These two times are equal when $n_P/n_D \sim 0.01^{7/6} \sim 0.005$, or $n_P \sim 1\,\text{cm}^{-3}$. Because this is the average density of all interstellar matter in the solar neighborhood, the filling factor of puffed up gas would be about unity if all clouds were puffed up at the same time. In general, the filling factor for gas at low density depends on the fraction of the evolution cycle that is spent there.

The basic point of this revised long-life scenario is that cloud destruction may be incomplete. The 95% residue from a 40 My old star-forming event could remain as a well-defined entity, ready for more star formation later, although perhaps removed in space from the previous event by some 200-500 pc distance. Note that there is no theoretical problem with long-term cloud support in this revised scenario because the cloud continuously tries to collapse as fast as internal dissipation and gravity allow; *the collapse is delayed only on average,* and the origin of this delay is star formation pressure during the active phase. This model differs from the turbulent-support models (see above), which delay the collapse somewhat uniformly at all times. Intuition suggests that such precise self-regulation may not be possible when only a few massive stars can disperse a whole cloud.

It seems reasonable that some clouds are not completely destroyed, i.e., converted back into unbound diffuse gas, after only one star formation event. A dense cloud like the M17 cloud may require several star formation events over a period of 100 My or more to completely disperse it. Other clouds may be destroyed after the first event, and then the gas may find its way back into a new cloud by one of the processes discussed in Section 4. The basic question to be addressed by observations is *what happens to a self-gravitating cloud after it forms stars and gets dispersed?* Does it recollapse into essentially the same cloud without much mixing with other clouds, or does it disperse completely and mix with other gas before returning to star-forming densities as a new cloud? The distinction between these two possibilities is important for theories of cloud formation because in the first case the initial conditions for cloud formation are special, having been set-up by a previous event of star formation in the same gas. In the second case, cloud formation begins fresh from the ambient medium each time. In the more likely combination of scenarios, some clouds will form from previous special conditions, and others will form fresh. The ratio of these two events is the important issue.

An example of cloud recycling might go as follows. An existing molecular cloud, formed either from the ambient gas or as residual from a previous epoch of star formation, becomes broken apart and puffed up because of pressures from internal star formation lasting 20 My. During this puffed-up stage it may have the form of a shell or other coherent structure, but what is important is that it does not completely evaporate into the intercloud medium or mix with a large number of other shells or ambient gas, or disperse by galactic tidal forces; it remains pretty much the same entity that it was before the star formation occurred (although with a different shape). And then, after 20 My more, it collapses spontaneously into the dense phase again, or collides with another puffed up cloud and is compressed into the dense phase. Then it begins to form more stars. A single cloud entity could, in principle, form several OB associations over a period of 100 My or more, each separated from the other by 40 My and several hundred parsecs (see discussion in Section IV of Elmegreen 1979). The observation of multiple and delayed star formation events in giant star complexes (Efremov and Sitnik 1988) supports this picture.

What other observations could help to distinguish between long-lived clouds with repetitive star formation events and short-lived clouds with single, highly disruptive events? The most obvious thing to do is to map in HI and CO around old OB associations and clusters (Leisawitz, *et al.* 1989, 1990) out to as far a distance from the clusters as necessary to include $\sim 10^5 M_\odot$, which is the likely residual mass from an OB association. If this mass is still in the form of a well-defined cloud, with boundaries that do not blend with ambient gas at the same velocity, and if the cloud is gravitationally bound or close to it, then the gas in that cloud could be left over from its previous star formation event and on its way (in a future 20 My) to another similar event.

A related observation would be to study large clouds without star formation (Pérault, Falgarone and Puget 1985; Maddalena and Thaddeus 1985) to see if they are within some 300 parsecs of a gas-free association that is older by ~ 300 pc $/ 10$ km s^{-1} = 30My. Here a likely drift velocity of 10 km s^{-1} has been chosen. Star-forming clouds could also be surveyed for middle-aged stars that might have been gravitationally dragged along with the gas during the disruption of the previous association. Only a small fraction of stars is likely to follow, but some might. There might also be evidence for metallicity variations from complex to complex if the gas did not mix well after each 40 My star formation event.

6. What Do We Do Next?

The problems to be solved before cloud formation can be understood are vast and varied. Many were discussed in the previous sections; they are summarized here.

1. What fraction of the interstellar gas has a clump/filament structure (e.g., spaghetti and meatballs) and what fraction has a hole/tunnel structure (e.g., a sponge)? Is there a general descriptive theory that can combine both topologies? What is the kinetic energy dissipation

rate and the magnetic diffusion rate for each type of structure? What is the origin of the small clumps and filaments in the first case, and the small holes and tunnels in the second?

2. What are the heating processes for the large scale motions of interstellar gas (the Γ in Section 4.1)? How do they depend on density and velocity dispersion? How patchy are they? And what are the cooling processes and their dependences on density and velocity dispersion? Is the gas ever in kinematic equilibrium for prolonged times (i.e., dissipation rate = stirring rate)? Is there a macroscopic thermal instability?

3. What is the origin of the cloud mass spectrum? Is it from an initial spectrum, determined by the mass-dependent growth rate of some instability, and then modified by cloud interactions? Does the spectrum tell us about cloud formation at all if several models give the same spectrum (which they do)? Is the spectrum a manifestation of turbulence?

4. What is the probability distribution function for the masses of the fragments that come off of a collision between two magnetic clouds (as a function of speed, impact parameter, mass, etc.)? How much cloud growth can be attributed to random coalescence (or are collisions usually shattering)?

5. What is the condition for the formation of secondary giant molecular clouds inside shells and chimneys that are swept up around OB associations? How does this condition depend on ambient density, shear, scale height, association pressure, lifetime, etc.? What is the fraction of molecular clouds that are formed by such stimulated processes, as opposed to those formed by spontaneous processes (e.g., instabilities and random coalescence)?

6. How much of the interstellar structure can be explained by turbulent motions and shocks that originally got their energy from processes operating on vastly different scales (i.e., from turbulent cascades up or down)? What does turbulence look like in a three-dimensional, supersonic, magnetic, self-gravitating medium?

7. What happens to the gas in a giant molecular cloud after star formation disrupts the cloud? Does it form another cloud without much mixing, or does it completely disperse and mix before cloud formation begins again?

Some of these questions may not be answered for decades. Some have their best answers in direct observation, and some require theory and computer modelling. There is a place for everyone in the study of cloud formation!

Acknowledgement: Many thanks to John Scalo for providing Figure 1 and for his stimulating company on the trek down Samaria Gorge.

7. References

Ames, S., and Heiles, C. 1970, *Ap.J.*, **160**, 59.
Balbus, S.A. 1986, *Ap.J.(Letters)*, **303**, L79.
Balbus, S.A. 1990, in *The Interstellar Medium in Galaxies,* H.A. Thronson, Jr. and J.M. Shull (eds.), Kluwer, Dordrecht, 305.
Bash, F.N., Green, E., and Peters, W.L. 1977, *Ap.J.*, **217**, 464.
Begelman, M.C. 1990, in *The Interstellar Medium in Galaxies,* H.A. Thronson, Jr. and J.M. Shull (eds.), Kluwer, Dordrecht, 287.
Beichman, C.A., Myers, P.C., Emerson, J.P., Harris, S., Mathieu, R., Benson, P.J., and Jennings, R.E. 1986, *Ap.J.*, **307**, 337.
Blaauw, A. 1964, *Ann.Rev.Astr.Ap.*, **2**, 213.
Blitz, L., and Shu, F.H. 1980, *Ap.J.*, **238**, 148.
Bonazzola, S., Falgarone, E., Heyvaerts, J., Pérault, M., and Puget, J.L. 1987, *Astr.Ap.*, **172**, 293.

Brand, P.W.J.L., and Zealey, W.J. 1975, *Astr.Ap.*, **38**, 363.
Chiang, W.H., and Prendergast, K. 1985, *Ap.J.*, **297**, 507.
Clark, B.G. 1965, *Ap.J.*, **142**, 1398.
Cox, D.P. 1990, in *The Interstellar Medium in Galaxies,* H.A. Thronson, Jr. and J.M. Shull (eds.), Kluwer, Dordrecht, 181.
Cox, D.P., and Smith, B.W. 1974, *Ap.J.(Letters)*, **189**, L105.
Dame, T.M., Elmegreen, B.G., Cohen, R.S., and Thaddeus, P. 1986, *Ap.J.*, **305**, 892.
DiFazio, A. 1986, *Astr.Ap.*, **159**, 49.
Efremov, Yu.N. 1989, *Stellar Complexes,* Harwood, London.
Efremov, Yu.N., and Sitnik, T.G. 1988, *Sov.Astr.Letters*, **14**, 347.
Elmegreen, B.G. 1979, *Ap.J.*, **231**, 372.
Elmegreen, B.G. 1983, *M.N.R.A.S.*, **203**, 1011.
Elmegreen, B.G. 1985, *Ap.J.*, **299**, 196.
Elmegreen, B.G. 1987a, in *IAU Symposium No. 115, Star Forming Regions,* M. Peimbert and J. Jugaku (eds.), Reidel, Dordrecht, 457.
Elmegreen, B.B. 1987b, in *Interstellar Processes,* D.J. Hollenbach and H. Thronson (eds.), Reidel, Dordrecht, 259.
Elmegreen, B.G. 1989a, *Ap.J.*, **338**, 178.
Elmegreen, B.G. 1989b, *Ap.J.*, **340**, 786.
Elmegreen, B.G. 1989c, *Ap.J.*, **344**, 306.
Elmegreen, B.G. 1989d, *Ap.J.*, **347**, 859.
Elmegreen, B.G. 1990a, in *The Evolution of the Interstellar Medium*, L. Blitz (ed.), San Francisco, Astron. Soc. of the Pacific, 247.
Elmegreen, B.G. 1990b, *Ap.J.(Letters)*, **361**, L77.
Elmegreen, B.G. 1991a, in *Protostars and Planets III,* E.H. Levy and M.S. Matthews (eds.), Univ. of Arizona, Tucson, in press.
Elmegreen, B.G. 1991b, submitted to *Ap.J.*
Falgarone, E., and Phillips, T. 1991, in *Fragmentation of Molecular Clouds and Star Formation*, E. Falgarone, F. Boulanger, and G. Duvert (eds.), Kluwer, Dordrecht, in press.
Ferriere, K.M., MacLow, M.M., and Zweibel, E.G. 1990, *Ap.J.*, in press.
Field, G.B. 1965, *Ap.J.*, **142**, 531.
Field, G.B., and Saslaw, W.C. 1965, *Ap.J.*, **142**, 568.
Field, G.B., Goldsmith, D.W., and Habing, H.J. 1969, *Ap.J.(Letters)*, **155**, L149.
Franco, J. 1990, in *Chemical and Dynamical Evolution of Galaxies,* F. Ferrini, J. Franco, and F. Matteucci (eds.), Giardini, Pisa-Lugano, in press.
Franco, J., Tenorio-Tagle, G., Bodemheimer, P., Rozyczka, M., and Mirabel, I.F. 1988, *Ap.J.*, **333**, 826.
Fukui, Y., and Mizuno, A. 1991, in *Fragmentation of Molecular Clouds and Star Formation,* E. Falgarone, F. Boulanger, and G. Duvert (eds.), Kluwer, Dordrecht, in press.
Gerola, H., and Seiden, P.E. 1978, *Ap.J.*, **223**, 129.
Goldreich, P., and Lynden-Bell, D. 1965, *M.N.R.A.S.*, **130**, 97.
Grabelsky, D.A., Cohen, R.S., Bronfman, L., and Thaddeus, P. 1987, *Ap.J.*, **315**, 122.
Heiles, C. 1967, *Ap.J.Suppl.*, **15**, 97.
Igumentshchev, I.V., Shustov, B.M., and Tutukov, A.V. 1990, *Astr.Ap.*, **234**, 396.
Kwan, J. 1979, *Ap.J.*, **229**, 567.
Kwan, J. 1988, in *Molecular Clouds in the Milky Way and External Galaxies,* R.L. Dickman, R.L. Snell, and J.S. Young (eds.), Reidel, Dordrecht, 281.
Lada, C.J. 1985, *Ann.Rev.Astr.Ap.*, **23**, 267.
Larson, R.B. 1981, *M.N.R.A.S.*, **194**, 809.
Larson, R.B. 1987, in *Starburst and Galaxy Evolution,* T.X. Thuan, T. Montmerle, and J. Tran Thanh Van (eds.), Gif sur Yvette, Editions Frontières, 467.

Larson, R.B. 1988, in *Galactic and Extragalactic Star Formation*, R.E. Pudritz and M. Fich (eds.), Kluwer, Dordrecht, 459.
Leisawitz, D., Bash, F.N., and Thaddeus, P. 1989, *Ap.J.Suppl.*, **70**, 731.
Leisawitz, D. 1990, *Ap.J.*, **359**, 319.
Léorat, J., Passot, T., and Pouquet, A. 1990, *M.N.R.A.S.*, **243**, 293.
Lioure, A., and Chièze, J.P. 1990, *Astr.Ap.*, **235**, 379.
Low, F.J., Bientema, D.A., Gautier, T.N., Gillett, F.C., Beichman, C.A., Neugebauer, G., Young, E., Aumann, H.H., Boggess, N., Emerson, J.P., Habing, H.J., Hauser, M.G., Houck, J.R., Rowan-Robinson, M., Soifer, B.T., Walker, R.G., and Wesselius, P.R. 1984, *Ap.J.(Letters)*, **278**, L19.
MacLow, M.M., McCray, R., and Norman, M.L. 1989, *Ap.J.*, **337**, 141.
Maddalena, R.J., and Thaddeus, P. 1985, *Ap.J.*, **294**, 231.
McCray, R., and Kafatos, M. 1987, *Ap.J.*, **317**, 190.
Mooney, T.J., and Solomon, P.M. 1988, *Ap.J.(Letters)*, **334**, L51.
Myers, P.C. 1987, *Star Forming Regions,* IAU Symposium No. 115, M. Peimbert and J. Jugaku (eds.), Reidel, Dordrecht, 33.
Norman, C.A., and Silk, J. 1980, *Ap.J.*, **238**, 158.
Norman, C.A., and Ikeuchi, S. 1989, *Ap.J.*, **345**, 372.
Nozakura, T. 1990, *M.N.R.A.S.*, **243**, 543.
Oort, J.H., 1954, *Bull.Astr.Inst.Neth.*, **12**, 177.
Palous, J., Franco, J., and Tenorio-Tagle, G. 1990, *Astr.Ap.*, **227**, 175.
Parravano, A. 1987, *Astr.Ap.*, **172**, 280.
Parravano, A. 1988, *Astr.Ap.*, **205**, 71.
Parravano, A. 1989, *Ap.J.*, **347**, 812.
Parravano, A., Rosenzweig, P., and Teran, M. 1990, *Ap.J.*, **356**, 100.
Passot, T., Pouquet, A., and Woodward, P. 1988, *Astr.Ap.*, **197**, 228.
Pérault, M., Falgarone, E., and Puget, J.L. 1985, *Astr.Ap.*, **152**, 371.
Pudritz, R.E., 1990, *Ap.J.*, **350**, 195.
Rand, R.J., and Kulkarni, S. 1990, *Ap.J.(Letters)*, **349**, L43.
Safronov, V.S. 1960, *Ann d'Ap.*, **23**, 979.
Scalo, J.M., 1985, in *Protostars and Planets II,* D.C. Black and M.S. Matthews (eds.), University of Arizona, Tucson, 201.
Scalo, J. 1990, in *Physical Processes in Fragmentation and Star Formation*, R. Capuzzo-Dolcetta, C. Chiosi, and A. Di Fazio (eds.), Kluwer, Dordrecht, in press.
Scalo, J., and Pumphrey, W.A. 1982, *Ap.J.(Letters)*, **258**, L29.
Schneider, S., and Elmegreen, B.G., 1979, *Ap.J.Suppl*, **41**, 87.
Scoville, N.Z., and Hersh, K. 1979, *Ap.J.*, **229**, 578.
Scoville, N., Sanders, D.B., and Clemens, D.P. 1986, *Ap.J.(Letters)*, **310**, L77.
Skillman, E.D. 1987, in *Star Formation in Galaxies,* C.J. Lonsdale (ed.), NASA Conference Publication 2466, 263.
Solomon, P.M., Rivolo, A.R., Barrett, J., and Yahil, A. 1987, *Ap.J.*, **319**, 730.
Spitzer, L. Jr. 1968, *Diffuse Matter in Space,* Wiley, New York.
Spitzer, L. Jr. 1978, *Physics of Interstellar Matter,* Wiley, New York.
Stark, A.A., and Brand, J. 1989, *Ap.J.*, **339**, 763.
Struck-Marcell, C., and Scalo, J.M. 1984, *Ap.J.*, **277**, 132.
Tenorio-Tagle, G., and Palous, J. 1987, *Astr.Ap.*, **186**, 287.
Tenorio-Tagle, G., and Bodenheimer, P.H. 1988, *Ann.Rev.Astr.Astrop.*, **27**, 145.
Tenorio-Tagle, G., Rozyczka, M., and Bodenheimer, P. 1990, *Astr.Ap.*, **237**, 207.
Tomisaka, K. 1987, *Pub.Astron.Soc.Japan*, **39**, 109.
Tomisaka, K. 1990, *Ap.J.(Letters)*, **361**, L5.
Toomre, A. 1964, *Ap.J.*, **139**, 1217.

Verschuur, G. 1970, *A.J.*, **75**, 687.
Vishniac, E.T. 1983, *Ap.J.*, **274**, 152.
Yuan, C. 1969, *Ap.J.*, **158**, 889.

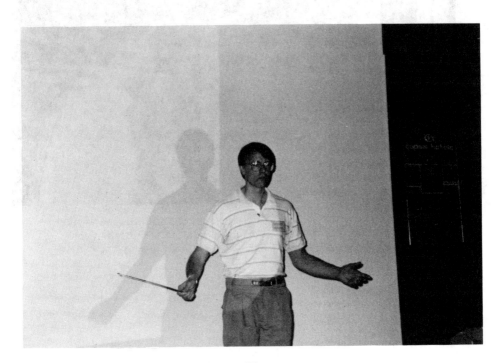

Bruce Elmegreen

Bruce Elmegreen

COSMIC MAGNETISM AND THE BASIC PHYSICS OF THE EARLY STAGES OF STAR FORMATION

TELEMACHOS CH. MOUSCHOVIAS
University of Illinois at Urbana-Champaign
Departments of Physics and Astronomy
1002 West Green street
Urbana, IL 61801, U. S. A.

ABSTRACT. Multifluid magnetohydrodynamics (nonideal MHD) is an indispensable tool for studying the physics of star formation. The physical origin of processes crucial to the formation of stars becomes clear, and the essence of their quantitative theoretical description becomes straightforward when a first-principles approach is taken. Progressively simpler forms of the equations are obtained and used to introduce concepts and describe processes such as ambipolar diffusion and the *self-initiated* formation and gravitational collapse of fragments (or cores); ohmic dissipation of magnetic flux; flux freezing in the electrons, ions, and neutral particles (through collisional coupling with ions and, possibly, charged grains); quasistatic and dynamical contraction of cloud cores; magnetic braking and the resolution of the thorny angular momentum problem; the magnetic Rayleigh-Taylor (or Parker) instability and formation of cloud complexes; equilibria of isothermal, magnetic clouds; the physical origin of the relation $B_c \propto (\rho_c T_c)^{1/2}$ between the magnetic field strength and the gas density and temperature in cores of contracting, not necessarily isothermal, self-gravitating clouds; and critical masses for the collapse of interstellar clouds. Improvements to the Virial Theorem are also presented. Close contact with observations is emphasized and maintained throughout the paper.

New results from analytical and numerical axisymmetric calculations on the formation and collapse of cloud cores and consequent refinements of a modern, basic theory of star formation are presented in the accompanying paper.

1. INTRODUCTION - OVERVIEW

Stars are the inevitable offsprings of the self-gravity of interstellar matter once it becomes comparable in magnitude to the disruptive forces due to magnetic fields, thermal pressure, hydromagnetic waves or, perhaps, turbulence, and rotation. Therefore, understanding star formation becomes a study of and a search for nature's delaying tactics through agents such as magnetic fields, thermal and wave or turbulent pressure, rotation, and even stars themselves (through winds and ionizing photons, and through the spectacular way in which some stars die as supernovae).

Stimulated by the pioneering ideas of Spitzer (1968) and Mestel (1965), and with early encouragement by and insightful discussions with G. B. Field, our work over the past fifteen years has established magnetic fields first as gravity's most faithful ally in reducing the otherwise potentially fierce centrifugal forces to near insignificant levels (through magnetic braking) and, later, as gravity's most formidable opponent in converting the would-be violent star formation process into a relatively quiescent one for most of its duration (see reviews by Mouschovias 1987a, b, 1989; hereafter referred to as TM87a, TM87b, etc.). Even as ambipolar diffusion, the means which gravity devises to win its battle against magnetic fields, allows the mass-to-flux ratio of cloud cores to increase, magnetic forces still regulate the contraction rate at least until the mass-to-flux ratio exceeds the critical value $(63G)^{-1/2}$, calculated by Mouschovias and Spitzer (1976). Gravitational forces have at this stage won a crucial battle, and the contraction of cold cores evolves into rapid collapse, which progresses until nature invents a new way to turn gravity against itself. This is done by the trapping of gravitational potential energy, released by contraction, and its conversion into thermal energy, which enforces relatively slow contraction all the way to the Main Sequence, where energy released by nuclear reactions (another strategic but unavoidable mistake of gravity) ensures a long but finite standoff between thermal-pressure and gravitational forces.

1.1. The Approach, and a Viewpoint

This paper is neither a detailed review of the literature nor a description of the most recent or most exciting theoretical or observational results relating to star formation --although very recent and very exciting results are discussed. (For reviews see Mestel 1965, 1977; Spitzer 1968a, b, 1978; Field 1970; Mouschovias 1978, 1981, 1983a, 1987a, b, 1991 (§§ 1 - 3, and 4.3); Nakano 1981, 1984; Shu, Adams, and Lizano 1987; McKee et al. 1991.) It focusses, instead, on providing the basic theoretical framework necessary to discover, describe and understand, at a basic physics level, nature's fundamental interstellar processes that lead to the birth of stars. These processes include the formation of cloud complexes, the equilibrium and contraction of individual clouds, the resolution of the thorny angular momentum problem by magnetic braking, the formation of fragments (or cores) by ambipolar diffusion, and the contraction or collapse of such fragments that leads to single and binary star formation. This approach almost guarantees that the paper will not become outdated in the near future.

The paper (and, therefore, the level of presentation) is aimed at graduate students beginning their research in astrophysics and at readers whose intellectual curiosity, unsatisfied by qualitative or order-of-magnitude descriptions or guesswork, desires either rigorous or simple but reliable analytical answers and explanations. The key theoretical discoveries of what precisely the role of the interstellar magnetic field is in the star formation process have been made not just because a worker has come up with a fundamentally new idea but also because that idea was used to introduce *appropriate* idealizations (or simplifications) in the formulation of the problem, not just to make its solution tractable, but to include in the formulation the *essential* physics of the problem. That is the way a theory is born, and that is the way of science. The young readers are forewarned that, occasionally, such approaches are termed "unrealistic" by some authors and are dismissed as such, only for these same authors to return to the problem years later, make inconsequential changes to its original formulation (presumably to make it more "realistic"), rediscover the same result(s) as the original solution, and then proceed to claim the result(s) as new. That is *not* the way of science, and it cannot be the way of theoretical astronomy or astrophysics either.

1.2. Key Developments: Elements of a Modern Theory of Star Formation

Complex as the structure of molecular clouds may be, particularly on small scales, those *individual* self-gravitating entities which have not yet given birth to stars are nearly isothermal ($T \simeq 10$ K) systems, typically characterized by a mean density $n \simeq 10^3$ cm^{-3} and a mass $\sim 10^2 - 10^4$ M$_\odot$, and are typically threaded by a mean magnetic field 10 - 100 μG (e.g., see reviews by Myers 1985; Heiles 1987). At the one extreme, unbound molecular clumps of mass $\sim 1 - 10$ M$_\odot$ are also observed (Blitz 1987); they are thought to be transient structures that form behind interstellar shocks. At the other extreme, cloud complexes containing more than 10^6 M$_\odot$ and having mean densities ≤ 50 cm^{-3} often represent the extended neighborhoods of star forming molecular clouds, and they are shown to form by the Parker (1966) instability (Mouschovias 1974; Mouschovias, Shu, and Woodward 1974; Mouschovias 1978; Blitz and Shu 1980; see also the review by Mouschovias 1981). Of direct interest to star formation are the individual self-gravitating clouds. If it were not for the presence of the magnetic field they would all be collapsing on essentially the free-fall time scale because the Bonnor-Ebert critical mass for a representative molecular cloud is only 5.8 M$_\odot$. In general, it is given by

$$M_{BE} = 1.2 \frac{C^4}{(G^3 P_{ext})^{1/2}}, \qquad (1a)$$

$$= 5.8 \left[\frac{T}{10 \text{ K}}\right]^{3/2} \left[\frac{10^3 \text{ cm}^{-3}}{n}\right]^{1/2} \text{ M}_\odot \qquad (1b)$$

(Bonnor 1956; Ebert 1955, 1957). The quantity C [$= (k_B T/\mu m_H)^{1/2}$] in equation (1a) is the isothermal speed of sound in the cloud, P_{ext} is the fixed external pressure, and G the universal gravitational constant; k_B is the Boltzmann constant, and μ the mean mass per particle in units of the atomic hydrogen mass, m_H. In order to account for the cosmic abundance of helium, we used $\mu = 2.33$ in obtaining equation (1b).

Subsonic turbulence can be accommodated as an effective increase of T (by at most a factor of 2). Thus the combination of thermal and turbulent pressure forces can support at most a 16 M$_\odot$ molecular clump. Supersonic (but subAlfvénic) small-scale turbulence dissipates too rapidly (by shocks along field lines and by ambipolar diffusion perpendicular to field lines) to be of relevance for any significant length of time, certainly in clouds which have not yet given birth to stars.

Early observational evidence revealed a strong spatial correlation between young stars and relatively dense concentrations of interstellar matter (Baade 1944). This led to the hypothesis, which prevailed until the mid-1970s, that star formation was the result of the collapse (and fragmentation) of an interstellar cloud *as a whole*. However, neither observations (e.g., see review by Zuckerman and Palmer 1974), which could not detect velocities characteristic of collapse, nor theoretical calculations (see Mouschovias 1976a, b, 1978; Mouschovias and Spitzer 1976), which showed that even relatively weak magnetic fields could provide very effective support against gravity, lent any support to that popular notion. The predicted, relatively weak fields turned out to be consistent with those subsequently measured (see reviews by Heiles 1987, and Mouschovias 1981, 1987a). A new picture emerged in which formation of fragments (or cores) and their subsequent collapse was initiated in cloud interiors by ambipolar diffusion in otherwise magnetically supported clouds (Mouschovias 1976b, p. 156; 1977; 1978; 1979b). *These papers were the first to demonstrate, both qualitatively and quantitatively, that ambipolar diffusion initiates the quasistatic (i.e., nonaccelerated) formation and, later, collapse of cores while the envelopes remain much better supported by magnetic forces.* [This process of *self-initiated collapse* of cores within magnetically supported envelopes was subsequently referred to as "inside-out collapse" by some authors (e.g., see Lizano and Shu 1987, p. 185).] The theoretical conclusion was reached that low-mass stars should preferentially form first, and perhaps only, in the cores of dense clouds. Soon thereafter, observations confirming that prediction began to accumulate (see extensive discussion in Vrba, Coyne, and Tapia 1981). In other words, the theoretical conclusion is reached that *the collapse of a molecular cloud as a whole is not necessary for stars to form in its interior --nor is it, in fact, commonplace in nature.*[1]

Observations also reveal that the topology of the magnetic field is too orderly (e.g., see Vrba *et al.* 1981; Moneti *et al.* 1984; Heyer *et al.* 1987) and its pressure too large (compared to the thermal pressure) to permit the inference that chaotic motions (or turbulence) play a significant role in the overall support of molecular clouds (see review TM87a, § 2.2.1). Interplay between theory and observations has led to more detailed calculations which represent improvements and refinements of the qualitatively novel theory of star formation. New observations were also undertaken. The updated, but still incomplete, theory is described in the papers TM87a and TM87b, which include a number of new predictions and a comparison with observations. Its key elements are summarized in Paper II. In the present paper, we focus on the theoretical description of the underlying physical processes, not necessarily in the order in which they occur in nature, but in an order revealing the strength(s) as well as the limitation(s) of the theoretical description, by quantifying the validity of the approximations used. This approach lends itself to straightforward generalization and refinement to include additional phenomena as the need arises.

1.3. Outline and Perspective

In general, interstellar matter, whether in clouds or in the diffuse intercloud medium, consists of neutral particles, ions, electrons, charged and neutral grains, and (charged) relativistic particles (cosmic rays). The bulk of this matter is in neutral particles and comprises approximately 10% of the mass of the Galaxy. All these particles are acted upon by the galactic gravitational field (due to stars) and, in the neighborhood or inside dense clouds, by the self-gravity of these clouds. Electromagnetic forces affect the charged particles directly, while collisional drag between charged and neutral particles transmits the electromagnetic forces to the neutrals. Since collisions between neutral and charged particles are hardly ever so perfect as to allow a theoretical description of the system as a single fluid, a multifluid magnetohydrodynamic (MHD) approach becomes, in general, necessary for the low-frequency (long time scale) phenomena relevant to star formation. These equations, referring to a system consisting of electrons, ions and neutrals, are presented and discussed in § 2 from a physical point of view, with particular emphasis on the conditions under which each term should be included or ignored. Although grains, especially small ones, play an important role in the evolution of molecular cloud interiors, they are not

---

[1]The notion that the contraction of a cloud in the presence of ambipolar diffusion can be quasistatic (i.e., the acceleration of the neutrals is negligible) was not invented by Mouschovias (1976b - 1979b), or Nakano (1976 - 1979), or Shu (1983), or Lizano and Shu (1989) --other claims in the literature notwithstanding. Spitzer (1968a) had the magnificent intuition to see that the quasistatic approximation could be used, and he did use it correctly to estimate the ambipolar diffusion time scale in a cylindrical cloud model which maintains a uniform density as it contracts. What TM contributed is the idea and an actual calculation that ambipolar diffusion *initiates* the quasistatic formation and, later, collapse of *cores* (and, therefore, star formation) *in otherwise magnetically supported envelopes* (Mouschovias 1976b, p. 156; 1977; 1978; 1979b); *hence, external triggers and/or the collapse of a cloud as a whole were no longer necessary, although both possible, for stars to form.* Shu (1982, p. 243) acknowledges the originality and significance of TM's contribution. In recent years, nevertheless, a number of papers are presenting both Spitzer's and TM's ideas and results as if they are now being discovered for the first time.

considered here. The goal is to keep the formalism as simple as possible in the limited space available and to introduce in an uncomplicated but rigorous way important phenomena (e.g., ambipolar diffusion) which change quantitatively but not qualitatively by the introduction of additional species. Although the three-fluid equations have much wider applicability than in the interiors of molecular clouds, we motivate them in § 2.1 in the context of the role of ambipolar diffusion in the collapse of self-gravitating clouds and star formation. The concept of flux freezing in the plasma (electrons and ions) is introduced in § 2.2, where it is shown that it is a concept applicable to a wide variety of interstellar conditions. This is then used in § 2.3 to simplify the formal description of the plasma and to arrive at the two-fluid (plasma and neutrals) MHD equations, which form the basis of theoretical studies of ambipolar diffusion (the bulk motion, or drift, of the plasma relative to the neutrals). General properties of the two-fluid equations directly related to the evolution of molecular clouds and star formation are also obtained. A simple but accurate and general analytical solution for the formation and quasistatic (i.e., nonaccelerated but not necessarily slow) contraction of cloud cores due to ambipolar diffusion is presented in § 2.3.1. The effect of the dynamical contraction of cloud cores on ambipolar diffusion is also studied analytically in § 2.3.2. The original, exact solution obtained by TM79b, which put on a rigorous footing the earlier ideas and approximate calculations (TM76b, TM77, TM78) that ambipolar diffusion *initiates* the formation and collapse of cores (and only of cores) of self-gravitating, magnetically supported clouds, is summarized in § 2.3.3.

The physical conditions under which the two-fluid MHD equations can be further reduced to the single-fluid equations, appropriate for the description of large-scale processes in the interstellar medium (such as the formation of interstellar clouds), are explored in § 3. A basic description of cosmic rays as a fluid is presented in § 3.1, and the linearized single-fluid equations are used in § 3.2 to study the possible formation of clouds through the magnetic Rayleigh-Taylor (or Parker) instability, after a brief review of other possible mechanisms is given. The transport of angular momentum (J) from a (subAlfvénically) contracting cloud (or core) to the "external" medium (or envelope) due to a frozen-in magnetic field (B) is studied in the case of aligned ($J \parallel B$) and perpendicular ($J \perp B$) rotators in turn in §§ 3.3.2 and 3.3.3 (magnetic braking). The results are explained in terms of exact analogies with transverse waves on strings. The study of the effect of ambipolar diffusion on magnetic braking of subAlfvénically contracting clouds requires the use of the two-fluid equations; it is given in § 4. The results are explained in terms of exact analogies with electrical resistors connected in series and in parallel. Angular velocities of clouds and the periods of binary stars from 10 hr to 100 yr are predicted or explained naturally (TM77; see, also, the review TM87a, §§ 3.5.4 and 3.6).[2]

Equilbria of clouds are studied by using a further reduced form of the single-fluid MHD equations, namely, the magnetohydrostatic equations (MHS). Unfortunately, these equations form an inherently open set (there is one more unknown quantity than there are equations). A self-consistent way to close this system of equations is shown in § 5. Solutions representing large-scale condensations (or cloud complexes) in valleys of field lines along spiral arms are presented in § 5.1. The key results of calculations of equilibria of self-gravitating, isothermal, magnetic clouds are described in § 5.2. A simple analytical derivation of the value of the exponent κ in the relation $B_c \propto \rho_c^\kappa$ between the magnetic field strength and the matter density in contracting cloud cores is also given; it is valid even in the presence of hydromagnetic waves. The scalar Virial Theorem is obtained, improved and employed to reexamine the equilibrium and criteria for collapse of magnetic clouds in § 6.

---

[2]Nakano (1991) claims this result as new!

Results of new analytical and numerical calculations on the formation and evolution of molecular cloud cores due to ambipolar diffusion are left for Paper II, in which a basic theory of star formation, refined by the new results, is presented. Surprises are in store.

The most serious omission in this paper, entirely due to space limitations, is a discussion of discontinuities (such as shock waves and ionization fronts) permitted by the MHD equations, occurring in the interstellar medium, and sometimes playing a role in star formation. The reader can find excellent reviews in Spitzer (1978) and in McKee and Hollenbach (1980).

2. ELEMENTS OF MULTIFLUID MAGNETOHYDRODYNAMICS

2.1. Motivation, and the Key Questions

The magnetic flux problem of star formation has traditionally been stated as follows: If the magnetic field of a collapsing cloud remained frozen in the matter all the way up to main-sequence densities, spherical isotropic collapse would result in typical stellar (mean) fields $B_* \sim n^{2/3}$ $\mu G \sim 10^{10}$ Gauss. These are much too large compared to observed values, which are $\lesssim 10^4$ Gauss. However, because the contraction of a magnetic cloud is neither spherical nor isotropic, it is much more precise to state the problem in a fashion independent of the details of the contraction: *The magnetic flux of a ~ 1 $M_\odot$ blob of interstellar matter at the mean density (~ 1 cm^{-3}) and magnetic field ($\simeq 3$ μG) of the interstellar medium is estimated to exceed the observed flux of magnetic stars by 2 - 5 orders of magnitude, depending on the assumed geometry; i.e., thin cylindrical magnetic flux tube or spherical blob, respectively* (see TM83a; TM87a, § 3.2; TM91, § 4.3). (Nakano [1983] strongly insists that the discrepancy is 4 - 5 orders of magnitude. More recently, however, he [1991] accepts 2 - 5 orders of magnitude as the proper discrepancy between stellar and corresponding interstellar magnetic fluxes.) In either case, it follows that the interstellar magnetic field must decouple from the matter that goes into stars at some stage during cloud collapse and star formation. Existing observations do not establish the stage by which the magnetic flux problem must be resolved. Measurements of the field strength in protostellar objects in the density range 10^5 - 10^{10} cm^{-3} could decide this issue (see TM87a, eq. [18]) and would, in any case, be invaluable as a guide to theoretical calculations. Ambipolar diffusion was proposed by Mestel and Spitzer (1956) as a means of resolving the magnetic flux problem. Regardless of whether it can do so or not and regardless of the stage at which such resolution may occur, ambipolar diffusion is an unavoidable process at least in the cores of self-gravitating, magnetically supported clouds and its onset has profound effects on the evolution of a cloud and star formation (TM76b, TM77, TM79b) even if the drift speed, v_D, is only comparable to, or even smaller than, the isothermal speed of sound, C.

Theoretical studies of ambipolar diffusion must ultimately answer at least the following questions. (i) At which stage of a cloud's evolution does it set in at a significant rate to affect the evolution, and precisely how rapidly does it progress? (ii) What is the size, mass, magnetic flux, and structure of the objects (fragments or cores) which form? (iii) Can it reduce the magnetic fluxes of protostellar fragments sufficiently to agree with observations and, if so, at which stage and in how long a time? (iv) By how much does it lengthen the time scale for magnetic braking of a contracting fragment and is that sufficient to render magnetic braking ineffective? If yes, at which stage does this happen, and is the leftover angular momentum what is required to explain binary periods and single star rotation rates? (v) Does it play a role in the determination of the initial (stellar) mass function (IMF), by introducing a length scale in the problem? (vi) What are its observational consequences?

2.2 Flux Freezing in the Plasma, and Reduction of the Three-Fluid Equations

We consider a system consisting of electrons, ions (each of charge $+ze$), and neutrals in the presence of electric (E), magnetic (B), and gravitational ($g = -\nabla\psi$) fields, with at least partial collisional coupling among the species. Assuming charge neutrality, i.e.,

$$n_i z e = n_e e ,\qquad (2)$$

(usually $z = 1$) so as to suppress high frequency (plasma) phenomena, and ignoring ionizations and recombinations (see § 2.3 below), the three force equations are written as

$$\rho_e \left[\frac{d}{dt}\right]_e v_e = -\nabla P_e - \rho_e \nabla\psi - n_e e \left[E + \frac{v_e}{c} \times B\right] + F_{ei} + F_{en} ,\qquad (3a)$$

$$\rho_i \left[\frac{d}{dt}\right]_i v_i = -\nabla P_i - \rho_i \nabla\psi + n_i z e \left[E + \frac{v_i}{c} \times B\right] + F_{ie} + F_{in} ,\qquad (3b)$$

$$\rho_n \left[\frac{d}{dt}\right]_n v_n = -\nabla P_n - \rho_n \nabla\psi + F_{ne} + F_{ni} ,\qquad (3c)$$

where

$$\left[\frac{d}{dt}\right]_k () \equiv \frac{\partial()}{\partial t} + (v_k \cdot \nabla)() ,\qquad (3d)$$

is the time derivative comoving with species k ($= e$, or i, or n). A quantity F_{12} ($= -F_{21}$ by Newton's third law) represents the force per unit volume on species 1 due to collisions with species 2, and is given by

$$F_{12} = n_1 \nu_{12} \left[\frac{m_1 m_2}{m_1 + m_2} (v_2 - v_1)\right] .\qquad (4)$$

The quantity ν_{12} is the collision frequency of a particle of species 1 with particles of species 2, and the quantity in the large parentheses in equation (4) is the mean momentum change upon collision.

The time evolution of the magnetic field can be determined from Maxwell's equation

$$\frac{\partial B}{\partial t} = -c (\nabla \times E) \qquad (5)$$

once E is known. (The displacement current has been ignored for the low frequency phenomena of interest here.) To obtain E, the small electron inertia is neglected in equation (3a), which is solved for E. Then the *curl* of this expression is taken and substituted in the right-hand side of equation (5) to find that

$$\frac{\partial B}{\partial t} - \nabla \times (v_e \times B) = -c \nabla \times \left[\frac{j}{\sigma_{ei}}\right] - \frac{c}{n_e^2 e} \nabla n_e \times \nabla P_e - c \nabla \times \left[\frac{F_{en}}{n_e e}\right] ,\qquad (6a)$$

where the electric current density j and electrical conductivity σ_{ei}, limited only by e-i collisions (the contribution of e-n collisions is included in the term involving F_{en} --see eq. [6b] below), are given by

$$j = n_e e\,(v_i - v_e) = (\sigma_{ei}/n_e e)\, F_{ei}\,, \qquad (7a, b)$$

$$\sigma_{ei} = (n_e e^2/m_e \nu_{ei}) \sim 10^7\, T^{3/2} \quad s^{-1} \qquad (8a, b)$$

(Spitzer 1962), and use has been made of equation (4) in the right-hand side of equation (7a) to express the electron-ion relative velocity in terms of F_{ei} and to obtain equation (7b).

The magnetic field is "frozen in" the electrons (in that the magnetic flux threading a volume which bounds always the same electrons as they move remains constant in time) if each term on the right-hand side of equation (6a) is negligible compared to the second term on the left-hand side, which represents the advection of magnetic flux by the motion of electrons.

The second term on the right-hand side ("Biermann's battery") vanishes if the electron isodensity contours and isobars coincide --this is expected in interstellar clouds, but not necessarily in stellar interiors.

The condition for the Ohmic dissipation term (first one on the right-hand side) to be negligible is

$$\frac{\text{Ohmic term}}{e\text{-advection term}} \sim \frac{\nu_{ei}}{\omega_{c,e}} \frac{|v_i - v_e|}{v_e} \sim \left(\frac{c}{4\pi \sigma_{ei} L}\right)\left(\frac{c}{v_e}\right) \ll 1\,, \qquad (9a)$$

where $\omega_{c,e} = eB/m_e c$ is the electron cyclotron frequency. To obtain the first part of equation (9a), equations (7a) and (8a) have been used, while to obtain the second part, use has been made of Ampere's law,

$$\nabla \times B = (4\pi/c)\, j\,, \qquad (10)$$

and $|\nabla|$ has been approximated with the inverse of a characteristic length L of the physical system. Estimating the above ratio for molecular cloud cores, for which $L \sim 10^{17}$ cm, $v_e \sim 10^4$ cm s^{-1}, and $T \sim 10$ K, we find that it is $\sim 10^{-10}$. Only when the electrons virtually disappear by attachment onto grains, at densities $\gtrsim 10^{10}$ cm^{-3}, would Ohmic dissipation come into play.

The last term, representing a drift of the electrons relative to neutrals, is negligible if

$$\frac{e\text{-}n \text{ drift term}}{e\text{-advection term}} \sim \frac{\nu_{en}}{\omega_{c,e}} \frac{|v_n - v_e|}{v_e} \ll 1\,. \qquad (9b)$$

Since $\nu_{en} \sim 10^2\, \nu_{in} \sim 10^{-7}\, n_n$ s^{-1} (see Mouschovias and Paleologou 1981, p.52 for a discussion of the i-n collision cross section and of a critical speed for breakdown of the Langevin approximation), we find that this ratio is $\sim 10^{-6}$ in dense molecular cloud cores, for which we take $(|v_n - v_e|)/v_e \sim 1$ and $B \sim 10^{-4}$ Gauss as representative, although even the strict lower limit of $B \sim 10^{-6}$ Gauss would not alter this conclusion.

Note that the e-n drift term in equation (6a) may be rewritten so as to bring out explicitly the contribution of electron-neutral collisions to the electrical resistivity $\eta = \sigma_c^{-1}$. The quantity F_{en} is eliminated by using equation (4), and v_e in this term is expressed in terms of j and v_i through equation (7a) to find that equation (6a) becomes

$$\frac{\partial B}{\partial t} - \nabla \times (v_e \times B) = -c\,\nabla \times \left[\frac{j}{\sigma_c}\right] - \frac{c}{n_e^2 e} \nabla n_e \times \nabla P_e - c\,\nabla \times \left[\frac{m_{en}}{e} \nu_{en} (v_n - v_i)\right] \qquad (6b)$$

where the total electrical conductivity σ_c is given by $\sigma_c^{-1} = \sigma_{ei}^{-1} + \sigma_{en}^{-1} = (m_e/n_e e^2)(\nu_{ei} + \nu_{en})$, and the electron-neutral reduced mass is $m_{en} = m_e m_n/(m_e + m_n) \simeq m_e$. In weakly ionized gases (e.g., molecular cloud cores with $n_n \gtrsim 10^4$ cm^{-3}), $\nu_{en} \gtrsim \nu_{ei}$ so that $\sigma_c < \sigma_{ei}$. Hence, electrical resistivity will be dominated by e-n collisions (at least up to the high densities at which n_e is reduced significantly due to attachment onto grains.

It still remains to examine whether the field is also frozen *in the ions* in molecular cloud cores. We use equation (7a) to eliminate v_e from the left-hand side as well of equation (6b) in favor of v_i and j to find that

$$\frac{\partial B}{\partial t} - \nabla \times (v_i \times B) = -c \nabla \times \left[\frac{j}{\sigma_c}\right] - \frac{c}{n_e^2 e} \nabla n_e \times \nabla P_e - c \nabla \times \left[\frac{m_{en}}{e} \nu_{en} (v_n - v_i)\right]$$

$$- \nabla \times \left[\frac{j}{n_e e} \times B\right]. \qquad (11)$$

Equation (11) differs from equation (6b) in that v_i has replaced v_e on the left-hand side, and a new term (the last one) has appeared on the right-hand side; this is known as the Hall term. It can be neglected if

$$\frac{\text{Hall term}}{i\text{-advection term}} \sim \frac{|v_i - v_e|}{v_i} \ll 1. \qquad (9c)$$

That this is so can be seen by estimating from equation (10) the relative speed $|v_i - v_e|$ needed to maintain the magnetic field in molecular cloud cores; we find that $(|v_i - v_e|) \sim cB/4\pi n_e eL \sim 1$ cm s^{-1} for $n_e \sim 3 \times 10^{-3}$ cm^{-3}. Since $v_i \sim 10^4$ cm s^{-1}, the Hall term may be neglected as well, and *one considers the magnetic flux as being frozen in the plasma*.

2.3. Physical Considerations and the Basic Two-Fluid Equations: Ambipolar Diffusion

Under extreme interstellar conditions (i.e., in the densest, cold cores of dark or molecular clouds and in hot supernova remnants), the speed of sound, the Alfvén speed, and material speed lie in the approximate interval 10^{-1} - 10^2 km s^{-1}. On the other hand, the relative speed ($v_i - v_e$) between ions and electrons needed to maintain the observed magnetic fields can be estimated from Ampere's law (eq. [10]), as done above; it is found to be in the respective range 10^{-5} - 10^{-10} km s^{-1}. It is therefore an excellent approximation to consider the ions and the electrons as a single fluid (the plasma), any element of which (of density ρ_i and velocity v_i) inside a cloud is acted upon by magnetic and gravitational forces, thermal pressure gradients of the plasma, and frictional forces due to collisions with neutral particles (of density ρ_n and velocity v_n). Since the electrons contribute negligibly to the mass density of the plasma ($m_e \ll m_i$ and $n_e \simeq n_i$), and since $v_i \simeq v_e$ for practical purposes, the mass density and velocity of the ions (ρ_i and v_i, respectively) will also refer to the total density and mean bulk velocity of a plasma element.

The neutral particles, which are much more numerous than ions in molecular clouds ($10^{-9} \lesssim n_i/n_n \lesssim 10^{-3}$), respond to gravitational forces, partial-pressure gradients, and frictional forces due to collisions with ions (electron-neutral momentum exchange is negligible by comparison). By virtue of the fact that $\rho_n \gg \rho_i$, the plasma contributes negligibly to the generation of the gravitational field. The magnetic field affects the neutrals only indirectly: if (with respect to the center or axis of symmetry of a cloud) the ions move outward under the action of magnetic forces, ion-neutral collisional drag imparts outward-directed momentum to the neutrals. If, on the other hand, the neutrals are contracting, collisional drag tends to pull the ions inward as well. Since the field is frozen in the plasma, this motion compresses and/or deforms the field lines, giving rise to magnetic forces which tend to retard the inward motion of the ions. The result in both situations is an increase in the mass-to-flux ratio in at least those flux tubes threading a cloud's core, although in the latter case the magnetic energy of the cloud can keep increasing despite the fact that ambipolar diffusion is in progress (TM78, TM79b).

If the above were the entire story, one could summarize the physical picture by stating that the neutrals affect the plasma through the gravitational field (generated by the neutrals) and through collisions; the plasma affects the neutrals only through collisions. Ionization and recombination alter this picture. These microscopic processes constitute a vehicle for mass and momentum exchange between ions and neutrals. The fact that $\rho_n \gg \rho_i$ and $|v_n| \sim |v_i|$ implies that such exchanges have a negligible effect on the neutrals, but may have a significant effect on the ions if ρ_i becomes very small. We denote by S_i the *net* rate at which the ion density *increases* due to microscopic processes (S_i carries an algebraic sign). Since cosmic rays, which constitute the dominant source of ionization (as long as the surface density of matter does not exceed about 60 g cm^{-2} [e.g., see Nakano and Tademaru 1972]) in a dense cloud which has not yet given birth to stars, are isotropic, ionization does not impart a net momentum to a fluid element. One may therefore assume that a newly created ion initially has the velocity of the parent neutral particle; hence, the rate at which momentum per unit volume is added to the ions is $G_i v_n$, where G_i is the rate of increase of ρ_i due to ionizations. On the other hand, when ions are destroyed by recombination, the rate at which momentum per unit volume is lost by the ions is $L_i v_i$, where L_i is the rate of decrease of ρ_i due to recombinations. (Note that $S_i = G_i - L_i$.)

With the above explanations/approximations, we may write the basic two-fluid (plasma plus neutrals) MHD equations in a general form as follows.

$$\frac{\partial \rho_i}{\partial t} + \nabla \cdot (\rho_i v_i) = S_i , \tag{12a}$$

$$\frac{\partial \rho_n}{\partial t} + \nabla \cdot (\rho_n v_n) = - S_i , \tag{12b}$$

$$\frac{\partial}{\partial t}(\rho_i v_i) + \nabla \cdot (\rho_i v_i v_i) = - \nabla\left[2P_i + \frac{B^2}{8\pi}\right] - \rho_i \nabla\psi + \frac{1}{4\pi}(B \cdot \nabla)B + F_{\text{in}} + F_{\text{inel}} , \tag{12c}$$

$$\frac{\partial}{\partial t}(\rho_n v_n) + \nabla \cdot (\rho_n v_n v_n) = - \nabla P_n - \rho_n \nabla\psi - F_{\text{in}} - F_{\text{inel}} , \tag{12d}$$

$$\frac{\partial B}{\partial t} = \nabla \times (v_i \times B) , \tag{12e}$$

$$\nabla^2 \psi = 4\pi G \rho_n , \tag{12f}$$

$$P_n = \rho_n C_n^2 , \quad \text{and} \quad P_i = \rho_i C_i^2 , \tag{12g, h}$$

where the frictional force (per unit volume) *on the ions* due to neutrals (atomic or molecular hydrogen, depending on the application) is given by (see eq. [4])

$$F_{\text{in}} = \rho_i \rho_H \frac{\langle \sigma w \rangle_{\text{in}}}{m_i + m_n} (v_n - v_i) \equiv \frac{1}{\tau_{ni}} \rho_H (v_n - v_i) , \tag{12i}$$

and the force (per unit volume) on the ions due to ionizations and recombinations is

$$F_{\text{inel}} = G_i v_n - L_i v_i . \tag{12j}$$

In addition, we have that

$$S_i \equiv G_i - L_i \, . \tag{12k}$$

Equation (12j) becomes important at small ion densities. The quantities ψ, C_n, and C_i signify, respectively, the gravitational potential, and the isothermal speeds of sound in the neutrals and ions. All other symbols have their conventional meaning. Under the usual assumption that the plasma and the neutrals are at the same temperature, and since $\rho_i \ll \rho_n$, equation (12h) drops out and so does the term involving P_i on the right-hand side of the force equation (12c) for the plasma. (The factor 2 in this term accounts for the contribution of the electrons.) We have not written down an energy equation because in the density regime of interest ($n_n \lesssim 10^9$ cm^{-3}) isothermality, described by the equation of state (12g), is a good approximation. (Draine 1980 considered the energy equation in the context of time-independent ambipolar diffusion in radiative shocks.) The mass of an ion is denoted by m_i and that of a neutral by m_n. The average collisional rate $\langle \sigma w \rangle_{in}$ is equal to 1.69×10^{-9} cm^3 s^{-1} for HCO$^+$-H$_2$ collisions and is very close to this number for Na$^+$-H$_2$ collisions as well (e.g., see McDaniel and Mason 1973) as long as the relative speed w does not exceed the critical value 10.24 km s^{-1} (see Mouschovias and Paleologou 1981, p. 52).

With P_i neglected compared to P_n, it remains to specify G_i and L_i in order to close the system of equations (12a) - (12k). One can easily show that F_{inel} can be neglected compared to F_{in} if ionization equilibrium takes several ion-neutral collision times to be established. This is usually the case. Detailed calculations of ionization equilibrium in static clouds showed that the ion number density n_i is determined by the neutral number density n_n, the cosmic ray ionization rate, and the extent to which ionized metals are depleted onto grains (Elmegreen 1979; Nakano 1979). We approximate Nakano's Figure 1 in the range of density $10^3 \le n_n \le 10^9$ cm^{-3} by the relation

$$n_i = K \left(\frac{n_n}{10^5 \text{ cm}^{-3}} \right)^k , \tag{13}$$

where the constant $K \simeq 3 \times 10^{-3}$ cm^{-3} and the exponent $k \simeq 0.5$. However, it should be borne in mind that K can easily differ from this value by a factor of 10 (or more) due to uncertainties in the rate of cosmic-ray ionization and depletion onto grains, and k is uncertain by about 20% - 30%. If one adopts equation (13), or any relation between n_i and n_n, it is straightforward to express S_i in terms of ρ_n, v_i, and v_n, and thus close the system of equations. Some special solutions of these equations are given in §§ 2.3.1 - 2.3.3.

With ionizations and recombinations having a negligible effect on the neutrals, the continuity equation (12b), with $S_i = 0$, can be used in the left-hand side of the force equation (12d) to write it in the more familiar form $\rho_n(d\mathbf{v}_n/dt)$. This approximation is employed in the remainder of this paper.

Given the above approximations (but retaining the thermal pressures for the moment), one may study the relative motion of the plasma (hence, of the magnetic field) and the neutrals (i.e., *ambipolar diffusion*) by adding the force equations (3a) and (3b) for the electrons and ions in order to eliminate F_{ei}, simplifying the resulting equation, and then subtracting from it the force equation (3c) for the neutrals, to find that

$$\frac{d\mathbf{v}_i}{dt} - \frac{d\mathbf{v}_n}{dt} = - \left[\frac{\nabla(P_e + P_i)}{\rho_i} - \frac{\nabla P_n}{\rho_n} \right] + \frac{1}{\rho_i} \frac{\mathbf{j}}{c} \times \mathbf{B} - \frac{1}{\rho_i} \mathbf{F}_{ni} , \tag{14}$$

where we have used the fact that $\rho_e \ll \rho_i \ll \rho_n$, and that $F_{ne} \ll F_{ni}$ because of the relatively small electron mass. The subscripts have also been dropped from the comoving time derivatives (see eq. [3d]) for simplicity. For comparable plasma and neutral temperatures, the thermal pressure terms in equation (14) are of order C^2 (C is the speed of sound), while the magnetic term is seen (by using eq. [10]) to be of order $v_{A,i}^2$ (where $v_{A,i}$ is the Alfvén speed in the ions, and is $> 10^3$ km s^{-1} in molecular cloud cores). Thus the pressure terms contribute negligibly to ambipolar diffusion. A drift of the plasma relative to the neutrals caused by the magnetic force is seen from equation (14) to reach a terminal (steady-state) velocity when the magnetic and collisional forces balance, namely,

$$v_D \equiv v_i - v_n = \frac{\tau_{ni}}{\rho_H} \frac{j}{c} \times B = \frac{\tau_{ni}}{\rho_H} \frac{1}{4\pi}(\nabla \times B) \times B, \tag{15a}$$

$$= \frac{\tau_{ni}}{\rho_H}\left[-\nabla\left(\frac{B^2}{8\pi}\right) + \frac{1}{4\pi}(B \cdot \nabla)B\right] = \frac{\tau_{ni}}{\rho_H}\left[-\nabla_\perp\left(\frac{B^2}{8\pi}\right) + \hat{n}\frac{B^2}{4\pi R_c}\right]. \tag{15b}$$

Equation (10) has been used to eliminate j in favor of B. The mean (momentum exchange) collision time (τ_{ni}) of a neutral in a sea of ions is given in the second part of equation (12i). In the last part of equation (15b), the subscript $\perp$ refers to a direction perpendicular to a field line, the unit vector $\hat{n}$ is normal to a field line and points toward the local center of curvature, and R_c is the local radius of curvature of the field line.

It is seen immediately from equation (15) that

$$v_D \sim \frac{v_A^2 \tau_{ni}}{L}, \tag{16a}$$

$$\sim 0.4 \left[\frac{v_A}{2 \text{ km s}^{-1}}\right]^2 \left[\frac{\tau_{ni}}{10^4 \text{ yr}}\right]\left[\frac{0.1 \text{ pc}}{L}\right] \text{ km s}^{-1}, \tag{16b}$$

where in equation (16b) all quantities have been normalized to values typical of molecular cloud cores (see TM87a, eq. [5a] for the Alfvén speed in the neutrals $v_A \equiv v_{A,n}$). Thus, sonic ($C \simeq 0.2$ km s^{-1}) but subAlfvénic drift speeds are to be expected in molecular cloud cores, as found by detailed dynamical calculations in a number of model clouds.

Equation (16) suggests a natural ambipolar diffusion time scale $L/v_D \sim \tau_A^2/\tau_{ni}$, so that

$$\tau_{AD} \sim \frac{\tau_A^2}{\tau_{ni}}, \tag{16c}$$

$$\sim 2 \times 10^5 \left[\frac{2 \text{ km s}^{-1}}{v_A}\right]^2 \left[\frac{10^4 \text{ yr}}{\tau_{ni}}\right]\left[\frac{L}{0.1 \text{ pc}}\right]^2 \text{ yr}, \tag{16d}$$

where $\tau_A \equiv L/v_A$ is the Alfvén crossing time. *Note that equation (16c) is similar to equation (12a) of TM87a, except that τ_A has replaced the free-fall time, τ_{ff}. The fact that these two equations yield almost identical numerical values for τ_{AD} is not an accident. For magnetically supported clouds, to which equation (12a) of TM87a refers, we have that $\tau_A \sim \tau_{ff}$* (see also eq. [5b] in TM87a). In § 2.3.1 we show from first principles that the expression $\tau_{AD} \sim \tau_{ff}^2/\tau_{ni}$ is indeed a geometry-independent, general result, valid during the quasistatic contraction phase of cold cloud cores (TM89, § III also accounts for the effect of thermal pressure on τ_{AD}).

The time evolution of the magnetic field threading a predominantly neutral fluid element, by virtue of frequent collisions of ions with neutrals within the fluid element, can be obtained from the flux-freezing equation (12e) by eliminating v_i in favor of v_D and v_n (from the definition of the drift velocity), and using equation (15) to find

$$\frac{\partial B}{\partial t} - \nabla \times (v_n \times B) = \nabla \times (v_D \times B), \tag{17a}$$

$$= \nabla \times \left\{ \frac{\tau_{ni}}{\rho_H} \left[\frac{1}{4\pi} (\nabla \times B) \times B \right] \times B \right\}, \tag{17b}$$

$$= \nabla \times \left\{ \frac{\tau_{ni}}{\rho_H} \left[-\nabla \left(\frac{B^2}{8\pi} \right) + \frac{1}{4\pi} (B \cdot \nabla) B \right] \times B \right\}. \tag{17c}$$

This is clearly a diffusion equation, albeit a nonlinear one, with a diffusion coefficient $\mathscr{D} \sim v_A^2 \tau_{ni}$. Hence, the diffusion time scale is $\tau_{\text{diff}} \sim L^2/\mathscr{D} \sim \tau_A^2/\tau_{ni}$, which is identical with τ_{AD} (see eq. [16c]).

2.3.1. Quasistatic Contraction: General Treatment. As justified above, the force equation (12c) for the plasma can be represented accurately by its two dominant terms (i.e., the magnetic and the frictional force), $(j/c) \times B + F_{in} = 0$. This can be used to eliminate the drag force F_{in} from the neutral-particle force equation (12d), which, *for quasistatic contraction of cold cores*, is thereby written as

$$\rho_n g_\perp + \frac{j}{c} \times B = 0, \tag{18}$$

where $g \equiv -\nabla \psi$, and the subscript $\perp$ denotes a direction perpendicular to the field lines. Thus the drift velocity (eq. [15a]) becomes

$$v_D = -1.4 \, \tau_{ni} g_\perp, \tag{19}$$

where the cosmic abundance of helium has been accounted for (i.e., $\rho_n/\rho_H = 1.4$ for $n_{He}/n_H = 0.1$), and only ion-hydrogen collisions have been considered. In the equatorial plane of the core, we have that $|g_\perp| \propto G\rho_n r \propto r/\tau_{ff}^2$. Substituting this in equation (19) and using the resulting expression for v_D in the definition of the ambipolar diffusion time scale in the core, $\tau_{AD} \equiv r/|v_D|$, we find that

$$\tau_{AD} = const \times \frac{\tau_{ff}^2}{\tau_{ni}}, \tag{20a}$$

$$\propto n_i/n_n \propto n_n^{-(1-k)}, \tag{20b}$$

where the definition of τ_{ni} (see eq. [12i]) and the n_i-n_n relation (13) have been used to obtain (20b) from (20a). Note that, since $k \lesssim 1/2$, τ_{AD} decreases upon contraction at least as rapidly as τ_{ff}, which varies as $n_n^{-1/2}$ (see, also, eq. [21]).

In one-dimensional (slab) geometry, the *const* on the right-hand side of equation (20a) is equal to 1/2; in two-dimensional (cylindrical) geometry, it is $(2/1.4\pi) = 0.45$; and in three-dimensional (axisymmetric) geometry, it is $(8/1.4\pi^2) = 0.58$ (see TM79b, TM89). Hence, for practical purposes, the *const* is always $\simeq 1/2$, and *equation (20a) for τ_{AD} is seen to be a general property of the equations describing ambipolar diffusion in quasistatically contracting, magnetically supported cores* (see, also, TM87a and TM87b).

As long as the drift speed remains smaller than the continuously decreasing (because of ambipolar diffusion) Alfvén speed in the neutrals in the core, τ_{AD} is expected to be a good measure of the rate of reduction of the core's flux. (Actually, the flux-loss time scale from the core, $\tau_\Phi \equiv \Phi_{B,c}/|d\Phi_{B,c}/dt|$, is exactly equal to $\tau_{AD}/2$; see TM90b, or TM91, eq. [31].)

Equation (20a) is written as

$$\frac{\tau_{AD}}{\tau_{ff}} = \frac{1}{2}\frac{\tau_{ff}}{\tau_{ni}} \simeq 7\left[\frac{n_i/(3 \times 10^{-3}\text{ cm}^{-3})}{(n_n/10^5\text{ cm}^{-3})^{1/2}}\right],$$

(21)

$$\simeq 7\left[\frac{K}{3 \times 10^{-3}\text{ cm}^{-3}}\right]\left[\frac{10^5\text{ cm}^{-3}}{n_n}\right]^{0.5-k},$$

where, to obtain the last form, we have used the relation (13) between the ion and neutral density, with $K = 3 \times 10^{-3}$ cm^{-3} and $k \simeq 0.5$ regarded as the "canonical" values (Elmegreen 1979; Nakano 1979). Some authors obtain $k = 0.4$ (see Falgarone and Pérault 1987), which has the effect of decreasing the ratio τ_{AD}/τ_{ff} by the factor $n_n^{0.1}$ (= 3.16 for $n_n = 10^5$ cm^{-3}). Equation (21) indicates that ambipolar diffusion may be important even in nearly free-falling clouds. As explained previously (TM77, TM87a), the free-fall time scale is not a relevant time scale for magnetically subcritical clouds; the evolutionary time scale *is* the ambipolar diffusion time scale. Even at the stage by which ambipolar diffusion has led to an increase of the mass-to-flux ratio of the core to the critical value necessary for collapse, the contraction of a magnetically critical and thermally supercritical core progresses on the magnetically diluted free-fall time, not on τ_{ff} itself. It is evident from equation (21) that a reduction of K by only a factor 3 - 5 (which is permitted by observations) below the value 3×10^{-3} cm^{-3} can make ambipolar diffusion effective even in a nearly free-falling cloud. The effect of the precise value of K (or, more generally, the free parameter $\nu_{ff} \equiv \tau_{ff}/\tau_{ni}$; see TM82) on τ_{AD} is therefore quite clear. Recent numerical calculations have exhaustively studied the dependence of the formation and collapse of molecular cloud cores on the free parameters of the problem (see summary in TM91, §§ 5 - 7).

After dynamical contraction begins and the expression (20a) for τ_{AD} is no longer valid, one may still calculate τ_{AD} analytically from first principles (see TM89, § IIIb) as follows.

2.3.2. Dynamical Contraction: Its Effect on Ambipolar Diffusion. We consider, for simplicity, an initially spherical core which has just exceeded the critical mass-to-flux ratio because of ambipolar diffusion during the earlier quasistatic contraction phase, described by equation (20a). The critical mass-to-flux ratio is (Mouschovias and Spitzer 1976; also § 6)

$$M_{crit} = \frac{0.53}{3\pi}\left[\frac{5}{G}\right]^{1/2}\Phi_B = \left[\frac{1}{63G}\right]^{1/2}\Phi_B \simeq \frac{0.13}{G^{1/2}}\Phi_B.$$

(22a)

(Recall that eq. [22a] is a necessary but not sufficient condition for collapse; see §§ 5.2.2 and 6.) In a strict sense, equation (22a) gives the critical *total* mass-to-flux ratio *of the cloud*. When applied to the central flux tube(s) *as a unit*, equation (22a) has a slightly different constant on the right-hand side, whose precise value depends on the differential mass-to-flux ratio, $dm(\Phi_B)/d\Phi_B$, i.e., on how the various flux tubes of the cloud are loaded with mass. For initially spherical, uniform clouds threaded by a uniform magnetic field, the total mass-to-flux ratio M/Φ_B is related to the central mass-to-flux ratio $(dm/d\Phi_B)_c$ by $M/\Phi_B = (2/3)(dm/d\Phi_B)_c$ (see TM76a, eq. [44]; or eq. [113a] below). Therefore,

$$\left[\frac{dm}{d\Phi_B}\right]_{c,crit} = 1.5\left[\frac{M}{\Phi_B}\right]_{crit}.$$

(22b)

For an initially cylindrical mass distribution, the central and total mass-to-flux ratios are identical. In applications of equation (22a) to cores (*many* cores within a cloud), we have not in the past and we do not presently attempt to distinguish among different possible mass distributions in the cloud's flux tubes, by multiplying the right-hand side of equation (22a) by a factor a ($1 \leq a \leq 1.5$), for several reasons: (1) The effect on the critical mass-to-flux ratio is usually less than 50%. (2) We are *not* considering the collapse of the central part of a massive cloud as a *single* unit, which would give birth to a supermassive star, in which case an increase of the critical mass-to-flux ratio by up to 50% might be appropriate. We envision, instead, a gravitational breakup of matter in the central flux tube of a massive cloud into *several* fragments (or cores) because, typically, we expect the central flux tube of a molecular cloud to be longer than a few times the thermal Jeans length (see Figs. 1 and 2 in Mouschovias and Morton 1985b, or Figs. 4 and 5 in TM87b, p. 514; see, also, the quantitative results in TM91, § 4). In such a case, the simple relation between the total mass-to-flux ratio of the cloud and the mass-to-flux ratio of a single core breaks down and, therefore, it makes no sense to pretend that we are improving it at the 10% level. Since the *local* distribution of matter is likely to be more important now in determining the gravitational field in and near the fragments (or cores), we continue to use equation (22a) while bearing in mind that it represents reality probably only to within a factor of two. Finally, (3) by the stage at which a core goes critical because of ambipolar diffusion, a significant velocity field has been established. Consequently, any critical mass-to-flux ratio obtained from a *static* calculation is only an estimate anyhow, and should be used only as an approximate measure of the stage at which dynamic contraction of a core may begin.

Using the definition of $\tau_{AD} \equiv r/v_D$ and the expression (15a) for v_D, we find that

$$\tau_{AD} \equiv \frac{r}{v_D(r)} = r \frac{\rho_H}{\tau_{ni}} \frac{4\pi r}{B^2}, \qquad (23)$$

where r is the characteristic size (radius) of the core. We multiply and divide the right-hand side of equation (23) by ρ_n and by the appropriate factors of π and r so as to introduce M^2/Φ_B^2, and we use the expression for the free-fall time

$$\tau_{ff} = (3\pi/32G\rho)^{1/2} \qquad (24)$$

and equation (22a) for the critical flux at a given mass to write equation (23) as

$$\tau_{AD} = 0.3 \frac{\tau_{ff}^2}{\tau_{ni}} \left[\frac{\Phi_{B,crit}}{\Phi_B}\right]^2, \qquad \text{for } \Phi_B \leq \Phi_{B,crit}; \qquad (25a)$$

$$= 1 \times 10^3 \left[\frac{n_i/n_{H_2}}{10^{-10}}\right] \left[\frac{\Phi_{B,crit}}{\Phi_B}\right]^2 \quad \text{yr} \qquad (25b)$$

(see TM89, § IIIb), where Φ_B is the actual flux of the core at any stage past the onset of dynamical contraction, and $\Phi_{B,crit}$ is uniquely determined by the mass of the core from equation (22a) as $\Phi_{B,crit} = (63G)^{1/2}M$. The normalization of n_i/n_{H_2} in equation (25b) refers to a neutral density $\sim 10^9$ cm^{-3}, above which n_i no longer increases with n_n (Spitzer 1963; Elmegreen 1979; Nakano 1979). Canonical values of K and k have been assumed.

It is emphasized that equation (25) refers to a core, not to the entire cloud. The cloud as a whole is magnetically supported. However, a core of mass ~ 1 M$_\odot$ has $M < M_{crit}$ or, equivalently, $\Phi_B > \Phi_{B,crit}$ for a relatively long time, i.e., until ambipolar diffusion during the quasistatic phase of contraction manages to reduce the flux of this (fixed) mass so as to make the mass-to-flux ratio equal to its critical value. *If* this marks the onset of accelerated contraction of the core (in that the core is both magnetically and thermally supercritical), then equation (25) shows clearly that this will also mark the stage at which τ_{AD} begins to increase above its quasistatic contraction value by the factor $(\Phi_{B,crit}/\Phi_B)^2$. It therefore becomes more difficult to lose flux during this late phase of accelerated contraction. Although ambipolar diffusion may have reduced the flux of such a core by a few orders of magnitude prior to this stage, equation (22a) yields a critical flux for a ~ 1 $M_\odot$ core which is larger than the observed fluxes of magnetic stars by a factor 10 - 50. It seems, therefore, that ambipolar diffusion may not resolve the entire magnetic flux problem of star formation. To complete the task, then, one may need to appeal to another dissipative process (such as ohmic dissipation; see eq. [9a] and associated discussion) that sets in at higher densities, at which the electron density becomes negligible and the negative charge is mainly carried by grains (Spitzer 1963). Ohmic dissipation at densities $\gtrsim 10^{12}$ cm^{-3} has been studied by Nakano and Umebayashi (1986), who conclude that it can reduce the protostellar flux to the desired level prior to flux re-freezing (originally studied by Pneuman and Mitchell 1965) due to thermal ionization. Detailed collapse calculations beyond the density $\sim 10^9$ cm^{-3}, accounting for the significant reduction of the electron density and the fact that collisions no longer permit the dominant charge carriers to remain attached to the magnetic field, are essential for understanding precisely how nature resolves the magnetic flux problem.

For the time being, equation (25) taken at face value implies that, if magnetic forces maintain quasistatic contraction up to a density $n_n \sim 10^9$ cm^{-3} (not an unreasonable proposition given the conclusions of TM76b and TM78 concerning the effective support against gravity by magnetic tension, confirmed also by Nakano's 1979 quasistatic calculation), it may be possible for ambipolar diffusion to reduce the core's flux by at least another order of magnitude prior to the stage $n_n \sim 10^{12}$ cm^{-3}. This is so because we now have that (i) $\tau_{AD}/\tau_{ff} \propto \rho_n^{-1/2}\Phi_B^{-2}$ (since $n_i \simeq const$ above $n_n \simeq 10^9$ cm^{-3}), and (ii) the charged particles themselves are not well attached to the magnetic field. Hence, ambipolar diffusion may still have its day in resolving the magnetic flux problem.

2.3.3. Self-Initiated Collapse due to Ambipolar Diffusion.
We consider for simplicity a cylindrically symmetric, self-gravitating, cold cloud threaded by a magnetic field parallel to its axis of symmetry, and embedded in a medium of constant pressure Π (thermal plus magnetic). If the cloud contracted from an initially uniform state of density ρ_0, radius R_0, and field strength B_0, its density and magnetic field *at equilibrium* are given by (see TM79b)

$$\rho(r) = \rho(0) J_0(k_0 r), \qquad (26a)$$

$$g(r) = -4\pi G\rho(0)k_0^{-1} J_1(k_0 r), \qquad (26b)$$

where r is the distance from the symmetry axis, J_0 and J_1 are Bessel functions of order zero and one, respectively, and k_0 is defined by

$$k_0 = 4\pi G^{1/2}\rho_0 B_0^{-1}, \qquad (26c)$$

while the density on the axis and the cloud radius are, respectively,

---

[3]Note that equation (25) remains unchanged if we use equation (22b), instead of equation (22a), as a (necessary) criterion for the collapse of a cylindrical central flux tube of radius equal to its half-height. This is so because its mass is 1.5 times greater than the mass of the spherical core (of the same radius and density) considered above. Hence, the factor by which the *actual* mass increases is exactly equal to the factor by which the *critical* mass of the core given by equation (22b) exceeds the critical mass given by equation (22a).

$$\rho(0) = \frac{\rho_0}{J_0(y_m)} \frac{(8\pi\Pi)^{1/2}}{B_0} , \qquad (26d)$$

$$R = k_0^{-1} y_m(a) , \qquad (26e)$$

where $y = k_0 r$. The dimensionless radius y_m is the only maximum of the function $y^a J_0(y)$ in the interval $0 \leq y \leq 2.4$, and the constant a is given by

$$a = \frac{1}{2}(k_0 R_0)^2 \frac{B_0}{(8\pi\Pi)^{1/2}} > 0 . \qquad (26f)$$

The above equations contain two free parameters, namely,

$$y_0 \equiv k_0 R_0 \quad \text{and} \quad \alpha_0 \equiv B_0^2 / 8\pi\Pi . \qquad (27a, b)$$

The first is the dimensionless radius of the cloud in the uniform initial state, and the second is the initial ratio of magnetic pressure in the cloud and total external pressure. Specific examples of an H I and a molecular cloud described by such a model are given in TM79b.

A model cloud would remain in the above equilibrium state indefinitely if it were not for ambipolar diffusion, which allows the neutrals in the deep interior of the cloud to begin to contract under their self-gravity. As they do so, they experience a drag due to the ions, which are prevented by the magnetic force from contracting at the same rate as the neutrals. We have already seen that the magnetic force is balanced by the drag force, and a terminal drift velocity is established. A quasistatic contraction will be maintained as long as the neutral-ion collisional coupling is good enough for the drag force to remain strong enough to balance the gravitational force on the neutrals, i.e.,

$$\rho_n g + F_{ni} = 0 . \qquad (28)$$

The drift speed is obtained from equations (28) and (12i), by considering only ion-hydrogen collisions (see eq. [35] in TM79b), as done in § 2.3.1:

$$v_D(r) = -1.4 \, \tau_{ni}(r) \, g(r) . \qquad (29)$$

This is then used in the definition of τ_{AD} along with equations (26b) and (26a) to find that

$$\tau_{AD}(r) \equiv \frac{r}{v_D(r)} = \frac{2}{1.4\pi} \frac{\tau_{ff}^2}{\tau_{ni}} \xi(y) , \qquad (30a)$$

$$= 1.8 \times 10^5 \left[\frac{n_i/n_{H_2}}{10^{-8}} \right] \xi(y) \quad \text{yr} , \qquad (30b)$$

where $y = k_0 r$. This equation, evaluated on the axis of symmetry, was applied to molecular cloud cores in TM79b; see, also, TM87a, § 2.2.5. It differs from Spitzer's (1968a) τ_{AD} by the multiplicative factor $\xi(y) \equiv y J_0(y)/2 J_1(y)$, and in that the density is a function of position inside the cloud. The function $\xi(y)$ is tabulated in TM79b (Table 1, column 2); it has a maximum equal to 1 at $y = 0$ (on the axis), and decreases monotonically with increasing y. For the molecular cloud model discussed therein, it has the value 3.32×10^{-2} at the surface, and remains greater than 10^{-3} for a wide variation of the physical parameters. In dense molecular cloud interiors, where the degree of ionization x ($\equiv n_i/n_{H2}$) is $10^{-9} \lesssim x \lesssim 10^{-7}$, equation (30b) yields $2 \times 10^4 \lesssim \tau_{AD} \lesssim 2 \times 10^6$ yr, a significantly short time to lead to core formation and contraction. In the envelope, however, where the degree of ionization is $x \sim 10^{-4}$, we find that $\tau_{AD} \gtrsim 10^8$ yr. Hence, *the envelope remains magnetically supported while stars can form in the core (and only in the core) because of ambipolar diffusion*, as originally suggested by TM76b. Observations have subsequently shown precisely that kind of core-envelope separation in massive molecular clouds (see Myers 1982; Myers and Benson 1983; Myers, Linke, and Benson 1983; see, also, review by Myers 1985).

3. SINGLE-FLUID MAGNETOHYDRODYNAMICS

The physical conditions in most regions of the interstellar medium are such that the considerations of § 2.2 yield that flux freezing in the plasma is a better approximation in these regions than it is in molecular clouds. Although the degree of ionization outside molecular clouds is $10^{-4} \lesssim x \leq 1$, one cannot use equation (20a) to estimate $\tau_{AD} \gtrsim 10^9$ yr and then conclude that ambipolar diffusion is negligible. Equation (20a) is valid only in self-gravitating, magnetically supported clouds (see eq. [18]). One must revert to equation (16), which is valid for nongravitating clouds as well. In the intercloud medium, in which $B \simeq 3$ μG and $n_n \simeq 0.1$ cm^{-3}, the Alfvén speed in the neutrals is $v_A \simeq 14$ km s^{-1}. Since the electron and ion density is $\simeq 0.03$ cm^{-3}, the neutral-ion collision time is $\sim 10^3$ yr. With a typical length scale being $L \gtrsim 10$ pc, equation (16d) yields $\tau_{AD} \sim 4 \times 10^8$ yr. For the large-scale phenomena of interest here (e.g., formation of interstellar clouds, motion of clouds through the intercloud medium, etc.), which occur over time scales $\sim 10^7$ yr, ambipolar diffusion may be neglected.[4]

In a more intuitive fashion, the conditions that allow us to treat the three-component interstellar gas as a single compressible fluid, can be stated as follows. For long-wavelength (or low frequency) hydromagnetic disturbances, for which an electron collides many times with ions and neutrals before it is forced to reverse its direction of motion by the oscillating electric field of the disturbance, charge neutrality (see eq. [2]) is an excellent approximation and, therefore, collective plasma effects, being relatively high-frequency phenomena (the plasma frequency is $\omega_p \sim 10^3$ s^{-1}), can be ignored. Although collision frequencies ($\omega_{ei} \sim 10^{-7}$ s^{-1} and $\omega_{in} \sim 10^{-9}$ s^{-1}) are relatively large compared to the frequency of the hydromagnetic wave ($\omega \sim 10^{-14}$ s^{-1}), they are nevertheless much smaller than gyrofrequencies ($\omega_c \sim 10$ s^{-1} for electrons). Hence, diffusion across the magnetic field, which would result from collisions between opposite charges or between a charged and a neutral particle before a gyration is completed, may be neglected (see, also, eqs. [9a] and [9b]). In summary, we have the inequalities

$$\omega_p \gg \omega_c \gg \omega_{coll} \gg \omega, \tag{31}$$

where ω_{coll} denotes collision frequencies. For length scales much larger than collision mean free paths, viscosity and thermal conduction may also be neglected. (The issue of energy transport did not arise in considering the two-fluid equations [12a] - [13] because of the assumed/observed near-isothermality in cold cloud interiors.)

We thus consider a conducting fluid of density ρ, thermal pressure P, and temperature T, threaded by a frozen-in magnetic field B, and affected by a gravitational field g derivable from a potential ψ. Both the gas and stars, of density ρ_*, may contribute to ψ. The velocity of a fluid element is denoted by v. An electric current density j maintains the magnetic field, which is derivable from a vector potential A. The entropy per gram of

---

[4]However, for magnetic disturbances with characteristic length scales significantly smaller than 1 pc (e.g., interstellar hydromagnetic shock waves), ambipolar diffusion is not negligible, and may produce interesting phenomena, such as magnetic precursors (e.g., see Draine 1980). Because of ambipolar diffusion, a magnetic disturbance (forerunner) forewarns the pre-shock medium of the shock's arrival.

matter is denoted by S, and $\mathcal{L}$ represents the net rate of energy loss (losses minus gains) per gram of matter. The MHD equations, then, are

mass conservation
$$\frac{d\rho}{dt} + \rho \nabla \cdot v = 0 , \tag{32a}$$

force equation
$$\rho \frac{dv}{dt} = -\nabla P - \rho \nabla \psi + \frac{j}{c} \times B , \tag{32b}$$

energy equation
$$\rho T \frac{dS}{dt} = -\rho \mathcal{L}(\rho, T) , \tag{32c}$$

flux freezing
$$\frac{\partial B}{\partial t} = \nabla \times (v \times B) , \tag{32d}$$

Ampere's law
$$\nabla \times B = \frac{4\pi}{c} j , \tag{32e}$$

Poisson equation
$$\nabla^2 \psi = 4\pi G (\rho_* + \rho) , \tag{32f}$$

ideal-gas law
$$P = \frac{\rho}{m} k_B T , \tag{32g}$$

definitions
$$g = -\nabla \psi , \tag{33a}$$

$$B = \nabla \times A , \tag{33b}$$

$$S = \frac{1}{\gamma - 1} \frac{k_B}{m} \ln\left(\frac{P}{\rho^\gamma}\right) + \text{const} . \tag{33c}$$

where m is the mean mass per particle ($m = \mu m_H$), and γ is the ratio of specific heats.

A hydromagnetic disturbance in the interstellar medium which occurs over a relatively short time scale may be considered as an adiabatic process, in which case the energy equation becomes simply

adiabatic process
$$\frac{d}{dt}\left(\frac{P}{\rho^\gamma}\right) = 0 . \tag{34a}$$

At the other extreme, a slow process may allow thermal equilibrium to be maintained, in which case an isothermal description is appropriate and the energy equation reduces to $T = \text{const}$ or, by using equation (32g),

isothermal process
$$\frac{d}{dt}\left(\frac{P}{\rho}\right) = 0 . \tag{34b}$$

Note that, if one views equation (34a) as the operational definition of the exponent $\gamma \equiv (\partial \ln P / \partial \ln \rho)_\theta$, where θ is a thermodynamic variable (not necessarily the entropy) kept fixed, then equation (34a) can be used to describe processes in addition to adiabatic ones. For example, an isothermal process will be characterized by $\gamma = 1$, while an adiabatic process in an ideal monoatomic gas will have $\gamma = 5/3$.

3.1. Cosmic Rays as a Fluid

The dynamical effects of cosmic rays in the Galaxy have been the subject of intensive investigation for many years. A series of papers has examined the conditions under which cosmic rays, considered as a very hot, collisionless plasma, may be described in the MHD approximation. Three excellent reviews point out what phenomena are excluded when such a description is adopted, but they argue that the MHD description is the proper one for cosmic rays in the Galactic environment (Parker 1968b, 1969; Lerche 1969). Here we give a physically motivated, brief but adequate for our purposes, MHD description of the cosmic-ray gas, so that the conspiracy between magnetic fields and cosmic rays to destabilize the interstellar medium and lead to the formation of cloud complexes can be studied.

The interstellar magnetic field, being frozen in the matter, couples the cosmic rays to the interstellar matter because the relativistic-particle gyroradii are much smaller than typical length scales in the interstellar medium. For example, the gyroradius of a 10^{11} eV proton is only about 10^{14} cm. In the absence of a magnetic field, the cosmic-ray pressure is maintained isotropic to within $\sim 1\%$ by various rapidly growing ($\tau \sim 10^3$ yr) relativistic microinstabilities (e.g., see Lerche 1969). In introducing the magnetic field, we restrict our attention to very long-wavelength (much greater than gyroradii) hydromagnetic waves. In this regime and for slow bulk motions ($v_{rel}^2 \ll c^2$), Parker argues that an isotropic cosmic-ray pressure is a fair approximation in most astrophysical situations even in the presence of a magnetic field. In the extreme relativistic case, the relation between the cosmic-ray pressure P_{rel} and density ρ_{rel} is simply

$$P_{rel} = \tfrac{1}{3}\rho_{rel} c^2 \equiv \rho_{rel} C_{rel}^2 \,. \tag{35}$$

Because the cosmic rays and the thermal gas are tied to the magnetic field, their bulk velocities v_{rel} and v (v_{rel}^2, $v^2 \ll c^2$) will have equal components in a direction perpendicular to the field; i.e.,

$$v \times B = v_{rel} \times B \,. \tag{36a}$$

Thus, an equation identical to (32d), with v replaced by v_{rel}, holds for the cosmic-ray gas as well. Note that, as in the case of the thermal gas, P_{rel} will, in general, respond to changes in the volume of a flux tube; it is *not* a constant of the motion. Hence, the right-hand side of the equation

$$\frac{dP_{rel}}{dt} = ? \,, \tag{36b}$$

will not, in general, vanish. Under the assumptions that (i) the motion of cosmic rays along field lines is completely uncoupled from that of the thermal gas (cf. Kulsrud and Pearce 1969), (ii) the cosmic-ray gas is not subject to the galactic gravitational field (because the ratio of gravitational and cosmic-ray-pressure forces $\sim v_{esc}^2/C_{rel}^2 \ll 1$; the quantity v_{esc} is the escape speed in the galactic gravitational field), and (iii) inertial effects of the relativistic gas are negligible (in a typical H I region $\rho_{rel}/\rho = 3P_{rel}/\rho c^2 \sim 10^{-10}$), one may write the force equation for cosmic rays in a direction parallel to the field simply as

$$\nabla_\parallel P_{rel} \equiv \frac{B}{B} \cdot \nabla P_{rel} = 0 \,. \tag{36c}$$

Physically, the nearly instantaneous communication of cosmic rays along field lines establishes pressure equilibrium in the cosmic-ray gas over a distance L in a time $L/C_{rel} \simeq L/c$. Even if L is as large as 1 kpc, this time is 10^{11} s, which is much smaller than the time scale of the hydromagnetic phenomena of interest to us here ($\simeq 10^{14}$ s). Hence, the inertial effects of the cosmic-ray gas can indeed be ignored. *Equation (36c) states that the cosmic-ray pressure is constant on a field line, but it does not determine its value, which is different for different field lines.* (TM75a shows how to calculate P_{rel} at the position of any field line in a quasi-equilibrium state.) Equation (36c) will be exact at equilibrium insofar as the hot ($T_{\text{rel}} \sim 10^{13}$ K) and tenuous ($n_{\text{rel}} \sim 10^{-10}$ cm^{-3}) cosmic-ray gas is not affected by the galactic gravitational field ($g \sim 10^{-9}$ cm s^{-2}).

It remains to specify how cosmic rays will modify the force equation (32b) in a direction normal to B. Under the approximations discussed above, the main contribution of the cosmic rays to the thermal-gas force equation (32b) is to add a term $-\nabla_\perp P_{\text{rel}}$ to the right-hand side. The electric current density j may, in principle, have a contribution from the cosmic rays, but as long as one uses the *total* current density on the right-hand side of Ampere's law (32e) and denotes it by j, the magnetic force term in equation (32b) does not change. All together, then, the combined force equation becomes

$$\rho \frac{d\mathbf{v}}{dt} = -\nabla P - \nabla P_{\text{rel}} - \rho \nabla \psi + j \times B/c, \tag{37}$$

where j is the total current density, which appears on the right-hand side of equation (32e). Note that, because of equation (36c), the term $-\nabla_\perp P_{\text{rel}}$ has been replaced by $-\nabla P_{\text{rel}}$ in equation (37).

In summary, the presence of cosmic rays adds the term $-\nabla P_{\text{rel}}$ on the right-hand side of the equation of motion (32b) of the thermal gas. Hence, another equation for dP_{rel}/dt must, in general, be specified in order to close the system of the single-fluid MHD equations. In the case of the interstellar gas, which is supported by thermal-pressure, magnetic, and cosmic-ray-pressure forces against the vertical component of the galactic gravitational field, Parker (1966) considered perturbations which leave the volume of each magnetic flux tube unchanged. For such disturbances, the cosmic-ray pressure is a constant of the motion, so that the equation needed to close the system is simply

$$\frac{dP_{\text{rel}}}{dt} \equiv \frac{\partial P_{\text{rel}}}{\partial t} + (\mathbf{v} \cdot \nabla) P_{\text{rel}} = 0, \tag{38}$$

where d/dt is the time derivative comoving with the thermal gas. Ames (1973) showed that equation (38) is indeed valid (to first order) for the perturbations considered by Parker.

3.2. Formation of Giant Clouds and Cloud Complexes

The mechanisms proposed for the formation of interstellar clouds were reviewed a decade ago (TM81). Although the old mechanisms have been applied, sometimes with new twists, to new situations, no fundamentally new physics has been introduced. The reader can find more details and specific references in TM81 and in Elmegreen's article in this volume. Here, after a very brief outline of the Jeans and the thermal instabilities as they relate to the formation of clouds, we focus on the basic physics of the most promising mechanism, namely, the magnetic Rayleigh-Taylor (or Parker) instability.

3.2.1. The Jeans Instability.
This is the oldest available mechanism (Jeans 1928; Chandrasekhar and Fermi 1953), and refers to the development of self-gravitating condensations in an infinite, uniform medium (of density ρ_0), threaded by a uniform magnetic field, B_0. Only length scales λ greater than the Jeans wavelength,

$$\lambda_J \equiv \lambda_{J,\parallel} = C_a \left[\frac{\pi}{G\rho_0}\right]^{1/2} = 1.23 \times 10^3 \left[\frac{T}{6000 \text{ K}}\right]^{1/2} \left[\frac{1}{n_H}\right]^{1/2} \text{ pc}, \tag{39a}$$

parallel to the field lines, and longer than

$$\lambda_{J,\perp} = \lambda_J (1 + b)^{1/2}, \qquad b \equiv 0.47 \left[\frac{(B/3 \ \mu G)^2}{(T/6000 \ K)}\right] \left[\frac{1 \ cm^{-3}}{n_H}\right] \qquad (39b, c)$$

perpendicular to the field lines can become gravitationally unstable. A 10% helium abundance has been accounted for. The temperature 6000 K is chosen because it is near the equilibrium temperature of a gas with $n_H \simeq 1$ cm^{-3} ($\simeq$ the mean density of interstellar matter), $n_e \simeq 10^{-2}$ cm^{-3}, which is heated by cosmic-ray ionization at the rate 10^{-17} s^{-1} and cooled by e-H collisions. The e-folding time of the instability along field lines is given by

$$\tau_J = [4\pi G \rho_0 (1 - h^2)]^{-1/2}, \qquad h \equiv \lambda_J/\lambda \leq 1; \qquad (40a, b)$$
$$= 2.3 \times 10^7 \, [n_H (1 - h^2)]^{-1/2} \quad yr. \qquad (40c)$$

The requirement that τ_J be less than 3×10^7 yr (otherwise stars formed by the collapse of such clouds would appear too far downstream from a galactic shock, contrary to observations --see Roberts 1969) implies that $h \leq 0.64$; i.e., $\lambda \geq 1.56 \lambda_J \simeq 1.9$ kpc. To gather matter from such long distances into a region even as large as 100 pc (a large cloud indeed) within a time 3×10^7 yr, speeds of about 60 km s^{-1} are required. They much exceed typical free-fall velocities, and they are certainly not observed over such extended regions in the Galactic plane. Even in the absence of the magnetic field, in which case the instability can grow in three dimensions for $\lambda > \lambda_J$, the mass inside a (marginally unstable) sphere of diameter equal to λ_J is

$$M_J \equiv \frac{4\pi}{3} \rho_0 \left(\frac{\lambda_J}{2}\right)^3 = \frac{\pi^{5/2}}{6} \frac{C_a^3}{(G^3 \rho_0)^{1/2}}, \qquad (41a)$$

$$\simeq 3 \times 10^7 \left[\frac{1 \ cm^{-3}}{n_H}\right]^{1/2} \left(\frac{T}{6000 \ K}\right)^{3/2} M_\odot. \qquad (41b)$$

Hence, only superclouds could form by the Jeans instability.

3.2.2. Thermal Instability. The isobaric nature of the condensation mode of the thermal instability (see Field 1965) implies an upper bound on the fastest-growing wavelengths of a perturbation. It is approximately that distance within which a sound wave can establish pressure equilibrium in a time not exceeding the cooling time ($< 10^6$ yr) of the medium. Since the sound speed is ≤ 10 km s^{-1}, the wavelengths which can grow at a rate near maximum are ≤ 10 pc. The final size of the resulting condensation is, of course, much smaller than this (< 0.1 pc), and involves only a fraction of a solar mass (see review TM78, Appendix A). Wavelengths much larger than 10 pc (i) grow nearly isochorically and, therefore, cannot explain the observed cloud densities; (ii) grow at a rate slower than the magnetic Rayleigh-Taylor instability (see below), in which the magnetic field is instrumental, rather than a nuisance, in the formation of large condensations. (Field [1965] has shown that a magnetic field as weak as 1 μG suppresses the thermal instability, except in a direction parallel to the field.) Altogether, then, only cloudlets could form by the thermal instability.

3.2.3. Magnetic Rayleigh-Taylor (or Parker) Instability. A "light" (low density) fluid can support a "heavy" (denser) fluid against a vertical (downward) gravitational field (assumed to be constant for simplicity) if the interface is perfectly horizontal. However, deformations of the interface grow, as fingers of heavy fluid protrude downward into the light fluid, thus reducing the energy of the system. Shorter wavelengths along the interface tend to grow faster than longer wavelengths. This is a classical Rayleigh-Taylor instability. It can also develop if the downward gravitational field is replaced by an upward acceleration of the heavy fluid by the lighter one. This instability changes if a frozen-in, horizontal magnetic field plays the role of the light fluid in supporting the gas against the gravitational field. The light fluid (the field) and the heavy fluid (the gas) now coexist in the same region of space, and analogies with the nonmagnetic case break down. The nature of the magnetic Rayleigh-Taylor instability has been worked out by Parker (1966) in the context of the interstellar medium. Is it a viable mechanism for the formation of interstellar clouds? In other words, are the unstable wavelengths of the proper size (at least a few hundred parsecs), and are the corresponding growth times short enough ($\lesssim 3 \times 10^7$ yr)? In addition, what final densities are achieved by the nonlinear development of the instability?

3.2.3(a). The Zero-Order State. Observations show that the interstellar magnetic field is predominantly parallel to the Galactic plane (Mathewson and Ford 1970), although arches of magnetic field lines rising high above the plane are also revealed. Parker (1966) assumed a zero-order equilibrium state which, for simplicity, has field lines exactly parallel to the galactic plane. The galactic gravitational field (due to stars) was taken, again for simplicity, to be constant and to reverse its direction across the galactic plane; i.e., $\mathbf{g} = -\hat{z}g(z)$, where $g(z) = -g(-z) = const > 0$. The assumption was also made that the pressure ratios

$$\alpha \equiv B_0^2/8\pi P_0 \quad \text{and} \quad \beta \equiv P_{\text{rel},0}/P_0 \qquad (42\text{a, b})$$

are constant *in the zero-order, equilibrium state*. With the magnetic field in the zero-order state taken parallel to the galactic plane and written as $\mathbf{B}_0 = \hat{y}B_{0,y}(z) \equiv \hat{y}B_0(z)$, the MHS force equation (eq. [32b] with the inertial term set equal to zero) becomes

$$(1 + \alpha + \beta)\frac{dP_0}{dz} = -\frac{g}{C^2}P_0 . \qquad (43)$$

Because of the symmetry of the problem about the plane $z = 0$, we need only be concerned with the upper half-plane $z \geq 0$. The solution is

$$P_0(z) \equiv \beta^{-1}P_{\text{rel},0}(z) \equiv \frac{B_0^2(z)}{8\pi\alpha} \equiv \frac{[-A_0(z)]^2}{32\pi\alpha H^2} = \rho_0(0)\,C^2 \exp\left(\frac{z}{H}\right), \qquad (44\text{a})$$

where

$$H \equiv (1 + \alpha + \beta)\,C^2/g \qquad (44\text{b})$$

is the total scale height of the gas in the zero-order state, and A_0 is the only nontrivial (x-)component of the magnetic vector potential (see eq. [33b], with $\mathbf{B}$ having only a y-component), namely, $\mathbf{A}_0(z) = \hat{x}A_0(z)$. The thermal pressure contribution to the scale height, for $T \simeq 6000$ K and $g \simeq 3 \times 10^{-9}$ cm s^{-2}, is $C^2/g \simeq 42$ pc. This is smaller than the observed scale height by at least a factor of 3, possibly 4. With $\alpha \simeq \beta \simeq 1$, a value consistent with observations, one finds that $H \simeq 126$ pc --not an unreasonable value.

3.2.3(b). Stability Analysis. Parker showed, through a linear stability analysis, that the above one-dimensional equilibrium state is unstable with respect to deformations of the field lines in the (y, z)-plane. The vertical gravitational field acquires a component along a deformed (nonhorizontal) field line, thus causing gas to slide along the field line from a raised into a lowered portion. The unloading of gas from the raised portion leaves magnetic and cosmic-ray pressure gradients unbalanced in that region, thereby causing further inflation of the already raised portion of a field line. The component of gravity along the now more vertical field line is greater, with the result that gas can be unloaded more effectively into the "valley" of the field line. The process will stop only when field lines have inflated enough for their tension to balance the expansive magnetic and cosmic-ray pressure gradients (TM74, TM75a). In this picture, the matter which accumulates in the valleys of the field lines represents interstellar clouds. Or, does it?

We first obtain Parker's instability criterion for two-dimensional (y, z) perturbations, discuss the physics of the instability more quantitatively, and then address the question of whether the "final" states of the system in which the instability has developed resemble interstellar clouds (see, also, § 5.1).

The MHD equations (32a) - (32g), with the energy equation (32c) replaced by (34a) and the force equation (32b) modified as in equation (37) so as to accommodate the cosmic-ray gas, are written in an expanded form in terms of the magnetic vector potential $A \equiv \hat{x}A(y, z)$ and then linearized to find

$$\frac{\partial \rho_1}{\partial t} + (v_1 \cdot \nabla)\rho_0 + \rho_0 \nabla \cdot v_1 = 0 , \qquad (45a)$$

$$\rho_0 \frac{\partial v_1}{\partial t} + \nabla(P_1 + P_{\text{rel},1}) - \rho_1 g + \frac{1}{4\pi}[(\nabla^2 A_0)\nabla A_1 + (\nabla^2 A_1)\nabla A_0] = 0 , \qquad (45b)$$

$$\rho_0 \frac{\partial P_1}{\partial t} + \rho_0(v_1 \cdot \nabla)P_0 - \gamma P_0 \frac{\partial \rho_1}{\partial t} - \gamma P_0(v_1 \cdot \nabla)\rho_0 = 0 , \qquad (45c)$$

$$\frac{\partial A_1}{\partial t} - v_1 \times B_0 = 0 , \qquad (45d)$$

$$\frac{\partial P_{\text{rel},1}}{\partial t} + (v_1 \cdot \nabla)P_{\text{rel},0} = 0 . \qquad (45e)$$

Ampere's law (32e) has been used to eliminate j from the force equation, and then the definition of A (eq. [33b]) to eliminate B. Although equation (38) was taken as an adequate description of the cosmic-ray gas in order to close the system of the MHD equations, it is emphasized that this step is legitimate only because the equations have been linearized --as explained at the end of § 3.1, equation (38) is valid only to first order for the perturbations considered by Parker. We choose to work with A in this geometry, instead of

$$B = \nabla \times A = - \hat{x} \times \nabla A , \qquad (46)$$

because $v \times B = - (v \cdot \nabla)A$, so that the flux-freezing equation (32d) becomes

$$\frac{\partial A}{\partial t} - v \times B = \frac{\partial A}{\partial t} + (v \cdot \nabla)A \equiv \frac{dA}{dt} = 0 , \qquad (47a)$$

which states that A is *a constant of the motion*. Since equation (46) implies directly that

$$B \cdot \nabla A = 0 , \qquad (47b)$$

A is also constant on a field line (but has different values on different field lines). One may therefore calculate A at some convenient time, for example, in the initial equilibrium state, and then identify different values of A with different field lines with the assurance that those values will always label the corresponding field lines no matter how the field lines are deformed during the evolution of the system --as long as the magnetic field remains frozen in the matter.

We look for a solution of equations (45a) - (45e) having the time dependence $\exp(nt)$, where n is the *growth rate* and is, in general, complex. A real, positive n implies instability, a real negative n exponential decay of the perturbation, and an imaginary n corresponds to waves. The time derivatives in equations (45a) - (45e) are thus replaced by n. Since the first-order part of the $\mathbf{v} \cdot \nabla$ is $v_z(\partial/\partial z)$, and $\mathbf{v}_1 \times \mathbf{B}_0 = -\hat{x}v_z B_0$, it is seen that the coefficients of all the perturbed quantities in these equations do not depend on y. We may therefore look for solutions of the form (i.e., fourier analyze in y)

$$a(y, z, t) = f(z) \exp(nt + ik_y y) , \tag{48}$$

which, in addition, has the effect of replacing $\partial/\partial y$ with ik_y. Because the coefficients of the perturbed quantities in the resulting equations still depend on z, it is not yet possible to fourier analyze in z. To get constant coefficients, one changes variables to[5]

$$\xi \equiv \frac{B_0}{\rho_0} \rho_1 , \qquad \eta \equiv B_0 v_y , \tag{49a, b}$$

which then permits fourier analysis in z as well. The resulting set of algebraic equations is then reduced as follows. Equation (45d) is solved for the perturbed velocity v_z (note that the subscript 1 has been dropped from the velocities since v_0 vanishes identically), equation (45c) is solved for P_1, equation (45e) for $P_{\text{rel},1}$, and the resulting expressions are substituted in the rest of the equations to obtain a homogeneous system of three algebraic equations for the three unknowns ξ, η, and A_1. By setting the determinant of the coefficients equal to zero, one obtains a dispersion relation for motions in the (y, z)-plane:

$$n^4 + C^2[(\gamma + 2\alpha)(k^2 + k_0^2/4)]n^2 + k_y^2 C^4 \{2\alpha\gamma k^2 + k_0^2[\gamma(1 + \beta + 3\alpha/2) - (1 + \alpha + \beta)^2]\} = 0, \tag{50}$$

where we have set

$$k_0 \equiv H^{-1} , \tag{51}$$

and we have let the magnitude of the wave number be denoted by k; i.e.,

$$k^2 = k_y^2 + k_z^2 . \tag{52}$$

Clearly, equation (50) has the form $n^4 + c_1 n^2 + c_2 = 0$, which is a quadratic for n^2, with the constant c_1 being always positive, and the constant c_2 being capable of having either sign.

One can show (Parker did not) that all roots of equation (50) are real --a tedious algebraic proof. Here we are content with finding a criterion that one root $n^2 > 0$, so that one root $n > 0$, corresponding to instability. Let the two roots of the quadratic be denoted by $n_{(1)}^2$ and $n_{(2)}^2$. From the properties of quadratic equations, the sum of the two roots is

$$n_{(1)}^2 + n_{(2)}^2 = -c_1 < 0 . \tag{53a}$$

Therefore, at least one root (say, $n_{(1)}^2$) is < 0, and the corresponding modes are waves. Hence, *at most* one root (the second one, $n_{(2)}^2$) corresponds to an unstable mode. To determine whether this is so, we look at the algebraic sign of the product of the two roots, which is equal to the constant term of the quadratic, namely,

$$n_{(1)}^2 n_{(2)}^2 = c_2 . \tag{53b}$$

---

[5] To the best of our knowledge, this transformation was first pointed out by Field (1970±1).

It follows, since we have already found that $n_{(1)}^2 < 0$, that the second root $n_{(2)}^2$ has the same algebraic sign as $-c_2$. It is therefore sufficient for instability to demand that $c_2 < 0$, so that $n_{(2)}^2$ will be positive. All together, *the criterion for having one unstable mode is*

$$\left[\frac{k}{k_0/2}\right]^2 < \frac{2(1 + \alpha + \beta - \gamma)(1 + \alpha + \beta) - \alpha\gamma}{\alpha\gamma}, \qquad \alpha \neq 0, \qquad (54)$$

which is the instability criterion obtained by Parker (1966).

In terms of wavelengths, the criterion (54) is equivalent to the following two conditions, which must be satisfied simultaneously,

$$\lambda_y > \Lambda_y \equiv 4\pi H \left[\frac{\alpha\gamma}{2(1 + \alpha + \beta - \gamma)(1 + \alpha + \beta) - \alpha\gamma}\right]^{1/2}, \qquad (55a)$$

$$\lambda_z > \Lambda_z(\lambda_y) \equiv \Lambda_y \left[1 - \left(\frac{\Lambda_y}{\lambda_y}\right)^2\right]^{-1/2} > \Lambda_y. \qquad (55b)$$

For the interstellar gas, $\gamma \equiv d\ell n P/d\ell n \rho \simeq 1$. If $\lambda_y < \Lambda_y$, the radius of curvature of a typical, deformed field line is small, hence the tension is large and straightens the field line out by propagating the perturbation away as a wave. If $\lambda_z < \Lambda_z(\lambda_y)$, the system is stable even though λ_y may exceed its critical value Λ_y. The physical reason lies in the fact that the volume available for the field lines to expand in, and thereby decrease the magnetic energy of the system, is limited. The increase in the field strength in the valleys and the pileup of field lines near the first undeformed field line, which forms a natural "lid" to the system below, represent an increase in magnetic energy which suppresses the instability (see TM74, TM75b for details). For a fixed λ_y, the growth rate of the perturbation increases monotonically as λ_z ($> \Lambda_z$) increases. For a fixed $\lambda_z > \Lambda_z$, the growth rate first increases and then decreases as λ_y increases. Equation (55a) shows that the horizontal critical wavelength for $\alpha \simeq \beta \simeq \gamma \simeq 1$ is $\Lambda_y = 1.2\pi H \simeq 477$ pc. The maximum growth rate occurs at $\lambda_y = 1.8\Lambda_y = 2.2\pi H \simeq 868$ pc and $\lambda_z = \infty$, and its inverse (the e-folding time) is given by

$$\tau_{\min} = 1.1 \frac{H}{C} = 2.2 \times 10^7 \left(\frac{T}{6000 \text{ K}}\right)^{1/2} \left(\frac{3 \times 10^{-9} \text{ cm s}^{-2}}{g}\right) \text{ yr}. \qquad (56a, b)$$

This growth time is short enough to be relevant for cloud formation behind a spiral density shock wave. In fact, it may be smaller than the value given in equations (56a, b) because, in a strict sense, the quantity H is the scale height in the *initial* state; not its value today. It has been shown by exact determination of final equilibrium states for the Parker instability that, in the valleys of field lines, $H_{\text{final}} \simeq 1.7 H_{\text{initial}}$ (TM74, Fig. 2b). This implies that $\tau_{\min} \simeq 1.3 \times 10^7$ yr and $\lambda_y(\tau_{\min}) \simeq 511$ pc. A further decrease in $\tau_{\min}$ can take place because of the fact that the instability is externally *driven* by a spiral density shock wave (TM75b, p. 73). There is yet another reason for which $\tau_{\min}$ can decrease further. Giz and Shu (1980) took into consideration the actual variation of g with z, and found that the value of g which enters equation (56b) is greater than the one given above by a factor of 3. The amount of matter involved in a cylinder (along a spiral arm) of length 511 pc and diameter 250 pc (the approximate thickness of a galactic shock, as well as the thickness of the galactic disk in which most of the gas is found) is 8.6×10^5 M$_\odot$. Thus the Parker instability is most suitable for the formation of large-scale condensations (or cloud complexes), rather than individual interstellar clouds (TM74; Mouschovias, Shu, and Woodward 1974). The implosion by shock waves of individual clouds within these complexes can give rise to OB associations and giant H II regions, all aligned along spiral arms "like beads on a string" and separated by regular intervals of 500 - 1000 pc, in agreement with observations both in our galaxy and in external galaxies (Westerhout 1963; Kerr 1963; Hodge 1969; Morgan 1970).

In the direction $g \times B$ (the "third direction"), wavelengths ranging from a very small fraction of the vertical scale height H to many times H can grow with almost identical growth rates (Parker 1967a, b). If some mechanism (other than a galactic shock) selects wavelengths $\lambda_x \sim 10$ pc in this direction, then the mass involved would be only slightly larger than 10^4 $M_\odot$. This begins to approach masses of *individual* clouds. Whether in fact individual clouds can form by the Parker instability will be decided only when nonlinear three-dimensional calculations are carried out. Although there is a wealth of ideas on how phase transitions (e.g., see Field, Goldsmith, and Habing 1969) and conversion of atomic to molecular hydrogen in the valleys of field lines can convert the nongravitating clumps of gas into dense molecular cloud complexes (Field 1969; TM75b, TM78; Blitz and Shu 1980), no quantitative calculation has been produced yet. Initial perturbations of the field lines which have an odd symmetry about the Galactic plane are more likely to initiate the necessary phase transitions. Such perturbations allow field lines originally coinciding with the Galactic plane to deform, and they therefore can lead to a gas density (and pressure) in the plane significantly greater (a necessary condition for phase transitions) than its value in the initial (unstable) equilibrium state. Perturbations with even symmetry about the Galactic plane can lead to such phase transitions only under special circumstances (TM75b, pp. 55 - 57).

3.3. Redistribution of Angular Momentum by Magnetic Braking

The angular momentum problem is one of the two central dynamical problems of star formation. Its resolution by magnetic braking is discussed for a single aligned rotator (fragment, or cloud, or core), a single perpendicular rotator, and a series of magnetically linked, aligned rotators; the field is frozen in the matter in all cases. Exact analogies with transverse waves on strings are invaluable for communicating the results of these calculations. Then, in § 4, the first detailed quantitative study of the effect of ambipolar diffusion on magnetic braking is described, and the results are explained by analogy with electrical resistors connected in series or in parallel.

Since no calculation concerning dynamical processes in the interstellar medium can be complete at the present time, the only way to distinguish the important from the unimportant effects is to introduce (guided by observations when available) one effect at a time, study it well, and then further complicate the model by relaxing another assumption.

It is often useful to solve a theoretical problem under two opposite assumptions because, when the results of both calculations are taken together, they bracket all other situations expected to arise in nature. In this spirit we have studied analytically for two different geometries ($J \perp B$ and $J \parallel B$) the loss of angular momentum, via torsional Alfvén waves, by a cloud (or fragment) of density ρ_{cl} in the shape of a disk (or cylinder) of radius R and half-thickness Z (not necessarily small), threaded by a frozen-in magnetic field and surrounded by an "external" medium of constant density ρ_{ext}. In the case of a cloud, the quantity ρ_{ext} refers to the density outside the cloud but within the extended cloud complex. In the case of a fragment (or core) within a cloud, ρ_{ext} refers to the envelope density.

3.3.1. Observational Constraints. The angular momentum problem, as traditionally stated (e.g., see Spitzer 1968a, p. 231), expresses the impossibility of forming a single star from a cloud (or a fragment) of typical interstellar density, even if the cloud spins about an axis only once per galactic rotation. Since over 70% of all stars occur in binary systems (Heintz 1967, 1969), and the orbital angular momentum of any binary system exceeds the spin angular momentum of its components by several orders of magnitude (Ambartsumian 1956), one may ask whether it is possible to form a typical *binary* system from a collapsing cloud (or fragment) while angular momentum is conserved. TM77 quantified and addressed this problem. The answer was still discouraging: A blob (fragment) of interstellar matter of mass 2 $M_\odot$ and density 1 cm^{-3} (the approximate mean density of the interstellar medium) possesses an amount of angular momentum $J_{fr}(2M_\odot, 1\text{cm}^{-3}) \sim 10^{55}$ g cm^2 s^{-1}. A relatively wide binary system of the same mass and a period of 100 yr is characterized by an angular momentum $J_b(2M_\odot, 100\text{yr}) \sim 10^{53}$ g cm^2 s^{-1}. The Sun possesses $J_\odot \sim 10^{49}$ g cm^2 s^{-1}, but the angular momentum of the solar system is almost entirely in the orbital motion of Jupiter (and Saturn) and is $10^2 J_\odot$. The angular momentum of the solar system is not atypical of that of single protostars. Observational evidence therefore defines the angular momentum problem: *The interstellar parent blob must lose at least two and four orders of magnitude of angular momentum for a relatively wide binary system and a single star or planetary system, respectively, to become dynamically possible.*

Since observations also show that stellar clusters are spherical, rather than highly flattened (disklike), systems with no appreciable rotation, it follows immediately that the original angular momentum of the parent cloud could not possibly be stored in the orbital motion of individual stars in a cluster. If angular momentum were to be conserved during the spherical contraction of a cloud possessing the minimum amount of angular momentum associated with its participation in the general galactic rotation, the angular velocity would increase with number density as

$$\Omega_{J=\text{const}} \simeq 10^{-15} \, n^{2/3} \quad \text{rad s}^{-1} \, . \tag{57}$$

For the density range $10^4 - 10^6$ cm^{-3}, typical of individual, dense molecular clouds, one would then commonly observe angular velocities in the corresponding range $5 \times 10^{-13} - 1 \times 10^{-11}$ rad s^{-1}. *Observations actually reveal angular velocities at least one, and (depending on the density) usually more than two or three, orders of magnitude smaller than those expected from equation (57).* In this context, even the fastest rotating clouds, the globule B 163 SW (Martin and Barrett 1978) and a fragment in the dark cloud B 213 NW (Clark and Johnson 1978), are *slow* rotators; and so are the fastest rotating fragments known to date, the two H_2CO fragments observed by Wadiak *et al.* (1985) in ρ Ophiuchi B which are characterized by $\Omega \simeq 10^{-12}$ rad s^{-1} but $n > 5 \times 10^6$ cm^{-3} (and $M \simeq 2$ $M_\odot$). (Note in the same context that the mean density of an open cluster is comparable with the mean density of an individual molecular cloud.) *The conclusion, therefore, seems inescapable that an efficient angular-momentum redistribution mechanism must operate at the early, relatively diffuse stages of cloud contraction.* Detailed calculations have shown that magnetic braking is such a mechanism. No other mechanism has been proven *by calculations*, as opposed to qualitative ideas, capable of resolving the angular momentum problem. This does not, of course, mean that nature employs no other such mechanism. However, if it does, it also does an excellent job hiding it from us.

That torsional Alfvén waves in a medium with a frozen-in magnetic field carry angular momentum has been demonstrated by Ebert, Hoerner, and Temesvary (1960). The issue, nevertheless, of whether magnetic braking can resolve the angular momentum problem of star formation is a *quantitative*, not a qualitative one: *How long does it take for magnetic braking to reduce the angular momentum of a contracting blob of interstellar matter by a sufficient factor (typically $10^{-2} - 10^{-4}$) for stars, ranging from wide binaries to single, to become dynamically possible?* We summarize below the (analytical) calculations which provided unambiguous, quantitative answers and we discuss their most important conclusions, with emphasis on observational consequences. (A review of the literature through 1980 is given in TM81. In this paper we present only the underlying physics of the resolution of this problem.) In order of increasing complexity, a single, *rigid*, aligned rotator (fragment, or cloud, or core) embedded in an "external" medium (or envelope) is considered; then the assumption of rigid rotation is relaxed and the propagation of the torsional Alfvén waves inside (as well as away from) the rotator is accounted for (Mouschovias and Paleologou 1980b); results from the two cases are discussed and compared. A single, rigid, perpendicular rotator is considered next; the results differ both quantitatively and qualitatively from those of the aligned rotator (Mouschovias and Paleologou 1979, 1980a; hereafter MP79, MP80a, etc.). Finally, the effect of fragment-fragment interaction through torsional Alfvén waves on the resolution of the angular momentum problem is considered for a system of N (magnetically linked) aligned rotators (Mouschovias and Morton 1985a, b); unexpected results emerge. The exact analogy with transverse waves on strings is made explicit; it was employed by MP80b in interpreting the results physically.

3.3.2. Single, Aligned Rotator. (a) Rigid Rotator. We consider a conducting gaseous disk (cloud, or fragment, or core) of uniform density ρ_{cl}, radius R, and half-thickness Z (not necessarily small) threaded by a magnetic field B_{cl}, which is initially uniform, parallel to the disk's axis, and equal to B_0. The surrounding ("external") medium (or "envelope") has a constant density ρ_{ext} and field B_{ext} and is also electrically conducting. By continuity of the field across the cloud surface, the external field in the neighborhood of the cloud must be equal to B_{cl}. We therefore introduce the simplification $B_{ext} = B_{cl}$ in this geometry. [This assumption implies that the magnetic braking time scale to be obtained from the solution of this problem is a strict *upper* limit on the time scale that one obtains in a more realistic geometry in which the field lines "fan out" away from the cloud --see § 3.3.2(d).] We choose cylindrical polar coordinates (r, ϕ, z) such that the initial field has only a z-component and the disk's equatorial plane coincides with $z = 0$. The cloud is then imparted a uniform initial angular velocity Ω_0 with respect to the surrounding medium. Assuming axial symmetry, negligibly small velocities v_z (compared to the Alfvén velocity) in the z-direction, and no sources or sinks of matter on the z-axis, one can show that velocities v_r in the radial direction are also negligible. It then follows from the flux-freezing equation (32d) and the initial uniformity of the field that $B_z(z, t) = B_0$ and $B_r(r, z, t) = 0$ at all times. Hence, the condition $\nabla \cdot B = 0$ is satisfied identically. The single-fluid MHD equations in the external medium imply a wave equation for the angular velocity Ω_{ext}, namely,

$$\frac{\partial^2 \Omega_{ext}(z, t)}{\partial t^2} = v_{A,ext}^2 \frac{\partial^2 \Omega_{ext}(z, t)}{\partial z^2}, \qquad z > Z, \tag{58}$$

and that the equation of motion for a rigidly rotating cloud, responding to the instantaneous external (magnetic) torque, is

$$\frac{\partial^2 \Omega_{cl}(t)}{\partial t^2} = v_{A,ext}^2 \frac{\rho_{ext}}{\rho_{cl}} \frac{1}{Z} \frac{\partial \Omega_{ext}(z, t)}{\partial z}\bigg|_{z=Z} \tag{59}$$

(MP80b). The quantity $v_{A,ext}$ is the initial Alfvén velocity in the external medium. By symmetry about the plane $z = 0$, only the region $z \geq 0$ is considered. [Note that $B_{\phi,ext}(r, z, t)$ also satisfies the wave equation (58) and that it has the form $rf(z, t)$.]

An exact solution of the coupled system of equations (58) and (59) will describe uniquely the evolution of the angular velocity of the cloud in time and the response of the external medium to the propagation of the torsional Alfvén waves along the z-axis. We solve these equations under the following boundary and initial conditions.

We require that the angular velocity vanish at infinity, be continuous at the disk surface, and be an even function of z; i.e.,

$$\Omega_{\text{ext}}(z = \infty, t) = 0, \quad \Omega_{\text{ext}}(Z, t > 0) = \Omega_{\text{cl}}(t > 0), \quad \Omega(z, t) = \Omega(-z, t). \tag{60a, b, c}$$

We may then consider only the region $z \geq 0$. The external medium is assumed to be at rest initially, and the disk to be given an angular velocity Ω_0:

$$\Omega_{\text{ext}}(z, t \leq 0) = 0, \quad \Omega_{\text{cl}}(t < 0) = 0, \quad \Omega_{\text{cl}}(t = 0) = \Omega_0. \tag{61a, b, c}$$

The initial uniformity of the field and the ϕ-component of the force equation imply that

$$\frac{\partial \Omega_{\text{ext}}(z, t \leq 0)}{\partial t} = 0 = \frac{\partial \Omega_{\text{cl}}(t \leq 0)}{\partial t}. \tag{62a, b}$$

We write the basic equations (58) and (59), and the boundary and initial conditions (60a) - (62b) in dimensionless form by choosing as units of length, speed, density, and magnetic field the quantities Z, $v_{A,\text{ext}}$, ρ_{ext}, and B_0, respectively. (Note that this choice specifies uniquely the units of time and mass.) The basic equations become

$$\frac{\partial^2 \Omega}{\partial \tau^2} = \frac{\partial^2 \Omega}{\partial \zeta^2}, \quad \zeta > 1; \tag{63a}$$

$$\frac{\partial^2 \Omega_d}{\partial \tau^2} = \frac{1}{\rho} \left. \frac{\partial \Omega}{\partial \zeta} \right|_{\zeta=1}, \tag{63b}$$

where

$$\tau = t/(Z/v_{A,\text{ext}}), \quad \zeta = z/Z, \quad \rho = \rho_{\text{cl}}/\rho_{\text{ext}}. \tag{64a, b, c}$$

In equations (63a, b) the quantity Ω without a subscript refers to the envelope, while the quantity Ω_d refers to the disk. The arguments of the angular velocities have been omitted for simplicity. It is clear that *ρ is the only dimensionless free parameter in the equations.* The dimensionless form of the boundary and initial conditions is obtained simply by replacing z with ζ, Z with 1, and t with τ in equations (60a) - (62b).

The equations (63a, b) are identical with those describing small-amplitude transverse waves on a flexible string which extends from $-\infty$ to -1 and from $+1$ to $+\infty$ along the ζ-axis and which joins to a rigid string extending from $\zeta = -1$ to $\zeta = +1$ as shown in the figure below --simply replace Ω with y, the transverse displacement. The string analogue of the initial value problem set up above is the problem in which (only) the rigid portion of the string is given a small initial displacement y_0 and then released (at time $t = 0$). The only difference between the two problems is that, under the assumptions of the magnetic braking problem, the angular velocity of the disk, unlike the displacement of the string, is not restricted to be small.

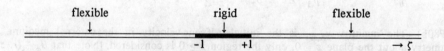

Figure 1. *The string analogue of the rigid rotator. The shaded, inflexible, dense portion of the string corresponds to the rigidly rotating cloud (or fragment, or core). The lower-density, flexible string corresponds to the external medium (or envelope). The transverse displacement of the string plays the role of the angular velocity Ω.*

3.3.2(b) Differential Rotator. The assumption of rigid rotation made above is synonymous with an instantaneous response of any interior fluid element to magnetic torques exerted on the surface of the model cloud. In reality, it would take a finite time for the torsional Alfvén waves to propagate inward from the surface and convey the message to an interior fluid element that it has to slow down because of the application of an external torque. It would therefore seem that, if the assumption of rigid rotation is relaxed, magnetic braking may take longer to remove the same angular momentum from the cloud. As we explain in § 3.3.2(c), this is not so.

We now relax the assumption of rigid rotation for the cloud, which is still imparted an initially uniform angular velocity Ω_0 with respect to the external medium. The boundary and initial conditions (60a) - (62b) remain in effect. However, the coordinate z must now appear explicitly as an argument of Ω_{cl}. In addition, one must impose continuity of B_ϕ across the cloud boundary at all times. This and the flux-freezing equation imply that $\partial \Omega / \partial z$ must also be continuous; i.e.,

$$\frac{\partial \Omega_{ext}(z, t)}{\partial z} = \frac{\partial \Omega_{cl}(z, t)}{\partial z}, \qquad z = Z. \tag{62c}$$

By symmetry, B_ϕ must always vanish on the equatorial plane; hence,

$$\frac{\partial \Omega_{cl}(z = 0, t)}{\partial z} = 0. \tag{62d}$$

The cloud as a whole can no longer be described by a single equation of motion. A derivation identical with the one that led to the wave equation (58) for the external medium yields a wave equation for the disk as well, namely,

$$\frac{\partial^2 \Omega_{cl}(z, t)}{\partial t^2} = \left[v_{A,ext}^2 \frac{\rho_{ext}}{\rho_{cl}} \right] \frac{\partial^2 \Omega_{cl}(z, t)}{\partial z^2}, \qquad z < Z. \tag{65}$$

The quantity in parentheses on the right-hand side of equation (65) is the square of the initial Alfvén speed *in the cloud*, $v_{A,cl}^2$. With the definitions (64a, b, c), equation (65) obtains the dimensionless form

$$\frac{\partial^2 \Omega(\zeta, \tau)}{\partial \tau^2} = \frac{1}{\rho} \frac{\partial^2 \Omega(\zeta, \tau)}{\partial \zeta^2}, \qquad \zeta < 1. \tag{66}$$

The subscript "cl" has been dropped from Ω_{cl} since the range of validity ($\zeta < 1$) of equation (66) is specified explicitly. (The azimuthal component of the magnetic field, B_ϕ, satisfies an identical wave equation inside the cloud.) Equations (63a) and (66), instead of (63b), must be solved simultaneously under the (dimensionless form of the) initial and boundary conditions (60a) - (62d). *These equations are identical with those describing small-amplitude, transverse waves on the string of Figure 1, but with the shaded, denser portion now being flexible too* --just replace, as before, Ω with y, the transverse displacement. The present initial value problem for a differentially rotating disk corresponds to giving an initial rigid displacement (only) to the shaded portion of the string (hence, the term "discontinuous initial condition") and releasing it at $t = 0$, with continuity at the string interfaces imposed at all times $t > 0$. Waves now propagate inside the shaded, denser string as well, and suffer numerous partial reflections (and transmissions) at the interfaces --corresponding to trapping (and leakage) of torsional Alfvén waves within (from) a cloud (or fragment, or core). This string analogy made it tempting to refer to the torsional Alfvén waves and their numerous reflections at density discontinuities (or density gradients) within clouds as *cosmic music*. It always precedes the birth of stars --a cosmic festivity that leaves no participant untouched!

MP80b also studied a more realistic, continuous initial condition across the disk surface; namely,

$$\Omega(z, t = 0) = (\Omega_0/2)[1 + \cos(\pi z/\beta Z)], \qquad z \leq \beta Z; \qquad (67a)$$
$$= 0, \qquad z \geq \beta Z, \qquad (67b)$$

where β is a constant greater than unity. The two wave equations, the boundary conditions and the rest of the initial conditions are identical with those describing the differential rotator with discontinuous initial Ω given above.

3.3.2(c) Results and Applications. **Rigid Rotator:** The solution of the rigid rotator problem, set up in § 3.3.2(a) and represented by the dimensionless equations (63a), (63b) and the associated initial and boundary conditions, is obtained either by a Laplace transform technique or by inspection. It is a simple exponential, namely,

$$\Omega(\zeta, \tau) = \Omega_0 \exp[-(\tau - \zeta + 1)/\rho], \qquad \text{if } \tau - \zeta + 1 \geq 0; \qquad (68a)$$
$$= 0, \qquad \text{if } \tau - \zeta + 1 < 0, \qquad (68b)$$

for $\zeta \geq 1$. By assumption, $\Omega(\zeta < 1, \tau) = \Omega(\zeta = 1, \tau)$. The meaning of the conditions on $(\tau - \zeta + 1)$ becomes clear when one realizes that the position of the original wavefront propagating in the region $\zeta \geq 1$ is given by $\tau - \zeta + 1 = 0$ at any time. Thus equation (68b) expresses the physically obvious fact that no disturbance exists ahead of the wavefront, which originates at $\zeta = 1$ at $\tau = 0$ and propagates outward with dimensionless speed equal to unity --recall that $v_{A,ext}$ has been chosen as the unit of speed (see discussion preceding eq. [63a]).

The ϕ-component of the magnetic field in the region $\zeta > 1$ is given by

$$B_\phi(\xi, \zeta, \tau) = -\xi\Omega_0 \exp[-(\tau - \zeta + 1)/\rho], \qquad \text{if } \tau - \zeta + 1 \geq 0; \qquad (69a)$$
$$= 0, \qquad \text{if } \tau - \zeta + 1 < 0, \qquad (69b)$$

where $\xi = r/Z$. At a time τ, a fluid element in the external medium at location ζ (> 1) has rotated through an angle ϕ given by

$$\phi = \rho\Omega_0 \{1 - \exp[-(\tau - \zeta + 1)/\rho]\}, \qquad \text{if } \tau - \zeta + 1 \geq 0; \qquad (70a)$$
$$= 0, \qquad \text{if } \tau - \zeta + 1 < 0. \qquad (70b)$$

By flux-freezing, this is also the angle through which a point on a field line at ζ has rotated in the same time τ; i.e., equations (70a, b) describe the field lines.

The angular velocity of the disk for positive times is obtained by evaluating Ω at $\zeta = 1$:

$$\Omega_d(\tau) = \Omega_0 \exp(-\tau/\rho). \qquad (71)$$

It is clear that the dimensionless e-folding time for magnetic braking of a rigid rotator is equal to ρ. This is put in dimensional form with the aid of equations (64a) and (64c), i.e.,

$$\tau_\parallel = \frac{\rho_{cl}}{\rho_{ext}} \frac{Z}{v_{A,ext}}. \qquad (72)$$

The term $Z/v_{A,ext}$ on the right-hand side of equation (72) is the time it takes an Alfvén wave in the *external* medium to traverse a distance along field lines equal to the half-thickness of the cloud; we refer to it as *the Alfvén crossing time*, since the internal Alfvén crossing time for a rigid rotator is exactly zero [see beginning of § 3.3.2(b)]. The ratio of densities multiplying the Alfvén crossing time is exactly the number of crossing times required for the waves to propagate along the (straight-parallel) field lines far enough from the disk so as to set into rotational motion an amount of external matter with moment of inertia I_{ext} equal to that of the cloud, I_{cl}. In fact, equation (72) can be recovered from a simple mechanical argument (see MP80b). Since the condition $I_{ext} = I_{cl}$ is satisfied when the waves reach a distance $Z_I = Z [1 + (\rho_{cl}/\rho_{ext})]$ from the equatorial plane, the time necessary for this to happen is

$$\tau_I = \int_Z^{Z_I} \frac{dz}{v_{A,ext}} = \frac{\rho_{cl}}{\rho_{ext}} \frac{Z}{v_{A,ext}} = \tau_\parallel \ . \tag{73}$$

This physical interpretation of the characteristic time makes it intuitively clear why magnetic braking is much more efficient for a perpendicular than for an aligned rotator. In the case of a perpendicular rotator, the waves propagate away from the axis of rotation and thereby affect external matter with larger and larger moment of inertia. That is, the equality $I_{ext} = I_{cl}$ is achieved much faster in this case and, therefore, $\tau_\perp \ll \tau_\parallel$. This simple argument, however, misses a very important feature of the exact solution for the perpendicular rotator, namely, the oscillatory approach to corotation with the background which leads naturally to retrograde rotation of some fragments in clouds and of some stars in stellar systems (see § 3.3.3 below).

The dotted curves in Figures 2a and 2b, marked Ω_{SB} for "Solid Body" rotation, represent the angular velocity of the cloud, normalized to its (arbitrary) initial value, as a function of time, normalized to the crossing time

$$\tau_0 = Z_0/v_{A_0} \tag{74}$$

of an Alfvén wave across the *original* half-thickness of the cloud $Z = Z_0 \simeq R_0$; i.e., the value of Z when the cloud, or fragment, began forming out of a background medium. At that initial stage, $\rho = 1$ and, therefore, $v_{A,cl} = v_{A,ext} \equiv v_{A0}$. Figures 2a and 2b refer to the cases $\rho \equiv \rho_{cl}/\rho_{ext} = 10$ and 100, respectively. The angular velocity falls to 1% its initial value in a time less than $5\tau_0$. For the representative parameters $n_{ext} \simeq 1$ cm^{-3}, $B_0 \simeq 3$ μG, $R_0 \simeq 20$ pc (corresponding to $M_{cl} \simeq 10^3$ $M_\odot$), we find that $\tau_0 \simeq 3 \times 10^6$ yr. Thus the angular momentum can decrease to a low enough value for binaries to begin forming in a time $\lesssim 10^7$ yr. This conclusion does not depend on the density ratio ρ_{cl}/ρ_{ext} because $\tau_\parallel$ is independent of the stage of contraction in this model with straight-parallel field lines. This is so because ρ_{ext} is fixed, $B_{ext} \simeq B_{cl}$ for $r \leq R$, and, by mass and flux conservation, $\rho_{cl} \propto R^{-2}Z^{-1}$ and $v_{A,ext} \propto R^{-2}$, respectively. Substituting these relations in equation (72), it follows that

$$\tau_\parallel = const = \tau_0 \ , \tag{75}$$

independent of the stage of contraction. The physical meaning of this result is that, although ρ_{cl} increases upon contraction, tending to make magnetic braking less effective, the Alfvén speed also increases and the thickness of the cloud decreases, which together tend to reduce the Alfvén crossing time and thereby increase the effectiveness of magnetic braking. We show in § 3.3.2(d) that, in a more realistic geometry in which the field lines fan out away from the cloud (i.e., $B_{ext} < B_{cl}$), the characteristic time for magnetic braking $\tau_\parallel$ actually *decreases* upon contraction. *Thus the above conclusion, that magnetic braking can resolve the angular momentum problem at least for binary stars, is much stronger in reality than it appears here.* In fact, we show that even the angular momentum problem for single stars can be resolved by magnetic braking of cloud cores. We therefore bear in mind that $\tau_\parallel$ given by equation (72) is a strict upper limit on the actual braking time of an

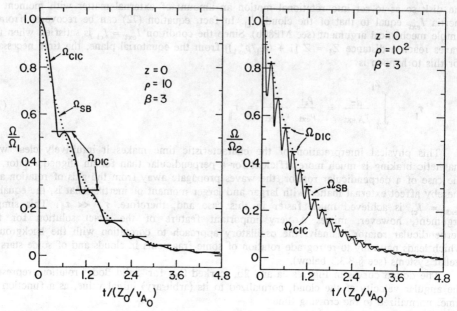

Figure 2a (*left*), **2b** (*right*). *The angular velocity of the equatorial plane $z = 0$ of the cloud as a function of time for the density ratios $\rho \equiv \rho_{cl}/\rho_{ext} = 10^1, 10^2$. The curves labeled Ω_{SB}, Ω_{DIC}, and Ω_{CIC} refer to solid-body rotation of the cloud, and differential rotation with a discontinuous initial condition, and a continuous initial condition on Ω across the cloud surface, respectively. The latter two cases include the multiple internal reflections of torsional Alfvén waves on the disk surfaces. In each case, Ω is normalized to its (arbitrary) initial value (maximum initial value in the case of Ω_{CIC}). The unit of time ($\tau_0 = Z_0/v_{A0} \lesssim 3 \times 10^6$ yr for $M_{cl} \lesssim 10^3 \, M_\odot$) is the same in both graphs and is equal to the Alfvén crossing time across the half-thickness (or polar radius) of the cloud at the time of its formation, when $\rho = 1$. A larger density ratio represents a later stage of contraction. The parameter β (= 3) specifies the point z (= βZ), in units of the half-thickness of the disk when the density ratio was ρ, beyond which the initial Ω_{CIC} and its first z-derivative vanish. Note that the rigid rotator approximates well the mean behavior of the cloud.*

aligned rotator.

It should be emphasized that our requirement that the angular momentum of the cloud or fragment be reduced by at least two orders of magnitude *at each and every value of the density ratio ρ* (i.e., at each and every stage of contraction) represents a severe test of magnetic braking. It is as if magnetic braking is expected to resolve the entire angular momentum problem over and over again, at each and every stage of contraction.

Using equation (5a) of TM87a for the Alfvén speed in the envelope of a molecular cloud and assuming the same mean mass per particle in the core and envelope, equation (72) is written in a form appropriate for applications to cloud cores as

$$\tau_\parallel = 5.7 \times 10^5 \left[\frac{n_n/n_{env}}{10}\right] \left[\frac{Z}{0.1 \text{ pc}}\right] \left[\frac{B_0}{3 \text{ }\mu G} \frac{M_{env}}{10^4 \text{ M}_\odot}\right]^{-1/4} \text{ yr}, \qquad (76)$$

where all quantities have been normalized to representative values. Although the core density n_n is expected to join smoothly to the envelope density n_{env} (unless there are phase transitions), the ratio n_n/n_{env} appearing in equation (76) should not be taken as small as unity because n_n and n_{env} represent *mean* densities.

Differentially Rotating Disk: The formal solution was obtained by MP80b through a Laplace transform technique. It is mathematically tedious, involving the evaluation of infinite sums. Partly for this reason, MP80b (§ Va[ii]) regenerated the solution in a much simpler way, by making use of the (exact) analogy with waves on a string and the *reflection coefficient*

$$\mathcal{R} = \frac{\rho^{1/2} - 1}{\rho^{1/2} + 1}, \qquad (77)$$

where $\rho \equiv \rho_{cl}/\rho_{ext}$. Here we describe the main results physically.

The curves labeled Ω_{DIC} (DIC standing for "Discontinuous Initial Condition") in Figures 2a and 2b represent the normalized angular velocity of the equatorial plane ($z = 0$) as a function of time, measured in units of the *initial* Alfvén crossing time τ_0 (i.e., when ρ was equal to unity), in the cases $\rho = 10^1$ and 10^2, respectively. In each case, the angular velocity maintains its initial value until the two wavefronts generated at the surfaces arrive (simultaneously) for the first time, when $t = Z/v_{A,cl} = (\rho_{cl}/\rho_{ext})^{1/2}(Z/v_{A,ext}) \equiv \rho^{1/2}(Z/v_{A,ext}) = \rho^{-1/2}\tau_0$ (see eqs. [74] and [75]). It then decreases discontinuously by a factor $\mathcal{R}$ ($\simeq 0.52$ and 0.82 for the two values of ρ), and retains its new value until the reflected waves arrive, a time $2\rho^{-1/2}\tau_0$ later. At this time Ω decreases by another factor $\mathcal{R}$, and so forth.

A comparison with the respective curves Ω_{SB} reveals that, except for short-lived departures, *the rigid rotation is a good approximation insofar as loss of angular momentum by the equatorial plane of the cloud is concerned. The later the stage of contraction, the better this approximation becomes. This is so because the internal Alfvén crossing time decreases upon contraction, thus tending to better fulfill the assumption of instantaneous response (of the cloud interior to a retarding torque exerted at the surface) which is implied by rigid rotation. The same conclusion holds for any interior point.*

The case with a continuous initial condition (CIC) on Ω across the cloud boundary (see eqs. [67a, b]) eliminates the unphysical discontinuous changes of Ω and the existence of wavefronts as such. Yet the formally messy exact solution can be understood physically in terms of the Huygens principle (see MP80b, § Va[ii]). The decrease of the angular velocity of the equatorial plane as a function of time is shown as a solid curve, labeled Ω_{CIC}, in Figures 2a and 2b for $\rho = 10^1$ and 10^2, respectively. The constant β, which defines how far (in units of Z) from the disk boundary the initial angular velocity is nonzero (see eqs. [67a, b]), has the value 3 for the solution shown in these figures. A comparison of the curves Ω_{SB} and Ω_{CIC} reveals that rigid rotation is still a good approximation, except for short-lived transients; and so is it for any other interior point. It is also clear from a comparison of Figures 2a and 2b that the rigid rotator becomes a better approximation for the differentially rotating cloud as ρ increases, i.e., at later stages of contraction. Ω_{SB} bounds from above Ω_{CIC}, and goes exactly through the middle of the straight-line segments representing Ω_{DIC}.

The behavior of B_ϕ (recall that the component B_z is a constant of the motion) for the above three rotators (except inside the rigid rotator, where $B_\phi = 0$ at all times) is discussed in detail in MP80b. Its oscillations inside the differential rotators are relatively more violent than those of Ω, but its mean value decreases in time like that of Ω.

The above results for the magnetic braking of an aligned rotator can be extended in a straightforward fashion to apply to an aligned oblate rotator. In the latter case, the internal Alfvén crossing time decreases from the value $Z/v_{A,cl}$ on the axis of symmetry to zero on the equator. Thus the outermost (near the equator) parts of an oblate cloud will slow down very rapidly. It is clear that differential rotation sets in not only along the axis of symmetry --which is the case for a disk-- but in a direction perpendicular to the axis as well. The magnetic braking of an ellipsoidal spheroid is obtained from the results for a disk by simply replacing the half-thickness of the disk Z by $Z[1 - (r/R)^2]^{1/2}$; where R is the equatorial radius of the spheroid, Z its polar radius, and r the distance from the axis ($\leq R$).

3.3.2(d) The Effect of Field Lines "Fanning Out" Away from Aligned Rotator. In a geometry more realistic than the one with straight-parallel field lines discussed above, that is, one in which the field strength decreases from its value in the cloud (or fragment), through a transition region, to that of the "background" field, the time scale $\tau_\parallel$ acquires a multiplicative factor ≤ 1 whose precise value depends on the moment of inertia of the transition region. If the transition region, which is assumed to have the same density ρ_{ext} as the background, is characterized by a moment of inertia $I_{tr} \leq I_{cl}$, it is the external medium (with a uniform field B_{ext}) beyond the transition region that accepts the cloud's angular momentum. In such a case, $\tau_\parallel$ for a disk cloud is given by the time at which the torsional Alfvén waves have propagated far enough to satisfy the condition $I_{ext}(t) = I_{cl}$; i.e.,

$$\tau_\parallel = \frac{\rho_{cl}}{\rho_{ext}} \frac{Z}{v_{A,ext}} \left(\frac{R_{cl}}{R_0}\right)^4 \tag{78}$$

(TM83b), where R_0 is the original radius of the flattened cloud, when its density was $\rho_0 > \rho_{ext}$ (e.g., see eq. [2] of TM87a), due to its formation by motions along field lines, and its magnetic field was equal to that of the external medium (see Mouschovias and Morton 1985b, Fig. 2, which shows this geometry, but for several magnetically connected fragments). (Note that $\rho_{cl}Z = \sigma_{cl}/2$, where σ_{cl} is the present surface density of the cloud.) It is clear that, for a cloud of fixed mass, $\tau_\parallel$ decreases as R_{cl}^2 upon contraction.

A comparison of equation (78) with equation (72) reveals that magnetic braking in this, more realistic geometry is much more effective in removing angular momentum from the cloud than in the case with straight-parallel field lines. Equation (72) or, equivalently, equation (76) is useful nevertheless because it gives an *upper limit* on $\tau_\parallel$. If application of equation (72) in a particular problem leads to the conclusion that enough angular momentum is lost for any given purpose, it will be the case that a more precise or more realistic calculation will strengthen that conclusion.

3.3.3. Single, Perpendicular Rotator. (a) Formulation of the Problem. In view of the above conclusion, that a rigid rotator is an excellent approximation for a differential rotator, we consider a rigidly rotating cloud (or fragment) in the shape of a cylinder or disk of uniform density ρ_{cl} and radius R. A frozen-in field links the cloud with an external medium of constant density ρ_{ext}. To tackle the problem with analytical means we adopt a two-dimensional geometry with cylindrical symmetry. The z-axis of a cylindrical polar coordinate system (r, ϕ, z) is chosen to coincide with the axis of symmetry, and the magnetic field to have only r- and ϕ-components. In this geometry, the assumption that ρ_{ext} is constant results in no loss of generality. MP79 derived a wave equation for the external medium and an equation of motion for the cloud as outlined below.

The assumption of cylindrical symmetry, flux-freezing, and the equation $\nabla \cdot B = 0$, which must be satisfied at all times, imply that the radial component of the magnetic field is independent of time and given by

$$B_r = B_0 R/r, \quad r \geq R. \tag{79}$$

The quantity B_0 is the value of the field at the cloud surface at $t = 0$. The azimuthal component of the field is assumed to vanish at $t \leq 0$. Assuming that there is no exchange of matter across the cloud surface, the continuity equation and the constancy of ρ_{ext} imply that only rotational motion can take place, with a velocity $v_\phi(r, t) = r\Omega_{ext}(r, t)$ for $r \geq R$.

Since v and B lie in the (r, ϕ)-plane and there is no ϕ-dependence, it follows that only the ϕ-component of the flux-freezing equation (32d) is nontrivial. It is

$$\frac{\partial B_\phi(r, t)}{\partial t} = RB_0 \frac{\partial \Omega_{ext}(r, t)}{\partial r}, \tag{80}$$

where we have made use of equation (79) and of the relation $v_\phi = r\Omega_{ext}$.

The ϕ-component of the force equation (32b) in the external medium can be written out in terms of Ω_{ext} and B_ϕ, by eliminating j and B_r with the aid of equations (32e) and (79), respectively. We find that

$$\frac{\partial \Omega_{ext}}{\partial t} = \frac{RB_0}{4\pi \rho_{ext} r^3} \frac{\partial}{\partial r}(rB_\phi). \tag{81}$$

Taking the partial derivative of equation (81) with respect to time and using equation (80) to eliminate B_ϕ, we obtain a wave equation for Ω_{ext}, namely,

$$\frac{\partial^2 \Omega_{ext}}{\partial t^2} = \frac{B_0^2}{4\pi \rho_{ext}} \frac{R^2}{r^2} \left[\frac{1}{r} \frac{\partial}{\partial r} \left(r \frac{\partial \Omega_{ext}}{\partial r} \right) \right]. \tag{82}$$

Clearly, the waves propagate with a local Alfvén speed $v_A(r) = v_{A0}R/r$, where $v_{A0} = B_0/(4\pi\rho_{ext})^{1/2}$ is the initial Alfvén speed just outside the cloud; their profile changes in time.

Once equation (82) is solved for $\Omega_{ext}(r, t)$ under appropriate initial and boundary conditions, the azimuthal component of the field can be obtained from equation (80) upon integration over time. Then the radial component of the force equation (32b) yields the pressure required to maintain hydrostatic equilibrium in the r-direction at all times.

We obtain the equation of motion for the cloud by setting the rate of change of its angular momentum equal to the torque exerted on it by the external magnetic field:

$$I_{cl} \frac{\partial \Omega_{cl}(t)}{\partial t} = N_m. \tag{83}$$

The quantities I_{cl} and N_m are the moment of inertia of the cloud and the torque exerted on the cloud, respectively. Both are expressed per unit length along the axis of symmetry. So, $I_{cl} = \rho_{cl}\pi R^4/2$, and N_m can be obtained from Maxwell's stress tensor. At the cloud surface, the tangential component of the magnetic force per unit surface area is $B_\phi(R, t)B_r(R)/4\pi$. Since there is no ϕ-dependence, the tangential force per unit length along z is equal to $RB_\phi(R, t)B_r(R)/2$, from which we conclude that $N_m = R^2 B_\phi(R, t)B_r(R)/2$. Equation (83) is then written as

$$\frac{\partial \Omega_{cl}(t)}{\partial t} = \frac{B_0}{\rho_{cl}\pi R^2} B_\phi(R, t), \tag{84}$$

where equation (79) has been used. We eliminate B_ϕ from equation (84) in favor of Ω_{ext} by differentiating with respect to time and using equation (80). We find that

$$\frac{\partial^2 \Omega_{cl}(t)}{\partial t^2} = \frac{B_0^2}{\rho_{cl}\pi R} \left.\frac{\partial \Omega_{ext}(r, t)}{\partial r}\right|_{r=R}. \qquad (85)$$

The boundary and initial conditions are obtained in a fashion similar to that of the rigid, aligned rotator (see eqs. [60a] - [62b]). They are

$$\Omega_{ext}(r = \infty, t) = 0, \qquad \Omega_{ext}(R, t > 0) = \Omega_{cl}(t > 0), \qquad (86a, b)$$

$$\Omega_{ext}(r, t \le 0) = 0, \qquad \Omega_{cl}(t < 0) = 0, \qquad \Omega_{cl}(t = 0) = \Omega_0, \qquad (86c, d, e)$$

$$\frac{\partial \Omega_{ext}(r, t \le 0)}{\partial t} = 0 = \frac{\partial \Omega_{cl}(t \le 0)}{\partial t}. \qquad (86f, g)$$

3.3.3(b) The Dimensionless Equations. We choose as the unit of length the radius of the cloud R, and as the unit of density the external density ρ_{ext}. This determines uniquely the unit of mass. A natural unit of time is the crossing time of an Alfvén wave across the cloud radius with a speed equal to the Alfvén speed just outside the cloud surface; i.e., $R/v_{A_0} \equiv R/[B_0/(4\pi\rho_{ext})^{1/2}]$. The unit of magnetic field is taken to be B_0, the initial value of the field at the cloud surface. With the change of variables

$$\xi = \left(\frac{r}{R}\right)^2, \qquad \tau = \frac{2t}{R/v_{A_0}}, \qquad (87a, b)$$

the basic equations (82) and (85) become, in dimensionless form,

$$\frac{\partial^2 \Omega(\xi, \tau)}{\partial \tau^2} = \frac{1}{\xi}\frac{\partial}{\partial \xi}\left[\xi \frac{\partial \Omega(\xi, \tau)}{\partial \xi}\right], \qquad \xi > 1; \qquad (88a)$$

$$\frac{\partial^2 \Omega(\xi, \tau)}{\partial \tau^2} = \frac{2}{\rho}\frac{\partial \Omega(\xi, \tau)}{\partial \xi}, \qquad \xi = 1. \qquad (88b)$$

In equation (88b) we have used the boundary condition (86b) to replace $\Omega_{cl}(\tau)$ with $\Omega_{ext}(\xi = 1, \tau)$. The subscripts "cl" and "ext" have been dropped from Ω since the range of validity of equations (88a) and (88b) is specified explicitly. As in the case of the aligned rotator, the dimensionless density $\rho \equiv \rho_{cl}/\rho_{ext}$ is the only free parameter in the equations. The dimensionless form of the boundary and initial conditions (86a) - (86g) is obtained by simply replacing r with ξ, t with τ, and R with 1.

3.3.3(c) Results and Applications. An exact solution of equations (88a) and (88b), under the above boundary and initial conditions, is given in MP79 in terms of Bessel functions (I_0, I_1, K_0, K_1), exponential and trigonometric functions. Here we give a physical discussion of the solution as it relates to the effectiveness with which magnetic braking redistributes angular momentum and, therefore, to the resolution of the angular momentum problem.

For a fixed ρ_{ext}, we have that $\rho_{cl} \propto R^{-2}$ and $v_A(R) \propto R^{-1}$, so that the Alfvén crossing time varies as $R^2 \propto \rho_{cl}^{-1}$. Hence, it decreases upon contraction. Since the moment of inertia of the cloud (or fragment) also decreases, *magnetic braking redistributes angular momentum more efficiently as contraction progresses.*

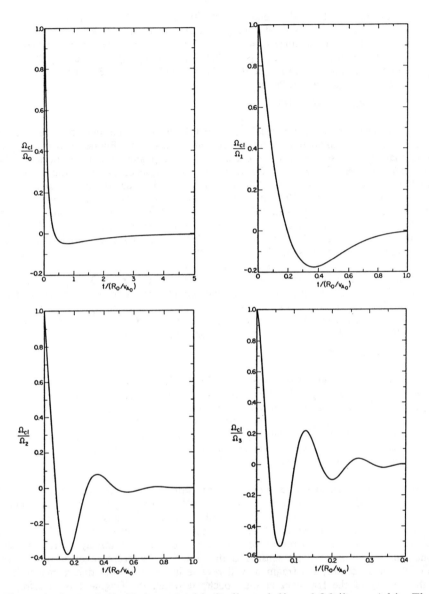

Figure 3a (*upper left*), **3b** (*upper right*), **3c** (*lower left*), and **3d** (*lower right*). *The angular velocity of the cloud as a function of time for the density ratios* $\rho \equiv \rho_{cl}/\rho_{ext}$ = 1, 10, 10^2, 10^3, *respectively.* In each case, Ω_{cl} is normalized to its (arbitrary) initial value. The unit of time is the same in all four figures and is equal to τ_0, the Alfvén crossing time across the cloud radius R_0 at the time of its formation, when $\rho = 1$, with a speed equal to the Alfvén speed just outside the cloud surface. A larger density ratio represents a later stage of contraction.

Figures 3(a-d) show the angular velocity of the cloud, normalized to its initial (arbitrary) value, as a function of time, normalized to the *initial* Alfvén crossing time (i.e., when the cloud began forming) $\tau_0 = R_0/v_{A0}$. For the same representative values used as an example in the aligned rotator case [see § 3.3.2(c)], namely, $n_{ext} \simeq 1$ cm^{-3}, $B_0 \simeq 3$ μG, and $R_0 \simeq 20$ pc (corresponding to $M_{cl} \simeq 10^3$ M$_\odot$), we find that $\tau_0 \simeq 3 \times 10^6$ yr. Since the free-fall time (eq. [24]) can be written as

$$\tau_{ff} = 4.35 \times 10^7 \, n_H^{-1/2} \quad \text{yr} \tag{89}$$

for the canonical helium abundance $n_{He}/n_H = 0.1$, magnetic braking has at least 15 crossing times to resolve the angular momentum problem at this stage. Figure 3a for $\rho \equiv \rho_{cl}/\rho_{ext} = 1$ shows that Ω_{cl} drops rapidly to zero (in a time $0.2\tau_0$) and then turns negative, implying *retrograde rotation* of the cloud (or fragment) with respect to its surroundings. Within 5 initial Alfvén crossing times, Ω_{cl} falls to 1% its initial value. Figure 3b exhibits a similar behavior for the case $\rho = 10$, but evolution occurs on a shorter time scale. On the other hand, for the cases $\rho = 10^2$ and 10^3 (Figs. 3c and 3d, respectively) the angular velocity of the cloud actually oscillates about zero with an ever decreasing amplitude. The number of oscillations increases but the time within which Ω_{cl} falls to 1% its initial value decreases as the density ratio ρ increases. In the case $\rho = 10^3$, the 1% level is reached in a time less than $0.4\tau_0$, or about 10^6 yr for the example given above.

In the case of the aligned, rigid rotator we found that $\tau_\parallel \simeq 3 \times 10^6$ yr for the above example, independent of the value of ρ. One should bear in mind, however, that $\tau_\parallel$ is the *e*-folding time for magnetic braking --not the time within which Ω_{cl} falls to 1% of its initial value. The initial *e*-folding time in the present case, $J \perp B$, is $< 6 \times 10^{-2} \, \tau_0 \simeq 1.8 \times 10^5$ yr for $\rho = 10^2$, and $< 2.2 \times 10^{-2} \, \tau_0 \simeq 6.6 \times 10^4$ yr for $\rho = 10^3$. Hence,

$$\tau_\parallel = (19 - 45) \times \tau_\perp \tag{90}$$

for the cases $\rho = 10^2$ and 10^3, respectively. Evidently, *magnetic braking removes angular momentum from a perpendicular rotator much more efficiently than it does from an aligned rotator* with identical physical parameters. This relative efficiency increases as the density ratio $\rho \equiv \rho_{cl}/\rho_{ext}$ increases. The effectiveness of magnetic braking in removing angular momentum will ultimately be limited by ambipolar diffusion and/or rapid (Alfvénic or superAlfvénic) collapse, as discussed in TM77 and TM87a (§ 3), respectively, and as further quantified by MP86 (see § 4 below).

3.3.3(d) Application to a Fragment - Retrograde Spin. A remarkable conclusion follows if the results exhibited in Figures 3(a-d) are applied to a fragment within a cloud at some distance from the axis of rotation. We identify as *direct* the sense of rotation of a cloud at the moment a fragment within it begins to separate out. By definition, the fragment will contract more rapidly than the cloud will as a whole. In a relatively short time, while the cloud still rotates in a direct sense and the fragment therefore revolves in a direct sense about the cloud center, the fragment will reverse its sense of spin due to magnetic braking. If the density of the fragment at this epoch, at which the fragment is in retrograde spin with respect to the cloud, is appropriate for ambipolar diffusion to lead to a magnetically and thermally supercritical state, the retrograde spin angular momentum will be conserved during the subsequent collapse of the fragment. *Therefore, retrograde spin of an object (star or planet) in a multicomponent system (stellar or planetary) may be a natural consequence of magnetic braking.*

The above results imply that a retrograde revolution of a fragment with either retrograde or direct spin is also possible. Which and how many fragments will acquire retrograde spin will depend on the stage at which fragmentation takes place and on the magnetic history of the cloud. Although the last question cannot be quantified at present, it is significant that retrograde rotation can be regarded as a *natural* consequence of a purely magnetic phenomenon. MP79 concluded that retrograde rotation of a fragment in a molecular cloud, "if it is observed, would constitute strong evidence for magnetic braking in action." Retrograde rotation has since been observed in at least two independent studies (Clark and Johnson 1981; and Young et al. 1981).[6] For example, Clark and Johnson observed the 6-cm and 2-cm lines of H_2CO in the cloud L134 and plotted the radial velocity v_{rad} as a function of offset from the center of the cloud along the major axis. The 6-cm line shows a systematic gradient in v_{rad} --the signature of rotation. The 2-cm line, which samples a more compact, higher density region (presumably a fragment, or core) shows a reversal in the algebraic sign of the *slope* of v_{rad}; i.e., the angular velocity of the high density fragment (or core) and that of the surrounding envelope have opposite algebraic signs.

3.3.4. Alternative Forms of the Magnetic Braking Time Scales. The characteristic time for magnetic braking of both aligned and perpendicular, as well as of oblique, rotators can be calculated, without regard to the details of torsional wave propagation, from a simple mechanical argument: *it is that time at which the waves have propagated far enough from the cloud (or core) in the external medium (or envelope) to affect an external moment of inertia equal to that of the cloud* (MP79, MP80b; see, also, reviews TM85 (§ 3), TM87a, TM87b).[7] For example, the mean behavior of a perpendicular rotator, as far as loss of angular momentum is concerned, is thus found to be described by (MP80b)

$$\tau_\perp = \frac{1}{2}\left[\left(1 + \frac{\rho_{cl}}{\rho_{ext}}\right)^{1/2} - 1\right]\frac{R}{v_A(R)}, \tag{91a}$$

$$\simeq \frac{1}{2}\left(\frac{\rho_{cl}}{\rho_{ext}}\right)^{1/2}\frac{R}{v_A(R)}, \quad (\rho_{cl}/\rho_{ext}) \gg 1, \tag{91b}$$

where the quantity $v_A(R) = B(R)/(4\pi\rho_{ext})^{1/2}$ is the Alfvén speed *just outside* the cloud (or core) surface. Although useful, this expression masks the oscillatory nature of the approach to corotation with the background, which, as we have seen in §§ 3.3.3(c) and 3.3.3(d), has the important consequence of retrograde rotation.

The expressions (72), (78), and (91) for $\tau_\parallel$ and $\tau_\perp$ can be rewritten so as to reveal different features of the magnetic braking phenomenon and its effectiveness under different

---

[6]Nakano (1991, § 3.4) presents the retrograde rotation of a fragment (or core) as a new prediction, although Nakano (1981, § 7) has already credited Mouschovias and Paleologou 1979 (MP79) for that prediction while criticizing their calculation on an inconsequential matter, and expresses the hope that this phenomenon may be observable!

[7]This result also is presented as new in a paper by Nakano (1989), which, in addition, follows step-by-step the formulation and solution of the magnetic braking problem by MP79 but without reference. The equation of (rotational) motion of the cloud (or core) in the form $\partial^2\Omega/\partial\tau^2 \propto \partial\Omega/\partial\zeta$ (see Nakano's eq. [18]), where τ denotes time and ζ a spatial coordinate, and the particular way of coupling it to the wave equation describing the propagation of torsional Alfvén waves in the external medium (or envelope), was first derived by MP79 (see their eq. [20]) for a perpendicular rotator and by MP80b (see their eq. [12b]) for an aligned rotator.

conditions. For example, one may introduce the mass-to-flux ratio of the cloud (or core) M/Φ_B, or the critical mass M_{crit} necessary for collapse against a given magnetic flux Φ_B, and/or τ_{ff}, etc., depending on the intended application (see MP80b; TM83b; especially TM85, § 3; also, TM87b, § 2.4). In the case of straight-parallel field lines, one finds that

$$\tau_\parallel = \frac{\rho_{cl}}{\rho_{ext}}\frac{Z}{v_{A,ext}} \equiv \frac{\sigma_{m,cl}}{2\rho_{ext}v_{A,ext}} \equiv \left(\frac{\pi}{\rho_{ext}}\right)^{1/2}\frac{M}{\Phi_B} \equiv 0.4\left(\frac{\rho_{cl}}{\rho_{ext}}\right)^{1/2}\frac{M}{M_{crit}}\tau_{ff},$$
(92a, b, c, d)

where R and Z are the equatorial and polar radius, respectively, of a cloud (or fragment) rotating about its (z-) axis of symmetry which is aligned with the magnetic field. The quantity $\sigma_{m,cl} \equiv M/\pi R^2$ is the column density of matter along field lines, M_{crit} the critical mass given by equation (22a), and τ_{ff} the free-fall time given by equation (24). For an aligned totator with field lines fanning out away from it, equation (78) yields

$$\tau_\parallel = \frac{\rho_{cl}}{\rho_{ext}}\frac{Z}{v_{A,ext}}\left(\frac{R_{cl}}{R_0}\right)^4 \equiv \frac{\sigma_{m,cl}}{2\rho_{ext}v_{A,ext}}\left(\frac{R_{cl}}{R_0}\right)^4 \equiv \left(\frac{\pi}{\rho_{ext}}\right)^{1/2}\frac{M}{\Phi_B}\left(\frac{R_{cl}}{R_0}\right)^2.$$
(93a, b, c)

Finally, for a perpendicular rotator with $\rho_{cl}/\rho_{ext} \gg 1$, $\tau_\perp$ becomes

$$\tau_\perp = \frac{1}{2}\left(\frac{\rho_{cl}}{\rho_{ext}}\right)^{1/2}\frac{R}{v_A(R)} \equiv 2\left(\frac{\pi}{\rho_{cl}}\right)^{1/2}\frac{M}{\Phi_B} \equiv (2/3)^{1/2}\frac{M}{M_{crit}}\tau_{ff},$$
(94a, b, c)

where $v_A(R)$ is, as in equation (91), the Alfvén speed just outside the cloud (or core) surface. It is indeed very clear from equations (92) - (94) that, in general, an aligned rotator in a geometry with straight-parallel field lines yields the least effective magnetic braking, a perpendicular rotator the most effective, and an aligned rotator with field lines "fanning out" (as a result of the contraction of the cloud; e.g., see the quantitative figures in TM76b) intermediate between the other two. It is for this reason that we concluded early on that *most clouds and fragments tend to become aligned rotators in time*.

It follows from equations (92c) and (93c) that, as a cloud (or core) contracts at constant M/Φ_B in an environment whose properties (ρ_{ext}) do not change much as a result of the cloud's (or core's) contraction, $\tau_\parallel$ remains constant in the geometry with straight-parallel field lines, but it decreases as R_{cl}^2 in the more realistic case in which the field lines fan out significantly. A similar contraction for the perpendicular rotator considered by MP79 yields that $\tau_\perp$ decreases as $\rho_{cl}^{-1/2}$ and does not depend on ρ_{ext} (see eq. [94b]).

TM79a (eq. [9]) has also accounted for the balance of forces (gravity and thermal pressure) along field lines for oblate clouds (or cores) contracting quasistatically and isothermally due to self-gravity (TM76b), and found that, in the case of an aligned rotator with straight-parallel field lines, $\tau_\parallel$ depends only on the temperature and the background magnetic field. That result was disputed by Mestel and Paris (1984) (see comments following their eq. [3.12]) because of two oversights on their part (see TM85).

3.3.5. Magnetically Linked, Aligned Rotators.

Observations have shown by now that clouds (or fragments, or cores) often exist in close proximity in molecular cloud complexes (e.g, see Myers 1982; Myers and Benson 1983; Myers, Linke, and Benson 1983; review by Myers 1985; Cernicharo 1991, in this volume). Polarization observations also show that large-scale, orderly (as opposed to tangled) magnetic fields permeate these complexes (e.g., see Vrba, Coyne, and Tapia 1981; Moneti et al. 1984; Heyer et al. 1987). Accounting for such observations, Mouschovias and Morton (1985a, b; hereafter MM85a, MM85b) raised the possibility and studied analytically the interaction of magnetically linked fragments by means of torsional Alfvén waves. They found that these waves, which are generated by the rotation of a fragment and propagate into the surrounding ("external") medium, get reflected off a neighboring fragment and return to the original fragment, thus delaying or, in principle, altogether preventing the resolution of the angular momentum problem for this fragment. If, by the time the waves return, a fragment contracts significantly with the magnetic field still nearly frozen in the matter, the waves can cause reexpansion of the rotating fragment. (*Such fragments would be observed to rotate near or even above centrifugal breakup --because of, not despite, magnetic braking!*) In addition, as these waves reach other fragments, they recreate an angular momentum problem for these fragments *after* those fragments have managed to resolve their original angular momentum problem. These phenomena were described as trapping and sharing of rotational kinetic energy (but not necessarily *net* angular momentum), regardless of its origin, among magnetically connected fragments (or clouds). For typical parameters of cloud cores and the surrounding envelopes, it was found that a typical core spends most of its lifetime in low spin states. Mechanical analogies (e.g., with waves on a string loaded by point masses separated by a given distance, or with oscillations of metal disks joined by rubber bands) were employed to interpret the results physically.

3.3.6. Conclusion.
For subAlfvénically contracting clouds with frozen-in magnetic fields, magnetic braking is very effective in removing angular momentum. It can resolve the angular momentum problem of star formation in $< 10^7$ yr for aligned rotators, and much faster for perpendicular rotators (the ratio of time scales $\tau_\perp/\tau_\parallel$ is less than 10^{-1} for $\rho_{cl}/\rho_{ext} = 10^2$, and becomes smaller at larger density ratios). This conclusion remains valid even when fragment-fragment interaction via torsional Alfvén waves is accounted for; the bounces of the waves among fragments cause each fragment to go through a succession of relatively high- and low-spin states (once every $\simeq 10^6$ yr, for typical core and envelope parameters) before each fragment irreversibly resolves its angular momentum problem.

Alfvénically or superAlfvénically contracting *clouds*, which should be relatively rare (see § 1.2), trap the torsional Alfvén waves inside and therefore contract at a nearly constant angular momentum (TM85, § 3.2; TM87a, § 3). Centrifugal forces stop the collapse perpendicular to the axis of rotation at very early, diffuse stages, and further contraction can proceed only as rapidly as magnetic braking removes angular momentum. Thus, even for these clouds, magnetic braking seems to play a crucial role in star formation.

The ability of magnetic braking to remove angular momentum from a contracting *core or fragment* is ultimately impaired by ambipolar diffusion and/or Alfvénic or superAlfvénic collapse. Certainly, as just explained, if/when the contraction speed of a thermally and magnetically supercritical core approaches the Alfvén speed in the neutrals, v_A, then magnetic braking ceases to be effective. However, by that stage of the evolution of a core, its angular momentum has been reduced by magnetic braking to manageable levels, and protostar formation can proceed with very little or no interference from centrifugal forces. The effect of ambipolar diffusion on magnetic braking is summarized in § 4 below. We find that, typically, significant redistribution of angular momentum by magnetic braking can still occur in molecular cloud cores at densities $n_n \sim 10^7$ cm^{-3} (see MP86). This is sufficient for the resolution of the angular momentum problem even for single stars.

Finally, it is emphasized that the conclusions of our calculations on magnetic braking do not depend on the magnitude or origin of the initial angular momentum of a cloud --the latter could equally well be galactic rotation of local interstellar turbulence.

4. THE EFFECT OF AMBIPOLAR DIFFUSION ON MAGNETIC BRAKING

A rigorous study of the effect of ambipolar diffusion on magnetic braking requires the use of the two-fluid equations of § 2.3. A clue as to what this effect may be can be obtained from equation (93c), applied to a core. The greatest effect would occur in the (unrealistic) case in which the degree of ionization is so small that a core of fixed mass contacts through the field lines (and the plasma) while the latter are held in place by magnetic forces. This implies that the flux of the core decrease as R_c^2, which exactly cancels the factor R_c^2 in the numerator of equation (93c) (whose origin is the reduction of the core's moment of inertia upon contraction); hence, $\tau_\parallel$ remains constant upon contraction. In a more realistic situation, in which the plasma (and the field lines) are dragged in by the contracting neutrals, the flux of a core decreases only as R_c^ϵ, where $0 \leq \epsilon < 2$. Thus $\tau_\parallel$ is expected to *decrease* upon contraction as $R_c^{2-\epsilon}$. The lower limit of zero on ϵ corresponds to flux freezing (or near free fall), while the upper limit of 2 corresponds either to a neutral-ion collision time $\tau_{ni} \gtrsim \tau_{ff}$ or to an unrealistically strong magnetic field (maintained by the envelope) that prevents the compression of the plasma (and the field lines) by the self-gravity of the contracting core. It seems clear, then, that magnetic braking of typical molecular cloud cores will be operative even in the presence of ambipolar diffusion, at least until the stage at which the redistribution of mass in the central flux tubes leads to magnetically and thermally supercritical conditions and thereby to accelerated contraction. Unfortunately, although these considerations make the qualitative effect of ambipolar diffusion on magnetic braking clear, they cannot be used to quantify this effect because equation (93c) and all other expressions obtained for $\tau_\parallel$ and $\tau_\perp$ thus far are valid only under strict flux freezing and subAlfvénic contraction.

MP86 reverted to first principles and reconsidered the rigid, aligned rotator of § 3.3.2(a) by relaxing the assumption of flux freezing. A two-fluid, analytical calculation yielded the key result that, as long as the contraction of a core remains subAlfvénic, ambipolar diffusion simply lengthens the magnetic braking time scale by the neutral-ion collision time, i.e.,

$$\tau_B \equiv \tau_{\parallel,AD} = \tau_\parallel + \tau_{ni} . \qquad (95)$$

The physical origin of this result is the fact that the collisional coupling of the neutrals to the magnetic field and the magnetic braking of the rotation of the neutrals are processes occurring *in series*; i.e., a typical neutral particle must wait for a time τ_{ni} before it is informed that it must slow down because of the presence of a magnetic torque. The two processes resemble electrical resistors connected in series. (On the other hand, the collisional coupling of a typical *ion* with neutrals and the magnetic braking of ions are processes occurring *in parallel*. Hence, the magnetic braking time scale of the ions is the harmonic mean of $\tau_{\parallel,i}$ and τ_{in} --just like resistors connected in parallel.) For typical molecular cloud cores, $\tau_{ni}/\tau_\parallel \lesssim 5 \times 10^{-2}$ (see MP86, eq. [46b]). It is therefore more likely that, for typical low-mass cores, magnetic braking will become ineffective only when supercritical conditions are attained. *It is precisely this effectiveness of magnetic braking up to relatively high densities* ($\sim 10^7$ cm^{-3}) *that makes the formation of low-mass stars possible.*

Since, as we have seen in § 2.3, the contraction time scale (τ_{contr}) of a subAlfvénically contracting core is essentially the ambipolar diffusion time scale (τ_{AD}), one may obtain a quantitative estimate of the stage at which contraction itself will begin to render magnetic braking less effective by comparing the magnetic braking time scale (τ_B; see eq. [95]) in the presence of ambipolar diffusion and τ_{AD}. This comparison was made and applied to molecular cloud cores before they were established observationally (TM77), and was refined to include the result $\tau_{contr} \simeq 10\tau_{ff} \simeq \tau_{AD}$ of detailed collapse calculations (TM83b). The result is written as (see MP86, eq. [49])

$$\frac{\tau_B}{\tau_{AD}} \simeq 0.2 \left[\frac{3 \times 10^{-3} \text{ cm}^{-3}}{K}\right] \left[\frac{n_c}{10^5 \text{ cm}^{-3}}\right]^{1/6} \left[\frac{n_c/n_{env}}{10}\right] \left[\frac{M_c}{M_\odot}\right]^{1/3} \left[\frac{B_0}{3 \ \mu G} \frac{M_{env}}{10^4 \ M_\odot}\right]^{-1/4},$$
(96)

where M_c is the mass of the core, M_{env} that of the envelope, and n_c the density of neutrals in the core; B_0 is the field strength at which the whole cloud became self-gravitating so that further contraction enhances the field strength as $B \propto \rho^{1/2}$ (see § 5.2.3 below). (The constant 0.2 on the right-hand side of eq. [96] changes to $\simeq 0.1$ if one considers a spherical, rather than a disklike, core.) Since the geometry with straight-parallel fields lines considered here results in a strict *upper limit* on τ_B (see § 3.3.4), it follows from equation (96) that *magnetic braking in nature is indeed expected to be effective in subAlfvénically contracting cores, whose gravitational contraction is controlled by ambipolar diffusion.*

Note the remarkable insensitivity of the ratio τ_B/τ_{AD} on n_c, M_c, and M_{env}, and the fact that it is directly proportional to the ratio of mean densities n_c/n_{env} and inversely proportional to K. The ratio n_c/n_{env} is, in this geometry with straight-parallel field lines, a measure of the distance that the torsional Alfvén waves must travel in the envelope before they affect a moment of inertia equal to that of the core. The parameter K directly affects τ_{AD}, as shown in equation (21).

An earlier calculation on the effect of ambipolar diffusion on magnetic braking obtained typical equatorial (rotational) velocities $v_{eq} \simeq 120$ km s^{-1} for single main-sequence stars of spectral type O5 to F5 (TM83b). This compares well with observed rotational velocities of 100 - 160 km s^{-1} (see Vogel and Kuhi 1981; Wolff et al. 1982).

A recent numerical calculation followed the quasistatic contraction of a cloud to an enhancement of the central density from its initial (equilibrium) value by a factor $\simeq 100$ (Tomisaka, Ikeuchi, and Nakamura 1990). They find that the loss of angular momentum occurs over MP80b's time scale $\tau_\parallel$ given by equation (92), and that the core, having started with a mass-to-flux ratio not far from critical, reaches the stage of dynamical contraction in a time comparable to τ_{AD} given by equation (20a).

In summary, the effect of ambipolar diffusion on magnetic braking in typical molecular cloud cores is to lengthen the magnetic braking time scale $\tau_\parallel$ obtained under strict flux-freezing by only a few percent. However, for values of K smaller than the "canonical" value 3×10^{-3} cm^{-3}, equation (96) shows that it may compete with magnetic braking at $n_n \sim 10^5$ cm^{-3}. We recall that the stage at which ambipolar diffusion tends to render magnetic braking ineffective (either by decoupling the field from the neutral matter or by inducing Alfvénic contraction; see TM77 and TM87a) determines the amount of angular momentum remaining in a contracting fragment (or core) destined to give birth to stars. The entire range of periods of binary stars from 10 hr to 100 yr has been explained in this fashion (TM77), and the theory was refined to specifically account for Alfvénic or superAlfvénic collapse (TM87a, § 3.6) and to explain the angular momenta of planetary systems and single stars (TM83b, MM85a, b, TM87a, b).[8] It seems that nature's way of choosing between the

---

[8]Nakano (1990, eq. [41]) derives in an identical fashion with TM77 an expression for the period of binary stars (in terms of the density at which angular momentum conservation begins) identical with equation (5) of TM77 (or eq. [20] of TM87a), and then misquotes TM77 in an effort to distinguish his result from that of TM, while ignoring TM87a (§ 3.6), which leaves no room for such nonexistent distinctions. Then Nakano (1991, eq. [7]) ignores both TM77 and TM87a and lays claim on that one of TM's results, too!!

formation of binary (or multiple) stellar systems and single stars (or planetary systems) may be through the microscopic physical processes that determine the ion density and its variations (and hence the coupling of the magnetic field to the neutral matter) from fragment to fragment in self-gravitating clouds.

5. MAGNETOHYDROSTATICS: EQUILIBRIA OF INTERSTELLAR CLOUDS

The magnetohydrostatic (MHS) equations are obtained from the single-fluid MHD equations (32a) - (32g) by dropping the terms containing time derivatives and velocities:

$$-\nabla P - \rho \nabla \psi + j \times B/c = 0 , \qquad (97a)$$

$$\nabla \times B = (4\pi/c) j , \qquad (97b)$$

$$P = \rho C^2 , \qquad (97c)$$

where $C = const$ for an isothermal gas. If the self-gravity of the gas is negligible, the gas responds to the (known) gravitational field of the stars, and $g \equiv -\nabla\psi$ in equation (97a) is not an unknown. If the self-gravity of the gas is important, then the Poisson equation,

$$\nabla^2 \psi = 4\pi G \rho \qquad (97d)$$

must also be considered. The above system of four equations is an inherently open set — it contains five unknowns, namely, P, ρ, ψ, j, and B. *Ad hoc* assumptions are usually made to close the system and obtain equilibrium solutions. In what follows, we return, once again, to first principles and close the system in a *self-consistent* manner. Then we use the equations to determine (final) equilibrium states for the Parker instability (see § 3.2.3) and for self-gravitating, isothermal, magnetic clouds. Two important results of the latter set of calculations will be the critical mass-to-flux ratio (eq. [22a]), already used extensively in our discussion of ambipolar diffusion, and the $B_c \propto \rho_c^{1/2}$ relation in cloud cores contracting quasistatically along field lines and only as rapidly as magnetic forces allow perpendicular to field lines.

It may appear as a puzzle that, although the MHD equations form a closed system, the MHS equations, which are obtained from the former, are inherently open. One may also think that adding the condition $\nabla \cdot B = 0$ would close the system. It does not. What is required is a relation between one of the electromagnetic variables (j or B) and at least one of the two variables P, ρ, and, possibly, the gravitational potential, ψ. In order to find a clue as to how the system may be closed self-consistently, we look for the physical principle(s) that must have gotten lost in going from the MHD to the MHS equations. Equations (32a) and (32d), representing mass and flux conservation, respectively, are the culprits. Since all their terms contain either time derivatives or velocities, they all drop out of the MHS equations. In other words, the problem of closing the system of MHS equations is reduced to one of finding a way to build into them the lost physical principles of conservation of mass and magnetic flux. *It is necessary to discover a new equation.* This was done by TM74 for the two-dimensional cartesian geometry relevant to the determination of final equilibrium states for the Parker instability, and by TM76a for the three-dimensional axisymmetric geometry relevant to the study of equilibria of self-gravitating, magnetic clouds. We outline the method and the results in §§ 5.1 and 5.2, respectively.

We first point out that Cowling's integral of the continuity and flux-freezing equations, $B/\rho = [(B_0/\rho_0) \cdot \nabla] r'$, which relates the ratio B/ρ in a fluid element at some time t to its value B_0/ρ_0 in the same fluid element at a previous time t_0 through the displacement vector r' of the fluid element, does not close the system of MHS equations because r' is not known — in fact, to know r' requires a knowledge of the solution.

5.1. Large-Scale Condensations or Cloud Complexes

5.1.1. Motivation. Since it has been argued that, once the Parker instability develops in the interstellar medium, the field lines will inflate forever, it is necessary first to give at least a plausibility argument why this is not so. We consider two neighboring field lines labeled by A and $A+\delta A$ (recall eqs. [47a, b], which show that the only nontrivial component of the magnetic vector potential in this two-dimensional geometry is both a constant of the motion and constant on a field line), which have inflated to a distance above the galactic plane (z = 0) comparable to the (unstable) horizontal wavelength of the perturbation, λ_y. Let h be the (small) separation between the above two field lines in the "wings" or "mountains" of the field lines, where inflation is taking place. Then the ratio of the magnitudes of the confining magnetic-tension force and the expansive magnetic-pressure force is (see terms in brackets in last part of eq. [15b])

$$\frac{B^2/4\pi R_c}{B^2/8\pi h} \propto \frac{1/h^2 R_c}{1/h^3} \propto \frac{1/\lambda_y}{1/h} \propto h, \tag{98a}$$

where we have used $B \simeq \delta A/h \propto 1/h$, $|\nabla_\perp| \sim 1/h$, and $R_c \sim \lambda_y \simeq const$. Since the separation h between neighboring field lines increases upon expansion, the magnetic-tension force will eventually overwhelm the magnetic-pressure force and the expansion will not continue indefinitely. After all, this is the reason for which the field lines of a bar magnet do not all recede to infinity: the tension and pressure forces exactly balance.

The expansive cosmic-ray-pressure force in the region where inflation occurs is

$$|\nabla_\perp P_{rel}| \propto h^{-1} n_{rel}^{4/3} \propto h^{-1} V^{-4/3} \propto h^{-7/3}, \tag{99}$$

where the relativistic relation $P_{rel} \propto n_{rel}^{4/3}$ between pressure and density has been used, and the quantity V is the volume of the flux tube $(A, A+\delta A)$, which was conservatively taken to increase only as h upon expansion. It therefore follows that the ratio of the magnitudes of the magnetic-tension force and the cosmic-ray-pressure force varies as

$$\frac{B^2/4\pi R_c}{|\nabla_\perp P_{rel}|} \propto \frac{h^{-2}}{h^{-7/3}} \propto h^{1/3}, \tag{98b}$$

which increases upon expansion. Hence, magnetic tension will also overwhelm the expansive tendencies of cosmic rays, and equilibrium becomes possible as long as the number of cosmic rays in each flux tube of the system remains quasi-steady (see TM75a for a more complete picture and for an explanation of a Galactic fat radio disk of half-thickness $\sim$ 1 kpc in a quasi-steady state).

5.1.2. Self-Consistent Formulation of the Problem. First we reduce the system (97a) - (97c) to one equation. We define the scalar function of position $q(y, z)$ by[9]

$$q = P \exp(\psi/C^2). \tag{100}$$

By taking the gradient of both sides of equation (100) and using equation (97c), we find that the thermal-pressure force and the gravitational force in equation (97a) combine into a

---

[9]For simplicity, we ignore cosmic rays here although their presence can be accounted for in a straightforward fashion: a second function, q_{rel}, is defined which is exactly equal to P_{rel} (see TM75a) because the cosmic-ray gas is not affected by the galactic gravitational field; i.e., $\psi/C_{rel}^2 \simeq (v_{esc}/c)^2 \ll 1$ (see § 3.1).

single term if written in terms of q, namely,

$$-\nabla P - \rho\nabla\psi = -\exp(-\psi/C^2)\,\nabla q\,. \tag{101a}$$

The magnetic force is written in terms of the magnetic vector potential, by using Ampere's law to eliminate j in favor of B and eq. [46] to eliminate B in favor of $A \equiv A_x(y, z)$, as

$$\frac{j}{c} \times B = -\frac{1}{4\pi}\nabla^2 A \times (-\hat{x} \times \nabla A) = -\frac{1}{4\pi}(\nabla^2 A)\,\nabla A\,, \tag{101b}$$

where in the last step we have used the fact that $\nabla^2 A$ is in the $\hat{x}$-direction, and $\hat{x} \cdot \nabla A = 0$. Hence, the MHS equations reduce to

$$-\exp(-\psi/C^2)\,\nabla q - \frac{1}{4\pi}(\nabla^2 A)\,\nabla A = 0\,. \tag{102}$$

To reduce equation (102) further, we look at the properties of the function q. Since A is constant on a field line (see eq. [47b]), we may take the dot product of B and equation (102) to find that $B \cdot \nabla q = 0$; i.e.,

$$q \equiv P \exp(\psi/C^2) = \text{constant on a field line} = q(A)\,. \tag{103a}$$

We may therefore write in equation (102) $\nabla q = \nabla A\,[dq(A)/dA]$ and, since ∇A does not vanish identically, finally find that

$$\nabla^2 A(y, z) = -4\pi\,\frac{dq(A)}{dA}\exp(-\psi/C^2)\,. \tag{104}$$

Note that, since from Ampere's law written in terms of A in this geometry we have that $\nabla^2 A = -(4\pi/c)j$, in deriving equation (104) we have also determined j:

$$\frac{j}{c}\exp(\psi/C^2) = \frac{dq(A)}{dA} = \text{constant on a field line}\,. \tag{103b}$$

The physical meaning of what we have done in order to reduce the MHS equations (97a) - (97c) to one equation (104) is as follows. (i) Equation (103a) states that, at hydrostatic equilibrium, the gravitational forces balance the thermal-pressure forces *along* field lines (since in eq. [101a] $\nabla_\parallel q = 0$). (ii) Equation (103b) gives the current density j for which forces balance *perpendicular* to field lines if q is such that forces balance along field lines [i.e., if $q = q(A)$]. However, j and B will not, in general, be consistent with each other unless they satisfy equation (104).

One should note that the gravitational potential in equation (104) has been left unrestricted. It can be due to stars (and therefore externally specified), as we shall assume in this section, or due to the self-gravity of the gas itself, in which case Poisson's equation (97d) must also be considered (see § 5.2.1).

In order to determine final equilibrium states for the Parker instability, we must solve equation (104) for a pair of *unstable* wavelengths (λ_y, λ_z) with $q(A)$ determined in such a manner that a final state can be reached (by the nonlinear time evolution of the system) from the corresponding initial, one-dimensional state through continuous deformations of the field lines.

What is $q(A)$? In the case of a cold gas ($T = 0$), Parker (1968a, b) assumed that q is proportional either to A or to A^2, so that equation (104) becomes linear in A and can easily be solved. However, it follows from the definition of q (eq. [103a]) and from the solution of the zero-order equilibrium state (eq. [44a] with $P_{\text{rel}} = 0$) that

$$q_0(A) = \rho_0(0)\, C^2 \left[-\frac{2HB_0(0)}{A} \right]^{2\alpha}, \tag{105}$$

which clearly is an *inverse power of A* (since $\alpha > 0$).

Although $q = q(A)$ at equilibrium, $dq/dt \neq 0$; i.e., q is *not* a constant of the motion. Therefore, one cannot calculate q once and for all from the initial equilibrium state and assume that a final state is characterized by the same $q(A)$.

Calculation of the Function $q(A)$. With $Y \equiv \lambda_y/2$, the mass (δm) in a flux tube between field lines A and $A+\delta A$ is, by definition,

$$\delta m(A) = \int_{-Y}^{+Y} dy \int_{z(y, A)}^{z(y, A+\delta A)} dz(y, A)\, \rho[y, z(y, A)]. \tag{106}$$

It is natural to consider y and A as the independent variables. Since the integration over z in equation (106) is performed keeping y fixed, we may write $dz = (\partial z/\partial A)dA$ and effect the change of variables from z to A. We eliminate ρ in favor of A by using equations (103a) and (97c), perform the trivial integration over A, and then solve the resulting equation for $q(A)$ to find that

$$q(A) = \frac{C^2}{2} \frac{dm(A)}{dA} \left\{ \int_0^Y dy\, \frac{\partial z(y, A)}{\partial A} \exp\left[-\frac{\psi(y, A)}{C^2} \right] \right\}^{-1}, \tag{107}$$

where the quantity $z(y, A)$ refers to the z-coordinate of the field line A at y.

The function $dm(A)/dA$ on the right-hand side of equation (107) is the mass-to-flux ratio in a flux tube about field line A in one period of the system in the y-direction and having a unit length in the "third" (x-) direction. By conservation of both mass and flux, dm/dA *is a constant of the motion.* If it is given/known, then $q(A)$ can be calculated for any equilibrium configuration characterized by the *same dm/dA*.

Note that $q(A)$ depends on the shape of the field lines, which, for the final equilibrium state, are not known. Hence, in general, one must solve equations (104) and (107) simultaneously for any given dm/dA. The initial state of the interstellar gas and magnetic field system is not known in reality, for it depends on the mechanism which creates the magnetic flux. Here we take it to be the plane-parallel system proposed by Parker (1966) [see § 3.3.2(a)]. We emphasize, however, that the only information needed in order to determine an equilibrium state is the mass-to-flux ratio in each flux tube. As discussed by TM74, if it becomes possible for the distribution of mass among the various flux tubes to be obtained from observations, we can determine an equilibrium state without reference to any particular initial state.

5.1.3. Determination of Equilibrium States. The system of (self-consistently) closed MHS equations (103) and (107) was solved by a successive underrelaxation method (see TM74, Appendix C) with the differential mass-to-flux ratio calculated from the zero-order state of § 3.2.3(a). For an isothermal gas, a pair of initially unstable wavelengths $\lambda_y \equiv 2Y = 24C^2/g$ and $\lambda_z \equiv 2Z = 50C^2/g$, and $\alpha_0 = 1$ (see eq. [42a]), the final equilibrium state is shown in Figure 4. The unit of length is the thermal-pressure scale height, C^2/g, where $g = -\hat{z}g(z)$, with $g(z) = -g(-z) = $ a positive constant. *Solid curves* represent field lines chosen such that the increment of magnetic flux (per unit length in the x-direction), ΔA, is constant between consecutive field lines; hence, the spacing between consecutive field lines is inversely proportional to the mean strength of the magnetic field in that region. The *dashed curves* represent isodensity contours at which the density decreases to e^{-1}, e^{-2}, and e^{-3} its value on the plane $z = 0$. The number on each curve is the z-coordinate of that curve in the initial state, in which all curves were horizontal.

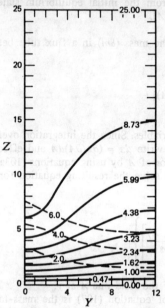

Figure 4. *An equilibrium state of the interstellar gas and magnetic field system in a vertical, uniform galactic gravitational field g.* Distance is measured in units of C^2/g, where C is the isothermal speed of sound in the gas. Only half a wavelength in the y- and half a wavelength in the z-direction are shown. Half the critical wavelength in the y-direction is equal to 7.26. Field lines (*solid curves*) are chosen so that the magnetic flux between any two consecutive ones is constant. The isodensity contours (*dashed curves*) at which the density decreases to e^{-1}, e^{-2}, and e^{-3} its value on the y-axis are shown. The number labeling each curve is the z-coordinate of that curve in the initial state, in which $\alpha = 1$.

The particular horizontal wavelength λ_y chosen for this state is that which corresponds to the maximum growth rate at the chosen (unstable) λ_z. The critical horizontal wavelength is $\Lambda_y = 14.52 C^2/g$ (see eqs. [55a] and [44b], with $\alpha = \gamma = 1$ and $\beta = 0$). The increase of the gas density in the "valleys" of the field lines is due primarily to efficient drainage along field lines, rather than to compression perpendicular to the galactic plane. A parameter study found that, for any given λ_z ($> \Lambda_z$), this drainage is more efficient the larger λ_y ($> \Lambda_y$) is. This makes the magnetic field a nearly vacuum field at high z in the region where inflation of field lines has taken place. The gas density in the galactic plane $\rho(y, z = 0)$ remains constant because g has no horizontal component and, therefore, no thermal-pressure gradients can be sustained (at equilibrium) along field lines in the plane. (Perturbations with odd symmetry about the galactic plane have been studied recently by Mouschovias, Basu, and Paleologou 1991, who found typical final enhancements of the density in the plane by a factor of 2.) Observations of spiral arms of a galaxy seen face-on were predicted to show a column density of gas through the center of a condensation (along the z-axis of Fig. 4) typically twice as large as the column density observed through the "mountains" of the field lines. The galaxy M81 shows precisely this kind of variation of column density along its spiral arms (Rots 1974).

For a detailed discussion of the physics of the Parker instability and the energetics of final equilibrium states, see TM74, where an energy principle for an isothermal gas is also obtained. The effect of cosmic rays on final equilibrium states is discussed in TM75a, while the explanation of the alignment of cloud complexes, OB associations, and giant H II regions "like beads on a string" along spiral arms is given by Mouschovias, Shu, and Woodward

(1974). The nonlinear development of the Parker instability, with a gravitational field simulating that due to a central source in accretion disks, has recently been followed numerically by Matsumoto et al. (1988), with emphasis on the behavior of the gas at large z and the appearance of shocks in that region.

5.2. Self-Gravitating Clouds

5.2.1. Self-Consistent Formulation of the Problem. The open set of MHS equations (97a) - (97d), referring to isothermal, self-gravitating, magnetic clouds, can be closed in a manner similar to that of § 5.1.2 for nongravitating clouds (see TM76a). We consider a conducting, axisymmetric cloud [cylindrical polar coordinates (r, ϕ, z)] embedded in a hot and tenuous external medium, of pressure P_{ext} and threaded by a frozen-in magnetic field aligned with the (z-)axis of symmetry. Far from the cloud the field is uniform. We choose the plane $z = 0$ to be the equatorial plane of the cloud, so that the symmetry about this plane allows us to consider only the half-plane $z \geq 0$. In this geometry, only the ϕ-component of the magnetic vector potential, $A_\phi(r, z)$, is nontrivial; i.e., A is a toroidal and B, related to A through $B = \nabla \times A$, a poloidal vector. The magnetic field may be written as $B = -r^{-1} \hat{\phi} \times \nabla\Phi$, where Φ is a scalar function defined by

$$\Phi(r, z) \equiv rA_\phi(r, z) \equiv rA(r, z) . \tag{108a}$$

Hence, $B \cdot \nabla\Phi = 0$, so that Φ *is constant on a magnetic surface.* [The intersection of a magnetic surface with the (r, z)-plane is referred to as a field line.] The flux-freezing equation (32d), written in terms of Φ, becomes $\partial\Phi/\partial t + (v \cdot \nabla)\Phi \equiv d\Phi/dt = 0$, so that Φ *is also a constant of the motion.* This is the familiar statement that the magnetic flux (Φ_B) through a surface comoving with the fluid is conserved, for it is easily shown directly that the flux through a contour of radius r is given by

$$\Phi_B = 2\pi\Phi . \tag{108b}$$

We may therefore use Φ to label the field lines once and for all.

As done in § 5.1.2, we define the scalar function of position, $q(r, z)$, in the cloud as in equation (100), where ψ is now the gravitational potential due to the self-gravity of the cloud. Then an identical procedure with that of § 5.1.2 shows that, at equilibrium, $q = q(\Phi)$; i.e., q *is constant on a field line.* For the hot and tenuous external medium, $q = P_{ext} =$ constant, the same on all field lines; i.e., the external medium is both force-free and current-free. Ampere's law, the equation corresponding to (104), now becomes

$$\frac{\partial}{\partial r}\left[\frac{1}{r}\frac{\partial\Phi}{\partial r}\right] + \frac{1}{r}\frac{\partial^2\Phi}{\partial z^2} = -4\pi r \exp(-\psi/C^2) \frac{dq(\Phi)}{d\Phi} , \qquad \text{inside;} \tag{109a}$$

$$= 0 , \qquad \text{outside;} \tag{109b}$$

where "inside" and "outside" stand for "inside the cloud" and "outside the cloud," respectively. Equation (109) must be considered simultaneously with the Poisson equation (97d), from which ρ is eliminated in favor of q to find that

$$\frac{1}{r}\frac{\partial}{\partial r}\left[r\frac{\partial\psi}{\partial r}\right] + \frac{\partial^2\psi}{\partial z^2} = \frac{4\pi G}{C^2} q(\Phi) \exp(-\psi/C^2) , \qquad \text{inside;} \tag{110a}$$

$$= 0 , \qquad \text{outside.} \tag{110b}$$

Equations (109) and (110) are coupled, nonlinear differential equations for Φ and ψ; the quantity q has yet to be determined as a function of Φ. We calculate q, as we did in § 5.1.2, in a manner consistent with conservation of mass and flux.

Calculation of the Function $q(\Phi)$. The cloud boundary may be specified uniquely by the function $Z_{cl}(\Phi)$, which represents the projections onto the z-axis of the intersections of field lines and the cloud boundary. This amounts to choosing a coordinate system (z, Φ). Then, the *total* mass (δm) in a flux tube between field lines characterized by Φ and $\Phi+\delta\Phi$ is

$$\delta m(\Phi) = 2 \int_0^{Z_{cl}(\Phi)} dz \int_{r(z, \Phi)}^{r(z, \Phi+\delta\Phi)} dr\, 2\pi r\, \rho(r, z)\,. \tag{111}$$

Since the integration over r is performed keeping z fixed, we may write $dr = (\partial r/\partial \Phi)d\Phi$ and change variables from r to Φ. Then, once again, we eliminate ρ in favor of Φ by using equations (97c) and (100), we perform the trivial integration over Φ in equation (111), and we solve for $q(\Phi)$ to find that[10]

$$q(\Phi) = \frac{C^2}{4\pi} \frac{dm(\Phi)}{d\Phi} \left\{ \int_0^{Z_{cl}(\Phi)} dz\, r(z, \Phi)\, \frac{\partial r(z, \Phi)}{\partial \Phi} \exp\left[-\frac{\psi(z, \Phi)}{C^2}\right] \right\}^{-1}. \tag{112}$$

The quantity $r(z, \Phi)$ refers to the r-coordinate of the field line Φ at z. If the differential mass-to-flux ratio, $dm/d\Phi$, were known either through observations or through a complete theoretical understanding of the mechanism which creates the interstellar magnetic flux, a unique equilibrium configuration for a dense cloud could be calculated by solving equations (109), (110), and (112) simultaneously, subject to appropriate boundary conditions. Since neither observations nor theory provide $dm/d\Phi$, it remains a free function in the equations. *Stability considerations imply that $dm/d\Phi$ must be a decreasing function of Φ.*

TM76b solved these equations by obtaining $dm/d\Phi$ from a reference ("initial") state corresponding to a uniform, spherical cloud of density ρ_i, radius R_i, threaded by a uniform magnetic field B_i. It was emphasized that clouds do not necessarily start their lives from such conditions; the choice was deliberately made to permit comparison of the solution of the self-consistent problem, on the one hand, and the equilibria of nonmagnetic (spherical), isothermal clouds (Bonnor 1956; Ebert 1955, 1957) and Mestel's (1966) magnetic but (by construction) spherical model, on the other hand. For such reference states, $\delta m = \rho_i\, 4\pi r \delta r$ $(R_i^2 - r^2)^{1/2}$ and $\delta \Phi \equiv \delta \Phi_B/2\pi = B_i r \delta r$. It therefore follows that

$$\frac{dm(\Phi)}{d\Phi} = \frac{3}{2} \frac{M_{cl}}{\Phi_{cl}} \left[1 - \frac{\Phi}{\Phi_{cl}}\right]^{1/2}, \qquad \Phi \leq \Phi_{cl}; \tag{113a}$$

$$= 0, \qquad \Phi > \Phi_{cl}; \tag{113b}$$

where we have used the fact that $4\pi\rho_i R_i/B_i = 3M_{cl}/2\Phi_{cl}$. Note that equation (113a) still holds if Φ (and Φ_{cl}) is replaced by the actual magnetic flux Φ_B (and $\Phi_{B,cl}$) everywhere it occurs in

---

[10] Nakano (1979) credits equation (112) to TM76a (eq. [18]). In subsequent papers, however, he meticulously attributes this equation to Nakano 1979 (e.g., see Nakano 1981, eq. [47] and associated credits).

this equation (see eq. [108b]). (We made such a replacement in deriving eq. [22b], which relates the critical central and critical total mass-to-flux ratios.)

5.2.2. Equilibrium States and Gravitational Collapse. When boundary conditions appropriate to a hot and tenuous external medium are imposed and equations (109), (110), and (112) (with the mass-to-flux ratio specified by eq. [113]) are put in dimensionless form by choosing as units of length, speed, density, and magnetic field the quantities $C/(4\pi G\rho_i)^{1/2}$, C, ρ_i, and B_i, respectively, the problem contains three dimensionless free parameters, referring to the reference state; namely, the (dimensionless) radius R_i, the ratio of magnetic and thermal pressure α_i, and the ratio of thermal pressures just outside and just inside the cloud surface P_0 (P_0 always becomes equal to unity when an equilibrium state is reached). The parameters R_i and α_i are measures of the magnitude of the initial gravitational (W_g) and magnetic (W_m) energies of the cloud relative to the thermal energy (U). In fact, one finds that (see TM76a, § IV)

$$|W_g|/U = (2/15)R_i^2 , \qquad W_m/U = (2/3)\alpha_i . \qquad (114a, b)$$

The parameter P_0 describes the effect that the external pressure *by itself* would have on the contraction ($P_0 > 1$) or expansion ($P_0 < 1$) of the cloud boundary --contraction of the cloud boundary while $P_0 < 1$ is entirely due to self-gravity.

Equilibrium states, including ("critical") states on the verge of gravitational collapse were obtained for a wide range of values of the free parameters. For our present purposes, we show one such state (see Fig. 5) and we summarize two key results relating to observable predictions of magnetic field strengths in self-gravitating clouds and to criteria for the collapse of clouds and star formation.

In order to demonstrate the effect that even a weak magnetic field can have on the structure of a self-gravitating cloud, we choose $\alpha_i = 1$, so that the magnetic forces do not dominate the thermal-pressure forces. Moreover, to avoid compression due to the external pressure (rather than due to self-gravity), we choose $P_0 = 0.5$. Then R_i (and, therefore, the mass-to-flux ratio $\propto R_i$) is increased until no equilibrium is possible. The critical state obtained in this fashion is shown in Figure 5; it was found to have $R_i = 3.20$. One notes immediately that, despite the relatively small compression of the magnetic field lines (*solid curves with arrows*), the isodensity contours (*solid, oblate curves*), particularly in the core, have a ratio of major to minor axis of about 3, and that a much greater ratio of central and surface densities has been achieved (23.0) than that permitted by nonmagnetic, isothermal spheres (14.3). An extended, relatively low-density envelope is also evident. The contraction has been highly nonhomologous and nonisotropic.

A detailed parameter study concluded the following. (i) For a fixed α_i and P_0, increasing R_i leads to equilibria with higher central concentration, more oblate isodensity contours, and more compressed magnetic field lines, until a value of R_i (or, equivalently, the mass-to-flux ratio) is reached that no further equilibrium is possible. (ii) For fixed R_i and α_i, provided that R_i is large enough, increasing P_0 also leads to more compression, more flattening, and eventual collapse. (iii) Decreasing α_i while R_i and P_0 are kept fixed leads to more spherical equilibria, as magnetic forces become less important compared to thermal-pressure forces. The critical mass-to-flux ratio and the critical external pressure are given in § 6, in which the virial theorem is improved so as to account for the results of the exact equilibrium calculations.

By studying sequences of equilbrium states, a relation between the magnetic field strength and the gas density in cloud cores was found:

$$B_c \propto \rho_c^\kappa , \qquad 1/3 \lesssim \kappa \lesssim 1/2; \qquad (115)$$

where *the value $\kappa = 1/2$ is the consequence of establishment of force balance along the field* lines of isothermal, magnetic, self-gravitating clouds, while lateral contraction proceeds only as rapidly as magnetic forces permit. Actually, the result proposed by TM76b is that α tends to remain constant once such force balance is established. In its latter form, this conclusion is valid even *for nonisothermal contraction*, in which case

$$B_c \propto (\rho_c T_c)^{1/2} \tag{116}$$

(e.g., see review TM83a, eq. [2]). This implies that, *if the temperature of a cloud core decreases during contraction, the enhancement of the magnetic field strength will be smaller than that expected for isothermal contraction*. The predicted decrease is equal to the square root of the factor by which the temperature has decreased.

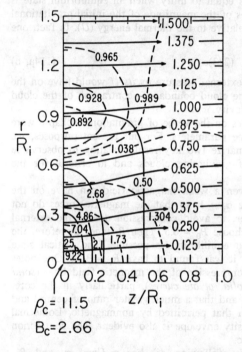

Figure 5. *A critical equilibrium state.* The free parameters have the values $\alpha_i = 1.0$, $P_0 = 0.5$, and $R_i = 3.20$. Both axes are labeled in units of the initial radius of the cloud, R_i. The curves bearing arrows represent field lines; each is labeled by its r-coordinate in the initial state, in which field lines are equidistant and parallel to the z-axis. The solid, oblate curves are isodensity contours; they are labeled by the value of the density in units of the (uniform) density of the initial state. The dashed curves are contours of equal magnetic field strength (isopedion contours); they are labeled by the magnitude of the field in units of the (uniform) field of the initial state. From the isodensity and isopedion contours, one may estimate α at equilibrium by using the expression $\alpha_{eq} = \alpha_i B_{eq}^2/\rho_{eq}$.

Relatively recent comparisons of the theoretical predictions with observations of magnetic fields in H I and molecular clouds concluded that the theory is consistent with observations (e.g., Troland and Heiles 1986; TM87a, § 2.1). A brief review of the issue of what is the proper value of the proportionality constant in the relation (115) is given by TM85, § 2.3. Further discussion is given below (§ 5.2.3) and in Paper II.

5.2.3. Physical Origin and Validity of the Relation $B_c \propto \rho_c^{1/2}$. We consider the contraction of an oblate (or disklike) core, with minor axis along B, balance of thermal-pressure and gravitational forces maintained along field lines, and lateral contraction taking place only as rapidly as magnetic forces allow. Force balance along field lines implies that $|\rho_c g_\parallel| = |\nabla_\parallel P_c|$ (see eq. [97a]). By integrating over z (along field lines) and denoting the half-thickness (or polar radius) of the core by Z, we find that

$$2\pi G \rho_c Z^2 \simeq C^2 , \tag{117a}$$

where we have used the fact that $g_\parallel \simeq -4\pi G\rho_c z$, and we have ignored the external pressure compared to the central pressure. For contraction with the mass M_c of the core fixed, ρ_c is eliminated in favor of M_c, R (the equatorial radius), and Z to find that

$$Z/R^2 \simeq C^2/GM_c \, . \tag{117b}$$

For isothermal contraction, equation (117a) yields that

$$\rho_c \propto Z^{-2} \, . \tag{117c}$$

Moreover, mass and flux conservation imply

$$\rho_c \propto (R^2 Z)^{-1} \, , \quad \text{and} \quad B_c \propto R^{-2} \, , \tag{117d, e}$$

respectively. Hence, we use equations (117d) and (117c) in equation (117e) to find that

$$B_c \propto R^{-2} \propto \rho_c Z \propto \rho_c^{1/2} \, , \tag{118}$$

which is the relation obtained by the exact calculations for isothermal contraction.

It is clear from this simple derivation that, for nonisothermal contraction, the relation (117c) is no longer valid; it is replaced by equation (117a), which now yields $\rho_c \propto (C/Z)^2$. Tracing the effect of this change through the rest of the derivation, we find that

$$B_c \propto \rho_c^{1/2} C \propto (\rho_c T_c)^{1/2} \, , \tag{119}$$

as originally found by TM76b from the constancy of α argument (see eq. [116] above).

This derivation also reveals that *even the presence of hydromagnetic waves or turbulence does not change the $B_c \propto \rho_c^{1/2}$ relation, provided only that the wave pressure can be written as $P_w = \rho_c u^2$, where u^2 is a mean-square wave speed*. The essence of this B-ρ relation lies not so much in what provides the support against gravity along field lines, but in the establishment of force balance along field lines in the first place, accompanied by gravitational contraction against the magnetic forces perpendicular to the field lines. As a corollary, different B-ρ relations may be obtained by assuming different P-ρ relations; e.g., if $P_c \propto \rho_c^\gamma$, we find that

$$B_c \propto \rho_c^{\gamma/2} \, . \tag{120}$$

For a contracting core, the proportionality constant $B_0/\rho_0^{1/2}$ in the relation $B_c \propto \rho_c^{1/2}$ is determined by the values of B_c and ρ_c at which (1) force balance along field lines is established (hence, by T_c [or effective T_c] and the column density $\sigma_{m,c} \equiv M_c/\pi R^2$ of the core; see eqs. [117a, b]) *and* (2) self-gravity becomes strong enough to initiate contraction perpendicular to the field lines. In general, but especially for *individual* low-mass cores, we expect condition 2 to be satisfied at a higher density than condition 1. This is so because of the gravitational breakup we envision along the central flux tube of a cloud (see discussion following eq. [22b]), which produces fragments (or cores) with mass-to-flux ratios *smaller* than that of the parent flux tube before breakup. It is only because of ambipolar diffusion in *each* core, after force balance along field lines is (re)established, that an individual core's mass-to-flux ratio increases toward the critical value (see eq. [22a]), beyond which the $B_c \propto \rho_c^{1/2}$ relation is obeyed again, even in the presence of ambipolar diffusion, for typical core parameters. We discuss further the effect of ambipolar diffusion on the B-ρ relation in Paper II; see, also, TM91, Figure 4c.

We now turn briefly to a further reduction of the MHD equations, namely, the scalar virial theorem, and its application (in an improved form) to the equilibria of isothermal, self-gravitating, magnetic clouds. Critical states are revisited and quantified.

6. THE SCALAR VIRIAL THEOREM AND CLOUD COLLAPSE

Since it is often the case that the detailed structure [e.g., $\rho(r)$, $v(r)$, $T(r)$, $B(r)$, etc.] of an astrophysical system is not known, it is useful to have at hand an equation that washes out the details of the system and relates instead the mean values of certain physical quantities. The scalar virial theorem is obtained from the single-fluid force equation (32b) by taking its inner product with the position vector r and integrating over a volume V, bounded by a surface S, that contains the physical system of interest (e.g., see Chandrasekhar and Fermi 1953; Spitzer 1978). One finds that

$$\frac{1}{2}\frac{d^2 I}{dt^2} = 2\mathcal{T} + 3(\gamma - 1)U + \mathcal{M} + W_g + \frac{1}{4\pi}\int_S (r \cdot B) B \cdot dS - \int_S \left(P + \frac{B^2}{8\pi}\right) r \cdot dS , \qquad (121)$$

where dS is the directed (outward normal) surface element, and the quantities I, $\mathcal{T}$, U, $\mathcal{M}$, and W_g are given by the following volume integrals:

$$I = \int dV \, \rho r^2 , \qquad \mathcal{T} = \frac{1}{2}\int dV \, \rho v^2 , \qquad U = \frac{1}{\gamma - 1}\int dV \, P , \qquad (122a, b, c)$$

$$\mathcal{M} = \int dV \, \frac{B^2}{8\pi} , \qquad W_g = -\int dV \, \rho r \cdot \nabla \psi . \qquad (122d, e)$$

The quantity I is referred to as the "generalized moment of inertia;" it should be noted that r in the integrand is the *spherical* polar coordinate, not the distance from the axis of rotation. The kinetic energy of the system is denoted by $\mathcal{T}$, the internal energy by U, and the magnetic energy within the surface S by $\mathcal{M}$. The quantity W_g is the gravitational potential energy of the system only if the system is isolated, in which case W_g becomes

$$W_g = \frac{1}{2}\int dV \, \rho \psi . \qquad (122f)$$

If a system is in an external gravitational field, then W_g should be calculated from its original form (122e). The cosmic-ray gas can be included by replacing P everywhere it occurs with $P + P_{rel}$ and noting that $\gamma_{rel} = 4/3$ (see § 3.1).

For the equilibrium of a spherical, uniform, isothermal, self-gravitating, magnetic cloud, of mass M and radius R, bounded by a fixed external pressure P_{ext}, and with a magnetic field (B_0) uniform inside and dipolar outside (but with the normal component of B being continuous across the cloud surface), we calculate the integrals to find that

$$P_{ext} = \frac{3MC^2}{4\pi R^3} - \frac{1}{4\pi R^4}\left[\frac{3}{5}GM^2 - \frac{\Phi_B}{4\pi^2}\right] , \qquad (123a)$$

$$\equiv \frac{3MC^2}{4\pi R^3}\left\{1 - \frac{1}{5}\frac{GM}{RC^2}\left[1 - \left(\frac{M_{cr,vir}}{M}\right)^2\right]\right\} , \qquad (123b)$$

$$M_{cr,vir} = \left(\frac{5}{12\pi^2 G}\right)^{1/2} \Phi_B . \qquad (124)$$

The virial critical mass, $M_{\text{cr,vir}}$, is the minimum mass necessary for the gravitational potential energy to exceed in magnitude the magnetic energy at the given flux Φ_B $(= B_0 \pi R^2)$.[11]

Equation (123b) is viewed as giving P_{ext} as a function of R at a given mass, magnetic flux, and temperature. Calculating $\partial P_{\text{ext}}/\partial R$ and setting it equal to zero, one finds that P_{ext} has a maximum, namely,

$$P_{\text{cr,vir}} = P_{\text{cr,th}} \left[1 - \left(\frac{M_{\text{cr,vir}}}{M} \right)^2 \right]^{-3}, \tag{125a}$$

where $P_{\text{cr,th}}$ is the critical external thermal pressure in the case $\Phi_B = 0$ and is given by

$$P_{\text{cr,th}} = 3.15 \, C^8/G^3 M^2. \tag{125b}$$

(Eq. [125b] is the virial analogue of the exact Bonnor-Ebert result, eq. [1a].) The maximum $P_{\text{cr,vir}}$ of the external pressure occurs at a radius

$$R_{\text{cr,vir}} = \frac{4}{15} \frac{GM}{C^2} \left[1 - \left(\frac{M_{\text{cr,vir}}}{M} \right)^2 \right]. \tag{125c}$$

If the critical states are determined from exact equilibria (see § 5.2), one expects to find a different (smaller) critical mass and critical external pressure because of the inevitable flattening and development of a central concentration in the exact equilibrium states, both of which tend to increase the relative strength of the gravitational forces perpendicular to the field lines. We therefore improve the virial theorem by introducing two "correction factors," which are then determined from comparison with the exact critical states (see Mouschovias and Spitzer 1976; Spitzer 1978). We let

$$\beta_M \equiv \frac{M_{\text{crit}}}{M_{\text{cr,vir}}}, \quad \text{and} \quad \beta_P \equiv \frac{P_{\text{crit}}}{P_{\text{cr,vir}}}, \tag{126a, b}$$

where the subscript "crit" denotes exact critical values. Then the corrected equation (125a) is

$$\frac{P_{\text{crit}}}{P_{\text{cr,th}}} = \beta_P \left[1 - \left(\beta_M \frac{M_{\text{cr,vir}}}{M} \right)^2 \right]^{-3}, \tag{127a}$$

or, equivalently,

$$\beta_P^{1/3} \left(\frac{P_{\text{cr,th}}}{P_{\text{crit}}} \right)^{1/3} = 1 - \beta_M^2 \left(\frac{M_{\text{cr,vir}}}{M} \right)^2. \tag{127b}$$

---

[11]The number 12, instead of 9, appears in the parentheses on the right-hand side of equation (124) because in calculating the magnetic contribution (last term on right-hand side of eq. [123a]) we have included the magnetic surface integral and accounted for the continuity of the normal component of B across the cloud surface. This results in somewhat different values for β_M and β_P (see eqs. [128a, b]) from those of Mouschovias and Spitzer (1976), but the final expressions for $(M/\Phi_B)_{\text{crit}}$ and P_{crit} (eqs. [129a, b]) remain unchanged to two significant-figure accuracy.

Since M, Φ_B, C, and P_{ext} are known in the critical states of the exact calculations, the quantities in parentheses in equation (127b) are also known, and, therefore, the parameters β_P and β_M can be adjusted so as to obtain a best fit to the exact critical states. The result is

$$\beta_M = 0.64, \quad \text{and} \quad \beta_P = 0.59, \tag{128a, b}$$

yielding a critical mass-to-flux ratio and a critical external pressure (to two significant-figure accuracy)

$$\left[\frac{M}{\Phi_B}\right]_{crit} = \frac{0.13}{G^{1/2}}, \tag{129a}$$

$$P_{crit} = 1.9 \frac{C^8}{G^3 M^2} \left[1 - \left(\frac{M_{crit}}{M}\right)^2\right]^{-3}, \quad M \geq M_{crit}. \tag{129b}$$

For a cloud to collapse, both conditions must be satisfied. (Eq. [129a] is the same as eq. [22a], used earlier in the paper.) By solving equation (129a) for M_{crit} and substituting in equation (129b), we obtain a single sufficient condition for the collapse of self-gravitating clouds supported by magnetic and thermal-pressure forces:

$$M \geq M_{TM} = 1.4 \left\{1 - \left[\frac{0.13}{G^{1/2}(M/\Phi_{B,cl})}\right]^2\right\}^{-3/2} \frac{C^4}{(G^3 P_{ext})^{1/2}}, \tag{130a}$$

where $\Phi_{B,cl}$ and M are the total flux and mass of the cloud, respectively. Since the total mass-to-flux ratio $M/\Phi_{B,cl}$ of initially spherical clouds threaded by a uniform magnetic field is related to the central mass-to-flux ratio $(dm/d\Phi_B)_c$ by $M/\Phi_{B,cl} = (2/3)(dm/d\Phi_B)_c$ (see eq. [113a]), equation (130a) can equivalently be written in terms of $(dm/d\Phi_B)_c$ as

$$M_{TM} = 1.4 \left\{1 - \left[\frac{0.19}{G^{1/2}(dm/d\Phi_B)_c}\right]^2\right\}^{-3/2} \frac{C^4}{(G^3 P_{ext})^{1/2}}. \tag{130b}$$

The form (130b) brings out the role of the central mass-to-flux ratio in the support and collapse of clouds, as emphasized earlier by TM78 (p. 218). Recent numerical determination of critical masses by Tomisaka, Ikeuchi, and Nakamura (1988) is completely consistent with equations (130a) and (130b). Other useful forms of the critical mass-to-flux ratio, for example, in terms of the mean column density of matter or the visual extinction and the mean magnetic field, can be found in the review TM87a, in which the role of the critical mass-to-flux ratio in fragmentation is also discussed.

7. CONCLUSION - SUMMARY

The labyrinth of evolutionary paths that a blob of matter (mean density ~ 1 cm^{-3}) can take in the restless interstellar environment, with all its violent as well as seductive diversions, before, if ever, it emerges as a star (mean density $\sim 10^{24}$ cm^{-3}) or a stellar cluster does not imply a hopelessly complex theoretical description. The single-mindedness of the gravitational force and the opposition to it, led half-heartedly by the cosmic magnetic force, are for theorists what Ariadne's ball of thread was for Theseus some three millenia ago, almost at the very location of this conference (e.g., see Hamilton 1942, p. 215). Among the otherwise bewildering theoretical possibilities, there is surprising simplicity. It is this simplicity that we have tried to convey in motivating, developing, and using basic theoretical tools necessary for the discovery and description of the key elements of the early, diffuse stages of star formation, whose importance lies in the seeming fact that they determine the ultimate nature of a forming stellar system (single, binary, or multiple) and in that observations continuously check the theoretical predictions pertaining to these (molecular cloud evolution) stages.

If we have managed to demystify the formalism and to show how relatively simple the solution of apparently complicated theoretical problems of star formation can be once we separate and focus on the *essence* of a problem, and if some young minds are attracted by and to cosmic magnetism (we need all the help we can get!), this paper will have served its purpose. In the process, we have examined the proposed mechanisms for the formation of cloud complexes; the equilibrium structure of clouds; the relation between the magnetic field strength and the gas density in self-gravitating cloud cores (even for nonisothermal contraction, and in the presence of waves); the self-initiated quasistatic (or subAlfvénic) formation and (later, dynamic) collapse of cores due to ambipolar diffusion; the resolution of the angular momentum problem by magnetic braking; the effect of ambipolar diffusion (and consequent rapid contraction) on magnetic braking; and the role of these processes in a quantitative but still incomplete theory of star formation.

Recent results of numerical studies, which follow the formation and evolution of axisymmetric cores to densities ~ 3×10^9 cm^{-3} (Fiedler and Mouschovias 1991a, b; Morton and Mouschovias 1991a, b), are left for the accompanying paper; see, also, TM91. Characteristic sizes of protostellar cores, due to effective magnetic support of envelopes (as originally found by TM76b), are discovered both analytically and numerically and are incorporated into a detailed, modern (albeit incomplete) theory of star formation (TM87a, TM90a, TM90b, TM91), summarized and refined in Paper II. The effect of ambipolar diffusion on the $B_c \propto \rho_c^{1/2}$ relation is also quantified.

Acknowledgements. This work was supported in part by the National Science Foundation. All computations and graphics were performed at the National Center for Supercomputing Applications, at the University of Illinois, Urbana-Champaign.

8. REFERENCES

Ambartsumian, V. A. 1956, in *Vistas in Astronomy*, ed. A. Beer (London: Pergamon Press), 2, 1708
Ames, S. 1973, *ApJ*, **182**, 387
Baade, W. 1944, *ApJ*, **100**, 137
Blitz, L. 1987, in *Physical Processes in Interstellar Clouds*, ed. G. E. Morfill and M. Scholer (Dordrecht: Reidel), 35
Blitz, L., and Shu, F. H. 1980, *ApJ.*, **238**, 148
Bonnor, W. B. 1956, *MNRAS*, **116**, 351
Cernicharo, J. 1991, *in this volume*
Chandrasekhar, S., and Fermi, E. 1953, *ApJ*, **118**, 116
Clark, F. O., and Johnson, D. R. 1978, *ApJ*, **220**, 550
_____. 1981, *ApJ*, 247, 104
Draine, B. T. 1980, *ApJ*, **241**, 1021
Ebert, R. 1955, *Zs. Ap.*, **37**, 217
_____. 1957, *Zs. Ap.*, **42**, 263

Ebert, R., Hoerner, S. von, and Temesváry, S. 1960, *Die Entstehung von Sternen durch Kondesnation diffuser Materie* (Berlin: Springer-Verlag)
Elmegreen, B. G. 1979, *ApJ*, 232, 729
———. 1991, in this volume
Falgarone, E., and Pérault, M. 1987, in *Physical Processes in Interstellar Clouds*, ed. G. E. Morfill and M. Scholer (Dordrecht: Reidel), 59
Fiedler, R. A., and Mouschovias, T. Ch. 1991a, *ApJ*, to be submitted
———. 1991b, *ApJ*, to be submitted
Field, G. B. 1965, *ApJ*, 142, 531
———. 1969, in *Interstellar Gas Dynamics*, ed. H. J. Habing (Reidel: Dordrecht), 51
———. 1970, in *Proc. 16th Liege Astrophys. Symp.* (Liege: Inst. d' Astrophysique), 29
———. 1970±1, *unpublished notes*
Field, G. B., Goldsmith, D. W., and Habing, H. J. 1969, *ApJ*, 155, L149
Giz, A., and Shu, F. H. 1980, *private communication*
Hamilton, E. 1942, *Mythology* (Boston: Little, Brown & Co.)
Heiles, C. 1987, in *Physical Processes in Interstellar Clouds*, ed. G. E. Morfill and M. Scholer (Dordrecht: Reidel), 429
Heintz, W. D. 1967, in *On the Evolution of Double Stars*, ed. J. Dommanget, *Commun. Obs. Roy. Belgique*, Ser. B., No. 17, 49
———. 1969, *JRAS Canada*, 63, 275
Hodge, P. W. 1969, *ApJ*, 156, 847
Heyer, M. H., Vrba, F. J., Snell, R. L., Schloerb, F. P., Strom, S. E., Goldsmith, P. F., and Strom, K. M. 1987, *ApJ*, 321, 855
Jeans, J. H. 1928, *Astronomy and Cosmogony* (Cambridge: Cambridge Univ. Press)
Kerr, F. J. 1963, in *The Galaxy and the Magellanic Clouds*, ed. F. J. Kerr (Dordrecht: Reidel), 81
Kulsrud, R., and Pearce, W. 1969, *ApJ*, 156, 445
Lada, C. J. 1985, *ARA&A*, 23, 267
Lerche, I. 1969, in *Adv. Plasma Phys.*, ed. A. Simon and W. B. Thompson (New York: Interscience), 2, 47
Lizano, S., and Shu, F. H. 1987, in *Physical Processes in Interstellar Clouds*, ed. G. E. Morfill and M. Scholer (Dordrecht: Reidel), 173
———. 1989, *ApJ*, 342, 834
Martin, R. N., and Barrett, A. H. 1978, *ApJ Suppl.*, 36, 1
Mathewson, D. S., and Ford, V. L. 1970, *Mem. RAS*, 74, 143
Matsumoto, R., Horiuchi, T., Shibata, K., and Hanawa, T. 1988, *PASJ*, 40, 171
McDaniel, E. W., and Mason, E. A. 1973, in *The Mobility and Diffusion of Ions in Gases* (New York: Wiley)
McKee, C. F., and Hollenbach, D. J. 1980, *ARA&A*, 18, 219
McKee, C. F., Zweibel, E. G., Goodman, A. A., and Heiles, C. 1991, in *Protostars and Planets III*, preprint
Mestel, L. 1965, *QJRAS*, 6, 265
———. 1966, *MNRAS*, 133, 265
———. 1977, in *Star Formation*, ed. T. de Jong and A. Maeder (Dordrecht: Reidel), 213
Mestel, L., and Paris, R. B. 1984, *A&A*, 136, 98
Mestel, L., and Spitzer, L., Jr. 1956, *MNRAS*, 116, 503
Moneti, A., Pipher, J. L., Helfer, H. L., McMillan, R. S., and Perry, M. L. 1984, *ApJ*, 282, 508
Morgan, W. W. 1970, in *The Spiral Structure of Our Galaxy*, ed. W. Becker and G. Contopoulos (Dordrecht: Reidel), 9
Morton, S. A., and Mouschovias, T. Ch. 1991a, *ApJ*, in preparation

———. 1991b, *ApJ, in preparation*
Mouschovias, T. Ch. 1974, *ApJ*, **192**, 37
———. 1975a, *A&A*, **40**, 191
———. 1975b, *Ph.D. Thesis*, University of California at Berkeley
———. 1976a, *ApJ*, **206**, 753
———. 1976b, *ApJ*, **207**, 141
———. 1977, *ApJ*, **211**, 147
———. 1978, in *Protostars and Planets*, ed. T. Gehrels (Tucson: Univ. of Arizona Press), 209
———. 1979a, *ApJ*, **228**, 159
———. 1979b, *ApJ*, **228**, 475
———. 1981, in *Fundamental Problems in the Theory of Stellar Evolution*, ed. D. Sugimoto, D. Q. Lamb, and D. N. Schramm (Dordrecht: Reidel), 27
———. 1982, *ApJ*, **252**, 193
———. 1983a, *Adv. Space Res.*, Vol. 2, No. 12, 71
———. 1983b, in *Solar and Stellar Magnetic Fields: Origins and Coronal Effects*, ed. J. O. Stenflo (Dordrecht: Reidel), 479
———. 1985, *A&A*, **142**, 41
———. 1987a, in *Physical Processes in Interstellar Clouds*, ed. G. E. Morfill and M. Scholer (Dordrecht: Reidel), 453
———. 1987b, in *Physical Processes in Interstellar Clouds*, ed. G. E. Morfill and M. Scholer (Dordrecht: Reidel), 491
———. 1989, in *The Physics and Chemistry of Interstellar Molecular Clouds*, ed. G. Winnewisser and J. T. Armstrong (Berlin: Springer-Verlag), 297
———. 1990a, in *Galactic and Intergalactic Magnetic Fields*, ed. R. Beck, Kronberg, and R. Wielebinski (Dordrecht: Reidel), 269
———. 1990b, in *Physical Processes in Fragmentation and Star Formation*, ed. R. Capuzzo-Dolcetta, C. Chiosi, and A. di Fazio (Dordrecht: Kluwer), 117
———. 1991, *ApJ, May 20*
Mouschovias, T. Ch., Basu, S., and Paleologou, E. V. 1991, *ApJ, in preparation*
Mouschovias, T. Ch., and Morton, S. A. 1985a, *ApJ*, **298**, 190
———. 1985b, *ApJ*, **298**, 205
———. 1991a, *ApJ*, **371**, 296
———. 1991b, *ApJ, submitted*
———. 1991c, *ApJ, submitted*
Mouschovias, T. Ch., and Paleologou, E. V. 1979, *ApJ*, **230**, 204
———. 1980a, *The Moon and the Planets*, **22**, 31
———. 1980b, *ApJ*, **237**, 877
———. 1981, *ApJ*, **246**, 48
———. 1986, *ApJ*, **308**, 781
Mouschovias, T. Ch., Paleologou, E. V., and Fiedler, R. A. 1985, *ApJ*, **291**, 772
Mouschovias, T. Ch., Shu, F. H., and Woodward, R. 1974, *A&A*, **33**, 73
Mouschovias, T. Ch., and Spitzer, L., Jr. 1976, *ApJ*, **210**, 326
Myers, P. C. 1982, *ApJ*, **257**, 620
———. 1985, in *Protostars & Planets II*, ed. D. C. Black and M. S. Matthews (Tucson: University of Arizona Press), 81
Myers, P. C., and Benson, J. 1983, *ApJ*, **266**, 309
Myers, P. C., Linke, R. A., and Benson, P. J. 1983, *ApJ*, **264**, 517
Nakano, T. 1979, *PASJ*, **31**, 697
———. 1981, in *Progress in Theoretical Physics*, No. 70, 54
———. 1983, *PASJ*, **35**, 87

———. 1984, *Fundam. Cosmic Phys.*, **9**, 139
———. 1989, *MNRAS*, **241**, 495
———. 1990, *MNRAS*, **242**, 535
———. 1991, in *Fragmentation of Molecular Clouds and Star Formation*, IAU Symp. No. 147, *in press*
Nakano, T., and Tademaru, T. 1972, *ApJ*, **173**, 87
Nakano, T., and Umebayashi, T. 1986, *MNRAS*, **218**, 663
Paleologou, E. V., and Mouschovias, T. Ch. 1983, *ApJ*, **275**, 838
Parker, E. N. 1966, *ApJ*, **145**, 811
———. 1967a, *ApJ*, **149**, 517
———. 1967b, *ApJ*, **149**, 535
———. 1968a, *ApJ*, **154**, 57
———. 1968b, in *Stars and Stellar Systems*, Vol. 7, *Nebulae and Interstellar Matter*, ed. B. Middlehurst and L. H. Aller (Chicago: Univ. of Chicago Press), 707
———. 1969, *Space Sci. Rev.*, **9**, 651
Pneuman, G. W., and Mitchell, T. P. 1965, *Icarus*, **4**, 494
Roberts, W. W. 1969, *ApJ*, **158**, 123
Rots, A. H. 1974, *Ph.D. Thesis*, Groningen University
Scalo, J. M. 1977, *ApJ*, **213**, 705
Shu, F. H. 1982, *The Physical Universe* (Mill Valley: Univ. Science Books)
———. 1983, *ApJ*, **273**, 202
Shu, F. H., Adams, F. C., and Lizano, S. 1987, *ARA&A*, **25**, 23
Spitzer, L., Jr. 1962, *Physics of Fully Ionized Gases*, 2nd ed. (New York: Interscience)
———. 1963, in *Origin of the Solar System*, ed. R. Jastrow and A. G. W. Cameron (New York: Academic Press), 39
———. 1968a, *Diffuse Matter in Space* (New York: Interscience)
———. 1968b, in *Stars and Stellar Systems*, Vol. 7, *Nebulae and Interstellar Matter*, ed. B. Middlehurst and L. H. Aller (Chicago: Univ. of Chicago Press), 1
———. 1978, *Physical Processes in the Interstellar Medium* (New York: Wiley-Interscience)
Tomisaka, K., Ikeuchi, S., and Nakamura, T. 1988, *ApJ.*, **335**, 239
———. 1990, *ApJ*, **362**, 202
Troland, T. H., and Heiles, C. 1986, *ApJ*, **301**, 339
Vogel, S. N., and Kuhi, L. V. 1981, *ApJ*, **245**, 960
Vrba, F. J., Coyne, G. V., and Tapia, S. 1981, *ApJ*, **243**, 489
Wadiak, E. J., Wilson, T. L., Rood, R. T., and Johnson, K. J. 1985, *ApJ*, **295**, L43
Westerhout, G. 1963, in *The Galaxy and the Magellanic Clouds*, ed. F. J. Kerr (Dordrecht: Reidel), 78
Wolff, S. C., Edwards, S., and Preston, G. W. 1982, *ApJ*, **252**, 322
Young, J. S., Langer, W. D., Goldsmith, P. F., and Wilson, R. W. 1981, *ApJ*, **251**, L81
Zuckerman, B., and Palmer, 1974, *ARA&A*, **12**, 279

II. THE PHYSICS OF STAR FORMATION

OB ASSOCIATIONS AND THE FOSSIL RECORD OF STAR FORMATION

ADRIAAN BLAAUW
Kapteyn Laboratory
P.O.Box 800
9700 AV Groningen
The Netherlands

Abstract. Properties of OB Associations within 1.5 kpc are reviewed as "fossils" of local star formation, and as a reference frame for the more detailed discussions, elsewhere in this volume, of star formation occurring within the sample. Special attention is paid to the process of sequential star formation. For the nearest associations this can be traced back to the formation of the Gould Belt between 30 and 50 million years ago. A few remarks on future research conclude the review.

1. Introduction, Definition, Nomenclature

The present chapter deals with "fossil" properties such as structures, motions and stellar content of the OB Associations in the domain within about 1.5 kpc from the sun, and in some more detail with those within 1 kpc. Ages range from a few to about 30 million years, and dimensions from a few parsec upward. Emphasis will be on morphology more than on interpretation.

Increasing observational resolution and sensitivity allows modern research on star formation to focus more and more on the detailed phenomena, approaching the sub-parsec scale, and aiming at revealing the formation of the individual star or compact cluster. However, virtually all of these local birthplaces have a physical history as parts of larger structures, and perhaps there might be a tendency for investigators to not see the wood for the trees. For example, the star forming ISM in the Ophiuchus clouds has a history over the past 30 or 40 million years within which during the last 5 million years or so it has been exposed to supernova explosions and stellar winds from the rapidly evolving massive stars in the neighbouring, youngest subgroup of the Ophiuchus-Scorpius-Centaurus Association, whereas this latter subgroup was formed at an earlier stage under the influence of its neighbouring still older subgroup on the opposite side. The formation of this oldest subgroup, in turn, probably was the result of star formation processes that have to be linked to the origin of the Gould Belt system. Similarly, the current active star formation in Orion, including that in and around the Trapezium Cluster, must be linked genetically to the earlier generations of massive stars in adjacent subgroups of Ori OB1.

We distinguish three main themes:

Section 2 describes the Ophiuchus-Scorpius-Centaurus Association, henceforth referred to as OSCA. This will serve a double purpose: because of its proximity to the sun (distance about 150pc) it allows more complete analysis than most of the other associations, and as it exhibits many of the features characteristic for OB Associations in general, it is a good example to illustrate the model of sequential star formation. The equally close, and evolutionary more advanced association Cas-Tau is next dealt with in this section, and we also deal briefly with some features of the three times more remote, but intensively studied association Ori-OB1.

Section 3 describes the pattern of space distribution, ages, association with molecular clouds, and some statistics of intrinsic properties of the ensemble of associations surrounding the sun up to distances of 1500 pc.

Section 4 pays attention to the Gould Belt System. This subsystem in the local ensemble shows features pointing to a common, very early, formation process that may be related to the Cas-Tau Association, and of which the formation of the (younger) associations within the Gould Belt may be a consequence.

The study of stellar associations is somewhat complicated by the lack of precise definition of their stellar content. Whereas in most cases it is not difficult to point to the group of stars that form the association's main body, in its outer parts the gradual transition to the field star population renders the assignment of "membership" uncertain, unless precise data are available on the motion of the star which allows tracing back its location of origin.

As to nomenclature, for most objects the IAU system applies, like for instance for Mon OB1 and Mon OB2, i.e. reference to the constellation in which the (main body of the) object is located followed by running numbers which are not necessarily in order of right ascension because of the inclusion of recent discoveries. However, for such nearby groups as Cas-Tau and Oph-Sco-Cen this system did not work to begin with, because they did not stand out clearly as clusterings against the general background population; in fact, as a consequence they did not even occur in the earliest listings of associations. On the other hand, we now list as OB Associations some objects, like Collinder 121, that were classified originally as open clusters.

2. The Nearest Associations OSCA and CAS-TAU; ORI-OB1

2.1. THE OPHIUCHUS-SCORPIUS-CENTAURUS ASSOCIATION (OSCA)

Properties of the Ophiuchus-Scorpius-Centaurus Association (henceforth called OSCA) are shown in Figure 1 and Table 1. The association spans an angle of about 70° in galactic longitude, from 290° to 360°. At the low longitudes it is centred slightly above the galactic circle, but between 310° and 360° around latitude +17°, well detached from the plane. The largest linear dimension is about 200 pc, and the lateral size about 40 pc.

The association can be subdivided into subgroups of different ages. The principal ones are denoted OPH (Ophiuchus), US (Upper Scorpius), UC-L (Upper Centaurus - Lupus), and LC-C (Lower Centaurus - Crux), the existence of which was recognized long ago (Blaauw 1960). Recent investigation (de Geus et al. 1990, de

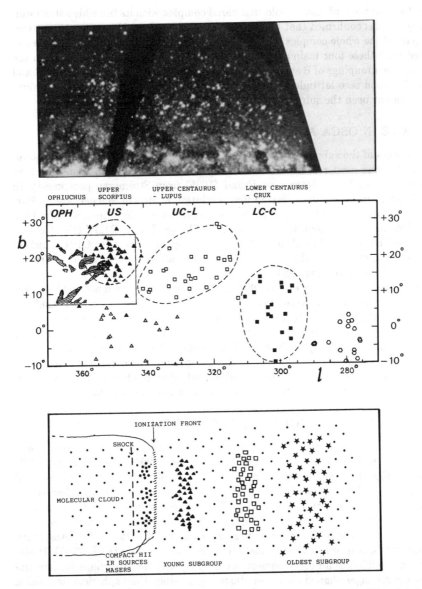

Fig. 1. Top: Section of a wide-angle blue photograph of the southern Milky Way showing the Oph-Sco-Cen Association (OSCA), published by Code and Houck (1955). Middle: Plot of the bright B-type stars in the region of the photograph, based on de Geus et al (1986), with outline of the subgroups and sketch of the Ophiuchus dark clouds. Bottom: Schematic model of sequential star formation, as it may apply to OSCA, adapted with certain modifications (see text) from Thaddeus (1977).

Geus 1988) of the Ophiuchus molecular cloud complex with its imbedded star forming regions has confirmed that OPH must be considered as the youngest, but integral part of the whole complex OSCA. Loosely connected, but probably genetically related with these four main subgroups, and at about the same distance from the sun, are two groupings of B stars: at negative latitudes between longitudes 325° and 5°, and around zero latitude between longitudes 270° and 290°. These two groupings have not been the subject of recent investigation.

2.2. AGES IN OSCA AND PROGRESSION OF STAR FORMATION

Nuclear ages of the subgroups US, UC-L and LC-C: 7, 13 and 11 million years, respectively, were most recently determined by de Geus et al (1989) from Walraven photometry and by de Zeeuw and Brand (1985) from Strömgren photometry, in combination with isochrones derived from evolution models of Maeder (1981). Earliest main sequence spectral types of the stars used in these determinations are B0V (Tau Sco) in US (not counting Zeta Oph, O9.5V, see § 3.8); B1.5V (Eta Cen) in UC-L, and B1.5V (Xi-2 Cen) in LC-C. The zero-age in Table 1 for OPH is indicative for the properties of the extremely young imbedded infrared cluster (Vrba et al 1975, Wilking and Lada 1983).

Whereas an uncertainty of, say, 20% may well have to be assigned to these ages in absolute sense, the age ratios between the different subgroups must be considered sufficiently well established that we may conclude that star formation has progressed over the body of the association, beginning at UC-L or/and LC-C, to US to OPH (where it still is manifest), with steps of the order of 2 to 6 million years.

The relative number contents, the "richness" given in Table 1 for the three oldest subgroups as judged from the numbers of unevolved stars of types B5 and earlier, is not widely different, nor is this the case for their global linear sizes.

2.3. OSCA AND THE MODEL OF SEQUENTIAL STAR FORMATION

Whereas the general occurrence of a sequential age order among association subgroups was noticed in the early 1960s (Blaauw 1960,1964), understanding of its physical origin grew only after the discovery of the intimate relation between OB Associations and molecular clouds. This soon led to a model for the mechanism involved, due to Elmegreen and Lada (1977) of which we reproduce the schematic presentation in the lower section of Figure 1. This has been taken from Thaddeus (1977), but was adapted to include the more uniformly distributed low mass stars, supposedly formed prior to the formation of the massive stars (Lada 1987). It also has been modified in an other respect: we assume that the subgroups in their initial stage are cigar shaped chains of clusterings, rather than spherical or disklike units, for reasons given in § 3.6.

2.4. STELLAR CONTENT OF OSCA

For the subgroups UC-L and LC-C our knowledge of the stellar content does not reach beyond about spectral type B5, i.e. about 5 solar masses, notwithstanding the proximity of the association to the sun. Principal clue to membership would be the proper motions; however, for the stars fainter than about visual magnitude 6, available proper motions (derived from meridian observations) do not have the required accuracy of ".003 per year. Considerable gain in our knowledge of membership of low massive stars may be expected from the extensive programme of proper motions in OB Associations that forms part of the Hipparcos project now in operation (de Geus et al 1986).

The situation is more favourable for subgroup US for which a special effort in determining proper motions of faint stars by means of the meridian positions collected over the years was made by Bertiau (1958), and photometric and spectroscopic follow-up by Garrison (1967) and Glaspey (1972). As a result, we know that for this subgroup membership extends to at least spectral type A7 (2 solar masses). For the OPH subgroup, we know of the occurrence of intermediate and low mass stars in the infrared imbedded cluster (Wilking and Lada 1983).

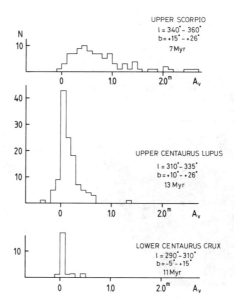

Fig. 2. Statistics of visual absorption, Av, in the regions of the three subgroups of the Oph-Sco-Cen Association.

2.5. RELATION OF OSCA TO THE INTERSTELLAR MEDIUM

In dealing with the relation of OSCA with the interstellar medium (ISM) we distinguish between a) the primordial ISM, i.e. the remnants of the molecular cloud from which the association was formed, and b) the envelopes or loops in the ISM of which the structure has been determined by the stellar winds and supernovae in the association. First, however, a general remark on the degree of association with the ISM is useful.

In Figure 2 we present statistics of the visual absorption, Av, for the three groups US, UC-L and LC-C, based on photometric data for B-type stars in de Geus et al (1989). The striking difference between US on the one hand, and UC-L and LC-C on the other hand is apparent: in the two latter subgroups values of Av peak around 0.2 and 0.1 mag., respectively, whereas in US Av ranges more or less uniformly between 0 and 1 mag. and in some cases reaches values up to 2 mag. and more. Clearly, UC-L and LC-C, with ages around 12 million years have been efficiently "cleaned", whereas this is by no means the case for US at the age of 7 million years. These statistics suggests that the time scale for cleaning of an OB-Association subgroup is about 10 million years.

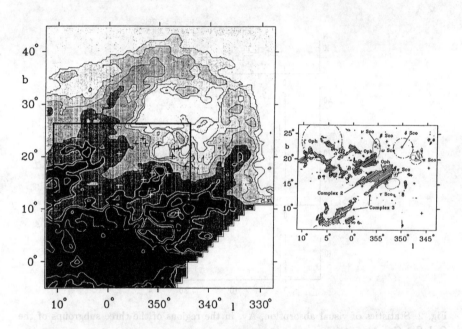

Fig. 3. Left: Shell structure in HI related to the young subgroup Upper Sco, with indication of the area of the Ophiuchus dark cloud star-forming complex, the latter shown separately in the figure at right (based on de Geus, 1988).

The process of breaking up the ambient ISM in subgroup US is nicely demonstrated in Figure 3, taken from a recent discussion by de Geus (1988). HI is still present in a nearly complete shell structure with diameter of about 50 pc, inside of which HII is present, whereas the molecular cloud remnant with its elongated "streamer" structures, now the site of current star formation, extends from the US OB stars toward higher galactic longitudes.

Shell structures in HI on a much larger scale, enveloping large parts or all of

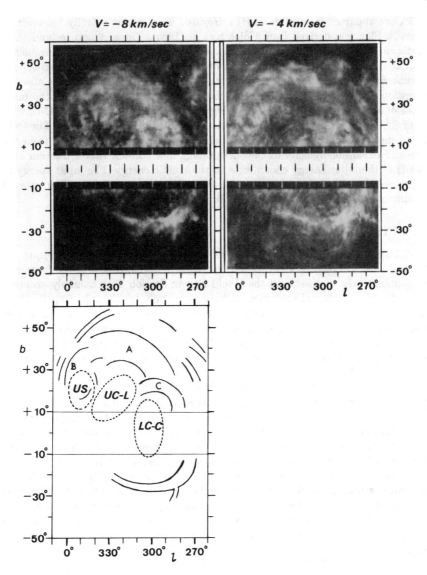

Fig. 4. Top: Large-scale shell structures in HI for the velocity intervals around -8 and -4 km/sec according to Colomb et al (1980). Bottom: schematic outline of principal shell structures with respect to the OSCA subgroups. Note that in these presentations the scale in galactic longitude has been contracted with respect to the latitude scale!

OSCA are apparent from plots of HI integrated over narrow velocity intervals as shown in Figure 4, derived from Colomb et al (1980). These structures and their kinetic energies can be reasonably well understood as the consequence of the energy input into the ISM over the time span of the age of the subgroups, taking into account what we know of their (initial) stellar content (de Geus 1988; see also the discussion of the Evaporation Function in § 3.4.) According to de Geus, the main source of energy input to the ISM, causing shell A, has been subgroup UC-L, with about 80% of this energy due to supernovae and 20% due to stellar winds. The total energy input compares reasonbly well with the kinetic energy estimated from the mass and expansion velocity of the shell, taking into account that a fraction only (20%?) of the input energy contributes to the kinetic energy. The input energy of subgroups US and LC-C (with about 50% and 75%, respectively, due to supernovae) accounts satisfactorily for the energies observed in shells B and C.

2.6. KINEMATIC PROPERTIES OF OSCA

We distinguish two aspects of the kinematics: the motion of OSCA with respect to the local standard of rest, and internal velocities. We shall return to the former in the context of the discussion of the Gould Belt in Section 4. Reasonably accurate knowledge of the internal velocities we have so far only for the subgroup US. It is based on proper motions determined especially for this object by means of available fundamental and meridian catalogues, with an accuracy of about ".0015 per year which corresponds to about 1.2 km/sec (Blaauw 1978). The resulting pattern of (projected) internal motions after elimination of common motion is shown in Figure 5. From it, we find:

a) the average relative velocity in one component is about 2.2 km/sec, corresponding to an average relative space velocity of 4.4 km/sec if an isotropic velocity distribution is assumed. Evidently, the accuracy of the proper motions just sufficed for establishing the internal velocity spread. Radial velocities of the required accuracy, 1.5 km/sec or better are not available yet for these B stars as a consequence of the high incidence of spectroscopic duplicity and fast stellar rotations.

b) tracing back the individual proper motions, as shown in Figure 5, leads to a "minimum-size configuration" at epoch about 4.5 million years ago, with dimensions of about 45 x 15 pc. We return to the latter aspect in § 3.6. The time scale mentioned here, 4.5 million years, is called the kinematic age of the subgroup. We note that it is close to, but somewhat shorter than, the nuclear age of 6-8 million years mentioned before.

2.7. DOUBLE STAR PROPERTIES AND RUN-AWAY STAR IN OSCA

We shall return to these two subjects in the context of the more general discussion of § 3.7 and § 3.8.

2.8. THE CASSIOPEIA-TAURUS ASSOCIATION (CAS-TAU)

The most fossil in character, and the nearest to full disintegration and dispersion into the general field population, is the Cas-Tau Association. It is located at the

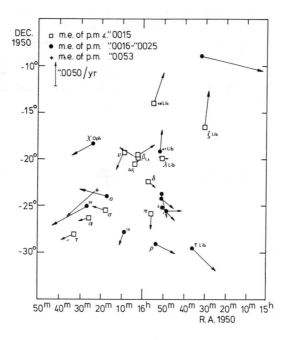

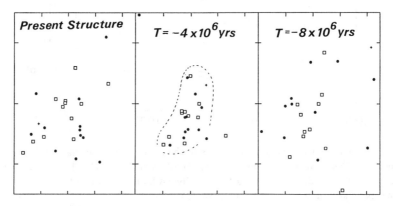

Fig. 5. Top: Internal motions in subgroup US of OSCA (relative to mean motion) as shown by the proper motions. Bottom: The present positions of these stars and those traced back for 4 and 8 million years, respectively, suggesting minimum-size configuration around -4 million years.

same distance from the sun as OSCA, about 140 pc, but on the opposite side, between galactic longitudes 110° and 200°. As shown in Figure 6, it is recognized only by the parallelism of the motions of its members with respect to the sun, and since the projected solar motion reflex amounts to about 20 km/sec, this implies that the internal velocities are much lower: the average velocity in one component was found to be 2.0 km/sec (Blaauw 1956), about the same as that in OSCA. The dimension of Cas-Tau also is close to that of OSCA.

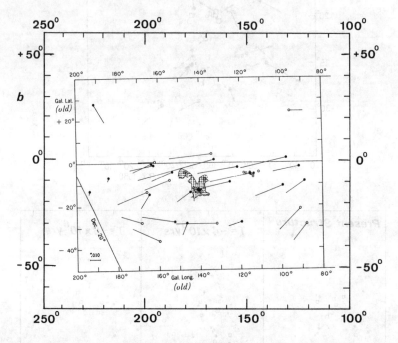

Fig. 6. Proper motions of early B-type stars in the region of Cassiopeia and Taurus, the parallelism of which reveals the existence of the old association Cas-Tau. In the centre, the Taurus dark clouds are marked schematically. (Adapted from an almost fossil paper, Blaauw 1956, still in "old" galactic coordinates.)

With two exceptions, earliest spectral types in Cas-Tau are B2V, and in accordance with this the photometric age estimate of de Zeeuw and Brand (1985) turned out to be about 25 million years. Nothing is known about membership later than B5, although this may be remedied once accurate proper motions will be available from Hipparcos. An interesting aspect is the probably close association of the - probably unbound - Alpha Persei Cluster with Cas-Tau. In agreement with the "cleaning time scale" of 10 million years for subgroups found from OSCA, there is no obvious association with interstellar matter. An interesting, and perhaps puzzling feature is the location of the Taurus molecular cloud complex near the centre of the association (see also Section 4).

2.9. THE ASSOCIATION ORI-OB1

Star formation in Ori-OB1 covers the whole range, from incipiency in a wide range of locations and particularly in the high density rims of Clouds A and B, tot the advanced stage of the Trapezium Cluster and its immediate neighbourhood. The star forming regions in the Orion molecular clouds are currently probably the ones most intensively studied. Moreover, there is the fortunate circumstance that the association is situated at the celestial equator, and hence accessible to both northern and southern observatories. For a recent review we refer to Genzel and Stutzki (1989). Features of the large-scale envelope of atomic hydrogen were most recently studied by Chromey et al. (1989); for an early study of "Orion's Cloak", the expanding shell centered on the association, see Cowie et al. (1979). A broad programme of photometry and analysis of the subgroups in Ori OB1 was carried out by Warren and Hesser (1977, 1978), to which we refer for more detailed study. The present review only describes some large-scale properties as they resulted from the consecutive stages of star formation, and to their relation to current star formation.

These past stages of star formation can be recognized in a series of subgroups of sequential age order. They are listed in Table 1, and shown in Figure 7 together with the main structures in the molecular clouds, taken from the study of Maddalena et al. (1986). Not included in Figure 7 is the star forming region around lambda Orionis at Decl. $+10°$ that is loosely connected with the main body of Ori OB1 (Murdin and Penston 1977, Duerr et al. 1982, Maddalena and Morris 1987), and that may have a more isolated history of formation.

There is interesting similarity to the association OSCA: in both, earliest star formation occurred about 12 and 7 million years ago and continues into the present: in OSCA in subgroup OPH, and in Orion most strongly in the region around the Trapezium Cluster in Cloud A, and less conspicuously in Cloud B. Age-wise in between the Trapezium Cluster and subgroup 1b is subgroup 1c which we see projected on the region around the Trapezium Cluster, but the degree of detachment is not clear. (Its age estimate seems uncertain: de Zeeuw and Brand mentioned 8-12 million years, but the estimate of Warren and Hesser, several million years seems more likely.) Subgroup 1c probably is the one most responsible for the inducement of star formation in the upper part of Cloud A.

There is conspicuous interaction between the oldest subgroup, Ori-OB1a, and molecular Cloud B. According to Table 1, over the past 12 million years around six or seven massive stars must have evolved away from this subgroup which is a particularly rich one. As its age is about the same as that of subgroup UC-L of OSCA and its stellar content about twice larger, the energy output over the years must have been about twice that of UC-L. Subgroup 1a probably is the main cause of the sharp "molecular ridge" on the western side of Cloud B that was pointed out by Baud and Wouterloot (1980) on the basis of OH observations, and that is also conspicuous in CO (Maddalena et al. 1986) and in CS (E.A.Lada 1990). The influence of subgroup 1a can, in fact, be traced all along the western edge of Cloud

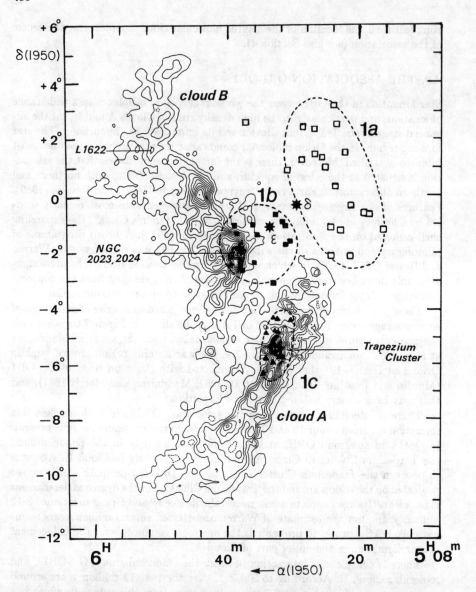

Fig. 7. The association Ori OB1 with the subgroups a, b and c outlined, in relation to the molecular clouds A and B, the latter according to Maddalena et al (1986). Note that this map does not show the other clouds occurring in this region - partly also related to Ori OB1 - as revealed by these authors and by, for instance, Kutner et al (1977).

TABLE 1. OB ASSOCIATIONS WITHIN 1 KPC

	Projected dimensions pc	Age Myr	Relative richness (U Sco =1)	Number evaporated
Within Gould Belt				
OPHIUCHUS	-	0	-	-
UPPER SCORPIUS	40 x 40	7	1.0	0.8
UPPER CEN - LUPUS	70 x 35	13	1.2	3.5
LOWER CEN - CRUX	50 x 40	11	0.6	1.3
CAS - TAU	200 x 50	25	1.6	15.0
PER OB2	60 x 40	7	0.3	0.2
ORI OB1a	45 x 25	12	2.6	6.5
ORI OB1b	25 x 20	7	1.0	0.8
ORI OB1c	20 x 10	3?	1.1	0.1?
ORI OB1d	-	0	-	-
Lambda ORI ASSOC.	12 x 8	4		
Outside Gould Belt				
CYG OB7	80 x 80	?	?	?
LAC OB1a	110 x 90	25	0.7	6.6
LAC OB1b	35 x 35	16	0.8	3.4
CEP OB2a	120 x 100	7	1.6:	1.3:
CEP OB2b		3	1.6:	0.2:
CEP OB3a	25 x 15	10	0.7	1.3
CEP OB3b	15 x 7	7	0.8	0.6
CEP OB4	60 x 60	2:	1.4	0.1
MON OB1	60 x 45	15	0.8	3.0
COLL 121	130 x 65	25	1.3	12.2

B and its extensions northward, for instance in the CO maps of the region around L1622 at declination +1°50' described by Reipurth in this Volume.

The centre of subgroup 1a lies approximately in the extension of the main body of Cloud A, and subgroup 1c perhaps in between these two. Relation between the supernovae and stellar winds of subgroup 1a and Cloud A is not as clear as that with Cloud B.

The energy output of subgroup 1b has been only 10-15% of that of subgroup 1a. However, due to its probably closer proximity to Cloud B this subgroup may well be the one mostly responsible for the star formation in the region around NGC 2023 and 2024. Its role seems comparable to that of subgroup US of OSCA in relation to the Ophiuchus clouds.

3. The Sample within 1500 pc

We review properties of the ensemble of OB Associations within 1500 pc, with attention to both features of the sample as a whole and, for the individual associations, properties we encountered in the description of the nearest associations.

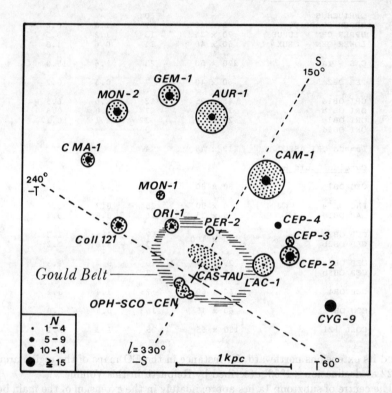

Fig. 8. Schematic presentation of the positions of the OB Associations within 1.5 kpc, projected on the galactic plane. The Sun is located at the origin of the coordinate system S,T of which the T axis was chosen in the direction of the ridge line of the "Orion Arm" (galactic longitude 60°,240°). The sizes of the central dots mark the current or recent star formation activity (N being the number of the O-type stars plus those more luminous than visual absolute magnitude -5.0). The diameters of the sorrounding circles correspond to the projected dimensions of the associations. (Adapted from Blaauw 1985a.)

3.1. GLOBAL PROPERTIES

Overall properties of the sample are exhibited in Figures 8 and 9 showing, respectively: the projection on the galactic plane, and on the two planes perpendicular to this and to the axes S and T marked in Figure 8 (from Blaauw 1985a). The axis T has been chosen roughly coinciding with the ridge line of the Orion Arm as it appears in plots on still larger scale of associations and clusters (See, f.i. Becker and Fenkart 1970). In the present context, the precise choice of the axes S and T is not important. As explained in the legend of the Figure 8, the size of the central dot for each association is a measure of the rate of ongoing or very recent star formation. The diameter of the circle surrounding the dot equals the estimated projected dimension of the association.

It must be realized that these presentations are somewhat biased by selection effects, in spite of the fact that we are dealing with relatively nearby objects. An association like Cas-Tau would not be identified as such at distances beyond, say, five hundred pc, because of the absence of very luminous stars, its lack of spatial concentration, and the present lack of accurate proper motions for stars fainter than 6th magnitude. To a lesser degree this also holds for an association like OSCA. Accordingly, at large distances there is a bias in favour of the youngest associations containing very luminous stars. Also, note that for the more distant associations it is difficult to discriminate subgroups of different ages and their differing degree of association with the ISM.

The most striking feature of the overall distribution is perhaps its flatness, apparent from Figure 9. The average distance of the associations from the galactic plane is only 3 pc, and the average without regard to sign only 50 pc. Within this extreme flatness, there is a disturbance; the Gould Belt System marked in Figures 8 and 9, comprising the OB Associations within 500pc from the centre of Cas-Tau. We return to this in Section 4.

Except for deviating kinematics in the Gould Belt System, the ensemble is kinematically very quiet: the average residual velocity in one component as derived from radial velocities amounts to only 4.5 km/sec, a value corroborated by the residual velocities of the molecular clouds with which the associations are connected (Blaauw 1985a).

3.2. FORMATION OF MASSIVE STARS IN ASSOCIATIONS AS COMPARED TO CLUSTERS

For the fraction of massive stars formed in open clusters as compared to associations, a canonical figure encountered often in the literature is 10%, due to Roberts (1957). This percentage is confirmed if we count the B stars within 1.5 kpc occurring in clusters and in OB Associations. However, a more relevant question would seem to be, what percentage is formed in bound clusters, i.e. systems with negative total energy. This probably is much lower, for the former statistics includes among the clusters objects (like, for instance, NGC 1502) which probably are in a state of disintegration and eventually will disperse into the general field, just like the association stars.

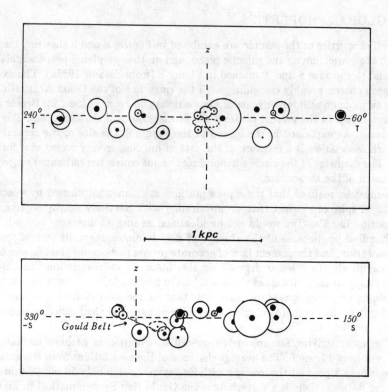

Fig. 9. The same as Figure 8, but now projected on planes perpendicular to the axes S (top) and T (bottom).

3.3. AGES AND SUBGROUPS; STELLAR CONTENT

Data on ages for the subgroups in the associations of Figure 8 are shown in Figure 10 and, for those within 1 kpc, in Table 1. Asterisks in the Figure indicate those associations - the majority - where star formation is still going on. In spite of the selection effects mentioned before, it seems justified to state that there is little evidence of systematic progression of star formation over the volume of space considered here.

Ages listed range from zero to about 25 million years, this upper limit being determined by detectability and stellar evolution: at that age the most luminous stars have evolved into anonymity, and the remaining association stars gradually disperse into the general field population: the average internal space velocity of 4 km/sec, mentioned before, corresponds to a path length of 100 pc after 25 million years, so that no spatial concentration is left except for the stars with lowest relative

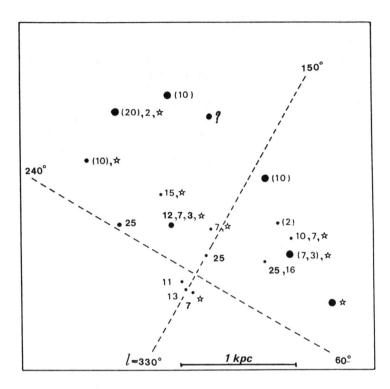

Fig. 10. Ages, in million years, of the (subgroups of) associations plotted in Figure 8. Asterisks mark objects with ongoing star formation. (Adapted from Blaauw 1985a).

velocities.

A not yet satisfactorily answered question concerns the "membership" of stars of spectral types later than B. Useful objective prism classification surveys have been made in the regions of several associations; examples are the work on Lac-OB1, Ori-OB1, and Per-OB2 by Guetter (1976,1977). These have established excesses of stars down to middle A-types and evidence for a pre-main sequence population, in global agreement with what would be expected at the age of the association. An interesting problem remains, however: were these stars formed at the same location and at the same epoch as the more massive ones? Probably the question of mass dependent location and/or epoch of formation can be taken up only when accurate proper motions will allow back tracing of the past path of these stars.

3.4. EVAPORATION FUNCTION AND EVOLVED STARS

The generally accepted theory that massive stars evolve away from the associations via the supernova phase allows an estimate of the rate at which the subgroups "evaporate" their most massive members. The resulting Evaporation Function $E(t)$ is shown in Figure 11 (Blaauw 1985b). It applies to a "standard subgroup" which is assumed to contain 24 stars of masses exceeding 6 solar masses, that initially follow the standard Initial Mass Function (IMF), and it gives the number of stars that evaporate per million years as a function of age. The shape of $E(t)$ is rather sensitive to the assumed IMF and to the evolution tracks adopted; a somewhat different shape was arrived at by de Geus (1988).

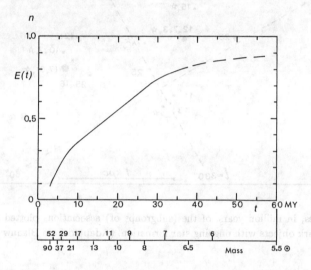

Fig. 11. Evaporation Function $E(t)$, expressed in number of stars evolving away per million years, for a standard association subgroup starting out with 24 stars of masses exceeding 6 solar masses.

The numbers of "evaporated" stars for the (sub)groups within 1 kpc are given in the last column of Table 1. They are of considerable interest when it comes to understanding the influence the evolved stars may have exerted on the neighbouring ISM through their supernova explosion and, preceding these, by their stellar winds. We referred to this when dealing with the shell structures around OSCA in § 2.5, and with the rim structures observed in the molecular clouds associated with Ori-OB1 in § 2.9. The evaporation function also plays an essential role in estimates of the rate of generation of pulsars (Blaauw 1985b).

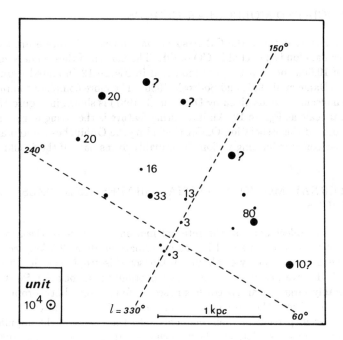

Fig. 12. Masses of the molecular clouds (in units of 10,000 solar masses) connected with the associations of Figure 8.

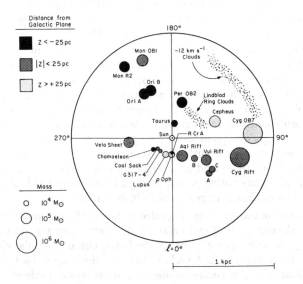

Fig. 13. Map of the molecular clouds as given by Dame et al (1987).

3.5. ASSOCIATED MOLECULAR CLOUDS

With very few exceptions, the OB Associations shown in Figure 8 are associated with molecular clouds detected in CO or OH. The masses of these clouds, expressed in units of 10,000 solar masses, are indicated in Figure 12 (updated from Blaauw 1985a with Dame et al 1987, and Sodroski 1990). The more complete compilation of the local molecular clouds given by Dame et al (1987) is shown in Figure 13, on the same linear scale as Figure 12. An interesting feature is the string of clouds within which is located the association OSCA (including the Ophiuchus clouds) as well as the Chamaeleon star-forming region. It probably forms part of the Gould Belt.

3.6. INTERNAL MOTIONS; INITIAL SHAPES AND SIZES OF SUBGROUPS

In Table 2 we collect data on the internal motions for some of the associations. For most of them, a mean speed in one coordinate of about 2.0 km/sec is found, hence a mean space velocity of about 4.0 km/sec if isotropic velocity distribution is assumed. However, there is at least one exception: the association Per-OB2 with a considerably larger mean speed, 6 km/sec as found from both proper motions and radial velocities. The association is in a state of rapid disintegration. As the time scale for dispersion of their members into the general field population for associations like this is relatively short, such cases tend to soon escape detection.

The velocity of about 2 km/sec found for most of the OB Associations is the same as that found for various categories of objects of very recent formation: 1.5 to 2.3 km/sec in the Orion Trapezium Cluster (Jones and Walker 1988, van Altena et al. 1988); 1.5 km/sec for Taurus Clouds TT stars (Jones and Herbig 1979), and 2.0 km/sec for the Lambda Ori Association (Mathieu 1986, 1987).

In § 2.6 we noted that back tracing of the positions of stars in the subgroup US of OSCA led to the conclusion that its initial, oblong shape probably cannot have had a largest dimension smaller than 45 pc, and excludes initial cluster size of the order of several pc. It is important to note that evidence from other cases, for which a reasonably accurate estimate of the limiting initial size can be derived from proper motions, leads to the same conclusion. Thus, a minimum size of about 50 pc was found for the strongly elongated initial configuration of Per OB2 (Blaauw 1983) and unpublished estimates by the author based on proper motion studies in the Ori OB1a subgroup by Lesh (1968) lead to a minimum largest initial dimension of 35 pc. Preliminary analysis of existing proper motions in subgroup UC-L of OSCA by the author leads to a minimum largest initial dimension of 50 pc.

The interpretation of these findings is guided by what is found from observations of those, most condensed, parts of molecular clouds where a new generation of stars is about to emerge. A striking case is that of the L1630 Orion molecular cloud as it results from CS observations by E.A.Lada (1990); see the chapter by C.J.Lada in this Volume. A linear string of clumps is observed at the western side of this cloud over a total length of about 25 pc, that may give rise to nearly coeval emergence of new stars that after several million years will show the typical features of subgroups observed now in the youngest associations. Somewhat similar evidence occurs in the CO contours observed in the W3-W4-W5 region by Lada et al (1978).

As a preliminary conclusion we infer from these cases, that in the model of sequential star formation elongated initial structures, consisting of a series of small clusterings of pre-main sequence stars, should be assumed to figure, as illustrated in the (revised) model at the bottom of Figure 1.

TABLE 2. INTERNAL VELOCITIES IN ASSOCIATIONS
(Average velocity in one coordinate)

	From proper motions (km/sec)		From radial velocities (km/sec)	
SCO-CEN (all subgroups)	1.7	(1)	≤ 2	(1)
UPPER SCORPIO	2.2	(2)		
CAS-TAU	2.0	(3)	3.0	(3)
PER OB2	6.5	(4)	6.0	(5)
ORI OB1a	≤ 6	(6)		
LAC OB1			2.0	(7)

(1) Bertiau (1958). (2) Blaauw (1978). (3) Blaauw (1956).
(4) Derived from Lesh, J.R. 1969, Astron.J. 74,891.
(5) Blaauw, A. 1952, Bull.Astron. Inst. Netherlands, 11,405. (6) Derived from Lesh (1968). (7) Blaauw, A. and Delhaye, J., unpublished.

3.7. STATISTICS OF CLOSE BINARIES

Duplicity and multiplicity properties of newly born stars are among the most important clues to understanding the process of star formation. We shall deal here only with duplicity and limit ourselves to the statistics of separations in the associations of the local sample. Accordingly, the properties observed are those occurring in the interval between several and, say, 30 million years after star birth, so that by necessity we skip changes in the elements during the first few million years. As we shall see, even during the subsequent 25 million years, evolutionary effects can be recognized in the statistics.

The frequency of spectroscopic, i.e. the closest, binaries among B-type stars has been investigated by many authors; for a summary we refer to van Albada (1985), from whose paper also the data below have been taken. There appears to be a concensus of opinion that the frequency of duplicity is about 30%.

Useful statistics is obtained by plotting the semi-amplitude of the radial velocity variation of the primary component against the (logarithm of the) period; both

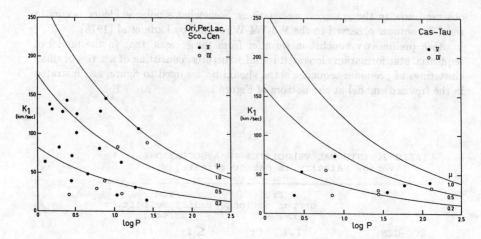

Fig. 14. Semi-amplitude of the radial velocity variation of the primary components of early-type spectroscopic binaries, plotted against Log Period; left: for a number of the young associations or their subgroups; right: for the older Cas-Tau group. From van Albada (1985).

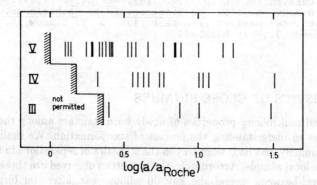

Fig. 15. Distribution of the ratio a/a(Roche) for the binaries of Figure 14, for the luminosity classes V, IV and III.

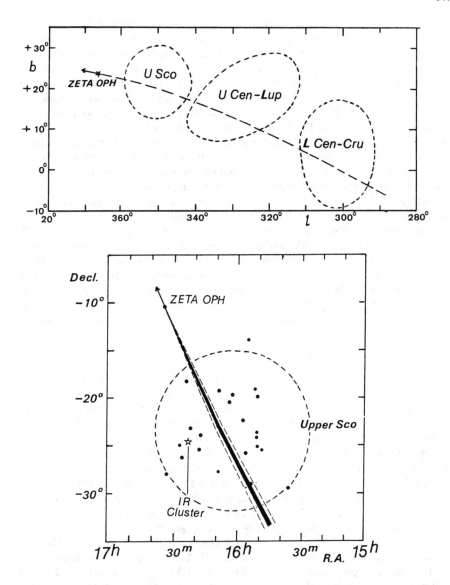

Fig. 16. The Run-Away star Zeta Ophiuchi (O9.5V) in relation to the association OSCA (relative space velocity about 42 km/sec). Top: The projection on the sky of the past path; Zeta Oph originated most likely in subgroup U Sco about 1 Myrs ago, but origin in U Cen-Lup, about 2.5 Myrs ago cannot be excluded. Bottom: Close-up of the situation with respect to U Sco; dashed lines indicate the uncertainty in the direction of the projected motion corresponding to the probable error of the proper motion.

quantities are directly obtained from the observations. [Note that bias favouring systems with large velocity amplitudes was avoided by specially designed observing programmes as referred to by van Albada (1985).] Fig.14 shows such plots for a) stars in the associations, respectively subgroups, with ages in the interval 5 to 15 million years, and b) those in the older association Cas-Tau, with age about 25 million years. We note the lack of binaries with high velocity amplitudes and short periods in the latter, and thus recognize the effect of the evaporation of the most massive stars from the Cas-Tau association as estimated in Table 1. Since for systems with a given mass of the primary component the period is primarily determined by the separation of the two components, it follows from plots like these that evolution effects have to be taken into account in determining the initial frequency distribution of the separations of the components.

The effect is also illustrated in Fig. 15, showing for the same stars statistics of the ratio between the actual separation of the components and the Roche radius of the primary component, separately for the luminosity classes V, IV and III. Clearly, already for classes IV, and even stronger for class III, as a consequence of the evolution of the primary the peak in the distribution of the ratios has shifted, indicating that for the estimate of the initial distribution of the separations only luminosity class V should serve. A median initial value for the separation in terms of the Roche radius appears to be about 3. For a discussion of the effects in the distribution of the eccentricities, reference is also made to van Albada's paper.

3.8. RUN-AWAY OB STARS (RAOBS)

Run-away OB stars, henceforth denoted as RAOBs, are those OB stars that with moderately to high space velocities, i.e. from about 40 to more than 100 km/sec, run away from the young (subgroups of) OB Associations that can be identified as their place of origin. This definition has to be relaxed somewhat for those (older) objects that must be considered to fall in the same physical category, but for which the origin cannot be so well established. The run-away phenomenon seems to be a regular feature, inherent to the evolution of every OB Association.

Typical cases are those illustrated in Figures 16 and 17 (Blaauw 1988): the star Zeta Oph moving away from OSCA and most likely originating from its subgroup US, and the stars AE Aur, Mu Col and 53 Ari moving away from Ori OB1 but for which it is less well possible to identify the subgroup within this association from which they came. For these four examples, the velocities with respect to the generating associations are estimated to be 42, 137, 141, and 55 km/sec, respectively, and the times elapsed since they left their origin 1.0 (or perhaps 2.5), 2.6, 2.8, and 6.0 million years. Principal aspects of current investigations of the run-away phenomenon are: the question, how to explain it, and the implication for the inducement of star formation far outside the parent association of the RAOB.

As to the first, at this moment there are two hypotheses. The one assumes an explosive event in close, massive double stars, most likely a supernova explosion of one of the components, as a consequence of which the other component is released from the original gravitational binding and escapes with a velocity close to its orbital velocity (Blaauw 1961; see also Stone 1985). In this scenario, the run-away

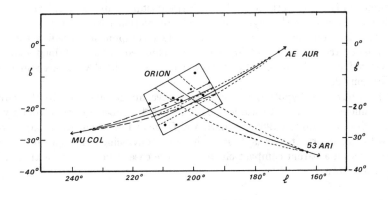

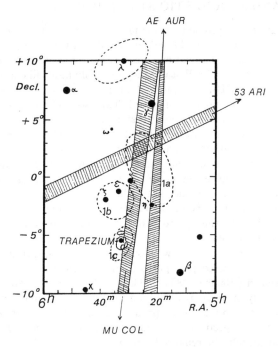

Fig. 17. Top: The Run-Away stars AE Aur (O9.5V), Mu Col (O9.5V) and 53 Ari (B2.5) in relation to the association Ori OB1. Relative space velocities are about 137, 141 and 55 km/sec, respectively and "kinematic ages" about 2.6, 2.8 and 6.0 Myrs. Bottom: close-up indicating that 53 Ari most likely originated in subgroup 1a or 1b, but that for AE Aur and Mu Col the generating subgroup cannot be well established from the projected past path.

phenomenon is not directly related to the circumstances of star formation. An obstacle encountered by this hypothesis is the prediction that, as a consequence of mass transfer from the primary to the secondary component preceding the explosion, a large fraction of the RAOBs would be expected to have retained a companion; yet radial velocity observations show that most of the RAOBs are single (Gies and Bolton 1986).

The alternative hypothesis, which does have direct relevance to the circumstances under which RAOBs are formed, assumes dynamical ejection from a young cluster (Gies and Bolton 1986, Leonard and Duncan 1990, Leonard 1990). A problem with this interpretation is, that we have no convincing evidence yet of the existence of such a parent compact cluster in those cases where the past track of the RAOB is known.

3.9. RAOBS AND STOCHASTIC STAR FORMATION

An aspect relevant to the problem of star formation, and independent of the mechanism of their origin, is the possible role of RAOBs in the propagation of star formation outside the parent association. We have no reason to assume that RAOBs are different from "ordinary" OB stars; they probably also will end their life in supernova explosions. RAOBs, however, will normaly do this far outside their region of origin. With run-away velocities of 50 to 100 km/sec and life times of 5 to 10 million years, the explosion will happen at distances of the order of 500 pc. We must expect this to give rise to stochastic induced star formation if the explosion occurs inside, or in the neighbourhood of, an other molecular cloud.

The efficiency of this mechanism, to be investigated further, depends on the frequency at which the inducement occurs. It will suffice here to state that obviously, for a plane parallel layer of clouds, the efficiency of the process goes with the square of the projected density of the clouds on the galactic plane, as both the production of RAOBs and the reciprocal of the projected mean free path of RAOBs are proportional to this density.

Figure 18 shows the positions of well established RAOBs projected on the galactic plane, and their projected past paths from the parent association in so far as these are well known (Blaauw 1985a). The fact that the best established RAOBs occur preponderately in the solar neighbourhood is an effect of observational selection. Only with much more accurate proper motion data in a fundamental, absolute system, as may be produced by Hipparcos, may we expect the areas void of RAOBs in the diagram to be reliably filled in.

4. The Gould Belt System

The Gould Belt derives its name from B.A.Gould who was the first to describe in some detail the belt of bright stars along a great circle including the constellations Scorpius, Centaurus, Crux, Orion and Perseus. At closer inspection it turns out to be marked especially by the bright B-type stars, part of which belong to the nearby associations OSCA, Ori-OB1, and Per-OB2. Early research on the Gould Belt System has been reviewed by Lindblad (1974). It comprises young stars and interstellar matter - dust as well as molecular clouds and neutral Hydrogen - and is characterized not only by its peculiar space distribution but also by its internal kinematics. It is a flat subsystem among the local OB population and the interstellar medium with largest dimension about 700 pc, tilted about 15° with respect to the galactic plane, and with low central density; see the sketch in Figure 8. The system is in a state of expansion, with the spatial centre of the system nearly at rest with respect to the local standard of rest. The motions of the associations participating in this expansion are revealed in their radial velocities by positive excesses, up to 10 km/sec and more, when compared to the velocities expected on the basis of the standard solar motion and regular differential galactic rotation.

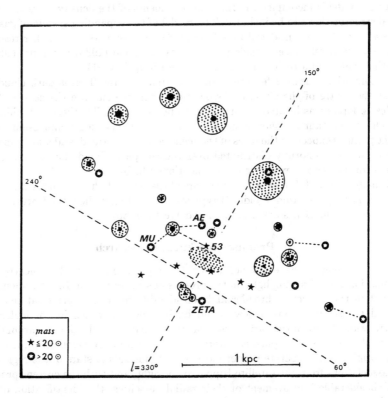

Fig. 18. The sample of best established RAOBs in relation to the OB Associations presented in Figure 8, projected on the galactic plane. Dashed lines connect RAOBs with their parent association in so far as the connection seems well established. The four cases of Figures 16 and 17 are marked individually. The lack of RAOBs at large distances from the sun must be due to observational incompleteness.

A suggestive, but not yet quite satisfactory model of the evolutionary history of the Gould Belt System starts from the assumption that between 25 and 40 million years ago, expansion set in in the ISM in the area where now the Cas-Tau association is situated, and identifies the cause of the expansion with the supernovae and stellar winds inherent to the early stage of this association. The expanding, swept-up ISM, settling at velocities of the order of 20 km/sec, was subsequently subject to the general galactic field of force, leading to a pattern of motions which has certain similarity to the one now observed. Meanwhile, secondary star formation took place in the swept-up ISM, leading to the formation of the OB Associations with ages of 15 million years and lower in the oval shaped configuration marked in Figure 8. The fact that the observed radial velocities of these associations with respect to their local standard of rest are not all in satisfactory agreement with predictions implies that the model is incomplete. Perhaps also elements of the density wave theory of spiral structure have to be included. The model of the expanding system of gas and stars was worked out in detail by Olano (1982). For reviews and further research on the subject, reference is made to papers by Lindblad (1983) and Lindblad and Westin (1985) and a review in preparation by Pöppel (1991).

A somewhat puzzling feature is the occurrence of the Taurus dark clouds at about the centre of the Cas-Tau Association where we surmise the centre of the explosive forces was located. Contrary to cloud structures like those in Ophiuchus and Orion, the Taurus clouds do not show evidence of one-sided compression as one might perhaps expect on the basis of the scenario just described, and star formation is more or less uniformly distributed over the complex. Could the Taurus clouds have formed after the explosive events mentioned before?

Finally, it is useful to note that the stars of the Gould Belt System are immersed in the general population of older B-type stars that do not show the features (tilt, expansion) of the Belt stars (Lindblad and Westin 1985).

5. Promises for Future Research

In the foregoing we have scanned a variety of prospects, offered by nearby OB Associations, for throwing light on the process of star formation. In view of current interest in this process a broader and considerably more in depth-treatment, far beyond the scope of this review would be in order, in which in particular the interrelations between the various sites of star formation in the domain around the sun would deserve to be pursued. Moreover, the future of research on these fossils looks bright. With accurate proper motions of thousands of stars in the regions of the associations from the satellite Hipparcos in sight, and with the current promise of a considerable improvement of their radial velocities, the identification of the member stars and the early spatial and kinematic history of the associations and their subgroups should bring us a step closer to their formation stage.

References

Baud, B., Wouterloot, J.G.A. 1980, *Astron. Astrophys.* , **90**,297.
Becker, W., Fenkart, R. 1970, in IAU Symp. **38** *The Spiral Structure of Our Galaxy*, W.Becker and G.Contopoulos (eds), p.205.
Bertiau, F.C. 1958, *Ap.J.* , **128**,533.
Blaauw, A. 1956, *Ap.J.* , **123**,408.
Blaauw, A. 1960, in *Present Problems Concerning the Structure and Evolution of the Galactic System* (Nuffic Intern. Summer Course), The Hague, p.III.1.
Blaauw, A. 1961, *Bull. Astron. Inst. Netherlands* , **15**,265.
Blaauw, A. 1964, *Ann. Rev. Astron. and Astroph.* , **2**,213.
Blaauw, A. 1983, *Irish. Astron. J.* , **16**,141.
Blaauw, A. 1985a, in IAU Symp. **106** *The Milky Way Galaxy* , H. van Woerden, R.J.Allen and W.B.Burton (eds), 335.
Blaauw, A. 1985b, in *Birth and Evolution of Massive Stars and Stellar Groups* , W.Boland and H.van Woerden (eds), Reidel, Dordrecht, p.211.
Blaauw, A. 1978, in *Problems in Physics and Evolution of the Universe* , L.Mirzoyan (ed.), Yerevan USSR, p.101.
Blaauw, A. 1989, *Astrophysics* , **29**,417.
Chromey, F.R., Elmegreen, B.G., Elmegreen, D.M. 1989, *Astron.J.* **98**,2203.
Code, A.D., Houck, T.E. 1955, *Ap.J.* , **121**,553.
Colomb, F.R., Pöppel, W.G.L., Heiles, C. 1980, *Astron. Astrophys. Suppl. Ser.* , **40**,47.
Cowie, L.L., Songaila, A., York, D.G. 1979, *Ap.J.* , **230**,469.
Dame, T.M., Ungerechts, H., Cohen, R.S., de Geus, E.J., Grenier, I.A., May, J., Murphy, D.C., Nyman, L.-A., Thaddeus, P. 1987, *Ap.J.* , **322**,706, Fig.7.
de Geus, E.J. 1988, *Stars and Interstellar Matter in Scorpio Centaurus* , Thesis Leiden University.
de Geus, E.J., Bronfman, L., Thaddeus, P. 1990, *Astron. Astrophys.* , **231**,137.
de Geus, E.J., Lub, J., de Zeeuw, P.T. 1986, *The Messenger - El Mensagero* (ESO), **43**,7.
de Geus, E.J., de Zeeuw, P.T., Lub, J. 1989, *Astron. Astrophys.* , **216**,44.
de Zeeuw, P.T., Brand, J. 1985, in *Birth and Evolution of Massive Stars and Stellar Groups* , H. van Woerden and W.Boland (eds), Reidel, Dordrecht, p.95.
Duerr, R., Imhoff, C.L., Lada, C.J. 1982, *Ap.J.* , **261**,135.
Elmegreen, B.G., Lada, C.J. 1977, *Ap.J.* , **214**,725.
Garrison, R.F. 1967, *Ap.J.*, **147**,1003.
Genzel, R., Stutzki, J. 1989, *Ann. Rev. Astron. Astrophys* ., **27**,41.
Gies, D.R., Bolton, C.T. 1986, *Ap.J.Suppl.* , **61**,419.
Glaspey, J.W. 1972, *Astron.J.* , **77**,474.
Guetter, H.H. 1976, *Astron.J.* , **81**,537.
Guetter, H.H. 1976, *Astron.J.* , **81**,1120.
Guetter, H.H. 1977, *Astron.J.* , **82**,598.
Jones, B.F., Herbig, G.H. 1979, *Astron.J.* , **84**,1872.
Jones, B.F., Walker, M.F. 1988, *Astron.J.* , **95**,1755.
Kutner, M.L., Tucker, K.D., Chin, G., Thaddeus, P. 1977, *Ap.J.* , **215**,521.
Lada, C.J. 1987, in IAU Symp.**115**, *Star Forming Regions* , M. Peimbert and J.Jugaku (eds), Reidel, Dordrecht, p.1.
Lada, C.J., Elmegreen, B.G., Cong Hong-Ih, Thaddeus, P. 1978, *Ap.J.* , **226**,L39.
Lada, E.A. 1990, *Global Star Formation in the L1630 Molecular Cloud* , Thesis Univ. of Texas, Austin.
Leonard, P.J.T. 1990, *Journ.R.A.S.Canada* , **84**,216.

Leonard, P.J.T., Duncan, M.J. 1990, *Astron.J.* , **99**,608.
Lesh, J.R. 1968, *Ap.J.* , **152**,905.
Lindblad, P.O. 1974, in *Stars and the Milky Way System*, Proc. First Europ. Astron. Meeting, Vol.2, L.M.Mavridis (ed.), Springer, Berlin, p.65.
Lindblad, P.O. 1983, in *Kinematics, Dynamics and Structure of the Milky Way* , W.L.H.Shuter (ed.), Reidel, Dordrecht, p.55.
Lindblad, P.O., Westin, T. 1985, in *Birth and Evolution of Massive Stars and Stellar groups* , H. van Woerden and W.Boland (eds), Reidel, Dordrecht, p. 33.
Maddalena, R.J., Morris, M. 1987, *Ap.J.* , **323**,179.
Maddalena, R.J., Morris, M., Moscowitz, J., Thaddeus, P. 1986, *Ap.J.* , **303**,375.
Maeder, A. 1981, *Astron. Astrophys.* , **93**,136; **99**,97; **102**,401.
Mathieu, R.D. 1986, in *Highlights of Astronomy* , Vol.7, J.P.Swings (ed.), Reidel, Dordrecht, p.481.
Mathieu, R.D. 1987, in IAU Symp. **115**, *Star Forming Regions* , M.Peimbert and J.Jugaku (eds), p.61.
Murdin, P., Penston, M.V. 1977, *Monthly Not.R.A.S.* , **181**,657.
Olano, C.A. 1982, *Astron. Astrophys.* , **112**,195.
Pöppel, W.G.L. 1991, Private communication.
Roberts, M.S. 1957, *Publ. Astron. Soc. Pac.* , **69**,59.
Sodroski, T.J. 1990, *LASP Preprint* No.90-8, NASA/Goddard Space Flight Center.
Stone, R. 1985, in *Birth and Evolution of Massive Stars and Stellar Groups* , H.van Woerden and W.Boland (eds), Reidel, Dordrecht, p.201.
Thaddeus, P. 1977, in IAU Symp. **75** *Star Formation* , T. de Jong and A.Maeder (eds), Kluwer Acad. Publ., Dordrecht, p.37.
van Albada, T.S. 1985, in *Birth and Evolution of Massive Stars and Stellar Groups* , H. van Woerden and W.Boland (eds), Reidel, Dordrecht, p.89.
van Altena, W.F., Lee, J.T., Lee, J.-F., Lu, P.K., Upgren, A.R. 1988, *Astron.J.* , **95**,1744.
Vrba, F.J., Strom, K.M., Strom, S.E., Grasdalen, G.L. 1975, *Ap.J.* , **197**,77.
Warren Jr., W.H., Hesser, J.E. 1977, *Ap.J.Suppl.Ser.* , **34**,115 and 207.
Warren Jr., W.H., Hesser, J.E. 1978, *Ap.J.Suppl.Ser.* , **36**,497.
Wilking, B.A., Lada, C.J. 1983, *Ap.J.* , **274**,698.

PHYSICAL CONDITIONS AND HEATING/COOLING PROCESSES IN HIGH MASS STAR FORMATION REGIONS

REINHARD GENZEL
Max-Planck-Institut für extraterrestrische Physik
Garching
FRG

1. Introduction

Massive stars have a profound impact on their environment. At a distance of a few pc from a massive O-star ($> 10^6$ L$_\odot$) the flux of UV photons is about 1000 times larger than in the average radiation field of the solar neighborhood. Massive stars also have supersonic winds with velocities up to several thousand km s^{-1} carrying between 0.1 and 1% of the star's luminosity in form of kinetic energy. Massive stars affect their environment very soon after a stellar core has formed as their thermal (Kelvin-Helmholtz) time scale is very short compared to all other time scales ($\tau_{KH} \leq 10^4$ y for an O-star). At the end of its lifetime (several 10^6 to 10^7 y) an O star explodes as a supernova, ejecting 10^{51} erg of energy into its surroundings.

Radiation and winds from nearby OB stars dramatically affect the physical conditions, structure and chemistry of dense molecular clouds. Discussion of observations and theoretical models of dense clouds in regions of massive star formation and the physical processes occuring there is the purpose of this review. Dense interstellar clouds primarily consist of molecules, atoms, and dust particles. Analysis of the infrared, submillimeter and radio line, and continuum radiation from these species are our primary tools for studying the physical conditions and processes in these clouds. Chapter III presents an overview of excitation and radiative transport of line and continuum emission in these wavelength bands and shows how physical parameters are derived from them. Chapter IV is a discussion of heating and cooling processes in clouds near newly-formed OB stars.

The most important global application of the effects of massive star formation on the interstellar medium as a whole is in *star burst galaxies*. The average flux of UV photons in the central 700 pc of the infrared luminous galaxy M82 (3×10^{10} L$_\odot$) is about the same as that 5 pc from an O-star. The entire interstellar medium of star burst galaxies can thus be expected to resemble a gigantic star forming cloud rather than cold clouds within our Galaxy. There are a number of interesting puzzles that are raised by the recent observations. From those chapter V selects two: the issue of finding "protostars", and the interpretation of warm quiescent molecular gas at the surfaces of clouds.

We begin with a brief overview of the *observational phenomena* in the case of the nearest cloud with massive star formation, the *Orion Molecular Cloud* (for a more detailed review see Genzel and Stutzki 1989).

2. An Example Of A Star Forming Cloud: Orion

2.1. GLOBAL STRUCTURE OF THE CLOUD

Fig. 1 (left) gives an overview of the Orion molecular cloud complex (adapted from Maddalena et al. 1986). The complex is made up of several sub-complexes (Orion A and B clouds etc.) and has a size of about 120 pc (15° at a distance of 450 pc). This massive interstellar cloud is mostly made up of molecular hydrogen molecules, a conclusion which is not based on direct measurement of H_2 itself (H_2 does not have "allowed" electric dipole, rotational or ro-vibrational transitions), but on extensive observations of dust and trace molecules (CO, NH_3, H_2CO, CS etc.). The complex's mass is estimated to be a few $10^5 \, M_\odot$ from observations of CO 1→0 line emission (Maddalena et al. 1986). This corresponds to a mean hydrogen density of 100 cm^{-3} or more.

In this cloud massive OB stars ($M > 10 \, M_\odot$) have been forming for the past 10^7 y or so. Four distinct OB associations can be distinguished in visible studies (Blaauw 1964). The three older ones are marked by hatched circles in Fig. 1. The youngest is associated with the Orion nebula. In addition to the associations which are in front of or near the cloud the most recent star forming activity (within the last 10^6 y) has occured within the cloud itself. Fig. 1 (right) shows an image of the 12 to 100 μm continuum emission that was obtained with the IRAS satellite (Beichman 1988). The emission comes from dust grains that are heated by the short-wavelength radiation from embedded stars. The dust has temperatures between 30 to 70 K and emits strongly in the far-infrared. The IRAS image marks the sites of the most recently formed massive stars and gives estimates of their luminosities. It is clear that star formation in the Orion cloud is taking place at a high rate.

The most prominent and luminous of these star forming regions is the Orion nebula itself. It contains the θ^1 (Trapezium) OB association ($L = 10^5 \, L_\odot$) as well as a remarkable concentration of lower mass stars surrounding it (Herbig and Terndrup 1986, McCaughrean et al. 1991). About 0.1 pc behind the HII region that is ionized by the Trapezium stars is the Orion-KL region, containing a luminous ($10^5 \, L_\odot$) cluster of compact infrared sources near the center of the densest core of the molecular cloud. The size of this core is about 0.1 pc, its mass 200 $M_\odot$ and hydrogen volume densities range between 10^5 and a few 10^7 cm^{-3}.

2.2. INFRARED AND SUBMILLIMETER EMISSION IN STAR FORMING CORES

Fig. 2 shows a schematic infrared/submillimeter spectrum toward the central 0.1 pc, the Trapezium OB cluster and the Orion-KL star forming region. Most of its luminosity emerges between 1 μm and 1 mm. In addition to intense infrared continuum this region emits also very bright infrared line radiation in a large number of atomic, ionic, and molecular transitions. They have all been detected for the first time within the last decade as the result of dramatic progress in infrared instrumentation (see Genzel and Stacey 1985 for a review). Hydrogen and ionic line emission from the Orion HII region in the visible range has of course been studied for a long time.

Quite remarkable is the intense far-infrared and submillimeter emission from species with ionization potentials <13.6 eV (OI, CI, CII). They indicate the presence

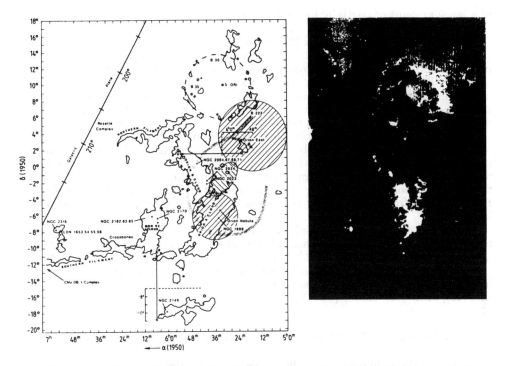

Figure 1. Large scale distribution of clouds and star formation in Orion. *Left:* Outlines of the molecular cloud complexes in Orion and Monoceros, from CO mapping by Maddalena et al. (1986), along with the standard designations of clouds and individual star forming regions. The hatched circles are the Orion Ia, Ib, and Ic OB associations (NW to SE). The Orion Id (θ^1) association is located at the Orion nebula. The shaded arc is Barnard's Loop. *Right:* Composite 12 to 120 μm IRAS image of the same region (Beichman 1988).

of a large amount of partially ionized or atomic gas along the line of sight through the Orion Trapezium/Orion-KL region.

Detailed observational and theoretical studies during the past 10 years show that the atomic and molecular infrared line emission shown in Fig. 2 can plausibly be accounted for by two main processes that are direct results of the star forming activity: first, photodissociation of the dense quiescent molecular cloud by the intense far-ultraviolet radiation coming from the Trapezium OB cluster and second, shock excitation of molecular gas by mass outflow from the center of the Orion-KL cluster. These two processes will be subjects of more detailed discussions below.

There are also many molecular emission lines. First, there is intense quadrupole line emission from vibrationally and rotationally excited levels of molecular hydrogen. The v = 1 level of H_2 is 6500 K above ground, so detection of near-infrared H_2 line emission thus provides evidence for a component of hot (a few thousand K) molecular gas. High spectral resolution measurements show that the H_2 lines come from dynamically active gas with a velocity range of several tens of km s^{-1}.

Second, there is far-infrared and submillimeter/mm rotational line emission from CO, OH, NH_3, and H_2O which also require high temperatures (100 to 1000 K) and high (hydrogen) densities (10^5 to 10^8 cm^{-3}) for their excitation. The gas kinetic temperature of the bulk of the molecular gas close to the newly formed massive stars is between 50 and 200 K, significantly greater than in the rest of the Orion cloud (20 to 50 K), or in dark clouds (10 K).

2.3. CLOUD STRUCTURE AND CLUMPINESS

While these findings already suggest substantial changes of the physical conditions of molecular cloud material near newly formed stars, there is also a dramatic impact of star formation on the spatial structure and dynamics of molecular clouds. Fig. 3 shows the structure of the molecular cloud over about 4 orders of magnitude in spatial scale. The data in Fig. 3 very clearly indicate that the molecular cloud is highly filamentary or clumpy, with many bubbles and cavities. On the largest scales (1–30 pc) sampled by the ^{13}CO survey of Bally et al. (1987) clumps have volume densities of 10^4 hydrogen molecules per cm^3 or more and a volume filling factor less than 10%. The cloud has an overall V-shape with its apex pointing toward the Orion OB associations.

Bally et al. conclude that these features are evidence for strong dynamical effects of the OB associations on the cloud structure. It is likely that also the lower mass stars contribute to this process, as T-Tau stars and mass outflow sources are present throughout the cloud (Fukui 1989). In a scenario originally proposed by Norman and Silk (1980), and further discussed, for instance, by Fukui (1989) outflows and winds of the young embedded stars stir up the cloud, thereby maintaining or contributing to the cloud's "turbulence" that prevents large scale collapse.

Structure and clumpiness can be traced all the way to the smallest scales ($\sim 3 \times 10^{15}$ cm or 10^{-3} pc) that are observable (Fig. 3). The molecular gas near the interface to the Orion A HII region forms a narrow ridge with embedded massive clumps of size ≤ 0.1 pc ($M \sim 10 \rightarrow 10^2 M_\odot$, Fig. 3 middle, Mundy et al. 1988). The densest and most massive of these clumps (the BN-KL "hot core") again can be resolved into a large number of $\leq 1''$ diameter sub-clumps of density $\sim 2 \times 10^7$ cm^{-3}

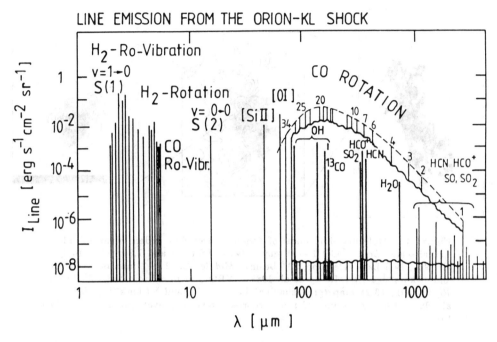

Figure 2. Schematic infrared, submillimeter and millimeter spectrum toward the central 0.1 pc of the Orion A/BN-KL star forming region (from Watson 1982).

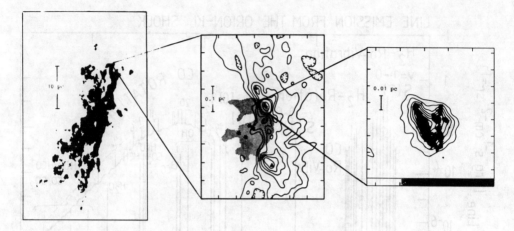

Figure 3. Molecular line maps of Orion A over four orders of magnitude in spatial scale. *Left:* ^{13}CO $1 \to 0$ map between LSR 6.5 and 7.5 km s^{-1} (Bally et al. 1987, Bell Labs telescope, 90″ beam). *Middle:* CS $2 \to 1$ map between -0.2 and 18 km s^{-1} LSR (Mundy et al. 1988, OVRO mm interferometer, 7.5″ beam). *Right:* NH$_3$ (3,2) map (grey shading) between 7.4 and 8.6 km s^{-1} (Migenes et al. 1989, VLA, 1.2″ beam). Contours represent a velocity averaged map, smoothed to 2″.

(Fig. 3 right, Migenes et al. 1989).

2.4. THE RELATION BETWEEN DENSE GAS AND STAR FORMATION

Star formation has already taken place in several of the 0.1 pc size clumps, suggesting that the large external pressure exerted by radiation, (expanding) ionization fronts and/or winds of the θ^1 OB stars have triggered a further generation of massive star formation (see below, Elmegreen and Lada 1977, Klein et al. 1987). The total mass and density of stars in the vicinity of the θ^1 association, including the large numbers of lower mass stars contained in the Orion-nebula or Trapezium Cluster ($> 10^3$ M$_\odot$ pc^{-3}) correspond to a mean density of 10^5 hydrogen nuclei per cm^3 or greater, quite comparable to the mean density in the nearby molecular cloud (Herbig and Terndrup 1986). Star formation in the vicinity of the θ^1 association thus must have been proceeding with reasonably large efficiency ($\eta_* = M_*/M_{\rm gas}$ = several tens of percent). In contrast the star formation efficiency averaged over large spatial scales in clouds in the disk of our Galaxy is known to be no more than a few percent (Myers et al. 1986).

E. Lada (1990) has recently presented a survey of molecular and near-infrared emission in the Orion B cloud that represents a key result for the understanding of star formation in molecular clouds. Her study shows that the number of 2 μm sources (as a measure of content of young embedded stars) has four peaks coincident with or very close to the locations of four of five prominent CS mm-emission peaks (as a measure of column density of dense gas) (Fig. 4). As in Orion A, the high density (> a few hundred M$_\odot$ pc^{-3}) of the young stellar associations suggests efficient stellar formation there. There is no other concentration of stars elsewhere in the Orion B cloud. The number of 2 μm sources found outside the major concentrations in fact is consistent with the background level not associated with the cloud. Star formation in the Orion B molecular cloud appears to proceed essentially only in the dense CS clumps. One obvious possible interpretation of this important result by Lada is a strong density dependence of both star formation efficiency and rate. It is tempting to speculate that star burst galaxies form efficiently stars exactly because they have (somehow) managed to form very dense molecular clouds at their nuclei. The four CS/2 μm peaks are also the locations of young OB-stars. This brings up the question whether the density peaks formed first, resulting in efficient star formation of all stellar types including the OB stars; or whether the OB stars formed first, compressing in a second step the gas in their vicinity leading to further star formation; or whether we have to consider a combination of both.

Molecular line mapping (Blitz and Stark 1986, Stutzki and Güsten 1990) and observations of the 158 μm [CII] fine structure line (Stutzki et al. 1988, Howe et al. 1991) suggest that clouds near OB star formation sites have a large clump to interclump density contrast (> 10). UV radiation may thus penetrate several parsec into these clumpy clouds, heat and compress the gas, create photodissociated "skins" of warm gas and perhaps destroy small dust grains (Genzel et al. 1989).

Another interesting question is the relationship between the mass spectrum of clumps and the initial mass function of stars. Stutzki and Güsten (1990) find that the C^{18}O 2→1 emission in the dense molecular ridge at the interface to the M17 HII region can be well described by the sum of about 180 gaussian shaped, approximately virialized clumps of size $\sim$ 0.1 pc. The clumps have a mass spectrum $dN/dM \propto M^{-\alpha}$ with $\alpha \sim 1.7$. The mass spectrum in M17 is similar to that found

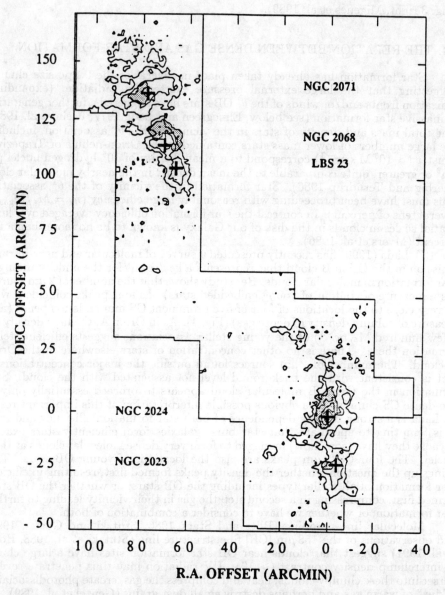

Figure 4. Locations of the embedded infrared clusters and dense CS cores in the L1630 molecular cloud (from Lada 1990). The locations and extents of the embedded stellar clusters are shown as shaded regions and the distribution of dense gas is presented as intensity contours of CS $(2 \to 1)$ emission. In addition, the peak intensity positions of the 5 most massive CS cores ($M > 200\,M_\odot$) are represented by crosses.

for the Rosette molecular cloud by Blitz (1987) and for ρ Oph by Loren (1989). A power law spectrum with $\alpha \approx 1.5$ is the plausible result of an equilibrium between coagulation and fragmentation (Spitzer 1982). It may also be consistent with a Salpeter type stellar mass spectrum ($\alpha = -2.35$) as the eventual outcome of that fragmentation if the fraction of the clump's mass that ends up in the star decreases with increasing mass (Zinnecker 1989).

The M17 clumps are massive enough that magnetic fields may not be able to support them from gravitational collapse on a free-fall time scale ($\tau_{\rm ff} \approx (3G\, m({\rm H}_2)\, n({\rm H}_2))^{-1/2} \leq 10^5$ y in M17). The maximum mass that can be supported by magnetic fields is $M_{\rm cr} \approx 7\, M_\odot\, [n({\rm H}_2)/2 \times 10^5\, {\rm cm}^{-3}]^{1/2} [R/0.1\, {\rm pc}]^2$ (cf., Shu et al. 1987), assuming that the magnetic field scales with density as $B = 2\,\mu{\rm G}\, [n({\rm H}_2)]^{1/2}$ (Myers and Goodman 1988). The M17 clumps of Güsten and Stutzki (size $\leq 0.1\,{\rm pc}$, mean hydrogen densities $2 \times 10^5\,{\rm cm}^{-3}$, local densities $\leq 10^6\,{\rm cm}^{-3}$) have masses of 10 to $10^3\, M_\odot$, larger than $M_{\rm cr}$. One way out may be if the clumps actually have substructure on much smaller spatial scales but also much larger density. The field can then support the clumps against collapse. If most of the M17 clumps are "supercritical" the M17 molecular ridge may be the current formation site of a very dense (bound?) stellar cluster.

2.5. THE MAGNETIC FIELD

Optical polarization measurements of magnetically aligned dust grains indicate that the magnetic field component in the plane of the sky is within 30° of the long axis of the Orion A molecular cloud (Vrba et al. 1988). Hence, the Orion A cloud cannot have contracted to its present elongated or flattened shape along magnetic field lines. Rather, the magnetic field may support the elongated structure of the Orion A cloud. Heiles and Stevens (1986) find that the line of sight magnetic field component, as determined from radio OH and HI Zeeman observations, reverses its direction across the cloud, pointing toward the Sun on the side of the cloud towards the Galactic plane and away from the Sun at lower Galactic latitudes. A possible scenario is a helical geometry in which the magnetic field wraps around the cloud (Bally 1989) with its largest component along the cloud's long axis. As indicated by the large scatter of polarization position angles, the orientation of the magnetic field on small scales does not necessarily trace the large-scale average orientation discussed above. Near-IR and far-IR polarization measurements in BN-KL indicate that the magnetic field is approximately along the long axis of the local outflow (Dyck and Lonsdale 1979, Dragovan 1986). On the other hand, the directions of several optical jets/outflows are aligned with the large-scale field (Strom et al. 1986) which suggests that the field plays an important role in the presently ongoing star formation. Line of sight field strengths for several positions range from 50 to 125 μG in the large beam OH/HI observations (see Genzel and Stutzki 1989 for references). The field strength inferred for the OH masers in BN-KL is a few mG (Hansen et al. 1977) and 40 mG in the H_2O maser spots (Fiebig and Güsten 1988).

2.6. A CLOSE-UP VIEW OF THE ENVIRONMENT OF MASSIVE YOUNG STARS:

The BN-KL Area

The BN-KL region is one of the best studied regions of recent massive star formation. There is not enough space to do justice to the many beautiful measurements of the last decade and the reader may want to turn to Genzel and Stutzki (1989) for a recent detailed review. Briefly, the observations show

- the presence of at least two *embedded, massive stars* (the Becklin-Neugebauer object [BN] and IRc2) with a combined luminosity of $\sim 10^5$ $L_\odot$, probably along with a compact cluster of lower luminosity/lower mass stars. The BN-IRc2 or BN-KL cluster emits most of its energy at $\lambda \geq 10\,\mu m$.
- *several components of dense molecular gas*, ranging from quiescent gas that is probably the remnant of the cloud out of which the BN-KL infrared cluster has formed to hot, dynamically active material in outflowing and shock-excited gas clumps. The different gas components are not only distinct in their physical and kinematic characteristics but also show clearly separate chemical compositions (Blake et al. 1987). The chemical characteristics of the different gas components are probably a direct consequence of their physical parameters; for instance, gas in the immediate vicinity of IRc2 shows evidence for recent ($\sim 10^4$ y) evaporation of saturated molecules off grain mantles (Walmsley et al. 1987).
- the existence of *very dense* ($n \sim 10^{10}\,cm^{-3}$) *circumstellar environments* around BN and IRc2. Both BN as well as IRc2 have a compact circumstellar HII region indicating that BN and IRc2 have hot ($\sim 30,000\,K$) central stars that are fueled by nuclear burning. Very dense molecular gas at $\sim 10\,AU$ from BN may come from a circumstellar disk (Scoville et al. 1983).
- the great *importance of interaction* between the stars, their circumstellar environment and the surrounding cloud. The outflows have cleared out a "cavity" of diameter $\sim 0.03\,pc$ surrounded by very dense clumps. The clumps have very high column densities ($10^{24}\,cm^{-2}$) so that the appearance of the infrared nebula is strongly affected by the distribution of the surrounding cloud material. As the outflows hit the cloud they create CH_3OH, H_2O, and OH masers and a zone of shock excited gas that emit intense infrared and submillimeter line emission.

3. Tools for Studying Dense Interstellar Clouds

3.1. EXCITATION AND INTERPRETATION OF INFRARED AND RADIO SPECTRAL LINES

In the following section, we will discuss the basic physics of radiative transport, excitation and interpretation of infrared/submillimeter and millimeter line, and continuum radiation.

a) Simple Radiative Transport

Consider radiative transport of a line at frequency ν (wavelength λ) in a plane-parallel cloud. The change of specific line intensity I_ν [ergs s^{-1} cm^{-2} Hz^{-1} sr^{-1}] along an infinitesimal element dz along the line of sight through the cloud is given by

$$\frac{dI_\nu}{dz} = \frac{h\nu}{4\pi}\phi(\nu)n_u A_{ul} - \frac{h\nu}{4\pi}\phi(\nu)I_\nu(n_u B_{ul} - n_l B_{lu})$$
$$= \epsilon_\nu - \kappa_\nu I_\nu. \qquad (1)$$

n_u, n_l are the volume densities of molecules in the upper and lower states of the transition, the former with statistical weight g_u and latter g_l. The Einstein Coefficient A_{ul} [s^{-1}] is the rate of decay of n_u by spontaneous radiative transitions. The line shape function $\phi(\nu)$ gives the probability per frequency interval that a photon is emitted at ν ($\int \phi(\nu)\,d\nu = 1$). It is easy to see that the first term in equation (1) represents the volume emissivity ϵ_ν, that is, the energy emitted per cm^3, sec, frequency interval and solid angle element through spontaneous emission. The second term takes into account the combined effects of absorption and stimulated emission. $B_{ul}I_\nu$ and $B_{lu}I_\nu$ [s^{-1}] give the corresponding rates per sec per molecule for stimulated radiative processes ($B_{ul} = \frac{2h\nu^3}{c^2}A_{ul}$ and $B_{ul}/g_l = B_{lu}/g_u$). The stimulated processes are characterized by the *absorption coefficient* κ_ν [cm^{-1}], or *optical depth* τ_ν, given by

$$\tau_\nu(z) = -\int_0^z \kappa_\nu\,dz = \int_0^z \frac{\phi(\nu)A_{ul}c^2}{8\pi\nu^2}\left(\frac{g_u n_l}{g_l n_u} - 1\right)n_u\,dz. \qquad (2)$$

Another important quantitity is the source function Σ_ν, defined as

$$\Sigma_\nu = \frac{\epsilon_\nu}{\kappa_\nu} = \frac{2h\nu^3}{c^2}\left[\frac{g_u n_l}{g_l n_u} - 1\right]^{-1} = \frac{2h\nu^3}{c^2}\left[\exp\left(\frac{h\nu}{kT_{\rm ex}}\right) - 1\right]^{-1}. \qquad (3)$$

The so defined *excitation temperature* $T_{\rm ex}$ is an equivalent Planck temperature that describes the population of states l and u through a thermal (Boltzmann) population. With (3) and (2) and an assumed spatially *constant* population (constant Σ_ν, $T_{\rm ex}$) equation (1) can be easily integrated to give the familiar equation

$$I_\nu({\rm observed}) = I_\nu({\rm background})\,e^{-\tau_\nu} + \Sigma_\nu(1 - e^{-\tau_\nu}). \qquad (4)$$

τ_ν is now the optical depth at ν through the cloud. A standard application is the case when the background source emits independent of frequency (continuum intensity I_B) and the emission of the cloud is pure line radiation. In that case the line intensity in excess of the continuum ΔI_{line} is

$$\Delta I_{\text{line}} = I_\nu(\text{observed}) - I_B = [\Sigma_\nu(T_{\text{ex}}) - I_B][1 - \exp(-\tau_\nu)]. \tag{5}$$

In real astronomical measurements one often deals with sources that do not fill the telescope beam. In that case equation (5) is modified by beam area filling factors Φ_{AL}, Φ_{AB} for line and background continuum respectively,

$$\langle \Delta I_{\text{line}} \rangle_{\text{beam}} = (\Phi_{\text{AL}}\Sigma_\nu(T_{\text{ex}}) - \Phi_{\text{AB}}I_B)\langle(1 - \exp(-\tau_\nu)\rangle. \tag{6}$$

Brackets $\langle\rangle$ indicate averages over the beam. The line appears in emission, as long as $\Phi_{\text{AL}}\Sigma_\nu > \Phi_{\text{AB}}I_B$. Otherwise it is in absorption. For optically thin ($\tau_\nu \ll 1$) emission, $1 - e^{-\tau_\nu} \to \tau_\nu$; for optically thick emission $1 - e^{-\tau_\nu} \to 1$.

In the radio region one usually has $h\nu \ll kT$ so that the *Rayleigh-Jeans approximation* ($\exp(h\nu/kT) \approx 1 + h\nu/kT$) is applicable. In that case it is convenient to express intensities in terms of *Rayleigh-Jeans radiation temperatures* $\widetilde{T}_R$ [K],

$$I_\nu = \frac{2h\nu^3}{c^2}\left(\exp\left(\frac{h\nu}{kT_R}\right) - 1\right)^{-1} \approx \frac{2k\nu^2}{c^2}T_R \quad \text{and,}$$

$$\Sigma_\nu \approx \frac{2k\nu^2}{c^2}T_{\text{ex}} \quad \text{for } h\nu/k \ll T_R, T_{\text{ex}}, \tag{7}$$

so that equation (6) turns into

$$\langle \Delta T_{\text{line}} \rangle = (\Phi_{\text{AL}}T_{\text{ex}} - \Phi_{\text{AB}}T_B)\langle 1 - \exp(-\tau_\nu)\rangle. \tag{8}$$

Equation (2) becomes

$$\tau_\nu = \frac{hc^2}{8\pi\nu^2 k}A_{\text{ul}}\left(\frac{\nu}{\Delta\nu_0}\right)\left(\frac{N_u}{T_{\text{ex}}}\right) \quad \text{for } h\nu \ll kT_{\text{ex}}. \tag{9}$$

$N_u = \int_0^z n_u\, dz$ is the *column density* [cm^{-2}] of particles in the upper state through the cloud, $\Delta\nu_0$ is the equivalent width of the line, $\phi(\nu) = 1/\Delta\nu_0$.

b) Optical Depths of Infrared and Radio Lines

The Einstein A_{ul} coefficient and the transition matrix element μ_{ul} of a transition u→l are related through

$$A_{ul} = \frac{64\pi^4 \nu^3}{3hc^3} \mu_{ul}^2. \tag{10}$$

Because of the increasing number of phase space modes with increasing ν, A_{ul} increases proportional to ν^3 making shorter wavelength lines typically stronger than longer wavelength lines. Another important aspect is whether transitions are allowed or forbidden in terms of electric dipole selection rules. Note that the transition moment of rotational transitions is proportional to the permanent electric dipole moment and the proportionality constant is of order unity. Many rotational lines in the millimeter and submillimeter range of molecules with permanent electric dipole moments have relatively large transition moments of about 1 Debye (10^{-18} cgs). One exception is CO which has an electric dipole moment of only 0.1 Debye. CO thus radiates about 100 to 1000 times slower than most other abundant heavy top molecules. Homonuclear molecules such as H_2, the most abundant interstellar molecule, have only electric quadrupole transitions that are more than 10 orders of magnitude slower than electric dipole transitions at the same wavelength. Rovibrational transitions of molecules in the mid-IR have transition moments of ~ 0.1 Debye, again leading to at least 100 times slower radiation rates than pure rotational transitions at the same wavelength. Finally, fine structure transitions of atoms and ions in the infrared and submillimeter range are magnetic dipole transitions that are typically 10^5 times slower than corresponding electric dipole transitions.

While magnetic dipole and electric quadrupole transitions radiate more slowly, they also have lower optical depths ($\tau \propto A/\nu^2$, equation (2)) and – as will be discussed below – they are easier to excite by collisions.

In addition to the Einstein coefficient, the populations of the upper and lower levels are important for the evaluation of the optical depth. We can write equation (2) as

$$\tau = \frac{c^2}{8\pi} \left(\frac{A}{\nu^3}\right) \left(\frac{\nu}{\Delta\nu_0}\right) (\exp[h\nu/kT_{ex}] - 1) N_u \tag{11}$$

In thermal equilibrium at temperature T the column density N_u can be expressed as

$$N_u = g_u \exp(-E_u/kT) N/Q(T). \tag{12}$$

$Q(T) = \sum_{\text{all levels } i} g_i \exp(-E_i/kT)$ is the partition function. $Q(T) = kT/hB$ for linear molecules of rotational constant B in *thermal equilibrium*. E_i is the energy of level i and N the total number of molecules (atoms) per cm^2. In astronomical applications high-J states are usually subthermally populated. Realistic partition functions thus increase more slowly with temperature than in thermal equilibrium. $Q(T) = $ const is often a good approximation. For most astronomical applications the line width $\Delta\nu_0$ is determined by Doppler shifts of velocity width $\Delta v = c\Delta\nu_0/\nu$. For infrared transitions $h\nu > kT_{\rm ex}$ and $N_{\rm u} \ll N_{\rm l} \approx N$ so that

$$\tau_{\rm IR} \propto \mu_{\rm ul}^2 N/(\Delta v Q(T)), \tag{13a}$$

while for radio transitions ($h\nu/kT_{\rm ex} \ll 1$)

$$\tau_{\rm rad} \propto \mu_{\rm ul}^2 \frac{h\nu_{\rm rad}}{kT_{\rm ex}} N/(\Delta v Q(T)). \tag{13b}$$

The ratio of infrared to radio optical depths of a given molecule (atom) then is simply

$$\tau_{\rm IR}/\tau_{\rm rad} \approx (\mu_{\rm IR}/\mu_{\rm rad})^2 kT_{\rm ex}/h\nu_{\rm rad}. \tag{14}$$

For typical warm interstellar clouds $kT_{\rm ex}/h\nu_{\rm rad} \sim 10\rightarrow 100$. Rotational transitions of molecules in the infrared/submillimeter range have thus one to two orders of magnitude greater optical depths than millimeter and centimeter transitions. For abundant molecules, such as CO, HCN, NH_3, H_2O etc., far-infrared and submillimeter lines in dense molecular clouds can be very optically thick ($\tau \sim 10^2\rightarrow 10^3$). For ro-vibrational transitions in the mid-IR $(\mu_{\rm IR}/\mu_{\rm rad})^2 \approx 10^{-2}$ so that they have similar optical depths as radio transitions.

c) Rotational and Ro-Vibrational Emission of Simple Linear Molecules

Diatomic and linear multiatomic molecules have a simple ladder of rotational energy states E_J depending on rotational quantum number J as (to first order)

$$E_J = hBJ(J+1), \tag{15}$$

where B [Hz] is the rotational constant. B is inversely proportional to the [reduced] mass of the molecular system. Electric dipole transitions in molecules with a permanent electric dipole moment are allowed between adjacent states ($J \to J-1$) and occur at frequency

$$\nu_J = 2BJ. \tag{16}$$

In the Rayleigh-Jeans regime, the optical depth at ν_J is proportional to $(\exp[h\nu_J/kT_{\rm ex}] - 1) N_J \sim \nu_J g_J \sim J^2$. The integrated intensity and brightness temperature of spectral lines at ν_J in the optically thin and optically thick limits then are

$$\left. \begin{array}{l} I_J = \dfrac{2k\nu^2}{c^2} T_{\rm ex}\tau(\nu_J)\Delta\nu \propto J^5 \\ T_J = T_{\rm ex}\tau(\nu_J) \propto J^2 \end{array} \right\} \quad \tau(\nu_J) \ll 1$$

and
$$\left.\begin{array}{l} I_J = \dfrac{2k\nu_J^2}{c^2}T_{ex}\Delta\nu \propto J^3 \\ T_J = T_{ex} \end{array}\right\} \quad \tau(\nu_J) \gg 1$$

[for $h\nu_J/kT_{ex} < 1$]. (17)

The maximum emission in a rotational ladder populated at temperature T_{rot} occurs approximately at $E_J = hBJ(J+1) \approx kT_{rot}$, or

$$J_{max} \approx \sqrt{\dfrac{T_{rot}}{hB/k}}. \tag{18}$$

The abundant CO molecule ($hB/k = 2.76$ K), for instance, has its strongest line emission in the 1.3 mm $J = 2 \to 1$ transition for a cold cloud of $T = 10$ K. A warm cloud ($T = 50$ to 100 K), on the other hand, emits strongest in the mid-J ($J = 4$ to 7) transitions at submillimeter wavelengths. Finally, for CO line emission to be strongest at far-IR wavelengths ($\lambda \approx 50$ to $200\,\mu$m, $J = 20$ to 40), temperatures of about 1000 K are required.

In the case of ro-vibrational transitions of linear molecules energy states are labeled by the vibrational (v) and rotational (J) quantum numbers

$$E(v, J) = \hbar\omega_0(v + \dfrac{1}{2}) + hBJ(J+1) + \text{corrections}(v, J). \tag{19}$$

The correction terms are due to *centrifugal stretching* and Coriolis coupling of the rotational and vibrational motions. In the simplest approximation ro-vibrational lines in electric dipole transitions ($v \to v - 1$, $J \to J \pm 1$) are found at

$$\nu_{v, J \to J\pm 1} = \omega_0 \mp 2hB(J+1). \tag{20}$$

Overtone transitions at $2\omega_0$, $3\omega_0$ are also possible due to deviations from a pure harmonic potential, but have substantially weaker transition moments. As mentioned before the important molecular hydrogen molecule (H_2) does not have any electric dipole, but only the much weaker *electric quadrupole transitions* with selection rules ($v \to v - 1$, $J \to J, J \pm 2$). Other molecules, such as OH and CH have non-zero electronic angular momentum or spin that makes their spectra more complicated.

While the principles are still the same as for linear molecules, non-linear molecules (asymmetric and symmetric tops) have more complicated spectra due to the presence of more than one rotation and vibration axis. The larger number of levels also results in a richer line spectrum. Simple cases are symmetric tops such as NH_3 and CH_3CN, where at least two moments of inertia are identical. Energy levels are labeled by the total rotational quantum number J and its projection onto the molecular symmetry axis K. Rotational transitions then occur at $\Delta J = \pm 1$ and $\Delta K = 0$. Table I gives a selected list of interstellar molecules, along with their rotational constants, permanent electric dipole moments and vibrational frequencies. Further details of molecular spectra can be found in the classical references of molecular spectroscopy (Herzberg 1948, Townes and Schawlow 1955).

Table I. Selected list of molecular lines

Molecule	Lowest rotational transition	Rotational dipole moment [Debye][a]	Vibrational transitions [μm]
H_2	28.1 μm	0	2.2
CO	2.6 mm	0.11	4.8
OH	119 μm	1.7	2.7
CS	6.11 mm	2	7.8
HCN	3.4 mm	3	3, 4.9, 14
H_2O	540 μm	1.85	2.7, 6.3
NH_3	524 μm	1.5	2.9, 6.1, 10.5

Notes: a 1 $Debye = 10^{-18}$ e.s.u. cm

d) Fine Structure Lines of Atoms and Ions

Atoms and ions with 1, 2, 4 and 5 p electrons in their outermost shell have fine structure in their electronic ground state as a result of the magnetic coupling between orbital angular momentum and electronic spin. The splitting is typically 0.01 to 0.1 eV, resulting in fine structure transitions in the 10 to 100 μm region. They are magnetic dipole transitions with transition probabilities typically $\alpha^2 = (1/137)^2$ or 4 orders of magnitude smaller than those of electric dipole transitions. Fine structure lines are, therefore, usually optically thin ($\tau \approx 10^{-2}$ to a few) and are relatively easily excited in dense interstellar clouds by collisions with hydrogen atoms and molecules (in neutral clouds), or protons and electrons (in ionized or partially ionized regions). Table II is a list of wavelengths of the most important infrared and submillimeter fine structure lines, along with their Einstein A coefficients, ionization potentials and critical densities (see below). In the optically thin limit the integrated line flux (erg s^{-1} cm^{-2}) of a fine structure line can be written as

$$F_{\text{line}} = \frac{h\nu}{4\pi} A_{ul} \iint_{\text{Beam}} d\Omega \left(\int_{\text{Line of sight}} n_u \, dl \right). \tag{21}$$

Table II. Infrared fine structure lines in HI/H$_2$ regions[a]

Species	Excitation potential [eV]	Ionization potential [eV]	Transition	λ [μm]	A [s^{-1}]	$n_{\text{crit}}^{\text{b}}$ [cm^{-3}]
C		11.26	$^3P, J=1\to 0$	609.1354	7.9×10^{-8}	4.7×10^2
						1×10^4 for H$_2$
			$^3P, J=2\to 1$	370.415	2.7×10^{-7}	1.2×10^3
C$^+$	11.26	24.38	$^2P, J=\frac{3}{2}\to\frac{1}{2}$	157.7409	2.4×10^{-6}	2.8×10^3
						5×10^3 for H$_2$
						50 for electrons
O		13.62	$^3P, J=1\to 2$	63.18372	8.95×10^{-5}	$4.7\times 10^5\, T_{300}^{-1/2}$ [c]
						$7\times 10^5\, T_{300}^{-1/2}$ [d]
			$^3P, J=0\to 1$	145.52547	1.7×10^{-5}	$9.5\times 10^4\, T_{300}^{-1/2}$
						$\geq 1\times 10^5\, T_{300}^{-1/2}$ [d]
Si		8.15	$^3P, J=1\to 0$	129.68173	8.25×10^{-6}	2.4×10^4
			$^3P, J=2\to 1$	68.473	4.2×10^{-5}	8.4×10^4
Si$^+$	8.15	16.35	$^2P, J=\frac{3}{2}\to\frac{1}{2}$	34.816	2.1×10^{-4}	3.4×10^5
S		10.36	$^3P, J=1\to 2$	25.249	1.4×10^{-3}	1.9×10^6
			$^3P, J=0\to 1$	56.309	3.0×10^{-4}	7.2×10^5
Fe		7.87	$^5D, J=3\to 4$	24.04	2.5×10^{-3}	3.1×10^6
			$^5D, J=2\to 3$	34.71	1.6×10^{-3}	3×10^6
Fe$^+$	7.87	16.18	$^6D, J=\frac{7}{2}\to\frac{9}{2}$	25.99	2.1×10^{-3}	2.2×10^6
			$^6D, J=\frac{5}{2}\to\frac{7}{2}$	35.352	1.6×10^{-3}	3.3×10^6

Notes: a for additional fine structure lines see Schmid-Burgk (1982), see also Tielens and Hollenbach (1985a).
b if not otherwise specified, critical density for collisions with atomic hydrogen: $n_{\text{crit}} = A_{ul}/\gamma_{ul}$; collisional rates with H$_2$ may be different; see Monteiro and Flower (1987) for CI and OI excitation
c T_{300} = kinetic temperature in units of 300 K
d critical density for collisions with H$_2$

e) Collisional Excitation

i) Two Level System. Consider a simple system consisting of only two levels, l (lower) and u (upper), spaced by $\Delta E_{ul} = h\nu$. Assume further that transitions l$\to$u and u$\to$l are triggered by collisions with a collision partner of volume density n.

The collisions are described by rates per sec per molecule/atom C_{ul} and C_{lu},

$$C_{ul} \stackrel{\frown}{=} C_{u \to l} = n\gamma_{ul} = n \langle \sigma_{ul} v \rangle \quad \text{and}$$
$$C_{lu} \stackrel{\frown}{=} C_{l \to u} = C_{ul}\, g_u/g_l \exp(-h\nu/kT_{\text{kin}}). \qquad (22)$$

σ_{ul} is the cross section (cm^2) for a collision u→l at velocity v, γ_{ul} [cm^3 s^{-1}] is the overall collisional rate coefficient u→l. It can be expressed as an average over all possible collision energies weighted by the distribution function that is assumed to be a Maxwell-Boltzmann distribution at kinetic temperature T_{kin}.

$$\gamma_{ul} = \langle \sigma_{ul} v \rangle = \frac{4}{\sqrt{\pi}} \left(\frac{\mu}{2kT_{\text{kin}}} \right)^{\frac{3}{2}} \int_0^\infty dv\, (\sigma_{ul}(v) v)\, v^2 \exp\left(-\mu v^2/(2kT_{\text{kin}})\right), \qquad (23)$$

where μ is here the reduced mass of molecule and collision partner. Typical values for γ_{ul} for neutral-neutral collisions range between 10^{-11} and 10^{-10} cm^3 s^{-1}, while neutral-charged particle collisions have collisional rate coefficients of about 10^{-9} cm^3 s^{-1}. The average of equation (23) is expressed by the brackets in equations (22) and (23).

The rate coefficients γ are obtained by detailed quantum mechanical calculations (see Flower 1987 for a review). The second line of equation (22) follows from detailed balance. Rate coefficients are typically uncertain by a factor of 2 for atoms and sometimes as much as an order of magnitude for collisions of complex atoms with H$_2$.

Given rate coefficients γ_{ul}, Einstein coefficient A_{ul}, kinetic temperature T_{kin} and density n it is then straightforward to calculate the populations n_u and n_l from detailed balance,

$$n_l (\gamma_{lu} n) = n_u (\gamma_{ul} n + A_{ul}) \quad \text{or,}$$

$$\frac{n_u}{n_{\text{tot}}} = \frac{g_u/g_l \exp(-h\nu/kT_{\text{kin}})}{1 + g_u/g_l \exp(-h\nu/kT_{\text{kin}}) + n_{\text{crit}}/n}. \qquad (24)$$

Here $n_{\text{tot}} = n_u + n_l$ and $n_{\text{crit}} = A_{ul}/\gamma_{ul}$. Equation (24) is a key formula. It shows that for very high density ($n \gg n_{\text{crit}}$) the population of the levels u,l is thermalized at kinetic temperature T_{kin}: $(n_u/n_l)_{\text{thermal}} = g_u/g_l \exp(-h\nu/kT_{\text{kin}})$. Below the *critical density* n_{crit}, however, spontaneous radiation rates u→l are much faster than collisions so that each collision l→u leads to photon emission; the population is subthermal: $n_u/n_l = (n/n_{\rm crit})(n_u/n_l)_{\rm thermal}$. With (24) and (21) the line flux of a simple two level system in the optically thin limit can then be easily expressed as

$$F_{\rm line} = \frac{h\nu}{4\pi} A_{ul}\Omega \int_{\text{line of sight}} n_u\, dz$$

$n < n_{\rm crit}$:
$$= \frac{h\nu}{4\pi}\gamma_{ul}\Omega g_u/g_l \exp(-h\nu/kT_{\rm kin}) \int_{\rm los} n_{\rm tot} n\, dz \propto n^2$$

$n > n_{\rm crit}$:
$$= \frac{h\nu}{4\pi}A_{ul}\Omega g_u/g_l \exp(-h\nu/kT_{\rm kin}) \int_{\rm los} n_{\rm tot}\, dz \propto N_{\rm tot}. \quad (25)$$

Ω is the solid angle of source or beam. In the subthermal regime line fluxes are proportional to (density)^2 [$\hat{=}$emission measure] while they are proportional to column density in the thermalized regime. In this way optically thin line fluxes can be easily calculated from the information given in Table II. The necessary corrections to (24) for optically thick lines are discussed below in IIIh.

ii) Multilevel System. The extension to a multilevel system is straightforward. The rate equation equivalent to (24) for level u then becomes

$$n_u\left[\sum_{k\neq u}(n\gamma_{uk} + R_{uk}) + \sum_{k<u} A_{uk}\right] = \sum_{u\neq k} n_k\left[n\gamma_{ku} + R_{ku}\right] + \sum_{k>u} A_{ku}n_k. \quad (26)$$

For the purpose of generality we have added in addition to collisional excitation, deexcitation and spontaneous emissions also rates due to stimulated emission and absorption, R_{uk}, R_{ku} [s^{-1}], where

$$R_{uk} = B_{uk}\int_{4\pi}\frac{d\Omega}{4\pi}(I_{\rm line} + I_{\rm cont}). \quad (27)$$

$I_{\rm line}$, $I_{\rm cont}$ are the wavelength/frequency integrated intensities of line and continuum radiation the molecule/atom sees. This extension takes care of radiative excitations as well. The set of rate equations (26) for u = 1...u$_{\rm max}$ can be solved for given n, $T_{\rm kin}$, $I_{\rm cont}$ through a matrix inversion. As soon as the line intensity itself contributes to the excitation, however, the rate equations become nonlinear and have to be solved iteratively.

f) Spectral Lines as Probes of Physical Conditions in Interstellar Clouds

Simple considerations involving the upper state energies, wavelengths, and critical densities of various molecular and atomic transitions at radio and infrared wavelengths as given in Tables I and II, already give fairly good first order impressions

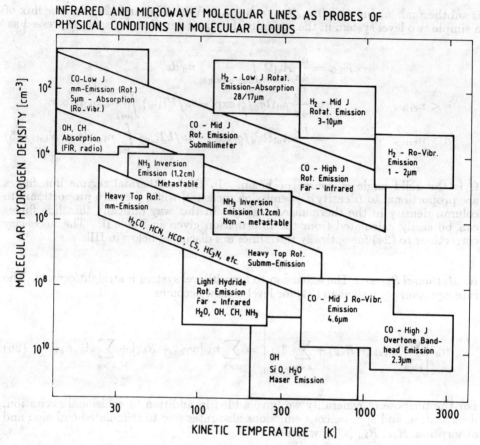

Figure 5. Molecular lines as probes of physical conditions in molecular clouds

which lines are probes of what astrophysical environments (see also discussion by Scoville 1984). Fig. 5 shows how the phase space of different physical conditions is probed by different molecular transitions. Rotational lines of CO, for instance, require molecular hydrogen densities of about $4000\,J^3$ [cm^{-3}] and temperatures of $2.7\,J(J+1)$ [K] for producing "strong" line emission in the $J \to J-1$ transition. The $1\to 0$ and $2\to 1$ millimeter CO transitions are excited in gas of molecular hydrogen density 10^3 to 10^4 cm^{-3} and temperatures 10 to 20 K. Mid-J CO submillimeter lines sample gas of hydrogen density a few 10^5 cm^{-3} and temperature ~ 100 K. Those values span the typical range of physical parameter in molecular clouds. Far-infrared CO lines require a few 10^6 cm^{-3} and 1000 K for thermal population, conditions that are found in interstellar shocks. The same parameter space as rotational emission lines is sampled by ro-vibrational lines in absorption against a hot background continuum source (e.g. the line of sight toward a newly formed star). CO ro-vibrational *emission* lines, on the other hand, require densities of 10^{10} to 10^{11} cm^{-3} and temperatures of a few 10^3 K, conditions that are typical of stellar atmospheres and envelopes. Other heavy top molecules, such as HCN, CS, SiO and HCO$^+$ are similar to CO but require typically 10^2 to 10^3 times greater densities for collisional excitation of their rotational states because of their greater electric dipole moments. As mentioned above, the H$_2$ molecule is a special case as it emits only through electric quadrupole transitions. As a result its rotational transitions are thermalized at very low densities $\sim 10^2 J^3$ cm^{-3} and even the 2 μm ro-vibrational transitions require only densities of about 10^5 cm^{-3}. As the H$_2$ transitions occur in the mid- and near-IR, only hot molecular gas with temperatures between a few hundred and a few thousand K emit observable line radiation. Such physical conditions are typical again of shocks or of UV heated gas at the surfaces of molecular clouds. Submillimeter and far-infrared transitions of light hydrides (H$_2$O, OH, CH, NH$_3$) are ideal probes of warm ($\sim 10^2$ K) and very dense gas (10^8 to 10^{10} cm^{-3}) that may sample the envelopes and disks of newly formed stars as well as the cooler downstream zones of shock fronts.

The critical densities of most infrared and submillimeter fine structure lines range between 10^2 and 10^6 cm^{-3}, depending mainly on wavelength ($n_{\text{crit}} \sim \lambda^{-3}$). Neutral species, such as O^0, Si^0, C^0, S^0, and ionized species with ionization potentials less than that of atomic hydrogen, such as Si$^+$, C$^+$ and Fe$^+$ sample UV-excited or shock excited, warm atomic gas at cloud surfaces between ionized (HII) zones and fully molecular regions.

g) Radiative Excitation

In addition to excitation by collisions, the rate equations (26) contain radiative excitations. It is immediately clear that radiative "pumping" (by a background continuum source, for instance) creates only net line photons when pump frequency and line frequency are different. Pumping at the line frequency itself is a scattering process that can only redistribute the continuum radiation in frequency space (e.g., P-Cygni profiles), special geometries not considered. Radiative pumping can thus be important for molecules with many interconnecting transitions (such as asymmetric rotors) and molecules that have strong mid- and far-infrared transitions where continua are intense (e.g., light hydrides).

As an example, consider a molecular cloud illuminated by a thermal background source of temperature T_d and continuum optical depth τ_d that subtends a solid angle Ω_d as seen from the cloud. The solid angle averaged radiative rate in equation (27) then becomes

$$R_{ul} = \frac{B_{ul}}{4\pi} \int d\Omega \, I_{back} = \frac{B_{ul}}{4\pi} \left[\frac{2h\nu^3}{c^2} \Omega_d \frac{(1-\exp(-\tau_d))}{(\exp(h\nu/kT_d)-1)} \right]$$
$$\widehat{=} B_{ul} \frac{2k\nu^2}{c^2} \widetilde{T}_{\text{eff}}. \tag{28}$$

$\widetilde{T}_{\text{eff}}$ is the equivalent Rayleigh-Jeans radiation temperature of the continuum source averaged over all angles. In the limit of the two level system of e,i) the line source function then is

$$\Sigma_{\text{line}} = \frac{2h\nu^3}{c^2} \left[\frac{g_u n_l}{g_l n_u} - 1 \right]^{-1} = \frac{2h\nu^3}{c^2} \left[\frac{g_u}{g_l} \left(\frac{A_{ul} + R_{ul} + C_{ul}}{R_{lu} + C_{lu}} \right) - 1 \right]^{-1} \tag{29}$$

Defining an equivalent Rayleigh-Jeans excitation temperature via $\Sigma_{\text{line}} = 2k\widetilde{T}_{\text{ex}}\nu^2/c^2$ we have

$$\widetilde{T}_{\text{ex}} = \frac{\widetilde{T}_{\text{eff}} + (h\nu/k)(n/n_{\text{crit}})}{1 + (n/n_{\text{crit}}) h\nu / \left(k\widetilde{T}_{\text{kin}} \right)}, \tag{30}$$

where $\widetilde{T}_{\text{kin}} = h\nu/k[1-\exp(-h\nu/kT_{\text{kin}})]^{-1}$. Equation (30) is a general formulation of excitation through radiation and collisions. In the Rayleigh-Jeans limit ($h\nu/kT \ll 1$) $\widetilde{T}_{\text{ex}} \to T_{\text{ex}}$, $\widetilde{T}_{\text{eff}} \to T_{\text{eff}}$ and $\widetilde{T}_{\text{kin}} \to T_{\text{kin}}$.
For pure radiative excitation ($n \to 0$)

$$\widetilde{T}_{\text{ex}} = \widetilde{T}_{\text{eff}}, \tag{31}$$

a well known result that states that without internal collisions the excitation temperature equilibrates to the effective radiation temperature (e.g., the 3 K cosmic background temperature). Note that for $n = n_{\text{crit}}$ and $h\nu/k > \widetilde{T}_{\text{eff}}$

$$\widetilde{T}_{\text{ex}} = \frac{\widetilde{T}_{\text{eff}} + h\nu/k}{1 + h\nu/k\widetilde{T}_{\text{kin}}} = \left((h\nu/k)^{-1} + \left(\widetilde{T}_{\text{kin}}\right)^{-1} \right)^{-1} \tag{32}$$

This means that at $n = n_{\text{crit}}$ an infrared transition has a relatively higher excitation temperature than a radio/mm transition.

Examples of radiative pumping for line emission in massive star formation regions are the excitation of far-IR transitions of OH (Melnick et al. 1990) and of vibrationally excited HCN (Ziurys and Turner 1986) in Orion-KL.

h) Line Trapping

The discussion up to now has not taken into account the effect of line optical depth and trapping on the excitation that is also implicitly contained in equation (27).

In order to qualitatively understand the effect of optical depth, let us consider the simple case of a cloud (volume V) emitting line photons in a thermalized transition. If the cloud is optically thin, a photon emitted anywhere within the volume of the cloud will escape without being reabsorbed. In this case the cloud's total line emission per sec is proportional to the number of molecules in the cloud and the spontaneous emission rate, $I_{\text{thin}} \propto n_u V A$ [s^{-1}]. n_u is again the volume density of molecules in the upper state of the transition under consideration, and A the Einstein coefficient. If the optical depth of the cloud is greater than one as the result of larger molecular volume or column densities or smaller velocity/frequency gradients through the cloud, the probability for a once emitted photon to escape without reabsorption will be less than one. Roughly speaking, the escape probability β will scale inversely proportional to the optical depth τ, $\beta \approx 1/\tau$. Once a photon is reabsorbed in our simple model, it will be "destroyed" in a collisional deexcitation and not contribute to the net photon emission. The total line emission rate in the optically thick case is then $I_{\text{thick}} \propto n_u V A \beta$. So the effect of "line trapping" in essence scales the spontaneous emission rate by β,

$$A \to A\beta(\tau) \approx A/a\tau, \tag{33}$$

where $a \sim 1...5$ is a numerical factor. More accurate formulations of the dependence of β on τ depend on the velocity field and have been calculated by a number of authors (e.g., Goldreich and Kwan 1974, de Jong et al. 1980).

Another effect of line trapping is that it drives a level toward thermalization that would be subthermal in the optically thin case. This can be immediately seen from the fact that for a fixed collision rate C, the ratio of effective spontaneous emission rate to collision rate $A\beta/C = n_{\text{crit}}\beta/n$ is smaller for the optically thick ($\beta < 1$) than for the optically thin ($\beta = 1$) case. The generalization of the equation for line emission from a collisionally excited two level system (equations (24) and (25)) for the optically thick case thus is

$$F_{\text{line}} = h\nu/4\pi \, A\,\beta(\tau)\, N_{\text{mol}} \frac{g_u/g_l \exp(-h\nu/kT)}{1 + g_u/g_l \exp(-h\nu/kT) + n_{\text{crit}}\beta(\tau)/n}, \tag{34}$$

N_{mol} is the total column density of molecules (or atoms, as equation (34) applies to fine structure lines as well). For optically thick, thermalized emission ($n_{\text{crit}}\beta/n \ll 1$) the line flux scales with $A\beta$, as discussed above. In the case of subthermal, optically thick emission ($n_{\text{crit}}\beta/n > 1$ and $\beta < 1$), however, the line flux keeps growing linearly with N_{mol} since the β's in the numerator and denominator cancel each other. So an optically thick, but subthermal line behaves much like an optically thin and subthermal line. The difference is that the line trapping results in a greater population in the upper state and a higher excitation temperature T_{ex}. The situation is somewhat more complicated for multilevel systems.

i) Line profiles

Line profiles contain important physical information as well. Let us first consider the simple case of a homogeneous cloud with a Gaussian profile of FWHM width Δv_{local} representing the thermal line width or local random motions (microturbulence). The line profile as a function of velocity shift from line center v can be written as

$$I_{\text{line}} = I_0(T_{\text{ex}})[1 - \exp(-\tau(v))], \tag{35}$$

where $\tau(v) = \tau_0 \exp(-(v/\Delta v_{\text{local}})^2)$.

Increasing the line center optical depth τ_0 to greater than 1 then changes the profile from Gaussian to flat-topped with a width that depends logarithmically on τ_0 (cf., Phillips et al. 1979)

$$\Delta v_{\text{FWHM}} = \frac{\Delta v_0}{\sqrt{\ln 2}} \left[\ln \left\{ \frac{\tau_0}{\ln(2/(1 + \exp(-\tau_0)))} \right\} \right]^{1/2} \tag{36}$$

A line center optical depth of 10, for instance, results in a FWHM line width 1.96 times as large as in the optically thin case, $\tau_0 = 100$ results in 2.7 and 1000 in 3.24 times the width. With increasing optical depth the lines become more and more flat-topped and rectangular shaped.

The simple single Gaussian model often is not a good description of molecular cloud kinematics. The clouds are highly clumped (section II.3), with large scale motions between clumps significantly larger than the local motions within each clump. This will tend to decrease the dependence of line width on optical depth by approximately the factor $\Delta v_{\text{cloud}}/\Delta v_{\text{local}}$, where Δv_{cloud} is the FWHM width of the large scale motions. Martin et al. (1984) have calculated models of the line profiles in clumpy clouds in the simplified case that large scale motions (macroturbulence) and local line profile (microturbulence) can both be described by Gaussians. They have shown that such models can plausibly fit the ratios of line intensities and widths of various CO isotopic lines in the warm molecular cloud core M17. One of the basic results is that an observed ratio Y of an optically thicker to an optically thinner line with an abundance ratio Z can imply an optical depth of the optically thicker line *much greater* than Z/Y. The reason is that the filling factor of the thicker line is larger than that of the thinner. For instance, in most massive, warm molecular clouds in the disk of our Galaxy one finds a ^{12}CO/^{13}CO 1→0 line ratio of ∼ 5. With an assumed ^{12}CO/^{13}CO fractional abundance ratio of 50 (Wannier 1980), the simple single Gaussian model of eq. (35) implies $\tau(^{12}$CO 1→0$) \sim 10$ and $\tau(^{13}$CO$) < 1$. In the corresponding clumpy cloud model one finds a $\tau(^{12}$CO$) \sim 10^2$ and $\tau(^{13}$CO$) \geq 1$. One also expects that optically thick line profiles are much more rounded than for the single Gaussian model. This is in fact observed (Martin et al. 1984).

3.2. INFRARED AND SUBMILLIMETER CONTINUUM EMISSION

The main processes of infrared and submillimeter continuum emission are thermal radiation from small dust grains, free-free radiation in hot plasmas and synchrotron radiation. Of those the first process is by far the most important in star formation regions and will be briefly discussed here (for more extensive reviews, see Hildebrand 1983, Cox and Mezger 1989). Discussions of the latter two processes can be found in chapters 1 and 2 of the book "Galactic and Extragalactic Radio Astronomy" (Verschuur and Kellerman 1988).

Small (diameter $a \sim 0.01$ to $1\,\mu m$) dust particles in interstellar clouds are efficient absorbers of short-wavelength ($\lambda \leq a$) radiation. In equilibrium of heating and cooling they reemit a continuous spectrum that for $a > 0.01\,\mu m$ closely resembles a thermal spectrum characterized by a temperature T_d and a smoothly varying absorption/emission efficiency $Q(\nu)$ $Q(\nu) = \sigma(\nu)/\pi a^2$ where $\sigma(\nu)$ is the radiation absorption crosss section at ν. The observed flux density S_ν at ν of such a cloud with a line of sight optical depth $\tau_d(\nu)$ and solid angle Ω_ν is given by

$$S_\nu = \frac{2h\nu^3}{c^2}\left[\exp\left(\frac{h\nu}{kT_d}\right) - 1\right]^{-1}[1 - \exp(-\tau_d(\nu))]\,\Omega_\nu. \quad (37)$$

Optical depth $\tau_d(\nu)$ and $Q(\nu)$ are directly proportional to each other. The proportionality factor is the line of sight column density of dust particles. For moderately large grains ($\lambda \sim 0.1\,\mu m$) $Q(\nu)$ approaches order of unity in the ultraviolet and decreases steadily with increasing wavelength in the infrared and submillimeter range (with the exception of a few resonances in the near- and mid-IR, see Draine 1989). The dust temperature at any position in the cloud depends on the angle-averaged radiation energy flux J_ν and is given by an energy balance equation

$$\int_0^\infty Q(\nu)J_\nu\,d\nu = 4\pi\int_0^\infty Q(\nu)B_\nu(T_d)\,d\nu. \quad (38)$$

B_ν is the Planck function. In the very simplified but interesting case of an optically thin cloud at distance R from a short-wavelength source of luminosity L (so that $Q(\nu_{abs}) \approx O(1) = $ const) and smoothly varying $Q(\nu) = Q_0(\nu/\nu_0)^\beta$ at $\nu \ll \nu_{abs}$ one can express T_d as

$$T_d(R) = T_0\left(\frac{L}{L_0}\right)^{\frac{1}{4+\beta}}\left(\frac{R}{R_0}\right)^{-\frac{2}{4+\beta}}. \quad (39)$$

Observations at far-infrared ($\lambda \sim 100\,\mu m$) wavelengths are consistent with $\beta \approx 1$ and $T_0 \approx 70\,K$ for $L_0 = 10^5\,L_\odot$, $R_0 = 3 \times 10^{17}$ cm (e.g., Scoville and Kwan 1976). Equation (39) demonstrates the well established fact that embedded, newly-formed OB stars heat the dust in a $\sim 1\,pc$ region to temperatures of 30 K or more. These dust particles radiate predominantly in the 30 to 300 μm band.

The data also indicate that optical depth effects ($\tau_d \geq 1$) set in at $\lambda \approx 30\,\mu m$ for hydrogen nuclei column densities of $N(H + 2H_2) \geq 3 \times 10^{23}\,cm^{-2}$ and at $\lambda \approx 400\,\mu m$ for $N \geq 10^{25}\,cm^{-2}$ (Hildebrand 1983, Draine 1989). Excepting the densest regions the submillimeter dust emission is thus optically thin and can be used as a tracer of interstellar gas mass once its temperature is known (see below). The $\lambda \geq 100\,\mu m$ emission can, however, become optically thick within massive cloud cores and more detailed radiative transport calculations have to be performed to obtain dust temperature and emergent flux. Such calculations for massive star formation regions have been discussed by Scoville and Kwan (1976), Yorke (1988), and Natta et al. (1981).

The assumption of an average equilibrium dust temperature breaks down for very small dust particles ($\lambda < 0.01\,\mu$m). Their heat capacity is so small that absorption of a single UV photon leads to a large fluctuation in lattice temperature. Evidence for such non-equilibrium heating in very small dust particles, or large molecules, has been found from observations of the 3 to 10 μm continuum and in particular from the $3-10\,\mu$m "unidentified" emission features (Sellgren 1984, Puget and Leger 1989).

3.3. ESTIMATES OF CLOUD COLUMN DENSITIES AND MASSES

The main fraction of the mass of dense interstellar clouds is in form of molecular hydrogen. Atomic hydrogen is the dominant species in diffuse and translucent clouds, but contributes only a small fraction to the mass of dense molecular clouds. While atomic hydrogen can be easily detected via its 21 cm hyperfine transition, molecular hydrogen can currently be observed directly only in special circumstances, such as through its infrared, ro-vibrational emission in shocks and photodissociation regions (see section IV), or via UV-absorption in diffuse clouds. In most cases it is thus necessary to determine the mass of a cloud by less direct indicators, such as the submillimeter/millimeter dust emission, by line emission of CO, the most abundant molecule after H_2, and by MeV γ-ray emission. We now briefly discuss these methods in turn.

a) Submillimeter dust emission

About one percent of the mass of interstellar matter is in form of dust grains. Thermal emission of interstellar dust may be used to determine the dust optical depth and then infer a gas column density. As discussed above submillimeter dust emission is usually optically thin and the flux density of a source of temperature T_d at frequency ν can be written as

$$S_\nu = 2h\nu^3/c^2[\exp(h\nu/kT_d) - 1]^{-1} \langle \tau_d(\nu) \rangle \Omega_\nu, \qquad (40)$$

where $\langle \tau_d(\nu) \rangle$ is the source/beam averaged dust optical depth at ν. Ω_ν is the smaller of either the source or the telescope beam solid angle. Equation (40) can be further simplified in the Rayleigh-Jeans $(h\nu/kT_d \ll 1)$ limit which is often appropriate for the dust emission from dense clouds at wavelength longward of 300 μm,

$$S_\nu = 2kT_d\nu^2/c^2 \langle \tau_d(\nu) \rangle \Omega_\nu. \qquad (41)$$

In this limit the flux density is linearly dependent on temperature and dust optical depth. Measurement of dust optical depth in the submillimeter or millimeter range also has the advantage that it likely is a measure of *both* warm *and* cold dust

(Mezger et al. 1986). The optical depth depends on the gas to dust ratio and the dust extinction efficiency at ν. The dust extinction efficiency in the submillimeter range is unfortunately only poorly known (see Draine 1989 for a discussion). Whitcomb et al. (1981) have related the dust emission at 400 µm to A_V in the reflection nebula NGC 7023 giving an absolute scaling from $\tau(400\,\mu m)$ to A_V. It is also typically assumed that the extinction efficiency in the submillimeter scales with wavelength as $\lambda^{-\beta}$ with $\beta \approx 1.5$ to 2 (see Hildebrand 1983, Draine 1989, Cox and Mezger 1989). A_V can then be related to the nuclei of hydrogen atoms $N(H + 2H_2)$ with the calibration of Bohlin, Savage, and Drake (1978: $N(H + 2H_2) = 1.9 \times 10^{21} A_V$) from UV measurements in diffuse clouds. Assuming a gas to dust mass ratio of 100 is equivalent to the Bohlin et al. calibration. The resulting "best" relationship near 400 µm then is (Hildebrand 1983, Cox and Mezger 1989, Draine 1989)

$$N(H + 2H_2) = 1.2 \times 10^{25} \, Z/Z_\odot \, \tau_d(\lambda) \, (\lambda/400\,\mu m)^\beta \quad [\text{cm}^{-2}]. \qquad (42)$$

Here $Z/Z_\odot$ is the abundance of heavy elements in units of the solar abundance.

b) CO Line Emission

The CO molecule has transitions in the UV (electronic transitions), at 2.3/4.6 µm (ro-vibrational) and between 50 µm and 2.6 mm (rotational). It is also the most abundant molecule after H_2. Observations in dense clouds made at infrared, submillimeter and millimeter wavelengths all suggest that the CO/H_2 fractional abundance is close to 10^{-4} (about 30% of carbon in CO). The most common method of determining CO column densities uses measurements of the low-J ($1 \to 0$, $2 \to 1$) transitions in emission. ^{12}CO transitions in dense molecular clouds are very optically thick ($\tau \approx 100$).

i) Tracing mass with CO isotopes.
One way of avoiding this problem is to observe the less abundant (optically thinner) ^{13}CO, $C^{18}O$ and $C^{17}O$ lines and then to multiply by the appropriate abundance ratio. If only one line is measured, an extrapolation of the column density in the measured levels to all other levels is necessary. Often this is done by assuming LTE population at some temperature. However, the critical density of the excited states increases rapidly with J ($n_{\text{crit}} \approx J^3$), so that typically only the lowest three or four levels are populated. Excitation/radiative transport calculations show that the integrated intensities of the $2 \to 1$ transitions of ^{13}CO (if optically thin) or $C^{18}O$ are a good measure of column density and more or less independent of temperature over a reasonably wide parameter range (a few $10^3 < n(H_2) <$ a few $10^6\,\text{cm}^{-3}$ and $15 < T < 80\,\text{K}$). With $[C^{18}O]/[H_2] = 1.7 \times 10^{-7}$ (Frerking et al. 1982) one then finds

$$N(H_2) = 4 \times 10^{21} \, I(C^{18}O \quad 2 \to 1) \quad [\text{cm}^{-2}/\text{K km s}^{-1}]. \qquad (43)$$

Similar relationships hold for optically thin ^{13}CO or $C^{17}O$ lines; one has to just choose the appropriate abundance ($^{13}CO/C^{18}O \approx 8$, $C^{18}O/C^{17}O \approx 4$). The $1 \to 0$

transitions can also be used, although they are more sensitive to temperature; in that case multiply the coefficient on the right by a factor of four.

ii) Tracing mass with ^{12}CO $1 \to 0$. In some cases the isotopic lines are too weak for measurement. A column density is then inferred from the optically thick ^{12}CO line(s) by assuming either that the integrated line flux is proportional to the filling factor of clouds of a constant flux, or by applying the virial theorem (e.g., Dickman et al. 1986, Solomon et al. 1987, Scoville and Sanders 1987). This method is widely used for external galaxies, for instance. In the following, the physical basis and possible pitfalls of the "virial" method are discussed.

Following van Dishoeck and Black (1987), let us consider an ensemble of clouds (or clumps) moving around each other, or around the center of a galaxy with an overall velocity width Δv_{source} (FWHM). Assume that each cloud has an intrinsic velocity width Δv_{cloud} that is determined by virial equilibrium (kinetic energy stabilizes clouds against gravitational collapse). Assume further that at each velocity the (beam) area filling factor of clouds $\Phi(v)$ is smaller than 1 (no shadowing). Each cloud emits optically thick ^{12}CO $1 \to 0$ lines with an excitation temperature T_{ex}, so that the integrated line flux [K km s^{-1}] is given by

$$I(^{12}\text{CO}) = 1/\Omega_B \iiint (\widetilde{T}_{\text{ex}} - \widetilde{T}_{\text{BG}})(1 - \exp[-\tau(v)]) \, dv \, d\Omega. \tag{44}$$

Ω_B is the solid angle of the beam. The integration is over the solid angle of the beam and the profile. $\widetilde{T}_{\text{ex}}$ and $\widetilde{T}_{\text{BG}}$ are the Rayleigh-Jeans temperatures corresponding to the excitation and 3 K background temperatures,

$$\widetilde{T} = h\nu/k \left[\exp(h\nu/kT) - 1\right]^{-1}. \tag{45}$$

For $T_{\text{ex}} \gg 3$ K and for $\tau \gg 1$, equation (41) becomes

$$I(^{12}\text{CO}) = \widetilde{T}_{\text{ex}} \Phi(v) \Delta v_{\text{source}} = \widetilde{T}_{\text{ex}} \Phi \Delta v_{\text{cloud}}, \tag{46}$$

where $\Phi = \Phi(v) \Delta v_{\text{source}}/\Delta v_{\text{cloud}}$ is the filling factor of clouds irrespective of velocity. The assumption of virialization of each cloud of radius R, mass $M(R)$ and mean (volume averaged) density $\langle n(\text{H}_2) \rangle$ means

$$\frac{G M(R)}{3R} = \sigma^2(\text{cloud}) + kT_{\text{kin}}/\langle m \rangle$$
$$\approx \sigma^2(\text{cloud}) = (\Delta v_{\text{cloud}}/2.35)^2. \tag{47}$$

In this equation G is the gravitational constant. We have assumed that the intrinsic cloud motions are significantly larger than thermal. The beam averaged H_2 column density can be expressed as

$$\langle N(H_2)\rangle = \Phi\,[N(H_2)]_{\text{cloud}} = \Phi\,2R\,\langle n(H_2)\rangle. \tag{48}$$

With $M(R) = (4\pi/3)\,R^3\,\langle n(H_2)\rangle\,\langle m\rangle$ ($\langle m\rangle = 4.3\times 10^{-24}$ g is the mean mass per hydrogen molecule, including heavier elements), equations (42), (43), and (44) can be combined to give the conversion factor, X, of

$$X_{CO} = \langle N(H_2)\rangle/I(^{12}CO) = 3\times 10^{20}\,(8\,K/\widetilde{T}_{ex})(\langle n(H_2)\rangle/200)^{1/2}. \tag{49}$$

The values for $\widetilde{T}_{ex}$ and $\langle n(H_2)\rangle$ chosen in equation (45) are those quoted by Scoville and Sanders (1987) to be representative for giant molecular clouds in the Galaxy. The conversion factor derived in equation (49) is the same within a factor of 2 as the ones obtained for galactic disk clouds by several other methods (isotopes, empirical correlation, γ-rays). This forms the basis of the common assumption that the above relationship holds *universally*.

Given equation (49) and the evidence, presented in sections III.4 and V, of much higher excitation temperatures of the ^{12}CO emitting gas near OB star forming regions (and in the nuclei of starburst nuclei) than in the average disk gas, this assumption has to be questioned (Maloney and Black, 1988). Yet the fact that mean densities also strongly increase in star forming clouds may "save" the relationship as a first order estimate, albeit with increased uncertainty. For instance, the integrated $^{12}CO\ 1\to 0$ flux in the $+20/+40\,\mathrm{km\,s^{-1}}$ clouds near Sgr A is about $1600\,\mathrm{K\,km\,s^{-1}}$. From $X = 3\times 10^{20}$ one infers an H_2 column density of $4.8\times 10^{23}\,\mathrm{cm^{-2}}$, remarkably close to the estimates from submillimeter dust continuum and $C^{18}O\ 2\to 1$ flux (3 to 6×10^{23}). This agreement must be fortuitous, as the ^{12}CO line brightness temperature is 47 K (6 times higher than $\widetilde{T}_{ex} = 8\,K$) and since the ^{12}CO line width is 40% broader than ^{13}CO and $C^{18}O$, indicating that the line width is not solely determined by bulk cloud motion. To understand the result, one has to assume that the mean density in these clouds is 70 higher than the $200\,\mathrm{cm^{-3}}$ assumed above. This is in fact quite reasonable in the environment of the Galactic Center. This lucky coincidence does not always work, however. In M82, for example, mass estimates via the standard conversion factor come out 3 to 5 times larger than from other lines of evidence, consistent with the greater kinetic temperature of the gas ($\approx 40\,K$). In the core of M17 it is the other way around. This uncertainty has to be kept in mind when applying the relationship to external galaxies.

c) Tracing mass with MeV γ-rays

High energy γ-rays (30 to 1000 MeV) in interstellar clouds are produced when cosmic rays collide with hydrogen nuclei. The π-mesons produced in this collision decay into γ's. If the cosmic ray density inside clouds in constant, the γ-ray flux thus measures the number of hydrogen nuclei in a given volume (see Bloemen 1989 for a review). Bloemen et al. (1984) and Strong et al. (1988) have used the COS-B galactic plane survey to determine masses and the conversion rate between ^{12}CO flux and H$_2$ column density. The average conversion factor for the Galactic disk corresponds to $X = 2.4 \times 10^{20}$. The conversion factor in the Galactic Center may be 10 times lower (Blitz et al. 1985), supporting cautionary remarks of the last paragraph.

3.4. MEASURING TEMPERATURE AND DENSITY IN MOLECULAR CLOUDS

All basic tools necessary for deriving these basic physical parameters have been discussed in the preceding sections. Different methods for extracting temperature and density information have been applied to the investigation of clouds with massive star formation; the most important ones are listed in Table III in increasing order of complexity (see also Goldsmith 1987). All have advantages and disadvantages. The simplest tools in each category (temperatures from line brightness temperatures and densities from column densities and cloud sizes) are very rough measures only and should be applied with great caution. Even the most sophisticated methods (fitting a non-LTE cloud model to observations of a wide variety of transitions) typically suffer from problems of simplification (single component models), large number of ill-defined parameters (for models with density and temperature gradients and/or complex velocity fields) and uncertainties in cross sections. If only a small set of lines is considered one has to be aware that they typically sample only a limited range of n and T.

A good recent review of the observational situation has been given by Goldsmith (1987), a somewhat older is that of Evans (1981). The basic result is that molecular clouds have a wide range of temperatures and densities, depending on scale size and environment (distance to OB stars, Galactic disk vs. Galactic center). Kinetic temperatures vary from $6 \rightarrow 10\,\mathrm{K}$ in cold, dark clouds to $\geq 50\,\mathrm{K}$ in warm cloud cores close to OB star forming sites. Submillimeter spectroscopy indicates that in these warm cloud cores there is a component of quiescent gas $\geq 100\,\mathrm{K}$ with H$_2$ column densities $10^{22} \rightarrow 10^{23}\,\mathrm{cm}^{-2}$ (Harris et al. 1987, Jaffe et al. 1987, Schmid-Burgk et al. 1989, Graf et al. 1990). This warm component represents a significant fraction ($10 \rightarrow 50\%$) of the mass in star forming cores (see more detailed discussion below in V.2.). Gas with temperatures $\geq 50\,\mathrm{K}$ is also common and probably dominant throughout the Galactic center molecular clouds (Güsten et al. 1985) and in the nuclei of starburst galaxies (Wild et al. 1991, Harris et al. 1991).

Typical molecular hydrogen densities in molecular cores range from about $10^4\,\mathrm{cm}^{-3}$ in dark clouds (Myers 1986) to $10^5 \rightarrow 10^6\,\mathrm{cm}^{-3}$ in massive star formation regions (Snell et al. 1984, Stutzki and Güsten 1990). The "hot core" clump (mass $\geq 10 M_\odot$, size $\sim 0.01\,\mathrm{pc}$) in the BN-KL region even has densities of $1 \rightarrow 3 \times 10^7\,\mathrm{cm}^{-3}$ (Genzel and Stutzki 1989).

4. Heating and Cooling Processes In Star Forming Clouds

A dense cloud or cloud core in a region of massive star formation is exposed to the intense ultraviolet radiation from the newly-formed OB star that in part

Table III. Temperature and density probes in molecular clouds

METHOD	STRENGTH	WEAKNESS
Temperature		
Brightness of optically thick mm-lines (e.g., ^{12}CO 1→0: $T_R = T_{ex} \sim T_{kin}$)	easy to measure	filling factor of emission $\Phi_{AL} < 1$ (eq. 8) and method may underestimate T_{kin}
Level population as a function of energy from metastable NH$_3$ inversion lines at ~ 1.2 cm (also other symmetric tops, such as CH$_3$CN, CH$_3$C$_2$H)	many energy levels available at about the same wavelength	optical depth and density correction not easy, only applicable to dense regions where NH$_3$, CH$_3$CN etc. are abundant
Level population as a function of energy from rotational lines of molecules in mm/submm (emission, e.g., CO)	very sensitive for $E_{ul} > kT_{kin}$ (eq. 18), gives also density information	laborious measurements need to be made at very different wavelengths, must solve the entire excitation problem (eq. 26) and take into account optical depth effects
.... from ro-vibrational lines in near-IR (absorption)	same as above, many energy levels at about the same wavelength	only line of sight toward bright continuum sources
Density		
Average volume density from column density and cloud size (e.g., ^{13}CO or C^{18}O)	easy to measure	underestimes local densities as volume filling factors $\ll 1$ (clumpiness), strongly depends on abundances and T_{ex}
Detection of "density tracers", such as mm-lines of CS, HCN with *high* n_{crit}	easy to measure	rough first order indication, depends strongly on optical depth (trapping, eq. 34)
Measurement of non-metastable NH$_3$ inversion lines at 1.2 cm	many energy levels available at about the same wavelength	optical depth effects and FIR radiative pumping need to be taken into account
Level population as a function of energy from rotational lines above n_{crit} (subthermal regime), especially for optically thin species (C^{18}O, H^{13}CN, C^{34}S etc.)	very sensitive to $n(H_2)$, gives also temperature information	laborious measurements need to be made at very different wavelengths, cross sections for all molecules but CO uncertain, must solve the entire excitation problem (eq. 26)

ionizes the surrounding cloud and forms an *HII region*. Beyond the penetration depth of Lyman-continuum ($E_{\text{Lyc}} \geq 13.6\,\text{eV}$) photons there is a zone where far-UV ($6 \rightarrow 13.6\,\text{eV}$) radiation ionizes species with ionization potentials less than that of hydrogen (C,S,Si,Fe), dissociates molecules and triggers a photon-dominated chemistry. These *photo-chemical* or *photo-dissociation regions* (PDR's) are of particular interest to the following discussion as in these surface zones most of the stellar luminosity is absorbed and reradiated in form of infrared/submillimeter line and continuum radiation.

Another important set of heating processes includes the conversion of *mechanical supersonic energy sources (stellar winds and outflows, supernova explosions)*, into thermal energy via *shocks*. Like radiative heating one should expect shock heating to predominantly affect the cloud/clump surfaces.

Energetic particles (cosmic rays) and *hard X-rays* penetrate much deeper into clouds and significantly contribute to the heating there. If magnetic fields are present and move (slip) relative to the bulk material of the cloud (e.g., during gravitational collapse), the resulting *frictional (ambipolar diffusion) heating* needs to be taken into account.

The just mentioned *heating sources* and connected *heating mechanisms* (such as photoionization, photoelectric effect, shocks etc.) contribute a net *heating rate* per volume element $\Gamma\,(n, T_{\text{kin}})$ [$\text{erg}\,\text{s}^{-1}\,\text{cm}^{-3}$]. The cloud cools at a rate $\Lambda(n, T_{\text{kin}})$ that again is a sum of a number of processes (atomic and molecular line cooling, dust continuum cooling etc.). In *equilibrium* the cloud reaches a kinetic temperature T_{kin} to just balance heating and cooling

$$\sum_i \Gamma_i(n, T_{\text{kin}}) = \sum_i \Lambda_i(n, T_{\text{kin}}).$$

The cooling rate also controls the *cooling time* during which the cloud can radiate away a significant fraction of its energy

$$\tau_{\text{cool}} = \frac{E_{\text{therm}} + E_{\text{kin}} + E_{\text{turb}}}{\sum_i \Lambda_i(n, T_{\text{kin}})}.$$

The cooling time is in most cases very small compared to other time scales of the system so that one can in fact expect the cloud to reach thermal equilibrium.

We will now briefly discuss in turn the most important heating and cooling mechanisms for dense, *neutral* clouds (see also Spitzer 1978, Goldsmith and Langer 1987, Black 1987). We will not address fully ionized clouds (HII regions) and refer to the books by Spitzer (1978) and Osterbrock (1989) for an in depth discussion. We begin with the processes affecting the surfaces of dense clouds (PDR's, shocks).

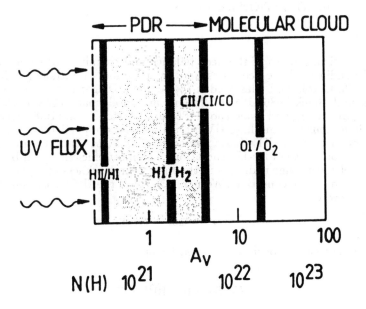

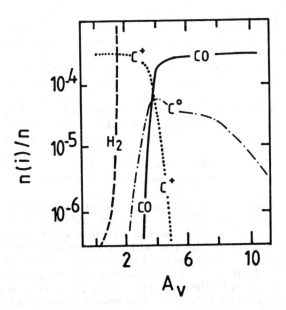

Figure 6. Schematic of a PDR (top), and chemical abundances of the major species (bottom) in the models of Tielens and Hollenbach (1985a)

4.1. PHOTODISSOCIATION REGIONS

a. Theory

Theoretical models of the heat balance and chemistry of clouds illuminated by far-UV radiation have been presented by a number of authors (Walmsley 1975, Langer 1976, Gerola and Glassgold 1978, de Jong, Dalgarno and Boland 1980, van Dishoeck and Black 1986, 1988). Tielens and Hollenbach (1985), Hollenbach et al. (1991), Sternberg and Dalgarno (1989) and Burton et al. (1990) treat the cases of high cloud densities and high UV intensities that are particularly interesting for the present discussion. The basic size scale of the dissociation region is given by the penetration depth of far-UV radiation. The most efficient broad-band absorber in this wavelength range is dust, and the maximum thickness of a photodissociated region is given by the condition (Werner 1970) that the UV dust *absorption* optical depth be a few, or

$$d_{\rm PDR} \stackrel{\frown}{=} \tau_{\rm abs} \approx {\rm a \ few},$$

corresponding to a hydrogen column density of

$$N_0({\rm H}) \approx 6 \times 10^{21} \, {\rm cm}^{-2}. \tag{50}$$

The chemical structure in such a cloud interface that is common to all models is displayed in Fig. 6. At $A_V \approx 1$ to 2 from the surface there is a sharp transition from atomic to molecular hydrogen. This is a direct consequence of the fact that photodissociation of H_2 requires UV absorption from the $X^1\Sigma_g$ ground electronic state to the $B^1\Sigma_u$ and $C^1\Pi_u$ excited states, followed by a transition into the unbound continuum of the ground state. There is no direct continuum dissociation process. The UV transitions of H_2 (the Lyman and Werner bands) rapidly become optically thick and self-shielding sets in (cf., Jura 1974). In this outer layer almost all of the gaseous carbon and oxygen is in form of C^+ and O^0. The transition between singly ionized carbon and CO occurs significantly deeper into the cloud ($A_V \approx 5$) with a layer of atomic carbon arising between $A_V \approx 3$ and 8. In Fig. 6 it is assumed that all carbon is in CO as soon as CO can form. Oxygen thus remains atomic deep into the cloud ($A_V \approx 20$) and OI may be the most important carrier of oxygen in interstellar clouds (see Dalgarno 1990). In addition to these most abundant species, the photodissociation zone contains many less abundant ions, such as S^+, Mg^+, Si^+, and Fe^+ whose abundances depend heavily on depletion.

Photoelectric heating and UV pumping of H_2 are likely the most important gas heating mechanisms in dense PDR's. These mechanisms are summarized in Fig. 7. The first mechanism involves the absorption of a UV photon by a dust grain, followed by ejection of an electron which then collisionally heats the gas (Draine 1978, de Jong 1977). Per incoming 10 eV photon, the energy per electron available for heating is about 1 eV. The remaining 9 eV go into overcoming the binding energy of the electron to the grain (the "work function", about 6 eV) and the grain's positive electrostatic potential. The potential energy which has to be overcome increases with increasing UV energy density (higher positive charge) and decreases with increasing density (higher rate of recombination). With a "yield" of conversion of incoming photons into photoelectrons of about 0.1 the photoelectric

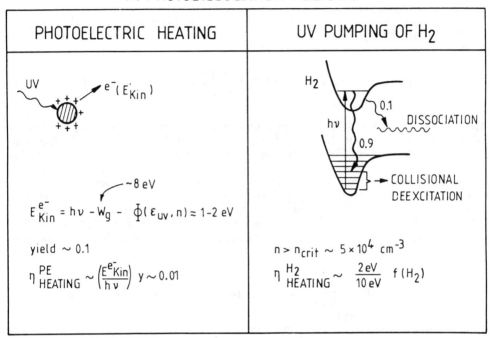

Figure 7. Heating mechanisms in PDR's (see text)

mechanism can plausibly convert about 1% of the UV photons into gas heating ($\eta_{pe} \sim 0.01$ to 0.03). The numerical values are, of course, grain-model dependent and also depend on carbon abundance and grain size distribution. Of great interest for future study is the importance of small dust grains and large molecules (PAH's) for this mechanism.

The net photoelectric heating rate can thus be expressed as (de Jong et al. 1980, Black 1987)

$$\Gamma_{pe} = n_d \sigma_d \eta_{pe} \chi \approx 4 \times 10^{-26} \left(\frac{\chi}{\chi_0}\right) n_H \quad [\text{erg cm}^{-3} \text{ s}^{-1}]. \quad (51)$$

Here we have used that the product of volume density n_d and cross section σ_d of dust grains is about $1.5 \times 10^{-21} n_H$ (Spitzer 1978). χ is the UV flux and $\chi_0 = 2.5 \times 10^{-3}$ erg cm^{-2} s^{-1} is the far-UV flux in the solar neighborhood (Draine 1978). χ is of course dependent on the depth into the cloud and scales like $\exp(-\tau_d)$.

Photoelectric heating by the average interstellar radiation field can plausibly explain the observed temperatures of moderate to low density HI clouds in the solar neighborhood ($T_{kin} \sim 70 \to 100$ K). There the main cooling line is the 158 μm $^2P_{3/2} \to {}^2P_{1/2}$ transition of C$^+$. Balancing its cooling rate for C$^+$/H $= 10^{-4}$ at low densities ($n < n_{crit} \sim 3 \times 10^3$ cm^{-3}) with the photoelectric heating rate from equation (51) gives

$$\Gamma_{pe} = 4 \times 10^{-26} \left(\frac{\chi}{\chi_0}\right) n_H = 2 \times 10^{-27} \exp\left(\frac{-91 \text{ K}}{T_{kin}}\right) n_H^2 \quad \text{and,}$$

$$T_{kin} = \frac{-91 \text{ K}}{\ln(20(\chi/\chi_0)n_H)} \approx 80 \text{ } K \quad (52)$$

for $n_H = 30$ cm^{-3} and $\chi = (1/2)\chi_0$ (solar neighborhood). Within the uncertainties, the photoelectric heating of equation (51) also is consistent with the C$^+$-cooling in the *diffuse* ISM inferred from UV spectroscopy by Pottasch et al. (1979), $\Lambda \sim 10^{-25}$ erg s^{-1} per hydrogen atom.

UV pumping of H$_2$ molecules can convert a similar fraction of the UV if the hydrogen density is sufficiently high (Sternberg 1988). In this case, electronic excitation of H$_2$ molecules in the Lyman and Werner bands at 1000 Å leads to subsequent radiative decay into a vibrationally excited level of the electronic ground state (2 to 3 eV) in 9 out of 10 excitations (Fig. 6). At densities greater than the critical density for collisional deexcitation of the vibrational states ($n(H_2)_{crit} \approx 7 \times 10^4$ cm^{-3}), that energy can again be converted into gas heating. In the UV transtions of H$_2$, the conversion efficiency of a single absorbed photon into kinetic energy can thus be as high as 30%. The total efficiency of converting far-UV photons into heating depends on the fraction of incident radiation absorbed by H$_2$ molecules throughout the cloud. This fraction depends on the dust to gas ratio and on the ratio of gas density to intensity of the UV field and is usually $\leq 10\%$ (Sternberg 1988, Sternberg and Dalgarno 1989).

At visual extinctions *greater than a few* the gas is predominantly heated through *collisions with warm dust particles* that transfer a fraction of their internal energy in the process. As this heating process involves a collision between two particles it scales proportional to (density)² and depends on the composition of the gas through the collision cross sections (Hollenbach and McKee 1979). It also depends on the difference between dust temperature (T_d) and gas kinetic temperature ($T_{kin} \leq T_d$) so that for molecular gas

$$\Gamma_{d-gas} = 2 \times 10^{-33} T_{kin}^{\frac{1}{2}} (T_d - T_{kin}) n_H^2 \quad [\text{erg s}^{-1} \text{cm}^{-3}]. \tag{53}$$

Clearly the gas is only coupled well to the dust at high densities where the cooling time scale is sufficiently small

$$t_{d-g} = 1.5 \times 10^4 \left(\frac{n_H}{10^5 \text{ cm}^{-3}}\right)^{-1} \left(\frac{T_{kin}}{50 \text{ K}}\right)^{-\frac{1}{2}} \quad [\text{y}]. \tag{54}$$

Gas cooling of the photodissociation zone at $A_V \leq 5$ is primarily by the fine structure lines of [O I] (63 μm), [C II] (158 μm) and [Si II] (35 μm). For very dense gas the 26 and 35 μm fine structure lines of [Fe II] and the forbidden lines of [O I] and [C I] in the visible may also contribute. Of lesser importance are the H₂ ro-vibrational and rotational lines and the submm fine structure lines of [C I] largely because of the small column densities in the upper state and because of their small A-coefficients.

The cooling rate at point z in the cloud is given by

$$\Lambda = \sum_{\substack{\text{transitions}(i),\\ \text{species}(j)}} h\nu_{ij} A_{ij} \beta(\tau_{ij}) n_{ij}(z). \tag{55}$$

For the fine structure lines the cooling rate can be simplified to

$$\Lambda_{fs} \simeq (h\nu A) \beta n_{tot} \frac{g_u/g_l \exp(-E_{ul}/kT)}{1 + g_u/g_l \exp(-E_{ul}/kT) + (n_{crit}/n)\beta}, \tag{56}$$

where the values for ν, A and n_{crit} can be taken from Table II.

The major coolants and the resulting thermal structure are depicted in Fig. 8 as a function of depth from the cloud surface, for the "standard" model of Tielens and Hollenbach (1985a,b): $\chi_{UV} \approx 10^5$ times the UV flux χ_0, $n(H_2) \approx 2 \times 10^5$ cm^{-3}, "cosmic" abundances. The standard model is representative for dense photodissociation regions within a few tenths of a pc of a massive O star, such as the HII region/cloud interfaces in Orion and M17 or the circum-nuclear disk in the Galactic center (cf., Genzel et al. 1989). The gas temperature in the PDR ranges between about 1000 K at the surface to about 100 K at the C$^+$/CO transition.

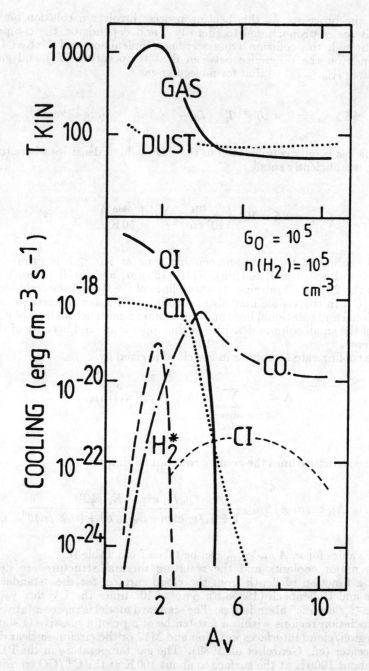

Figure 8. Distribution of gas/dust temperatures (top) and cooling lines (bottom) through a PDR, for the specific case of the "standard model" of Tielens and Hollenbach (1985a).

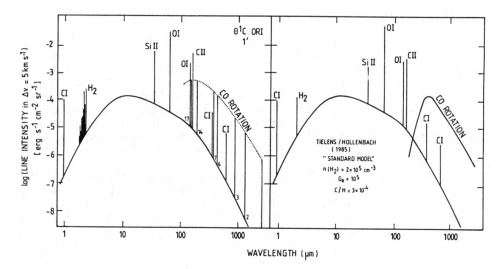

Figure 9. Line emission from the Orion A PDR. The left diagram shows the observations, the right diagram shows the line emission predicted by the "standard model" of Tielens and Hollenbach (1985a).

b. OBSERVATIONS OF THE ORION PHOTODISSOCIATION REGION

Fig. 9 gives a direct comparison of Tielens and Hollenbach's (1985a) standard model with the observed infrared and submillimeter line and continuum emission toward the Trapezium region in Orion. Observations and theory agree fairly well. It seems quite convincing that the infrared measurements are accounted for by a PDR model of a dense molecular cloud at a distance of about 0.1 pc from the Trapezium OB association. About 0.6% of the impinging UV radiation is converted into line emission. Although the standard model is not a fit to the Orion data, the good agreement with the data should, of course, not come as a complete surprise, as the theoretical work was in part motivated by the observations of the intense infrared line emission.

Another key test of the PDR model is the measurement of the isotopic [^{13}C II] fine structure line that gives – in combination with the [^{12}C II] line – a fairly direct estimate of column density and excitation temperature. Recently Stacey et al. (1991b) were able to detect the F = 1→0 [^{13}C II] line in the Orion PDR. The results are shown in Fig. 10. The large ratio of [^{12}C II]/[^{13}C II] F = 1→0 fluxes (~ 100) clearly shows that the [^{12}C II] line does not have a large optical depth ($\tau([^{12}\text{C II}]) \leq 1$) and must come from gas with temperature ~ 200 K or more, in excellent agreement with the theoretical models.

The standard model does not predict enough [Si II] emission which may indicate that the assumed gas phase abundance of silicon was too low. More serious is the failure of the model to account for the observed submm and far-infrared CO line emission. Again this is not too surprising, as the models discussed above predict very little CO at temperatures > 100 K (Fig. 7) that is required to explain the Orion CO data (Stacey et al. 1991a). The model *does*, however, predict fairly intense (50 to 100 K) ^{12}CO 1→0 emission originating just behind the transition region from C$^+$ to CO. The mid- and high-J CO emission in Fig. 9 comes from warm, quiescent gas and not the high-velocity shock excited gas associated with Orion-KL (see section V.2.).

4.2. SHOCKS IN DENSE INTERSTELLAR CLOUDS

Interstellar shock waves are generated by supersonic mass motions, such as fast moving clouds, outflows from young stars, stellar winds and supernovae. The kinetic energy of the supersonic motions is converted into thermal energy. In the process, shocks compress and accelerate gas and dust. Shocks radiate primarily in lines from the cooling gas behind the front where the flow motion is converted into random thermal motions (McKee and Hollenbach 1980). With the exception of high velocity shocks ($v_s > 50$ km s^{-1}) in dense ($n(\text{H}_2) > 10^6$ cm^{-3}) clouds, dust emission is small compared to line emission from hot gas. Shocks in dense molecular clouds often radiate mostly in the infrared and submillimeter lines of molecules, atoms and ions. High velocity, gas dynamic (J)-shocks dissociate most molecules and ionize atoms. Magneto-hydrodynamic (C)-shocks create large column densities of moderately warm molecular gas.

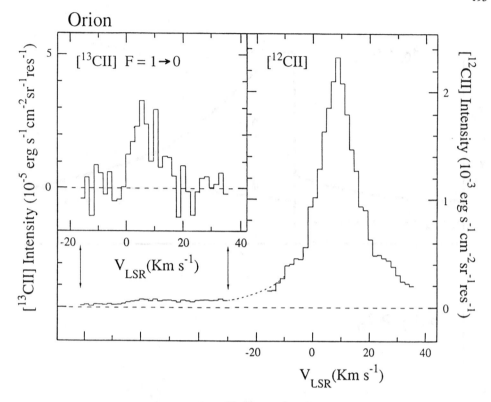

Figure 10. First detection of the $[^{13}C\,II]$ $F = 1 \to 0$ hyperfine component toward the center of the Orion A HII region (from Stacey et al. 1991b)

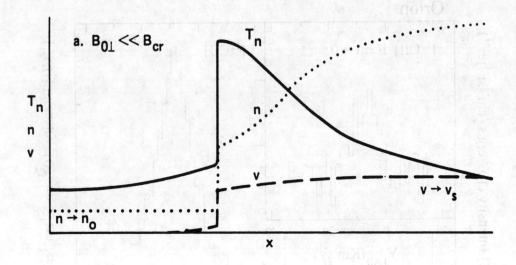
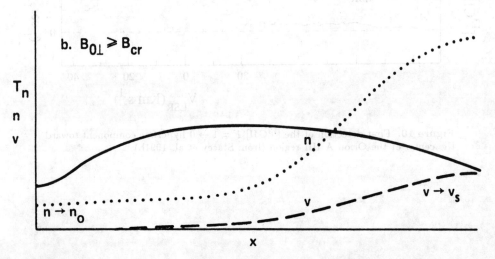

Figure 11. Shock structure in J- and C-shocks (from Hollenbach et al. 1989, see text).

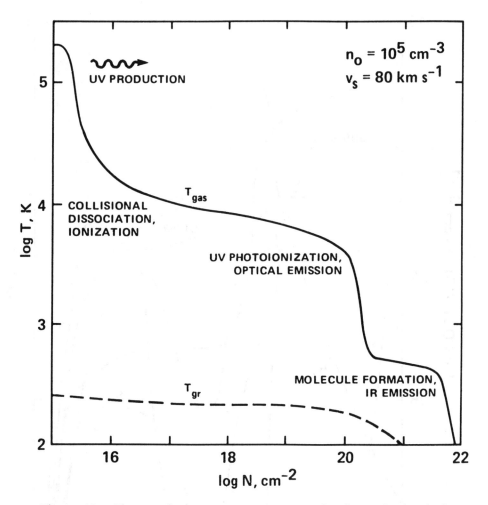

Figure 12. The post-shock temperature structure of a fast molecular shock ($n_0 = 10^5 \,\text{cm}^{-3}$, $v_s = 80 \,\text{km s}^{-1}$ shown here). Three regions are delineated: (1) the hot, $T \sim 10^5$ K, immediate post-shock region, where gas is collisionally dissociated and ionized and UV photons are produced which affect both the pre-shock and post-shock gas; (2) the "recombination plateau" where the Lyman continuum photons are absorbed, maintaining $T \sim 10^4$ K; and (3) the recombining and molecule-forming gas downstream, where chemical energy of H_2 formation can maintain a lower temperature plateau. The column densities of the first two regions are nearly independent of n_0. Note that the grains are weakly coupled to the gas, so that $T_{\text{gr}} \ll T_{\text{gas}}$ (adapted from Hollenbach and McKee 1989).

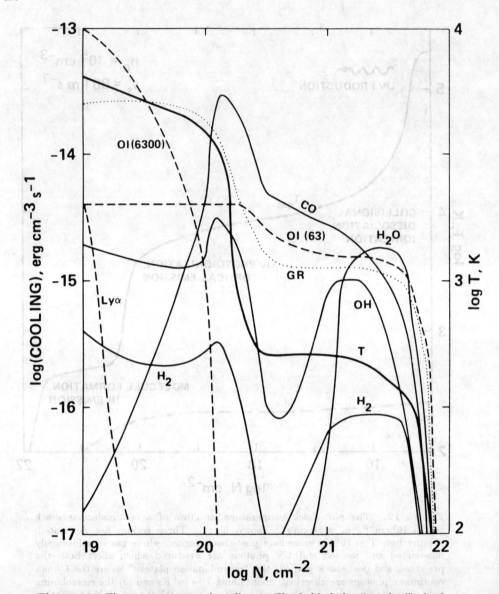

Figure 13. The temperature and cooling profiles behind the "standard" shock of Fig. 12. GR denotes the gas cooling by collisions with the cooler grains. The molecular cooling includes vibrational and rotational contributions (adapted from Hollenbach and McKee 1989).

a. Theory

i) J-Shocks. Shock waves in dense interstellar clouds have been recently reviewed in some detail by Shull and Draine (1987) and Hollenbach et al. (1989) who also refer to the many detailed calculations of the last 10 years or so. We summarize here the main results of these theoretical studies. In cold gas (sound speed $< 1\,\mathrm{km\,s^{-1}}$) with no or weak magnetic fields, a fast ($> 50\,\mathrm{km\,s^{-1}}$) pressure disturbance creates a *J-shock*, where temperature, density and flow velocity suffer a virtually discontinuous jump across the shock front. Figure 11 (top) gives a schematic of the change of the variables across the front. Conservation of mass and momentum (and energy for an adiabatic shock) yield relations between preshock variables and downstream variables. These Rankine-Hugoniot relations are, for instance, discussed in Spitzer's (1978) book. The thermal structure behind the front is shown in Fig. 12, for the case of a $50\,\mathrm{km\,s^{-1}}$ shock into a $n = 10^5\,\mathrm{cm^{-3}}$ pre-shock density gas (from Hollenbach and McKee 1989). Immediately behind a fast shock in a molecular cloud the temperature is high,

$$T_s \approx 1.5 \times 10^5 \, (v_{\mathrm{shock}}/100\,\mathrm{km\,s^{-1}})^2 \quad [\mathrm{K}]. \tag{55}$$

Consequently the gas dissociates and ionizes. It radiates in the UV and visible wavelength bands in resonance, semi-forbidden and forbidden lines of hydrogen, helium and the ions of oxygen, carbon, sulfur, and iron. Further downstream ($N_H \geq 10^{16}\,\mathrm{cm^{-2}}$), the hydrogen ionizing photons are absorbed, creating a $10^4\,\mathrm{K}$ temperature plateau. Hydrogen recombination lines and [Ne II] fine structure lines originate there. Further downstream again ($N_H \approx 10^{20}\,\mathrm{cm^{-2}}$), the gas recombines and cools rapidly. Molecular formation sets in at temperatures of a few hundred K. As a result of the high temperatures, neutral-neutral chemical reactions proceed rapidly. Carbon is efficiently converted to CO. The remaining oxygen is converted to OH and H_2O. In at least moderately dense gas ($n(H_2) \geq 7 \times 10^4\,\mathrm{cm^{-3}}$), the H_2 formation energy is converted into gas heating and creates another temperature plateau at $\approx 400\,\mathrm{K}$ (Neufeld and Dalgarno 1989a). The calculations show that much of the infrared emission comes from this plateau and from the temperature region $< 10^4\,\mathrm{K}$ where the gas density is about 100 times greater than in the pre-shock gas. Typically 1 to 10% of the shock's energy emerges in infrared lines. Hollenbach and McKee (1989) and Neufeld and Dalgarno (1989a,b) have calculated the chemical and thermal structure and cooling of dissociative shocks with speeds between 40 and $150\,\mathrm{km\,s^{-1}}$ at preshock densities between 10^3 and $10^6\,\mathrm{cm^{-3}}$. Fig. 13 shows the cooling for the shock of Fig. 12 (Hollenbach and McKee 1989). At medium densities ($\approx 10^4\,\mathrm{cm^{-3}}$) and velocities, the dominant coolants are [O I] $63\,\mu\mathrm{m}$, [O I] $6300\,\mathrm{\AA}$ and [C I] $9849\,\mathrm{\AA}$, followed by H_2 and CO rotational and ro-vibrational line emission. At densities of $10^6\,\mathrm{cm^{-3}}$ or greater the rotational line emission of H_2O and OH dominates and grain cooling becomes important. The [Fe II] lines at $1.3/1.6\,\mu\mathrm{m}$, along with vibrational transitions of H_2, are the most prominent emission lines in the near-infrared range. Infrared hydrogen recombination lines of atomic hydrogen is also detectable from J-shocks, but may easily be confused with HII regions along the line of sight. Depending on the gas phase abundance of silicon, [Si II] $35\,\mu\mathrm{m}$ emission may also be easily detectable. Submillimeter emission of highly excited rotational states of SiO, HCN, CN, SO and NO may be characteristic tracers of J-shocks in dense clouds (Neufeld and Dalgarno 1989b).

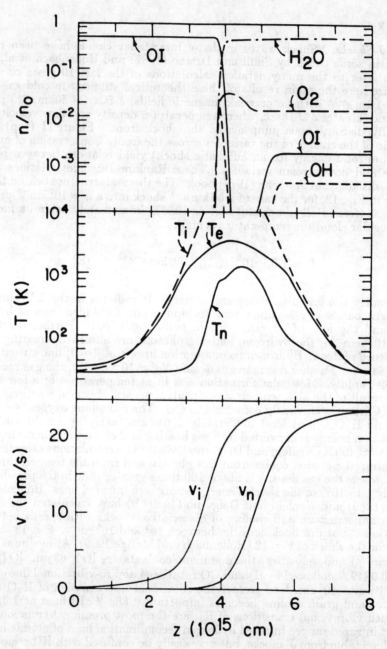

Figure 14. Structure of a $v_s = 25\,\mathrm{km\,s^{-1}}$ shock in a dense molecular cloud with $n_H = 10^6\,\mathrm{cm^{-3}}$, $x_e = 10^{-8}$, and $B_0 = 1.0\,\mathrm{mG}$ (adapted from Draine et al. 1983).

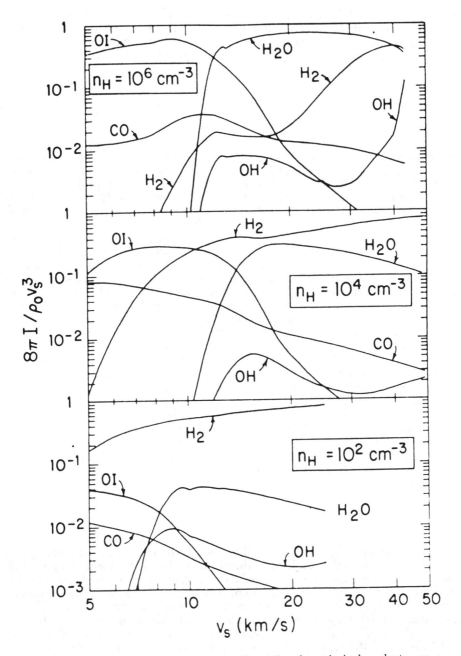

Figure 15. The normalized power radiated by the principal coolants, as a function of the shock speed v_s, in molecular gas with pre-shock density with $n_H = 10^2\,\text{cm}^{-3}$, $x_e = 10^{-4}$ and $B_0 = 10\,\mu\text{G}$; $n_H = 10^4\,\text{cm}^{-3}$, $x_e = 10^{-7}$, $B_0 = 100\,\mu\text{G}$; and $n_H = 10^6\,\text{cm}^{-3}$, $x_e = 10^{-8}$, $B_0 = 1\,\text{mG}$ (adapted from Draine et al. 1983).

ii) C-Shocks. A different type of shock occurs if shock velocities are less than about 40 km s^{-1}, if there is a moderately strong magnetic field and if the ionization fraction in the cloud is moderately low (x_e = [electrons]/[hydrogen] $\leq 10^{-6}$). In the presence of a magnetic field disturbances propagate at the magnetosonic, or Alfvén speed v_A, given by

$$v_A = \frac{B}{\sqrt{4\pi\rho}} = 22\,(B/1\,\text{mG})\,(n/10^4\,\text{cm}^{-3})^{-\frac{1}{2}} \quad [\text{km s}^{-1}], \tag{56}$$

where ρ and n are the mass and number density of the gas. Shocks are possible for $v_A < v_{\text{shock}}$, otherwise the pressure disturbance is "communicated" and damped by Alfvén waves. In partially ionized gas, the ions react rapidly to changes in the magnetic field and then communicate those more slowly to the neutrals by ion-neutral collisions. Because of this difference in reaction speed, it is possible to transmit damped Alfvén waves in the ion "fluid" at the ion magnetosonic speed, $v_A(\text{ion}) = (\rho/\rho(\text{ion}))^{1/2} v_A \gg v_A$. For small ion fraction one may then have the situation $v_A < v_{\text{shock}} < v_A(\text{ion})$. In that case magnetic field and ion density $\rho(\text{ion})$ must vary *continuously* through the shock front. If the neutrals (because of their interaction with the ions) also vary continuously, the shock is called a *"C-shock"* (for continuous, Draine 1980). This situation is depicted in the lower part of Fig. 11. The shock sends a message to the upstream gas via the ions and magnetic field. The ions begin to compress and accelerate, so that they drift relative to the neutrals and heat and accelerate them. The neutral gas can rapidly radiate its thermal energy away. In practice this happens if molecules are not dissociated ($T <$ a few 10^3 K) and the efficient cooling of H_2, CO, OH and H_2O is available. The detailed structure of a 25 km s^{-1} C-shock into a 10^6 cm^{-3} pre-shock density cloud at a magnetic field of 1 mG is shown in Fig. 14 (from the models of Draine et al. 1983). C-shocks radiate nearly the entire energy of the shock in many molecular and atomic infrared lines. Models of Draine et al. (1983) are shown in Fig. 15 for three different pre-shock densities. [O I] 63 µm emission dominates the cooling for slow (10 km s^{-1}) shocks. Rotational and ro-vibrational line emission of H_2 is the main coolant for higher velocities ($10 < v_{\text{shock}} < 50$ km s^{-1}), but moderate pre-shock densities. For densities of 10^6 cm^{-3} or greater, H_2O and OH rotational emission become more and more important.

b. Observations of shocks

The best investigated region is the shock in the Orion-KL star forming region. Since the first detection of H_2 ro-vibrational emission by Gautier et al. (1976), many infrared lines have been inferred to come from shock excited gas. Fig. 16 is a composite of the observed line emission. The shocked zone has a size of about one arcminute (0.15 pc) and is centered on the Orion-KL infrared cluster. The shock is created as a (clumpy) molecular outflow emanating from the young stars in the cluster (IRc2, BN) strikes the surrounding molecular cloud. Chernoff et al. (1982) and Draine and Roberge (1982) have successfully modeled most of the emission as coming from a 40 km s^{-1} C-shock, with a magnetic field of about 1 mG and a pre-shock hydrogen density of 10^5 to 10^6 cm^{-3}. The momentum flux required to drive the shock ($\dot{M}v_{\text{shock}} \approx 10^{30}$ g cm sec^{-2}) is quite consistent with other estimates of

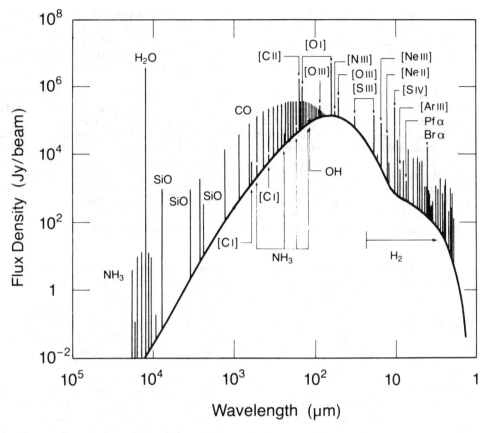

Figure 16. Line emission from the Orion-KL shock (dereddened and for a 30″ beam)

the mass outflow from IRc2. It is apparent from Fig. 16 that the qualitative features of the shocked gas in Orion-KL in fact closely match the theoretical expectations for a C-shock, as they were described in the last section. H_2 and CO infrared line emission are the dominant coolants ($L(H_2 + CO) \approx 500 L_\odot$). The H_2 emission comes from thermalized gas at a temperature near 2000 K. There is little ionization.

While much of the infrared line emmission in Orion-KL can be fit quite plausibly by a single C-shock, there remain several questions. The C-shock models cannot explain the strong [Si II] (and probably also the [O I]) emission and an additional J-shock (perhaps the "wind-shock", Hollenbach 1989) may be required. Brand et al. (1988) have recently argued that the H_2 lines can be better explained by a J-shock. Finally the simple "cloud" C-shock models can also not account for the high velocity wings in the H_2 lines (e.g., Geballe et al. 1986). These high velocity wings may be created in collisions of high velocity cloudlets in the wind. Finally and perhaps most worrysome Wardle (1990) has recently concluded from a theoretical stability analysis that C-shocks are *unstable* to a buckle (Parker) instability of the magnetic fields and the previous time-independent models may not be applicable.

i) Chemical Probes. Fig. 16 shows intense millimeter line emission from many sulfur- and oxygen-rich molecules. Fig. 17 is a comparison of the resulting abundances derived for the shock (the "plateau") relative to those in the quiescent cloud (the "ridge"). This figure is adapted from the paper by Blake et al. (1987) who present an analysis of the Caltech 200 GHz line survey of the Orion region. The measurements clearly indicate a significant change in the chemical composition for the different physical kinematic components in the Orion-KL star formation region. The results for the "plateau" are quite consistent with high temperature shock chemistry. Saturated molecules (HCN, H_2S) or molecules which are preferentially formed at high temperatures (SiO) are one to two orders of magnitude more abundant than in the quiescent cloud where ion-molecule chemistry dominates (see Irvine 1990). Melnick et al. (1990) derive an OH/H_2 abundance of 2×10^{-5} from observations of far-infrared OH lines from the shock, again about 2 orders of magnitude greater than in the quiescent cloud and consistent with the theoretical predictions. One important remaining test is the observation of infrared H_2O lines which ought to be very bright as well ($\geq 10^{-1}$ erg s^{-1} cm^{-2} sr^{-1}: Neufeld and Melnick 1987).

ii) J-Shocks vs. C-Shocks vs. PDR's. How can J- and C-shocks be observationally distinguished and how can shocks be distinguished from photodissociation regions? In comparison to J-shocks, C-shocks have essentially no infrared emission from ionized species ([Ne II], [Fe II], [Si II], hydrogen recombination lines). Furthermore, C-shocks have much larger column densities of moderately high temperature (10^3 K) molecular gas, resulting in a relatively high ratio of far-infrared CO to near-infrared H_2 cooling. Another potential way of distinguishing J- from C-shocks is from the line profiles. In J-shocks, the cooling gas has been fully accelerated and thus emits a narrow line displaced from the velocity of the quiescent cloud (by the shock velocity). In C-shocks, there is a smooth, broad emission profile centered on the velocity of the quiescent cloud, as most of the emission occurs when the gas is just being accelerated.

Perhaps the most definite indicators of a photodissociation region vs. an (ionizing) J-shock are the absolute and relative intensities of the [C II] 158 µm and

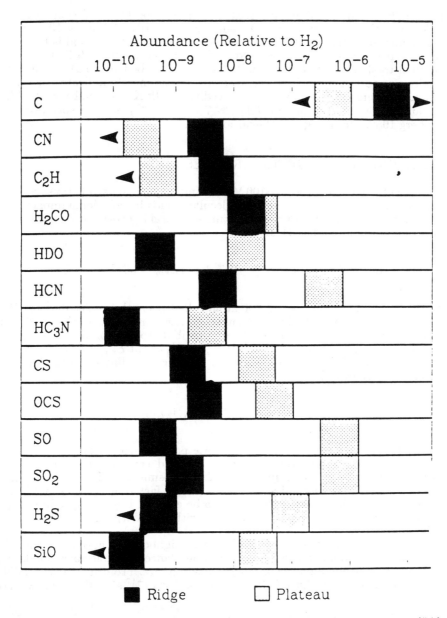

Figure 17. Molecular abundances in the shocked ("plateau") and quiescent ("ridge") gas components of Orion-KL, derived from the 200 GHz Caltech spectral survey (adapted from Blake et al. 1987).

[O I] 63 μm fine structure lines. In shocks one finds weak [C II] emission (or no [C II] emission in C-shocks), but strong [O I] emission ($I([\text{O I}])/I([\text{C II}]) \gg 10$), while the reverse tends to be true in PDR's. Another commonly used method is the ratio of H_2 v = 2→1 to v = 1→0 S(1) line fluxes. If the ratio is small (≤ 0.1), the emission is ascribed to shocks. If the ratio is ≈ 0.5, it is ascribed to a PDR. This is not generally a good indicator, however, as a small ratio merely indicates thermalized emission at a temperature not much exceeding $\approx 10^3$ K. A small v = 2→1/1→0 ratio can also occur in a dense ($n(H_2) \gg 7 \times 10^4 \text{ cm}^{-3}$) photodissociation region (Sternberg 1988, Sternberg and Dalgarno 1989).

4.3. HEATING BY ENERGETIC PARTICLES

Energetic particles (e.g., 1→100 MeV cosmic rays) and hard X-rays (> 250 eV) contribute to the heating throughout molecular clouds by collisional impact ionization. A cosmic ray (usually a proton) hitting a neutral hydrogen atom, for instance, produces a proton and a fast electron,

$$\text{H} + \text{CR}\,(1\to 100\,\text{MeV}) \to \text{H}^+ + e^-\,(E_{\text{kin}} \sim 35\,\text{eV}) + \text{CR}. \tag{57}$$

In the case of molecular instead of atomic hydrogen the molecular ion H_2^+ is produced, the key ion in the ion-molecule chemistry chain (cf., Herbst 1987). The primary electron can ionize again or it can contribute to non-ionizing heating of the gas. In dense neutral gas ($x = n(\text{H}^+)/n(\text{H}) \leq 10^{-3}$) about $E_{\text{nh}} = 7$ eV (Cravens and Dalgarno 1978) are fed into non-ionizing gas heating including secondary processes. The total heating rate then is

$$\Gamma_{\text{cr}} = \zeta_{\text{cr}} E_{\text{nh}}\, n(\text{H}, \text{H}_2), \tag{58}$$

where ζ_{cr} (s^{-1}) is the primary ionization rate per hydrogen atom/molecule. The primary ionization rate in the solar neighborhood can in principle be directly determined from in situ measurements of the cosmic ray flux. The principal difficulty of this direct method so far has been the fact that the measured cosmic ray flux in the vicinity of the Earth needs to be corrected by a large and uncertain factor involving the interaction of cosmic rays with the interplanetary magnetic field and the solar wind.

The best determination of ζ_{cr} in the solar neighborhood currently comes from an analysis of the ionization and chemical abundances in molecular clouds; both being strongly dependent on the primary ionization rate. From an analysis of diffuse clouds van Dishoeck and Black (1986) find $\zeta_{\text{cr}} \approx 4 \times 10^{-17}$ s^{-1}. With this we have

$$\Gamma_{\text{cr}} = 4.7 \times 10^{-28} \left(\frac{\zeta_{\text{cr}}}{4 \times 10^{-17}\,\text{s}^{-1}}\right) \left(\frac{E_{\text{nh}}}{7\,\text{eV}}\right) n(\text{H}, \text{H}_2) \quad [\text{erg s}^{-1}\,\text{cm}^{-3}]. \tag{59}$$

Heating by cosmic rays with $\zeta_{\text{cr}} \sim 10^{-17}$ to 10^{-16} s^{-1} can plausibly account for the kinetic temperatures of the bulk of cold non-star forming molecular clouds,

$T \sim 10\,\text{K}$ (e.g., Solomon et al. 1987). A simplified but fairly reasonable assumption is that the cooling in these clouds is dominated by CO rotational line emission so that

$$\Lambda_{\text{CO}} = n_{\text{CO}} \sum_{J=1} h\nu_J A_J \beta_J(\tau_J) \left(\frac{n_J}{n_{\text{CO}}}\right). \tag{60}$$

Numerical calculations of CO excitation by H_2 with the CO-H_2 cross-sections of Flower and Launay (1985) and radiative transport taken into account by an escape probability formalism (de Jong et al. 1980) indicate for molecular hydrogen densities between 10^3 and $10^4\,\text{cm}^{-3}$ and kinetic temperatures between 5 and 30 K a combined ^{12}CO and ^{13}CO cooling function scaling as

$$\Lambda_{\text{CO}} \approx 10^{-24} \left(\frac{T_{\text{kin}}}{10\,\text{K}}\right)^{2.75} \left(\frac{n(H_2)}{10^3\,\text{cm}^{-3}}\right)^{0.3} \left(\frac{\Delta v/\Delta l}{1\,\text{km s}^{-1}\,\text{pc}^{-1}}\right) \quad [\text{erg s}^{-1}\,\text{cm}^{-3}]. \tag{61}$$

Earlier calculations by Goldsmith and Langer (1978) find a similar total cooling rate (including O_2 and C I). The heating and cooling rates of equations (59) and (61) balance each other at

$$T_{\text{kin}} \approx \left(\frac{2.1\,(\zeta_{\text{cr}}/4 \times 10^{-17})}{(\Delta v/\Delta l)/1\,\text{km s}^{-1}\,\text{pc}^{-1}}\right)^{0.36} n(H_2)^{0.25}. \tag{62}$$

At the default values for velocity gradient, cooling rate and cosmic ray ionization rate and for local molecular hydrogen densities between 10^3 and $10^4\,\text{cm}^{-3}$ equation (62) implies kinetic temperatures between 7 and 13 K, quite consistent with the observations. Cosmic ray heating can also explain higher kinetic temperatures in the vicinity of strong cosmic ray sources, such as supernova remnants.

X-rays photoionize and heat gas in much the same manner as cosmic rays so equations like (58) or (59) are similarly applicable (Krolik and Kallman 1983, Lepp and McCray 1983). In the solar neighborhood the X-ray ionization rate is much smaller than the ionization rate due to cosmic rays suggesting that X-ray heating plays only a minor role in the average interstellar medium. The X-ray luminosity of OB stars is about $10^{-7}\,L_{\text{tot}}$ and T-Tau stars emit about $10^{-4}\,L_{\text{tot}}$ in X-rays (Rosner et al. 1985). The total X-ray luminosity of the central region around θ^1 Ori is about 3×10^{33} to $10^{34}\,\text{erg s}^{-1}$ (Ku and Chanan 1979). Assuming that the stellar X-ray spectrum is thermal with $kT_* \sim 500\,\text{eV}$, Krolik and Kallman (1983) find an overall X-ray heating as a function of hydrogen column density N_H into the cloud and distance R from the X-ray sources of

$$\Gamma_X \approx 3 \times 10^{-26} \left(\frac{N_H}{10^{22}\,\text{cm}^{-2}}\right)^{-1.6} \left(\frac{L_X}{10^{34}\,\text{erg s}^{-1}}\right) \left(\frac{R}{\text{pc}}\right)^{-2} n(H_2) \quad [\text{erg s}^{-1}\,\text{cm}^{-3}]. \tag{63}$$

The assumed X-ray optical depth scales with photon energy E approximately as $\tau_X \sim N_{22} E_{\text{keV}}^{-2.5}$ (Cruddace et al. 1974) where $N_{22} = (N_H/10^{22}\,\text{cm}^{-2})$. Equation (63) shows that near OB and T-Tau associations X-ray heating may significantly contribute to cloud heating and ionization (Krolik and Kallman 1983).

4.4. HEATING BY AMBIPOLAR DIFFUSION

In addition to shocks, "mechanical" heating processes involving the interaction of the gas with surrounding and embedded magnetic fields or magnetic waves may make a significant contribution to cloud heating (Black 1987). In the following we want to consider in particular the heating that results from a slow drift between magnetic field and neutral gas. This "ambipolar diffusion" heating may occur during cloud collapse in dense gas. Following Spitzer (1978) and Zweibel (1987) let us consider a magnetic field of strength B drifting through a neutral cloud of density n_n and ionization fraction x_c ($n_c = x_c n_n$). Recall that the charged particles in the cloud are closely coupled to the magnetic field while the neutrals interact with the field only indirectly via collisions with the charged particles on the "slow down" time scale

$$t_{cn} = (n_n \langle \sigma u \rangle_{cn})^{-1} \approx \frac{10^9}{n_n} \quad [\text{sec}]. \tag{64}$$

The change in momentum of the drift motion due to these collisions is

$$\frac{dp_d}{dt} = -\frac{\mu n_c w_{drift}}{t_{cn}}. \tag{65}$$

w_{drift} is the drift velocity and $\mu = \frac{m_c m_n}{(m_c + m_n)}$ is the reduced mass. For a constant drift velocity to occur this momentum change must be balanced by the force of a magnetic field gradient (magnetic field assumed to be straight and perpendicular to the drift velocity) over scale length L,

$$\frac{dp_d}{dt} = -\left|\frac{\vec{\nabla} B^2}{8\pi}\right| \approx -\frac{B^2}{8\pi L}. \tag{66}$$

The drift velocity can then be expressed as

$$w_{drift} \approx \frac{B^2}{8\pi L n_n^2 x_c \mu \langle \sigma u \rangle_{cn}} \approx 8 \frac{B_{\mu G}^2}{L_{pc} n_n^2 x_c (\mu/m_H)} \quad [\text{cm s}^{-1}]. \tag{67}$$

$B_{\mu G}$ is the magnetic field in units of μG and L_{pc} is in units of pc. The resulting heating rate then is

$$\Gamma_{amb} = \frac{\mu n_c w_{drift}^2}{t_{cn}} \approx 10^{-31} \frac{B_{\mu G}^4}{L_{pc}^2 (\mu/m_H) x_c n_n^2} \quad [\text{erg cm}^{-3} \text{ s}^{-1}]. \tag{68}$$

If B and L were free variables, t_{amb} would scale as n_n^{-2} since the drift velocity scales as n_n^{-2}.

From observations of clouds over a wide range of density and scale size, mean density and scale size appear to be correlated inversely to each other ("Larson's" relation, Larson 1981),

$$n_n L_{\text{pc}} \approx N_0 \approx 2 \times 10^3 \tag{69}$$

(see, for example, Scalo 1990). Furthermore it is expected theoretically that magnetic fields increase with increasing density, $B \sim n_n^\alpha$ with $\alpha \approx 1/2$ (Mouschovias 1987). From the measured values of magnetic fields in diffuse clouds and dense dark clouds (see Heiles 1987, Myers and Goodman, 1988) a reasonable estimate may be

$$B \approx 200 \left(\frac{n_n}{10^4 \text{ cm}^{-3}}\right)^{\frac{1}{2}} [\mu G] = 2 \left(\frac{B_0}{2\mu G}\right) n_n^{\frac{1}{2}}. \tag{70}$$

With (69) and (70) substituted into eq. (68) one finds for molecular hydrogen gas ($\mu = 0.67 m_H$)

$$\Gamma_{\text{amb}} \approx 10^{-29} \left(\frac{B_0}{2\mu G}\right)^4 \left(\frac{10^{-7}}{x_c}\right) \left(\frac{5 \times 10^{21}}{N_0(H_2)}\right)^2 n^2(H_2) \quad [\text{erg cm}^{-3}\text{ s}^{-1}]. \tag{71}$$

Γ_{amb} now appears to increase with n_n^2. Ambipolar heating is of particular importance in dense, very low ionization fraction clumps of small size (or low column density). Ambipolar heating can be neglected in the outer layers of clouds where the ionization fraction may approach a few 10^{-3} due to the far-UV photoionization of carbon.

An interesting question is whether ambipolar heating can contribute significantly to the heating of dense ($n(H_2) \geq 10^5 \text{ cm}^{-3}$), warm ($T \geq 40 \text{ K}$) cloud cores with column densities of $N(H_2) \geq 5 \times 10^{22} \text{ cm}^{-2}$. For these conditions we use the total line cooling function given by Goldsmith and Langer (1978)

$$\Lambda_{\text{tot}} \approx 7 \times 10^{-30} T^{2.2} n(H_2) \quad [\text{erg cm}^{-3}\text{ s}^{-1}]. \tag{72}$$

With equations (71) and (72) we find

$$T_{\text{kin}} \approx 27 \text{ K} \left(\frac{B_0}{2\mu G}\right)^{1.8} \left(\frac{x}{10^{-7}}\right)^{-0.45} \left(\frac{N(H_2)}{5 \times 10^{22}}\right)^{-0.9} \left(\frac{n(H_2)}{10^5}\right)^{0.45} \tag{73}$$

Given the considerable uncertainties of all the parameters going into eq. (73), ambipolar heating may well be an interesting mechanism for explaining temperatures near 50 K. It is less likely, however, that this mechanism can account for substantial column densities at temperatures $\geq 100 \text{ K}$ (see V.2.).

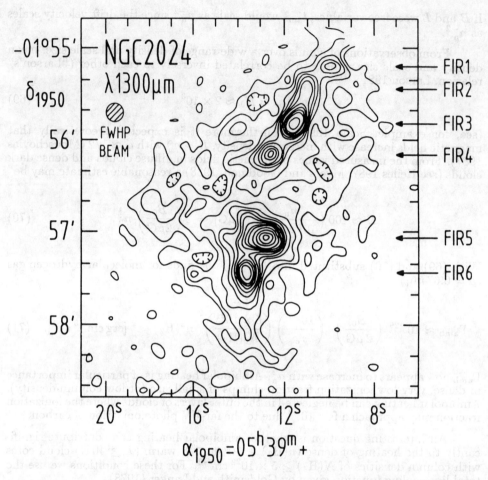

Figure 18. 1.3 mm continuum map of NGC 2024 (Mezger et al. 1988, IRAM 30 m, 13″ beam).

5. Two Puzzles

5.1. HOW TO FIND MASSIVE PROTOSTARS

Finding protostellar condensations after gravitational collapse has formed a stellar core but before nuclear reactions have set in has always been a major goal of millimeter and infrared astronomy. Attempts based on observations of the mid-infrared dust emission in massive star forming sites have largely been unsuccessful since the dust condensations turned out to already contain hot, central stars (Wynn-Williams 1982). Observations in the millimeter and submillimeter range until recently either did not have the necessary spatial resolution, or were hampered by optical depth and chemical effects. Recently Mezger and colleagues (Mezger et al. 1987, 1988, and 1990) have found very compact (size $< 10''$) condensations of 1 mm dust emission from high spatial resolution observations with the IRAM 30 m telescope toward a number of OB star forming regions (Orion-KL, NGC 2024, S255, S106). Fig. 18 shows the map of NGC 2024 (Mezger et al. 1988). There are about half a dozen compact dust emission knots along the north-south dust band across the face of the NGC 2024 HII region. Mezger et al. deduce very high column densities ($N(H_2) \approx$ a few 10^{24} cm^{-2} beam averaged) and low dust temperatures (16 K) from the 1 mm fluxes and a decomposition of the overall millimeter, submillimeter and far-infrared spectrum into 3 different components (Table IV). A key assumption is that the submm/mm absorption efficiency of dust scales like λ^{-2}. Mezger et al. infer masses for the individual condensations of 10 to 60 $M_\odot$ and interpret them as massive isothermal protostars without luminous central stellar cores. Mezger et al. also conclude that the knots were not recognized before in molecular line observations because at the low temperatures molecules are frozen out on dust grains and are significantly depleted in the gas phase.

Table IV. Column densities toward NGC 2024 FIR 5/6[a]

Measurement	$N(H_2)$ [cm^{-2}]	Comments
1 mm dust emission (Mezger et al. 1988)	$2...4 \times 10^{24}$	multicomponent fit $T_d = 16$ K, $Q_d \propto \lambda^{-2}$
dust spectrum (Moore et al. 1989)	$3...6 \times 10^{23}$	single component fit $T_d = 47$ K, $Q_d \propto \lambda^{-1.6}$
^{13}CO 6→5/3→2[b] (Graf et al. 1990)	1.3×10^{23}	$T_{gas} = 95$ K [^{13}CO]/[H$_2$] = 1.2×10^{-6}
C^{17}O/C^{18}O 2→1[c] (Graf et al. 1991)	2×10^{23}	$T_{gas} \approx 67$ K, $n(H_2) = 10^6$ cm^{-3} [C^{17}O]/[H$_2$] = 4.0×10^{-8}

Notes: a 10 to 15" beam
  b $\tau(^{13}$CO 2→1$) \approx 1$ for $N(H_2) \approx 2 \times 10^{23}$ cm^{-2} at $\Delta v \approx 2.5$ km s^{-1}
  c $\tau(C^{17}O) \approx 1$ for $N(H_2) \approx 6 \times 10^{24}$ cm^{-2}

If correct, the observation by Mezger et al. clearly is a key result in understanding protostellar evolution and shows the way to future searches for protostellar candidates from observations of small scale structure in molecular cloud cores.

A detailed comparison with molecular data is necessary in assessing the conclusions of Mezger et al. Moore et al. (1989) have presented a 15″ resolution JCMT map of CS 7→6 emission of the NGC 2024 ridge and Graf et al. (1991) have made 14″ $C^{17}O$ and $C^{18}O$ 2→1 maps with the IRAM 30 m telescope. The molecular maps are in good agreement with each other and in reasonable agreement with the submillimeter dust continuum map. They clearly show condensations in the ridge close to but not identical with the submillimeter peaks. It appears that molecular line data of optically thin or "thinnish" species, if taken at about the same resolution as the dust observations, reproduce the gross features found by Mezger et al. (1988), but leave some significant differences. In particular, there is only one major peak in between FIR 5 and 6 and the density contrast between clumps and the extened emission is less in the molecular maps than in the 1 mm dust emission. While it is possible that some of the CS 7→6 emission is optically thick (Moore et al. 1989), the optical depth of the $C^{17}O$ 2→1 line is low (Table IV). The ratio of the main to satellite hyperfine emission in the $C^{17}O$ emission toward FIR 5 and the ratio of $C^{17}O$ to $C^{18}O$ line intensities across the source are fully consistent with optically thin emission (Graf et al. 1991).

The temperature of the gas emitting in CS and the CO isotopes is fairly high and inconsistent with the 16 K derived for the dust emission. From an analysis of various CS lines Moore et al. (1989) and Schulz et al. (1991) derive a lower limit to the gas temperature of about 28 K and a most probable value near 45 K (see also Evans et al. 1987). Graf et al. (1990) find a gas temperature near 100 K from observations of the ^{13}CO 6→5 and 3→2 lines. Graf et al. (1991) find no evidence for a cold ($T_{ex} < 20\,K$) component in available molecular line data, including the optically thin $C^{17}O$ measurements. Column densities toward FIR 5/6 derived from the optically thin $C^{17}O$ data for any assumed temperature greater than 20 K are within a factor of 2 of the column densities derived from the ^{13}CO measurements that measure the warm gas. This is shown in Table IV. On the other hand, Moore et al. (1989) have shown that a single component model of temperature 47 K can be fit to all dust measurements if the assumption of the λ^{-2} wavelength dependence of the dust emissivity is relaxed (Table IV). The resulting column densities of the one component model are then quite comparable to the column densities derived from the istopic CO data (Table IV). While a single component fit to the entire dust emission spectrum is probably not physically reasonable (Mezger et al. 1988), the argument of Moore et al. is nevertheless a "proof of feasibility" of a solution without a large amount of cold dust. There is other evidence suggesting a weaker than λ^{-2} dependence of submm/mm dust emissivity. Wright and Vogel (1985) and Woody et al. (1989) derive $Q_d \propto \lambda^{-1.2...-1.7}$ from 3 mm to 250 μm continuum observations in several star forming regions. Tielens and Allamandola (1987) comment that layered, amorphous dust particles with a large surface to volume area are theoretically expected to have a λ^{-1} emissivity dependence in the far-infrared.

Several of the FIR peaks of Mezger et al. (1988) may be associated with stellar activity. Richer et al. (1990) find a compact outflow associated with FIR 6. Moore and Chandler (1989) report a 2 μm stellar source at the position of FIR 4.

The argument about the nature of the dust clumps in NGC 2024 and elsewhere is certainly not closed. Molecules are expected to freeze out on dust grains at low temperatures and the resulting near-infrared absorptions by grain mantles have been seen toward a number of embedded compact near-infrared sources. It is suspicious, however, that no trace of any cold molecular material has been seen in the vicinity of the NGC 2024 condensations, or that no compact dust peaks have yet been reported at some distance from luminous OB stars which are likely to substantially heat up the material in their vicinity (Graf et al. 1991).

5.2. WARM QUIESCENT GAS AT THE SURFACES OF MOLECULAR CLOUDS

There is now a substantial amount of evidence for the presence of ≥ 100 K atomic and molecular gas at the surfaces of molecular clouds. The warm atomic gas traced in the bright fine structure lines of [O I], [C II], [Si II] and [C I] has been recently reviewed by Genzel et al. (1989) and will not be discussed here in detail. The far-infrared atomic line emission almost certainly comes from dense photodissociation regions (PDR's) excited by the far-UV radiation impinging on the cloud's surface (Tielens and Hollenbach 1985 and Sternberg and Dalgarno 1989). Harris et al. (1987) found intense submillimeter CO emission (CO 7→6 brightness temperatures near 100 K) of moderately narrow velocity width ($\Delta v \approx 5$ to $10 \, \mathrm{km \, s^{-1}}$) toward S106 and the M17 interface region. Observations in Orion, M17, and S106 show that fluxes increase in intensity all the way to the far-infrared transitions with $J \geq 14$ (Harris et al. 1987, Stacey et al. 1991, Schmid-Burgk et al. 1989 and Boreiko et al. 1989). This requires temperatures in excess of 100 K and densities of at least $10^6 \, \mathrm{cm^{-3}}$. The derived gas temperatures are high enough above the temperatures of the dust in the same regions that heating of the gas by gas-dust collisions can be excluded with some certainty (Harris et al. 1987, Graf et al. 1990). Spatial distributions of the warm molecular gas are similar to that of the warm atomic material (Stutzki et al. 1988, Jaffe et al. 1990, Schmid-Burgk et al. 1989 and Stacey et al. 1991). It is thus highly likely that the warm molecular gas is a surface phenomenon that requires an external heating mechanism.

The warm molecular gas at the cloud surfaces is not just relevant when observing star forming regions like Orion A. Castets et al. (1989) and Gierens (1990), for example, show that an externally heated warm cloud surface is a key assumption in a quantitative interpretation of the ^{12}CO and ^{13}CO emission from the *entire* Orion molecular cloud. The warm molecular gas found in starburst galaxies may also be a direct consequence of the enhanced heating rate at the cloud surfaces (e.g., Genzel 1991).

How much material is associated with the warm gas component? Estimates based on brightness of the mid-J ^{12}CO lines give a lower limit of $\approx 10^{18} \, \mathrm{cm^{-2}}$ (($\tau(^{12}$CO 7→6) ≈ 1) to the CO column density in gas at temperatures of at least 100 K. This limit corresponds to between 5 and 20% of the mass in these regions. The ^{12}CO mid-J lines are likely optically thick. Recently Graf et al. (1990) detected bright ^{13}CO 6→5 emission with the JCMT toward the Orion Bar and NGC 2024. Together with ^{13}CO 3→2 data obtained in the same positions, they were able to derive better column densities, as well as an independent estimate of the kinetic temperature of the isotopic CO emission. Toward the HII region/molecular cloud interfaces in the Orion Bar and NGC 2024 Graf et al. estimate CO column densities between 2.5 and $10 \times 10^{18} \, \mathrm{cm^{-2}}$ ($N(\mathrm{H}_2) \approx 3$ to $13 \times 10^{22} \, \mathrm{cm^{-2}}$) for gas at temperature ≥ 70 K. This is about a factor of 3 to 5 more than obtained from the ^{12}CO data and corresponds to at least 30% of the total molecular column densities in these regions.

The temperatures and column densities of the warm, quiescent molecular gas are very difficult to explain with any reasonable heating mechanism. The high temperatures almost certainly exclude gas-dust collisions, as mentioned above. Slow shocks consistent with the line width fail to explain the line intensities by several orders of magnitude (section IV.2.). Ambipolar diffusion heating (section IV.4.) cannot account for temperatures ≥ 100 K *and* large column densities (eq. (73)). The spatial correlation with the atomic emission suggests an interpretation in terms of photodissociation regions (section IV.1.). The "standard" models of Tielens and Hollenbach (1985) and Sternberg and Dalgarno (1989) fail to explain the submm and far-infrared ^{12}CO emission in Orion by an order of magnitude and the ^{13}CO emission by more. The basic reason is that in the standard models ($n(H_2) \geq 10^5$ cm^{-3}) CO is mostly dissociated in the zones that have temperatures ≥ 100 K. A similar and possibly related problem is the failure of current models to account for the CO abundance in diffuse clouds (van Dishoeck and Black 1986, 1988).

Burton et al. (1989) have proposed that a very dense ($n(H_2) \approx 10^6$ to 10^7 cm^{-3}) photodissociation region accounts for the measurements. In this case the combination of accelerated CO formation rate, enhanced heating rate and H$_2$/CO self-shielding can keep a significant column of molecular gas at temperatures ≈ 100 K. The models of Sternberg (1990) and Burton et al. (1991) with $n(H_2) = 10^7$ cm^{-3} can plausibly fit the ^{12}CO data, but still don't account for the large column densities inferred from the ^{13}CO 6→5 measurements ($I(^{13}$CO 6→5) $\approx 30\%$ to 100% of $I(^{12}$CO 6→5)). If the observed ^{13}CO 6→5 emission is formed in PDR's, the effective far-UV dust absorption cross secton per hydrogen nucleus must be 5 to 10 times smaller than presently assumed ($\approx 10^{-21}$ cm^{-2}) in order to heat the gas to an H$_2$ column density of a few times 10^{22} cm^{-2} (Sternberg 1990). An interesting speculation is whether the clumpy (fractal?) structure of molecular clouds increases the surface to volume ratio so that the UV penetration is significantly enhanced (Boisse 1990).

Acknowledgements. I am grateful to S. Harai, B. Schnelle, and P. van der Werf, for their help in preparing, and A. Harris for a critical reading of this long manuscript. I would also like to thank the organizers, N. Kylafis and C. Lada, for the invitation to attend this most enjoyable summer school.

References

    Bally, J., Langer, W.D., Stark, A.A. and Wilson, R.W. 1987, *Astrophys. J.* **312**, L45

    Bally, J. 1989, in *Low Mass Star Formation and Pre-Main Sequence Objects*, ed. B. Reipurth, ESO Conf.Proc. 33, 1

    Blaauw, A. 1964, *Ann. Rev. Astron. Astrophys.* **2**, 213

    Black, J.H. 1987, in *Interstellar Processes*, eds. D.J. Hollenbach and H.A. Thronson, Reidel (Dordrecht), p. 731

    Black, J.H. and Willner, S.P. 1984, *Astrophys. J.* **279**, 673

    Blake, G.A., Sutton, E.C., Masson, C.R. and Phillips, T.G. 1987, *Astrophys. J.* **315**, 621

    Blitz, L., Bloemen, J.B.G.M., Hermsen, W. and Bania, T.M. 1985, *Astron.*

Astrophys. **143**, 267
Blitz, L. and Stark, A.A. 1986, *Astrophys. J.* **300**, L89
Blitz, L. 1987, in *Physical Processes in Interstellar Clouds*, eds. G. Morfill and M. Scholer, Reidel (Dordrecht), 35Bloemen, J.B.G.M. et al. 1984, *Astron. Astrophys.* **139**, 37
Bloemen, J.B.G.M. 1989, *Ann. Rev. Astron. Astrophys.* **27**, 469
Bohlin, R.C., Savage. B.D. and Drake, J.F. 1978, *Astrophys. J.* **224**, 132
Boisse, P. 1990, preprint
Boreiko, R., Betz, A. and Zmuidzinas, J. 1989, *Astrophys. J.* **337**, 332
Brand, P.W.J.L., Moorhouse, A., Burton, M.G., Geballe, T.R., Bird, M. and Wade, R. 1988, *Astrophys. J.* **334**, L103
Burton, M., Hollenbach, D. and Tielens, A.A.W.G. 1989, in *IR Spectroscopy in Astronomy*, ed. B.H. Kaldeich, ESA-SP-290, 141
Burton, M., Hollenbach, D. and Tielens, A.G.G.M. 1991, *Astrophys. J.* in press
Castets, A., Duvert, G., Bally, J., Wilson, R.W. and Langer, W.D. 1989, in *The Physics and Chemistry of Interstellar Molecular Clouds*, eds. G. Winnewisser and J.T. Armstrong, Springer (Berlin), 133
Chernoff, D.F., Hollenbach, D. and McKee, C.F. 1982, *Astrophys. J.* **259**, L97
Cox, P. and Mezger, P.G. 1989, *Astron. Astrophys. Rev.* in press
Cravers, T.E. and Dalgarno, A. 1978, *Astrophys. J.* **219**, 750
Cruddace, R. Paresce, F., Bowyer, S. and Lampton, M. 1974, *Astrophys. J.* **187**, 497
Dalgarno, A. 1991, in *Interstellar Processes*, eds. D.J. Hollenbach and H.A. Thronson, Reidel (Dordrecht), p. 71
de Jong 1977, *Astron. Astrophys.* **55**, 137
de Jong, T., Dalgarno, A. and Boland, W. 1980, *Astron. Astrophys.* **91**, 68
Dickman, R.L., Snell, R. and Schloerb, F.P. 1986, *Astrophys. J.* **309**, 326
Dragovan, M. 1986, *Astrophys. J.* **308**, 270
Draine, B.T. 1978, *Astrophys. J. Suppl. Ser.* **36**, 595
Draine, B.T. 1980, *Astrophys. J.* **241**, 1021
Draine, B.T. and Roberge, W.G. 1982, *Astrophys. J.* **259**, L91
Draine, B.T., Roberge, W.G. and Dalgarno, A. 1983, *Astrophys. J.* **264**, 485
Draine, B.T. 1989, in *IR Spectroscopy in Astronomy*, ed. B.H. Kaldeich, ESA-SP-290, 93
Dyck, H.M. and Lonsdale, C.J. 1979, *Astrophys. J.* **84**, 1339
Elmegreen, B.G. and Lada, C.J. 1977, *Astrophys. J.* **214**, 725
Evans, N.J., Mundy, L.G. Davis, J.H. and Vanden Bout, P. 1987, *Astrophys. J.* **312**, 344
Falgarone, E. and Puget, J.L. 1988, in *Galactic and Extragalactic Star Formation*, eds. R. Pudritz and M. Fich, Kluwer (Dordrecht), 195
Fiebig, D. and Güsten, R. 1989, *Astron. Astrophys.* **214**, 333
Flower, D.R. and Launay, J.M. 1985, *Monthly Notices Roy. Astron. Soc.* **214**, 271
Frerking, M.A., Langer, W.D. and Wilson, R.W. 1982, *Astrophys. J.* **262**, 590
Fukui, Y. 1989, in *Low Mass Star Formation and Pre-Main Sequence Objects*, ed. B. Reipurth, 95
Gautier, T., Fink, U., Treffers, R. and Larson, H. 1976, *Astrophys. J.* **207**, L129
Geballe, T.R., Persson, S.E., Simon, T., Lonsdale, C.J. and McGregoe, P.J.

1986, *Astrophys. J.* **302**, 500
Genzel, R. and Stacey, G.J. 1985, *Mittg. Astr.G.* **63**, 215
Genzel, R., Harris, A.I. and Stutzki, J. 1989, in *Infrared Spectroscopy in Astronomy*, ed. B.H. Kaldeich, ESA SP-290, 115
Genzel, R. and Stutzki, J. 1989, *Ann. Rev. Astron. Astrophys.* **27**, 41
Genzel, R. 1991, in *Space Chemistry*, eds. J.M. Greenberg and V. Pirronello, Kluwer (Dordrecht), 123
Gerola,H. and Glasssgold,A. 1978, *Astrophys. J. Suppl. Ser.* **37**, 1
Gierens, T. 1990, Ph.D. Thesis, Univ. Cologne
Goldreich, P. and Kwan, J. 1974, *Astrophys. J.* **189**, 441
Goldsmith, P.F. 1987, in *Interstellar Processes*, eds. D.J. Hollenbach and H.A. Thronson, Reidel (Dordrecht), p. 51
Goldsmith, P.F. and Langer, W.D. 1978, *Astrophys. J.* **222**, 881
Graf, U.U. et al. 1990b, in prep.
Graf, U.U., Genzel, R., Harris, A.I., Hills, R.E., Russell, A.P.G. and Stutzki, J. 1990, *Astrophys. J.* **358**, L49
Graf, U.U. et al. 1991, in prep.
Güsten, R., Walmsley, C.M., Ungerechts, H. and Churchwell, W. 1985, *Astron. Astrophys.* **142**, 381
Hansen, S.S., Moran, J.M., Reid, M.J., Johnston, K.J., Spencer, J.H. and Walker, R.C. 1977, *Astrophys. J.* **218**, L65
Harris, A.I. et al. 1991, in prep.
Harris, A.I., Stutzki, J., Genzel, R., Lugten, J.B., Stacey, G.J. and Jaffe, D.T. 1987, *Astrophys. J.* **322**, L49
Heiles, C. 1987, in *Interstellar Processes*, eds. D.J. Hollenbach and H.A. Thronson, Reidel (Dordrecht), p. 171
Heiles, C. and Stevens, M. 1986, *Astrophys. J.* **301**, 331
Herbig, G.H. and Terndrup, D.M. 1986, *Astrophys. J.* **307**, 609
Herbst, E. 1987, in *Interstellar processes*, eds. D.J. Hollenbach and H.A. Thronson, Reidel (Dordrecht), p. 611
Herzberg, G. 1950, *Molecular Spectra and Molecular Structure, I. Spectra of Diatomic Molecules*, Van Nostrand (Toronto)
Hildebrand, R.H. 1983, *Quart. J. Roy. Astron. Soc.* **24**, 267
Hollenbach, D., McKee, C.F.: 1979, *Astrophys. J. Suppl. Ser.* **41**, 555
Hollenbach, D., Chernoff, D.F. and McKee, C.F. 1989, in *IR Spectroscopy in Astronomy*, ed. B.H. Kaldeich, ESA-SP-290, 245
Hollenbach, D. and McKee, C.F. 1989, *Astrophys. J.* **342**, 306
Hollenbach, D., Tielens, A.A.W.G. and Takahashi, T. 1991, *Astrophys. J.*, in press
Howe, J.E., Jaffe, D.T., Genzel, R. and Stacey, G.J. 1990, *Astrophys. J.* in press
Irvine, W.M. 1991, in *Chemistry in Space*, eds. J.M. Greenberg and V. Pirronello, Kluwer (Dordrecht), p. 89
Jaffe, D.T., Harris, A.I. and Genzel, R. 1987, *Astrophys. J.* **316**, 231
Jaffe, D.T., Genzel, R., Harris, A.I., Howe, J., Stacey, G.J. and Stutzki, J. 1990, *Astrophys. J.* **353**, 193
Klein, R.I., McKee, C.F., Sandford, M.T., Whitaker, R. and Ho, P.T.P. 1987, in *Star Forming Regions*, eds. M. Peimbert and J. Jugaku, Reidel (Dordrecht), p. 435
Krolik, J.H. and Kallman, T.R. 1983, *Astrophys. J.* **267**, 610

Ku, W.H.M. and Chanan, G.A. 1979, *Astrophys. J.* **234**, L59
Lada, E. 1990, Ph.D. Thesis Univ. of Texas
Langer, W. 1976, *Astrophys. J.* **206**, 699
Larson, R.B. 1981, *Monthly Notices Roy. Astron. Soc.* **194**, 809
Lepp, S. and McCray, R. 1983, *Astrophys. J.* **269**, 560
Loren, R.B. 1989, *Astrophys. J.* **338**, 902
Maddalena, R.J., Morris, M., Moscovitz, J. and Thaddeus, P. 1986, *Astrophys. J.* **303**, 375
Maloney, P. and Black, J.H. 1988, *Astrophys. J.* **325**, 389
Martin, H.M., Sanders, D.B. and Hills, R.E. 1984, *Monthly Notices Roy. Astron. Soc.* **208**, 35
McCaughrean, M.J., McLean, I.S., Rayner, J.T. and Aspin, C. 1991, in prep.
McKee, C.F. and Hollenbach, D.I. 1980, *Ann. Rev. Astron. Astrophys.* **18**, 219
Melnick, G., Stacey, G.J, Genzel, R., Lugten, J.B. and Poglitsch, A. 1990, *Astrophys. J.* **348**, 161
Mezger, P.G., Chini, R., Kreysa, E. and Gemuend, H.P. 1986, *Astron. Astrophys.* **160**, 324
Mezger, P.G., Chini, R., Kreysa, E. and Wink, J.E. 1987, *Astron. Astrophys.* **182**, 127
Mezger, P.G., Chini, R., Kreysa, E., Wink, J.E. and Salter, C.J. 1988, *Astron. Astrophys.* **191**, 44
Mezger, P.G., Wink, J.E. and Zylka, R. 1990, *Astron. Astrophys.* **228**, 95
Migenes, V., Johnston, K.J., Pauls, T.A. and Wilson, T.L. 1989, *Astrophys. J.* **347**, 294
Monteiro, T. and Flower, D. 1987, *Monthly Notices Roy. Astron. Soc.* , 228, 101
Moore, T.J.T., Chandler, C.J., Gear, W.K. and Mountain, C.M. 1989, *Monthly Notices Roy. Astron. Soc.* **237**, 1P
Moore, T.J.T. and Chandler, C.J. 1989, *Monthly Notices Roy. Astron. Soc.* **241**, 19p
Mouschovias, T. 1987, in *Physical Processes in Interstellar Clouds*, eds. G. Morfill and M. Scholer, Reidel (Dordrecht), p. 453
Mundy, L.G., Cornwell, T.J., Masson, C.R., Scoville, N.Z., Baath, L.B. and Johansson, L.E.B. 1988, *Astrophys. J.* **325**, 382
Myers, P. 1986, in *Star Forming Regions*, eds. M. Peimbert and J. Jugaku, Reidel (Dordrecht), 33
Myers, P.C., 1987, in *Chemistry in Space*, eds. J.M. Greenberg and V. Pirronello, Kluwer (Dordrecht), p. 71
Myers, P.C., Darne, T.M., Thaddeus, P., Cohen, R.S., Silverberg, R.F., Dwek, E. and Hauser, M.G. 1986, *Astrophys. J.* **301**, 398
Myers, P.C. and Goodman, A.A. 1988, *Astrophys. J.* **329**, 392
Natta, A., Palla, F., Panagia, N. Preite-Martinez, A. 1981, *Astron. Astrophys.* **99**, 289
Neufeld, D.A. and Melnick, G.J. 1987, *Astrophys. J.* **322**, 266
Neufeld, D.A. and Dalgarno, A. 1989a, *Astrophys. J.* **340**, 869
Neufeld, D.A. and Dalgarno, A. 1989b, *Astrophys. J.* **344**, 251
Norman, C. and Silk, J. 1980, *Astrophys. J.* **238**, 158
Osterbrock, D.E. 1989, *Astrophysics of Gaseous Nebulae and Active Galactic Nuclei*, Univ. Science Books (Mill Valley)

Phillips, T.G, Huggins, P.J., Wannier, P.G. and Scoville, N.Z. 1979, *Astrophys. J.* **231**, 720
Pottasch, S.R., Wesselius, P. and van Duinen, R. 1979, *Astron. Astrophys.* **74**, L15
Puget, J.L. and Leger, A. 1989, *Ann. Rev. Astron. Astrophys.* **1989 27**, 161
Richer, J. Hills, R. and Padman, R. 1990, in prep.
Rosner, R., Golub, L. and Vaiana, G.S. 1985, *Ann. Rev. Astron. Astrophys.* **23**, 413
Scalo, J. 1990, in *Physical Processes in Fragmentation and Star Formation*, eds. Capuzzo-Dolcetta, R., Chiosi, C. and DiFazio, A., Kluwer (Dordrecht), in press
Schmid-Burgk, J. 1982, Landolt-Boernstein, VI,2c, p. 115
Schmid-Burgk, J. et al. 1989, *Astron. Astrophys.* **215**, 150
Schulz, A., Zylka,R. and Gueston, R. 1991, *Astron. Astrophys.* in press
Sellgren, K. 1984, *Astrophys. J.* **277**, 623
Scoville, N. and Kwan, J. 1976, *Astrophys. J.* **206**, 718Scoville, N.Z., Kleinmann, S.G., Hall, D.N.B. and Ridgway, S.T. 1983, *Astrophys. J.* **275**, 201
Scoville, N.Z. 1984, in *Galactic and Extragalactic Infrared Spectroscopy*, eds. M. Kessler and J. Phillips, AP-SP-108, Reidel (Dordrecht), p. 167
Scoville, N.Z. and Sanders, D.B. 1987, in *Interstellar Processes*, eds. D. Hollenbach and H. Thronson, Reidel (Dordrecht), p. 21
Shu, F.H., Adams, F.C. and Lizano, S. 1987, *Ann. Rev. Astron. Astrophys.* **25**, 23
Shull, M. and Draine, B.T. 1987, in *Interstellar Processes*, eds. D. Hollenbach and H. Thronson, Reidel (Dordrecht)
Snell, R.L. et al. 1984, *Astrophys. J.* **276**, 625
Solomon, P.M., Rivolo, A.R., Barrett, J. and Yahil, A. 1987, *Astrophys. J.* **319**, 730
Spitzer, L. 1978, *Physical Processes in the Interstellar Medium*, Wiley (New York)
Spitzer, L. 1982, *Searching Between Stars*, Yale Univ.Press, ch. VII, 148
Stacey, G., Viscuso, P., Fuller, C. and Kurtz, N. 1985, *Astrophys. J.* **289**, 803
Stacey, G.J., Geis, N., Genzel, R., Harris, A.I., Jaffe, D.T., Poglitsch, A. and Stutzki, J. 1991a, *Astrophys. J.* to be submitted
Stacey, G.J., Geis, N., Genzel, R. Jackson, J.M., Madden, S. Poglitsch, A. and Townes, C.H. 1991b, *Astrophys. J.* submitted
Sternberg, A. 1988, *Astrophys. J.* **332**, 400
Sternberg, A. and Dalgarno, A. 1989, *Astrophys. J.* **338**, 197
Sternberg, A. 1991, in prep.
Strom, K.M., Strom, S.E., Wolff, S.C., Morgan, J. and Wenz, M. 1986, *Astrophys. J. Suppl. Ser.* **62**, 39
Strong, A.W. et al. 1988, *Astron. Astrophys.* **207**, 1
Stutzki, J. and Güsten, R. 1990, *Astrophys. J.* **356**, 513
Stutzki, J., Stacey, G.J., Genzel, R., Harris, A.I., Jaffe, D.T. and Lugten, J.B. 1988, *Astrophys. J.* **332**, 379
Tielens, A.G.G.M. and Allamandola, L.J. 1987, in *Interstellar Processes*, eds. D. Hollenbach and H.A. Thronson, Kluwer (Dordrecht), p. 397
Tielens, A.A.W.G. and Hollenbach, D. 1985a, *Astrophys. J.* **291**, 722
Tielens, A.A.W.G. and Hollenbach, D. 1985b, *Astrophys. J.* **291**, 747

Townes, C.H. and Schawlow, A.L. 1955, *Microwave Spectroscopy*, McGraw Hill (New York)
Ungerechts, H. and Thaddeus, P. 1987, *Astrophys. J. Suppl. Ser.* **63**, 645
van Dishoeck, E. and Black, J. 1986, *Astrophys. J. Suppl. Ser.* **62**, 109
van Dishoeck, E. and Black, J. 1987, in *Physical Processes in Interstellar Clouds*, eds. G. Morfill and M. Scholer, NATO ASI 210, p. 241
van Dishoeck, E. and Black, J. 1988, *Astrophys. J.* **334**, 711
Verschuur, G.L. and Kellermann, K.I. 1988, *Galactic and Extragalactic Radio Astronomy*, Springer (Berlin)
Vrba, F.J., Strom, S.E. and Strom, K.M. 1988, *Astrophys. J.* **96**, 680
Walmsley, C.M. 1975, in *HII Regions and Related Topics*, eds. T. Wilson and D. Downes, Springer (New York), p. 17
Walmsley, C.M., Hermsen, W., Henkel, C., Mauersberger,R., and Wilson,J.L., 1987, *Astron. Astrophys.* **172**, 311
Wannier, P.G. 1980, *Ann. Rev. Astron. Astrophys.* **18**, 399
Wannier, P.G., Penzias, A.A. and Jenkins, E.B. 1982, *Astrophys. J.* **254**, 100
Wardle, M. 1990, *Monthly Notices Roy. Astron. Soc.* **246**, 98
Werner, M. 1970, *Astrophys. J.* **6**, L81
Whitcomb, S.E., Gatley, I., Hildebrand, R.H., Keene, J., Sellgren K. and Werner, M.W. 1981, *Astrophys. J.* **246**, 416
Wild, W. et al. 1991, in prep.
Woody, D.P., Scott, S.L., Scoville, N.Z., Mundy, L.G., Sargent, A.I. et al. 1989, *Astrophys. J.* **337**, L41
Wright, M.C.H. and Vogel, S.N. 1985, *Astrophys. J.* **297**, L11
Wynn-Williams, C.G. 1982, *Ann. Rev. Astron. Astrophys.* **20**, 587
Yorke, H.W. 1988, in *Radiation in Moving Gaseous Media* (18th Saas-Fee Course), eds. Y. Chmielewski and T. Lanz, Geneva Observatory, p. 193
Zinnecker, H. 1989, in *Evolutionary Phenomena in Galaxies*, ed. J. Beckman, Cambridge Univ. Press
Ziurys, L.M. and Turner, B.E. 1986, *Astrophys. J.* **300**, L19
Zweibel, E.G. 1987, in *Interstellar Processes*, eds. D. Hollenbach and H.A. Thronson, Kluwer (Dordrecht), p. 195
Zweibel, E.G. 1987, in *Chemistry in Space*, eds. J.M. Greenberg and V. Pirronello, Kluwer (Dordrecht), p. 195

Reinhard Genzel, Telemachos Mouschovias, Ed Churchwell

NEWLY FORMED MASSIVE STARS

ED CHURCHWELL
University of Wisconsin
Astronomy Department
475 North Charter Street
Madison, WI 53706
U. S. A.

1. Introduction

Massive O stars are responsible for most of the spectacular phenomena occurring in galaxies. There is a dramatic difference in the appearance of galaxies with few and with many O stars. The visual appearance of spiral galaxies is determined by the number and distribution of O stars. O stars emit most of their light in the ultraviolet (UV), giving rise to large HII regions. They have extremely powerful winds which deposit momentum and mechanical energy into the interstellar medium that, over their lifetime, are comparable to those in supernovae explosions. They play an important role in the heating and destruction of molecular clouds. They modify the chemistry in their neighborhood by providing energy for gas phase endothermic reactions, by heating dust which may evaporate volatile grain mantles thus enriching certain molecules in the gas phase, and by dust destruction close to the star. They largely determine the nature of a galaxy's interstellar medium, which controls the rate of formation and composition of the next generation of stars. Finally, at the end of their brief lifetimes (a few times 10^6 yr), they explode as supernovae, thereby enriching the interstellar medium with heavy elements, causing violent shocks which accelerate cosmic rays, destroy dust grains, and propel gas and dust to large distances from the galactic plane. Because of their dominant effects on both local and galactic scales, it is important that we understand the conditions and processes that give rise to the formation and early evolution of O stars and the consequences of this for the structure and evolution of galaxies.

Ultracompact (UC) HII regions are manifestations of newly formed massive stars that are still embedded in their natal molecular clouds. They can be distinguished from classical HII regions produced by O and B stars that are no longer fully embedded in their natal molecular clouds by their small sizes (diameter $\leq 10^{17}$ cm), high densities ($n_e \sim 10^5$ cm^{-3}), and high emission measures (EM $> 10^7$ pc cm^{-6}). They are among the brightest and most luminous objects in the Galaxy at 100 μm due to their warm dust envelopes. Typically, the IR emission peaks at about 100 µm and is 3.5 to 4 orders of magnitude above the free-free emission at this wavelength. In order to produce a detectable UC HII region, the total luminosity must be equivalent to a B3 or hotter main sequence or ZAMS star. By contrast, classical HII regions, such as the Orion Nebula, have diameters typically $\geq 10^{18}$ cm, $n_e \leq 10^4$ cm^{-3}, and EM $\leq 10^6$ pc cm^{-6}. UC HII regions are often seen in the direction of large classical HII region complexes such as W49, W51, NGC7538, etc. because these are the regions where current massive star formation is occurring. Classical diffuse HII regions are produced by a previous generation of massive stars that have either moved out of or managed to destroy the natal molecular cloud in its neighborhood. The UC HII region(s) in the direction of large, diffuse HII regions are located in the molecular cloud which gave birth to both the previous and current generation of massive stars.

In this review, I will discuss our current knowledge of newly formed, massive stars and the impact they have on their environment. In particular, I will briefly summarize in §II the evolution of massive stars up to the UC HII region phase. It is important to establish the evolutionary phase occupied by UC HII regions since this will provide the basis for understanding the physics of these objects. This will be followed by a review of their observed and physical properties in §III and §IV. The bow shock hypothesis for UC HII regions is reviewed in §V. Finally, future programs which might make key contributions to our understanding of this class of objects are discussed in §VI.

2. Evolutionary Perspective

Most of what we know about O stars pertains to relatively advanced evolutionary states after they have either moved out of, or destroyed, the natal molecular cloud out of which they formed. The most interesting phases of an O star's life, its birth and earliest evolutionary phases, are very poorly understood. This is, in part, because they are obscured at optical and ultraviolet (UV) wavelengths by dust in the molecular cloud in which the star formed. They are generally only observable at infrared and radio wavelengths, primarily with low spatial resolutions (single dish observations mostly), and they do not have the benefit of a long history of careful optical study. Until about 10 years ago the only clearly identified pre-main-sequence (PMS) objects were T Tauri stars (masses less than ~2 solar masses). Recognition that luminous, compact IR and radio continuum sources in molecular clouds (such as the BN object in Orion) may also be PMS or ZAMS (zero age main sequence) stars has allowed us to extend the study of formation and early evolution to massive O stars.

Recent reviews of theoretical and observational progress in star formation and early stellar evolution are included in the following compendia: "Birth and Infancy of Stars", edited by Lucas, Omont, and Stora (1985); "Galactic and Extragalactic Star Formation" edited by Pudritz and Fich (1988); and "Workshop on Star Formation" edited by Wolstencroft (1983). Here, I will summarize the most salient steps in the formation of massive stars, as they are presently understood, to set the stage for an in-depth discussion of newly formed O stars. Once a molecular cloud, or some part of it, manages to become gravitationally unstable, three major phases of star formation have been predicted theoretically. The earliest is the **isothermal collapse phase** in which nonhomologous collapse proceeds at the free-fall rate. In this phase, the velocity of collapse is $\propto R^{-1/2}$ where R is the radius, so infall velocities increase toward the center. This results in the formation of a small, dense core. The heat of compression is freely radiated away through an optically thin envelope so the collapsing cloud is isothermal. The time scale for this phase is

$$\tau_{f\text{-}f}(yr) = \left(\frac{3\pi}{32G\rho_o}\right)^{1/2} = 2.1 \times 10^5 \, \rho_o^{-1/2} \tag{1}$$

where ρ_o is the initial density in units of 10^{-19} g cm^{-3}. This phase continues until a small, star-like, optically thick, hydrostatic core is formed containing a few tenths of a solar mass inside a free-falling outer envelope of diameter ~0.1 pc. Since the core is optically thick and cannot cool efficiently, its temperature increases, temporarily halting collapse of the core because of the increase in central pressure. At this point the **accretion phase** begins. During this phase the core grows in mass via accretion from the surrounding protostellar envelope which continues to collapse at the free-fall rate. The core slowly contracts due to accretion toward main sequence temperatures and densities on a Kelvin-Helmholtz time scale given by:

$$\tau_{KH}(yr) = \frac{GM_*^2}{R_*L_*} \approx 3 \times 10^7 \frac{M_*^2}{R_*L_*} \tag{2}$$

where M_*, R_*, and L_* are the mass, radius, and luminosity of the core in solar units. The core evolves at a rate much slower than the envelope. The luminosity of the core is produced both by accretion shocks and by possible thermonuclear reactions. Prior to reaching the main sequence, the luminosity is primarily produced by accretion shocks given by:

$$L_{core} \approx \frac{GM_*\dot{M}}{R_*} \approx 300 L_\odot \frac{M_*\dot{M}}{R_*} \tag{3}$$

where $\dot{M}$ is the accretion rate in units of 10^{-5} M$_\odot$ yr^{-1}. As the core grows in mass, it finally reaches densities and temperatures high enough to support nuclear burning. Ultimately, accretion is halted by a process(es) not yet fully understood (perhaps due to radiation pressure on dust, the on-set of a wind, all available matter used, etc.) and the star settles down on the main sequence at the position appropriate for its mass. It is possible and indeed likely that the core will reach the main sequence before it stops accreting; it will thereafter move up the main sequence until accretion is stopped. During the accretion phase, the infalling matter may prevent the central star from forming a detectable HII region, but as accretion is turned off stars more massive than about 10 solar masses (B3 or hotter) produce enough UV photons to form <u>detectable</u> HII regions. This is the **UC HII region** phase of evolution for massive stars. It is **believed** that this stage includes the ZAMS and earliest stages of evolution on the main sequence. All stars of this class are young relative to stellar evolutionary time scales. They are young even in comparison to the lifetimes of the most massive O stars. Even so, there is now clear evidence that this subclass of young stars spans a significant range of ages far in excess of the dynamical ages one would infer from the compactness of their HII regions (an issue I will return to in §V). Since all massive stars have revealed a wind when observed with high enough sensitivity, I will **assume** that the central stars in UC HII regions have energetic winds, although it is not known when the wind phenomenon begins in massive stars. The morphologies of UC HII regions provide strong indirect evidence for stellar winds (see the discussion in §V). Figure 1 is a schematic illustration of the main stages of massive star formation from its earliest phase of gravitational instability and free-fall collapse to the UC HII region stage. The size of the initial condensation is uncertain because the fraction of the initially collapsing cloud that is finally incorporated into the star is unknown. An initial volume of diameter ~0.5 to ~1 pc is reasonable for a star of final mass 30 $M_\odot$ to form out of a cloud of density $10^4 \, cm^{-3}$.

This review will deal only with the UC HII region phase of massive star evolution. It will be implicitly assumed that the ionizing star has stopped accretion and is on the main sequence or in the final stages of settling onto it. A precautionary note, however, is in order. It has recently been found that several UC HII regions are apparently at the center of luminous, bipolar molecular outflows. Such outflows are thought to be driven by the accretion process and are integrally associated with PMS evolution. The issue for UC HII regions is whether the outflows are still being driven or are relics of an earlier accretion phase. If they are still being driven, the implication is that the stellar core is probably still accreting matter. Resolution of this issue could change our ideas about the evolutionary status of the ionizing stars in UC HII regions.

3. Observed Properties

A. THE CONTINUOUS SPECTRUM

The continuous spectrum of UC HII regions is produced by two mechanisms. At wavelengths ≥ 3 mm, the spectrum is dominated by free-free emission from the HII region. At wavelengths < 3 mm, the spectrum is dominated by thermal emission of warm circumstellar dust which lies mostly outside the HII region. The free-free flux density is:

$$S_\nu = \int B_\nu(T_B) \, d\Omega \quad (4)$$

where $B_\nu(T_B)$ is the Planck function at frequency n and T_B is the brightness

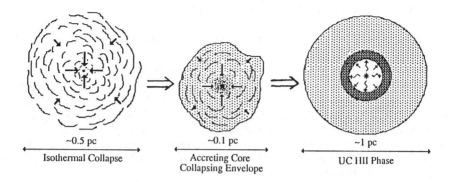

Figure 1
A schematic of the major phases of star formation with approximate size scales. Effects produced by rotation, magnetic fields, clumping, etc. are ignored here.

Left: During the isothermal collapse phase, the cloud is in free-fall and remains cool because the heat generated by infall freely escapes due to the cloud's low optical depth.

Middle: During the accretion phase, a small, dense, optically thick, star-like core has formed at the center of the cloud which contracts on a Kelvin-Helmholtz time scale. The outer envelope continues to contract at the free-fall rate.

Right: After the core finally stops accreting and settles on the main sequence, it forms a wind-blown central cavity surrounded by a dense HII shell which is enshrouded in a large warm molecular envelope. This is the UC HII region phase of evolution.

temperature of the nebula at frequency n. The brightness temperature is related to the kinetic temperature of the electrons, T_e, and the optical depth of the nebula, τ_ν, at frequency ν as follows

$$T_B(\nu) = T_e(1-e^{-\tau_\nu}) \tag{5}$$

In the Raleigh-Jeans regime (a good approximation at radio wavelengths), the flux density of an isothermal nebula becomes:

$$S_\nu \approx \frac{2k}{c^2}\nu^2 T_e \tau_\nu \Omega_s \text{ for } \tau_\nu \ll 1 \tag{6a}$$

and

$$S_\nu \approx \frac{2k}{c^2}\nu^2 T_e \Omega_s \text{ for } \tau_\nu \gg 1 \tag{6b}$$

The free-free optical depth at radio frequencies is

$$\tau_\nu \approx 0.08235 T_e^{-1.35} \nu^{-2.1} EM \qquad (7)$$

where T_e is in K, ν is in GHz, the source solid angle Ω_s is in steradians, and the emission measure $EM \equiv \int f n_e^2 dl$ is in pc cm^{-6} (the integral is taken over the line of sight through the nebula and f is the filling factor). Thus in the optically thin regime (high frequencies), the free-free flux density is almost constant with frequency, $S_\nu \propto \nu^{-0.1}$, and in the optically thick regime (low frequencies) $S_\nu \propto \nu^2$. The emission measure determines the "turn-over" frequency at which a nebula changes from the optically thick to optically thin regime; it is also a measure of the brightness of the nebula when $\tau_\nu < 1$.

At wavelengths shortward of about 3 mm, the thermal emission of circumstellar dust heated by the central star begins to dominate free-free emission. As the stellar UV and optical radiation propagates outward, it is converted to increasingly longer IR wavelengths by repeated absorption and reemission by circumstellar dust. The dust grains radiate like black bodies characterized by the temperature of the grains. The grain temperatures decrease with increasing distance from the star because of increasing dilution of the stellar radiation field and decreasing ability of the dust to absorb the increasingly longer wavelength radiation. Finally, at large enough distances from the star, the radiation is shifted to FIR wavelengths where the dust is an ineffective absorber and the radiation escapes mostly as FIR emission. The dust temperature for a given grain size and composition at a given distance from the star is determined by the equilibrium between the rate of energy absorbed and the rate it is emitted (erg cm^{-3} s^{-1}) which is given by:

$$\int_0^\infty J_\lambda(r) Q_\lambda d\lambda = \int_0^\infty B_\lambda(T_d) Q_\lambda d\lambda \qquad (8)$$

$J_\lambda(r)$ is the mean radiation intensity at distance r from the star at wavelength λ and Q_λ is the absorption/emission efficiency of dust at wavelength λ. The left side of this equation is primarily determined by the stellar UV radiation at the inner surface of the circumstellar dust shell, but rapidly shifts to longer wavelengths at greater distances from the star. $J_\lambda(r)$ is determined by the transfer of radiation through the dust shell out to radius r. Q_λ depends on the grain size and composition and is different at UV and IR wavelengths; on the right side of the equation emission occurs mostly at IR wavelengths because the dust cannot reach temperatures high enough to emit significantly at visible and UV wavelengths. The emergent IR spectrum is determined by the transfer of radiation through the warm dust shell when the whole range of dust sizes and compositions is present; that is, one has to integrate over both the size distribution and composition of the dust at each distance from the star as well as integrate over the line of sight to calculate the emergent spectrum.

The combined radio and IR flux density distribution of W3 (a complex of UC HII regions) was first published by Wynn-Williams, Becklin, and Neugebauer (1972) and extended by Wynn-Williams and Becklin (1974). Prior to the IRAS mission, only a few UC HII regions had spectra available which spanned radio to near infrared (NIR) wavelengths. In the meantime, the IRAS database has become available and the spectra of more than 50 UC HII regions are known from ~6 cm to ~1 μm (see Wood and Churchwell 1989, hereafter WCa). The most striking feature of the spectra shown by WCa is their similarity. The **shape** is essentially the same for all UC HII regions, independent of the spectral type of the ionizing star and of the morphology of the HII region. Figure 2 shows a montage of typical spectra, taken from WCa, which dramatically illustrates this point.

The general features of the IR part of these spectra are: (1) they peak at ~100 μm; (2) the 100 μm peak lies 3.5 to 4 orders of magnitude above the extrapolated free-free flux density at 100 μm; (3) none can be accurately represented by a single-temperature Planck function; and, (4) for wavelengths shortward of ~5 μm, the observed flux densities lie below the extrapolated NIR free-free flux densities. Several objects have a strong 9.7 μm silicate absorption feature. Although no survey has been made to systematically search for the 9.7 μm silicate feature toward UC HII regions, it is likely to be a general feature of their spectra (Jourdain de Muizon, Cox, and Lequeux 1990). In the NIR ($\lambda < 5$ μm), the spectra of UC HII regions do not agree as well from source to source as they do at longer wavelengths. There could be several reasons for this, but not enough work has been done on this to merit speculation at this time.

Because of their high emission measures, UC HII regions have rather high turn-over frequencies. For example, for EM = 5×10^8 pc cm^{-6} and T_e = 8000 K (i.e. typical values for UC HII regions), $\tau_\nu \geq 1$ for $\nu \leq 13$ GHz. Thus, searches for, or attempts to study the properties of, UC HII regions at frequencies significantly below ~5 GHz will exclude most of them, especially those with the highest emission measures which are the ones of highest interest.

The relationship between the flux density of an HII region, the UV photon flux of the ionizing star, and the distance to the HII region is:

$$S_\nu = 2.10 \times 10^{-49} F\, N_c^* a(\nu, T_e) \nu^{-0.1} T_e^{0.45} D^{-2} \tag{9}$$

where S_ν is in Jy, F is the fraction of stellar ionizing photons absorbed by the gas, N_c^* is the number of ionizing photons produced by the central star per second, $a(\nu, T_e)$ is a slowly varying function of frequency and temperature of order unity tabulated by Mezger and Henderson (1967), ν is in GHz, T_e is in K, and D is the distance to the HII region in kpc. As pointed out above, an effective search for UC HII regions should be done at a frequency high enough that the most compact objects will not be missed due to large optical depths. At 23 GHz (1.3 cm), an HII region would have to have an emission measure $>3 \times 10^9$ pc cm^{-6} before optical depth effects would become important; this is also a frequency at which most observatories have good receivers. In Figure 3 the maximum distance to which an HII region ionized by a single star can be detected at 23 GHz as a function of spectral type and telescope sensitivity is plotted. It is assumed here that T_e = 8000 K, F = 0.5,

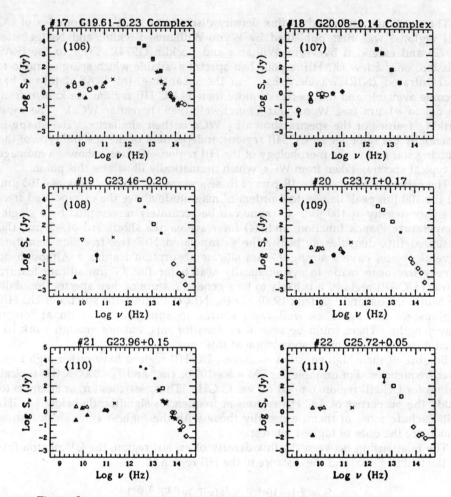

Figure 2
The flux density distribution from 6 cm to about 1 mm is shown for 6 typical UC HII regions. Explanation of the symbols and references are given in WCa. It is noteworthy that the shape of the infared part of these spectra are similar for all UC HII regions independent of morphology and spectral type of the central star.

and the relationship between N_c^* and spectral type is that given by Panagia (1973).This illustrates why surveys with detection limits higher than 100 mJy are all biased toward HII regions ionized by the hottest O stars; HII regions ionized by O6 or hotter stars can be detected in the entire Galaxy, but those ionized by B0.5 or cooler can only be detected within a few hundred parsecs of the Sun.

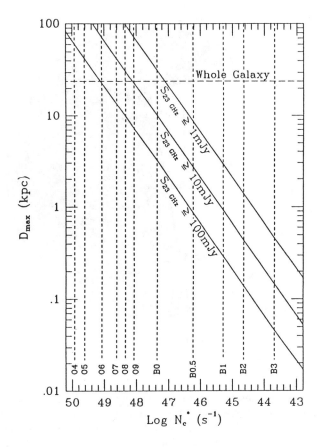

Figure 3 a
The maximum distance to which a UC HII region ionized by a single star with UV photon flux N_c^* can be detected with detection limits of 1, 10, and 100 mJy. It is assumed that dust absorbs 50% of the stellar UV photon flux and that the electron temperature is 8000 K. Assuming the galactic disk has a radius of 15 kpc, one sees that B0 and hotter stars can be detected anywhere in the Galaxy at 23 GHz with a detection limit of 1 mJy.

3.2 SPECTRAL TYPES

One might think that the spectral types of the ionizing stars could be accurately determined for those objects with known distances when the radio continuum

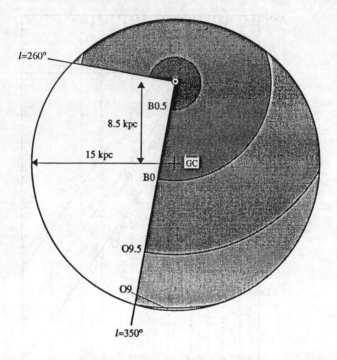

Figure 3 b
This illustrates the fraction of the galactic disk within which a star of a given spectral type can be detected at 23 GHz with the VLA (i.e. $\delta \geq -40°$). A detection limit of 10 mJy is assumed. The scale is indicated. Stars O9 and hotter can be detected anywhere in the galaxy at this detection limit.

and IR flux densities are known. This is not the case for three reasons: a) large uncertainties still exist in the model atmospheres of O stars; b) poor spatial resolution at FIR wavelengths includes emission from stars other than those responsible for ionization; and, c) the fraction of ionizing radiation absorbed by dust within HII regions is still uncertain. The usual technique of converting the radio continuum flux density to a UV photon flux from which the central type of the ionizing star is inferred via model atmospheres suffers both from uncertainties in the models and from the fact that the radio free-free emission only counts the ionizing photons absorbed by the gas. If a significant fraction of the ionizing photons is absorbed by dust in the HII region or escape (i.e. density bounded), the radio continuum emission will **under-estimate** the spectral type of the central star. All stellar UV radiation is likely absorbed within the UC HII region, but the question of absorption by dust **within** the ionized gas is potentially a serious problem. In §IV it will be argued on the basis of models of the circumstellar dust emission that the dust probably lies mostly outside UC HII regions. Integration of the IR flux densities would be a good technique to determine the spectral type of the ionizing star if the total stellar luminosity is converted to IR radiation close to the star and the FIR emission originates from a single star. The first of these requirements appears to be generally satisfied for UC HII regions (see the discussion in §IV below). Resolutions at FIR wavelengths longward of ~60 μm, however, have been typically ≥1 arc min. Toward many regions of massive star formation, this would include several stars in an emerging cluster or association. Even if they are not hot enough to contribute to ionization, they will contribute to the FIR flux densities. Therefore, the single star assumption for the production of the FIR emission will usually estimate too early a spectral type for the ionizing star if only low resolution FIR flux densities such as the IRAS data are available. A third technique, intermediate between the lower limit inferred from the radio continuum emission and the upper limit inferred from the single star assumption for the FIR emission, is to assume an initial mass function (IMF) to estimate the most massive star in a cluster that produces the luminosity derived from the observed FIR flux densities. The problem with this technique is that the IMF for O stars is very poorly known (Scalo 1986). All three techniques have been applied to more than 30 UC HII regions by WCa. They found that the infrared emission using the single star assumption predicts ionizing stars that are 1.5 to 2 subclasses hotter than that inferred from use of an IMF and 2 to 4 subclasses hotter than than those inferred from the radio continuum emission (see Table 18 of WCa). Higher resolution FIR flux density measurements could make a key contribution to the determination of accurate spectral types of the ionizing stars in UC HII regions.

3.3 MORPHOLOGIES

Radio continuum measurements of the structure of UC HII regions began soon after radio interferometers achieved resolutions commensurate with their angular diameters which, as we will see below, are generally ≤ 5". The early observations of DR21 (Ryle and Downes 1967; Webster and Altenhoff 1970; Wynn-Williams 1971; and Harris 1973) and W3 (Webster and Altenhoff 1970; Wynn-Williams 1971; Wynn-Williams, Becklin, and Neugebauer 1972) represent the approximate period when high resolution radio continuum studies of UC HII regions began. In the meantime, with increasing telescope sensitivity and resolution, many more UC HII regions have been identified and imaged at radio and infrared wavelengths. Most of the interferometric radio observations of UC HII regions prior to 1989 have been tabulated by Churchwell (1990).

With few exceptions, prior to the survey of WCa, most UC HII regions that had been imaged with high spatial resolution had a ring or shell morphology. A few were either unresolved or showed a central core surrounded by a more extended halo. One object, G34.26+0.15, was observed to have a cometary morphology (Benson and Johnston 1984; Reid and Ho 1985; Garay, Rodriguez, and van Gorkom 1986). The majority had shell type morphologies. Of the 6 objects imaged by Turner and Matthews (1984), 4 had shell morphologies and 2 had core/halo structures. Felli, Churchwell, and Massi (1984) tabulated 7 UC HII regions with known shell type morphologies up to 1984. Dreher *et al.* (1984) argued that the cluster of UC HII regions they observed toward W49N required shell structures to understand the sizes of the HII regions.

The perception of shells dominating UC HII morphologies changed dramatically with the survey of WCa. They identified five morphological types and found that only 4% of their sample of 75 UC HII regions had shell-type morphologies. Fully 20% of the sample had cometary shapes, 16% had core/halo morphologies, 17% had irregular or multiply peaked brightness distributions, and 43% were either unresolved (resolution 0.4") or could be fitted by gaussian distributions. The latter category was designated as a spherical morphology. An example of three morphological classes is shown in the montage of UC HII regions displayed in Figure 4. Figure 5 is a schematic illustration of the brightness distributions which define each morphological class. The frequency of occurrence of each class is also indicated in Figure 5. Both figures are from WCa.

Since none of the criteria used to select the WCa sample should favor one morphological type over another, it is believed that this distribution is probably representative of UC HII regions. Preliminary results of an independent follow-up survey by Kurtz, Churchwell, and Wood (1990, hereafter KCW) of 64 additional fields at 3.6 and 2 cm support this conclusion. With better resolution, some of the unresolved sources in the spherical category will be found to belong to one of the other morphological types. Also, as pointed out by WCa, their classification scheme is subjective so agreement on classification will not be unanimous. Classification may differ depending on the response of the telescope to different size scales. In particular, objects classified as irregular, or multiply peaked, or spherical are especially prone to be classified differently when observed with telescopes having responses to size scales different from that used by WCa. More precise definitions of the morphological types are given by WCa; the names are reasonably descriptive of the projected radio brightness distribution. Regardless of the name one chooses to give the observed brightness distributions, the important point is that a small number of projected shapes occur often. The interesting question is why. This will be explored in some detail in §V.

3.4 CLUSTERING AND LOCATION

Many UC HII regions appear to be located in clusters. For sometime, regions such as W3, DR21, NGC6334, W49, W51, and NGC7538 have been known to contain clusters of UC HII regions. The VLA surveys of WCa, Garay *et al.* (1990), and KCW have revealed more fields with similar clusters (e.g. G19.61, G20.08, G30.54, G34.26, G10.47, etc.). Clustering is illustrated in Figure 6 which shows a 6 cm image of the CEPH A East region obtained by Hughes (1988).

Extreme examples of compact radio source clumping are found in Orion and Sgr B2. Roughly 30 very compact radio sources have been observed in an area of about 1.5' x 1.5' near the center of the Orion nebula (Garay, Moran, and Reid 1987; Churchwell *et al.* (1987). Not all of these are UC HII regions; some have been identified with stars, some are stellar nonthermal emitters (e.g. θ^1 Ori A), some are

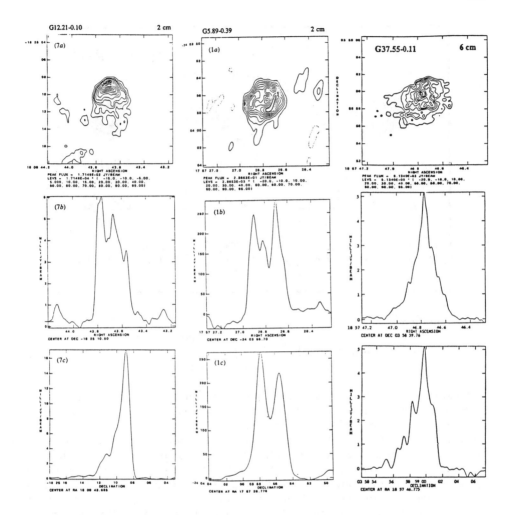

Figure 4
Examples of 3 morphological types of UC UII regions found by WCa. The left panels show a contour map of the shell type UCHII region G5.89-0.39 at λ2 cm. The profiles below are crosscuts through the source in RA and Dec. The middle panels show the cometary UC HII region G12.21-0.10 at 2 cm with crosscuts in RA and Dec under the contour map. The right panels show the core/halo UC HII region G37.55-0.11 at 6 cm with crosscuts in RA and Dec below. The resolution in all cases is 0.4".

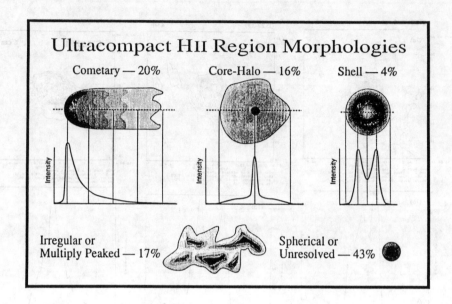

Figure 5
A schematic representation of the morphological types found in the VLA survey of WCa. The percentage of the sample of 75 UC HII regions that show each morphological type is indicated.

embedded neutral globules with thin but dense, ionized, outer envelopes, and some are objects whose emission mechanism(s) have not yet been identified. Gaume and Claussen (1990) have identified at least 20 UC HII regions in an area of about 2' x 1.1' in the core of the Sgr B2 molecular cloud. The mean separation of UC HII regions in the main cluster of this cloud is ~0.06 pc. This is about half that found by Beichman (1979) for 31 Trapezium like clusters. However, proximity of Sgr B2 to the Galactic center probably plays an important role in this issue. Sgr B2 is much denser than molecular clouds of similar size lying further out in the galactic disk; it is the only molecular cloud known to be optically thick even at 100µm (Gatley et al. 1978). The higher densities and resulting deeper gravitational potentials may be the primary determinant of the mean separations in forming clusters. Fragmentation during gravitational collapse has been proposed as the best explanation of such clustering (Beichman 1979).

UC HII regions **do not** appear to be preferentially located near the edge of molecular clouds. They generally have large visual extinctions (e.g. Wood and Churchwell 1990; hereafter WC90), which is consistent with them being deeply embedded in molecular clouds. Further, comparison of the positions of UC HII regions with molecular cloud maps and submillimeter continuum maps shows that UC HII regions are located preferentially in molecular cloud cores. Beichman

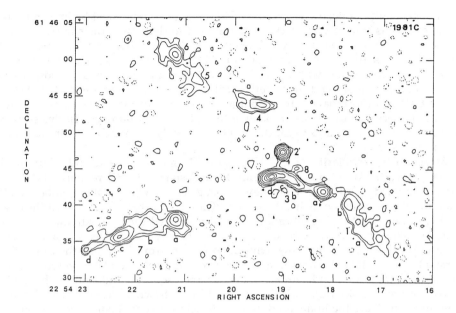

Figure 6
A 6 cm image to the CEPH A East region obtained with the VLA by Hughes (1988). The resolution is ~1". The contour units and other details are given by Hughes (1988). This illustrates the tendency for UC HII regions to be found in clusters.

(1979) reached the same conclusion for IR sources presumed to be protostars. Regions where this conclusion is particularly well illustrated are W3, DR21, and Sgr B2.

3.5 GALACTIC POPULATION AND DISTRIBUTION

If the number, spatial distribution, and ages of newly formed O stars in the Galaxy were known, they would permit a direct determination of the current rate of massive star formation in the Galaxy (independent of uncertainties in the IMF of O stars), provide estimates of the mechanical and radiative energy and momentum input to molecular clouds, allow a determination of the contribution of massive stars to the global energetics of the Galaxy, show the locations of massive star formation relative to spiral arms, and presumably delineate the spiral structure of the Galaxy better than gas tracers such as HI or CO. To estimate the contribution of O stars to the global energetics, one must know, in addition, what fraction of the total O star population resides in molecular clouds where the stars are not generally optically observable. Distances, of course, must be known to use these objects as tracers of galactic structure and to determine their impact on the global energetics of the Galaxy. Obviously, the number and distribution of embedded O stars are fundamental quantities. The availability of the IRAS database in conjunction with recent VLA radio surveys have made it possible to determine the number and **projected** distribution on the sky of embedded O stars in the Galaxy; unfortunately, their **spatial** distribution in the Galaxy is not as well determined because distances are known for only a small number. In this section, a technique to identify and count the number of newly formed O stars in the entire Galaxy is described.

Based on the similarity of UC HII region IR spectra, Wood and Churchwell (1989b; hereafter WCb) noted that embedded O stars (UC HII regions) should be tightly confined in FIR color-color plots. Using the known sample of embedded O stars in WCa to calibrate their FIR colors, WCb showed that this is indeed the case. Perhaps of even greater significance, WCb also showed that the extremely red FIR colors of embedded O stars are shared by few other types of objects. Figure 7 is a color-color plot which shows the embedded O stars from WCa (i.e. the calibration sample) as solid dots. Sources in a $2^o \times 2^o$ box in the galactic plane centered at the arbitrarily chosen longitude 40° are shown as open squares, and sources that lie in a strip around the sky from $13^h \ 00^m$ to $13^h \ 10^m$ are shown as crosses. The dashed lines in Figure 7 are the limits used by WCb to select potential embedded O stars.

All the sources from the WCa calibration sample lie within the dashed box. WCb, therefore, assumed that all IRAS point sources with flux density ratios $\log(F_{60}/F_{12}) \geq 1.30$ and $\log(F_{25}/F_{12}) \geq 0.57$ are potential embedded O stars; the subscripts refer to the wavelength band in μm. Sources with upper limits at either 60 or 25 μm were rejected; those with upper limits at 12 μm, however, were included because this would only shift a source further to the red (diagonally to the upper right within the dashed box in Fig. 7). A search of the entire IRAS Point Source Catalog (PSC) located 1717 objects with the above range of colors. All IRAS sources identified with known stars and galaxies were rejected. The distribution of the selected objects in galactic coordinates is shown in Figure 8.

Figure 8 demonstrates that the selected sample is tightly confined to the galactic plane. Only 7% (126 objects) of the sample have galactic latitudes greater 15°; of these, 58 are in the LMC and 13 are in the SMC. The overwhelming majority of the galactic objects lie within 1° of the plane. Also, most of the sources lie in the 1st and 4th quadrants as one would expect for a disk population observed from the Sun's location in the Galaxy. A histogram of the distribution of the sample in galactic latitude is reproduced in Figure 9. The distribution is centered at 0° to within a single bin ($< 0.2^o$) and falls off exponentially away from the plane with an angular scale height of $0.6^o \pm 0.05^o$ (represented by the dashed curve in Figure 9). WCb point out that this is consistent with the scale height of known O stars in the solar neighborhood.

The distribution of UC HII regions is essentially identical to the distribution of CO clouds projected onto the plane of the sky. This is illustrated in Figure 10 where the distribution of CO emission observed by Dame *et al.* (1987) is compared with that of potential UC HII regions from WCb. In every area where CO emission

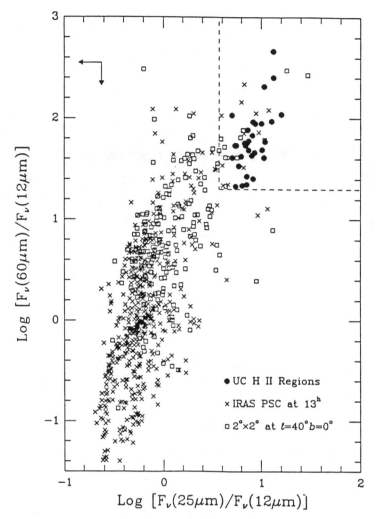

Figure 7
A comparison of the IR colors of known UC HII regions (filled dots) with IRAS point sources in a 2° × 2° box centered in the galactic plane at longitude 40° (open squares) and with IRAS point sources that lie between $13^h\,00^m$ and $13^h\,10^m$ in right ascension (crosses). This figure was taken from WCb and it illustrates that: 1) the FIR colors of UC HII regions are similar; 2) the FIR colors of UC HII regions are distinctly different from IRAS point sources chosen randomly either in the galactic plane or anywhere else on the sky. The limits used to select UC HII regions by FIR colors are indicates by the dashed lines in the upper right of the figure.

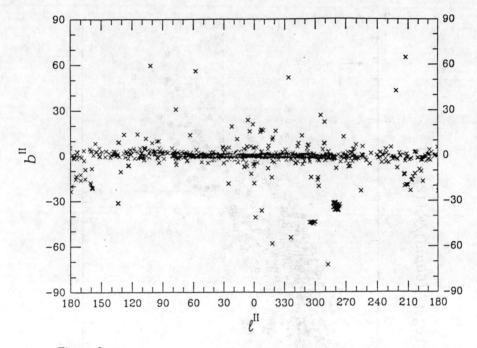

Figure 8

The galactic distribution of the ~1700 embedded OB star Candidates selected on the basis of their FIR colors by WCb. The distribution of the sources close to the galactic plane have an almost one-to-one correlation with the distribution of CO emission of Dame *et al.* (1987). Well known molecular cloud complexes located out of the galactic plane such as Orion, Ophiuchus and the Ceph OB association are easily identified with embedded OB stars in this figure. The LMC and SMC are also apparent at about (l,b)=(270°,-30°) and (300°,45°), respectively.

is concentrated, one also sees a concentration of FIR color selected objects ("embedded O stars"), and in every area in the plane where CO emission is absent, there is a corresponding lack of "embedded O stars". Essentially, one can identify every giant molecular cloud by a concentration of FIR color selected objects that are believed to be predominantly newly formed O stars still embedded in their natal molecular clouds. That we are dealing with a population of objects associated with molecular clouds is obvious from Figure 10. It is not obvious, however, what fraction of the sample is represented by embedded, newly formed, **low-mass** stars, which have FIR colors similar to their more massive counterparts. Because of their lower luminosities, embedded stars of only a few solar masses can only be detected out to a few hundred parsecs from the Sun by all the IRAS broadband

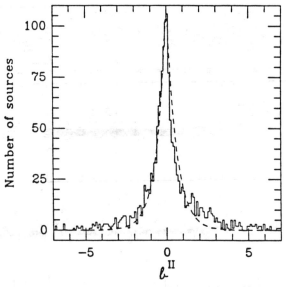

Figure 9
A histogram of the galactic latitude distribution of the embedded OB star candidates selected by the FIR color-color criteria of WCb. The solid line is the data and the dashed line is the function 112 EXP($|b|/0.6°$). This figure illustrates that the OB star candidates are well confined to the galactic plane with a scale height of 0.6° in latitude. At a distance of 8.5 kpc, this corresponds to a z-distance of ~90 pc, typical of the distribution of O-stars in the solar neighborhood.

detectors, but O stars can be detected anywhere in the Galaxy by all detectors. The galactic luminosity function is such that one expects the sample to be dominated by O stars. Contamination by embedded low-mass stars occurs in some of the closest regions such as the Taurus cloud where several sources with the correct FIR colors have been selected even though no O stars are present in this cloud. Since Taurus is one of the nearest molecular clouds to the Sun (~140 pc), it is not surprising that several embedded, low-mass stars were selected there. Clouds more distant than Taurus are likely to be little contaminated by low-mass stars in the color selected sample.

Hughes and MacLeod (1989) have used **optically identified** HII regions to locate IR HII regions from the IRAS PSC. Since their calibration sample was based on optically visible HII regions which are on average probably more evolved, and are associated with smaller amounts of natal dust and molecular gas than UC HII regions, their sample contains a wider variety of objects, including ~950 planetary nebulae. They examined various selection criteria and showed very clearly that planetary nebulae have both different average FIR colors and a distinctly different

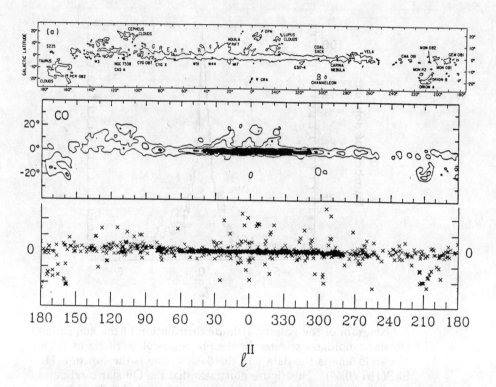

Figure 10
A comparison of the CO emission distribution in the galactic plane with the distribution of FIR color-selected potential embedded O stars. The top panel is a schematic of the CO emission distribution with the major molecular cloud complexes identified. The *middle* panel is a contour map of the CO(J=1-0) emission. The top and middle panels are from Dame *et al.* (1987). The *bottom* panel is the distribution of potential embedded O stars. It is noteworthy that everywhere the CO emission is most intense, the IR selected sources are also most numerous. Essentially every cloud complex identified by molecular line emission can be identified by embedded IR sources in the bottom panel of this figure.

distribution in the Galaxy than HII regions. Jourdain de Muizon, Cox, and Lequeux (1990) and van der Veen and Habing (1988) also find that planetary nebulae and UC HII regions have different FIR colors. The distribution of planetary nebulae above and below the galactic plane is substantially broader than HII regions, but the distribution of HII regions in galactic longitude is broader and more numerous than planetary nebulae. Both HII regions and planetary nebulae peak near longitude 0°. Using less restrictive FIR color selection criteria, Hughes and MacLeod (1989) identified 2298 HII regions to a confidence level of 80% in the IRAS PSC. Many of these are apparently more evolved HII regions that would not appear in the WCb sample. The distribution in galactic coordinates of those objects identified as HII regions in the Hughes and MacLeod sample looks similar to that found by WCb. There appears to be a large intersection of the two samples.

What fraction of the WCb sample is contaminated by objects other than embedded O stars? To answer this question, KCW imaged 64 fields toward IRAS point sources using the VLA in the snapshot mode at 2 and 3.6 cm with resolutions of 0.5" and 1.0", respectively. The sources were chosen entirely on the basis of the FIR color-color selection criteria given by WCb with no other knowledge of the sources. Objects in the WCa sample were not included. Although, the analysis of this survey is still in a preliminary state, it is already clear that a large percentage of the imaged fields contain UC HII regions; over 100 UC HII regions have been detected in this survey. It is too early to estimate the fraction of the imaged fields that contain UC HII regions, but it is clear that the FIR color-color selection criteria developed by WCb to locate UC HII regions is quite reliable.

Taking into account confusion and sources not counted because of upper limits at 25 and/or 60 μm, WCb found that the total number of embedded O stars is greater than 1650 and probably less than 3300. WCb also found that at least 10% of the total O star population resides in molecular clouds, indicating that O stars remain in their natal molecular clouds at least 10% of their main sequence lifetimes. Assuming that O stars remain embedded about 15% of the main sequence lifetime of an O6 star, WCb estimated a current O star formation rate in the Galaxy $>3 \times 10^{-3}$ O stars yr^{-1}. If the embedded massive star candidates found by WCb are assumed to have the same average luminosity as the IRAS sources associated with the UC HII regions in WCa (~$7 \times 10^5 L_\odot$), their total FIR luminosity would be ~$10^9 L_\odot$. This is ~30% of the total FIR luminosity of molecular clouds and ~8% of the FIR luminosity of the Galaxy as a whole (Sodroski 1988). These estimates can be improved when the IMF for massive stars and the current rate of star formation (for all masses) in the Galaxy are better determined and when a volume limited, rather than a flux limited, radio survey of UC HII regions becomes available.

4. Physical Properties

4.1 THE IONIZED GAS

The range of physical properties of UC HII regions are summarized in Table 17 of WCa. These data illustrate that UC HII regions are small (radius typically less than 10^{17} cm), dense (n_e> several times 10^4 cm^{-3}), and have high emission measures (EM > 10^7 pc cm^{-6}). The **rms** electron densities are, without exception, greater than 10^4 cm^{-3}. Several objects have been identified with EM > 10^9 pc cm^{-6} (G5.89, G10.62, G34.26, G45.07, G45.12, and W51d). Since most UC HII regions are not uniformly bright over their projected images (WCa; Garay *et al.* 1990; Gaume and Claussen 1990; and others), their **peak** densities are typically $\geq 10^5$ cm^{-3}. UC HII regions are among the highest emission measure objects in the Galaxy and, when observed with resolutions commensurate with their angular diameters, they are among the brightest IR and radio objects in the Galaxy.

Garay, Rodriguez, and van Gorkom (1986) have analyzed in some detail the structure and physical conditions of the ionized gas in the cometary nebula G34.3+0.2 and the shell structured nebula G45.07+0.13 from high resolution observations of the H76α line and 2 cm continuum emission. Wood and Churchwell (1990; hereafter WC90) have obtained similar data for the cometary nebula G29.96-0.02. Here, I will summarize two important properties of G29.96-0.02 found by WC90 that will be useful in the discussion of bow shocks in §V. First, there is a well ordered, systematic velocity gradient of the ionized gas from ~85 km s^{-1} at the leading edge of the cometary structure to ~110 km s^{-1} in the "tail". The gradient is about twice as steep in front of the cometary arc as behind it. Second, the H76α line widths are 10 to 15 km s^{-1} wider along the leading edge of the cometary arc than elsewhere in the nebula. These results are illustrated in Figure 11.

It is perhaps also of interest that the electron temperature in G29.96-0.02 cannot be higher than about 5000 K. This is consistent with the galactocentric distance of this nebula (4.2 kpc) and the gradient of heavy element abundances found by Shaver *et al.* (1983) with galactic radius. It is also consistent with direct abundance measurements of Herter *et al.* (1981), Lacy, Beck, and Geballe (1982), and Simpson and Rubin (1984; 1990) who find that G29.96-0.02 is over abundant in O, Ne, Ar, and S relative to those in the solar neighborhood.

4.2 THE WARM DUST COCOON

As already noted, UC HII regions are strong IR emitters whose flux density distributions are strikingly similar from object to object (Chini *et al.* 1986a,b; Chini, Krügel, and Wargau 1987; and WCa). The **shape** of the spectrum is independent of the morphology of the ionized gas and spectral type of the ionizing star. The IR emission is thermal radiation produced by circumstellar dust which has been heated by the central star. The similarity of the IR spectra of UC HII regions implies that this process is similar for all objects of this class. Therefore, a successful model of a single object should provide a general basis for understanding the underlying physics of the class as a whole. Other UC HII regions should basically only require modifications in scale length to accommodate differences in intrinsic stellar luminosity and ambient cloud density.

Yorke and Krügel (1977), Yorke (1979, 1980), Yorke and Shustov (1981), and others have calculated models of the expected spectrum of stars of various masses during their approach to the main sequence. Since these deal with the **premain** sequence phases of stellar evolution, I will not discuss them further here because the stars responsible for UC HII regions, although very young, are believed to be main sequence stars.

Krügel and Mezger (1975), Leung (1976), Tielens and de Jong (1979), and Tutukov

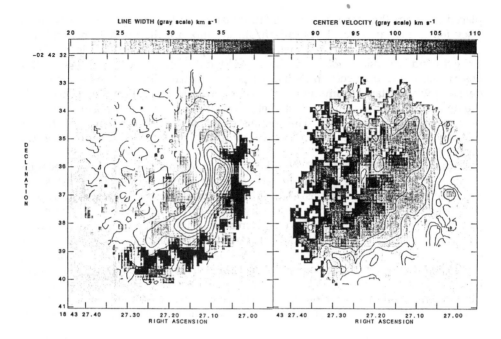

Figure 11
Left : The distribution of H76α line widths toward the cometary UC HII region G29.96-0.02 superimposed on the 2 cm continuum contours. The line widths are ~40 km s^{-1} along the leading edge of the I-front and they decrease to ~25km s^{-1} behind the I-front. The resolution is ~0.5". These data are from WC90.
Right : The distribution of H76α line velocities toward G29.96-0.02. The velocities change from ~85 km s^{-1} along the leading edge of the I-front to ~110 km s^{-1} in the "tail" behind the I-front. The solid curves are constant velocity contours.

and Shustov (1981) were among the first to develop models of the IR emission from warm dust surrounding O stars. Although these models were very informative, they suffered from a lack of FIR data to constrain them, the properties of interstellar dust were more poorly understood than they are today, and a single grain size was used rather than a distribution of sizes. Two recent models of the IR dust emission associated with UC HII regions (Chini, Krügel, and Kreysa 1986; Hoare *et al.* 1988) use the same grain size distribution and optical constants. Both are well constrained by measured IR flux densities. They arrive, however, at rather different conclusions, apparently because of the different density distributions used. Hoare *et al.* (1988) found that the graphite-silicate grain mixture and size distribution of Mathis, Rumpl, and Nordsieck (1977; hereafter MRN) can reproduce the observed FIR fluxes, but only if too much cold dust is placed far from the star when a free-fall density law (i.e. $\rho \propto R^{-3/2}$) is assumed. They found that an amorphous carbon/silicate mixture fits the data better, but it still requires a large outer dust shell radius to reproduce the observations. Chini, Krügel, and Kreysa (1986), on the other hand, obtained good fits to the observations with dust cocoon radii of ~1 pc by requiring, in most cases, increasing densities with distance from the ionizing star and large central cavities. Obviously, the density distribution in the dust shell, as well as the dust composition, are key parameters in such models. The models did not explore the range of density distributions and dust properties that would satisfactorily reproduce the observations.

Churchwell, Wolfire, and Wood (1990; hereafter CWW) have recently completed an extensive set of model calculations of circumstellar dust shells associated with UC HII regions. The primary aim of these models was to explore parameter space to establish the sensitivity of the calculated spectrum to the model parameters. In particular, they were interested in the range of parameter values that would satisfy the observational constraints. They used a mixture of graphite and silicate grains with the optical constants of Draine (1985) and Draine and Lee (1984, 1987; hereafter DL) and the grain size distribution of MRN. The models used 25 grain sizes in the range 0.005 to 0.25 μm for both the silicate and graphite grains. The spherical radiative transfer code of Wolfire and Cassinelli (1986) was used. The other assumptions, model parameters, and calculation techniques are discussed in detail by CWW.

Figures 12, 13, and 14 show the results of the "best fit model" found by CWW for the UC HII region G5.89-0.39. This object has an ionized spherical shell morphology with an outer radius of $\sim 9.3 \times 10^{16}$ cm, an inner radius of $\sim 4.6 \times 10^{16}$ cm, a maximum emission measure of 2.4×10^9 pc cm^{-6}, and an *rms* density of $\sim 8 \times 10^4$ cm^{-3}. If ionization is due to a single star, then an O6 ZAMS star is required. Figure 12 shows the calculated spectrum superimposed on the observations in the upper panel, the run of average dust temperatures with distance from the star in the middle panel, and the optical depth of the dust cocoon with frequency in the lower panel. Figure 13 shows the surface brightness as a function of distance from the star at 12, 25, 60, and 100 μm and Figure 14 shows the fraction of the flux density contained within angular radius θ at the same bands. This model produces a good fit to the observed spectrum (i.e. reproduces the 1.3 mm flux density, peaks at ~100 μm, matches the depth and breadth of the 9.7 μm silicate feature, and reproduces the slope of the NIR part of the spectrum); it predicts angular diameters consistent with inclusion in the IRAS PSC; and it is consistent with the NIR optical depths inferred from observed radio continuum and NIR hydrogen recombination lines. The most important general results of CWW are:

(1) The only models of dust cocoons around newly formed O stars that satisfied **all** the observational constraints required **large** but **thin** shells with about **half** the graphite/silicate abundance found for the **diffuse** interstellar medium by MRN and DL. Large outer radii (~1 pc for an O6 star) are required in order to have enough cool dust for the spectrum to peak at ~100 μm as observed. A large (inner radius ~10^{17} cm for an O6 star), dust-free cavity is required to prevent too much hot dust close to the star from producing more NIR emission than observed.

Figure 12

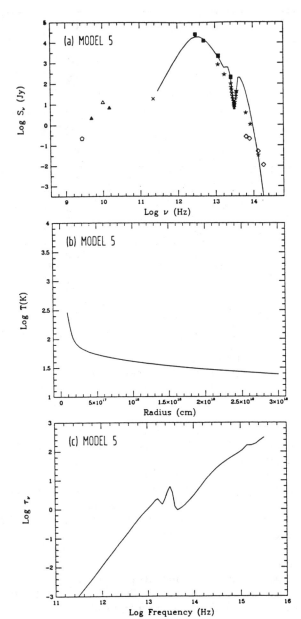

Top : The best fit dust model (solid curve) calculated by CWW superimposed on the observed data (symbols) of G5.89-0.39. References to the symbols are given by WCa, and the model parameters are given by CWW. The dust distribution is a constant density, evacuated shell of inner radius 9×10^{16} cm and outer radius 3×10^{18} cm. The dust has the size distribution of MRN ranging from 0.005 to 0.25 µm. An important feature of this model is that the graphite/silicate abundance ration of the dust had to be reduced by a factor of almost half that of the MRN-DL mixture appropriate for the diffuse interstellar medium.

Middle : The average dust temperature as a function of distance from the star for model 5. More than 99.9% of the volume of the dust shell has temperatures <100K. Paradoxically, the hottest main sequence stars appear to be among the coolest objects in the Galaxy at FIR wavelengths.

Bottom : The optical depth as a function of frequency for model 5. The dust cocoon is opaque at wavelengths shortward of ~7 µm.

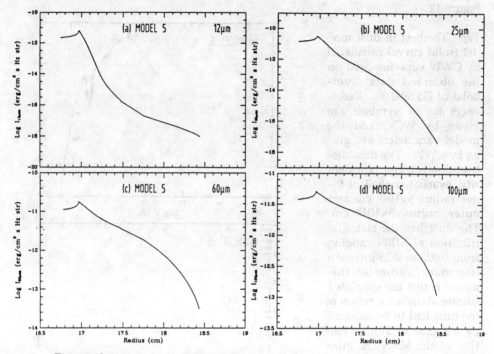

Figure 13
The surface brightness distribution with radius for model 5 at 12, 25, 60, and 100 μm. The peak in each panel is due to the shell geometry, it occurs where the line of sight is a maximum.

Chini, Krügel, and Kreysa (1986) also required large dust-free central cavities to achieve acceptable fits to the observations. It is also interesting to note that the inner dust radius is larger than the observed outer radius of the HII region, implying that absorption of stellar UV radiation by dust **within** the HII region may not be very important. A graphite/silicate abundance ratio of only about half that of the MRN-DL mixture (found for the diffuse interstellar medium) is required to fit the 9.7 μm silicate feature without the NIR optical depths becoming larger than implied by NIR hydrogen recombination lines. Lee and Draine (1985) came to a similar conclusion for the dust around the BN object in the Orion molecular cloud.

(2) Constant density models produce the best fit to the observations. Power law density distributions with indices ≤-1 produce too much warm dust to fit the observations with any combination of input parameters, and those with indices ≥+1 do not produce enough warm dust to fit the NIR part of the spectrum. Acceptable fits with indices between ±1/2 were found, but they were no better than the constant density models.

(3) The dust cocoons in the "best fit models" are optically thick at wavelengths

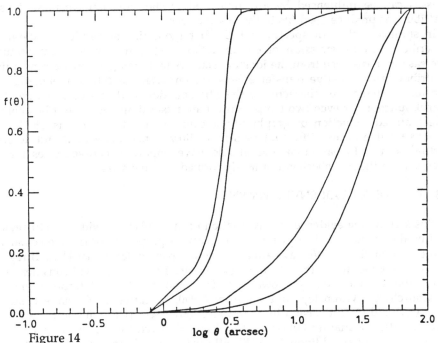

Figure 14

The fraction of the total flux density f(θ) contained within angular radius θ at 12, 25, 60, and 100 μm for model 5. Note that at 60 and 100 μm the half power radii are becoming significant (~20.3" and 32.2", respectively).

shortward of ~7 μm. Optical depths of 5-6 at 2.2 μm and 25-26 at 1 μm are predicted. Therefore, unless the dust is very clumpy, one should not expect to see the central star at wavelengths less than ~1 μm.

(4) The average dust temperatures drop very steeply at the inner boundary of the dust shell to <100 K in a distance < 0.1% of the outer cocoon radius. The average temperature implied by the peak of the FIR spectrum is only about 30 K, which makes embedded O stars among the coolest compact objects in the Galaxy at FIR wavelengths.

(5) Addition of water ice mantles to refractory grain cores produces a strong 3.07 μm absorption feature and weaker 12 and 45 μm absorption features, but does not significantly alter either the average dust temperatures with radius or the optical depths with frequency (except at ice resonances).

(6) The large dust shell radii and high densities required in the CWW models imply large masses in the warm dust cocoons. Total masses greater than 5000 $M_\odot$ are implied in all their models. The dust cocoon or the "sphere of influence" (i.e. the distance from the central star within which the dust temperatures are higher than the ambient cloud temperature) is larger than the volume involved in the formation of a massive star. Thus, one is not necessarily surprised that such large

masses can be influenced by an embedded O star nor that constant density distributions produce good fits to the observations.

In summary, the FIR spectra of UC HII regions (i.e. embedded O stars) are described well by emission of circumstellar dust grains whose temperatures decrease with distance from the ionizing star. A wide range of dust temperatures contribute and radiative transfer effects are important at NIR wavelengths and shorter. Attempts to characterize the FIR emission with a single temperature Planck function (or even two temperatures) are misleading and should be avoided. The required reduction of graphite-to-silicates in the dust cocoons of UC HII regions relative to the diffuse interstellar medium is not understood and deserves further study. In particular, this should have important consequences for the chemistry in the neighborhood of newly formed massive stars.

4.3 THE MOLECULAR ENVIRONMENT

In this section, the evidence for association of molecular gas with UC HII regions will be discussed and the dynamics and physical properties of the molecular gas will be summarized. We have already seen indirect evidence that UC HII regions are closely associated with CO clouds (see Figure 10). Here, I will consider direct observations of molecular gas toward UC HII regions and their interpretations.

Churchwell, Walmsley, and Cesaroni (1990, hereafter CWC) have observed NH_3 (1,1) and (2,2) inversion lines and H_2O (6_{16}-5_{23}) maser emission toward 84 sources with a spatial resolution of 40". They observed the UC HII regions in the WCa sample and an additional 20 UC HII *candidates* selected according to the FIR two-color selection criteria of WCb. This is the only extensive survey of molecular line emission toward UC HII regions presently available. There are several groups, however, currently involved with more extensive molecular line surveys: Bronfman, May, and Nyman (1990) have observed CS (2-1) line emission from more than 200 potential UC HII regions in the southern hemisphere; Henkel and Churchwell (1991) are observing CH_3CN emission for a subset of the sources for which CS lines have been observed to obtain estimates of the gas temperatures; and Kurtz *et al.* (1991) are observing CO emission toward several hundred UC HII regions observable from the Five College Radio Astronomy Observatory (FCRAO) in Massachusetts.

In the CWC survey, ammonia was detected in 70% of the sample, water in 67%, and water or ammonia in 83% of the sample. This high detection rate leaves no doubt that molecular gas is closely associated with UC HII regions and strongly supports the premise that they are enshrouded in natal molecular gas. The high percentage of detected H_2O masers implies that the conditions necessary to produce these masers must persist for most of the lifetime of the ultracompact stage of evolution. In addition to H_2O masers, OH masers (Gaume and Mutel 1987 and references therein) and CH_3OH masers (Menten 1987; Haschick, Menten, and Baan 1990 and references therein) are common features toward UC HII regions. Although no extensive survey in either molecule exists, enough detections toward UC HII regions have been made to suggest that both masers could be about as prevalent as H_2O masers.

The velocities of NH_3 and H_2O were used by CWC to estimate kinematic distances using a modern scale size and rotation curve of the Galaxy (Wouterloot and Brand 1989; Wouterloot et al. 1990). Even with great effort to resolve distance ambiguities, a plot of the UC HII regions projected onto the galactic plane does not delineate a clear-cut spiral arm pattern. This is presumably because distance errors are still too large to preserve the spiral pattern. The average NH_3 line width was 3.1 km s^{-1}. CWC reported column densities and kinetic gas temperatures, derived from the NH_3 data, for a number of objects. Several statistical relationships were also investigated.

What are the dynamics of the circumnebular molecular gas? There is strong evidence that the molecular gas surrounding UC HII regions is collapsing toward a central mass concentration. In a series of papers (Zheng et al. 1985; Ho and Haschick 1986; Keto, Ho, and Reid 1987; and Keto, Ho, and Haschick 1987, 1988) based on high resolution NH_3 (1,1) emission and absorption observations of G10.6-0.4 using the VLA, Ho and coworkers have argued that the molecular gas is gravitationally collapsing and spinning up as it approaches the central star or star cluster. In this picture, compactness of the HII region is maintained by infall of neutral matter. Forster et al. (1990) have interpreted their observations in terms of an HII region expanding into surrounding molecular gas which has a density gradient, somewhat reminiscent of a champagne flow. Forster et al. cannot account for the large number of UC HII regions. Ho et al. requires large infall rates for very long periods of time to be consistent with the number of UC HII regions. The Ho et al. picture would appear to be more appropriately applied to earlier evolutionary phases when the central star is still accreting matter. Reid et al. (1980) argued that the OH masers toward W3(OH) are consistent with infall of the molecular envelope toward the central star and inconsistent with a rotating disk or a shock compressed layer expanding outward with the HII region. Welch et al. (1987) observed inverse P Cygni profiles of HCO$^+$ toward W49A and interpreted these as evidence for infall of a cool molecular envelope. Rudolph et al. (1990) have found similar profiles toward four UC HII regions in the W51 complex; they also argue for in-falling molecular envelopes. The central masses inferred from the HCO$^+$ data toward both W49A and W51 are in excess of 10^3 M$_\odot$. This is much greater than the mass of the ionizing stars and their associated HII regions. The critical issue is whether the in-falling molecular envelope is responding to a dense, massive, molecular core or to the UC HII region and its associated stars. There may also be a contradiction between the predictions of the dust models discussed in §IV and the in-falling molecular envelope. The dust models require essentially a constant density with radius in the range ~0.1 to ~1 pc, but a free-falling envelope implies a strong dependence of density on radius ($\rho \propto R^{-3/2}$). If the infall is occurring over a size scale much larger than 1 pc in response to a massive, more extended, molecular core, then there may be no conflict between the dust models and the observed in-falling molecular gas. This issue is not yet resolved and will require more high resolution imaging and perhaps additional dust models to make further progress.

The detection of bipolar molecular outflows apparently centered on several UC HII regions introduces yet another complication. If bipolar molecular outflows turn out to be generally associated with UC HII regions, then we need to determine if they are remnants from an earlier accretion phase or if they are still being driven. In any case, outflows introduce a complication into the dynamics of the circumnebular gas that is not addressed by any of the proposed models.

The physical conditions of the molecular gas in the vicinity of the UC HII region, G10.6-0.4, have been studied by Keto, Ho, and Haschick (1987) via high resolution observations of NH_3. They found that the gas temperature scales with radius as $R^{-1/2}$, with T=140 K at 0.05 pc from the continuum peak and T= 54 K at 0.35 pc from the continuum peak. If the gas is heated by collisions with warm dust, then these temperatures are quite consistent with those predicted by the dust models of CWW discussed in §IV B. The NH_3 column density is centrally peaked and declines with radius as R^{-1}; at 0.05 pc the NH_3 column density is 1×10^{17} cm^{-2}. They determined that the gas density depends on radius as R^{-2} with $n(H_2) \geq 6 \times 10^6$ cm^{-3} in absorption (i.e. near the I-front) and 1×10^5 cm^{-3} in emission. The densities could be $\geq 10^7$ cm^{-3} near the I-front when correction for a filling factor of ~0.3 is applied. Such densities are required for the OH masers and are in the range expected in the neutral bow shock.

There is growing evidence that the chemistry in the vicinity of UC HII regions (embedded O stars) is different from that in cold dark clouds. This is not entirely unexpected because the environment around embedded O stars provides several additional chemical processes not generally available in ambient molecular clouds. Shock heating and density enhancements both increase reaction rates and make certain endothermic reactions possible. Recall that the shocks experienced by molecular gas around embedded O stars are expected to have velocities < 10 km s^{-1}. Although the shocks have large Mach numbers, they are nondissociative for most molecular species. Higher radiation densities also increase photoreaction rates. Evaporation of grain mantles due to grain heating may enhance gas phase abundances of mantle constituents such as NH_3, H_2O, and others by large factors. Although it is difficult to separate excitation from abundance effects in the warm regions around embedded O stars, NH_3/H_2 appears to be enhanced by an order of magnitude or more near UC HII regions relative to that in cold molecular clouds (Wilson *et al.* 1983; Mauersberger *et al.* 1986; Henkel *et al.* 1987). This may also be true of other nitrogen bearing molecules such as HC_3N and C_2H_5CN (Blake *et al.* 1987). It has been suggested that evaporation of grain mantles may be important in the formation of deuterated molecular species observed toward "hot cores" Walmsley (1989, and references therein). CWW found no evidence for water ice mantles on grains in the dust cocoon around G5.89-0.39, perhaps indicating that the ice mantles have been destroyed in the warm cocoon.

In summary, UC HII regions are closely associated with molecular clouds. We have learned about some of the statistical properties of the molecular gas, but much more needs to be done. There is strong evidence for infall of molecular gas, but the scale size over which this occurs is still somewhat uncertain. Fundamental uncertainties remain regarding the dynamics of the molecular gas interior to a few tenths of a parsec. The relative importance of grain mantle destruction, surface chemistry, and shock chemistry in the vicinity of embedded O stars are still unclear, although we do know that the chemistry is different from that in cold dark clouds. Finally, it is important to obtain more high spatial resolution molecular line observations toward UC HII regions to determine the run of temperature, density, and velocity with position.

5. Bow Shocks

5.1 MOTIVATIONS FOR BOW SHOCKS

The large number of UC HII regions detected by WCa, Garay et al. (1990), and KCW in a sparsely sampled fraction of the galactic disk poses an interesting problem. There are far too many UC HII regions! One is forced to conclude that the rate of massive star formation is much greater than other indicators suggest or UC HII regions somehow remain compact much longer than dynamical evolution calculations suggest. For example, a UC HII region with an electron density of 10^5 cm^{-3} surrounding a **stationary** O6 ZAMS star located at a distance of 5 kpc from the Sun would expand from its initial Strömgren radius (1.0 x 10^{17} cm) to an angular diameter $\geq 5"$ (radius $\geq 1.9 \times 10^{17}$ cm) in < 3500 yr. Without revising the rate of massive star formation upward by a factor of ~10^2, such short dynamical time scales for the UC phase of HII regions are inconsistent with the number imaged with the VLA and way out of line with the >1600 potential UC HII regions found in the IRAS PSC. An increase of the massive star formation rate by this much is inconsistent with other independent data and does not appear to be a viable option.

A possible solution to this dilemma is suggested by the observed morphologies of UC HII regions. Cometary structures are suggestive of bow shocks. The high frequency of cometary structures among UC HII regions essentially demands a bow shock interpretation. In fact, WCa suggested on the basis of the large number and the observed morphologies of UC HII regions that embedded O stars are probably in supersonic motion relative to the ambient molecular cloud and have formed molecular bow shocks supported by stellar winds from the ionizing stars. Further observational support for this hypothesis has come from molecular absorption line studies, which show that ionized and molecular gas approach each other. In the cometary UC HII region G34.26+0.15, the ionized and molecular gas approach each other with speeds of about 10 km s^{-1}. This is probably an extreme velocity difference. Foster et al. (1990) and Churchwell et al. (1990) find velocity differences to be generally <5 km s^{-1}. Figure 15 shows the velocities of molecular and ionized gas toward G34.26+0.15. The OH maser spots observed by Gaume and Mutel (1987) are situated around the leading edge of the parabolic ionization front (I-front) of G34.26+0.15, precisely where the molecular bow shock is expected to be. This is the region where the densities and temperatures are expected to be high enough to produce OH masers. The distribution of OH masers relative to the I-front is shown in Figure 16.

The presence of bow shocks around embedded O stars fundamentally alters the impact of newly formed O stars on their molecular environments. First, the bow shock provides a very dense neutral medium which will usually trap the I-front in

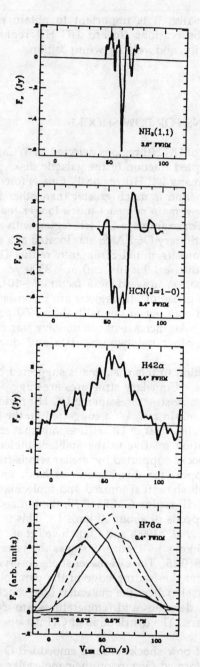

Figure 15

A comparison of the central line velocities and line profiles of NH_3 (1,1), HCN (J=1-0), H42α, and H76α toward G34.26+0.15 with resolutions $\leq 3.8"$. This figure is taken from Churchwell (1988). One sees that the molecular gas has a center velocity of about 60 km s^{-1} and the mean velocity of the HII region is about 50 km s^{-1}. Also, the HCN line is much broader than the NH_3 (1,1) line. The blending of the hyperfine structure components of HCN cannot account for all the additional broadening. The H76α lines are profiles from different positions in the nebula as indicated: this panel is from Garay, Rodriguez, and van Gorkom (1986).

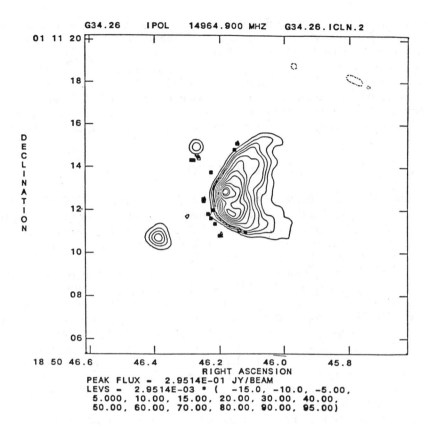

Figure 16
The position of OH masers (filled squares) from Gaume and Mutel (1987) superimposed on the 2 cm map of G34.26+0.15 from WCa. The distribution of the OH masers along the leading edge of the I-front is suggestive of a bow shock.

the forward and perpendicular directions of motion (see Van Buren et al. 1990), causing the HII region to remain small in these directions as long as the O star remains in its natal molecular cloud. Thus, the ambient molecular cloud is not dispersed or destroyed by ionizing radiation beyond a small distance from the star. Second, the stellar wind will be largely dissipated interior to the I-front (kinetic energy converted to radiation and momentum supports the bow shock) and therefore, it is not available to drive supersonic turbulence within molecular clouds. Although turbulence exists in molecular clouds, it is not clear if the winds from early evolutionary stages of O stars are primarily responsible. Third, populated highly excited energy levels of certain molecular species (eg.

vibrationally excited states and high rotational states) become explicable if they originate in the bow shocks associated with embedded O stars. For example, Walmsley et al. (1986) have observed OH $^2\Pi_{3/2}$, J=9/2 in absorption against 7 UC HII regions. This rotational level lies 511 K above ground. They inferred rotational excitation temperatures ranging from ~70 K to 550 K. In a sense, a bow shock acts like a glove which confines the ionizing stellar radiation, its wind, and energetic phenomena (such as excitation of high energy states) to a small region around the star, thereby insulating the rest of the molecular cloud from the destructive effects of the star.

5.2 STELLAR WIND SUPPORTED BOW SHOCKS

The reason bow shocks solve the problem with large numbers of UC HII regions is that a bow shock is a static configuration, neither expanding nor contracting with time as long as the star remains inside its natal molecular cloud. Forces are balanced; the ram pressure produced by the ambient molecular gas flowing around the bow shock (due to the star's supersonic motion through the molecular cloud) is balanced by the stellar wind pressure on the ionized gas. This balance of forces results in the characteristic parabolic shape of the I-front with strong edge-brightening in the direction of motion of the star and a slow tapering off of brightness in the direction from which the star came with the I-fronts becoming parallel to the velocity vector behind the star. Effectively, the star bores a hot ionized tunnel through the molecular cloud; the I-front does not continue to flare out away from the axis of symmetry behind the star in the bow shock model. The motion of the star through the ambient molecular gas need not be large, since the sound speed in molecular clouds is only ~0.2 km s^{-1}. Speeds of a few km s^{-1} are quite adequate to produce a strong shock in the molecular gas. The more fundamental question is how O stars achieve motions relative to their natal molecular gas. This is not clear, but it is an empirical fact that OB associations have velocity dispersions of several km s^{-1}. It is reasonable to assume that these motions will also be reflected in the interactions of newly formed stars with their environment. It is well known, for example, that the Trapezium stars associated with the Orion nebula have motions of ~3 km s^{-1} relative to the Orion molecular cloud (Zuckerman 1973).

Figure 17 shows a schematic of a stellar wind supported bow shock associated with an O star that is moving supersonically through the ambient molecular cloud with speed v_*. The theory of a stellar wind supported bow shock has been developed by Van Buren et al. (1990). They assume that the wind momentum flux supports the ram pressure of the ambient medium. The wind kinetic energy is believed to be converted to radiation in the shocked wind region and ultimately radiated away as IR emission. For a star with stellar wind terminal velocity v_w and mass loss rate $\dot{M}_*$ moving through an ambient medium with density nH = 2n(H$_2$) + n(HI) + n(HII), the terminal wind shock occurs at a distance l in front of the star (but behind the I-front, see Fig. 17) where the momentum flux in the wind equals the ram pressure of the ambient medium. This "stand off distance" is given by (Van Buren et al. 1990; Weaver et al. 1977):

$$l = 5.50 \times 10^{16} \sqrt{\frac{\dot{M}_* v_w}{\mu(H) n(H) v_*^2}} \tag{10}$$

where the units are l in cm, $\dot{M}_*$ in 10^{-6} $M_\odot yr^{-1}$, v_w in 10^8 cm s^{-1}, v_* in 10^6 cm s^{-1}, n(H) in cm^{-3}, and $\mu(H)$ is the mean mass per hydrogen particle in grams. This distance establishes a size scale for the bow shock which agrees very well with high resolution radio observations. A more precise comparison will be possible when the ionizing star can be localized by IR observations.

Mac Low et al. (1990) have also shown that the shape of the I-front near the vertex is closely described by the equation for a parabola

$$y = \frac{x^2}{3l} \tag{11}$$

where y is along the direction of motion on the sky with the origin at the vertex of the parabolic I-front. Others have also predicted a similar relation (see references in Van Buren et al. 1990). Van Buren et al. (1990) also showed that in most realistic cases, the I-front will be trapped behind the bow shock because of the high neutral particle densities achieved in a radiatively cooled bow shock. The thickness of the ionized shell in the vicinity of the stagnation point (i.e. at the stand off distance ahead of the star) can then be estimated from the requirement that it be in pressure equilibrium with the stellar wind momentum flux. This gives (Van Buren et al. 1990):

$$\delta_{HII} = 1.82 \times 10^{17} \frac{N_* T_e^2 \gamma_I^2}{(\dot{M}_* v_w n(H) \mu(H) \alpha v_*^2)} \tag{12}$$

where N_* is the stellar ionizing photon flux in units of 10^{49} photons s^{-1}, α is the recombination coefficient to all levels but the ground state in units of 10^{-13} cm^3 s^{-1}, γ_I is the ratio of specific heats in the neutral ambient medium. The units are δ_{HII} in cm, T_e in K, $\dot{M}_*$ in 10^{-6} $M_\odot$ yr^{-1}, v_w in 10^8 cm s^{-1}, n(H) in 10^5 cm^{-3}, and v_* in 10^6 cm s^{-1}. Essentially all the important parameters that one can compare with observations (eg. the stand off distance, the thickness of the ionized shell, and the shape of the bow shock) depend on the stellar wind strength ($\dot{M}_*$ and v_w), the velocity of the star relative to the molecular gas (v_*), and the ambient density (n_H). A schematic of the approximate variations of temperature and density expected with distance from the star along the direction of motion is illustrated in Figure 18. Note that the region between the star and the bow shock is isobaric.

In principle, with well chosen observations one should be able to unambiguously confirm or disprove the bow shock hypothesis. Van Buren et al. (1990) have suggested the following observational tests of the bow shock

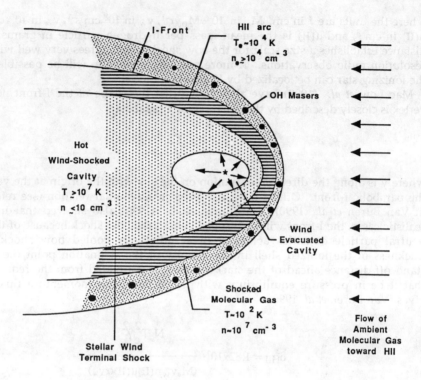

Figure 17
A schematic of a wind-supported bow shock associated with an O star moving supersonically through a molecular cloud. The rest frame is that of the star. The arrows on the right indicate the flow of ambient molecular gas toward the bow shock. The stellar wind velocity is typically ~2000 km s^{-1} and the velocity of the star through the cloud is typically < 10 km s^{-1}. Five regions are indicated. A small wind-evacuated cavity surrounds the star. This is enshrouded by a very hot, low density, wind shocked region bounded by the wind terminal shock. A dense, photoionized, HII shell lies between the wind terminal shock and the neutral bow shock. The bow shock is primarily molecular and probably contains the OH masers associated with UC HII regions. Outside the bow shock lies the ambient molecular cloud. The temperatures and densities are typical values.

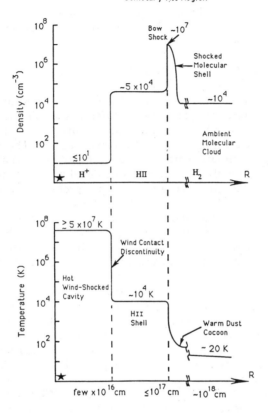

Figure 18
Schematic temperature and density profiles along the symmetry axis of a typical cometary UC HII region. The symmetry axis is along the velocity vector of the star relative to the molecular cloud in which it is embedded. The HII shell and bow shock are "ahead" of the star. The position of the star is indi-cated by the star symbol. The medium is isobaric out to the bow shock; this particular example assumes $nT=5 \times 10^8$ K cm^{-3}. The hot, low density, wind-shocked central cavity extends out to a few $\times 10^{16}$ cm from the star where the terminal wind shock occurs. The dense HII shell (observable at radio and mm waves) lies between the terminal wind shock and the neutral bow shock. The density of the bow shock is 2-3 orders of magnitude greater than the HII region and the ambient cloud because it can cool effectively via C^+, H_2, and CO lines. The precise shape of these profiles, particularly near the shock boundaries, are not well determined.

hypothesis: 1) measure the proper motions of UC HII regions relative to features in the molecular cloud; 2) measure the proper motions of OH maser spots relative to the bow shock pattern; 3) observe IR fine-structure lines of Fe VII (9.51 and 7.81 µm), which are expected to be produced in the central wind-shocked region with detectable strengths; and, 4) observe IR line emission from H_2 (9.665 µm) and O I (63 µm) from the outer molecular bow shock.

5.3 COMPARISON WITH OBSERVATIONS

The bow shock hypothesis must reproduce the observed morphologies with the correct relative frequencies if it is to be considered a viable theory for UC HII regions. Mac Low et al. (1990) have numerically calculated the equilibrium shape of wind-blown shells assuming momentum conservation. The shell was then illuminated by ionizing radiation from the central star; the radiation transfer of the free-free emission through the shell was calculated and projected onto the the sky from various aspect angles. The stellar wind properties were chosen to be consistent with the spectral type of the ionizing star which was inferred from the integrated FIR emission. The velocity of the star v_*, density of the ambient medium, and orientation of the bow shock with the line of sight were adjusted until an acceptable comparison with the observed brightness distribution of the free-free emission from the HII region was obtained. Figure 19 shows one such comparison.

Mac Low et al. (1990) found that all morphologies observed by WCa, except circularly symmetric shells and irregular shapes, could be reproduced by bow shocks. Also, the relative frequencies of the various morphological types found by WCa seem to be consistent with the bow shock hypothesis. The fact that only about 20% of all UC HII regions have cometary shapes is consistent with most of them being bow shocks when one takes into account different orientations of the line of sight, a range of optical depths, and the distribution of distances to UC HII regions (Mac Low et al. 1990). Recall that ~43% of the WCa sample were either unresolved or had gaussian shapes not much larger than their 0.4" beam. Magnetic fields have not been included in the models. This is a serious weakness and should be included in the next generation of bow shock models.

5.4 COMPARISON WITH CHAMPAGNE FLOWS

In a series of papers, Tenorio-Tagle and coworkers have presented numerical simulations of the evolution of HII regions in media with strong density gradients (Tenorio-Tagle 1979; Bodenheimer, Tenorio-Tagle, and Yorke 1979; Tenorio-Tagle, Yorke, and Bodenheimer 1979; Bedijn and Tenorio-Tagle 1981; and Yorke, Tenorio-Tagle, and Bodenheimer 1983). These models have been called "champagne flows" or "blisters" and seem to have been motivated by the belief prevalent in the 1970's that O stars form preferentially near the boundaries of molecular clouds; although true in a few well-known cases, this does not appear to be the general case (see § III D). Yorke et al. (1983) have presented model radio continuum maps of champagne flows seen from various aspect angles and at

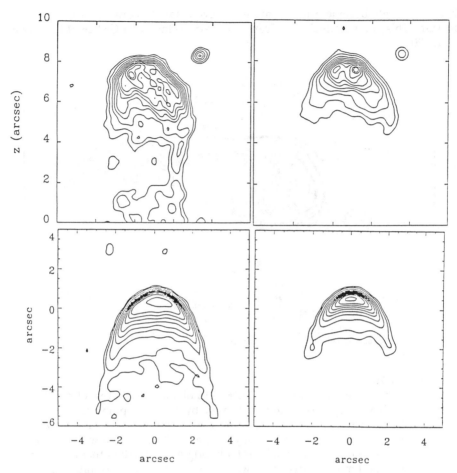

Figure 19
A comparison of the observed 6 cm and 2 cm maps of the cometary nebula G34.26+0.15 and the calculated bow shock models of LBWC. The orientation, velocity, and electron density have been adjusted to obtain an acceptable similarity with the observed object. This shows that a cometary morphology is consistent with the bow shock hypothesis, although it is not a sufficient proof.

different phases of their evolution. There are several serious discrepancies between the predictions of these models and real cometary UC HII regions. First, cometary UC HII regions are limb-brightened, but the models predict centrally peaked brightness distributions. Second, at the points where the ionized gas breaks out of the molecular cloud (near the nozzle), the models predict that the ionized gas should flare out away from the axis of symmetry, but the observations show that the I-fronts become almost parallel or even start to converge downstream of the star. This is illustrated very clearly in the deep image of G34.26+0.15 shown in Figure 20.

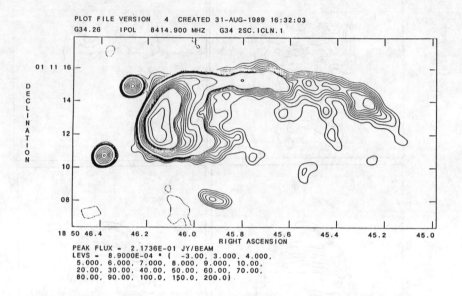

Figure 20

A deep 3.6 cm image of G34.26+0.15 obtained with the VLA. The HPBW is approximately indicated by the two point sources in front of the parabolic arc. The lower contours are emphasized here to show the morphology of the "tail". It is noteworthy that one side of the tail is substantially brighter than the other and that the tail tends to curve in toward the axis of symmetry rather than flare out away from it down stream from the star.

Third, the observed relative approach of the molecular and ionized gas is not explicable in the champagne model, which predicts that the ionized gas should be flowing away from the molecular gas toward lower density regions. Fourth, the velocities achieved should only approach the sound speed of the ionized gas near the nozzle, not near the coma as observed. Fifth, champagne flows are dynamical models which must evolve on dynamical time scales. Sixth, the increase in line width along the leading edge of the cometary arc in G29.96-0.02 and the change in radial velocities across this nebula (see Figure 11) are inconsistent with champagne models. Finally, the champagne models so far available have ignored the effects of stellar winds, which for O stars can be dramatic. Champagne flows do not account for the morphologies, gas dynamics, or the number of observed UC HII regions. This does not mean that champagne flows do not occur in nature; many of the specific objects that Tenorio-Tagle and coworkers had in mind when developing the models are well described by them, but cometary UC HII regions are not. S106, for example, appears to be a champagne flow, but it is not a UC HII region. Champagne flow models apparently apply to HII regions during their transition from the UC stage to a classical nebula as they exit their natal molecular cloud. A major improvement to the champagne flow models would be to include the effects of stellar winds.

The role of stellar winds is strongly implicated by cometary and shell morphologies. Such structures are unstable and would dissipate in a few thousand years without a force to maintain them. This is because over-pressure in the higher density shell would fill the inner cavity in about one sound crossing time. A stellar wind provides a natural mechanism to maintain the central cavity. In light of the model results discussed in §III, it does not appear that radiation pressure on dust is likely to be a viable mechanism to maintain the central cavity.

5.5 WARM MOLECULAR CLOUD CORES AND THE LIFETIME PROBLEM

We have postulated that bow shocks must increase the lifetimes of UC HII regions (over their dynamical time scale) by a factor of ~100 to be consistent with both the number of observed objects and the rate of massive star formation. That is, rather than ages of a few times 10^3 yr, they must remain in their UC state for periods of a few times 10^5 yr. WCb showed that O stars are embedded in their natal molecular clouds on average about 15% of the main sequence lifetime of an O6 star. This implies that UC HII regions could maintain their compact configuration for periods $\geq 10^5$ yr, rather than ~10^3 yr (for dynamical models). A consequence of this hypothesis is that most of the UC HII regions observed in the Galaxy must be bow shocks.

An important issue here is whether this is consistent with the properties of warm molecular cloud cores where O stars are formed. Stated another way, are warm molecular cloud cores where O stars are formed large enough to contain a supersonically moving O star for periods of a few times 10^5 yr as required? To address this issue it is useful to recall that: 1) the sound speed in molecular cores is only ~0.2 km s^{-1}; 2) the dispersion velocities of OB associations are generally <5 km s^{-1} with typical values of ~2 km s^{-1} (Blaauw 1964; Jones and Walker 1988); and, 3) the H$_2$ column densities in the directions of embedded massive stars are in the range of 10^{22} - 10^{25} cm^{-2} with a typical value of ~10^{23} cm^{-2}. If the ambient density is ~10^4 cm^{-3}, see Figure 18, then a typical massive cloud core radius would be ~1.6 pc. This size scale is also consistent with NH$_3$ maps of warm molecular cloud cores (Ungerechts 1980; Ungerechts, Walmsley, and Winnewisser 1986; and others). Thus a star traveling with a typical speed of 2 km s^{-1} could remain in the core for ~7.8x10^5 yr. Even if we take the extreme case in which the star has a speed of 10 km^{-1}, it could be embedded in the core for a period of ~1.6x10^5 yr. Warm massive molecular clouds are clumpy (see the review of Wilson and Walmsley 1989 and references therein). NH$_3$ analyses (Ungerechts 1980; Ungerechts, Walmsley, and Winnewisser 1986; and others) indicate typical beam filling factors of ~0.3 and clumps with density ~10^5 cm^{-3} and radii ~0.05 pc. Within a core of radius 1.6 pc, there would be about 300 such clumps with average separation ~0.47 pc which would fill only ~0.009 of the core volume and contain about 790 $M_\odot$ of material. The total mass in this core (with interclump density ~10^4 cm^{-3}) would be about 9000 $M_\odot$, a small fraction of a giant molecular cloud mass. A bow shock surrounding an O6 star in this core would have a cross section of ~0.064 pc (twice the Strömgren radius) and would collide with a clump on time scales of ~8×10^6 yr (v_*= 2 km s^{-1}) to ~2×10^6 yr (v_*=10 km s^{-1}). Thus it is unlikely that a bow shock will collide with an embedded clump before escaping from the core; it is, therefore, not surprising that observed morphologies are generally symmetric with almost no distortions that would imply a collision with a dense molecular clump. The warm cloud core properties used here can be adjusted somewhat in either direction and still satisfy the lifetime requirements. Difficulties arise only when the ambient core densities become $\geq 10^5$ cm^{-3} where the cores are either too small to contain a star long enough or the cores contain too much mass. Observational evidence seems to be against such high ambient densities over volumes of radius > 1 pc.

5.6 BIPOLAR MOLECULAR OUTFLOWS

Finally, I wish to mention a recent observational result which, if a common phenomenon, will have important consequences for our understanding of newly formed O stars and their impact on the molecular environment. A bipolar molecular outflow is associated with the UC HII region G5.89-0.39 (Harvey and Forvielle 1988). This is the most luminous outflow yet detected and it is the first one identified with a massive O star; thus it makes an important link with low-mass star formation.. Bipolar CO outflows are generally associated with low-mass star formation. The association of a massive molecular outflow with G5.89-0.39 raises many questions regarding the acceleration mechanism(s) and possible distortions of the ionized shell. Since the ionizing star has presumably long since stopped accreting matter, one wonders if the CO outflow is a relic from the accretion phase or if it is still being fed. NGC2024(FIR5) (Sanders and Willner 1985) and DR21 (Garden, Grolemund, and Carlstrom 1990) also apparently have associated bipolar molecular outflows: Guilloteau (1990) has also detected molecular outflows from several other UC HII regions. If molecular outflows turn out to be a general feature of UC HII regions, then major modifications may have to be made to our present understanding of these objects.

5.7 SUMMARY

In summary, the evidence that bow shocks dominate the structure of UC HII regions is very compelling. If this is the correct picture, then standard theoretical models of the evolution of newly formed O stars will require some fundamental modifications. The ionized zones around such stars are not spherical cocoons, nor are the O stars at the focus of champagne flows. Rather, the shape of the ionization zone is fundamentally determined by the supersonic motion of an O star through the molecular cloud in which the star is embedded. Bow shocks would drastically limit the effects of the stellar UV radiation and its wind to a small volume around the star, implying that their role in molecular cloud destruction is much less important than previously thought. These effects need to be included in theories of the early evolutionary phases of O stars.

6. Future Directions

6.1 GLOBAL ISSUES

An issue of primary importance is to verify that the FIR two-color selection criteria developed by WCb is a valid technique for selecting and counting embedded O stars in the Galaxy. The VLA survey of KCW, the sample for which was selected purely on the basis of FIR colors, should establish what fraction of the IRAS PSC FIR color-selected sample is contaminated by other types of objects. In the WCa survey about 35% of the fields observed had no detectable HII region but contained FIR sources with the appropriate colors. This may be because the HII regions were either too large (>10") to be detected by the VLA, or the star was too young to have formed a detectable HII region, or some of the fields contained warm cores heated by low mass stars. An additional criterion is needed to fine-tune the selection of embedded massive stars and to distinguish between other types of objects with similar FIR colors. A criterion that can be applied using existing large databases (such as IRAS) and requiring no additional information (such as distances, radio and IR images, etc.) will be difficult to find, but well worth investment of time.

Distances are key to the determination of the energetics, scale sizes, and galactic distribution of UC HII regions. Unfortunately, distances have been estimated for a very small fraction of the sample (less than 150 out of probably more than 1600). For those whose distances have been estimated using molecular and/or radio recombination line velocities, serious questions remain regarding resolution of distance ambiguities, departures from the simple galactic rotation model used, and the relationship of molecular line velocities to the systemic velocity of UC HII regions. The fact that the CWC survey did not show a well defined spiral arm pattern, as one would expect for newly formed O stars, leaves serious doubts about the adequacy of kinematic distances. More time should be invested in determining accurate distances. Possibilities are proper motion studies, larger samples of molecular line observations, and associations with clouds and/or clusters whose distances have been independently determined.

If reliable distances could be determined for a large enough sample of UC HII regions, then we could estimate the radiative and mechanical energies newly formed O stars contribute to the molecular clouds in which they are embedded and to the energetics of the Galaxy as a whole. It has been commonly assumed that O stars quickly destroy the molecular clouds out of which they were formed, however, the bow shock picture would drastically alter this assumption. These stars may not be as effective in ionizing and driving turbulence as previously thought. The radiant energy, of course, will ultimately escape in the form of IR emission and will contribute to the total IR luminosity of the Galaxy. The question is, what fraction of the total is contributed by embedded O stars? Further, the **total** contribution of O stars to the galactic energy balance could be better estimated if the fraction of O stars that reside in molecular clouds were known to better than a factor of two. So, the thrust here should be directed toward getting a better count of the number of embedded O stars, determining what fraction this represents of the total O star population, and finding the distribution of luminosities among embedded O stars.

The models of dust properties around embedded O stars all imply that dust will be heated to temperatures above ambient gas temperatures out to distances of almost a parsec from the star and that the total mass in this volume is large ($>10^3$ solar masses). The heated gas and dust in this volume may be sufficient to prevent formation of lower mass stars. Deep NIR imaging in the neighborhood of UC HII regions would be helpful to resolve this issue.

6.2 LOCAL PROPERTIES

The radiation field immediately surrounding UC HII regions is intense and the spectrum changes from one dominated by UV to FIR photons within a short distance from the star. This volume also contains a sharp ionization front with very dense, shocked, molecular gas just outside it. The dust temperatures change rapidly with distance from the star also (see Fig. 12). Although the density distribution on a large scale (~1pc) seems to be approximately constant with distance from the star, at radii $< 10^{17}$ cm from the star this is no longer likely to be the case. The velocity field is also likely to be strongly dependent on distance from the star. In all respects, the area in the immediate neighborhood of UC HII regions is characterized by rapidly changing conditions. For this reason, a carefully selected set of observational probes sensitive to a wide range of densities, temperatures, and radiation fields is required to determine the physical conditions and kinematics as a function of position.

High spatial resolution maps in the lines of many molecules are needed to trace the gas density, temperature, and changes of chemical abundances with position. Independent measurements of the gas and dust will tell us about the coupling of these two constituents, as well as the heating mechanisms of both. An important question is the generality of collapse of molecular gas toward the central star as implied by Keto, Ho, and Haschick (1987, 1988) from their study of G10.6-0.4. The presence of bow shocks should be established by observations of molecular transitions that can only be formed in the high densities and temperatures provided by a shock (eg. H_2 (v=1-0) S and Q branch lines in the 2.0 and 2.5 μm wavelength bands, and H_2O, OH, and CH_3OH masers). Are there molecular species whose gas phase abundances imply grain mantle destruction? Molecular line probes have not been utilized in proportion to the richness of the information they can reveal, but this is changing as new, large, single dish and interferometer mm-wave telescopes come on-line.

The models of CWW and Chini, Krügel, and Kreysa (1986) have made a number of important predictions regarding the properties of dust surrounding embedded O stars. Sensitive, high resolution images from NIR to FIR wavelengths would provide an important test of the predicted size of the dust cocoon as a function of wavelength. They would also indicate any structure that could provide important constraints for future, more refined models. The strength of the 9.7 μm feature provides an important constraint on the total amount of silicate grains in the dust cocoons. It also appears to be anticorrelated with the strength of the "PAH" features between 3.3 and 11.3 μm. A high resolution (spectral and spatial) survey of the 9.7 μm feature toward a large number of UC HII regions would be a very useful database both to constrain dust models and to use for correlation purposes with other IR spectral features. The 3.07 μm water ice feature has not been observed toward a UC HII region; much lower limits on the strength of this feature are necessary to establish possible destruction of grain mantles. Finally, good flux density measurements between 100 μm and 1 mm need to be obtained. Most UC HII regions have no data between 1.3 mm and 100 μm; thus, modeling the spectrum in this wavelength interval is mostly guess work.

High resolution dynamical studies of the ionized gas need to be obtained in infrared and radio recombination lines. Such studies will reveal systematic motions which will support or negate the bow shock hypothesis. If the HII shells are found to be expanding, for example, this would be a strong argument against the bow shock picture. Is there any evidence of bipolar flows inside the HII regions? Also the dynamics and ionization conditions on the inner side of the ionized shell where the stellar wind terminal shock occurs should become apparent by increased line widths and the appearance of high stages of ionization. The gas in the inner wind-shocked cavity must have very low density ($n_e < 10^2$ cm^{-3}) and be very hot ($T_e > 10^7$ K) with conditions varying widely between the inner part of the cavity and the wind contact discontinuity. This region must emit copious X-rays, but these are unlikely to be observable because of the optically thick dust cocoon which enshrouds the HII region. An abundant, highly ionized atom having transitions in the middle to far infrared wavelengths seems to be the only probe of this region. The problem is finding ions with middle infrared transitions that are abundant enough to be observed. Van Buren *et al.* (1990) have predicted that FeVII (9.51 and 7.81 μm) transitions should be observable in this region. There are almost certainly others. A measurement of very high importance would be a detection of the ionized stellar wind in several objects. This would establish the position of the star relative to the I-front; this is an important parameter in the bow shock models of Van Buren *et al.* (1990) and Mac Low *et al.* (1990). Finally, proper motion measurements could establish whether the UC HII regions are expanding and whether they are in motion relative to the ambient molecular gas in the sense expected in the bow shock picture. They could also give us a powerful independent estimate of distances.

There is still much to be done on these dynamically active, rapidly evolving regions that will challenge both the ingenuity of astronomers and the current and next generation of radio and infrared telescopes.

Acknowledgements

I wish to thank Doug Wood and Stan Kurtz who have worked with me on many aspects of the work reported on in this review. Dave Van Buren and Mordecai Mac Low have played pivotal roles in educating me about bow shocks and their significance for UC HII regions. Germination of the bow shock hypothesis for UC HII regions came from discussions with them. I especially thank Stan Kurtz for helping me prepare several of the figures in this review.

References

Beichman, C., 1979, Ph. D. Thesis, Univ. Hawaii.
Bedjin, P. J., Tenorio-Tagle, G. 1981, *Ast. Ap.*, **98**, 85.
Benson, J. M., Johnston, K. J. 1984, *Ap. J.*, **277**, 181.
Blaauw, A. 1964, *Ann. Rev. Astr. Ap.*, **4**, 213.
Blake, G. A., Sutton, E.C., Masson, C. R. Phillips, T. G. 1987, *Ap. J.*, **315**, 621.
Bodenheimer, P., Tenorio-Tagle, G., Yorke, H. W. 1979, *Ap. J.*, **233**, 85.
Bronfman, L., May, J., Nyman, L. 1990, in preparation.
Chini, R., Kreysa, E., Mezger, P. G., Gemünd, H.-P. 1986a, *Ast. Ap.*, **154**, L8.
Chini, R., Kreysa, E., Mezger, P. G., Gemünd, H.-P. 1986b, *Ast. Ap.*, **157**, L1.
Chini, R., Krügel, E., Kreysa, E. 1986, *Ast. Ap.*, **167**, 315.
Chini, R., Krügel, E., Wargau, W. 1987, *Ast. Ap.*, **181**, 378.
Churchwell, E. 1990, *Ast. & Ap. Reviews*, in press.
Churchwell, E., Felli, M., Wood, D. O. S., Massi, M. 1987, *Ap. J.*, **321**, 516.
Churchwell, E., Walmsley, C. M., Cesaroni, R. 1990, *Ast. Ap. Suppl.*, in press. (CWC)
Churchwell, E., Walmsley, C. M., Wood, D. O. S., Steppe, H. 1990, in "Radio Recombination Lines: 25 Years of Investigation", eds. M. A. Gordon and R. L. Sorochenko, Kluwer Acad. Pub., Dordrecht, p.83.
Churchwell, E., Wolfire, M. G., Wood, D. O. S. 1990, *Ap. J.*, **354**, 247. (CWW).
Dame, T. M., Ungerechts, R. S., Cohen, R. S., De Geus, E. J., Grenier, I. A., May, J., Murphy, D. D., Nyman, L.-Å., Thaddeus, P. 1987, *Ap. J.*, **322**, 706.
Draine, B. T. 1985, *Ap. J. Suppl.*, **57**, 587.
Draine, B. T., Lee, H. M. 1984, *Ap. J.*, **285**, 89.
Draine, B. T., Lee, H. M. 1987, *Ap. J.*, **318**, 485.
Dreher, J. W., Johnston, K. J., Welch, W. J., Walker, R. C. 1984, *Ap. J.*, **283**, 632.
Felli, M., Churchwell, E., Massi, M. 1984, *Ast. Ap.*, **136**, 53.
Forster, J. R., Caswell, J. L., Okumura, S. K., Hasegawa, T., Ishiguro, M. 1990, *Ast. Ap.*, **231**, 473.
Garay, G., et al. 1990, in preparation.
Garay, G., Moran, J. M., Reid, M. J. 1987, *Ap. J.*, **314**, 535.
Garay, G., Rodriguez, L. F., van Gorkom, J. H. 1986, *Ap. J.*, **309**, 553.
Garden, R., Grolemund, D., Carlstrom, J. E. 1990, Presented at the Centennial Meeting of the Ast. Soc. of the Pacific.
Gatley, I., Becklin, E. E., Werner, M. W., Harper, D. A. 1978, *Ap. J.*, **220**, 822.

Gaume, R. A., Claussen, M. J. 1990, *Ap. J.*, **351**, 538.
Gaume, R. A., Mutel, R. L. 1987, *Ap. J.*, Suppl., **65**, 193.
Guilloteau, S. 1990, private communication.
Harris, S. 1973, *Mon. Not. R. Ast. Soc.*, **162**, 5p.
Harvey, P. M., Forvielle, T. 1988, *Ast. Ap.*, **197**, L19.
Haschick, A. D., Menten, K. M., Baan, W. A. 1990, *Ap. J.*, in press.
Henkel, C., Churchwell, E. 1991, in preparation.
Henkel, C., Wilson, T. L., Mauersberger, R. 1987, *Ast. Ap.*, **182**, 137.
Herter, T., Helfer, H. L., Piper, J. L., Forrester, W. J., Mc Carthy, J., Houck, J. R., Willner, S., Puetter, R. C., Rudy, R. J., Soifer, B. T. 1981, *Ap. J.*, **250**, 186.
Ho, P. T. P., Haschick, A. D. 1986, *Ap. J.*, **304**, 501.
Hoare, M. G., Glencross, W. M., Roche, P. F., Clegg, R. E. S. 1988, in "Dust in the Universe", ed. M. E. Bailey and D. A. Williams, Cambridge Univ. Press.
Hughes, V. A. 1988, *Ap. J.*, **333**, 788
Hughes, V. A., MacLeod, G. C. 1989, *A. J.*, **97**, 786.
Jones, B. F., Walker, M. F. 1988, *A. J.*, **95**, 1755.
Jourdain de Muizon, M., Cox, P., Lequeux, J. 1990, *Ast. Ap. Suppl.*, **83**, 337.
Keto, E. R., Ho, P. T. P., Haschick, A. D. 1987, *Ap. J.*, **318**, 712.
Keto, E. R., Ho, P. T. P., Haschick, A. D. 1988, *Ap. J.*, **324**, 920.
Keto, E. R., Ho, P. T. P., Reid, M. J. 1987, *Ap. J.*, **323**, L117.
Krügel, E., Mezger, P. G. 1975, *Ast. Ap.*, **42**, 441.
Kurtz, S., Churchwell, E., Wood, D. O. S., 1990, in preparation. (KCW)
Kurtz, S. Wood, D. O. S., Myers, P., Bronfman, L., Churchwell, E. 1991, in preparation.
Lacy, J. H., Beck, S. C., Geballe, T. R. 1982, *Ap. J.*, **255**, 510.
Lee, H. M., Draine, B. T. 1985, *Ap. J.*, **290**, 211.
Leung, C. M. 1976, *Ap. J.*, **209**, 75.
Lucas, R., Omont, A., Stora, R. 1985, in "Birth and Infancy of Stars," EDS? North-Holland Pub., Amsterdam.
Mac Low, M.-M., Van Buren, D., Wood, D. O. S., Churchwell, E. 1990, *Ap. J.*, submitted.
Mathis, J. S., Rumpl, W., Nordsieck, K. H. 1977, *Ap. J.*, **217**, 425 (MRN).
Mauersberger, R., Henkel, C., Wilson, T. L., Walmsley, C. M. 1986, *Ast. Ap.*, **162**, 199.
Menten, K. M. 1987, Ph. D. Thesis, Universität Bonn
Mezger, P. G., Henderson, A. P. 1967, *Ap. J.*, **147**, 471.
Panagia, N. 1973, *A. J.*, **78**, 929.
Pudritz, R. E., Fich, M. 1988, in "Galactic and Extragalactic Star Formation," Kluwer Academic Pub., Dordrecht.
Reid, M. J., Haschick, A. D., Burke, B. F., Moran, J. M., Johnston, K. J., Swenson, G. W. Jr. 1980, *Ap. J.*, **239**, 89.
Reid, M. J., Ho, P. T. P. 1985, *Ap. J.*, **288**, L17.
Rudolph, A., Welch, W. J., Palmer, P., Dubrulle, B. 1990, *Ap. J.*, in press.
Ryle, M., Downes, D. 1967, *Ap. J.*, **148**, L17.
Sanders, D. B., Willner, S. P. 1985, *Ap. J.*, **293**, L39.
Scalo, J. M. 1986, *Fund. Cosmic Phys.*, **11**, 1.

Shaver, P. A., Mc Gee, R. X., Newton, L.M., Danks, A. C., Pottasch, S. R. 1983, *Mon. Not. R. A. S.*, **204**, 53.
Simpson, J. P., Rubin, R. H. 1984, *Ap. J.*, **281**, 184.
Simpson, J. P., Rubin, R. H. 1990, *Ap. J.*, **354**, 165.
Sodroski, T. J. 1988, Ph. D. Thesis, University of Maryland.
Tenorio-Tagle, G. 1979, *Ast. Ap.*, **71**, 59.
Tenorio-Tagle, G., Yorke, H. W., Bodenheimer, P. 1979, *Ast. Ap.*, **80**, 110.
Tielens, A. G. G. M., de Jong, T. 1979, *Ast. Ap.*, **75**, 326.
Turner, B.E. and Matthews, H.E. 1984, *Ap. J.*, **277**, 164.
Tutukov, A. V., Shustov, B. M. 1981, *Soviet Ast.*, **58**, 109.
Ungerechts, H. 1980, Ph. D. Thesis, Universität Bonn.
Ungerechts, H., Walmsley, C. M., Winnewisser, G. 1986, *Ast. Ap.*, **157**, 207.
Van Buren, D., Mac Low, M.-M., Wood, D. O. S., Churchwell, E. 1990, *Ap. J.*, **353**, 570.
Van der Veen, W. E. C. J., Habing, H. J. 1988, *Astr. Ap.*, **194**, 125.
Walmsley, C. M. 1989, in "Interstellar Dust," IAU Symp. No. 135, eds. L.J. Allamandola and A. G. G. M. Tielens, Kluwer Academic Pub.,Dordrecht, p. 263.
Walmsley, C. M., Baudry, A., Guilloteau, S. Winnberg, A. 1986, *Ast. Ap.*, **167**, 151.
Weaver, R., Mc Cray, R., Castor, J., Shapiro, P., Moore, R. 1977, *Ap. J.*, **218**, 377.
Webster, W. J., Altenhoff, W. J. 1970, *Ast. J.*, **75**, 896.
Welch, W. J., Dreher, J. W., Jackson, J. M., Tereby, S., Vogel, S. N. 1987, *Science*, **238**, 1550.
Wilson, T. L., Mausersberger, R., Walmsley, C. M., Batrla, W. 1983, *Ast. Ap.*, **127**, L19.
Wilson, T. L., Walmsley, C. M. 1989, *Ast. Ap. Rev.*, **1**, 141.
Wolfire, M. G., Cassinelli, J. P. 1986, *Ap. J.*, **310**, 207.
Wolstencroft, R. D. 1983, **Workshop on Star Formation,** Reports of the Royal `Observatory, Edinburgh, No. 13.
Wood, D. O. S., Churchwell, E. 1989a, *Ap. J. Suppl.*, **69**, 831 (WCa).
Wood, D. O. S., Churchwell, E. 1989b, *Ap. J.*, **340**, 265. (WCb)
Wood, D. O. S., Churchwell, E. 1990, *Ap. J.*, submitted. (WC90)
Wouterloot, J. G. A., Brand, J. 1989, *Ast. Ap. Suppl.*, **80**, 149.
Wouterloot, J. G. A., Brand, J., Burton, W. B., Kwee, K. K. 1990, *Ast. Ap.*, **230**, 21.
Wynn-Williams, C. G. 1971, *Mon. Not. R. Ast. Soc.*, **151**, 397.
Wynn-Williams, C. G., Becklin, E. E. 1974, *Pub. Ast. Soc. Pacific*, **86**, 5.
Wynn-Williams, C. G., Becklin, E. E., Neugebauer, G. 1972, *Mon. Not. R. Ast. Soc.*, **160**, 1.
Yorke, H. W. 1979, *Ast. Ap.*, **80**, 308.
Yorke, H. W. 1980, *Ast. Ap.*, **85**, 215.
Yorke, H. W., Krügel, E. 1977, *Ast. Ap.*, **54**, 183.
Yorke, H. W., Shustov, B. M. 1981, *Ast. Ap.*, **98**, 125.
Yorke, H. W., Tenorio-Tagle, G., Bodenheimer, P. 1983, *Ast. Ap.*, **127**, 313.
Zheng, X. W., Ho, P. T. P., Reid, M. J., Schneps, M. H. 1985, *Ap. J.*, **293**, 522.
Zuckerman, B. 1973, *Ap. J.*, **183**, 863.

MASERS AND STAR FORMATION

NIKOLAOS D. KYLAFIS
University of Crete
Physics Department
714 09 Heraklion, Crete
Greece

ABSTRACT. The physics of astronomical masers and their connection with star-forming regions are reviewed. For a better understanding of the subject, a qualitative discussion is given of the basic concepts about masers. These are: Amplification, saturation, thermalization, beaming, apparent size, variability, line width and polarization. The difference between laboratory and astronomical masers is discussed and a few examples of the usefulness of astronomical masers are given. The basic requirements for the construction of a maser model are presented and the accuracy with which the various inputs are known is commented on. A qualitative discussion is given of the most common models, which are collisional and radiative. Specific pumping mechanisms for OH and H_2O masers in star-forming regions are presented and criticized. The current status of the observations of these masers is reviewed and the implications on the theoretical models is discussed.

1. Introduction

The subject of astronomical masers is now more than thirty years old, and as a consequence, several good reviews have appeared in print. For the readers who want to enter the subject of astronomical masers in a serious way, I want to recommend two reviews (Elitzur 1982; Reid and Moran 1988). They contain a review of the observations as well as the basic ideas and equations that will enable the readers to comprehend the research papers on the subject.

In this review, I want to avoid the mathematical details but stress instead a qualitative understanding of astronomical masers. Needless to say, some overlap with the above reviews is unavoidable. Section 2 contains the basic concepts regarding masers. In § 3 the differences between laboratory and astronomical masers are briefly discussed and in § 4 a few examples which demonstrate the usefulness of astronomical masers are presented. Section 5 describes what it takes to build a maser model and how well we know the various input parameters. The observations and proposed pumping mechanisms are given in § 6 for H_2O and OH masers. The conclusions are given in § 7.

2. Basic Concepts

The energy level diagram of a molecule consists of an infinite number of levels. Nevertheless, one need only consider a finite and relatively small number of energy levels in order to perform an accurate model calculation regarding a specific molecule. Even better, for the purposes of a qualitative discussion of masers, one can consider only two energy levels and treat the rest of them as a reservoir. Figure 1 shows such a simplified model of a molecule.

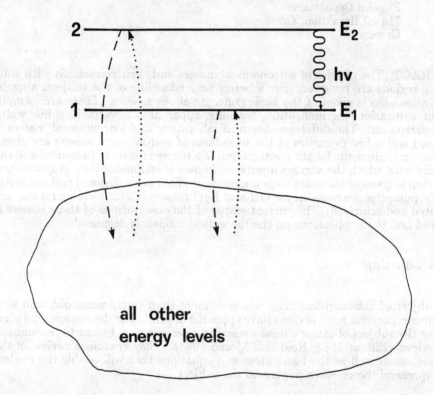

Figure 1. Schematic energy level diagram of a molecule. Only two levels are explicitly shown. The rest are lumped together and are considered as a reservoir. The dotted and dashed arrows indicate respectively the transitions from the reservoir to the two levels and the opposite.

Level 1 is the lower maser level and level 2 is the upper one. The energy difference between the two levels is $\Delta E \equiv E_2 - E_1 = h\nu$, where ν is the frequency of the maser. Collisional and radiative transitions cause the molecules to circulate between the reservoir and the maser levels. Thus, the dotted lines in Figure 1 indicate the transitions of molecules from the reservoir to the maser levels. The opposite transitions are indicated by dashed lines. Let n_1 and n_2 be the populations *per magnetic sublevel* of the two maser levels. The maser phenomenon occurs if, in steady state, the populations satisfy the condition $n_2 > n_1$. This is because, as a photon of frequency ν propagates, it is more likely to cause stimulated emission than to be absorbed.

2.1 AMPLIFICATION

Consider radiation of frequency ν and intensity I propagating through molecules whose energy level diagram is that of Figure 1. The change dI in the intensity that occurs as the radiation propagates by a length dl through the medium is

$$dI = -\kappa I dl , \qquad (1)$$

where $\kappa \propto (n_1 - n_2)$ is the absorption coefficient corrected for stimulated emission. Here it is assumed for simplicity that the molecules do not emit radiation by spontaneous emission. Defining the optical depth τ by $d\tau \equiv \kappa dl$, one finds that the intensity as a function of optical depth is given by

$$I = I_0 e^{-\tau} , \qquad (2)$$

where I_0 is the intensity of a background source that is incoming to the medium. If the populations of levels 1 and 2 in Figure 1 are such that $n_1 > n_2$, then $\tau > 0$ and the radiation is attenuated as it propagates through the medium. This situation is typical in astronomical environments and it leads to the formation of absorption lines. On the other hand, if $n_1 < n_2$, the optical depth is negative and the radiation is amplified. This is because a propagating photon is more likely to cause a stimulated emission than it is to be absorbed. Thus, one photon leads to two propagating in the same direction, two photons to four and so on. In other words, the amplification of the radiation is *exponential*.

As mentioned above, it was only for simplicity that spontaneous emission in the medium was neglected. In reality, such an emission always occurs and the radiation thus produced is also amplified. If there is no background source, as it is the case in most astronomical masers, it is only the spontaneous emission that gets amplified.

2.2 SATURATION

The exponential growth of the radiation intensity in a medium that exhibits inverted populations is truly fascinating. It is therefore important to investigate whether this exponential growth has any limits, assuming that the maser region is long enough. Since a stimulated emission event causes the transition of a molecule from the upper maser level to the lower, it is clear that stimulated emission tends to *reduce* the population inversion. At low intensities, the stimulated emission rate is relatively small and the population difference $n_1 - n_2$ remains essentially unaffected. As the intensity grows, so does the stimulated rate. Thus, a point is reached where the stimulated emission rate becomes comparable to the rate that causes the population inversion. At this point, *saturation* has set in because the rate of extraction of photons has reached its limiting value, which is the pumping rate (the rate of population inversion). Increasing the volume of the emission region under saturation conditions will only increase the photon emission rate in proportion to the volume.

It is important to remark here that the astronomical definition of *saturation* given above is exactly opposite to that used by experimental physicists for laboratory masers and lasers. Experimental physicists call a maser *saturated* when its intensity grows exponentially.

2.3 THERMALIZATION

The establishment of inverted populations in a medium containing a certain kind of molecules is not as difficult as one might think. In fact, it is not an exaggeration to say that if the molecules are abused, they will most likely exhibit inverted populations in *some pair* of energy levels. One cannot easily predict which pair it will be, but inversion in some pair is almost certain. Common methods of abuse of molecules in astronomical environments are a) collisions with other molecules, b) irradiation of the molecules and c) a combination of the above, which is the usual case. As an illustration, let us consider collisions with other molecules (e.g., H_2) as the cause of population inversion. If the density of H_2 molecules is low, the collision rate is low, the rate of population inversion is low and therefore the maser luminosity is correspondingly low. For low rates of abuse, the maser luminosity increases in proportion to the rate of abuse of the molecules. However, this proportionality has an upper limit. When the collision rate becomes very large, the collisional transitions of the molecules dominate all other transitions (see § 5.3 below), the molecules thermalize at the temperature of the H_2 molecules and the population inversion is destroyed. At thermalization, the level populations obey the Boltzmann distribution, namely

$$\frac{n_2}{n_1} = \exp(-\Delta E/kT) < 1 , \qquad (3)$$

where T is the temperature of the H_2 molecules.

2.4 BEAMING

The majority of the astronomical masers *do not* emit isotropically. This effect can be understood as follows: Consider a maser region which is not spherically symmetric but instead it is elongated in some direction, as shown in Figure 2. The gain $|\tau|$ along the direction of elongation is larger than in directions perpendicular to it. Consequently, the intensity of the maser is largest along the direction of elongation. Thus, a departure from spherical symmetry introduces an anisotropy in the intensity.

Once an anisotropy in the intensity is established, it is further enhanced through the following process. Consider the maser photons that reach a certain point in the maser region from all possible directions (see Figure 2). These beams of photons compete for the available pump photons there. The rate of stimulated emission caused by a given beam is proportional to its intensity. Therefore, the stronger beam wins and becomes even stronger (Alcock and Ross 1985a,b; 1986). As a result, a non-spherically symmetric maser region results in maser emission in two opposite cones along the direction of elongation.

Radio observers measure the flux of photons that reach their telescope and, not knowing the degree of beaming, make the assumption that the emission is isotropic. Thus, from the observed flux, an equivalent *isotropic* luminosity is quoted for the source. This procedure certainly overestimates the luminosity of a *maser spot*, but it gives the right value when one considers a source that contains many maser spots. This is because the cones of emission of the various spots have random orientations, but we only see the ones that are directed towards the earth.

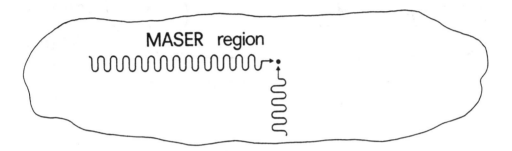

Figure 2. *Schematic representation of a maser region. The two wiggly arrows indicate the intensity of the maser in two directions.*

2.5 GEOMETRY AND APPARENT SIZE

The geometry of the maser region and its apparent size are intimately related. Furthermore, the apparent size of a maser region may be significantly smaller than its actual size. For the purposes of illustration we will consider two shapes of maser region, elongated and nearly spherical.

2.5.1. *Elongated Structures.* Consider a maser region which is more or less cylindrical as shown in Figure 3a. Let the length of the region be l and its characteristic width be d, with $l > d$. As discussed in § 2.4, most of the maser photons escape through the bases of the cylinder and into solid angles $\Omega \sim (d/l)^2$.

If the observational direction falls in this beaming angle Ω, then the observer sees the maser spot and determines an apparent size $\sim d$. If, on the other hand, the observational direction falls significantly outside Ω, the observer *does not* see the maser spot, because the intensity of the maser in the direction of the observer is below the detection limit.

2.5.2. *Nearly Spherical Structures.* Consider now a maser region which is nearly spherical with diameter D, as shown in Figure 3b. In this case, all observers see the maser spot as having the same apparent size. However, the apparent size is *less* than D because the rays that come out of the region have traversed different lengths and therefore the maser photons have experienced different gains. It is straightforward to estimate the apparent size of the maser spot once the detection limit of the telescope is given.

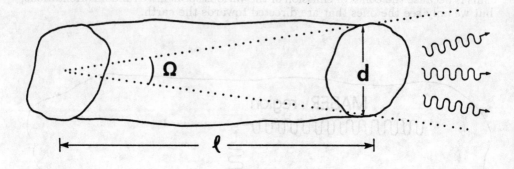

Figure 3a. Schematic of a maser region which is approximately cylindrical. The wiggly lines indicate the emergent maser intensity.

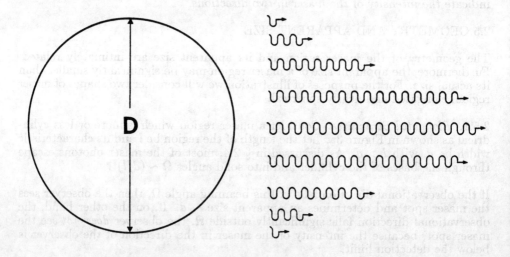

Figure 3b. Schematic of a maser region which is approximately spherical. The wiggly lines indicate the emergent maser intensity.

2.6 VARIABILITY

All masers, even the very strong ones, are variable with the shortest timescale of variability of order a day. This, of course, is not surprising because even a small change in the gain introduces a significant change in the intensity. Small changes in the gain can be caused either by changes in the aspect angle or by changes in the local physical parameters such as the density and the velocity field. For example, if the maser region is elongated, a small change in the aspect angle may result in a relatively large change in the intensity (see § 2.5.1).

Despite the short timescale of variability, some maser spots can be followed for years. The lifetime of individual spots can be as large as one to two years for H_2O and more than ten years for OH masers. This relatively long lifetime has allowed studies of proper motion of the maser spots and through this the determination of their distance (see § 4).

2.7 SPECTRA AND LINE WIDTHS

The maser spectra observed in star-forming regions are fairly complex (see, e.g., Genzel and Downes 1977 for a collection of such spectra). A sketch of a typical spectrum of H_2O emission in a region of star formation is shown in Figure 4.

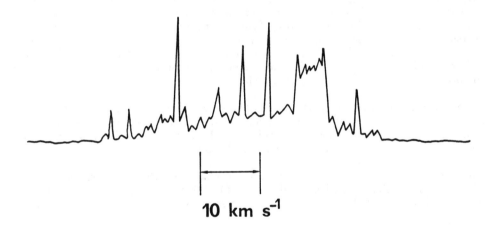

Figure 4. *A sketch of a spectrum of H_2O emission in a region of star formation.*

H_2O maser spectra in star-forming regions extend over a frequency range corresponding to a velocity difference of a few tens km s^{-1} (typically less than 10 km s^{-1} for OH) and consist of many spikes. VLBI maps which resolve the maser region have shown that a spike arises from one or more distinct spots. The line width of a strong spike is typically a few times smaller than the width expected from the thermal motions of the molecules. This is expected because unlike a thermal line, a maser line grows exponentially (in the unsaturated regime) and therefore the full width at half maximum is smaller than thermal. As the line center saturates, the line wings continue to grow exponentially and the line width approaches the thermal one.

The reader should not be left with the impression that unsaturated lines have narrow widths while saturated ones have thermal widths. It has been demonstrated by Goldreich and Kwan (1974) that the broadening of the line during saturation can be inhibited if trapping of the infrared radiation involved in the pumping occurs.

2.8 POLARIZATION

Polarization is fairly common in masers observed in star-forming regions. The OH masers are typically highly polarized while the H_2O ones show polarized features in about 25% of the sources. In order for a maser source to emit polarized radiation, the molecules must not only exhibit inverted populations but also unequal populations in the magnetic sublevels. It has been demonstrated (Goldreich, Keeley and Kwan 1973) that an ordered magnetic field in the maser region can establish unequal populations in the magnetic sublevels and therefore result in polarized emission. The magnetic field required is of order a few milligauss for OH and somewhat larger for H_2O.

The maser polarization would be an ideal way of measuring the magnetic field strength in maser regions if the magnetic field were the only cause of such polarization. As it has been demonstrated by Watson and his collaborators (Western and Watson 1983a,b,c, 1984; Deguchi and Watson 1986a,b; Deguchi, Watson and Western 1986; Deguchi and Watson 1990; Nedoluha and Watson 1990a,b,c), anisotropic radiation fields in the maser region can also cause polarization. Thus, the origin of the observed polarization is not always clear.

Despite the above ambiguity, the unmistakable Zeeman pattern has been seen by Moran et al. (1978) for OH and Fiebig and Güsten (1989) for H_2O. These observations have determined the magnetic field strength to be 2 to 9 milligauss in W3(OH) and of order 50 milligauss in Orion KL and W49N.

The Zeeman splitting of OH maser lines can also yield the line-of-sight direction of the magnetic field. Reid and Silverstein (1990) have in this way been able to infer the direction of the magnetic field in maser sources over large regions of the Galaxy.

3. Laboratory versus Astronomical Masers

Both laboratory and astronomical masers are based on the same principle, namely, some transition exhibits inverted populations. Nevertheless, there are significant differences in the size and the technical characteristics of the two.

In laboratory masers, the maser region is typically ~ 1 m long (to astronomical accuracy). In order to have significant gain, the radiation is bounced between two mirrors, one of which is semi-transparent to allow the amplified radiation to escape. On the other hand, the astronomical masers have significantly lower densities but gain lengths of order a few AU or larger. The absence of reflecting boundaries in astronomical masers makes them single pass as opposed to multi pass in the laboratory ones. Furthermore, the fixed size of the laboratory masers introduces a mode selection, i.e., not all wavelengths can fit in the length of the maser. On the other hand, the astronomical masers allow all possible wavelengths and they are broad band. Finally, laboratory masers emit coherent radiation, while the astronomical ones emit non-coherent. This is because in astronomical masers the beaming angle (see § 2.4) is fairly large and therefore a significant phase difference is introduced between two rays in the beam.

Because the radiative transfer in astronomical masers is relatively simple (see § 5), the main theoretical effort has concentrated on inventing efficient pumping mechanisms and understanding the origin of polarization.

4. Usefulness of Masers

The astronomical masers, because of their unique properties, provide very useful information. Below are listed some representative results that have been obtained from maser observations.

As mentioned in § 2.8, when a maser line shows the Zeeman pattern, the magnetic field in the maser region is determined unambiguously. Thus, we now have an accurate knowledge of the magnetic field strength in some OH and H_2O maser regions.

Observations of maser spots have provided detailed information on the dynamics of a region. As an interesting example, I mention the work of Chapman and Cohen (1986, see their Figures 5 and 6), who studied OH and H_2O maser emission from the circumstellar envelope of the supergiant star VX Sagittarius and inferred the velocity field there.

VLBI proper motion studies of H_2O maser spots have provided a means to determine distances across the Milky way. Thus, the distances to Orion KL and W51(Main) have been determined to be respectively 480±80pc (Genzel et al. 1981a) and 7.0 ± 1.5 kpc (Genzel et al. 1981b). The distance to W51(North) is 7 ± 2 kpc (Schneps et al. 1981) and the distance to Sgr-B2(North) is 7.1 ± 1.5 kpc (Reid et al. 1987). Ground based VLBI observations can extend to H_2O masers in nearby galaxies. In fact, the first such observation has already been performed (Greenhill et al. 1990). H_2O maser spots have been seen in a spiral arm of M33. The relative positions of some of these maser spots are accurately enough determined to provide first epoch measurements for proper motion studies. Space based VLBI may allow measurement of distances to galaxies out to 10 Mpc (Reid, Moran and Gwinn 1988). When the distance to even one galaxy is determined with the use of H_2O masers, we will have a crucial check on the calibration of Cepheids and of other distance indicators, and hence we will be able to remove significant sources of error in the determination of the Hubble constant.

Because of their high intensity and small angular size, the astronomical masers may be useful in probing the interstellar medium. Thus, irregularities in the electron density distribution in the interstellar medium may be detectable via scattering of the maser radiation. By using various maser lines, the scattering effects can be studied as a function of frequency.

The astronomical masers can also be useful on earth. The brightest maser source in the sky (Orion KL) has been used in a holographic technique to measure the figure of large radio telescopes (Moran 1989).

5. Maser Models

In what follows, the requirements for the construction of a maser model are discussed and the basic equations that need to be solved are given. Also, a qualitative discussion of how collisional and radiative pumping works is presented.

5.1 REQUIREMENTS

In order to construct a maser model one needs the following:

a) The energy levels of the molecule that exhibits maser emission. These are generally well known.

b) The einstein coefficients A_{ij} for the transitions from any upper level i to any lower level j. These also are well known, at least for the most common molecules.

c) The collision rates C_{ij} for transitions from any level i to any level j of the maser molecule. The colliding particles are the other molecules in the region (e.g., H_2). These rates are only approximately known. Therefore, when maser models are constructed, it should always be kept in mind that the currently published collision rates may be revised.

d) The velocity field in the maser region. This is unknown and it is only guessed in maser models. Two types of velocity field have generally been assumed. In one, the medium is static and the molecules obey a distribution of velocities (typically Gaussian). In the other, the medium exhibits a large velocity gradient, i.e., the macroscopic velocity differences are significantly larger than the thermal velocities of the molecules.

e) The geometry of the maser region is also unknown and in maser models a simple (typically spherical or cylindrical) geometry is assumed.

5.2 BASIC EQUATIONS

The set of equations that need to be solved consists of the statistical equilibrium equations for the level populations and the radiative transfer equations for the various line frequencies. For simplicity, we will assume that the molecules under study have only three, non-degenerate energy levels with energies E_1, E_2 and E_3 and corresponding populations n_1, n_2 and n_3. If only collisional and radiative transitions are taken into account, the steady state equations describing the level populations are

$$\frac{dn_3}{dt} = -(A_{32} + A_{31})n_3 + R_{13}(n_1 - n_3) + R_{23}(n_2 - n_3)$$

$$+ C_{13}n_1 + C_{23}n_2 - (C_{32} + C_{31})n_3 = 0 , \qquad (4a)$$

$$\frac{dn_2}{dt} = A_{32}n_3 - A_{21}n_2 - R_{23}(n_2 - n_3) + R_{12}(n_1 - n_2)$$

$$+ C_{12}n_1 + C_{32}n_3 - (C_{23} + C_{21})n_2 = 0 , \qquad (4b)$$

$$n_1 + n_2 + n_3 = N , \qquad (4c)$$

where N is the total number of maser molecules and R_{ij} are the radiative absorption rates corrected for stimulated emission. Generalization of these equations to any number of levels is straightforward.

If ν is the frequency of the transition from upper level j to lower level i and I_ν is the corresponding intensity, then the radiative transfer equation is given by

$$\frac{dI_\nu}{dl} = -\kappa_\nu I_\nu + \eta_\nu , \qquad (5)$$

where κ_ν is the absorption coefficient (which is a function of ν and $n_i - n_j$) and η_ν is the emissivity (which is a function of ν and n_j).

The system of equations (4) and (5) must be solved simultaneously. In the case where there is a large velocity gradient in the medium, equations (4) and (5) are decoupled (Sobolev 1960; Castor 1970). For a complete formalism in this case see, for example, Kylafis and Norman (1991). In the case of a static medium, approximate analytic solutions have been obtained by Goldreich and Keeley (1972) and Elitzur (1990a,b; 1991).

5.3 COLLISIONAL AND RADIATIVE PUMPING

Collisional pumping works as follows: The maser molecules are excited by collisions with the other particles (typically H_2) in the maser region. Their de-excitation is done both radiatively and collisionally. If the radiative de-excitations dominate the collisional ones, it is likely that some transition will exhibit inverted populations. However, when the collisions dominate both excitations and de-excitations, the molecules thermalize and the population inversion is destroyed. Equations (4), with only collision terms, yield the Boltzmann distribution for the level populations.

Radiative pumping works in a similar way. The excitation of the molecules is done with infrared radiation from an external source. The de-excitation is done via spontaneous and stimulated emission. In this "flow" of the molecules from level to level, it is likely that inverted populations will be established in some transition. If the external radiation dominates in the de-excitation, thermodynamic equilibrium is established at the temperature of the external radiation.

6. Masers in Star-Forming Regions

Maser emission is a common phenomenon in star-forming regions. Of the molecules that exhibit maser emission there, we will concentrate only on OH and H_2O. The SiO masers are rare (they have been seen only in Orion, W51 and Sgr B2). The methanol and formaldehyde ones have not as yet been studied well theoretically. Other molecules exhibit fairly weak maser emission.

Let's begin with some similarities and differences between OH and H_2O masers in star-forming regions.

The H_2O and OH masers are excellent signposts of massive-star formation. The OH masers are in most cases associated with compact HII regions. On the other hand, most H_2O masers are located near, but not coincident with, compact HII regions. The H_2O maser spectra are broad and span a frequency range that corresponds to velocities of few tens km/s. The OH spectra, on the other hand, typically span a range of 5 to 10 km/s. From theoretical models (e.g., Kylafis and Norman 1990; 1991) it has been determined that the OH molecules thermalize at densities $\gtrsim$ few $\times\ 10^7$ cm^{-3} and the H_2O at densities $\gtrsim$ few $\times\ 10^9$ cm^{-3}.

Forster and Caswell (1989) observed with the VLA the 1665 MHz line of OH (see § 6.2) and the 22 GHz line of H_2O (see § 6.1) in 74 star-forming regions and they concluded:

a) The OH and H_2O masers do not coincide.

b) The OH masers tend to occur in groups of diameter 5 - 30 mpc.

c) The H_2O masers tend to occur in more compact groups of size less than 10 mpc.

d) In cases where a compact H II region is detected at 22 GHz, OH and H_2O masers are generally found within $\sim$ 20 mpc of the peak of the continuum emission and typically extend to distances $\sim$ 65 mpc for OH and $\sim$ 130 mpc for H_2O.

e) OH and H_2O masers appear spatially intermixed but uncorrelated.

6.1 H_2O MASERS

The rotational energy levels of H_2O are characterized by three quantum numbers and are denoted by $J_{K_+ K_-}$, where J is the total angular momentum and K_+ and K_- are its projections on two molecular axes. For an energy level diagram of H_2O see, for example, Neufeld and Melnick (1991).

The first H_2O maser transition was detected more than twenty years ago (Cheung et al. 1969). It is the transition from level 6_{16} to 5_{23}, giving rise to radiation with frequency $\nu = 22$ GHz or wavelength $\lambda = 1.35$ cm. This transition lies 640 K above the ground state.

Up to 1990, the words H_2O masers were synonymous with radiation from this transition. In the last year, three more transitions were confirmed as masers. These are: $10_{29} - 9_{36}$ at 321 GHz (Menten, Melnick and Phillips 1990), $3_{13} - 2_{20}$ at 183 GHz (Cernicharo et al. 1990) and $5_{15} - 4_{22}$ at 325 GHz (Menten et al. 1990). Of these, the 321 GHz transition lies 1860 K above the ground state. Therefore, it is likely that this transition is excited in a warm (say, 1000 K) environment.

Radiative pumping of the H_2O masers can be ruled out because, even with 100% efficiency, one infrared pumping photon is needed for every maser photon emitted. However, the observed number of infrared pumping photons is significantly lower than the observed maser photons.

Collisions with H_2 molecules on the other hand *can* invert the 22 GHz transition (Elitzur, Holenbach and McKee 1989; Kylafis and Norman 1991) as well as the three recently discovered transitions (Neufeld and Melnick 1990; 1991). The detected transitions, as well as a few more that will undoubtedly be detected in the near future, will constrain the models significantly.

In my opinion, one persistent problem with H_2O masers is the following: While collisional pumping can, in principle, explain the medium and low power sources, the very strong ones (such as W49 and W51), which emit close to a solar luminosity and have brightness temperatures of order 10^{15} K or larger, *cannot* be explained. The reason is that the required densities are so high that the molecules thermalize.

Attempts to overcome the problem of thermalization led Strelnitskij (1984) to the interesting idea that the H_2O molecules collide with two kinds of particles (e.g., H_2 and electrons) at different kinetic temperatures, but with comparable collision rates. In other words, the H_2O molecules find themselves in contact with two heat baths, and not knowing which of the two Maxwellian distributions to obey, they obey none. Kylafis and Norman (1986; 1987) showed that, with the then available collision rates, collisional pumping of H_2O in environments where the temperature of the neutral particles is significantly less than that of the electrons, *can* sustain inverted populations. Such environments exist in the magnetic precursors of MHD shocks (Draine, Roberge and Dalgarno 1983).

This idea is now temporarily abandoned because the recently recalculated collision rates (Palma et al. 1988) are significantly different than the older ones. Anderson and Watson (1990) have shown that with the new rates inversion does not occur.

Elitzur, Hollenbach and McKee (1989) claim to have explained the powerful H_2O masers with conventional collisional pumping. They model the maser regions as static, in the shape of cylinders with aspect ratios (see Figure 3a) $10 \lesssim l/d \lesssim 100$. It is rather difficult to imagine how such long and narrow structures can exist in star-forming regions. Furthermore, if there is a somewhat large velocity gradient $\Delta v/l \gtrsim \Delta v_{thermal}/l$ along the axis of the cylinder (which is not unreasonable), then the gain is lower and one needs aspect ratios $l/d > 100$ to explain the powerful H_2O masers (Kylafis and Norman 1991).

An interesting idea regarding the powerful H_2O masers has been proposed by Deguchi and Watson (1989). Consider two cylindrical masers, of length l and base diameter d each, separated by a distance D and having their axes aligned. The observer is located on their common axis. If there were only one maser, its radiation would be beamed into a solid angle $\Omega \sim (d/l)^2$ (see § 2.4). However, the radiation of the back maser causes by stimulated emission the radiation of the front maser to be emitted in a much smaller solid angle $\Omega \sim (d/D)^2$. Thus, the observer detects radiation that would not be detected if the front maser emitted its radiation into $\Omega \sim (d/l)^2$. The aligned masers do not emit more than their sum, but *they give you the impression of a powerful maser*.

A question here may be, why don't we see the same phenomenon in OH masers? The answer is not clear.

6.2 OH MASERS

The energy levels of OH are grouped into two ladders, the $^2\Pi_{3/2}$ and the $^2\Pi_{1/2}$ (see, for example, Figure 1 of Elitzur 1982 for an energy level diagram). In each ladder, the energy levels are characterized by the total angular momentum J and are split into Λ-doublets. Each component in a given Λ-doublet is further split into two hyperfine levels. Thus, there are four allowed ground state transitions and many excited ones in the $^2\Pi_{3/2}$ and $^2\Pi_{1/2}$ ladders. Maser emission has been observed in all four allowed ground state transitions and also in several excited ones (Reid and Moran 1981, see their Table 1). With the recent detections of Wilson, Walmsley and Baudry (1990), microwave lines (not all maser) have been measured from all OH rotational levels up to 500 K above the ground state toward W3(OH).

The four transitions in the ground state are the best known ones. They have wavelengths approximately 18 cm and frequencies 1665, 1667 MHz (main lines) and 1720, 1612 MHz (satellite lines). We will concentrate on them and for simplicity we will refer to them as OH masers. For a review of the observations up to 1986 see Genzel (1986) and for an assessment of the theoretical understanding up to that time see Elitzur (1986).

Gaume and Mutel (1987) observed 11 star-forming regions which contained more than 30 centers of maser activity and concluded the following:

a) All four ground state transitions were detected.

b) A fraction of 10% of the sources exhibit emission at all four ground state transitions.

c) The ratio of the intensities of the two main lines in the observed sources varies from 1/4 to 10.

d) Most (but not all) OH masers are found at the projected edges of H II regions.

e) The observations *do not* support a simple shell model for the OH maser distribution.

f) No correlation was found between the fluxes of individual features and their distances from the center of the associated H II regions.

g) No correlation was found between the peak maser flux of a cluster and the peak or integrated flux of the associated H II region.

h) About 25% of the OH maser clusters do not appear to be closely associated with H II region emission. These masers, however, are much weaker than those associated with H II regions.

i) *No existing theoretical model explains the observations.*

After the last conclusion, it is important to give a brief account of the theoretical effort to explain the ground state OH masers.

The ground state OH masers *are not* pumped by collisions alone (Kylafis and Norman 1990).

An attractive suggestion by Andresen (1986) that the 18 cm OH masers are pumped by photodissociation of H_2O is also ruled out. This is because the H_2O has to be reformed at a rate which is larger than the collision rate. This is not possible.

Pumping by infrared radiation has always been a good candidate, but the question has been for years whether there are enough infrared photons in the maser regions (Elitzur 1986). Moore, Cohen and Mountain (1988) claim that there are more than enough.

A recent calculation by Cesaroni and Walmsley (1991) takes into account collisions, infrared radiation and line overlap. They attempt to model all the observed transitions of OH in W3(OH) and conclude that a suitable combination of a far infrared radiation field and overlap can qualitatively reproduce almost all the features observed in W3(OH).

7. Conclusions

From the above presentation I think it has become evident that the astronomical masers are very useful. They are particularly useful in probing the dense star-forming regions. We have now reached the point where several maser lines from the same molecule are observed. This places significant constraints on the theoretical models and for the first time it is reasonable to expect that maser observations will reveal the physical parameters in the dense parts of star-forming regions.

On the observational side, I anticipate lots of beautiful observations of masers with the VLBA. On the theoretical side, what is needed (at least for the OH masers) is *systematic studies* to completely understand the effects of collisions, infrared radiation, line overlap, etc.

In conclusion, I would like to encourage young theorists to enter the field of astronomical masers because it is ripe and waiting.

8. References

Alcock, C., and Ross, R. R. 1985a, *Ap. J.*, **290**, 433.
———. 1985b, *Ap. J.*, **299**, 763.
———. 1986, *Ap. J.*, **306**, 649.
Anderson, N., and Watson, W. D. 1990, *Ap. J. (Letters)*, **348**, L69.
Andresen, P. 1986, *Astr. Ap.*, **154**, 42.
Castor, J. I. 1970, *M.N.R.A.S.*, **149**, 111.
Cernicharo, J., Thum, C., Hein, H., John, D., Garcia, P., and Mattioco, F. 1990, *Astr. Ap. (Letters)*, **231**, L15.
Cesaroni, R., and Walmsley, C. M. 1991, *Astr. Ap.*, in press.
Chapman, J. M., and Cohen, R. J. 1986, *M.N.R.A.S.*, **220**, 513.
Cheung, A. C., Rank, D. M., Townes, C. H., Thornton, D. D., and Welch, W. J. 1969, *Nature*, **221**, 626.
Deguchi, S., and Watson W. D. 1986a, *Ap. J. (Letters)*, **300**, L15.
———. 1986b, *Ap. J.*, **302**, 750.
———. 1989, *Ap. J. (Letters)*, **340**, L17.
———. 1990, *Ap. J.*, **354**, 649.
Deguchi, S., Watson, W. D., and Western, L. R. 1986, *Ap. J.*, **302**, 108.
Draine, B. T., Roberge, W. G., and Dalgarno, A. 1983, *Ap. J.*, **264**, 485.
Elitzur, M. 1982, *Rev. Mod. Phys.*, **54**, 1225.
———. 1986, in *Masers, Molecules, and Mass Outflows in Star Forming Regions*, ed. A. D. Haschick (Haystack Observatory), p. 299.
———. 1990a, *Ap. J.*, **363**, 628.
———. 1990b, *Ap. J.*, **363**, 638.
———. 1991, *Ap. J.*, January 20.
Elitzur, M., Hollenbach, D. J., and McKee, C. F. 1989, *Ap. J.*, **346**, 983.
Fiebig, D., and Güsten, R. 1989, *Astr. Ap.*, **214**, 333.
Forster, J. R., and Caswell, J. L. 1989, *Astr. Ap.*, **213**, 339.
Gaume, R. A., and Mutel R. L. 1987, *Ap. J. (Suppl.)*, **65**, 193.
Genzel, R. 1986, in *Masers, Molecules, and Mass Outflows in Star Forming Regions*, ed. A. D. Haschick (Haystack Observatory), p. 233.
Genzel, R., and Downes, D. 1977, *Astr. Ap. (Suppl.)*, **30**, 145.
Genzel, R., Reid, M. J., Moran, J. M., and Downes, D. 1981a, *Ap. J.*, **244**, 884.
Genzel, R., Downes, D., Schneps, M. J., Reid, M. J., Moran, J. M., Kogan, L. R., Kostenko, V. I., Matveyenko, L. I., and Ronnang, B. 1981b, *Ap. J.*, **247**, 1039.
Goldreich, P., and Keeley, D. A. 1972, *Ap. J.*, **174**, 517.
Goldreich, P., Keeley, D. A., and Kwan, J. Y. 1973, *Ap. J.*, **179**, 111.
Goldreich, P., and Kwan, J. 1974, *Ap. J.*, **190**, 27.
Greenhill, L. J., Moran, J. M., Reid, M. J., Gwinn, C. R., Menten, K. M., Eckart, A., and Hirabayashi, H. 1990, *Ap. J.*, **364**, 513.
Kylafis, N. D., and Norman, C. 1986, *Ap. J. (Letters)*, **300**, L73.
———. 1987, Ap. J., **323**, 346.
———. 1990, Ap. J., **350**, 209.
———. 1991, Ap. J., May 20.

Menten, K. M., Melnick, G. J., and Phillips, T. G. 1990, *Ap. J. (Letters)*, **350**, L41.
Menten, K. M., Melnick, G. J., Phillips, T. G., and Neufeld, D. A. 1990, *Ap. J. (Letters)*, **363**, L27.
Moore, T. J. T., Cohen, R. J., and Mountain, C. M. 1988, *M.N.R.A.S.*, **231**, 887.
Moran, J. M. 1989, preprint to be published in *Handbook of Laser Science and Technology*, ed. M. J. Weber.
Moran, J. M., Reid, M. J., Lada, C. J., Yen, J. L., Jonston, K. J., and Spencer, J. H. 1978, *Ap. J. (Letters)*, **224**, L67.
Nedoluha, G. E., and Watson, W. D. 1990a, *Ap. J.*, **354**, 660.
———. 1990b, *Ap. J.*, **361**, 653.
———. 1990c, *Ap. J. (Letters)*, **361**, L53.
Neufeld, D. A., and Melnick, G. J. 1990, *Ap. J. (Letters)*, **352**, L9.
———. 1991, *Ap. J.*, **368**, 215.
Palma, A., Green, S., DeFrees, D. J., and McLean, A. D. 1988, *Ap. J. (Suppl.)*, **68**, 287.
Reid, M. J., and Moran, J. M. 1981, *Ann. Rev. Astr. Ap.*, **19**, 231.
———. 1988, in *Galactic and Extragalactic Radio Astronomy*, eds. G.L. Verschuur and K. I. Kellerman, (New York: Springer-Verlag), p. 255.
Reid, M. J., Moran, J. M., and Gwinn, C. R. 1988, *The Impact of VLBI on Astrophysics and Geophysics*, eds. M. J. Reid and J. M. Moran (Dordrecht: Kluwer), p. 169.
Reid, M. J., Schneps, M. H., Moran, J. M., Gwinn, C. R., Genzel, R., Downes, D., and Ronnang, B. 1987, in *Star Formation*, eds. M. Peibert and J. Jugaku, IAU Symposium 115 (Dordrecht: Reidel).
Reid, M. J., and Silverstein, E. M. 1990, *Ap. J.*, **361**, 483.
Schneps, M., Lane, A. P., Downes, D., Moran, J. M., Genzel, R., and Reid, M. J. 1981, *Ap. J.*, **249**, 124.
Sobolev, V. V. 1960, *Moving Envelopes of Stars*, (Cambridge: Harvard University Press).
Strelnitskij, V. S. 1984, *M.N.R.A.S.*, **207**, 339.
Western, L. R., and Watson, W. D. 1983a, *Ap. J.*, **268**, 849.
———. 1983b, *Ap. J.*, **274**, 195.
———. 1983c, *Ap. J.*, **275**, 195.
———. 1984, *Ap. J.*, **285**, 158.
Wilson, T. L., Walmsley, C. M., and Baudry, A. 1990, *Astr. Ap.*, **231**, 159.

Jochen Eislöffel with first place poster prize.

THE PHYSICAL CONDITIONS OF LOW MASS STAR FORMING REGIONS

J. CERNICHARO
*IRAM, Av. Divina Pastora 7, NC, 18012 Granada. Spain
and Centro Astronómico de Yebes. Apartado 148. 19080 Guadalajara. Spain*

ABSTRACT. The stellar population of dark molecular clouds and isolated Bok globules consist mainly of low mass stars. The physical conditions of these clouds from their largest to their smallest structures are reviewed and the different methods used to derive them are evaluated. The interaction of the newly born stars with the ambient gas is analyzed on the basis of new results obtained from high angular resolution molecular line observations.

1. Introduction

Within the standard classification of galactic molecular clouds, which is not always self-consistent (see Goldsmith 1987; Scalo, 1990), O and B stars, i.e., massive stars, form in massive clouds (giant molecular clouds; see Blitz in this volume) while low mass stars are more typically associated with dark clouds. The border line between low mass and massive star forming regions are not clearly defined since low mass stars can also form in clouds that also produce massive stars. However, the physical conditions of clouds forming solely low mass stars are actually very different from those clouds that form massive stars. Table 1 gives the typical physical conditions prevailing in interstellar clouds (adapted from Goldsmith 1987). Note, however, that the concept of cloud is a matter of observational definition (i.e., minimum column density defining the cloud, optical aspect, single entities, ...). The division shown in Table 1 is a practical one and does not imply a clear separation of these different objects. For example, dark cores are embedded in dark clouds, which are embedded in complexes of dark clouds, which could be considered as clouds of the GMCs, which are again associated with large HI superclouds, and so on (see e.g. Scalo, 1990). In this review I will analyze the physical conditions of the different "associations" of gas forming low mass stars (i.e., dark clouds and Bok globules. See Genzel in this volume for a review of the physical conditions in massive star forming regions; see also Blitz in this volume).

The capability of an interstellar cloud to form stars depends on its initial physical conditions, which are subjected to the influence of the external medium (nearby massive stars, ionization fronts, density waves, etc; see e.g. Myers, 1990). Consequently, the knowledge of the physical conditions in the parent clouds and those of the immediate regions around the newly born stars are a necessary step to study and to infer the physical processes that occur during their formation.

TABLE 1
PHYSICAL PROPERTIES OF INTERSTELLAR CLOUDS

	GIANT MOLECULAR CLOUD	DARK CLOUD
COMPLEX		
Size (pc)	20-60	6-20
Density (cm^{-3})	100-300	100-1000
Mass (M$_O$)	8 10^4-2 10^6	10^3-10^4
Line width (kms^{-1})	6-15	1-3
Temperature (K)	7-15	≈10
Examples	W51, W3, M17	Taurus, Perseus, ρ-Oph (See Figure 1)
CLOUD		
Size (pc)	3-20	0.2-4
Density (cm^{-3})	10^3-10^4	10^2-10^4
Mass (M$_O$)	10^3-10^4	5-500
Line width (kms^{-1})	4-12	0.5-1.5
Temperature (K)	15-40	≈8-15
Examples	Orion OMC1, W33 W3A	B227, HCL2, B5, L1495 (See Figure 2)
CORE		
Size (pc)	0.5-3	0.1-0.4
Density (cm^{-3})	10^4-10^6	10^4-10^5
Mass (M$_O$)	10-10^3	0.3-10
Line width (kms^{-1})	1-3	0.2-0.4
Temperature (K)	30-100	≈10
Examples	Orion (ridge)	TMC1, HCL2-C, TMC2, B1 (See Figure 3)
CLUMP		
Size (pc)	<0.5	
Density (cm^{-3})	>10^6	
Mass (M$_O$)	30-10^3	
Line width (kms^{-1})	4-15	
Temperature (K)	30-200	
Examples	Orion (Hot core), W3(OH)	

(Adapted from Goldsmith 1987)

As an example of interstellar clouds and of the difficult tasks related to the study of star formation, let us consider a representative complex of dark clouds. Figure 1 shows the integrated line emission of ^{13}CO (J=1-0) in the direction of the central region of the Taurus complex. The emission extends over several square degrees and various individual clouds can be distinguished. The area covered by one of these individual clouds is typically of 1-4x1-4 pc^2. The whole complex, as defined by the extent of the ^{12}CO emission, covers more than 1600 pc^2 (see upper left insert in Figure 1). Figure 2 shows the visual extinction map towards HCL2, one of the clouds shown in Figure 1. Figure 2 shows that individual clouds have some regions of high visual extinction where we can expect large gas column densities. These regions are also likely to have high volume densities and are potential sites for low mass star formation (see e.g. Myers, 1985). Figure 3 shows the emission of different molecules in HCL2-C, one of the dense cores in HCL2 (the cloud at the left in Figure 1, and shown in more detail in Figure 2). We can see that the cores, which have a typical size of 0.1 pc in the Taurus complex, are deeply embedded in more extended volumes of gas. In an attempt to derive the physical conditions of these cores we will need to known, without any doubt, the physical conditions of the gas surrounding them.

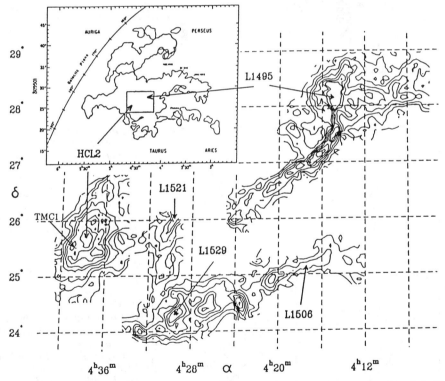

Figure 1. Map of the integrated intensity of the ^{13}CO (J=1-0) line in the central region of the Taurus complex. The data were collected with the Bordeaux 2.5-m telescope (POM I). The surveyed area is shown as a small rectangle against the ^{12}CO map of the Taurus-Perseus complex by Ungerechts and Thaddeus (1987) in the upper left insert. The lowest contour and the step are 1 K kms^{-1}. Data from Cernicharo and Guélin (1987), Duvert et al. (1986), and Nercessian et al (1988). (from Guélin and Cernicharo, 1988).

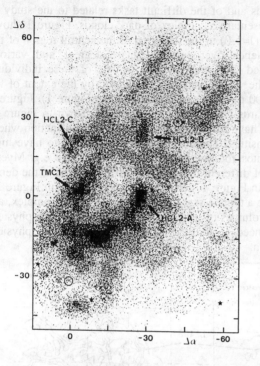

Figure 2. Visual extinction map towards HCL2 from star counts (Cernicharo et al. 1985). The densest regions inside this cloud are indicated by arrows. Indicated visual extinctions are only lower limits when $A_V>6.5$ mag (black contours). Circles indicate positions of reflection nebulae, black stars of T-Tauri stars. Offsets in arcmin. are relative to $\alpha=4^h 38^m 38.0^s$, $\delta= 25° 35' 45''$.

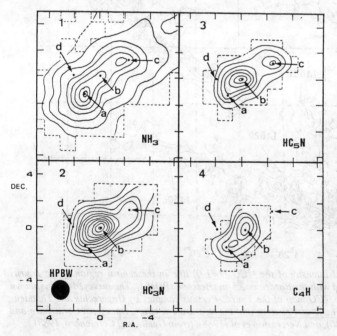

Figure 3. Contours of antenna temperature of the (1,1) main line of NH_3 (1), HC_3N $J=2-1$, $F=3-2$, (2), HC_5N $J= 8-7$ (3) and C_4H $N=2-1$, $J=5/2-3/2$ (4) in HCL2-C, one of the cloudlets shown in Figure 2. Offsets are in arcmin. Contour intervals are 0.1 K for NH_3 and HC_3N and of 0.05 for HC_5N and C_4H. First contour is 0.2 K for HC_3N and 0.1 for the other molecules. The positions a,b,c,d have a kinetic temperature of 8-9 K. Volume density is $5\,10^3$ cm^{-3} at position d, 10^4 cm^{-3} at positions b and c, and $4\,10^4$ cm^{-3} at position a. The core is surrounded by a low density envelope ($\approx 10^3$ cm^{-3}, see Figure 1 and Cernicharo and Guélin, 1987) of 2-3 mag. of visual extinction (from Cernicharo et al. 1984).

Looking at Figures 1, 2, and 3, we realize that the study of the physical conditions of star forming regions is a formidable challenge : the initial gas of a cloud distributed over several square parsecs (see e.g. the ^{12}CO limits of the Taurus complex in Figure 1), will contract and fragment into individual clouds of 1-4 pc size having a density an order of magnitude larger than the initial one (The ^{13}CO clouds of Figure 1). These clouds break up again into several fragments of higher density ($\approx 10^4$ cm^{-3}) and smaller size (≈ 0.1-0.2 pc; see Figures 2 and 3). Some fragments, similar to that shown in Figure 3, will continue to contract and will evolve until they reach the physical conditions necessary to activate the nuclear reactions that will support a new star. During this process, which could take a few 10^7 years, the cloud contracts by several orders of magnitude and its density increases from a few 10^{-23} g cm^{-3} to the typical stellar mean density of ≈ 1 g cm^{-3}.

In addition to this fantastic puzzle, the new formed stars will affect the surrounding medium through important mass loss (Lada 1985, 1988, and this volume). The subsequent evolution of the parent cloud and its potential to form more stars will be strongly affected by the newly born stars. Consequently, it will be very difficult to establish whether the physical conditions and the physical structure of the gas around the new stars reflect those of the parent cloud or those produced by the new stars.

The study of interstellar clouds can be carried out through the measurement of line emission radiated by their molecular gas components, and by the infrared continuum emission of their dust grains. However, while the determination of the large scale properties of these clouds can be derived using existing telescopes, the estimation of the physical conditions of the innermost regions where stars form is not an easy matter (the region of a few A.U. around the newly born star). Probably the next generation of radio and optical interferometers will open the way to such deep studies.

This review proceeds from the largest to the innermost cloud scales we can study with the present telescopes, i.e., the densest cloud regions where stars could form (Section 2 and 3). The physical conditions of isolated low mass Bok globules are analyzed in Section 4. The effects of star formation on the ambient gas will be analyzed in Section 5.

2. The Determination of the Physical Conditions of Low Mass Star Forming Regions.

Molecular dark clouds are characterized by low temperatures, narrow lines (see e.g. Guélin et al., 1982a), and low velocity gradients (see Table 1). For a given molecule, low temperatures will limit the number of rotational levels populated by collisions, thus favouring the increase in opacity of the low J rotational lines. Low velocity dispersion will limit the line broadening, while small velocity gradients will limit the doppler shift of the line center between different points of the cloud. These physical conditions imply that different regions of a molecular cloud are radiatively connected, i.e., photons emanating from the cloud's interior can be absorbed by the molecular gas in the intermediate and in the most external layers of the cloud. Consequently, the study of the molecular content of the densest regions of dark molecular clouds requires a detailed knowledge of the physical and chemical conditions of the gas surrounding these regions. This can be done by the measurement of different molecular tracers to discriminate the molecular emission of the cores from that arising in the less dense envelopes.

The parameters characterizing the physical conditions in molecular clouds are the temperature, the volume density, the density structure, the velocity field, the mass and the molecular abundances. In the following I will discuss the different methods currently used to derive these parameters, and to which degree they can be applied to low mass star forming regions. Another important parameter in the evolution of molecular clouds is the magnetic field (see Mouschovias in this volume) which now may be derived in molecular clouds through the observation of the Zeeman splitting of the 21-cm line of HI and of the OH lines (Crutcher, 1988; for a review see Heiles 1987).

Reviews on the determination of the physical conditions in molecular clouds can be found in Goldsmith (1984, 1987, 1988), Walmsley (1988), Wilson and Walmsley (1989), and Myers (1990).

2.1. TEMPERATURE

The temperature of an interstellar cloud depends on different heating and cooling processes operating throughout the cloud (see Goldsmith 1988). Since the gas cooling in a dark cloud is dominated mainly by the emission of the most abundant molecular species, its thermal structure is related to its density structure and to the molecular abundances. The thermal structure of the gas also depends on the presence of internal heating sources (IR sources, newly formed stars), and on the external heating (cosmic rays, UV photons). It is thus interesting to use the temperature distribution of the cloud to probe the physical processes occurring in it.

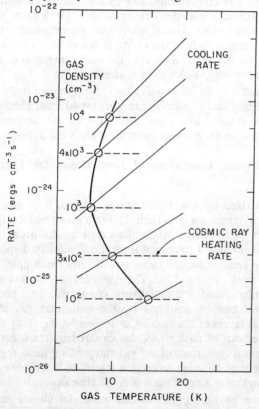

Figure 4. Expected kinetic temperature in dark clouds (X-axis) as a function of the cooling and heating rates, and of the density (from Goldsmith 1988; only heating by cosmic rays was considered).

The most commonly used tracer of the gas kinetic temperature is the ^{12}CO molecule. This molecule has a low dipole moment, and its low J rotational levels have an excitation temperature close to the cloud kinetic temperature under typical dark cloud conditions (the full thermalization of all the rotational levels, i.e., excitation temperature for all the energy levels equal to the kinetic temperature, will require a higher density). Its abundance in dark and molecular clouds assures its detection even in the external layers of the clouds. Due to the high CO abundance, its rotational transitions are optically thick and the observed line intensity of the lowest rotational transitions can be directly translated into the cloud's kinetic temperature. However, the ^{12}CO line opacities make the analysis of the observed emission difficult. We believe that the CO emission in dark clouds traces mainly the kinetic temperature of the external layers because photons emanating from the inner regions are absorbed and lost due to collisional de-excitation during their travel across the cloud. Only photons radiated from the external layers of the cloud are received at the telescope, and hence any temperature gradient or temperature structure in the cloud cannot be detected in the observed ^{12}CO emission.

The first systematic studies of dark clouds in the J=1-0 line of ^{12}CO were carried out by Dickman (1975, 1978) who found that these clouds have uniform kinetic temperatures, with T_K varying between 8 and 15 K from cloud to cloud. A complete study of HCL2 (see Figure 1 and 2) in the emission of several lines of different molecules by Cernicharo and Guélin (1987) and Cernicharo et al. (1984a) indicates a constancy of the kinetic temperature at least for the regions with more than 1 mag of visual extinction (the regions protected against the UV).

Figure 4 shows the results of thermal balance calculations made by Goldsmith (1988) assuming that heating is only produced by cosmic rays. The predicted kinetic temperature, for densities between 10^2 and 10^4 cm^{-3}, varies from 8 to 15 K which is in very good agreement with the observations of Dickman (1975). Other heating processes, like collisions with warm dust grains or photoelectric heating, obviously modifies this picture (see e.g. Falgarone and Puget, 1985; Hollenbach, 1988). Hence, these simple calculations apply to dark clouds which are well protected against the interstellar UV and do not have internal heating sources.

The observation of several transitions of ^{12}CO allows a more detailed study of the kinetic temperature (see Snell et al., 1981; Young et al., 1982; Cernicharo and Guélin, 1987). The variation of the line intensity ratios of the J=1-0 and J=2-1 transitions across the cloud can be used to derive a kinetic temperature profile. In B5, Young et al (1982) found that the line intensity ratio ^{12}CO(2-1)/^{12}CO(1-0) increases towards the cloud edges. Such variation can hardly be produced by a density gradient (the density should increase towards the cloud's edge in order to explain the observations), and Young et al concluded that the kinetic temperature increases from the cloud center, where $T_K \approx 15$ K, towards the cloud boundaries where it reaches a value of ≈ 30-40 K (see also Langer et al., 1989). This higher temperature of the cloud's external layers is probably due to photoelectric heating of the gas by the interstellar UV field. In case of B5, the Per-OB2 association enhances the UV field by at least one order of magnitude (Bachiller and Cernicharo, 1986a), rendering this effect more efficient than in dark clouds exposed to the standard UV field (the Dickman's (1975) clouds, or the clouds of Figure 1 and 2). More recently, Andersson et al. (1991) found evidence for general heating of the external layers of molecular clouds.

To derive the kinetic temperature of the inner cloud's regions, one has to observe a molecule less easy to excite and with optically thin transitions. Ammonia, NH_3, is such a molecule and because of its metastable levels, it is a good tracer of the gas kinetic temperature and density (Ho and Townes 1983; Walmsley and Ungerechts 1983). Its large dipole moment requires larger densities than ^{12}CO to produce a significant population of its energy levels. This molecule was extensively used by Myers and co-workers (see section 3.2) to derive the kinetic temperature and the density of an extensive sample of high visual extinction cores (selected from different dark clouds). Practically all the dense cores have a kinetic temperature of ≈ 10 K. A few of these cores were carefully mapped in the ammonia lines and the derived kinetic temperature across the core shows little fluctuations around the value of 10 K (Tölle et al., 1981; Myers et al. 1983; Cernicharo et al., 1984a, Bachiller and Cernicharo 1984,1986b). These cores are well protected against the interstellar UV field so that the only heating process is by cosmic rays (see Figure 4).

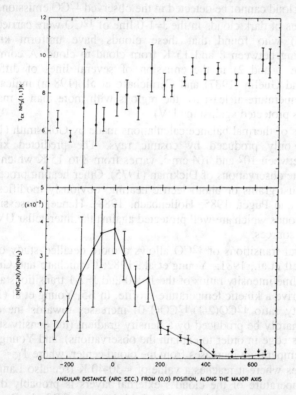

Figure 5. Excitation temperature of the (1,1) line of ammonia (top) as a function of the angular offset along the TMC1 major axis. This dense core in Taurus (see Figure 2) is the richest cyanopolyyne source in the sky and the offsets are relative to this source (offset 0,0). The bottom panel shows the variation of the ratio HC_7N to NH_3 along the same axis (from Olano et al., 1988).

One of the best studied dense cores is TMC1, an elongated object embedded in the Taurus cloud HCL2 (HCL2 is shown in Figures 1 and 2, and the position of TMC1 is indicated by an arrow in Figure 2). TMC1 is the richest cyanopolyyne source (HC_3N, HC_5N, ... $HC_{(2n+1)}N$), and in spite of the important molecular abundance variations across this dense core (see Bujarrabal et al., 1981; Tölle et al., 1982; Olano et al., 1988), its temperature seems to be almost constant (see Figure 5).

The situation may be different if the core were close to the cloud surface. Bachiller et al. (1987a) studied in detail one dense core in Perseus and found that the temperature varies from 12 K at the core center to 30 K at its edge. This core is submitted to the strong UV field existing in this region (Bachiller and Cernicharo, 1986b), and some UV photons can reach the external layers of the core. A similar behaviour of the kinetic temperature was found in the vicinity of o-Per by Bachiller et al. (1987b).

In addition to NH_3, rotational lines of other symmetric top molecules like CH_3CN or CH_3CCH can also be used as cloud thermometers (see e.g. Askne et al., 1984; Walmsley, 1988). However, the weakness of these lines and the limited regions towards which they can be detected, limit their usefulness for studies of selected cloud cores.

2.2. DENSITY

2.2.1. Density determinations. Among the different methods we may use to derive volume densities, there are mainly two currently applied by astronomers. The first is related to the determination of the cloud size and the average cloud gas column density (or the cloud mass). Gas column densities can be derived from the observation of the optically thin lines of ^{13}CO and $C^{18}O$, or from estimates of the visual extinction across the cloud (Encrenaz et al., 1975; Dickman 1978; Frerking et al., 1982; Cernicharo et al. 1985; Cernicharo and Guélin, 1987). Once the column density, or the cloud mass, is estimated, the average cloud density can be derived once the cloud size is known, i.e., if the distance to the cloud is available. It is also possible to derive a density profile from the observed variation of the gas column density across the cloud (see 2.2.2).

The second method to derive the gas density is based on the observation of several rotational transitions of a given molecule (see Walmsley 1988, and Wilson and Walmsley 1989). Since the cloud density can be derived from the observed line ratios, particular care must be taken to calibrate the lines correctly. An important instrumental problem is the different telescope beam size generally used to observe each transition. This problem can be overcome if the source size is known, but one should take into account that the source could have different sizes for each transition observed. Optically thin lines with excitation temperatures below the kinetic temperature are required in order to get some reliable information on the physical conditions of the cloud. These constrains arise because the observed transitions should be in the range over which line intensities are directly proportional to the volume density (which requires high dipole moment molecules if regions of high density are to be studied). If the lines are optically thick then the observed intensities, which are an integrated parameter of the cloud along the line of sight, will hardly reflect the physical conditions in the inner cloud regions.

Among the long list of molecules detected in the interstellar medium (see e.g. Irvine et al., 1988; Friberg and Hjalmarson, 1990) one can always select a molecule fulfilling the conditions quoted above. HC_3N, HC_5N, H_2CO, CS, and NH_3 have been extensively used to estimate densities in hot and cold cores (Bujarrabal et al., 1981; Myers and Benson, 1983; Benson and Myers, 1983, 1989; Cernicharo et al., 1984a; Snell et al., 1984; Mundy et al., 1986, 1987; Cernicharo and Guélin, 1987; Zhou et al., 1989,1990). Density estimates from different molecules for hot cores seem consistent with each other, but those of cold cores can be qualified to be somewhat disparate. Typical densities derived in cold dense cores from NH_3 and HC_3N are in good agreement and in the range of $5\,10^3$-$5\,10^4$ cm^{-3} (see e.g. Bujarrabal et al., 1981; Myers and Benson, 1983; Cernicharo et al., 1984a), while those derived from CS are between $\approx 10^4$-10^6 (see e.g. Zhou et al., 1989).

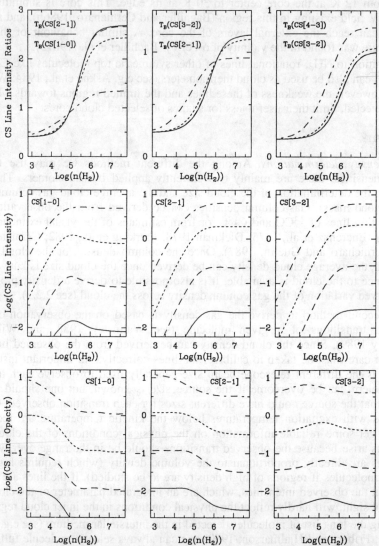

Figure 6. *LVG calculations for CS in a cloud with $T_K=10$ K. CS column densities are 10^{11} cm^{-2} (continuous line), 10^{12} cm^{-2} (- - - - - line), 10^{13} cm^{-2} (· · · · · line), and 10^{14} cm^{-2} (- · - · - line). The density varies between 10^3 and 10^8 cm^{-3}. The three top panels show the behaviour of the CS line intensity ratios as a function of the density and of the CS column density. The central panels show the $\log_{10}$ of the CS line brightness temperature. Bottom panels show the $\log_{10}$ of the CS line opacities (see text).*

The most popular method to analyze multi-line observations is the LVG approximation (Large Velocity Gradient, see e.g. Scoville and Solomon (1974) and Goldreich and Kwan (1974)). This approach is based on the hypothesis that the population of the molecular levels at a given point of the cloud depends only on the physical conditions at this point, i.e., there is no radiative connection between the different points of the cloud. This situation is achieved if Rdv/dr (where R is the cloud radius and dv/dr is the velocity gradient), is much larger than the local thermal line width. The parameters required to solve the radiative transfer problem within the LVG approximation are the cloud kinetic temperature, cloud size, molecular abundances, collisional cross sections, velocity gradients, and the observed line intensities.

Obviously, the extreme hypothesis of the LVG approximation is never reached in dark clouds, which are mainly characterized by very narrow lines (see above Guélin et al., 1982a), and special care should be taken in the interpretation of the observed line intensities when observing cold dense cores deeply embedded in dark clouds. In spite of this serious problem, the LVG approximation continues to be commonly used to derive densities in dark clouds from optically thick lines. The problem may be overcome by observing optically thin lines for which the LVG approximation gives reasonable results. As an example let us consider the case of CS, a high dipole moment molecule commonly used to derive the cloud density. Figure 6 shows CS line intensity ratios, line intensities and line opacities calculated with the LVG method. The calculations were done for a cloud with a kinetic temperature of 10 K and a line width at half intensity of 1 kms^{-1} (typical conditions in dark clouds; similar calculations for other values of T_K are available from Wilson and Walmsley, 1989; and from Walmsley, 1988). In Figure 6 one can see that while the CS[2-1]/CS[1-0] line intensity ratio depends strongly on the density and on the column density, the other line ratios depend mainly on density. For $N(CS)<10^{14}$ cm^{-2} and $n(H_2)$ larger than a few 10^4 cm^{-3}, the line opacities are moderate or weak, but they become larger for low densities. If the line of sight contains only gas with uniform density, then curves such as those of Figure 6 can be used to analyze the data, but if it contains gas of different physical conditions, then the derived densities should be taken with caution.

In order to illustrate the possible pitfalls that we may encounter in the LVG interpretation of multi-line data, I repeated calculations similar to those of Figure 6, but with a Monte Carlo radiative transfer code which takes into account the radiative connection between all the cloud layers. A "core+envelope" cloud structure was adopted (see Cernicharo and Guélin, 1987; see also Table 2). The core had a density of 10^5 cm^{-3} and a radius of 1.35 10^{17} cm (1' at the distance of the Taurus complex; see Figure 3). The envelope had a diameter four times larger than the core (see Figures 2 and 3), and its density was 5 10^3 cm^{-3}. The CS abundance was fixed in the core and in the envelope to 5 10^{-9}. Figure 7 shows the results in the direction of the core for the CS rotational transitions with $J_{upper}\leq 5$. We see that the line intensity from the core alone (continuous line) can be well reproduced by the calculations of Figure 6, but that the full CS profile emerging from the cloud (dashed line) is strongly affected by the low density envelope. The J=1-0, 2-1, and 3-2 lines of CS have an enhanced emission in the envelope due to radiative heating by the photons arising from the core. The emission of the J=1-0 and J=2-1 lines of CS will extend over a region much larger than the size of the core. In addition, their intensities and their line ratio will reflect the physical conditions prevailing in the core rather than those of the envelope. The line width, however, will reflect the turbulent motions in the envelope. When looking at the core, the observed line width will lead us to an erroneous interpretation of its dynamical state (see also Figure 8). Since the $C^{34}S$ emission profiles are not affected by the envelope, the emission of this molecule is a better tracer of the physical conditions of the core than $C^{32}S$.

If we try to fit the CS profiles of Figure 7 using a LVG model for a single component cloud model, then the physical conditions derived for the core will depend on the applied line ratios. From the J=1-0, J=2-1 and J=3-2 lines of CS, the density towards the core and across the envelope will be overestimated and the column density underestimated. Consequently, the source size and the cloud mass will be overestimated. The effect would be less important when comparing transitions of high J (the envelope is optically thin for these transitions).

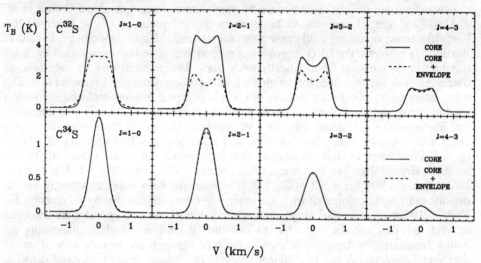

Figure 7. Radiative transfer calculations for a cloud consisting of a core with $n(H_2)=10^5$ cm^{-3} and an envelope with $n(H_2)=5\ 10^3$ cm^{-3}. The kinetic temperature and CS abundance for both components, core and envelope, are of 10 K and $5\ 10^{-9}$ respectively. Turbulent velocity in the core is 0.3 kms^{-1} and in the envelope 0.6 kms^{-1} (see text). The radius of the core is 0.04 pc. The envelope is four times larger than the core. Continuous lines represent the emission arising from the core alone, while dashed lines represent the emission arising in the core plus the envelope.

Another potentially interesting density tracer is the HCO+ molecule because of its high dipole moment ($\approx$ 4 Debye). However, the emission of HCO+ presents a very peculiar behaviour wherever it was observed in dark clouds : (i) HCO+ emission extends to regions of low extinction, hence, presumably, of low gas density; ii) the H^{13}CO+/HCO+ line intensity ratio is surprisingly large towards the cores ($\approx$ 1); and iii) HCO+ emission often peaks in the same direction as H^{13}CO+ (see Guélin et al., 1982b; Cernicharo and Guélin, 1987).

In order to explain the widespread HCO+ emission in HCL2 (the dark cloud shown in Figure 2), Cernicharo and Guélin concluded that the observed HCO+ radiation is essentially emission from the cores scattered by the low density envelope surrounding them. Figure 8 shows their calculations for a cloud model similar to that used above for CS. The effects of radiative scattering on the extent of the HCO+ emission, and on the HCO+ line intensities, are clearly substantiated in this Figure.

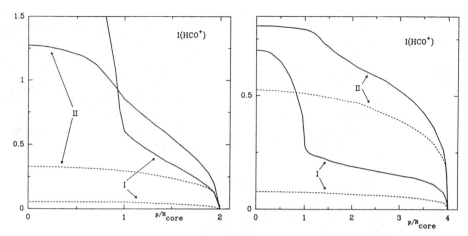

Figure 8. HCO^+ $J=1-0$ *integrated line brightness temperature distribution for spherical "core-envelope" cloud models, as predicted from radiative transfer calculations. Abcissa represents the offset from the center position (core direction) normalized to the core radius. The radius of the core is 10^{17} cm, its density is $3\ 10^4$ cm^{-3} and its kinetic temperature is 10 K. The "envelope" has a radius of $2\,R_c$ in* **a** *(left panel) and of $4\,R_C$ in* **b** *(right panel); its kinetic temperature is also of 10 K and its density is of 10^3 cm^{-3} in case I and of $3\ 10^3$ cm^{-3} in case II. Velocity dispersion is 0.3 kms^{-1} in the core and 0.7 kms^{-1} in the envelope. The large difference between the dashed lines (emission of the envelope alone) and the full lines (envelope illuminated by the core) shows the importance of diffusion in dark clouds. Even in case II where the opacity in the halo is large, we can see a blurred image of the core. (from Cernicharo and Guélin, 1987)*

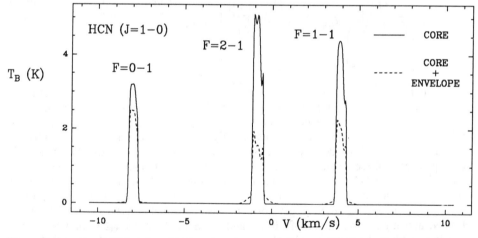

Figure 9. Emerging HCN $J=1-0$ profile from a high density core (continuous line) and a "core+envelope" model (dashed line). The parameters are the same than as for CS, except that a small velocity gradient was introduced and that the HCN abundance was 10^{-8}. Line overlap was included in the model (from González-Alfonso and Cernicharo, 1991a).

Finally, Figure 9 shows the HCN J=1-0 emerging profile from the same "core+envelope" model. The profile is characterized by a strong modification of the relative intensities of the HCN hyperfine components (see Walmsley et al. 1982 and Guilloteau and Baudry 1981, for a description of the HCN hyperfine anomalies). Figure

10 shows the observed HCN spectra towards TMC1 by Walmsley et al. (1982). These spectra look very similar to the calculated spectrum shown in Figure 9. The observed spectra show a considerable modification of the hyperfine line intensities, while the $H^{13}CN$ spectrum towards the same source observed by Irvine and Schloerb (1984), indicates normal hyperfine line intensity ratios. Thus, the cloud density structure and the HCN line opacities in the low density envelope are so important in determining the emergent profile, that any other effect, like selective collisional cross sections (as discussed by Guilloteau and Baudry, 1981), may be considered as irrelevant. A discussion of these effects can be found in Cernicharo et al., (1984b), Cernicharo and Guélin (1987), and González-Alfonso and Cernicharo (1991a).

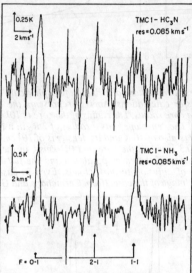

Figure 10. (Left) Observed HCN J=1-0 spectra towards the TMC1's HC_3N and NH_3 sources. The spectra look very similar to that of Figure 9. The relative intensities of the F=1-0, 2-1, and 1-1 hyperfine lines are in the ratio 1:5:3. However, the weak F=1-0 line is stronger than the other components. (from Walmsley et al., 1982).

Although the core-envelope model used for these radiative transfer calculations is still a pretty crude one, it accounts for several observational effects. The simulations shown in Figures 7 to 9 indicate that the emission from high dipole moment molecules, having also high molecular abundances, can be essentially dominated by the size and physical conditions of the low density envelopes that surround the cores. What are the consequences of these results? First, density and mass determinations in cold dense cores can be in error by 1-2 orders of magnitude if obvious pitfalls are not avoided ! Second, one could devote more telescope time to the observation of the less intense lines of the rare isotopes of CS in order to get correct density estimates. $H^{13}CO^+$, DCO^+, $C^{34}S$, $H^{13}CN$, and high J transitions of HC_3N and $C^{32}S$ (at least $J_{upper} \geq 3$ for CS) will be intense enough in dense cores to be observed with existing sensitive instruments. In some cases, the line intensity of the rare isotope molecule could be stronger than that of the main one (like $H^{13}CO^+$ and HCO^+, see e.g., Guélin et al., 1982b, Cernicharo and Guélin, 1987), and in addition, the rare isotope will trace the physical conditions of the cores more accurately.

2.2.2. The density profile of a self-gravitating cloud. The determination of the cloud's density is an indispensable step if one wants to study its dynamical, physical and chemical evolution. An isothermal self-gravitating cloud in equilibrium will have a density proportional to r^{-2} (Larson 1969). However, as shown by Shu (1977), an isothermal self-gravitating cloud will form a centrally condensed object with a spherical accretion flow following a $r^{-3/2}$ density law around it (see also Shu et al., 1987). Falgarone and Puget (1985), studied theoretically the density structure of thermally supported self-gravitating clouds including in their calculations cooling by molecular emission, heating by UV photons and cosmic rays, dissipation of turbulence and interaction of dust with gas. Their calculations indicate the presence of a central condensation having a radius $\approx 0.1\ R_T$, where R_T is the cloud radius. This condensation is composed of an isothermal core with a density profile $\propto r^{-2}$ and an envelope where the density changes sharply. For $r>0.1 R_T$ they found that the density changes slowly with a law roughly r^{-1}. It is interesting to note that any self-gravitating hydrostatic sphere having an ideal-gas equation of state and which is hotter at the outside than inside will have a density profile less abrupt than the r^{-2} characteristic of the isothermal case (Dickman and Clemens, 1983).

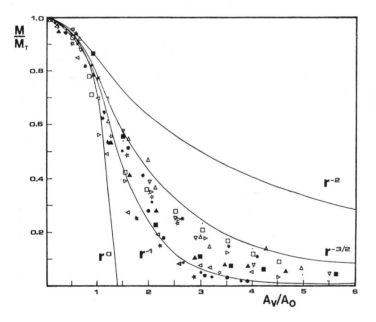

Figure 11. The mass within a given contour of A_V, normalized to the total mass of the cloud, M_T, as a function of the visual extinction normalized to the visual extinction at $0.7 R_T$, where R_T is the total radius of the cloud. These parameters are distance independent. The normalizing visual extinction has a value of ≈ 1-1.7 mag. The different symbols correspond to the different clouds in Taurus and Perseus. Solid lines represent spheres with a density law $n(r) \propto r^{-\alpha}$ for $\alpha=0$, 1, 3/2, and 2 (from Cernicharo et al., 1985).

Cernicharo et al. (1985) used visual extinctions derived from star counts to estimate the variation of the density with the cloud radius. Assuming that the density follows a law $\propto r^{-\alpha}$, we derived a density profile $n(r) = n_0 r^{-(1.3 \pm 0.2)}$ for the whole complex of dark clouds in Taurus and Perseus (see Figure 11). A similar result was obtained by Myers et al. (1978) from their study of ρ-Oph. This method can be used to derive a cloud average density and to estimate the density profile for regions of moderate gas column densities. However, it will give only very approximate results for the cores. The radial density structure of some cold and dense cores was determined by Westbrook et al. (1976), Cheung et al. (1980) and Walker et al. (1990) from observations of their dust continuum radiation. Their results indicate that the density distribution of the dust, $n_d(r)$, follows as a $r^{-(1.5\pm 0.5)}$ law. In the case that the dust-to-gas ratio is constant in these cores, the gas density distribution is also given by the same law. Many other density determinations, including multi-line observations of dense cores (Zhou et al., 1990), are consistent with $\alpha \approx 1.0-2.0$, with a definite tendency to values lower than 2 for the large scale density profile, and a value close to 2 for the inner core regions (Yamashita et al., 1990). However, such density determinations could be affected by the presence of different objects in the line of size, or by temperature gradients, or simply by selection effects. In case of Figure 11, Cernicharo et al. (1985) used distance free parameters and discussed the influence of cloud alignment in the line of sight on the derived density profile. However, they assume that clouds are spherical which, in view of Figures 1 to 3, can be considered as very crude approximation to the cloud geometry. This Figure clearly reflects the mass distribution inside the clouds and gives some indication that the density profile of dark clouds is less abrupt than r^{-2}.

It is obvious that the complex structure of dark clouds cannot be interpreted as a single cloud component. Cernicharo and Guélin (1987) modelled the dark cloud HCL2 as consisting of four "onion-like" layers of different physical conditions (see Table 2; the values given in these Table applies to the two different clouds appearing in the HCL2 line of sight). The derived physical structure for HCL2 follows the spatial distribution of the emission of the different molecular tracers used in this study. The halo corresponds to the large region of ^{12}CO emission seen in Figure 1 (see also Ungerechts and Thaddeus, 1987). The external envelope corresponds to the region where ^{13}CO is easily detected (see Figure 2), while the inner one corresponds to the region where $C^{18}O$ emission is strong. Finally, the cores are the different high visual extinction fragments embedded in the cloud (see Figures 2 and 3).

TABLE 2
A MODEL FOR THE DARK CLOUD HCL2

	T_K (K)	Size (pc)	$n(H_2)$ (cm^{-3})	Δv (kms^{-1})	Mass (M_O)	Mass Fract.
Halo	?	>3	≈30	2-3	>230	>0.38
Ext. Env.	15	1.5	300	1.0	290	<0.48
Int. Env.	10	0.4	1-3 10^3	0.7	55	<0.09
Cores	10	0.15	1-3 10^4	0.30	28	<0.05

Adapted from Cernicharo and Guélin (1987).

2.3. MOLECULAR ABUNDANCES AND EXOTIC MOLECULES IN DENSE CORES

Many important molecular species cannot be observed in interstellar clouds because of their lack of detectable radio transitions (O_2, C_2, C_2H_2, CH_4,...). H_2, the most abundant molecule in interstellar clouds, can be observed only in the line of sight of a few bright infrared sources. Some abundant molecules have optically thin lines and some physical information can be extracted through the observation of only one rotational line ($C^{18}O$ and ^{13}CO for example). These molecules can be used as indirect tracers of the molecular hydrogen abundance (see e.g. Cernicharo and Guélin, 1987). However, these determinations assume that the $N(H+2H_2)/A_V$ relation of Bohlin et al. (1978), which has been well established only for $A_V<3$ mag, applies also for large A_V. Many other interesting molecules have optically thick transitions and the determination of their abundance through complex radiative transfer models can be affected by important errors (see section 2.2.1). Thus, we can say that the determination of molecular abundances in cold and dense clouds is a difficult task, and that taking into account the possible problems quoted in the previous section, it is not surprising to find some inconsistency between the molecular abundances derived by different authors for the same molecule in the same source. Nevertheless, clear differences were established between the chemical processes occurring in quiescent dark clouds and those of massive star formation regions (see Friberg and Hjalmarson (1990) for a review). This result is not unexpected because of the different physical conditions of these clouds (see Table 1). However, the behaviour of the molecular abundances within a given dark cloud is even more amazing. TMC1 is probably the best example. The emission from carbon chains is enhanced towards the South-East of the TMC1 ridge (the HC_3N source), while other molecules, like NH_3, peak towards the North-West (see Figure 2 and 5b). Similar differences were found in L134N by Swade (1989). Some of these variations can be explained by the presence of density gradients in the cloud (see e.g. Bujarrabal et al., 1981), but the different spatial distribution for carbon chains (C_2H, C_4H, HC_5N, HC_7N, ...) and for ammonia, among other molecules, appears now to be well established in TMC1 (see Olano et al., 1988). These important variations can probably be related to a different evolutionary chemical stage across the TMC1 ridge (Prasad et al., 1987) or to slightly different physical conditions across it (see the slight variation of T_K in Figure 5a).

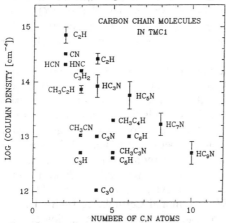

Figure 12. Column density of carbon-rich species in $TMC1(HC_3N)$ as a function of their number of heavy atoms. Column densities for ^{13}CO, and $C^{18}O$ in this cloud are $\approx 2\ 10^{16}$ and $\approx 3\ 10^{15}$ cm^{-2} respectively. The total gas column density is $\approx 10^{22}$ cm^{-2}. For comparison C_3O has also been included in this Figure (from Cernicharo et al., 1987b).

Figure 12 shows the molecular column densities derived for various molecules in TMC1(cyanopolyyne source). The total gas column density in this direction is $\approx 10^{22}$ cm^{-2} (Cernicharo and Guélin, 1987). We can see that molecules as exotic as C_5H, C_6H or HC_9N have a relatively high abundance, $\approx 5\ 10^{-10}$-10^{-9} (see e.g. Cernicharo et al., 1987a, 1987b; see Herbst, (1991), for a chemical model for the formation of heavy molecules). A new molecular species has been recently detected in this source : linear H_2C_3 (Cernicharo et al., 1991a). Although TMC1 can be considered, from the chemical point of view, as a very particular source, Cernicharo et al. (1984a,1986) showed that carbon chain molecules can be observed in practically all the dense cores of the Taurus complex.

Due to the chemical differentiation between clouds and within the same cloud we can say that, in addition to radiative transfer, chemistry can play an important role in modifying the spatial distribution of the emission from different molecules (see Prasad et al., 1987; Olano et al., 1988; Swade, 1989).

Reviews on the chemistry of dark clouds can be found in Herbst (1987), Prasad et al. (1987), and Millar (1990).

2.4. MASS

The difficulty to observe H_2 directly in dense interstellar clouds forces us to take recourse at indirect methods for the estimation of the mass of molecular gas. These methods use the visual extinction of dust grains (Dickman, 1978; Cernicharo et al., 1985), far-IR dust emission (see e.g. Boulanger and Pérault, 1988; Langer et al., 1989; Jarret et al., 1989; Snell et al., 1989), diffuse gamma ray emission (see e.g. Bloemen (1988) and references therein), and the CO integrated brightness temperature as gas tracers, or simply assume virial equilibrium (see Maloney (1990) for a discussion of this assumption). The most popular method, the CO brightness, assumes that the integrated intensity of the ^{12}CO, ^{13}CO, and $C^{18}O$ J=1-0 rotational line is proportional to the H_2 column density, the proportionality "constant" being determined from the comparison of the integrated intensity with the gas column density derived from star counts. This calibration method is based on the constancy of the gas/dust ratio, which has only been established for low visual extinctions ($A_V < 3$ mag.; see Bohlin et al., 1978). Important efforts have been directed to verify the constancy of this ratio for large values of the visual extinction, and it seems now well establish that the relations CO/A_V are valid for values of Av as large as 10 mag (Encrenaz et al., 1975; Dickman 1978; Frerking et al., 1982; Bachiller and Cernicharo, 1986b; Duvert et al., 1986; Cernicharo and Guélin, 1987; Nercessian et al., 1988).

The adequate method to derive the mass depends of the object under study. Thus, to derive the mass of a complex of dark clouds (see e.g. Guélin and Cernicharo, 1988), the integrated intensity of ^{12}CO will be the best tool (through the empirical relation $N(H_2) \approx 2\text{-}4\ 10^{20}\ W(^{12}CO)$ cm^{-2}, where $W(^{12}CO)$ is the integrated intensity of the J=1-0 line of ^{12}CO; it is also possible to show that $N(H_2) = 3\ 10^{20}$ $(10/T)(n(H)/1000)^{1/2} W(^{12}CO)$ -see van Dishoeck and Black, 1987). For the mass of an individual cloud defined by the regions having more than 1 mag of visual extinction, ^{13}CO will be a good mass tracer. Finally, for the mass of the darkest regions we will need to observe the emission of $C^{18}O$ because even the ^{13}CO lines will be optically thick (see e.g. Cernicharo and Guélin, 1987; Guélin and Cernicharo, 1988). To derive precise masses of the dense cores, it will be necessary to observe the emission of a good density tracer like NH_3, $H^{13}CO^+$, or $C^{34}S$ (see section 2.2.1).

As an illustration of the typical mass of dark clouds, Table 3 gives those of the largest dark clouds in Taurus and Perseus (as derived from star counts by Cernicharo et al., 1985). Although it is always difficult to define a cloud as an entity, the Taurus clouds (see Figure 1) are adequately defined. The area covered by each cloud, the minimum visual absorption defining the cloud, and the number of high obscuration fragments ($A_V > 4$ mag) in each cloud, are also given in this Table.

TABLE 3
MASS OF THE TAURUS AND PERSEUS CLOUDS

Cloud	Area (pc^2)	A$_V$(min) mag	Mass M$_O$	Total Mass M$_O$	N. Fragments
HCL2	15.6	1.0	640	1000	10
L1517	16.6	0.5	278	457	4
L1536	7.4	1.0	250	410	3
L1506	8.5	1.0	242	430	4
L1529	14.0	1.0	440	740	6
L1495	38.7	0.5	1480	1900	10
L1544	2.4	0.5	57	83	1
L1489	3.0	0.5	58	92	1
B5	14.3	0.5	210	360	1
Perseus	180	0.5	5400	7450	20

N. Fragments is the number of fragments in each cloud having A$_V$> 4 mag. Some of these fragments are fragmented again (see Figure 14).
Mass is the cloud mass inside the region of A$_V$(min) defining the cloud. The estimated total cloud mass is given in column 4.
Adapted from Cernicharo et al (1985)

Reviews on the different methods for mass determination in interstellar clouds can be found in Dickman (1988), Guélin and Cernicharo (1988), Bloemen (1988), and Wilson and Walmsley (1989).

3. The Mass Distribution and the Physical Structure of Low Mass Star Forming Clouds.

3.1. MASS DISTRIBUTION AND FRAGMENTATION

We know that a complex of dark clouds typically has a mass of several 10^3 M$_O$. Each individual cloud in the complex could have a mass ranging from 10^2 to 10^3 M$_O$ and the densest cores a mass of 1-10 M$_O$ (see Table 1 and 2). The free fall time, t$_{ff}$, of a cloud is given by t$_{ff}$=3.4 10^7 n$^{-1/2}$ yr, where n is the density in cm^{-3}. For the typical densities in dark clouds t$_{ff}$ is of a few 10^6 yr, and of $\approx 10^5$ yr for the dense cores. If the mass of a complex of dark clouds is completely processed to form low mass stars with a 100% efficiency and if there were no internal support against gravity, then the star formation rate will be several orders of magnitude larger than observed. We have to ask, what is the dynamical state of the dark clouds containing dense cores ? What is the dynamical state of the dense cores ? And finally, what is the actual star formation rate in dark clouds ?

Due to their proximity, local dark clouds are excellent laboratories to study the smallest details of the interstellar cloud structure. The Taurus complex and some of its individual clouds are shown in Figures 1 to 3. Figure 13 shows a visual extinction map of the whole Perseus complex. An overall picture of the ρ-Oph complex can be found in de Geus et al. (1990). The general characteristic of these complexes of dark clouds is their fragmentary aspect and the enormous variety of structure they offer (see for example the rich structure exhibited by the ^{13}CO observations of Orion A by Bally et al. (1987) and Bally (1989), or the OrionB cloud which contains 42 dense cores that represent only a small fraction of the total cloud mass (Lada et al., 1991)).

The mass distribution in Taurus and Perseus is shown in Figure 11, which shows that in these complexes 50% of the cloud mass is concentrated in the regions of $A_V<3$ mag, while only $\approx 10\%$ is concentrated in regions of large gas column density ($A_V>5$ mag). The cloud mass distribution can be interpreted in terms of a standard cloud density profile $n(r) \propto r^{-(1.3\pm 0.2)}$ (see section 2.2.2). Furthermore, all these large clouds in Taurus and Perseus seem to be in virial equilibrium (see e.g. Ungerechts and Thaddeus, 1987), i.e., the gravitational energy is balanced by the internal cloud pressure (thermal and/or magnetic). Under virial equilibrium and with a cloud density profile $\propto R_T^{-(1.3\pm 0.2)}$, it is straightforward to derive $M_T \propto R_T^{1.7\pm 0.2}$, $\Delta v \propto R_T^{(0.3\pm 0.2)}$, and $<n> \propto R_T^{-(1.3\pm 0.2)}$, where M_T is the total mass of the cloud, Δv is the cloud velocity dispersion, and $<n>$ is the average cloud density. Thus, dark clouds follow the Larson relations (1981), which were initially established for GMCs, but which also apply to dense regions of very small size (Myers 1983; see also Falgarone and Puget, 1986 and Scalo, 1987).

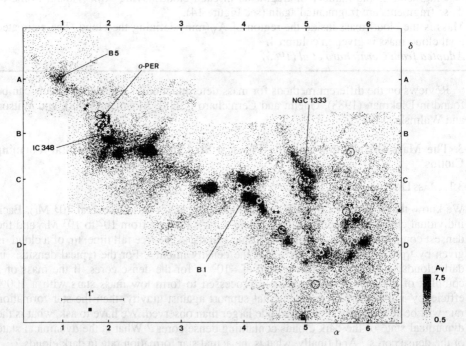

Figure 13. Visual extinction map of the Perseus complex derived from star counts with a resolution of 2.5' in the red prints of the Palomar Observatory Sky Survey. Asterisks represent the position of young stellar objects and circles those of reflection nebulae. The ticks on the axes indicate the reference positions of the counts. $\alpha(1)=3^h 44^m 22.6^s$, $\alpha(2)=3^h 40^m 50.7^s$, $\alpha(3)=3^h 35^m 19.9^s$, $\alpha(4)=3^h 30^m 39.8^s$, $\alpha(5)=3^h 25^m 56.0^s$, $\alpha(6)=3^h 21^m 15.5^s$; $\delta(A)=32° 42' 44''$, $\delta(B)=31° 52' 27''$, $\delta(C)=31° 10' 12''$, $\delta(D)=30° 10' 12''$ (from Cernicharo et al., 1985).

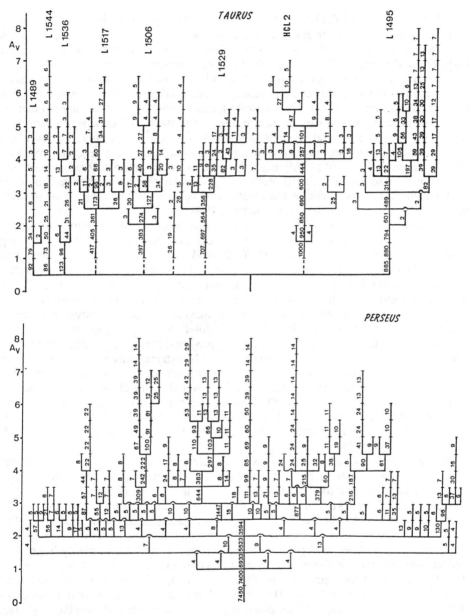

Figure 14. Hierarchical distribution of the mass in Taurus (top) and Perseus (bottom) complex of dark clouds. The vertical axis represents visual extinction in mag. Each cloud is defined starting at some value of visual extinction (see Table 2). The variation of the visual extinction across the cloud was used to characterize new well defined objects inside the original cloud. The parent cloud is fragmented for some value of A_V, and the new fragments are followed until the process stops or we reach the angular resolution limit (2.5'). The vertical numbers indicate the mass (in $M_\odot$) of each fragment inside the A_V contour defining it. Visual extinctions larger than 5.5 mag are only lower limits. The objects with A_V larger than this limit may also be fragmented, but higher angular resolution observations are necessary to check this.

If dark clouds follow the Larson relations, velocity dispersion in them could be part of a common hierarchy of interstellar turbulent motions : dark clouds appear to be supported against self-gravity by supersonic, but subalfvénic internal motions (see e.g. Falgarone and Puget, 1988). The turbulent pressure created by these supersonic internal motions can stabilize gravitationally unstable clouds, increasing their lifetime against collapse. This velocity field generates density fluctuations on all scales down to 0.015 pc (Falgarone and Pérault, 1988). The smallest objects created by these density fluctuations are the dense cores which may eventually become unstable and collapse to form stars (they have subsonic internal motions, Myers and Benson 1983). Note, however, that there are some dense cores that show important deviations from the general correlations (Martín-Pintado et al., 1985). Although these cores could be considered as particular because their association with HII regions, high angular resolution observations of the TMC1 cold "core" also indicate the presence of very small and dense "clumps" with supersonic line widths (Guélin and Cernicharo, 1988).

Another different school of thought assumes that magnetic forces dominate thermal-pressure, centrifugal forces, and the dynamical evolution of the clouds (see e.g. Mouschovias 1990 and this volume). With moderate values of the magnetic field strength, the clouds could be magnetically supported (see Mouschovias 1990; Myers and Goodman, 1988b). However, although the measured component of the magnetic field in the line of sight are in the required range to support the clouds, the observed cloud structure is so complex that in addition to the magnetic field other physical processes could be playing an important role in the cloud evolution (see Scalo, 1990).

It is worth noting that the different models proposed for the dynamical evolution of dark clouds (hierarchical turbulence, magnetic fields, external pressure,...) try to simulate the Larson relations. However, column densities or magnetic field estimates in molecular clouds cover a limited range compared with that covered by the cloud sizes (2-3 orders of magnitude), which could create correlations between these parameters and the cloud size where in fact there is none (Leisawitz, 1990; Maloney, 1990; Scalo 1990). Observational selection effects, important errors in mass and cloud size determinations could also be playing a role in the observed correlations.

Whatever the physical processes are, (see references quoted above), what is well accepted by the astronomers is that clouds are complex, that they appear nearly virialized, and that they form stars but, at a low rate. The study of the structure of the clouds and of their mass distribution can give important information about these physical processes. From Figure 11 and Table 2 we can conclude that only a small fraction of the cloud mass corresponds to regions of large column density. How is this mass distributed? We can see in Figures 2 and 13 that dark clouds are be highly fragmented, and an answer to the question can be found by studying their fragmentary state. In Figure 14, I have plotted the same data of Figure 11 but in a hierarchical path (data from Cernicharo and Bachiller 1984). This Figure shows the visual extinction at which the different fragments inside a given cloud begin to be well defined, together with their masses at different values of A_V (for a graphical description of such hierarchical trees see Scalo, 1990).

We can see from Figure 14 that the fraction of the cloud mass concentrated in regions of high A_V ($\approx 5\text{-}10\% M_T$), is in fact distributed among several fragments resulting from multiple steps of cloud fragmentation (for a description of hierarchical fragmentation see e.g. Elmegreen 1985a). The size of the smallest fragments in Taurus is 0.1 pc, while it is

≈0.5 pc in Perseus. Another interesting result from Figure 14 is that Taurus and Perseus clouds, with a total mass lower than a certain critical value, are non fragmented (to the limit of the angular resolution of the observations which is of 2.5 arcmin, i.e., ≈0.1 pc for Taurus and ≈0.2 pc for Perseus). From Figure 14 we can derive this critical value to be ≈100 M_O, which coincides with the typical Jeans mass of 32-100 M_O for the physical conditions of Taurus and Perseus clouds ($<n> \approx 10^2$-$5 \ 10^2$ cm^{-3}, $T_K \approx$ 10-20 K; the Jeans mass is 100 $(n(H)/100)^{-1/2} (T_K/20)^{3/2}$ M_O). In each fragmentation step there are 2-5 new fragments, with one or two of them much more massive than the others. The average of the mass-of-fragments to mass-of-the-parent-cloud(fragment) ratio is of 0.6±0.06. The total number of fragmentation steps in each cloud(fragment) with M>100 M_O is of ≈3-6. These values are only lower limits because they were derived without information about the velocity, i.e., a cloud appearing as not fragmented in the visual extinction maps may be composed of several fragments with different velocities.

Although there are similarities between the Taurus and Perseus fragmentation processes, the final fragments in Perseus are more massive and larger than those in Taurus. In addition, while individual clouds in Taurus begin to be well defined for $A_V \approx$ 0.5-1.0 mag, in Perseus the individual entities appear at $A_V \approx$ 2-3 mag. This difference in the visual extinction, at which the initial clouds begin to fragment, may be related to the different external conditions prevailing in both complexes. Star formation occurs when a cloud collapses under the force of its own gravity. The gravitational force threshold can be related to a minimum column density proportional to the square root of the external pressure, $A_V > 0.8 (P/3000)^{1/2}$ (see e.g. Elmegreen, 1985b,1989; Chièze, 1987). Thus, the external pressure should be ≈2-3 times larger in Perseus than in Taurus to explain the different A_V threshold in both complexes. This enhancement of the external pressure could be explained by the presence of several OB stars in the neighbourhood of the Perseus complex. Alternatively, if the clouds are magnetically supported, there is a critical column density below which the cloud cannot collapse, $A_V > 2$ [B/30 μG] (see e.g. Shu et al., 1987). If B≈8-15 and B≈30–50 μG for Taurus and Perseus respectively, then the different visual extinction thresholds for fragmentation in these complexes could be explained. Unfortunately, magnetic field measurements are only available for a few clouds, and it is not possible to establish a link between the magnetic field strength, B, and the visual extinction threshold for fragmentation (for magnetic field measurements see Heiles, 1987; Crutcher et al., 1987; Crutcher, 1988; Myers and Goodman, 1988a; Goodman et al., 1989). Nevertheless, the line of sight magnetic field intensities in Taurus seem to be low, ≈1-10 μG (see e.g. Crutcher, 1988), while in Perseus the derived values for B are ≈27 μG (Goodman et al. 1989).

Whatever the physical processes regulating the cloud fragmentation are (see e.g. Scalo, 1987,1988, 1990; Houlahan and Scalo 1990; Elmegreen, 1990; Mouschovias, 1990), the observational results in Taurus and Perseus indicate that a cloud (or a cloud complex), initially containing hundreds or thousands of solar masses of gas, will break, in the course of its evolution, into several fragments. The more massive fragments will be Jeans-unstable (see above) and will continue to collapse and fragment. The total mass of the smallest clumps produced after n fragmentation steps will be ≈$0.6^n M_T$, where M_T is the total mass of the cloud. For n=6 the total mass in fragments will be only 5% of the total cloud mass, and probably only a small fraction of this mass will be processed to form stars (for a review on small scale clumping see Wilson and Walmsley, 1989). Due to the reduction of the mass in fragments at each fragmentation step, the initial potential of the cloud to form stars is considerably reduced, and its life expectancy is increased. Consequently, the potential to form stars of a virialized and highly fragmented cloud against collapse is not defined by its total mass and total velocity dispersion, but by the actual physical conditions in each individual fragment (although gravitational interaction between fragments cannot be excluded. Note that the evolution of the cloud towards its actual state does obviously depend on its initial physical conditions).

Scalo (1990) analyzed the 100μm IRAS emission maps to study the fragmentary state of the Taurus clouds. He concludes that these clouds have a projected dimension fractal of ≈1.4, i.e., that the cloud perimeter, P (which depends strongly on the angular resolution of the observations and on the cloud definition), varies with cloud surface, S, as $P \propto S^{0.7}$ -note that for objects without irregularities the expected relation is $P \propto S^{1/2}$. From similarity with other well known physical processes (turbulent-nonturbulent interface of incompressible fluids, terrestrial clouds,...), he suggests that the cloud structure is perhaps connected with turbulence, but the physical processes connecting this interstellar cloud turbulence (supporting term) to cloud gravity (collapsing term) and cloud magnetic field (supporting term) are not yet well understood.

Will the fragmentation process stop after a given number of fragmentation steps or will it continue indefinitely ? Will stars form at some moment in the fragmentation process ? The only way to give an answer to these question is to study systematically, and with the highest available angular resolution, the dense cores of dark clouds (keeping in mind that the cores can also have a complex structure embedded in a large amount of gas).

3.2. Dense Cores : Low Mass Protostars ?

Most of the information we have about the physical conditions in cold dense cores stems, mainly, from the work of Myers and coworkers (see e.g. Benson and Myers, 1989 and references therein). They observed the line emission of NH_3, ^{13}CO, $C^{18}O$ and HC_5N in an extensive sample of cores selected from local dark clouds according to their optical appearance (i.e., high visual extinction). The main results of their work can be summarized as follows :

1) the typical source size from $C^{18}O$ measurements is ≈0.3 pc (inner envelope in Table 2) with a mass of 30 M_O and a line width of 0.6 kms^{-1}. The sources in Taurus are the coldest and have the narrowest lines in their survey (Myers et al., 1983).

2) dense cores in dark clouds have an average size of 0.1 pc, a density of 3 10^4 cm^{-3}, a kinetic temperature of ≈11 K, and Δv≈0.3 kms^{-1} (Myers et al., 1983).

3) the intrinsic line shapes are best fitted by either thermal and subsonic microturbulent motions, or thermal and collapsing motions. If the cores are supported only by thermal motions, they are either collapsing or in an unstable equilibrium. If they are supported by their internal turbulence as derived from the observed line widths, they are either collapsing or in a near-critical equilibrium (see Figure 14; Myers and Benson, 1983).

4) the cores follow the Larson relations that correlate velocity dispersion and average density with core size. These relations, initially established for supersonic regions, extend into the subsonic regime (Myers, 1983).

5) An alternative explanation for the dynamical state of the cores is given by Myers and Goodman (1988b) who suggest that a uniform, isothermal, self-gravitating sphere supported by thermal and nonthermal motions, with the nonthermal kinetic energy

density ≈ magnetic energy density, will simulate Larson's relations if the magnetic field strength is ≈15-40 μG (see Figure 15).

6) the ratio of nonthermal kinetic energy to gravitational potential energy is ≈1 for cores larger than 0.1 pc, but it is only of ≈0.2 for cores with R≈0.1 pc. Low mass cores seem to have little magnetic support and more thermal support against gravity than the larger clouds in which they are embedded (see Figure 16; Myers and Goodman, 1988b).

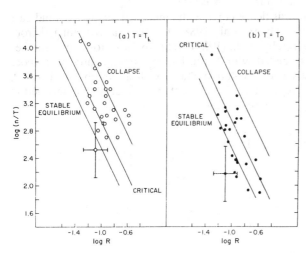

Figure 15. Comparison of cloud density, n (cm^{-3}), temperature (K), and size R (pc) with requirement for equilibrium and stability assuming that supporting motions are due only to the thermal part of line broadening (left panel) and to the entire line broadening (right panel). Solid lines are based on a model of an isothermal, pressure bounded equilibrium sphere. Points above the upper line represent clouds with no possible equilibrium. Points between the upper and the lower lines may or may not have possible equilibrium : if the cores are in equilibrium, it is unstable if the corresponding points are above the critical line and stable if they are below. Points below the lower line represent clouds in stable equilibrium (from Myers and Benson, 1983).

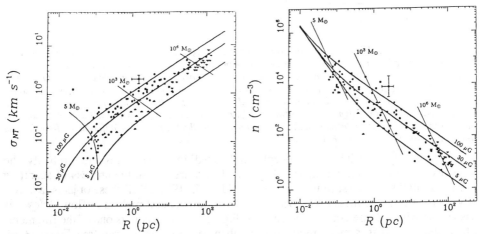

Figure 16. Nonthermal velocity dispersion (left panel) and density (right panel) versus size. Curves show predictions of virial equilibrium models : heavy curves correspond to constant magnetic field strength $B=5$, 30, and 100 μG and light curves to a cloud of constant mass $M=5$, 10^3, 10^6 $M_\odot$ (from Myers and Goodman, 1988b).

7) in a first search at 2 µm for embedded IR sources in dense cores, Benson et al., (1984) found that some of these cores are associated with optically invisible stars. The age of these stars is $\approx 10^5$ yr, and they were probably formed inside the cores. Subsequent studies by Beichman et al. (1986) showed that $\approx 50\%$ of the cores can be associated with IRAS sources. These objects were observed at optical, near IR, and submillimeter wavelenghts by Myers et al. (1987) and Ladd et al. (1991), who concluded that these sources are objects of stellar temperature and pointlike, confirming the idea that dense cores form low mass stars.

8) a search for CO outflows towards dense cores with embedded young stars indicate an outflow detection rate of 44% (Myers et al., 1988; this rate was reevaluated from interferometric observations by Tereby et al., (1989), to be $\approx 64\%$). However, searches for outflows in dense cores associated with T-Tauri stars (consequently, in a more advanced state of pre-main-sequence evolution than the heavily obscured stars), shows a smaller detection rate (Heyer et al., 1987), confirming that stars embedded in dense cores are in an early stage of their evolution.

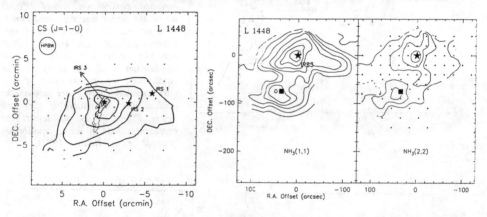

Figure 17. (left panel) Integrated line intensity of the CS J=1-0 line in L1448 from observations with the 13.7-m telescope of Centro Astronómico de Yebes (Spain). First contour and step are 0.3 K kms⁻¹. The star symbols indicate the position of the three IRAS sources associated with this cloud. The CO outflow associated with this source, as delineated by its red (dashed line) and blue (continuous line), is also shown. Ammonia (1,1) and (2,2) line observations (central and right panels) of the central region of L1448 indicate that the cloud is fragmented into two cores, one of them associated with IRS3 and the other with the exciting star of the molecular outflow (indicated as a black square). A different outflow is also associated with IRS3 (from Bachiller et al., 1990)

The optically invisible stars, deeply embedded in dense cores are probably the youngest known low mass stars ($t \approx 10^5$ yr), and they are still in the process of accreting mass via infall from the surrounding core (Ladd et al., 1991). Will these objects manifest any peculiarity in the mass loss process? One of the youngest molecular flows known is associated with a dense core in Perseus (see Figure 17). This core is one of the fragments of Figure 13, and it is much more massive than the cores in Taurus (see Figure 3). It shows a highly collimated CO jet with low mass bullets moving away from the exciting star (see Figure 18). The dynamical timescale of the outflow is $3 \cdot 10^3$ yr, and the ^{12}CO line wings extend over 140 kms⁻¹, with the highest velocities close to the exciting star.

The bullets are placed symmetrically with respect to the star suggesting that very young stellar objects, in addition to the normal neutral wind, have episodical events of mass loss (Bachiller et al., 1990; Bachiller and Cernicharo, 1990). From momentum considerations Bachiller and Cernicharo (1990) also conclude that the low velocity wind could be driven by the fast neutral wind (see Lada in this volume for a review of physical properties of molecular outflows). A high velocity outflow with a clumpy structure was also found in ρ-Oph by André et al. (1990; for a study of the physical properties of dense cores in ρ-Oph see Loren et al., 1990).

Finally, not all the cores are actually associated with low mass stars. TMC1 is one of the best studied cores and does not reveal the presence of any embedded stellar object. High angular resolution observations by Guélin and Cernicharo (1988) indicate that TMC1 is fragmented into several clumps near virial equilibrium. Their size is ≈ 0.03-0.04 pc and some of them have, in spite of their smallness, supersonic line widths. Cores without associated stellar objects and with subsonic internal motions will probably evolve in a free fall time, $\approx 10^5$ yr, towards low mass stars as those discussed above.

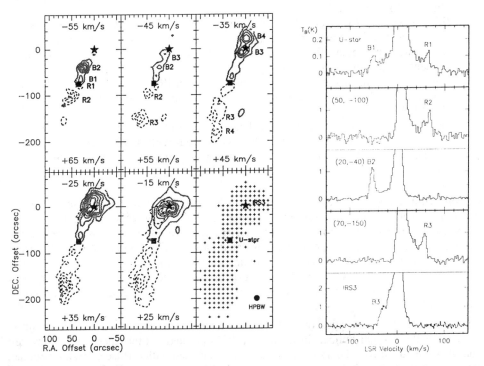

Figure 18. Maps of the ^{12}CO J=2-1 integrated line intensity towards L1448 for different velocity intervals (left panels). The ambient cloud velocity is ≈ 5 kms^{-1}. Crosses represent observed positions. First contour and step are 1.6 and 2 K kms^{-1} respectively, for all the maps except for that corresponding to the slower gas (emission at -15 and +25 kms^{-1}). Note, in the maps at extreme velocities, the existence of high velocity bullets which are placed at symmetric positions with respect to the exciting star, they are marked as Bn and Rn. The right panel shows different ^{12}CO J=2-1 spectra towards the positions indicated in the upper right corner of each spectrum. Note the broad wings and the existence of intense emission peaks that correspond to the different bullets shown in the left panel (from Bachiller et al., 1990)

4. Low Mass Stars and Low Mass Clouds : Bok Globules.

Forty years ago Bok and Reilly (1947) and Bok (1948) suggested that isolated dark clouds, known as Barnard objects or Bok globules, could be the sites of star birth. Bok globules are probably the nearest and simplest structures in the interstellar medium. Optically they appear as dark, roundish patches against the stellar background and have well defined edges and high visual extinctions. For many years, the only observational approach to study Bok globules was through optical star counts. Masses, distances, and sizes were derived for several globules (Bok and Cordwell 1973; Bok and McCarthy 1974), but no clear evidence for star formation in isolated globules could be obtained by this method. During the last ten years, however, infrared, optical, and radio observations have given evidence that low mass stars form in Bok globules :

(i) The number of emission line stars in the vicinities of Bok globules is higher by a factor of nearly two than in ordinary fields (Ogura and Hasegawa, 1983).
(ii) Several cometary globules in the Gum nebula are associated with Herbig-Haro objects (Bok 1978; Reipurth 1983).
(iii) Frerking and Langer (1982), Frerking et al (1987), and Cabrit et al. (1988), found that the CO emission from the large Bok globule B335 has moderate line wings with a bipolar flow structure, which indicates that there is a pre-main sequence object inside this globule.
(iv) Several cometary globules are associated with IRAS point sources which are powering bipolar outflows (Sugitani et al., 1989; Duvert et al., 1990; Cernicharo et al., 1991b).
(v) From an analysis of the IRAS data of 248 Bok globules, Yun and Clemens (1990) found that 23% of them are associated with IRAS point sources. A comparison of these sources with field sources indicates that stars inside the globules are young stellar objects.

Following Leung (1985), dark globules can be classified into four groups: elephant trunk and speck globules, cometary globules, globular filaments, and isolated dark globules. Their sizes vary from a few times 10^4 A.U., like the tear drops of the Rosette nebula (Herbig 1974), to ≈ 1 pc for large isolated globules like B335 (Bok and McCarthy 1974; Frerking et al., 1987). There are indications that the diverse kinds of globules could represent different stages in the evolution of these objects and an evolutionary path from cometary and elephant trunk globules towards isolated Bok globules was proposed by Reipurth (1983).

Among the different types of globules, cometary globules attracted early attention because of their bright optical rims. Struve (1937) pointed out that these bright rims are sharply defined, surround dark nebulae, and are associated with diffuse emission nebulae, not with reflection nebulae. Pottasch (1956, 1958) analyzed a significant number of bright rims and suggested that they are diffuse HII regions ionized by O stars located close to the neutral gas.

Although Bok globules are much smaller than normal dark clouds, their physical conditions are very similar to those of dark clouds (Frerking and Langer 1982; Frerking et al., 1987). However, the behaviour of the molecular emission in elephant trunks and cometary globules clearly indicated the presence of neighbouring massive stars (Schneps

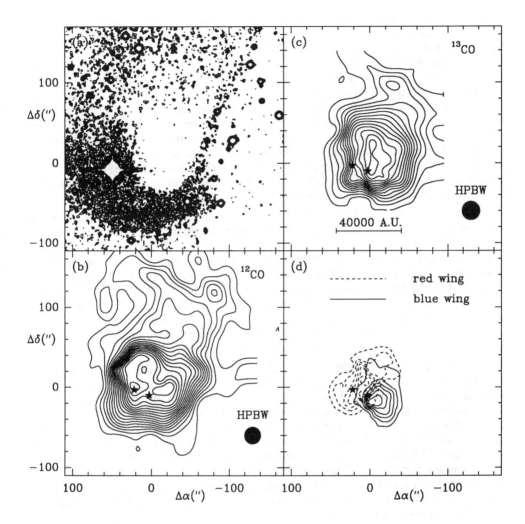

Figure 19. Optical aspect and distribution of the molecular emission in the cometary globule ORI-I-2 observed with the 30-m IRAM radio telescope. (**a**) Optical picture of ORI-I-2 obtained by digitizing the red plate of the Palomar Observatory. Center position is $\alpha = 5^h\ 35^m\ 33.0^s$, $\delta = -1°\ 46'\ 40''$. The stars apparently associated to the globule are plotted as two dark stars in panels b,c, and d. Star at Da=3 and Dd=-10 coincides in position with a point IRAS source. (**b**) 12CO main beam brightness integrated intensity contours. First contour and step are 1.5 and 2.3 K km s-1 respectively. (**c**) 13CO intensity contours. First contour and step are 0.75 and 1.2 K km s-1. (**d**) Red and blue 12CO integrated intensity wings. First contour and step are 1 and 0.75 K km s-1 respectively. (from Cernicharo et al., 1991b).

et al. 1980, Duvert et al., 1990; Harju et al., 1990; Cernicharo et al. 1991b). Cometary globules are submitted to the UV photons of nearby massive stars, and their lifetime against evaporation depends on the globules' density structures, masses and gas compositions (Reipurth, 1983). For an initial density of 10^5 cm^{-3}, these globules will survive at least for a few 10^6 yr, i.e., a time comparable to the lifetime of the stars responsible for their evaporation (Reipurth, 1983). For lower densities the lifetime of the globule will be considerably shorted. However, the ionization front can become dust-bounded in a short time (Sandford et al., 1982, 1984; see also Bertoldi, 1989), and during the succeeding evolution of the HII region, the ionization front will be shielded by the dusty HII region. Subsequent ionization will be affected by the absorbing dust, favouring the formation of a D-type shock front which will increase the density of the globule and will decrease its evaporation rate. The compressing shock front could modify the initial dynamical evolution of the globule and its potential to form stars.

Duvert et al. (1990) and Cernicharo et al. (1991) studied in detail, at optical and radio wavelenghts, two of these cometary globules. Figure 19 shows the results they obtained in the case of ORI-I-2, a cometary globule in Orion. The ^{12}CO emission shows red and blue wings with a clear bipolar spatial structure. These wings are centered on an IRAS source associated with the globule which has a FIR luminosity of 12 L_O (see also Sugitani et al., 1989). The line emission of ^{13}CO, CS, HCN, and C^{18}O, is concentrated in the darkest area of the globule. The projected size of the CS source is of 0.1 pc and its mass of »4 M_O. The total mass of the globule is of 12 M_O. Ha, SII, I, and red CCD images of the globule indicate unambiguously that the bright rim surrounding it is a HII region produced by the strong local UV field (Cernicharo et al., 1991b).

The physical structure of this cometary globule emerging from these radio and optical observations is depicted in Figure 20, and its physical properties are summarized in Table 4. The lower part of the globule is being illuminated by σ-Ori. The UV photons from this star are ionizing the gas in this part of the globule and producing a diffuse HII region. The ionization shock is preceded by a shock front which has compressed the globule (darkest area in Figure 20). The shocked gas has a kinetic temperature of 25 K while it is of 12 K for the gas not yet submitted to the UV photons. This gas is much less dense than the former. The CS emission arises from the thin compressed region. The density derived from different tracers indicate a compression factor of $\approx 10^2$. The shocked gas is blue shifted with respect to the low density gas suggesting that the globule is imploding (see also Sugitani et al., 1989).

Several cometary globules of the Gum nebula were also studied in detail (Reipurth, 1983; Harju et al., 1990; González-Alfonso et al., 1991b). They are more than 1 order of magnitude larger than the globule of Figure 19 and seem to be in an early stage of interaction with the UV field. In these cometary globules most of the mass is concentrated in the tail, while in the most evolved and compact cometary globules the mass is concentrated in the core and in the envelope surrounding it.

Globules submitted to a strong UV field and with an initial low density, or a low mass, will be completely evaporated (Reipurth, 1983). During the evaporation process their radii will decrease, becoming undetectable at some moment of their evolution. A good example of cometary globules in the last stages of their life are the tear drops of the Rosette nebulae, which could still survive to the UV field for 10^3-10^4 yr (Herbig, 1974).

TABLE 4

PHYSICAL PROPERTIES OF THE COMETARY GLOBULE ORI-I-2

DISTANCE	500 pc
PROJ. RADIUS (HEAD)	0.09 pc
PROJ. SIZE (HEAD+TAIL)	0.53 pc
EXCITING STAR	σ-Ori (O9.5V)
DISTANCE TO σ-ORI	16 pc
BRIGHT RIM	Diffuse HII region; Ionized shell of gas around a roundish globule.
SHELL THICKNESS	0.035 pc
SHELL INNER RADIUS	0.20 pc
SHELL DENSITY AND MASS	≈ 100 cm^{-3}, 0.1 M$_O$
GLOBULE LIFETIME AGAINST EVAPORATION	$\gg 2\ 10^6$ y
IRAS SOURCE FIR LUMINOSITY	12 L$_O$
DUST TEMPERATURE (12μm/25μm)	130 K
DUST TEMPERATURE (25μm/60μm)	53 K
DUST TEMPERATURE (60μm/100μm)	23 K
CORE GAS TEMPERATURE	25 K
CORE DENSITY (PEAK & AVERAGE)	$2\ 10^5$ cm^{-3}, $\approx 2\ 10^5$ cm^{-3}
CORE RADIUS, HEIGHT AND MASS	0.07 pc, 0.02 pc, 4 M$_O$
ENVELOPE GAS TEMPERATURE	12 K
ENVELOPE DENSITY (AVERAGE)	10^4 cm^{-3}
ENVELOPE RADIUS (^{13}CO) AND MASS	0.15 pc, 10 M$_O$
TAIL DENSITY (^{12}CO) AND MASS	few 10^2 cm^{-3}, ≈ 1 M$_O$
VELOCITY STRUCTURE	collapsing or imploding cloud; bipolar outflow centered on the IRAS source
OUTFLOW MASS	≈ 0.15 M$_O$
OUTFLOW DYNAMICAL AGE	$\approx 2.5\ 10^4$ yr
OUTFLOW MECHAN. LUMINOSITY	≈ 0.01 L$_O$

The smallest tear drops have a diameter of ≈ 0.02 pc (for a distance to the Rosette of 1.6 Kpc), a mass of $\approx 4.4\ 10^{-7} <n(H_2)>$ M$_O$, and a visual extinction of $A_V \approx 7\ 10^{-5} <n(H_2)>$ mag, where $<n(H_2)>$ is the average globule density in cm^{-3}. The Rosette's tear drops appear optically thick against the background of the HII region and probably have a few magnitudes of visual extinction. This implies that $<n(H_2)> \approx 10^4$-10^5 cm^{-3} and M $\approx 4.4\ 10^{-3}$-10^{-2} M$_O$, i.e., the Rosette's tear drops are probably the less massive objects of the interstellar medium. In spite of their smallness, these objects contain an important molecular component. Figure 21 shows the ^{12}CO J=2-1 emission towards two tear drops observed with the 30-m IRAM radiotelescope (from Cernicharo et al., 1991c). Note that the line widths in Figure 21 are clearly supersonic when

Larson's relations predict subsonic lines for these objects. Obviously, the external pressure and the local UV field are playing an important role in the evolution of the tear drops and of the cometary globules of the Rosette Nebulae. The high visual absorption in these objects protects the molecules against photodissociation by UV photons (see van Dishoeck and Black, 1988), allowing the study of these fascinating small objects, the last remnant gas of globules being destroyed by the action of the strong local UV field.

When the UV field will decline to standard values, i.e., at the end of the life of the exciting stars, some cometary globules will have a reduced mass and probably will then expand, others will be fully evaporated, but some of these cometary globules will probably keep inside the vestige of the action of the nearby massive stars : a new born star (Sugitani et al., 1989; Duvert et al., 1990; Cernicharo et al., 1991b).

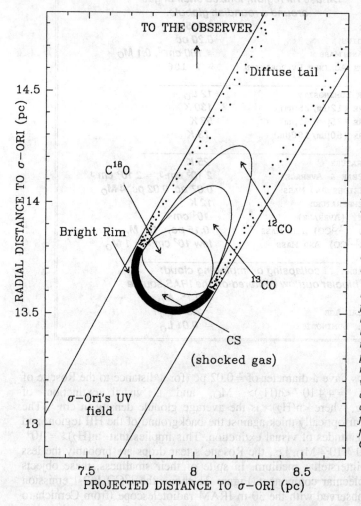

Figure 20. Schematic model of the physical structure of the cometary globule ORI-I-2. The globule's symmetry axis is tilted 60o with respect to the line of sight (counted from the West). The exciting star, s-Ori, is 16 pc from the globule and is behind it. The sizes of the different regions, as derived from different molecular tracers, are plotted according the parameters given in Table 4. The bright rim along the bottom of the globule, which is a HII region, is somewhat obscured by the globule itself, but the long tail is visible against the more tenuous neutral gas and dust far from the central condensation. In this model the CS and C18O emission regions have the same projected extent but represent different gas volumes. The flattened CS condensation is the consequence of the ionization front produced by s-Ori's UV field (see text; from Cernicharo et al. 1991b)

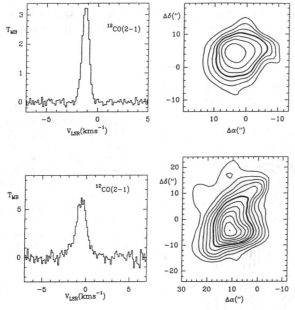

Figure 21. ^{12}CO J=2-1 integrated line intensity (right panels) and observed spectra (left panels) towards two tear drops of the Rosette nebulae. First contour and step are 0.5 Kkms^{-1} (right top panel) and 1 Kkms^{-1} (right bottom panel). Thick contours correspond to the half intensity emission level. The tear drops have a deconvolved size of 5±3"x10±2" and of 5±3"x18±1" (top and bottom respectively). The integrated intensity maps show clearly the cometary appearance of the tear drops. Note the broad wings in the spectrum shown in the left bottom panel (from Cernicharo et al., 1991c).

5. The Interaction of Newly Born Stars with the Ambient Gas.

There is increasing evidence that the primary winds from newly formed stars have extremely high velocities (EHV) and that they are mainly neutral. Lizano et al. (1988) detected HI and CO emission near HH7-11 spreading over ≈±170 kms^{-1} (see also Koo, (1989), Bachiller and Cernicharo, (1990), and Rodriguez et al., (1990)). Bachiller et al. (1990) detected similar winds in L1448 (see Figures 17 and 18), and in addition found that episodical events of mass loss could characterize the first evolutionary stages of the newly formed stars. Margulis and Snell (1989) found EHV CO winds in nine molecular outflows, and André et al., (1990) found similar results in a molecular outflow in ρ-Oph. EHV winds were also observed in the infrared lines of CO by Mitchell et al. (1988), Maillard and Mitchell (1988). All these authors suggest that the EHV winds may carry a considerable amount of mechanical energy and they have a momentum rate high enough to drive the low velocity and more extended winds. In addition, the enormous amount of mechanical power in the EHV wind can disrupt the core where the star was formed (see e.g. Mathieu et al., 1987). One of the most interesting feature of EHV outflows is the increase of collimation for the EHV winds. In all cases, the low velocity wind fills a more extended region than the EHV one, suggesting that important braking processes are taking place at the interface of the molecular cloud and the ambient gas (Bachiller and Cernicharo, 1990).

As an example of these possible effects, let us consider the Herbig Haro object HH34. HH34 is located in the Northern part of the L1641 cloud in Orion (see Bally, 1989, for a detailed picture of this cloud), and it was studied in detail by Reipurth (1989, and references therein; see also Reipurth in this volume for a nice picture of HH34). HH34 is associated with a conspicuous optical jet. The jet is well collimated, shows considerable structure and emanates from a red star (see Reipurth, 1989). The interaction of the jet with the ambient gas produces the HH34 object. We could expect that in the inner region of the shock the jet is decelerated while the ambient gas is being impelled or accelerated. As the shocked ambient gas is accelerated, a cavity, where internal pressure is higher than the ambient cloud pressure, is formed. Such a cavity appears clearly in the optical pictures of HH34 (see Reipurth, 1989).

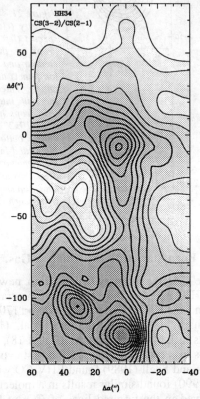

Figure 22. CS(3-2)/CS(2-1) integrated intensity ratio in HH34 observed with a 25" beam. First contour and step are 0.2 and 0.1 respectively. The ratio increases towards the walls of the optical cavity we can see in the pictures of this object by Reipurth (1989 and this volume). The maximum of this ratio, which indicates a density peak, is below the HH34 object. The cavity is expanding with a velocity of ≈ 1 kms^{-1} (from Cernicharo et al., 1991d).

Does the ambient gas show a similar cavity ? Figure 22 shows the CS(3-2) over CS(2-1) integrated intensity ratio as observed by Cernicharo et al. (1991c) with the 30-m IRAM radiotelescope. This ratio increases towards the walls of the optical cavity, and has a clear maximum below the HH34 object. This object is optically visible, thus having little material in front of it. Consequently, the CS lines are tracing regions of high density and are probably free of the problems quoted in section 2.2.1 (see the contrasted and sharp variation of the CS(3-2)/CS(2-1) ratio in Figure 22). The outflow from the exciting star of HH34 is thus introducing important modifications to the cloud density structure. The CS cavity seems to be incomplete, reflecting the inhomogeneous structure of the ambient gas. The CS velocity structure indicates clearly that the cavity is actually expanding with a velocity ≈ 1 kms-1. HH34 and its exciting jet are in the plane of the sky, but other molecular flows aligned in the line of sight may manifest the same density perturbations of the ambient gas. The interpretation of the data in these cases will be very difficult because emission from the core and from the perturbated regions will appear to be superposed.

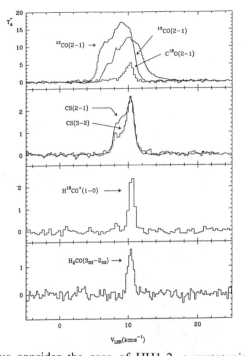

Figure 23. Observed spectra with the 30-m IRAM radiotelescope of the ^{12}CO, ^{13}CO, $C^{18}O$ $J=2-1$, CS $J=3-2$ and $J=2-1$, $H^{13}CO^+$ $J=1-0$ and H_2CO $3_{02}-2_{02}$ lines towards the exciting source of the HH1-2 system of Herbig-Haro objects. Note the presence of gas at three different velocities in the line of sight. These three components have very different physical conditions and only the gas at 10.5 kms^{-1} is associated with the core where the exciting star has been formed (from Cernicharo et al., 1991d).

Finally, let us consider the case of HH1-2, a prototypical system of Herbig-Haro objects. HH1-2 is also located in L1641 (see Bally, 1989; Reipurth 1989, and this volume) and was studied by many authors at optical and radio wavelengths (see e.g. Strom et al. (1985); Martín-Pintado and Cernicharo, 1987, and references therein). The presence of gas at different velocities in the line of sight and around the exciting star of HH1-2 has lead some authors to conclude that a toroidal expanding and rotating structure exists around the source, and that it is responsible for the collimation of the molecular outflow. Other authors prefered to talk about clumps at different velocities and with different physical conditions produced by the interaction of the molecular outflow with the ambient gas (see e.g. Martín-Pintado and Cernicharo, 1987, and references therein). Figure 23 shows the spectra of different molecular lines observed towards HH1-2 with the 30-m IRAM telescope (Cernicharo et al., 1991e). Gas at three different velocities is observed (6.5, 8 and 10.5 kms-1), but as concluded by Martín-Pintado and Cernicharo, they have very different physical conditions. The gas at 6.5 kms-1 is detected only in ^{12}CO and ^{13}CO and very marginally in $C^{18}O$. This gas has a moderate column density and a low density. Maps of this emission show it to be extended and unrelated to HH1-2. The gas at intermediate velocities has a larger density, $\approx$ a few 10^4 cm-3, but moderate column densities as it is not detected in $C^{18}O$ (note that this gas is not detected in the para-H_2CO $3_{03}-2_{02}$ line and only marginally detected in the J=1-0 line of $H^{13}CO^+$). Finally, the gas at 10.5 kms-1 has a very high density, $\approx 10^6$ cm-3, and a high column density. This gas is the only component that can be directly related to the exciting star of HH1-2. Maps of the line emission at this velocity (see Figure 24) indicate that it arises from a clump embedded in a long filament running from the star towards the South-West. The gas at intermediate velocity appearing in the line of sight of the exciting star of HH1-2 corresponds to the boundaries of another high density clump located at $\Delta\alpha \approx 40"$ and $\Delta\delta \approx 80"$. Figure 24 shows the spatial distribution of the CS J=2-1 emission around the HH1-2 system. The different clumps appearing at different velocities are clearly seen in this Figure. Some of these clumps seem to be produced by the interaction of the molecular outflow with the ambient gas and, in addition, these clumps are also associated with bipolar CO line wings indicating that star formation is a common process in the dense cores of the L1641 cloud (Cernicharo et al., 1990e).

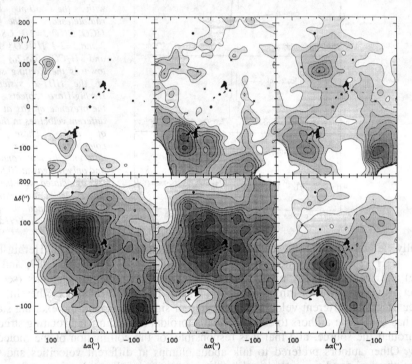

Figure 24. *Integrated intensity velocity maps of CS(2-1) for different velocity intervals (from left-top to right-bottom these intervals are from 5 to 11 by 1 kms^{-1}). Contour levels are 0.2 to 0.4 by 0.1 and from 0.4 to 1.2 by 0.2 kms^{-1}. The grey scale level is the same for all the panels. Dark contours outline the HH objects 1 and 2 and the Cohen-Schwartz star (filled circle at $\Delta\alpha \approx -20''$ and $\Delta\delta \approx 40''$). Asterisks show the position of the IR sources detected by Strom et al.(1985) at 2μm. The star at the center of the maps indicates the position of the exciting star of the HH1 and HH2 objects. The filled circle at $\Delta\alpha \approx 40''$, $\Delta\delta \approx 180''$ indicates the position of the star V380 Ori. Note that the clumps appearing at different velocities are typically associated with IR sources. Emission at intermediate velocities in the direction of the exciting star arises from different, unrelated clumps. Only the clump at 10.5 kms^{-1}, which is embedded in a large filament running from the exciting star towards the South-West, can be related to the exciting star of the Herbig-Haro objects HH1-2. The clumps immediately South of HH2 and that located West of HH1 (which appears at a velocity of ≈ 9.5 kms^{-1}), seem to arise from the interaction of the HH1-2 outflow with the ambient gas. The last clump is associated with an H$_2$O maser. Both clumps have their own bipolar molecular outflows. The clump located at $\Delta\alpha \approx 40''$ and $\Delta\delta \approx 80''$ is also associated with a bipolar molecular outflow, and its emission peaks right on an IR source detected by Strom et al (1985). Another clump located at $\Delta\alpha \approx 20''$ and $\Delta\delta \approx 40''$, and at a velocity of ≈ 9.5 kms^{-1}, is also associated with an IR source. The gas distribution around HH1-2 probably results from the action of the HH1-2 molecular outflow with the ambient gas (from Cernicharo et al., 1991e).*

For a detailed study of the molecular outflow of L1551 and of its shell-like structure see Moriarty-Schieven et al. (1987a,b), Moriarty-Schieven and Snell (1988), and Uchida et al. (1987). For the structure of the central object of L1551 see Sargent et al. (1988), Walmsley (1988), and Menten et al. (1989).

The cases of HH34 and HH1-2, as that of L43 (Matthieu et al., 1987), show that the energy input from the EHV star winds into the clouds, could modify their dynamical evolution. Consequently, in addition to magnetic field, external pressure, and interstellar turbulence, extremely high velocity winds should be considered in theoretical studies of the evolution of interstellar clouds (see e.g. Norman and Silk, 1980; see also Silk, 1985).

6. Conclusions.

Systematic high angular resolution observations of the fragmentary structure of dark cloud cores, with and without associated stars, is necessary to understand the physical processes occurring during star formation and to derive the physical conditions of the gas before and after gravitational collapse. Special care should be taken with the chosen molecular tracers in order to avoid important errors that could arise during the data interpretation. Molecular lines of high J, or those of optically thin emitters, are the best tools in determining the physical conditions of the dense cores deeply embedded in molecular cold clouds. Even when the obvious pitfalls are avoided, the interpretation of the physical structure around the cores associated with new stars is a difficult task in view of the considerable modifications produced in the ambient gas by the action of the newly born star through important mass loss : the cloud structure will be strongly affected by the stars it formed (see the case of HH1-2 and HH34 above).

The observation with good sensitivity and high angular resolution of cores close to critical equilibrium, could permit the study of the first stages of protostellar evolution.

Finally, the mapping with high angular resolution of the magnetic field strength in, and around, dense cores with, and without, embedded sources is necessary to evaluate its influence on the dynamical evolution of the clouds.

I would like to thank A. Greve, R. Bachiller, and C. Koempe for useful comments and suggestions, and to D. John, and G. España for a critical reading of the manuscript.

REFERENCES

Andersson, B.G., Wannier, P.G., Morris, M.:1991, Ap. J., *366*, 464.
André, P., Martín-Pintado, J., Despois, D., Montmerle, T.:1990, Astron. Astrophys., 236, 180.
Askne, J., Hoglund, B., Hjalmarson, A., Irvine, W.M.:1984, Astron. Astrophys., *130*, 311.
Bachiller, R., Cernicharo, J.:1984, Astron. Astrophys., *140*, 414.
Bachiller, R., Cernicharo, J.:1986a, Astron. Astrophys., *166*, 283.
Bachiller, R., Cernicharo, J.:1986b, Astron. Astrophys., *168*, 262.
Bachiller, R. Guilloteau, S., Kahane, C.:1987a, Astron. Astrophys., *173*, 324.
Bachiller, R. Cernicharo, J., Goldsmith, P., Omont, A.:1987b, Astron. Astrophys., *185*, 297.
Bachiller, R., Cernicharo, J.:1990, Astron. Astrophys., *290*, 276.

Bachiller, R., Cernicharo, J., Martín-Pintado, J., Tafalla, M., Lazareff, B.:1990, Astron. Astrophys., 231, 174.
Bally, J., Langer, W.D., Stark, A.A., Wilson, R.W.:1987, Ap. J. Letters, *312*, L45.
Bally, J.:1989, in "Low Mass Star Formation and Pre-Main Sequence Objects", ESO Conference and Workshop Proceedings No. 33, pag. 1, ed. Bo Reipurth.
Benson, P.J., Myers, P.C.:1989, Ap. J. Suppl., *71*, 89.
Benson, P.J., Myers, P.C., Wright, E.I.:1984, Ap. J. Letters, *279*, L27.
Bertoldi, F.:1989, Ap. J., *346*, 735.
Bloemen, J.B.G.M.:1988, in "Molecular clouds in the Milky Way" Lectures Notes in Physics, Vol 315, pag. 71, eds. R.L. Dickman, R.L. Snell, J.S. Young (Springer-Verlag).
Bohlin, R.C., Savage, B.D., Drake, J.F.:1978, Ap. J., *224*, 132.
Bok, B.J., and Reilly, E.F. :1947, Ap. J., *105*, 255.
Bok, B.J.:1948, Centennial Symposia, Harvard Obs. Monograph No. 7, 53.
Bok, B.J., Cordwell, C.S. :1973, in Molecules in the Galactic Environnement, Ed. by M.A. Gordon and L.F. Snyder (New York Interscience).
Bok, B.J., McCarthy, C.C. :1974, Astron. Journal, *79*, 42.
Bok, B.J. :1978, Pub. Astron. Soc. Pac., *90*, 489.
Boulanger, F., Pérault, M.:1988, Ap. J., *330*, 964.
Bujarrabal, V., Guélin, M., Morris, M., Thaddeus, P.:1981, Astron. Astrophys., *99*, 239.
Cabrit, S., Goldsmith, P.F., Snell, R.L.,:1988, Ap. J., *334*, 196.
Cernicharo, J., Bachiller, R.:1984, Astron. Astrophys. Suppl., *58*, 327.
Cernicharo, J., Guélin, M., Askne, J.:1984a, Astron. Astrophys., 138, 371.
Cernicharo, J., Castets, A., Duvert, G., Guilloteau, S.:1984b, Astron. Astrophys., 139, L13.
Cernicharo, J., Bachiller, R., Duvert, G.:1985, Astron. Astrophys., *149*, 273.
Cernicharo, J., Bachiller, R., Duvert, G.:1986, Astron. Astrophys., *160*, 181.
Cernicharo, J., Guélin, M.:1987, Astron. Astrophys., *176*, 299.
Cernicharo, J., Guélin, M., Walmsley, C.M.:1987a, Astron. Astrophys., *172*, L5.
Cernicharo, J., Guélin, M., Menten, M.N., Walmsley, C.M.:1987b, Astron. Astrophys., *181*, L1.
Cernicharo, J., Gottlieb, C.A., Guélin, M., Killian, T.C., Paubert, G., Thaddeus, P., Vrtilek, J.M.:1991a, Ap. J. Letters, 368, L39.
Cernicharo, J., Bachiller, R., González-Alfonso, E., Duvert, G., Gómez-González, J.:1991b, Astron. Astrophys., submitted.
Cernicharo, J., González-Alfonso, E., Gómez-Gonzalez, J., J. Masegosa.:1991c, in preparation.
Cernicharo, J., Lazareff, B., Bachiller, R., Martín-Pintado, J.:1991d, in preparation.
Cernicharo, J., González-Alfonso, E., Martín-Pintado, J.:1991e, in preparation.
Cheung, L.H., Frogel, J.A., Gezari, D.Y., Hauser, M.G.:1980, Ap. J., *240*, 74.
Chièze, J.P.:1987, Astron. Astrophys., *171*, 225.
Crutcher, R.M., Kázes, I., Troland, T.H.:1987, Astron. Astrophys., *181*, 119.
Crutcher, R.M.:1988, in "Molecular clouds in the Milky Way" Lectures Notes in Physics, Vol 315, pag. 105, ed. R.L. Dickman, R.L. Snell, J.S. Young (Springer-Verlag).
de Geus, E.J., Bronfman, L., Thaddeus, P.:1990, Astron. Astrophys., *231*, 137.
Dickman, R.L.:1975, Ap. J., *202*, 50.

Dickman, R.L.:1978, Ap. J. Suppl., *37*, 407.
Dickman, R.L., Clemens, D.P.:1983, Ap. J., *271*, 143.
Dickman, R.L.:1988, in "Molecular clouds in the Milky Way" Lectures Notes in Physics, Vol 315, pag. 55, eds. R.L. Dickman, R.L. Snell, J.S. Young (Springer-Verlag).
Duvert, G., Cernicharo, J., Baudry, A.:1986, Astron. Astrophys., 164, 349.
Duvert, G., Cernicharo, J., Bachiller, R., Gómez-González, J.:1990, Astron. Astrophys., 233, 190.
Elmegreen, B.G.:1985a, in "Birth and Infance of Stars", pag. 218, eds. A. Omont, R. Lucas, R. Stora, (Elsevier).
Elmegreen, B.G.:1985b, in Protostars and Planets II, eds. by D.C. Black, and M.S. Matthews, The University of Arizona Press, pag. 33.
Elmegreen, B.G.:1989, Ap. J., *338*, 178.
Elmegreen, B.G.:1990, Ap. J. Letters, *361*, L77.
Encrenaz, P.J., Falgarone, E., Lucas, R.:1975, Astron. Astrophys., *44*, 73.
Evans, N.J., II:1990, in "Frontiers of Stellar evolution", ed D.L. Lambert (San Francisco: Astronomical Society of the Pacific).
Falgarone, E., Puget, J.L.:1985, Astron. Astrophys., *142*, 157.
Falgarone, E., Puget, J.L.:1986, Astron. Astrophys., *162*, 235.
Falgarone, E., Puget, J.L.:1988, in "Galactic and Extragalactic Star Formation", NATO ASI Series, C232, pag. 195, eds. R. Pudritz, M. Fich, (Reidel).
Frerking, M.A., Langer, W.D.:1982, Ap. J., *256*, 523.
Frerking, M.A., Langer, W.D., Wilson, R.W.:1982, Ap. J., *262*, 590.
Frerking, M.A., Langer, W.D., Wilson, R.W.:1987, Ap. J., *313*, 320.
Friberg, P., Hjalmarson, A.:1990, in "Molecular Astrophysics", pag. 3, ed. T.W. Hartquist (Cambridge University Press).
Goldreich, P. Kwan, J.:1974, Ap. J., *189*, 441.
Goldsmith, P.F.:1984, in "Galactic and Extragalactic Infrared Spectroscopy" Astrophys. and Space Science Library, Vol. 108, eds. M.P. Kessler and J.P. Philips (Dordrecht : Reidel).
Goldsmith, P.F.:1987a, in "Interstellar Processes" Astrophys. and Space Science Library, Vol. 134, pag. 51, eds. D.J. Hollenbach, H.A. Thronson Jr., (Dordrecht : Reidel).
Goldsmith, P.F.:1988, in "Molecular clouds in the Milky Way" Lectures Notes in Physics, *Vol 315*, pag. 1, eds. R.L. Dickman, R.L. Snell, J.S. Young (Springer-Verlag).
González-Alfonso, E., Cernicharo, J.:1991a, submitted to Astron. and Astrophys.
González-Alfonso, E., Cernicharo, J., Radford, S., Greve, A.:1991b, in preparation.
Goodman, A.A., Crutcher, R.M., Heiles, C., Myers, P.C., Troland, T.H.:1989, Ap. J. Letters, *338*, L61.
Guélin, M., Friberg, P., Mezaoui, A.:1982a, Astron. Astrophys, *109*, 2.
Guélin, M., Langer, W.D., Wilson, R.W.:1982b, Astron. Astrophys, *107*, 107.
Guélin, M. Cernicharo, J.:1988, in "Molecular clouds in the Milky Way" Lectures Notes in Physics, *Vol 315*, pag. 81, ed. R.L. Dickman, R.L. Snell, J.S. Young (Springer-Verlag).
Guilloteau, S., Baudry, A.:1981, Astron. Astrophys., *97*, 213.
Harju, J., Sahu, M., Henkel, C., Wilson, T.L., Sahu, C.K., Pottasch, R.:1990, Astron. Astrophys., *233*, 197.

Heiles, C.:1987, in "Interstellar Processes" Astrophys. and Space Science Library, Vol. 134, pag. 171, eds. D.J. Hollenbach, H.A. Thronson Jr., (Dordrecht : Reidel).
Herbig, G.H.:1974, Pub. Astron. Soc. Pac., 86, 604.
Herbst, E.:1987, in "Interstellar Processes" Astrophys. and Space Science Library, Vol. 134, pag. 611, eds. D.J. Hollenbach, H.A. Thronson Jr., (Dordrecht : Reidel).
Herbst, E.:1991, Ap. J., 366, 133.
Heyer, M.H., Snell, R.L., Goldsmith, P.F., Myers, P.C.:1987, Ap. J., 321, 370.
Ho, P.T.P., Townes, C.H.:1983, Ann. Rev. Astron. Astrophys., 21, 239.
Hollenbach, D.:1988, in "Interstellar Matter", pag. 39, eds. J.M. Moran, P.T.P. Ho (Gordon and Breach Science Publishers).
Houlahan, P., Scalo, J.M.:1990, Ap. J. Suppl., 72, 133.
Irvine, W.M., Scholerb, F.P.:1984, Ap. J., 282, 516.
Irvine, W.M., Avery, L.W., Friberg, P., Matthews, H.E., Ziurys, L.M.:1988, in "Interstellar Matter", pag. 15, eds. J.M. Moran, P.T.P. Ho (Gordon and Breach Science Publishers).
Jarret, T.H., Dickman, R.L., Herbst, W.:1989, Ap. J., 345, 881.
Koo, B.C.:1989, Ap. J., 337, 318.
Lada, C.J.:1985, Ann. Rev. Astron. Astrophys., 23, 267.
Lada, C.J.:1988, in "Galactic and Extragalactic Star Formation", NATO ASI Series, C232, pag. 5, eds. R. Pudritz, M. Fich, (Reidel).
Lada, E., Bally, J., Stark, A.A.:1991, Ap. J., 368, 432.
Ladd, E.F., Adams, F.C., Casey, S., Davidson, J.A., Fuller, G.A., Harper, D.A., Myers, P.C., Padman, R.:1991, Ap. J., 366, 203.
Langer, W.D., Wilson, R.W., Goldsmith, P.F., Beichman, C.A.:1989, Ap. J., 337, 355.
Larson, R.B.:1969, MNRAS, 145, 271.
Larson, R.B.:1981, MNRAS, 194, 809.
Leisawitz, D.:1990, Ap. J., 359, 319.
Lizano, S., Heiles, C., Rodriguez, L. F., Koo, B., Shu, F.H., Hasegawa, T., Hayashi, S., Mirabel, I.F.:1988, Ap. J., 328, 763.
Leung, C.M. :1985, in Protostars and Planets II, eds. by D.C. Black, and M.S. Matthews, The University of Arizona Press, pag. 124.
Loren, R.B., Wooten, A., Wilking, B.A.:1990, Ap. J., 365, 269.
Maillard, J.P., Mitchell, G.F.:1988, 22nd ESLAB Symposium on "Infrared Spectroscopy in Astronomy", held at Salamanca (Spain).
Maloney, P.:1990, Ap. J. Letters, 348, L9.
Margulis, M., Snell, R.L.:1989, Ap. J., 343, 779.
Martín-Pintado, J., Wilson, T.L., Johnston, K.J., Henkel, C.:1985, Ap. J., 299, 386.
Martín-Pintado, J., Cernicharo, J.:1987, Astron. Astrophys., 176, L27.
Matthieu, R.D., Priscilla, J.B., Fuller, G.A., Myers, P.C., Schild, R.E.:1987, Ap. J., 330, 385.
Menten, K.M., Harju, J., Olano, C.A:, Walmsley, C.M.:1989, Astron. Astrophys., 223, 258.
Millar, T.J.:1990, in "Molecular Astrophysics", pag. 115, ed. T.W. Hartquist (Cambridge University Press).
Mitchell, G.F., Alenn, M., Beer, R., Dekany, R., Huntress, W., Maillard, J.P.:1988, Ap. J. Letters, 327, L17.

Moriarty-Schieven,, G.H., Snell, R.L., Strom, S.E., Grasdalen, G.L.:1987a, Ap. J. Letters, *317*, L95.
Moriarty-Schieven,, G.H., Snell, R.L., Strom, S.E., Schloerb, F.P., Strom, K.M., Grasdalen, G.L.:1987b, Ap. J., *319*, 742.
Moriarty-Schieven,, G.H., Snell, R.L.:1988, Ap. J., *332*, 364.
Mouschovias, T. CH.:1990, in "Physical Processes in Fragmentation and Star Formation", eds. R. Capuzzo-Dolcetta, C. Chiosi, A. Di Fazio, (Kluwer Academic Publishers).
Mundy, L.G., Scoville, N.Z., Booth, L.B., Masson, C.R., Woody, D.P.:1986, Ap. J., *304*, L51.
Mundy, L.G., Evans, N.J. II, Snell, R.L., Goldsmith, P.F.:1987, Ap. J., *318*, 392.
Myers, P.C., Ho, P.T.P., Schenps, M.H., Chin, G., Pankonin, V., Winnberg, A.:1978, Ap. J., *220*, 864.
Myers, P.C., Benson,, P.J.:1983, Ap. J., *266*, 309.
Myers, P.C., Linke, R.A., Benson, P.J.:1983, Ap. J., *264*, 517.
Myers, P.C.: 1985, in "Protostars and Planets II", pag 81., eds. D.C. Black, and M.S. Matthews, The University of Arizona Press.
Myers, P.C., Fuller, G.A., Mathieu, R.D., Beichman, C.A., Benson, P.J., Schild, R.E., Emerson, J.P.:1987, Ap. J., *319*, 430.
Myers, P.C., Goodman, A.A.:1988a, Ap. J. Letters, *326*, L27.
Myers, P.C., Goodman, A.A.:1988b, Ap. J., *329*, 392.
Myers, P.C., Heyer, M., Snell, R.L., Goldsmith, P.F.:1988, Ap. J., 324, 907.
Myers, P.C.:1990, in "Molecular Astrophysics", pag. 328, ed. T.W. Hartquist (Cambridge University Press).
Nercessian, E., Castets, A., Cernicharo, J., Benayoun, J.J.:1988, Astron. Astrophys., *189*, 207.
Norman, C., Silk, J.:1980, Ap. J., *238*, 158.
Olano, C.A., Walmsley, C.M., Wilson, T.L.:1988, Astron. Astrophys., *196*, 194.
Pottasch, S.:1956, Bull. of the Astron. Inst. of the Netherlands, *Vol XIII*, N. 471, pag. 77.
Pottasch, S.:1958, Bull. of the Astron. Inst. of the Netherlands, *Vol XIV*, N. 482, pag. 29.
Prasad, S.S., Tarafdar, S.P., Villers, K.R., Huntress Jr., W.T.:1987, in "Interstellar Processes" Astrophys. and Space Science Library, Vol. 134, pag. 631, eds. D.J. Hollenbach, H.A. Thronson Jr., (Dordrecht : Reidel).
Reipurth, B.:1983, Astron. Astrophys., *117*, 183.
Reipurth, B.:1989, in "Low Mass Star Formation and Pre-Main Sequence Objects", ESO Conference and Workshop Proceedings No. 33, pag. 247, ed. Bo Reipurth.
Rodriguez, L.F., Lizano, S., Cantó, J., Escalante, V., Mirabel, I.F.:1990, Ap. J., *365*, 261.
Sandford, M.T., Whitaker, R.W., Klein, R.I.:1982, Ap. J., *260*, 183.
Sandford, M.T., Whitaker, R.W., Klein, R.I.:1984, Ap. J., *282*, 178.
Sargent, A.I., Beckwith, S., Keene, J., Masson, C.:1988, Ap. J., *333*, 936.
Scalo, J.M.:1987, in "Interstellar Processes" Astrophys. and Space Science Library, Vol. 134, pag. 349, eds. D.J. Hollenbach, H.A. Thronson Jr., (Dordrecht : Reidel).
Scalo, J.M.:1988, in "Molecular clouds in the Milky Way" Lectures Notes in Physics, *Vol 315*, pag. 201, ed. R.L. Dickman, R.L. Snell, J.S. Young (Springer-Verlag).
Scalo, J.:1990, in "Physical Processes in Fragmentation and Star Formation", eds. R. Capuzzo-Dolcetta, C. Chiosi, A. Di Fazio, pag. 117, (Kluwer Academic Publishers).

Schneps, M.H., Ho, Paul T.P., Barret, A.H.:1980, Ap. J., *240*, 84.
Scoville, N.Z., Solomon, P.M.;1974, Ap. J. Letters, *187*, L67.
Shu, F.H.:1977, Ap. J., 214, 488.
Shu, F.H., Adams, F.C., Lizano, S.:1987, Ann. Rev. Astron. Astrophys., *25*, 23.
Silk, J.:1985, in "Birth and Infance of Stars", pag. 349, eds. A. Omont, R. Lucas, R. Stora, (Elsevier).
Snell, R.L.:1981, Ap. J. Suppl., *45*, 121.
Snell, R.L., Mundy, L.G., Goldsmith, P.F., Evans, N.J. II, Erickson, N.R.:1984, Ap. J., *276*, 625.
Snell, R.L., Heyer, M.H., Schloerb, F.P.:1989, Ap. J., 337, 739.
Strom, S.E., Strom, K.M., Grasdalen, G.L., Sellgren, K., Wolf, S., Morgan, J., Stocke, J., Mund, R.:1985, Astron. Journal, *90*, 2281.
Struve, O.:1937, Ap. J., *85*, 208.
Sugitani, K., Fukui, Y., Mizuno, A., Ohashi, N.:1989, Ap. J. Letters, *342*, L87.
Swade, D.A.:1989, Ap. J., *345*, 828
Tereby, S., Vogel, S.N., Myers, P.C.:1989, Ap. J., *340*, 472.
Tölle, F., Ungerechts, H., Walmsley, C.M., Winnewisser, G., Churchwell, E.:1981, Astron. Astrophys., *95*, 143.
Uchida, Y, Kaifu, N., Shibata, K., Hayashi, S.S., Hasegawa, T., Hamatake, H.:1987, Pub. Astron. Soc. Japan, *39*,907.
Ungerechts, H., Thaddeus, P.:1987, Ap. J. Suppl., *63*, 645.
van Dishoeck, E.F., Black, J.H.:1987, in "Physical Processes in Interstellar clouds", pag. 241, eds G. Morfill and M.S. Scholer (Reidel, Dordrecht).
van Dishoeck, E.F., Black, J.H.:1988, Ap. J., *334*, 711.
Walker, C.K., Adams, F.C., Lada, C.J.:1990, Ap. J., *349*, 515.
Walmsley, C.M., Churchwell, E., Nash, A., Fitzpatrick, E.:1982, Ap. J. Letters, *285*, L75.
Walmsley, C.M., Ungerechts, H.:1983, Astron. Astrophys., *122*, 164.
Walmsley, C.M.:1988, in "Galactic and Extragalactic Star Formation", NATO ASI Series, C232, pag. 181, eds. R. Pudritz, M. Fich, (Reidel).
Westbrook, W.E., Werner, M.W., Elias, J.H., Gezari, D.Y., Hauser, M.G., Lo, K.Y., Neugebauer, G.:1976, Ap. J., *209*, 94.
Wilson, T.L., Walmsley, C.M.:1989, Astron. Astrophys. Review, Vol 1, N2, 141.
Yamashita, T., Sato, S., Kaifu, N., Hayashi, S.S.:1990, Ap. J., *365*, 615.
Young, J.S., Goldsmith, P.F., Langer, W.D., Wilson, R.W., and Carlson, E.R.:1982, Ap. J., *261*, 513.
Yun, J.L., Clemens, D.P.:1990, Ap. J. Letters, *365*, L76.
Zhou, S., Yuefang, W., Evans II, N.J., Fuller, G.A., Myers, P.C.:1989, Ap. J., 346, 168.
Zhou, S., Evans II, N.J., Butner, H.M., Kutner, M.L., Leung, C.M., Mundy, L.G.:1990, Ap. J., *363*, 168.

THE FORMATION OF LOW MASS STARS: OBSERVATIONS

Charles J. Lada
Harvard-Smithsonian Center For Astrophysics
60 Garden Street
Cambridge, Massachusetts 02138
USA

ABSTRACT. In this chapter basic observational knowledge relevant to the formation and early evolution of low mass stars is reviewed. Particular emphasis is placed on describing: 1) the significance of the initial stellar mass function, 2) the importance of clusters in the scheme of low mass star formation, 3) the physical natures of embedded young stellar objects of low mass and 4) the critical importance of molecular outflows for the physics of star formation and early stellar evolution.

1. Introduction

Stars are the basic objects of the astronomical universe. Understanding their origin is one of the most interesting and fundamental problems of modern astrophysics. The study of the origins of dwarf or low mass stars generates particular interest not only because the sun, and for that matter most stars in the galaxy, are of low mass, but also because low mass stars in early stages of evolution offer a number of critical advantages for observational and theoretical investigation that high mass stars do not.

At the most basic level the entire evolution of a star from birth to death is driven by gravity, consequently the amount of mass a star contains plays a fundamental role in its life history. Stars form from cold, dense molecular cloud cores whose properties have been described in detail in earlier chapters of this volume. To create a star from scattered interstellar material, whose density is more that 20 orders of magnitude less than that of the star itself, clearly requires gravity to be at the heart of the overall process. The timescale for the gravitational collapse of a cloud core, the free-fall time, is determined largely by ρ, the density of the cloud:

$$\tau_{ff} = \sqrt{\frac{3\pi}{32G\rho}}$$

For the typical mean density ($n \approx 10^4$ cm^{-3}) of a cloud core (of either low or high mass) the free-fall time is about 4×10^5 years. However, once a stellar object of mass M_* forms at the center of a collapsing cloud the timescale for its (pre-hydrogen burning) evolution is initially given by the Kelvin-Helmholtz time:

$$\tau_{KH} \approx \frac{GM_*^2}{R_* L_*}$$

which is very rapid for a high mass star (i.e., $\approx 10^4$ years for $M_* = 50$ $M_\odot$) and relatively slow for a low mass star (i.e., $\approx 3 \times 10^7$ years for $M_* = 1$ $M_\odot$). More importantly for high mass stars $\tau_{KH} < \tau_{ff}$ and *these stars begin burning hydrogen and reach the main sequence before the termination of the infall or collapse phase of protostellar evolution*. On the other hand, for low mass stars $\tau_{KH} > \tau_{ff}$ and low mass stars have an *observable* pre-main sequence stage of stellar evolution.

Low mass stars are considerably less destructive to their natal environments than high mass stars. For example, stars with masses in excess of about 10 $M_\odot$ form HII regions which can quickly disrupt dense molecular gas. For stars with masses in excess of about 7 $M_\odot$ radiation pressure generated by the stellar radiation field dominates gravity in circumstellar material and consequently can hinder the process of collapse of the surrounding interstellar material and even blow it away. A final advantage offered by low mass stars is that they can form in relative isolation so that their immediate circumstellar environments are not influenced by the disruptive presence of nearby stars. High mass stars rarely form in isolation and observations of their environments are nearly always confused by the effects of other recently formed nearby stars.

For these reasons, regions of low mass star formation offer important laboratories for star formation research. Driven by advances in observational technology at infrared and millimeter wavelengths, the quest to decipher the mystery of star formation has met with considerable success, during the last decade. Indeed, with the new knowledge generated by observations obtained only within the last few years, a complete empirical picture of stellar origins is being synthesized and a more profound and penetrating understanding of the physical process of low mass star formation is beginning to emerge. In this chapter I will review some of the more interesting and important aspects of the basic observational foundation of our current understanding of star formation.

2. The Initial Mass Spectrum

The most important parameter which determines the evolution of a star is its mass. Consequently, one of the most fundamental and challenging tasks for a successful theory of stellar origins is to predict the frequency distribution of stellar masses which results from the process of stellar formation. Clearly then, observational determination of the mass spectrum of stars at birth is essential for a complete understanding of the physical process of star birth. Yet, in practice, determination of this initial mass spectrum has proved to be neither trivial nor straightforward. This is largely due to two reasons. First, the initial mass spectrum of stars at birth is an inherently global quantity and it is often not clear over what interval of time or space one has to integrate to obtain a meaningful spectrum. Second, it is stellar brightness that is directly observable not stellar mass. Therefore, the starting point of any attempt to determine a stellar mass spectrum is the determination of a stellar luminosity function. Then, an appropriate mass-luminosity relation is required to transform the luminosity function into a stellar mass function.

The first attempt to derive an empirical initial mass spectrum was carried out by Salpeter (1955) who used as a starting point the general (or van Rhijn) luminosity function determined for field stars within the neighborhood of the sun. The field star luminosity function, $\Phi(M_V)$, is the relative number of stars per unit magnitude interval per unit volume of space as a function of absolute magnitude, M_V, in the solar neighborhood. (The determination of the field star luminosity function is itself a difficult task and discussion of this is beyond the scope of this review, and the reader who desires more information on this topic is referred to the text by Mihalas and Binney (1968)). In principle, the field star luminosity function counts all presently visible field stars in the neighborhood of the sun. Because stars more massive than the sun have main sequence lifetimes shorter than that of the galaxy they are under represented in the field star luminosity function with respect to low mass stars, whose lifetimes are in excess of the age of the galaxy. Therefore, the present day field star luminosity function is biased toward low luminosity stars and must be corrected for the effects of post–main– sequence stellar evolution to obtain a luminosity function more representative of that produced at stellar birth.

Salpeter introduced the concept of an initial luminosity function, $\Psi(M_V)$, which is the relative number of stars per unit volume in each absolute magnitude interval originally produced by the star formation process. He derived $\Psi(M_V)$ from the field star luminosity function as follows:

$$\Psi(M_V) = \Phi(M_V) \quad for \quad \tau_{MS} \geq \tau_{MW}$$

and

$$\Psi(M_V) = \frac{\tau_{MW}}{\tau_{MS}} \Phi(M_V) \quad for \quad \tau_{MS} < \tau_{MW}$$

where τ_{MW} is the age of the galaxy and τ_{MS} is the main sequence lifetime of a star under consideration. Here, it has been assumed for simplicity that the stellar birthrate and the form of $\Psi(M_V)$ has been constant with time over the age of the galaxy. To derive an initial distribution of stellar mass we need to apply a mass luminosity relation to $\Psi(M_V)$. With the additional assumption that $\Psi(M_V)$ counts only main sequence stars, we can use the empirical mass- luminosity relation for main sequence stars (i.e., $L_* \sim m_*^p$, where p $\approx$ 3.45; Schwarzschild 1958; Allen 1973). Because $\Psi(M_V)$ is a distribution of stellar (absolute, visual) magnitudes and not luminosities, it is convenient to use a form of the mass-luminosity relation that relates M_V to stellar mass. Since $L_* \sim m_*^p$, and $M_V \sim \log L_*$, then $M_V \sim \log m_*$. Consequently it is useful to introduce the concept of the initial mass function $\xi(log m)$ which is the relative number of stars formed per unit volume per unit *logarithmic* mass interval and is straightforwardly related to $\Psi(M_V)$ as follows:

$$\Psi(M_V) dM_V = \xi(log m_*) d log m_*.$$

The shape of the initial mass function is usually characterized by a spectral index, $\beta \equiv \partial log \xi(log m_*)/\partial log m_*$, which is the slope of the function in a log-log plot. We note that the initial mass *spectrum*, $f(m_*)$, (i.e., the differential frequency distribution of stellar masses at birth) is related to the mass function by:

$$\xi(log m_*) = \frac{m_* f(m_*)}{0.434}.$$

The spectral index of the mass spectrum, γ, is equal to $\beta - 1$. The initial mass function is usually called the IMF, however in general it is convenient to also use this abbreviation when referring to the initial mass spectrum as well. Although this can sometimes lead to confusion, we will adopt this convention for the rest of this review.

Salpeter found that the initial mass function could be reasonably well represented by a simple power-law form viz:

$$\xi(log m_*) \sim m_*^{-1.35}$$

In other words, a constant spectral index over the range of stellar mass that was considered, between approximately 0.4 and 10 $M_\odot$. In addition, the value of the spectral index (being less than -1.0) indicated that more stellar mass was contained in low mass stars than in high mass stars. More recent studies, however, (c.f. Scalo 1986) suggest that the spectral index of the IMF is significantly flatter than this for low stellar masses (i.e., $m_* < 0.5\ M_\odot$) and perhaps significantly steeper for high mass stars (i.e., $m_* > 10\ M_\odot$). There may also be evidence for a turnover or peak in the function at about 0.3 $M_\odot$. In any event, it is clear that most stars that form in our galaxy are low mass objects (i.e., $m_* < 3\ M_\odot$).

In this regard, the mass spectrum of stars is both quantitatively and qualitatively different than that of molecular cloud cores and clumps. For molecular cloud cores the spectral index of the mass *spectrum* is approximately equal to -1.6 for cores between $10-10^3\ M_\odot$ (see chapter by Blitz, this volume). Therefore, massive cores contain more of the total available molecular mass than low mass cores. If stars from directly from molecular cores and clumps how is it that stars have such a different mass spectrum (i.e. $\gamma_* = -2.35$) ? The most likely explanation is that molecular cloud cores are themselves structured with subfragments and either the fragmentation is not self-similar in mass or the newly formed stars themselves affect the determination of their final mass (e.g., Lada and Shu 1990; and §5). It is interesting in this context to point out that the mass spectrum of open clusters determined by van den Bergh and LaFontain (1985) has been shown by Elmegreen and Clemens (1985) to have a spectral index of -1.5, similar to molecular cloud cores. This would seem to imply that it is in the physical process of the formation of individual stars that the final form of the stellar mass spectrum is determined.

The field star IMF, by itself, may not necessarily provide a very strong constraint for star formation theory. This is because the field star IMF is a globally averaged IMF, averaged over both the lifetime of the galaxy and a over a specific volume of space (i.e, the solar neighborhood). Although it is often assumed that the field star IMF is universal in both time and space, this has never been satisfactorily demonstrated (except, perhaps, in the general sense that more low mass stars have been produced than high mass stars over the lifetime of the galaxy). However, for the construction of a theory of star formation it is crucial to know whether the detailed form of the IMF is indeed universal. Are there spatial and/or temporal variations in the initial conditions and other important astrophysical parameters that can alter the process of star formation and the form of the IMF? Or is star formation such a robust process that the outcome is always an IMF of the same form?

In principle, it is possible to directly address at least one aspect of this problem: whether or not the IMF is constant in space. Observation of large enough groups of young or newly forming stars in different parts of the galaxy should be able to provide fundamental constraints on this question. The smallest spatial size

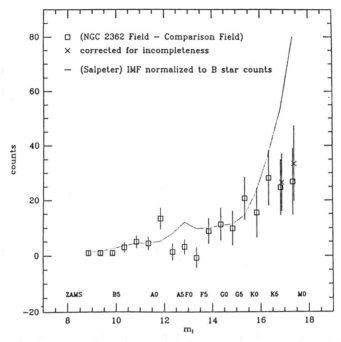

Figure 1. The I-band luminosity function of the young cluster NGC 2362 constructed by Wilner and Lada (1991). The dotted line shows the expected distribution of star counts assuming a Salpeter IMF normalized to the B star population of the cluster.

scale over which a meaningful determination of a luminosity function can be made is that characteristic of an open cluster. Clusters are important laboratories for studying the initial luminosity function as well as stellar evolution because they consist of statistically significant groups of stars who share the common heritage of forming from the same parental cloud at the same epoch in time. An example of the luminosity function of a young cluster is shown in Figure 1. Here the I band luminosity function of the young (7×10^6 yr.) open cluster NGC 2362 is plotted along with the I band luminosity function corresponding to the Salpeter IMF. As can be seen the cluster luminosity function significantly deviates from the Salpeter IMF at late spectral types (and low stellar masses). This is consistent with recent determinations of the field star IMF which suggest a flattening or even a turnover at low mass, although in NGC 2362 the deficit of low mass stars appears more severe than that determined for the field star IMF. How does the luminosity function of NGC 2362 compare to the luminosity functions of other clusters? Unfortunately, there are no clusters of a similar age for which luminosity functions have been determined that extend to such low masses. However, observations have been made of a few older clusters such as the Pleiades and the Hyades indicating similar trends (e.g., van Altena 1969; van Leeuwen 1980). However, these clusters are sufficiently old that dynamical mass segregation and evaporation of low mass stars makes meaningful evaluation of the shape of the low mass end of their initial

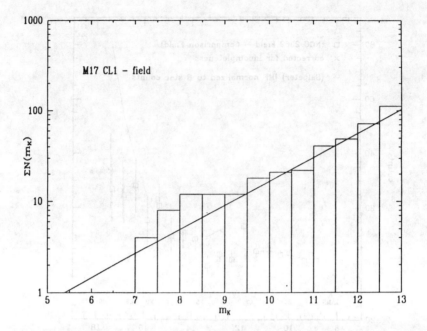

Figure 2. The k-band cumulative or integrated luminosity function i.e.:

$$\int^{M_k} \Psi(M_k) dM_k \text{ vs. } M_k$$

constructed by Lada et al. (1991) for the obscured cluster of OB stars in M17. The slope of the luminosity function corresponding to the Salpeter IMF is also plotted for comparison.

luminosity functions nearly impossible (van Leeuwen 1980). In general, such effects have made it difficult to confidently assess variations in the luminosity functions between most open clusters observed to date, and have resulted in contradictory conclusions in the literature (Scalo 1986).

Clearly, more observations of clusters as young as NGC 2362 would be extremely helpful. However, potentially even more useful would be the determination of the luminosity functions of even younger stellar systems, the embedded populations of molecular clouds. Molecular clouds cover large regions of space and often contain hundreds and thousands of Young Stellar Objects (YSOs) which have recently formed or are in the earliest stages of stellar evolution. Such populations are extremely young with ages probably between 1/2-5 million years. In particular, as discussed in §3.0, GMCs contain compact, often rich, embedded clusters. These clusters have small diameters ($\approx$ 1 pc) and can contain on the order of a few hundred YSOs within their boundaries. They provide excellent laboratories for luminosity function studies. Their embedded nature, however, necessitates that they be observed at infrared wavelengths.

In addition to minimizing the effects of extinction, infrared observations offer a number of advantages not available in other wavelength bands. For example, at these wavelengths we are in the Rayleigh-Jeans portion of the spectrum of the stellar photospheres. Therefore the observed flux is given by:

$$F_\nu = 2\pi k \frac{\nu^2}{c^3} T_*$$

where k is Boltzman's constant and T_* is the photospheric temperature. The luminosity, L_ν, is just $4\pi R_*^2 F_\nu$, consequently $L_\nu \sim R_*^2 T_*$. Empirically, for main sequence stars the photospheric temperature and stellar radius are related to the stellar mass by simple power laws, i.e., $\sim m_*^{0.55}$ and $\sim m_*^{0.75}$, respectively. Consequently, there is a straightforward relation between mass and luminosity at infrared wavelengths, $L_\nu \sim m_*^{2.0}$. As we have seen earlier, this means that the infrared luminosity function at some frequency or wavelength in the Rayleigh-Jeans regime can be straightforwardly related to a mass function viz:

$$\Psi(M_\lambda)dM_\lambda = \xi(logm_*)dlogm_*$$

Thus for a stellar mass function of spectral index γ, it follows that the spectral index of the infrared luminosity function is $\frac{\gamma}{5}$ which has a value of 0.26 for a Salpeter IMF. Therefore, for a cluster composed primarily of main sequence stars, infrared observations alone can be used to directly derive the form of the IMF.

Figure 2 shows the k band luminosity function of the famous obscured cluster in M17. In this diagram the integral form of the luminosity function (i.e., $\int \Psi(M_k)dM_k$ vs. M_k) has been plotted for the cluster and compared to that corresponding to the standard Salpeter IMF. In contrast to NGC 2362, the slope of the M17 cluster luminosity function agrees extremely well with that predicted for the IMF. However, it is important to note that in M17 we are observing a cluster of very massive stars. The range of masses represented in the figure is roughly 8-30 $M_\odot$ and has very little overlap with that of NGC 2362. Since such massive stars evolve very quickly we can safely assume that they are main sequence stars. However, this assumption is not a very good one for lower mass stars which have not yet been detected in the cluster or for the more typical observations of embedded clusters in which lower mass stars are readily detected. These clusters undoubtedly contain many pre- main sequence stars and even protostars. For these stars the mass- luminosity relation is not well understood, moreover, much of their infrared emission is not photospheric in origin (see discussion below). For such stars, significant (as yet undetermined) evolutionary corrections are needed to transform the infrared luminosity function into a mass function.

Nonetheless, useful information may still be derived from the infrared luminosity functions of embedded stellar populations and clusters which contain low mass stars or mixtures of low and high mass stars. The embedded clusters in the L1641 cloud (E. Lada et al. 1991, hereafter LDMG) provide a good example of the latter type. The infrared k band luminosity functions of the three rich clusters in this cloud have slopes which are (slightly) steeper than of the field star IMF. LDMG suggested that the steeper slopes might result in very young clusters whose underlying mass spectra are similar to the IMF. In these clusters the lower mass stars are pre-main sequence stars and are systematically more luminous for their mass than main sequence stars, producing an excess of stars at lower luminosities compared to that predicted by the IMF, while the high mass stars are on or much closer to the main sequence and have a distribution of luminosities closer to that predicted by the IMF.

Because of the concerns listed above it is clear that single color infrared luminosity functions for extremely young clusters cannot be straightforwardly transformed into stellar mass functions. In particular, for low mass YSOs the infrared luminosity at a particular wavelength may not be simply related to a total or bolometric luminosity for the object (see discussion below). Thus it is necessary to determine the bolometric luminosities of YSOs in order to construct meaningful luminosity functions for extremely young embedded populations of stars. Because the dust surrounding young stellar objects is capable of absorbing the bulk of the radiant luminosity of an embedded star, we expect that most if not all the luminosity of a YSO, however, will be radiated between 1-300 μm. As a result, observations across the infrared spectrum can be integrated to produce fairly accurate bolometric luminosities for the observed stars (e.g., Lada and Wilking 1984, Lada 1988a). Thus multi-wavelength near- and far-infrared observations can be used to *directly* construct the initial luminosity function (ILF) of a young embedded cluster. To date, the best determined ILF for an embedded cluster is that derived for the embedded population in the ρ Ophiuchi dark cloud (Wilking, Lada and Young 1989). In this cloud bolometric luminosities were estimated for more than 50 low luminosity YSOs using 1-100 μm observations. The luminosity function for this cluster appears to contain very few stars more massive than the sun. Its general shape is similar to that expected from a Salpeter IMF, however, there are so few intermediate to high mass stars in the cloud that it is hard to assess the significance of this result. Moreover, the cloud contains such high levels of extinction ($A_V \approx$ 50-100 mag) that even in the infrared bands numerous low luminosity embedded stars are likely to be invisible. Nonetheless existing observations provide evidence that this cluster is undergoing luminosity evolution which, if true, would imply that the cluster is deficient in stars of intermediate (2-10 $M_\odot$) mass compared to the field star IMF (Wilking, Lada and Young 1989).

Yet, even when armed with accurate bolometric luminosity functions of embedded clusters such as the Ophiuchi cluster, only limited information concerning the underlying stellar mass spectra can be derived. This is due to the fact that we as yet have no theoretical or clear observational understanding of the relation between stellar luminosity and mass for collections of pre-main sequence and protostellar objects. Additionally, even if we could determine the mass spectrum of an embedded population of a given molecular cloud, we could not be sure that the form of the spectrum wouldn't change as the cloud continued to evolve and produce more stars. Moreover, as objects evolve from protostars to pre-main sequence stars to main sequence stars within a young protocluster we do expect that the cluster luminosity function as a whole will also evolve, even if additional star formation is not taking place. For example, as a young cluster evolves, stars of progressively lower mass will arrive on the main sequence. At any given instant of time stars with pre-main sequence lifetimes less than the age of the cluster will be on the main sequence and will display a luminosity function with a form or slope appropriate for newly formed main sequence stars. Stars whose pre-main sequence lifetimes are greater than the age of the cluster will be brighter for their mass compared to main sequence stars and their luminosity function should be steeper than that which will they will ultimately display when they arrive on the main sequence. If the underlying mass spectrum is given by the standard IMF then the slope of the luminosity function will be that derived by Salpeter for the massive stars but will be steeper than Salpeter for the low mass stars. A break in the spectrum should occur at the luminosity or mass where the pre-main sequence lifetime of the stars is equal to the cluster age. Consequently, it may be possible that observations of embedded clusters in different clouds of differing evolutionary states will reveal significant differences in their luminosity functions, even if the underlying mass spectra of the stars they ultimately spawn are the same. Observations of such differences could provide important constraints for the development of a general theory of star formation and early stellar evolution in a similar way that differences in the HR diagrams of aging clusters provided important tests for, and confirmation of, stellar evolution theory more than four decades ago.

Unfortunately, from the perspective of trying to understand star formation, we must admit that the current state of knowledge concerning the IMF is poor. We do not yet know the extent to which the IMF is universal and our observational knowledge of the IMF in individual young stellar clusters and star forming regions has not yet been able to provide the critical tests of star formation theories that we desire. However, it is at least still clear from what we do know that low mass stars are the most typical outcome of the star formation process and understanding their origins must be an important first step toward the development of a general theory of star formation.

3. Clusters vs. Associations: Two Modes of Low Mass Star Formation

It has long been suspected that stars of all masses form predominately in clouds that give rise to unbound OB associations (see chapter by Blaauw, this volume; also Roberts 1957). OB associations are formed in giant molecular clouds (GMCs), therefore, in the current epoch of galactic history the vast majority of stars, of both low and high mass, must form in GMCs (see chapter by Blitz, this volume, also Lada 1987). With extents on the order of 100 parsecs and masses often in excess of 10^5 $M_\odot$, GMCs are the largest objects in the galaxy and rival globular clusters as the most massive objects in the Milky Way. The spectral index of the mass spectrum of GMCs is approximately equal to -1.6 indicating that most of the molecular (star-forming) mass in the galaxy is contained within the largest (M $\geq 10^5$ $M_\odot$) GMCs (see Blitz, this volume). In an earlier chapter of this book Chenicharo described the conditions in molecular clouds which spawn predominately low mass stars. In general these clouds are at the low mass end of the spectrum of GMCs and are likely precursors to T associations, loose groups of low mass stars first recognized as associations of pre-main sequence objects by Ambartsumian (1947). Within these clouds low mass stars sporadically form individually or in small groups of 2 or 3 stars from relatively isolated and well defined cloud cores (as discussed earlier in detail by Chenicharo). Such regions clearly represent important laboratories for studying the birth and early evolution of low mass stars since complications resulting from the disruptive presence of nearby, previously formed, stars are minimized. Such conditions constitute a "loosely-aggregated" mode of star formation. However, this mode may not be representative of the formation of all or even the majority of low mass stars. Indeed, based on the mass spectrum of GMCs we expect that most of the stars, both low and high mass, that formed in the galaxy were born in massive GMCs where conditions described in the chapter by Genzel could be more appropriate. Do low mass stars in these clouds form in a similar "loosely-aggregated" mode of star formation?

It has become apparent from recent studies, particularly the work of E. Lada and her colleagues (Lada 1991; LDMG), that a significant fraction of low mass stars that form in GMCs form within massive, dense cores in rich embedded star clusters. These constitute a second or "clustered" mode of star formation. The natal environment of a massive core which is capable of forming a cluster is quite different than that of an isolated core participating in the "loosely-aggregated" mode of star formation. Indeed, the physics of the two modes could differ in fundamental ways such as in the form of the IMF that is produced. Like isolated dark cloud cores, embedded clusters are also important laboratories for the study of star formation and early stellar evolution. Indeed, their main liability, a closely packed grouping of many young stars, also provides their main asset for star formation investigations. The close proximity of the members of an embedded cluster enables direct observational comparison of the physical natures of YSOs of the same parental heritage in differing stages of formation and early stellar evolution. Because of their potential significance, in the remainder of this section we will describe some of the recent and interesting observational developments pertaining to our knowledge of embedded clusters. Although, it is important to bear in mind that very little is presently known or understood about these interesting objects. The discussion below follows that recently presented in Lada and Lada (1991).

3.1 THE NATURE OF NEARBY EMBEDDED CLUSTERS

To determine the nature of star formation within a GMC or any molecular cloud requires a census of the embedded population of YSOs in the cloud. However, to obtain an accurate census of an embedded stellar population in a single GMC is a difficult task requiring extensive infrared observations. To date, only one GMC has been systematically and extensively surveyed in the near-infrared: the L1630 or Ori B molecular cloud. With the aid of an imaging array camera LDMG imaged a significant portion of the cloud obtaining approximately 3000 $1' \times 1'$ fields covering an area of ~ 0.8 square degrees. The survey was estimated to be complete to a K magnitude of 13 which at the distance of Orion corresponds to a 0.6 $M_\odot$ main sequence dwarf. Therefore the observations were sensitive enough to detect both high and low mass young stellar objects in the cloud and to investigate the overall spatial distribution of these embedded and obscured young sources. As a result of this survey, four spatially distinct, embedded clusters were identified. These clusters turn out to be associated with the well known star forming regions, NGC 2071, 2068, 2024 and 2023. The smallest cluster, associated with the reflection nebula NGC 2023, has a radius of 0.3 pc and contains only 21 sources ($m_K < 14$). The largest and most spectacular of the four clusters contains more than 300 objects within a radius of 1.0 pc and is associated with the HII region NGC 2024.

The basic physical properties of the four embedded clusters are summarized in Table 1 (from Lada and Lada 1991). All four clusters are centrally condensed. The stellar densities of the clusters are high, ranging from ~ 70 pc^{-3} to ~ 180 pc^{-3} for the regions contained within the cluster boundaries. These densities resemble the stellar densities of other young star forming regions such as the star forming core of the Rho Ophiuchi cloud. (e.g., Wilking and Lada 1983). In addition, two of the infrared clusters, NGC 2071 and NGC 2024, show evidence for spatial magnitude segregation, with the brighter sources displaying a tendency to be more

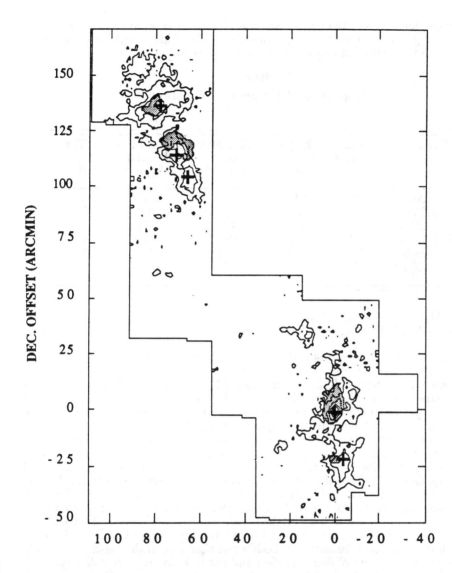

Figure 3. Locations of the embedded stellar clusters and dense cores in the L1630 Molecular Cloud (from E. Lada 1991). The shaded regions represent the location and extent of the embedded clusters. The distribution of dense gas is presented as intensity contours of $CS(2\rightarrow 1)$ emission. In addition, the peak intensity positions of the 5 most massive CS cores ($M>200\ M_\odot$) are represented by crosses.

centrally condensed than the fainter sources. This may represent a segregation in mass resulting from either the equipartition of stellar kinetic energies, the initial distribution of the star forming gas or some combination of these.

TABLE 1
Properties of Embedded Clusters

Cluster	# of Sources	Cluster Radius (pc)	Stellar Mass Density ($M_\odot pc^{-3}$)	Total Mass Density ($M_\odot pc^{-3}$)	SFE † %
NGC 2023[1]	21	0.30	185	2800	7
NGC 2071[1]	105	0.59	122	650	20
NGC 2068[1]	192	0.86	72	170	42
NGC 2024[1]	309	0.88	108	260	42
ρ Ophiuchi[2]	94	0.70	124	564	22
Trapezium[3]	142	0.44	1800	?	?

† SFE = M_{stars} / (M_{stars} + M_{gas})
(1) Lada 1990; (2) Wilking, Lada and Young 1989; (3) Herbig and Terndrup 1986

The most striking and surprising result of the 2.2 μm survey is that the vast majority of the detected sources believed to be embedded in the cloud are concentrated in the four well defined clusters. LDMG find that 58% of the objects they detected are contained within the four clusters. However, many of these sources are background/foreground field stars. After correction for the presence of field stars, LDMG estimate that approximately 96% of the total number of sources actually associated with the *entire* molecular cloud are contained within the four clusters! Moreover, the three richest clusters (see Table 1) contain the vast majority of these embedded stars. The total area covered by the four embedded clusters equals only 18% of the total region surveyed. From these results, LDMG conclude that star formation in L1630 is a highly localized process even for stars of low mass. Apparently in the L1630 cloud, the vast majority of stars that have formed in the cloud have formed in three rich embedded clusters. This leads to the astonishing conclusion, that if the L1630 GMC is typical of GMCs in the solar neighborhood, most star formation in GMCs and therefore in the galaxy may likely occur in the environment of dense clusters and not in isolated protostellar systems. Indeed, if we assume that the L1630 cloud is typical of other GMCs in its cluster forming properties, we can estimate the total number of rich embedded clusters, N_{EC}, within 2 kpc of the sun as follows:

$$N_{EC} \approx 3 \times N_{GMC} \approx 36$$

where N_{GMC} is the number of local GMCs (12; Blitz 1980). For a constant rate of cluster formation, we can predict that the number of visible (exposed) open clusters formed over the last 10^8 years should be:

$$N_{VC} \approx \frac{\tau_{VC}}{\tau_{EC}} \times N_{EC} \approx 720$$

where the lifetime of the embedded cluster phase, τ_{EC}, is estimated to be 5 million years. Clusters which reach an age of 10^8 years are most certainly bound since they have survived for a period equal to about ten internal crossing or dynamical times (for a cluster similar in density and size to the Pleiades). However, within 2 kpc of the sun only 94 exposed clusters are observed with ages of 10^8 years or less (Elmegreen and Clemens 1985). The ratio of observed to predicted clusters with ages 10^8 years or less is $\approx$ 0.13. Evidently, few clusters survive their emergence from molecular clouds as long-lived bound systems, such as the Pleiades (e.g., Lada, Margulis and Dearborn 1984).

Embedded clusters are distinguished from other types of clusters by their intimate association with the interstellar gas and dust from which they form. Therefore, to fully understand the nature of embedded clusters requires knowledge of *both* their stellar and gaseous contents. In the L1630 GMC an extensive survey for dense molecular gas (Lada, Bally and Stark 1991) has enabled the first systematic investigation of the relationship between dense gas and embedded clusters (Lada 1991). A substantial area of the L1630 cloud, including the regions surveyed at 2.2 μm, was mapped in the 2→1 transition of CS, in order to trace dense (n > 10^4 cm^{-3}) gas. The total area covered by the survey was $\sim$ 3.6 square degrees or $\sim$ 20% of the molecular cloud as measured in CO (Maddalena *et al.* 1986). CS emission was detected over $\sim$ 10 % of the area surveyed, revealing very clumpy structures. In fact, forty two individual dense cores were identified. These cores have masses ranging from < 8 $M_\odot$ to 500 $M_\odot$. Most have masses less than 100 $M_\odot$ and only 5 cores have masses greater than 200 $M_\odot$. The distribution of clump masses (for M > 20 $M_\odot$) can be described by the powerlaw, $dN/dM \propto M^{-1.6}$, where N equals the number of cores per solar mass interval. The spectral index, γ_{clumps} = -1.6 implies that a significant amount of the mass of the dense gas is contained in the most massive cores. Indeed, approximately 50% of the total mass of dense gas in L1630 is contained within the 5 most massive cores. These 5 cores cover a total area of $\sim$ 2 pc^2 or only 1% of the total area surveyed, indicating that the dense gas in the surveyed region is confined to a small area. Moreover, the cores appear to be distributed in groups with the small cores clustered around the most massive ones. This further indicates that the dense gas is highly localized within this cloud.

The CS and 2.2 μm surveys of the L1630 molecular cloud provide the most complete census of dense gas and young stellar objects within a single giant molecular cloud to date. Comparison of these surveys shows that the embedded clusters are coincident or nearly coincident with 4 of the 5 most massive CS cores (Lada 1991). These results are summarized in Figure 3 which displays the locations and extents of the embedded clusters and the dense molecular gas in the L1630 molecular cloud. Apparently, the formation of embedded clusters in L1630 occurs in regions having both high density and substantial gas mass. Further comparison of the stellar and dense gas components of the L1630 embedded clusters, reveals that in all cases the mass of the dense gas associated with (and presumably contained within) an embedded cluster is considerably larger than the total mass of the embedded stars (Table 1).

The observations of L1630 appear to confirm the notion that stars form almost exclusively in dense gas and that clusters form from massive cores. However, these conditions do not appear to be sufficient for cluster formation. The L1630 cloud was also found to contain a massive, dense core (LBS 23) that is not associated with a recognizable cluster. Moreover, another massive core (NGC 2023) was found to contain only a very poor cluster. In these two massive cores the level of star forming activity is considerably lower than that in the three other comparable mass cores which are producing rich clusters. The lack of a substantial embedded cluster in two of the five massive cores is intriguing, given the levels of star formation in the 3 most active cores in the cloud, which together account for 97% of the embedded sources but only 30% of the dense gas found in the L1630. This is reflected in the derived star formation efficiencies. For example, NGC 2024, NGC 2071 and NGC 2068 cores are associated with rich clusters and have star formation efficiencies on the order of 20-40%. In contrast, LBS 23 and NGC 2023 exhibit SFEs $\sim$ 7%. The vastly different levels of star forming activity in these cores could either be a result of fundamental differences in some basic physical property of the cores (e.g., structure of internal magnetic fields) or to evolutionary effects. Understanding the differences in activity between these cores should have important implications for understanding the process of star formation and it is likely that future investigation and comparison of the physical conditions in massive dense cores with and without rich clusters will provide important new insights into the processes of embedded cluster formation and stellar birth.

Another fully embedded cluster for which both the gas and stellar content have been extensively studied and which contains predominately low mass stars is the cluster embedded in the ρ Ophiuchi dark cloud. Molecular line observations towards this cluster reveal the presence of a large centrally condensed core (Wilking and Lada 1983). This core is a region of high visual extinction, with A_V as high as $\sim$ 100 magnitudes (Vrba et al. 1975, 1976; Chini et al. 1977; Chini 1981; Wilking and Lada 1983). Wilking and Lada (1983) have mapped the core in $C^{18}O$, which is a good tracer of gas column density. They find the core to be 1 pc x 2 pc in size and to contain $\sim$ 600 M$_\odot$ of gas. Estimates of the star formation efficiency (SFE = $M_{stars}/(M_{stars} + M_{gas})$) of this system produce a SFE $\gtrsim$ 20% (Wilking and Lada 1983; Wilking, Lada and Young 1989). In this example, as in the L1630 clusters, the stellar mass of the cluster is a small fraction of the mass of the dense molecular core within which it is embedded.

Besides the examples mentioned above numerous other embedded clusters are now known to exist in nearby molecular clouds and to account for a large fraction of the stars formed in them (e.g., the Trapezium, Herbig and Terndurp 1986; NGC 2264, Walker 1956; S255, McCaughrean et al. 1991; L1641N, Strom, Strom and Margulis 1990). Although the initial studies are suggestive, it still remains to be determined if the cluster mode of stellar formation contributes the bulk of low mass stars in the galaxy. Even if it doesn't, future studies of low mass YSOs in clusters should still provide important constraints concerning the nature of the star forming process in an environment quite different than that which gives rise to single, isolated stars.

4.0 The Nature of Low Mass Young Stellar Objects

4.1 INFRARED ENERGY DISTRIBUTIONS AND SPECTRAL CLASSIFICATION

The remainder of this review will be concerned with describing what is known about the nature of indivdual YSOs of low mass. Varying amounts of gas and dust surround YSOs embedded in molecular clouds and absorb and reprocess substantial amounts of the radiant energy emitted by the buried stars. Consequently, we expect YSOs to radiate a significant fraction, if not all, of their luminous energy in the infrared portion of the spectrum. Circumstellar dust associated with YSOs also has a spatial extent considerably greater than that of a stellar photosphere. Consequently, emitting circumstellar dust, which is in radiative equilibrium with the stellar radiation field of the YSO, will exhibit a wide range of temperatures and the emission that emerges will have a spectral distribution much wider than that of a single temperature blackbody. For this reason it can be very difficult, if not impossible, to meaningfully place a YSO on an HR diagram. To detect the bulk of the luminous energy radiated by an embedded YSO therefore requires observations over a broad range of infrared wavelengths, typically 1-100μm. The shape of the broad–band infrared spectrum of a young stellar object (YSO) depends both on the *nature* and *distribution* of the surrounding material. Clearly then, we expect that the shape of the spectrum will be a function of the state of evolution of a YSO. The earliest (protostellar) stages, during which an embryonic star is surrounded by large amounts of infalling circumstellar matter, should have a very different infrared signature than the more advanced (pre–main sequence) stages, where most of the original star forming material has already been incorporated into the young star itself. Recent observations of YSOs (e.g., Lada and Wilking 1984; Myers *et al.* 1987; Wilking, Lada and Young 1988) have largely confirmed this expectation and shown that YSOs can be meaningfully classified by the shapes of their optical–infrared spectral energy distributions (e.g., Lada 1987). Moreover, such classifications can be arranged in a more or less continuous evolutionary sequence which is well modeled by at least one physically self-consistent theory of star formation and early stellar evolution (Adams, Lada and Shu 1987).

It is convenient to employ energy distributions to analyze the emergent emission from a YSO. An energy distribution is the plot of observable quantities λF_λ vs λ (or equivalently, νF_ν vs ν) in log–log space. The quantity νF_ν is useful because it is proportional to the actual energy flux (as opposed to the spectral energy flux density, i.e., F_ν) radiated in each logarithmic frequency or wavelength interval. Thus a flat distribution of νF_ν when plotted against logν represents equal energy radiated in each logarithmic frequency interval of the spectrum. The peak of a source's energy distribution occurs at the frequency where the greatest emergent energy is located. Moreover, the blackbody function acquires certain useful properties when plotted as an energy distribution. For example, the shape of a blackbody curve is invariant to translations in energy flux and frequency on an energy distribution plot.

If one defines a spectral index $\alpha = -\text{dlog}\nu F_\nu/\text{dlog}\nu$ which is to be evaluated longward of 2.2 μm wavelength, then the spectral energy distributions (SEDs) observed between 1–100 μm of most known YSOs fall into three distinct morphological classes (e.g., Lada and Wilking 1984; Adams, Lada and Shu 1987; and Lada 1987).

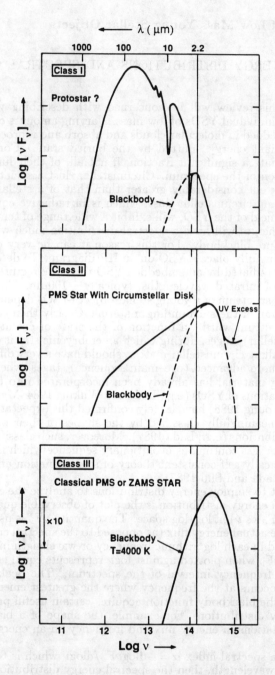

Figure 4. Classification scheme for YSO energy distributions

These are illustrated in Figure 4. Class I sources have SEDs which are broader than a single blackbody function and for which α is positive. These sources are often assumed to be protostellar in nature and are sometimes called "infrared protostars" (Wynn-Williams 1982). Class II sources have SEDs which are also broader than a single blackbody function but have values of α which are negative. Class III SEDs are also characterized by negative values of α but have widths that are comparable to those of single blackbody functions, consistent with the energy distributions expected from purely reddened photospheres of young stars. Class I sources derive their steep positive spectral indices from the presence of large amounts of circumstellar dust. These sources are usually deeply embedded in molecular clouds and rarely exhibit detectable emission in the optical band of the spectrum (e.g., Lada and Wilking 1984; Myers *et al*, 1987). However, nearly all known class II sources can be observed optically as well as in the infrared. When classified optically class II sources are usually found to be T Tauri stars or FU Ori stars, conversely, most all T Tauri and FU Ori stars exhibit class II SEDs (see Rucinski (1985) for some good examples of T-Tauri star SEDs). Their negative spectral indices indicate that class II YSOs are surrounded by considerably less circumstellar dust than class I sources. Class III sources are usually optically visible with no or very little detectable excess emission at near- and mid-infrared wavelengths, and therefore little or no circumstellar dust. Class III objects include both young main sequence stars and pre-main sequence stars, such as the so-called "post"-T Tauri stars (e.g., Lada and Wilking 1984) and the recently identified "naked"- T Tauri stars (e.g., Walter 1987). It is apparent from existing studies of YSOs that there is a more or less continuous variation in the shapes of SEDs from class I to class III (e.g., Wilking, Lada and Young 1989)

Although most YSO energy distributions can be classified into one of the three classes in Figure 4, there are a few sources which exhibit considerable, often well defined (e.g., double humped), structure in their SEDs and they cannot be characterized by a single spectral index in the infrared. Typically, a source with a double humped energy distribution will exhibit a high frequency hump at around $1\mu m$ wavelength and a low frequency hump at around $60\mu m$ wavelength. The energy distributions of such double humped sources can usually be classified as sub-classes of either class II or III (e.g., Wilking, Lada and Young 1989). A class II-D source has a double peaked energy distribution whose high frequency peak is characterized by a spectral index similar to that of a T-Tauri star. Such a source is likely intermediate between pure class I and pure class II. A class III-D source is characterized by a high frequency peak with a spectral index similar to that of a purely reddened photosphere. Early-type B stars which have reached the ZAMS often produce such energy distributions since they are luminous enough to heat relatively distant and cold dust to temperatures which produce observable far-infrared emission (e.g., Harvey, Thronson and Gatley 1980; Harvey, Wilking and Joy 1984). Low mass stars also produce class III-D SEDs, but usually the long wavelength excess is not strong enough to produce a separate peak in the SED; however, a clear break in the long wavelength slope or shape of the SED is apparent (see §4.2). In some situations it is possible that a complex energy distribution could result from confusion and the superposition of more than one source in an observer's beam. Such confusion is most likely to occur at long wavelengths and for distant sources where the effective spatial resolution may be relatively poor.

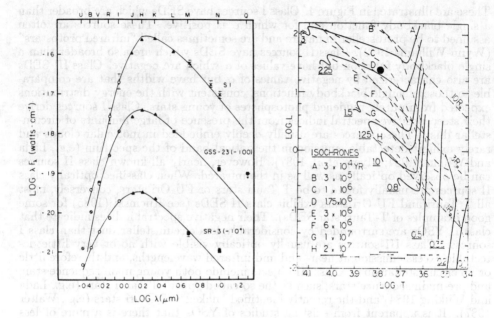

Figure 5. Energy distributions of three sources in the ρ Ophiuchi dark cloud which exhibit class III SEDs (from Lada and Wilking 1984). The HR diagram (from Cohen and Kuhi 1979) is shown with the position of GSS-23 indicated.

4.2 THE NATURE OF CLASS III SOURCES

Since their shapes are more or less similar to single temperature blackbodies, the energy distributions of class III sources are easily interpreted as arising from extincted or unextincted photospheres of young stars. By definition these stars display no infrared excess. However, their light still could be substantially extinguished by foreground dust. Extinction attenuates the spectrum of a YSO by a factor of $\exp(-\tau_\nu)$ and since $\tau_\nu \sim \nu$, for interstellar dust, then $F_\nu \sim \exp(-\nu)$ and it is difficult to distinguish such a spectrum from the high frequency side of a blackbody function where $F_\nu \sim \exp(-\frac{h}{kT}\nu)$. Thus an extincted star has an energy distribution similar to a blackbody but with the high frequency end exponentially extinguished and the peak displaced toward lower frequencies thus giving the appearance of being "reddened". Figure 5 shows the SEDs of three sources in the ρ Ophiuchi cluster. The line connecting the observations is that of a blackbody reddened by an amount derived directly from the infrared color excesses observed for these objects. Because these sources display no excess near-infrared emission their *color* excesses are exclusively produced by reddening. Moreover, since at these infrared wavelengths the Rayleigh-Jeans approximation is valid, the intrinsic infrared color indices (e.g., H_0-K_0) are essentially zero for a wide range of photospheric temperature. Thus the observed color indices give directly the extinction toward each object and the

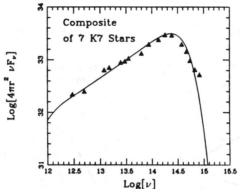

Figure 6. *The composite spectral energy distribution of seven class II sources together with a theoretical spectrum of a K7 star and circumstellar disk with a power-law temperature gradient of index n=0.6. From Adams, Lada and Shu (1987)*

blackbody fits give uniquely the effective temperatures of the stars. With knowledge of the distance to the cluster, we can also determine the total luminosities of these stars and place them on the HR diagram. The stars S1 and SR-3 were found in this way to be B stars, consistent with existing optical classifications, while the star GSS-23 was found to be a pre-main sequence star of approximately 2.25 $M_\odot$ as is seen from its position in the HR diagram in Figure 5. Clearly if a source has a class III spectrum it can be accurately placed on the HR diagram and its properties compared to classical evolutionary tracks. ZAMS stars, Naked T-Tauri Stars and Post T-Tauri stars usually display class III SEDs in the optical and near-infrared portions of the spectrum.

Inspection of the SEDs in Figure 5 also indicates that at the longer wavelengths the source SEDs appear to depart from the Planck curve. Although these sources have class III shapes in the near-infrared, they appear to have excesses in the mid- and far-infrared and are probably best classified as class III-D sources. Numerous sources have been detected within molecular clouds with this type of spectral signature (e.g., Harvey, Thronson and Gatley 1980; Strom et al. 1989).

4.3 THE NATURE OF CLASS II SOURCES

Class II sources have negative values of α similar to class III sources. The optical portion of their spectrum is blackbody-like and similar in shape to a cool stellar photosphere. However, the infrared portion of their spectrum departs from that of a normal stellar photosphere (see Figure 4.) producing "excess" emission at these wavelengths. For many class II sources the shape of the infrared portion of the SED is well described by a power-law form (e.g., Rucinski 1985; Rydgren and Zak 1987). This is illustrated in Figure 6 which shows the composite SED of seven class II sources. The dereddened and averaged data points are indicated by triangles and a theoretical fit to the spectrum (in which the infrared portion is modeled by a power-law slope) is indicated by a solid curve.

More than a decade ago Lynden-Bell and Pringle (1974) predicted that T

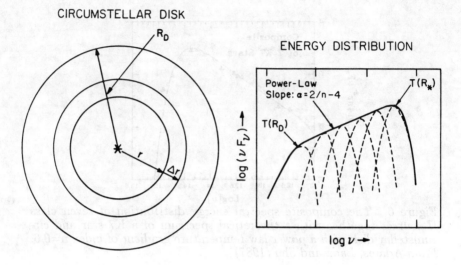

Figure 7. Schematic diagram of an optically thick disk and its emergent spectral energy distribution. The emergent energy distribution is produced by a superposition of blackbodies of appropriately varying temperature. The energy distribution consists of three parts. At high frequencies the emission falls off approximately exponentially appropriate for the Wien side of a Planck curve. At intermediate frequencies the emission follows a power-law form and at the lowest frequencies the emission falls with a slope given by the Rayleigh-Jeans Law or somewhat steeper if the emission becomes optically thin.

Tauri stars would display such energy distributions if surrounded by luminous accretion disks. Consider an optically thick and spatially thin disk that surrounds a young star and radiates everywhere like a blackbody. Imagine the disk to be composed of concentric annuli as illustrated in Figure 7 with radial dimension ΔR and area $2\pi R \Delta R$. Each annulus radiates as a blackbody of temperature $T(R)$. The emergent spectrum of the disk will then be the superposition of a series of blackbody curves of varying $T(R)$ as shown in Figure 7. Now if $T(R) \sim R^{-n}$, the Wien laws tells us that the frequency of maximum emission scales as $\nu \sim T(R) \sim R^{-n}$. The luminosity radiated in each annulus is given by:

$$L_\nu d\nu = 2\pi R dR \sigma T(R)^4 \sim R^{2-3n} d\nu \sim \nu^{3-\frac{2}{n}}$$

Therefore, if the temperature gradient in the disk is characterized by a radial power-law, the emergent spectrum will also be characterized by a power-law slope in frequency or wavelength. For an SED, $\nu L_\nu \sim \nu^{4-\frac{2}{n}}$ or $\alpha = \frac{2}{n} - 4$.

A viscous accretion disk is predicted to produce a temperature gradient characterized by n=0.75, corresponding to an α of -1.33 (Lynden–Bell and Pringle 1974). Recently Adams and Shu (1986) and Friedjung (1985) showed that a flat *passive* disk which derives all its luminosity from the reprocessing and re-radiation of light it has absorbed from the central star, also has an equilibrium temperature gradient characterized by n = 0.75. However, most class II sources have spectra characterized by slopes which are less steep than those predicted for either a passive disk or an accretion disk (e.g., see Figure 6, also Rucinski 1985; Beckwith *et al.* 1990). Kenyon and Hartmann (1987) have argued that a passive, reprocessing disk can produce a more shallow temperature distribution and a more shallow spectral slope if the disk is flared. Most observed class II energy distributions can probably be explained with such a model. However, it may also be possible that the shallow temperature gradients required for the disks arise from some nonviscous accretion process (Adams, Lada and Shu 1988). In particular, flat spectrum sources are not easily fit with flaring disks and the nature of their luminosity generation is unclear. However, it is likely that these sources, which include the star T Tauri itself (see Figure 8), are characterized by disks which are *active* and have an intrinsic luminosity source in addition to the reprocessed radiation from the central star. In any event, as many authors have argued (e.g., Rucinski 1985; Beall 1987; Adams, Lada and Shu 1987 and Kenyon and Hartmann 1987), the most likely interpretation of the nature of the class II sources is that they represent pre–main sequence stars which are surrounded by disks.

Apart from modeling spectral energy distributions, independent evidence for circumstellar disks has been obtained at optical and infrared wavelengths. For example, at high spectral resolution, T Tauri stars often have blue shifted forbidden-line emission. Because these lines probably form in stellar winds originating near the surface of the stars, the absence of corresponding red-shifted emission suggests the presents of occulting disks close to the stellar surfaces (Appenzeller, Jankovics and Ostreicker 1984; Edwards *et al.* 1987). In addition, the presence of relatively strong mid- infrared and even millimeter continuum radiation from class II sources is inconsistent with the relatively low visual extinctions toward these objects unless the emitting circumstellar material is distributed in a very non-isotropic fashion (Adams, Lada and Shu 1987; Beckwith *et al.* 1990). Perhaps the most compelling evidence to support the disk hypothesis comes from spectroscopic measurements of optical and infrared absorption lines in the spectrum of FU Ori by Hartmann and Kenyon (1987). Comparison of the velocity widths of the optical and infrared lines indicate that the infrared rotational velocity is smaller than the optical rotational velocity as would be expected for a differentially rotating structure such as a disk in Keplerian motion around a central star. Moreover, the energy distribution of FU Ori is class II and is well modeled by a disk with a temperature power-law index of n = 0.75 (Adams, Lada and Shu 1987).

Class II sources have been observed over more than three decades of wavelength from optical to millimeter. In general, their SEDs depart from a power–law form and turn over at low frequencies where they begin to become optically thin, most likely as a result of their finite masses. For a disk of infinite mass and finite radius (e.g. R_D, see Figure 7), the low frequency end of the SED would take the form of the Rayleigh–Jeans law, i.e., $\nu F_\nu \sim \nu^3$ and the frequency of the turn over could be used to determine R_D, the radius of the disk (see Figure 7 and Adams, Lada and Shu 1988). In reality, circumstellar disks have finite masses and because the low frequency emission becomes optically thin, the turnover occurs at higher

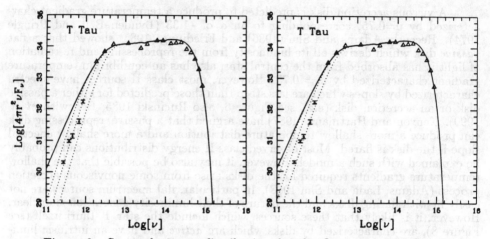

Figure 8. Spectral energy distribution for the flat spectrum class II source T Tauri. The left hand diagram also plots the theoretically derived fits to the SED assuming a circumstellar disk and an opacity law of the form $\kappa_\nu \sim \nu^2$ for disk masses of 0.01, 0.1, 1.0 $M_\odot$ (dotted lines) and for infinite disk mass (solid line). The right hand figure plots the same models but with a different opacity law, $\kappa_\nu \sim \nu^{1.5}$. With permission from Adams, Emerson and Fuller (1991).

frequency and can be used only to place a lower (but certainly useful) limit on the radius of the disk. Typical values derived from observations indicate that $R_D \gtrsim 100$ AU. On the other hand, the emission observed at optically thin frequencies can be used to determine the disk mass, provided the temperature and opacity law of the emitting material is known (Adams, Lada and Shu 1988; Beckwith et al. 1990).

Figure 8 shows observations and model fits to the spectrum of the flat- spectrum class II source T Tauri (Adams, Emerson and Fuller 1991) using two different opacity laws and a range of disk masses. The temperature of the inner disk was set equal to that of the stellar photosphere and a power-law radial temperature index n=0.55 was derived from the fits to the near to mid-infrared portion of the spectrum. As shown in the Figure the masses derived for the disk (i.e. 0.1 $M_\odot$ and 0.01 $M_\odot$ respectively) for opacities $\kappa_\nu \sim \nu^2$ and $\kappa_\nu \sim \nu^{1.5}$ differ by an order of magnitude. To date numerous class II sources have been observed and detected at millimeter and submillimeter wavelengths. Current dogma favors use of an opacity law with a spectral index between 1.5 and 2.0 and disk masses in the range between 0.01–0.1 $M_\odot$ have been derived for numerous sources (Adams, Emerson and Fuller 1991; Beckwith et al. 1990). The magnitude of the derived masses are very interesting since they are of the order believed to be necessary for planet formation.

Finally, in one of the more interesting developments to occur recently, a number of YSOs have been found to have excess emission at far-infrared and millimeter wavelengths but class III spectra at optical and near-infrared wavelengths (e.g., Strom et al. 1989; Beckwith et al. 1990; Montmerle and Andre 1989). It has been hypothesized that the long wavelength excesses observed toward these sources also arise in cold circumstellar disks. However, to be consistent with the lack of near-infrared excess emission the circumstellar disks must have holes or be optically thin in their inner regions. Such objects would clearly represent transition cases between pure class II and class III sources and their observation may provide fundamental insights to understanding how circumstellar disks evolve around young pre-main sequence stars. (e.g., Strom *et al.* 1989; Montmerle and Andre 1989).

4.4 THE NATURE OF CLASS I SOURCES

4.4.1 *Interpreting Spectral Energy Distributions*

Because of their deeply embedded nature, it has often and long been argued that class I sources are protostars, that is, embryonic stellar cores in the process of acquiring the bulk of the mass they will ultimately contain as main sequence stars (e.g., Becklin and Neugebauer 1967). Their luminosity is assumed to be derived almost entirely from accretion of infalling gas and dust. Although direct evidence for such a hypothesis is lacking, analysis of the energy distributions of these objects tends to lend strong support to such contentions. As mentioned earlier, the shape of the energy distribution of a YSO depends on the nature and distribution of surrounding circumstellar material. As a result, one should be able to derive some important general properties about the structure of such an object from its observed spectrum. As illustrated in Figure 9, the emergent intensity from a spherically symmetric extended source of radiation depends on the density and temperature distribution of the emitting material and the opacity law:

$$I_\nu(\mu, r) = \int_0^S n(r)\kappa_\nu B(T(r))e^{-\tau_\nu} ds.$$

With knowledge of the temperature distribution and the opacity law, one can in principle invert the observed emission spectrum and infer the density distribution or structure of the object. Let us assume that a class I source consists of a hot (i.e., 3000–5000 K) stellar–like central object, surrounded by an extended dust atmosphere which absorbs and reprocesses the radiant energy emitted by the central stellar object. In such a situation we expect the temperature distribution of the extended dust atmosphere to be determined by radiative equilibrium with the stellar radiation field. If we now assume an opacity law we can derive the derive the density distribution of the source from its SED. When such an analysis is performed on observed class I sources the following general picture of the structure of class I sources is obtained (e.g., Myers *et al.* 1987; Adams and Shu 1987; Adams, Lada and Shu 1987). A typical class I sources consists of three regions. On a large scale ($\approx 10^4$ AU) the density distribution is a power–law with a spectral index approximately equal to -1.5. However, the inferred power–law density distribution cannot continue all the way down to the stellar surface because if it did the extinction would be so great (on the order of 1000 visual magnitudes!) that near–infrared and mid-infrared emission would not be detectable in the spectrum. Therefore, the observed level of near-infrared emission constrains the amount of dust that can be present in the inner regions of the source. For typical sources, there must be a relatively evacuated (but not empty) cavity of roughly 200 AU in size around the central star (Myers *et al.* 1987; Adams and Shu 1986). However, there must be a third structural

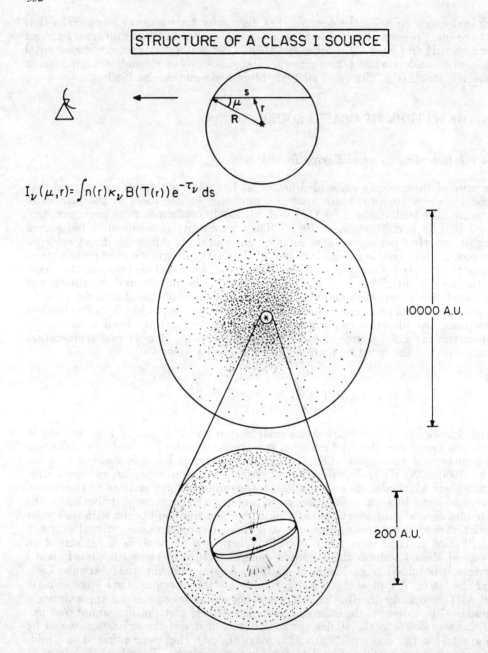

Figure 9. Schematic diagram illustrating the structure of the dust atmosphere or envelope of a typical class I source as derived from spectral modelling of well observed sources in a few nearby molecular clouds.

component to be able to explain the strong emission at mid–infrared wavelengths from class I sources. Radiative equilibrium demands for the typical luminosity of a class I source (5-50 $L_\odot$) that the mid–infrared emission must originate from regions within 200 AU of the star. The observed flux of such emission is so high that if distributed spherically around the source would cause too much extinction and a decrease in the near infrared flux to below what is observed. To reconcile the need for strong mid–infrared emission and the lack of excessive extinction, one invokes the presence of a circumstellar disk which will provide copious mid-infrared flux but only excessive extinction when observed edge on. The structure of a typical class I source is illustrated in Figure 9.

4.4.2 Protostellar Models

The physical characteristics of class I sources outlined above, interestingly, turn out to be predictions of a recent physically self–consistent theory of protostars developed by Shu and his collaborators (see chapter by Shu in this volume). In this picture the evolution of a magnetically supported molecular cloud leads to the formation of dense cores through the ambipolar diffusion of local fields (see Shu, Adams and Lizano 1987). These cores are assumed to have density distributions similar to that of an isothermal sphere, that is:

$$\rho(r) = \frac{a^2}{2\pi G} r^{-2}$$

where a is the speed of sound, G the gravitational constant and ρ the mass number density. Shu (1977) showed that the collapse of an isothermal sphere can proceed in a self–similar manner and be characterized by a single parameter, the mass infall rate, given simply by:

$$\dot{M} = \frac{.975 a^3}{G}$$

This collapse proceeds in a nonhomologous, inside–out manner with a flattening of the initial isothermal density gradient within the infalling region (i.e., $\rho(r) \propto r^{-1.5}$). In the center, the collapse is arrested at a dense protostellar core which grows in mass linearly with time, that is, $M(t) = \dot{M}t$. The protostellar core becomes increasingly luminous as it radiates away the gravitational potential energy lost by the infalling gas as it is arrested and thermalized in an accretion front at the protostar's surface, i.e.,

$$L_*(t) = \frac{GM(t)\dot{M}}{R_*}$$

Since this collapse solution specifies the density distribution of the collapsing core, one can calculate the emergent spectrum of the protostar with knowledge of the properties of the dust in the infalling envelope. Adams and Shu (1986) used such self–similar collapse models together with the best available calculation of dust opacities (i.e., Draine and Lee 1984) to predict the emergent spectra of protostars. However, because these particular collapse models did not naturally contain large cavities or disks in their central regions, they were unable to produce enough near and mid–infrared radiation to match the observations of class I sources.

By introducing a slight amount of rotation, Terebey, Shu and Cassen (1984) showed that collapsing isothermal spheres could, as a consequence of conservation of angular momentum, produce infalling density distributions with central cavities as well as flattened central disk–like structures. A schematic diagram of such a collapsing isothermal sphere is shown in Figure 10. In the outer regions the infall velocities are nearly radial and virtually similar to the non–rotating case. However, as material falls closer to the center, conservation of angular momentum produces curved infall trajectories and, in the equatorial plane, a centrifugal barrier whose radius is given by:

$$R_c = \frac{G^3 M^3 \Omega^2}{16 a^8}$$

here Ω is the rotation rate and M the mass of the central star. When R_c is larger than the radius of the central stellar core, a disk is formed as infalling material falls onto the equatorial plane. Moreover, the curved (parabolic) trajectories within R_c produce a region of significantly reduced density compared to the non–rotating solutions. Using the density distributions derived from such models and taking into account the luminosity which would be radiated by the central disks as well as the protostellar core, Adams and Shu (1986) and Adams, Lada and Shu (1987) were able to successfully model the observed spectra of numerous class I sources (see Figure 12 for an example). Although it is probably is not unique, this model provides a satisfying and self–consistent physical explanation of the simultaneous presence of a disk and a cavity of reduced dust density around the central source.

This theory has two additional virtues. First, the emergent spectrum or energy distribution of a model protostar is essentially completely specified by two parameters (a and Ω) which in principle can be *independently* determined from observation of molecular gas. Second, the model can successfully account for a wide variety of observed class I energy distributions. For example, the depth of the silicate absorption feature at 10 μm in class I SEDs is predicted to be inversely proportional to Ω and the breadth of the source energy distribution in frequency space. This is because the larger Ω and therefore R_c, the larger the inner cavity and the lower the observed extinction to the central protostar. Fits to the energy distributions of a variety of class I sources in Taurus and Ophiuchus appear to confirm this prediction (Adams, Lada and Shu 1987). The robust nature of this theory in its ability to fit the energy distributions of a variety of class I sources, appears to lend strong credence to the contention that class I objects are true protostars.

If class I sources are truly low mass protostars, then we expect them to derive a significant portion of their luminosity from accretion, in general they should be more luminous than similar mass class II or III objects. Evidence for the presence of such "excess" luminosity is suggested in the study of the YSOs in the embedded cluster in the ρ Ophiuchi dark cloud. Analysis of the luminosity function of that cluster indicated that class I sources were systematically more luminous than class II sources (Wilking, Lada and Young 1989) consistent with predictions of protostellar theory. Taken together the evidence of modern observation is strongly suggestive of a protostellar nature for class I objects of low mass.

4.4.3 *Extreme Class I Sources*

In concluding the discussion of the nature of class I sources and embedded YSOs I would like to briefly note an important new development in YSO research. Since 1986 a number of embedded sources have been discovered which display extremely

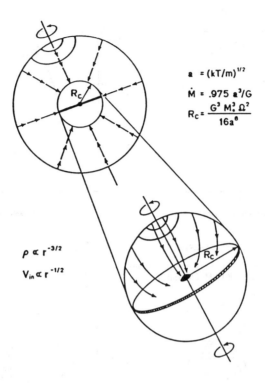

Figure 10. Schematic diagram of a rotating and collapsing protostar based on the models of Terebey, Shu and Cassen (1984) and Adams and Shu (1986).

steep infrared energy distributions. These energy distributions are so steep that they merit special treatment as either examples of a subclass of class I objects or a whole new YSO class altogether. However, for the time being I will refer to them as Extreme Class I sources. There are presently 5 examples of this class: IRAS 16293-2422, B335, VLA 1623, L1448U, and NGC 2264G. Nearly all these sources were discovered because they were the apparent driving sources of spectacular, well collimated outflows. They range in luminosity between about 3–30 $L_\odot$. None of the sources are detectable in the IRAS 12 μm band, and three, VLA 1623 (Andre et al. 1990, L1448U (Bachiller et al. 1990), and NGC 2264G (Margulis et al. 1990) are not detected in any of the four IRAS bands. All but B335 have been detected at the VLA at centimeter wavelengths and besides these VLA detections two sources, VLA 1623 and L1448U have been detected only at millimeter wavelengths. Indeed the only sources of this group that have been observed at enough frequencies to construct any kind of a SED are B 335 (e.g., Chandler et al. 1991) and IRAS

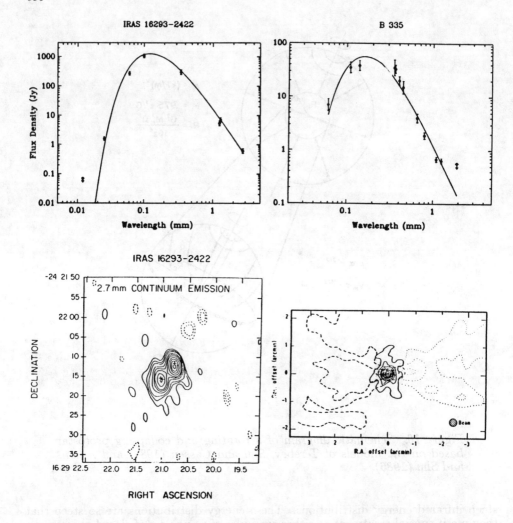

Figure 11. Top: The spectra of the Extreme Class I sources IRAS 162930-2422 (left) located in the Ophiuchi dark cloud complex and B335 (right). The spectra have been fit with single temperature (T=35 K and T=26 K, respectively) modified (i.e., $\kappa_\nu \sim \nu^{1.5}$) blackbody curves. Figures and fits courtesy of Bruce Wilking. Bottom: Interferometric map of IRAS 1629 (left) made at a wavelength of λ 3mm (from Mundy et al. 1991) and an 800 μm continuum map of B335 (right) with the outline of its associated bipolar flow (from Chandler et al. 1991).

1629 (e.g., Walker *et al.* 1986). Figure 11 shows the spectra (i.e., logF$_\lambda$ *vs.log*λ) for these two sources. Note the extremely steep slope in the mid to far–infrared portion of each spectrum. Along with the observed points, a modified blackbody function has been plotted as a solid line in each spectrum. Unlike a typical class I source, the shapes of the spectra of these objects are not marginally different than that of a single blackbody. As a result a temperature can be uniquely assigned characterizing the bulk of their emission. IRAS 1629 is fit somewhat better by a single temperature blackbody and it has a luminosity of approximately 30 L$_\odot$. To date it is the only class I source or protostar that can be placed reliably on the HR diagram! The emission at a wavelength of λ3mm toward IRAS 1629 has been resolved by interferometric measurements (see Figure 11) and the source has been found to have a size on the order of about 10 "corresponding to about 1400 AU at the distance of the Ophiuchi dark cloud in which it is located. It appears to be double (Wootten 1989). The observed size is similar to that derived from the modified blackbody fit to the spectrum. The steep nature of the slopes of the SEDs of these and other extreme class I objects and their association with powerful bipolar flows suggests that they may be the youngest known examples of class I sources. On the other hand, it is also possible that they are typical class I sources observed edge on so that their circumstellar disks and rotationally flattened infalling envelopes completely obscure them at all but the longest wavelengths observed. Their association with bipolar flows, whose high degree of collimation suggests that they are only slightly inclined out of the plane of the sky, may also suggest such a model. At the present time observations cannot distinguish between these two possibilities or some even more interesting explanation of the nature of these objects.

5.0 Molecular Outflows and the Spectral Evolution of YSOs

The variation in the shapes of YSO energy distributions from class I to III represents a variation in the amount and distribution of luminous circumstellar dust around an embedded YSO. It is natural therefore to hypothesize that the empirical sequence of spectral shapes is a sequence of the gradual dissipation of gas and dust envelopes around newly formed stars and is therefore an evolutionary sequence (see Lada 1987). Recently, Adams, Lada and Shu (1987) have been able to theoretically model this empirical sequence as a more or less continuous sequence of early stellar evolution from protostar to young main sequence star using the self-consistent physical theory of rotating collapsing isothermal spheres as the initial condition (see §4.4.3 and chapter by Shu). In this theoretical picture class I sources are the youngest and least evolved objects, indeed, true protostars, objects undergoing accretion and assembling the bulk of the mass they will ultimately contain when they arrive on the main sequence. They consist of central embryonic cores steadily gaining mass from accretion and infall of surrounding matter. As a result of rotation they are surrounded by luminous disks, which themselves are contained in cavities 10–100AU in extent. The cavities are surrounded by large shells of infalling matter which absorb and reprocess most of the radiation from the central star and disk. To evolve from a class I to a class II object requires the removal or dissipation of the infalling envelope.

It is expected that as the infalling envelope is depleted and removed the class I

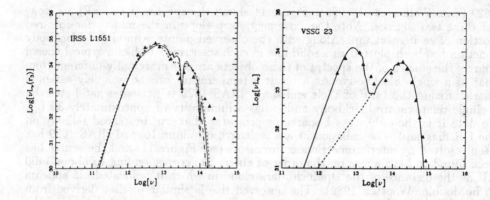

Figure 12. Theoretical model fits to the SEDs of two well known YSOs: on the left L1551 IRS5, a suspected protostar, and on the right VSSG-23, a Class II–D source (from Adams, Lada and Shu 1987).

energy distribution will become significantly modified. As the inner regions become excavated by the wind and the outer regions depleted by the infall, residual infalling gas and dust produce less and less extinction and contribute emission to the source energy distribution only at the longest infrared wavelengths. A double–humped or class II–D energy distribution should result, with a peak at short wavelengths (high frequencies) due to the less extincted central star and disk and a second peak at long wavelengths due to emission from the remnant infalling envelope.

Figure 12 displays the results of theoretical models calculated by Adams, Lada and Shu (1987; 1988) to fit observations of YSOs in two primary evolutionary states: class I and class II–D. The left panel shows the model fit to the well known class I source L1551 IRS 5. Here the solid curve represents a model which consists of a 1 $M_\odot$ accreting stellar core surrounded by a rotating luminous disk and infalling envelope. The mass infall rate needed to produce this fit was $\approx 10^{-5}$ $M_\odot$ per year. This source is therefore about one hundred thousand years old. The right panel shows the transition object, VSSG 23, which has been self-consistently fit by a reddened stellar photosphere surrounded by a luminous, spatially thin but optically thick circumstellar disk and an outer infalling shell which emits optically thin infrared radiation. The infalling shell has an inner radius of 80 AU and accounts for all the observed extinction to the star. However, its density has suffered a depletion of a factor of 100 compared to that of the initial cloud core. The luminous disk is passive in the sense that it derives its luminosity from purely reprocessing radiation absorbed from the central star. As the source evolves we expect the hump at low frequencies to gradually disappear as more of the infalling envelope is depleted or removed resulting in the transition to a pure class II spectrum. As discussed earlier such spectral energy distributions (class II) are well fit by disk models (Figure 6) without the need for any other circumstellar component of dust emission.

Class III sources represent the most evolved sources in the evolutionary sequence of YSOs. For these objects the vast majority of the original star forming material has been either all incorporated into the star or removed from its immediate vicinity. Therefore, to evolve from a class II to III source requires the further depletion and removal of the circumstellar disk. Recent observations and recognition of sources with a class III–D signature (e.g., Montmerle and Andre 1989; Strom et al. 1989) suggest that the removal of the circumstellar disks occurs from the inside out over a period of time of only a few million years (Strom et al. 1989).

As is evident from consideration of the spectral modelling described in this chapter, the simple theory outlined above (and in more detail in the chapter by Shu) appears to be able to account for the observed variation in the spectral shapes of YSO SEDs extremely well. It is likely that the vast majority of low mass stars in our galaxy pass through these stages (classes I–III) during their formation. However, it is not known if the duration of each of these stages is the same for all stars, including those of the same initial mass. It is possible that differing initial conditions could result in different rates of evolution through the various phases. On the other hand, observations of large samples of members of embedded populations in individual clouds and particularly clusters can be used to derive approximate estimates of the timescales or durations of the various stages. Such investigations (e.g., Beichman et al. 1986; Myers et al. . 1987; Wilking, Lada and Young 1989; Strom et al. 1989; Parker 1991) suggest that for low mass stars the class I phase has a duration of approximately $1-4 \times 10^5$ years, the class II–D phase a duration of roughly 10^5 years and the class II phase a duration of between $1-3 \times 10^6$ years.

The rate of evolution from the class I to III stages for a given star must depend on the physics of the clearing process of circumstellar material. In principle, the clearing of circumstellar dust could be accomplished by running out of material which can fall into the system. This possibility conflicts with the observation that star formation occurs with considerable inefficiency: the cores from which stars form contain much more mass than the stars themselves, and most of the surrounding material cannot end up inside the star. This indicates that at some point in early stellar evolution the cloudy material surrounding a YSO must be physically removed (blown to large distances) by some active agent which then drives the evolution of a YSO from the class I stage to the class II stage.

It is becoming increasingly apparent that the most likely agent for removing the large quantities of circumstellar dust from around a YSO is an energetic bipolar molecular outflow (e,g., Lada and Shu 1990 and Shu this volume). The basic physical and observational properties of bipolar flows are described in some detail in the chapter by Bally and Lane (this volume). In the present context, however, it is important to note two particular properties of bipolar outflows. First, bipolar molecular outflows are individually energetic enough to disrupt cloud cores and collectively powerful enough to have a significant impact on the dynamics and structure of an entire GMC. In fact, the molecular outflows generated by a population of embedded YSOs may be able to generate the turbulent pressure that keeps GMCs from global collapse, thereby solving one of the outstanding problems of cloud dynamics. Second, the large numbers of outflows discovered to date (e.g., Lada 1985; Fukui 1989) coupled with their short lifetimes indicates that nearly all stars must pass through an outflow phase sometime during their early evolution (Lada 1985; Terebey, Vogel and Myers 1989). We consider it interesting in this regard to determine by direct observation the nature of the embedded sources which drive cold molecular outflows. A growing body of observational data now clearly shows that molecular outflows are most frequently associated with class I type sources and only rarely with class II or III objects (e.g., Lada 1885, 1988b; Berrill et al. . 1989; Snell et al. 1988; Margulis, Lada and Young 1989). In fact survey observations of both embedded source populations within individual clouds (Margulis, Lada and Young 1989) and among all molecular clouds (Berill et al. 1989; Snell et al. 1989) indicate that at least half of all studied class I infrared objects drive detectable molecular outflows. In addition, recent studies with more sensitivity and angular resolution than those typical of the surveys have found outflows that would have otherwise escaped detection around additional embedded class I sources (Terebey, Vogel and Myers 1989). On the other hand, less than 10% of class II and III objects have associated molecular outflows, although many of these may still actively possess some type of stellar wind (e.g., Lada 1988b).

The high frequency of association between class I infrared sources and molecular outflows poses an apparent paradox. The statistics suggest that a class I object spends a significant fraction of the lifetime in the outflow phase. Yet, if class I sources represent true protostars, their evolution should have been characterized by the infall of surrounding material. Indeed, the observed luminosity excess of class I sources in Ophiuchus (Wilking, Lada and Young 1989) seems to suggest the presence of infall generated luminosity. How can an object simultaneously undergo both inflow and outflow? Alternatively, how can a star form by losing mass? The answer to this question holds the key to understanding the basic physics of star formation.

A possible solution to this paradoxical problem (for more details see Shu, this volume) contains two crucial ingredients: angular momentum and magnetic fields. The fact that disks are found around most YSOs implicates an important role for angular momentum. The formation of a disk around a young stellar object is the natural consequence of the initial presence of angular momentum (even in small amounts) and its conservation in dynamically evolving cloud cores. For a rotating protostar most of the mass that ends up on the star must be accreted from the surrounding disk. In order for material to flow through the disk and onto the protostar, the material must lose both energy and angular momentum. If the mass of the disk is not much larger than that of the central stellar object, the material in the disk should rotate differentially in Keplerian fashion. Gas falling through such a disk will reach the surface of the star with an orbital velocity and specific angular momentum which is relatively high compared to that in the star (Shu et al. 1989, 1991). If this material is added to the star it will spin up the star. The star will quickly reach breakup at which point material can no longer be added to it. A centrifugal barrier prevents further growth of the central protostellar core. Thus the process of star formation can proceed only if the star can somehow spin down while accretion is taking place.

Angular momentum can be carried away from a star by a stellar wind. Consequently, a protostar may be able to gain mass by generating a stellar wind which carries mass and angular momentum away. In this way the protostar can both gain and lose mass. To allow the net flux of mass to result in a gain for the protostar requires that the mass loss rate of the wind be a fraction of the accretion rate, i.e.,

$$\dot{M}_{wind} = f\dot{M}_{accretion}$$

where the fraction f is determined by the physics of the wind generation mechanism. The ideal protostellar wind carries away little mass but lots of angular momentum.

A number of recent investigations (as described in chapters by Pudritz *et al.* and Shu in this volume) have shown that centrifugally driven hydromagnetic winds are potentially capable of doing the job. Such winds could be driven form circumstellar disks or from the surfaces of the central protostars themselves. Thus, it may be that the natural outcome of the collapse of a rotating, magnetic cloud core is the formation of a protostar with a substantial circumstellar disk and a powerful outflow.

In conclusion, it is now clear that the generation of a powerful molecular outflow is of fundamental significance for any scenario or theory of star formation. The outflow is both necessary for star formation to proceed (by enabling accretion of material through a disk) and for providing a natural mechanism for the ultimate reversal of infall from the surrounding infalling envelope. The bipolar outflow limits the mass available to be accreted onto the protostar by clearing away the surrounding gas and dust and driving the evolution of a class I source to a class II source. Thus, the physics of wind formation and generation may be intimately coupled to the problem of the origin of the initial stellar mass spectrum.

Acknowledgements

I thank Elizabeth Lada and Bruce Wilking for providing data in advance of publication and Fred Adams, Elizabeth Lada and Bruce Wilking for assistance in preparation of the figures. I am grateful to Fred Adams, Leo Blitz, Elizabeth Lada, Frank Shu and Bruce Wilking for many stimulating and enlightening discussions during the last few years concerning the topic of this review. My participation in this ASI was funded in part by NSF grant AST # 8815753.

REFERENCES

Adams, F. C. and Shu, F.H. 1986 *Astrophys. J.*, **308**, 836.
Adams, F. C., Lada, C. J. and Shu, F. H. 1987 *Astrophys. J.*, **321**, 788.
Adams, F. C., Lada, C. J. and Shu, F. H. 1988 *Astrophys. J.*, **326**, 865.
Adams, F. C., Emerson, J. P. and Fuller, G. A. 1990 *Astrophys. J.*, **357**, 606.
Adams, M. T., Strom K. M. and Strom S. E. 1983 *Astrophys. J. Suppl. Ser.*, **53**, 893.
Allen, C. W. 1973 *Astrophysical Quantities*, (Athlone Press: London).
Ambartsumian, V. A. 1947 *Stellar Evolution and Astrophysics*, (Armenian Acad. of Science).
Andre, Ph., Martin-Pintado, J., Despois, D. and Montmerle, T. 1990 *Astron. Astrophys.*, **236**, 180.
Appenzeller I., Jankovics, I. and Ostreicker, R. 1984 *Astron. Astrophys.*, **141**, 108.
Bachiller, R. *et al.* 1990 *Astron. Astrophys.*, **231**, 174.
Beall J. H. 1987 *Astrophys. J.*, **316**, 227.
Beichman C. *et al.* 1986 *Astrophys. J.*, **307**, 337.
Becklin, E. E. and Nuegebauer, G. 1967 *Astrophys. J.*, **147**, 799.
Beckwith, S.V.W., Sargent, A.I., Chini, R.S., and Gusten, R. 1990 *Astron. J.*, **99**, 924
Berrill F. 1989 *Monthly Notices Roy. Astron. Soc.*, **237**, 1.
Blaauw, A. 1964 *Ann. Rev. Astron. Astrophys.*, **2**, 213.

Blitz, L. 1980 in *Giant Molecular Clouds in the Galaxy*, eds. P. M. Solomon and M. G. Edmunds, (Pergamon Press), pg. 1.
Chandler, C. J. et al. 1991 *Monthly Notices Roy. Astron. Soc.*, **243**, 330.
Chini, R. 1981 *Astron. Astrophys.*, **99**, 346.
Chini, R., Elasasser, H., Hefele, H. and Weinberger, R. 1977 *Astron. Astrophys.*, **56**, 323.
Cohen M. and Kuhi L. V. 1979 *Astrophys. J. Suppl. Ser.*, **41**, 473.
Draine, B. T. and Lee H. M. 1984 *Astrophys. J.*, **285**, 89.
Edwards, S. et al. 1987 *Astrophys. J.*, 321, 473.
Elmegreen, B. G. and Clemens, C. 1985 *Astrophys. J.*, **294**, 523.
Friedjung, M. 1985 *Astron. Astrophys.*, **146**, 336.
Fukui Y. 1989 in *Low Mass Star Formation and Pre-Main Sequence Objects*, ed. B. Reipurth, (ESO: Munich), p. 95.
Hartmann, L. and Kenyon S. J. 1987 *Astrophys. J.*, **312**, 243.
Harvey, P. M., Thronson, H. A. and Gatley I. 1980 *Astrophys. J.*, **213**, 115.
Harvey, P. M., Wilking, B. A. and Joy M. 1984 *Astrophys. J.*, **278**, 156
Herbig, G. H. and Terndrup, D. M. 1986 *Astrophys. J.*, **307**, 609.
Kenyon, S. J. and Hartmann, L. W. 1987 *Astrophys. J.*, **323**, 714.
Lada, C. J. 1987 in *I.A.U. Symposium No. 115: Star Forming Regions*, eds. M. Peimbert and J. Jugaku, (Dordrecht: Reidel), p. 1.
Lada, C. J. 1988a in *Formation and Evolution of Low Mass Stars*, eds. A.K. Dupree and M.T.V.T. Largo (Kluwer:Dordrecht), p 93.
Lada, C. J. 1988b in *Galactic and Extragalactic Star Formation*, eds., R. Pudritz and M. Fich, (Kluwer Academic Publishers: Dordrecht), p. 1.
Lada, C. J. and Lada, E. A. 1991 in *The Formation and Evolution of Star Clusters*, A.S.P. Conference Series No. 11, ed. K.A. Janes, (Chelsea: Michigan), in press.
Lada, C. J. and Wilking, B. A. 1984 *Astrophys. J.*, **287**, 610.
Lada, C. J., Margulis, M. and Dearborn, D. 1984 *Astrophys. J.*, **285**, 141.
Lada, C. J. and Shu, F. H. 1990 *Science*, **1111**, 1222.
Lada, C. J., Depoy, D. L., Merrill, M. and Gatley, I. 1991 *Astrophys. J.*, in press.
Lada, E. A. 1990 Ph. D. Thesis, University of Texas at Austin.
Lada, E. A., Bally, J. and Stark, A. A. 1991 *Astrophys. J.*, **368**, 432.
Lada, E. A., Depoy, D. L., Evans, N. J. and Gatley, I. 1991 *Astrophys. J.*, **371**, 171.
Lynden-Bell, D. and Pringle, J. 1974 *Monthly Notices Roy. Astron. Soc.*, **168**, 603.
Margulis, M. et al. 1990 *Astrophys. J.*, **352**, 615.
Margulis, M., Lada, C. J. and Young, E. T. 1989 *Astrophys. J.*, **345**, 906.
McCaughrean, M., Zinnecker, H., Aspin, C. and McLean I. 1991 Proc. Workshop on *Astophysics with Infared Arrays* ed. R. Elston, in press.
Mihalas, D. and Binney, J. 1981 *Galactic Astronomy*, (Freeman: San Francisco), p. 217.
Montmerle, T. and Andre Ph. 1989 in *Low Mass Star Formation and Pre-Main Sequence Objects*, ed. B. Reipurth, (ESO: Munich), p. 407.
Mundy, L. G., Wootten, A., Wilking B. A., Blake. G. A. and Sargent, A. I. 1991 *Astrophys. J.*, in press.
Myers, P.C. et al. 1987 *Astrophys. J.*, **319**, 340.
Parker, N. D. 1991 *Monthly Notices Roy. Astron. Soc.*, in press.
Roberts, M. S. 1957 *Publ. Astron. Soc. Pacific*, **69**, 59.
Rucinski, S. M., 1985 *Astron. J.*, **90**, 2321.

Rydgren, A. E., and Zak, D. 1987 *Publ. Astron. Soc. Pacific*, **99**, 141.
Salpeter, E. E. 1955 *Astrophys. J.*, **121**, 161.
Scalo, J. 1986 in *Fundamentals of Cosmic Physics*, **11**, 1.
Schwarzchild, M. 1958 *Structure and Evolution of Stars*, (Princeton University Press: Princeton).
Shu, F. H. 1977 *Astrophys. J.*, **214**, 488.
Shu, F. H., Adams, F. C. and Lizano, S. 1987 *Ann. Rev. Astron. Astrophys.*, **25**, 23.
Shu, F. H., Lizano, S., Ruden, S. P. and Najita J. 1988 *Astrophys. J. Letters*, **328**, 19.
Shu, F. H., Ruden, S. P., Lada, C. J. and Lizano, S. 1991 *Astrophys. J. Letters*, **370**, 31.
Snell R.N., Huang, Y. L., Dickman, R. L. and Claussen, M. 1988 *Astrophys. J.*, **325**, 853.
Spitzer, L. 1958 *Astrophys. J.*, **127**, 17.
Strom, K. M., Margulis, M., and Strom, S. E. 1989 *Astrophys. J. Letters*, **346** L33.
Strom, K. M., Strom, S. E., Edwards, S., Cabrit, S. and Skrutskie, M. 1989 *Astron. J.*, **97**, 1451.
Terebey, S., Shu, F. H. and Casssen, P. 1984 *Astrophys. J.*, **286**, 529.
Terebey, S., Vogel, S. N. and Myers P. C. 1989 *Astrophys. J.*, **340**, 472
van Altena, W.F. 1969 *Astron. J.*, **74**, 2.
van den Bergh S. and Lafontaine A., 1984 *Astron. J.*, **89**, 1822.
van Leeuwen, F. 1980 in *IAU Symp. No. 85, Star Clusters*, ed. J.E. Hesser, (Reidel: Boston), p. 157.
Vrba, F. J., Strom, K. M., Strom, S. E. and Grasdalen, G. L. 1975 *Astrophys. J.*, **197**, 77.
Walker, C. K., Lada, C. J., Maloney, P. Young, E.T. and Wilking B. A. 1986 *Astrophys. J. Letters*, **309**, L47.
Walker, M. F. 1956 *Astrophys. J. Suppl. Ser.*, **2**, 365.
Walter, F. M. 1987 *Publ. Astron. Soc. Pacific*, **99**, 31.
Wilner, D. and Lada C. J. 1991 *Astron. J.*, in press.
Wilking, B. A. and Lada, C. J. 1983 *Astrophys. J.*, **274**, 698.
Wilking, B. A., Lada, C. J. and Young, E. T., 1989 *Astrophys. J.*, **340**, 823.
Wootten, A. 1989 *Astrophys. J.*, **337**, 858.
Wynn-Williams, C. G. 1982 *Ann. Rev. Astron. Astrophys.*, **20**, 587.

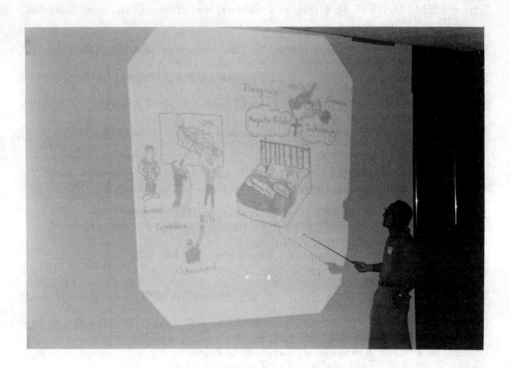

Frank Shu

# THE FORMATION OF LOW MASS STARS: THEORY

Frank H. Shu
Astronomy Department
University of California
Berkeley, CA 94720
USA

ABSTRACT. We consider the interrelationships among the structure of molecular clouds; the collapse of rotating cloud cores; the formation of stars and disks; the origin of molecular outflows, protostellar winds, and highly collimated jets; the birth of single and binary stars; and the dynamics of star/disk/satellite interactions. Our discussion interweaves theory with the results of observations that span from millimeter wavelengths to X-rays.

1. Overview

The chapters by Blaauw and by Elmegreen in this volume make clear that the history of the births and deaths of high mass stars probably interlock in a complicated way with the question of the origin and evolution of giant molecular clouds. High mass stars, when they form, tend to be born in rich groups and associations, and their large ultraviolet fluxes tend to make them much more destructive of their natal environments than their low mass counterparts (see the chapter by Genzel). The chapters by Churchwell and by Kylafis suggest, however, that many compact H II regions and maser regions share unifying characteristics that may lend greater systematics to theories of such objects than previously thought possible.

It remains unclear whether the bulk of the formation of low mass stars in the Galaxy takes place in the same environment as high mass stars. We do know that the *nearest* well-studied regions of low-mass star formation yield the birth of *individual* objects (either single stars or binaries). These regions appear to be much simpler in structure and kinematics than the large dense cores that give rise to stellar clusters and OB associations (see the chapters by Blitz and by Cernicharo). This relative simplicity has made possible a deeper development of the theory, and the mode of star formation that produces individual systems has correspondingly occupied most of the theoretical effort in this field since the pioneering studies by Larson (1969a, see also §5 below). It constitutes the subject that we undertake to review in the present chapter. Parts of this review also appear in Shu et al. (1991).

We start in §2 with a possible theoretical explanation for the existence of two different modes of starbirth: a "closely-packed" mode characterized by the more or less simultaneous formation of a tight group of many stars from large dense clumps of molecular gas and dust, and a "loosely-aggregated" mode in which an unbound association of individual systems (some of which may be binaries) form sporadically from well-separated, small, dense, cloud cores embedded within a more rarefied common envelope. We introduce the working hypothesis, adopted currently by many workers in the field, that the formation of sunlike stars by the second mode

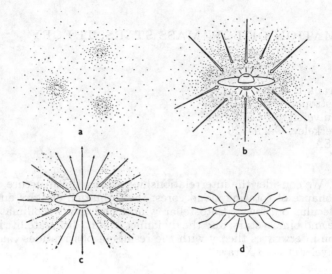

Figure 1. The four stages of star formation. (a) Cores form within molecular cloud envelopes as magnetic and turbulent support is lost through ambipolar diffusion. (b) Protostar with a surrounding nebular disk forms at the center of a cloud core collapsing from inside-out. (c) A stellar wind breaks out along the rotational axis of the system, creating a bipolar flow. (d) The infall terminates, revealing a newly formed star with a circumstellar disk. (From Shu, Adams, and Lizano 1987.)

occurs in nearby dark clouds like the Taurus region in four conceptually distinct stages (Fig. 1).

Of the four stages a-d outlined in Figure 1, the most surprising – the one totally unanticipated by prior theoretical developments – is c, the *bipolar outflow phase*. In §3, we review the observational discoveries and interpretations that have led to our current empirical understanding of this fascinating phenomenon. We pay particular attention to those aspects that set tight constraints on possible theories. In §§4-8, we give a theoretical discussion that indicates why, in retrospect, we should perhaps have known all along that the formation of stars would necessarily involve the heavy *loss* of mass. Given the angular momentum difficulty likely to be faced by any object which forms by contraction through many orders of magnitude from an initially extended, rotating state, the birth of stars (and, perhaps, galactic nuclei) through the accretion of matter from the surroundings may simply not be possible without the *simultaneous* accompaniment of powerful outflows.

Under the working hypothesis, gas and dust in the first stage (Fig. 1a) slowly contract under their own gravitation against the frictional support provided by a background of ions and magnetic fields via the process of ambipolar diffusion (§4; see also Mouschovias, this volume). The principal feature in this stage involves a *quasistatic evolution toward a $1/r^2$ density configuration* appropriate for a singular isothermal sphere. The relative statistics of cores with and without embedded infrared sources, as well as the theory of ambipolar diffusion, suggests that the timescale over which the configuration is observable as a quiescent ammonia core before it enters dynamical collapse (i.e., before it contains an embedded infrared source) should span about 10^6 yr.

When the contracting configuration becomes sufficiently centrally concentrated, it enters stage b, wherein the cloud core gravitationally collapses from *inside-out*. In such a situation, the inner regions form an accreting but otherwise secularly evolving protostar plus nebular disk (see also the chapter by Tscharnuter). An infalling envelope of gas and dust that rains down from the overlying (slowly rotating) molecular cloud core covers the growing star plus disk. The visual extinction to the central star measures from several tens to a thousand or more, so the embedded source during this stage does not appear as an optically visible object, but must be studied principally by means of the infrared, submillimeter, and millimeter radiation produced by dust reprocessing in the surrounding envelope (see the chapter by Lada).

At some point during this phase of the evolution (see §8), a powerful wind breaks out along the rotational poles of the system reversing the infall and sweeping up the material over the poles into two outwardly expanding shells of gas and dust (see the chapters by Bally and by Reipurth). The main source of momentum input in low mass protostars appears not to be an *ionized* wind, but rather a *neutral* wind (see the chapters by Panagia and by Natta). This stage (stage c) corresponds to the *bipolar outflow* phase observed spectroscopically at radio wavelengths by CO observers. Theory suggests that this stage features *combined* inflow (in the equatorial regions) and outflow (over the poles). Current consensus in the field holds that magnetohydrodynamic forces drive the wind sweeping up the molecular outflow. The main debate concerns whether the wind originates from the *star*, or from the *disk*, or from their *interface* (see also Pudritz, this volume). From the mass infall rate as well as the statistics of numbers of embedded sources compared to revealed ones (T Tauri stars), we can estimate the combined time spent in stages b and c (when the system still gains mass in net) as roughly 10^5 yr, almost independent of mass.

As time proceeds, we envisage the angle occupied by the outflow to open up (like an umbrella) from the rotation axes and to spread, halting even the rain of infalling matter over the equator. At this point, stage d, the system becomes visible, even at ultraviolet, optical, and near-infrared wavelengths as a star plus disk to all outside observers (see the chapter by Bertout). Among other signatures, the disk makes its presence known via scattered starlight, which introduces measurable polarization in the observed radiation (see the chapter by Bastien). The location of the star in the Hertzsprung-Russell diagram yields a constraint on the accretion timescale, and the numbers estimated by this method agree well with the dynamical calculations based on the concept of inside-out collapse from a molecular cloud core modeled as a (rotating) singular isothermal sphere. The principal questions during this stage then concern the mechanisms by which mass, angular momentum, and energy transport take place within the disk, and the nature of any companions (stellar or planetary) that may condense from such a disk (see also the chapter by Pringle). Intimately tied to these issues are the estimates of disk masses that can be obtained from observational measurements at submillimeter and millimeter wavelengths. Another important question concerns whether disk accretion mainly occurs episodically during FU Orionis outbursts or during the relatively quiescent (normal) states as well (see the chapter by Hartmann).

The observational evidence indicates that the presence of an inner disk constitutes the crucial distinction between the classical T Tauri systems that drive extraordinary winds and the weak or "naked" T Tauri stars that do not. The absence of an absorbing wind (and inner disk) makes the magnetic activity on the surfaces of the latter objects observable as X-ray sources (see Montmerle, this volume).

The above summary gives a sample of the rich diversity of physical processes that confronts the research worker in the field of star formation. This variety arises for a simple and basic reason: the problem spans physical conditions ranging from the depths of interstellar space to the interiors of stars, involving all the known states of matter and forces of nature, with observational diagnostics available across practically the entire electromagnetic spectrum. To attack its problems, the field has developed a broad range of technical tools – observational and theoretical – and the full exploitation of this range of tools during the past decade has led to rapid and impressive progress.

2. Bimodal Star Formation

The empirical notion that the birth of low- and high-mass stars may involve separate mechanisms has a long and controversial history (Herbig 1962, Mezger and Smith 1977, Elmegreen and Lada 1977, Gusten and Mezger 1982, Larson 1986, Scalo 1986, Walter and Boyd 1991). The name "bimodal star formation" usually attaches to this concept, but, more recently, the emphasis has changed from "low-mass versus high-mass" to "loosely-aggregated versus closely-packed." The latter phrasing roughly coincides with the older one of "associations versus clusters," except that it need not carry the connotation of "gravitationally unbound versus gravitationally bound."

The distinction between "loosely-aggregated" and "closely-packed" refers theoretically to whether gravitational collapse occurs independently for individual small cores to form single stars (or binaries); or whether it involves a large piece of a giant molecular cloud to produce a tight group of stars created more-or-less simultaneously. This tight group need not form a bound cluster if the winds or other violent events that accompany the formation of stars expel a large fraction of the gas not directly incorporated into stars (Lada et al. 1984, Elmegreen and Clemens 1985).

The occurrence of two separate modes of star formation has a natural theoretical explanation (Mestel 1985; Shu, Lizano, and Adams 1987) if we adopt the point of view that magnetic fields provide the primary agent of support of molecular clouds against their self-gravity (Mestel 1965, Mouschovias 1976, Nakano 1979). The inclusion of a conserved magnetic flux Φ threaded by an electrically conducting cloud introduces a natural mass scale (cf. Mouschovias and Spitzer 1976; Tomisaka et al. 1988, 1989),

$$M_\Phi \equiv 0.13\, G^{-1/2} \Phi, \tag{1}$$

that is analogous to Chandraskhar's limit M_{Ch} in the theory of white dwarfs. A critical mass arises whenever we try to balance Newtonian self-gravity by the internal pressure of a fluid which varies as the 4/3 power of the density (because of ultrarelativistic electron degeneracy pressure in the case of a white dwarf, because of the pressure of a frozen-in magnetic field in the case of a molecular cloud). White dwarfs with masses $M > M_{Ch}$ cannot be held up by electron degeneracy pressure alone, but must suffer overall collapse to neutron stars or black holes. Molecular clouds with masses $M > M_\Phi$ (supercritical case) cannot be held up by magnetic fields alone (even if perfectly frozen into the matter), but, in the absence of other substantial means of support, must collapse as a whole to form a closely-packed group of stars. The only trick in this case concerns how to get a supercritical cloud or clump from an initially subcritical assemblage (or they would have all collapsed by now). Theory and observation both suggest a natural evolutionary course: the agglomeration, with an increase of the mass-to-flux ratio, of the discrete cloud clumps that comprise giant molecular complexes (Blitz and Shu 1980; Blitz 1987, 1990; Shu 1987).

The analogy between white dwarfs and molecular clumps breaks down for the subcritical case. White dwarfs with $M < M_{Ch}$ can last forever as uncollapsed degenerate objects (if nucleons and electrons are stable forms of matter) because quantum principles never weaken. Magnetic clouds with $M < M_\Phi$ initially cannot last forever because magnetic fields do weaken, and the local loss of magnetic flux from unrelated dense regions allows loosely-aggregated star formation. In particular, in a lightly ionized gas, the fields can leak out of the neutral fraction by the process of *ambipolar diffusion* (Mestel and Spitzer 1956, Nakano 1979, Mouschovias 1978, Shu 1983). In §4, we shall examine in more detail the production of individual molecular cloud cores by this process. For the present, we merely note that theory predicts that self-gravitating cores sustained by magnetic fields will inevitably evolve to a state of spontaneous gravitational collapse. This prediction seemingly flies in the face of observations, which almost always find young stellar objects associated with outflows rather than inflows.

3. The Bipolar Outflow Phase: Observations

At the heart of the central paradox concerning star formation lies the problem of *bipolar outflows* (for reviews, see Lada 1985, Welch et al. 1985, Bally 1987, Snell 1987, Fukui 1989, Rodriguez 1990). Astronomical visionaries (see, e.g., the reminiscences of Ambartsumian 1980) have long worried that forming astronomical systems frequently exhibit *expansion*, rather than the *contraction* that would be naively predicted by gravitational theories. Astronomical conservatives have long persisted in ignoring such warnings, doggedly pursuing the theoretical holy grail that bound objects, such as stars and planetary systems, should form from more rarefied precursors by a process of gravitational contraction. The reconciliation of this fundamental dichotomy remains the central challenge facing theorists and observers alike (see the review by Lada and Shu 1990).

In 1979, Cudworth and Herbig discovered that two Herbig-Haro objects in a nearby dark cloud L1551 (Lynds 1962) exhibit very high proper motions (corresponding to ~ 150 km s^{-1}) that trace back to an apparent point of origin near to the location of an embedded infrared source IRS 5 found by Strom, Strom, and Vrba (1976). Knapp et al. (1976) had earlier found that CO millimeter-wave spectra in this region possess linewidths of the order 10 km s^{-1}, too large according to Strom et al. to correspond to gravitational collapse. The latter authors suggested instead that the disturbance in the ambient molecular cloud material might be produced by a powerful outflow from IRS 5. Snell, Loren, and Plambeck (1980) mapped the CO emission in the L1551 region and verified this conjecture in a dramatic fashion. They found that the high-velocity CO surrounds the tracks of the fast H-H objects,

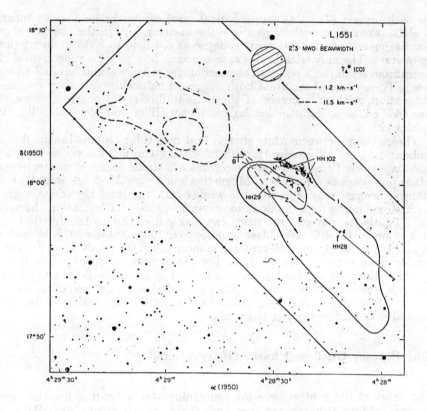

Figure 2. Contour map of the antenna temperature of the $J = 1-0$ transition of ^{12}CO at high velocities, superposed on an optical photograph of the L1551 dark cloud. The cross indicates the position of IRS 5; also shown are the directions of the proper motions of two Herbig-Haro objects, HH28 and HH 29. (From Snell, Loren, and Plambeck 1980.)

taking the form of two lobes of gas moving in diametrically opposed directions from IRS 5 (see Fig. 2). Putting together all of the empirical clues, Snell et al. made the prescient proposal (Fig. 3) that a stellar wind must blow at 100 - 200 km s^{-1} from IRS 5, in directions parallel and anti-parallel to the rotation axis of a surrounding accretion disk, and that this collimated wind sweeps up ambient molecular cloud material into two thin shells, which manifest themselves as the observed bipolar lobes of CO emission. The thin-shell nature of the CO lobes of L1551 has since received empirical validation in the investigations of Snell and Schloerb (1985) and Moriarty-Schieven and Snell (1988). Except for the further interpretation that IRS 5 represents a *protostar*, which, apart from suffering mass loss in the polar directions, accretes matter from *infall* occurring in the equatorial regions (see panel c in Fig. 1), the proposal of Snell, Loren, and Plambeck corresponds in every detail to the theoretical model that we shall pursue in §8.

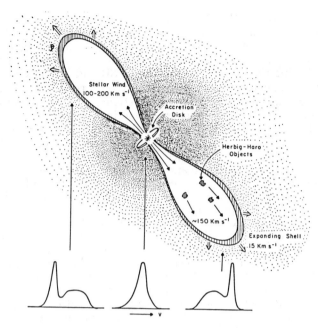

Figure 3. Schematic model for the bipolar flow in L1551 driven by a stellar wind emanating from IRS 5. At the bottom is depicted expected CO line profiles for different line of sights across the source. (From Snell, Loren, and Plambeck 1980.)

Shortly after the original suggestion, however, events moved to a different interim conclusion. In common with the central source of several other bipolar outflows (see the reviews of Cohen 1984 and Schwartz 1983), IRS 5 lies along a chain of Herbig-Haro objects (Strom, Grasdalen, and Strom 1974; Mundt and Fried 1983) that may just constitute the brighter spots of a more-or-less continuous optical jet (Mundt 1985, Reipurth 1989). The emission knots probably arise as a result of the interactions of a highly-collimated ionized stellar wind with the surrounding medium (possibly another [neutral] wind; see Stocke et al. [1988] and Shu, Lizano, Ruden, and Najita [1988, hereafter SLRN]). The free-free thermal emission from the ionized stellar wind resolves itself in VLA radio-continuum measurements as a two-sided jet centered on IRS 5 (Bieging, Cohen, and Schwartz 1984). It was natural to suppose that this ionized wind represents the driver for the bipolar molecular outflow. Unfortunately, the momentum input provided by the ionized wind, integrated over the likely lifetime of the system, falls below the value required to explain the moving CO lobes by one to two orders of magnitude (for the likely case of momentum-driven rather than energy-driven flows), a conclusion that holds as well for all other well-studied bipolar flow sources (Bally and Lada 1983, Levreault 1985). By comparing the required momentum input to the photon luminosity divided by the speed of light, the same authors argued persuasively against the possible importance of radiation pressure in the outflow dynamics.

The claim by Kaifu et al. (1984) of the detection of an extended, rotating, molecular disk encircling IRS 5, contributed to the puzzle. This result, and the inability of ionized stellar winds to drive the observed molecular flows, motivated Pudritz and Norman (1983, 1986) and Uchida and Shibata (1985) to suggest that bipolar outflows originate, not as shells of molecular cloud gas swept up by a stellar wind, but, following the work of Blandford (1976) and Lovelace (1976), as gaseous material driven magneto-centrifugally directly off large ($\sim 10^{17}$ cm) and massive ($\sim 10^2$ $M_\odot$) circumstellar disks. This suggestion seemingly received support from reports that one limb of the blue-shifted lobe in L1551 exhibits rotation (Uchida et al. 1987). However, subsequent observational studies have failed to confirm the presence of large and massive disks around young stellar objects with the properties required by the theoretical models (see, e.g., Batrla and Menten 1985, Menten and Walmsley 1985, Moriarty-Schieven and Snell 1988). Moreover, the claim of rotation in the blue-shifted lobe may have been contaminated by confusion with another bipolar flow source in the same field of view (Moriarty-Schieven and Wannier 1991). Finally, from a theoretical point of view, the original large-disk models have severe energetic problems (Shu, Adams, and Lizano 1987; Pringle 1989).

Recently, a number of different groups have attempted to salvage the disk model by postulating smaller and less massive disks, with the bulk of the wind emerging from radii much closer to the central star. This revised picture, however, suffers from the lack of a plausible injection mechanism. The paper by Blandford and Payne (1982) illuminates the basic difficulty.

Blandford and Payne consider a thin disk threaded by a poloidal magnetic field **B** in which rotation balances the radial component of gravity. If **B** has sufficient strength and a proper orientation, it can fling to infinity electrically conducting gas from the top and bottom surfaces of the disk. A freely sliding bead on a rigid wire anchored at one end to a point ϖ in a disk rotating at the angular velocity $\Omega_{\rm disk}(\varpi)$ provides a useful analogy (Henriksen and Rayburn 1971). In such an analogy, the termination of the wire at the free end corresponds to the Alfven surface, beyond which the magnetic field can no longer enforce corotation, even approximately, and the super-Alfvenic fluid motion becomes essentially ballistic. For the flinging effect to take place in a Keplerian disk (one in which all of the gravitational attraction comes from a central mass point), the poloidal field must enter or leave the disk at an angle larger than 30° from the normal. (The component of the centrifugal force parallel to the field [wire] in the meridional plane has no excess compared to gravity if the field [wire] makes too small an angle with respect to the rotation axis, e.g., if it points vertically through the disk.) However, the magnetic field on the two sides of a disk cannot *both* bend outward by a nonzero angle without generating a large kink across the midplane of the disk, one that would result in a very large $\nabla \times \mathbf{B}$, and which would yield, by Ampere's law, a very large current (infinitely large in a disk of infinitesimal thickness). The Lorentz force per unit volume,

$$\mathbf{f}_{\rm L} = \frac{1}{4\pi}(\nabla \times \mathbf{B}) \times \mathbf{B}, \qquad (2)$$

needed to produce order-unity departures from Keplerian motion above the disk (in order to drive a wind) must then have even bigger values inside the disk. The tendency for the field lines to want to straighten vertically (thereby shutting off the magneto-centrifugal acceleration) can be offset only by a heavy radial inflow through the disk, maintained, for example, by an bipolar diffusion in the presence of sub-Keplerian rotation inside the disk (Konigl 1989). Sub-Keplerian rotation everywhere in the disk necessitates the continuous removal of angular momentum

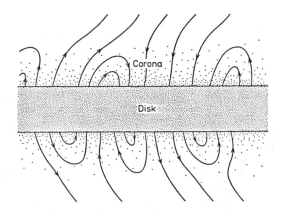

Figure 4. A schematic representation of a possible field geometry close to the disk. (From Blandford and Payne 1982.)

from it (the role of the wind), but the postulated deficit of centrifugal support in the disk then works against lift-off of the gas in the first place. Indeed, no one has yet succeeded in constructing a self-consistent cool model of this type, demonstrating a smooth connection from the region of the sonic transition (near the disk surface) to the region of the Alfvenic transition (far from the disk surface).

Blandford and Payne themselves adopt another strategy. They accept the conclusion that the symmetry of a disk geometry naturally forces any poloidal magnetic field to thread vertically through the midplane of the disk, from where it *gradually* bows outward (see Fig. 4). To lift the material off the disk to heights where the magnetic field bends over sufficiently for the magneto-centrifugal mechanism to operate, they invoke the thermal pressure of a hot corona above (and below) the disk. In other words, their final model starts off as a *thermally driven wind*. A hot gas with thermal speeds comparable to the virial values characteristic of the depth of the gravitational potential in the inner parts of a protostellar disk, however, suffers tremendous radiative losses at the injection densities needed to supply the observed mass outflows. While astrophysical systems with highly luminous compact objects at their centers, such as active galactic nuclei or binary X-ray sources, might well provide sufficient energy to sustain analogous radiative losses, we cannot expect the same bounty from the central machine of protostellar systems (cf. the discussion of DeCampli [1981] that even the relatively low-powered T Tauri winds cannot be thermally driven).

In summary, at radii greater than $\sim 10^{17}$ cm, disks *can* maintain thermal speeds that compete with the local virial values, but such regions lack enough rotational energy to drive the observed flows. Greater centrifugal power exists at smaller radii, but lift-off from the disk then requires an unidentified nonthermal injection mechanism. Compounding these difficulties is the unlikelihood of strong magnetic fields in thin disks. Recall that the energy density of the field at the Alfven surface must equal the kinetic energy density of a wind flowing close to its terminal velocity. But magnetic buoyancy effects probably limit the generation of fields by dynamo mechanisms acting within disks to energy densities comparable to or less than the prevailing *thermal* values (Stella and Rosner 1984). The inability of disks to retain large *toroidal* fields (Parker 1966, Matsumoto et al. 1988) limits the viability of mechanisms like those of Lovelace et al. (1991) that rely on the pressure gradient of such fields in the z-direction for the initial vertical acceleration of the gas.

Other theoretical arguments suggest that disk fields must be weak. If accretion disks owe their viscosity to the powerful MHD instability recently discussed by Balbus and Hawley (1991) and Hawley and Balbus (1991), the *poloidal* component of the field cannot have energy densities much higher than thermal gas values. Indeed, the instability itself may serve as one of the ingredients of a self-consistent dynamo mechanism responsible for generating the poloidal component of the disk field in the first place. In addition, dynamically strong fields (energy densities comparable to that contained in the *rotation* of the part of the disk supplying the matter for the outflow) cannot have been dragged in and amplified by lightly ionized gas that collapsed from interstellar dimensions by ambipolar diffusion (see §4 below), since mass infall at rates and speeds (free-fall) comparable to the inferred outflow values had to overcome the interstellar field to form the disk in the first place. Finally, even if we could surmount the above difficulties and drive a disk wind, we would *still* be left with an angular-momentum problem for the accreting central star (cf. the discussion in §7). For all of these reasons, we believe that Snell et al. (1980) intrinsically had the right idea when they speculated that *stellar winds* must constitute the basic driver for bipolar outflows in young stellar objects.

The issue then reverts to the original question: How can stellar winds drive the observed molecular outflows when the *ionized* component lacks the requisite power? Many researchers arrived independently at the obvious answer: perhaps young stellar objects possess *neutral* winds that supply the missing momentum input. Detection of fast neutral winds would require the spectroscopic observation of broad and shallow radio emission lines, a difficult task since so little mass resides within the telescope beam at any particular time. Nevertheless, the search for *atomic* hydrogen winds has now succeeded in at least three sources: HH 7-11, L1551, and T Tau (Lizano et al. 1988, Giovanardi et al. 1991, Ruiz et al. 1991). Fast neutral winds are also indicated through the detection of CO moving at velocities of $\pm$ 100 - 200 km s^{-1} in HH 7-11 and several other sources (Koo 1989; Margulis and Snell 1989; Masson, Mundy, and Keene 1990; Bachiller and Cernicharo 1991; Carlstrom, private communication). The mass-loss rate measured in HH 7-11 is especially impressive, $\dot{M}_w \approx 3 \times 10^{-6}\ M_\odot$ yr^{-1}, more than sufficient to drive the known bipolar molecular outflow associated with this source. Thus, the discovery of massive neutral winds from (low-mass) protostars solves one of the principal remaining mysteries concerning bipolar outflows, namely, the ultimate source for their power. However, this empirical discovery pushes to the fore another puzzling question: How and why do stars get born by *losing* mass at such tremendous rates?

4. Rotating, Magnetized, Molecular Cloud Cores

A major breakthrough occurred in the field when Myers and Benson (1983) identified the sites for the birth of individual sunlike stars in the Taurus cloud as small dense cores of dust and molecular gas that appear as especially dark regions in the

Palomar sky survey or as regions of NH_3 and CS emission (cf. the references in Evans [1991]). One theory for the formation of such cores begins with the view that large molecular clouds are supported by a combination of magnetic fields and turbulent motions (see the review of Shu, Adams, and Lizano 1987; see also Myers and Goodman 1988a,b). Because magnetic fields directly affect only the motions of the charged particles, magnetic support of the neutral component against its self-gravity arises only through the friction generated when neutrals slip relative to the ions and field via the process of ambipolar diffusion (Mestel and Spitzer 1956). This slippage occurs at enhanced rates in localized dense pockets where the ionization fraction is especially low. As the field diffuses out relative to the neutral gas, the Alfven velocity $B/(4\pi\rho)^{1/2}$ drops, and the turbulence must also decay if fluctuating fluid motions are to remain sub-Alfvenic. Thus, both processes – field slippage relative to neutrals and the lowering of turbulent line widths in regions of increasing mass-to-flux – reinforce the tendency for quiet dense cores to separate from a more diffuse common envelope.

Assuming axial symmetry (no φ dependence) in spherical polar coordinates (r, θ, φ) as well as reflection symmetry about the midplane $\theta = \pi/2$, Lizano and Shu (1989) performed detailed ambipolar diffusion computations, using the ionization balance calculations of Elmegreen (1979) and including empirically the effects of turbulence with the scaling laws found by radio observers. They find that the time to form NH_3 cores, $\sim 10^6 - 10^7$ yr, and the additional time to proceed to gravitational collapse, a few times 10^5 yr, agree roughly with the spread in ages of T Tauri stars in the Taurus dark cloud (Cohen and Kuhi 1979) and with the statistics of cores with and without embedded infrared sources (Fuller and Myers 1987). Toward the end of its quiescent life, a molecular cloud core tends to acquire the density configuration of a (modified) singular isothermal sphere:

$$\rho(r,\theta) = \frac{a_{\text{eff}}^2}{2\pi G r^2} Q(\theta), \qquad (3)$$

where a_{eff} is the effective isothermal sound speed, including the effects of a quasi-static magnetic field $\mathbf{B}_0$ and (as modeled) an isotropic turbulent pressure. The dimensionless function $Q(\theta)$, greater than 1 toward the magnetic equator and less than 1 toward the poles, yields the flattening that occurs because the magnetic field contributes no support in the direction along $\mathbf{B}_0$. We define $Q(\theta)$ so that it has a value equal to unity when averaged over all solid angles, i.e.,

$$\int_0^{\pi/2} Q(\theta) \sin\theta \, d\theta = 1; \qquad (4)$$

hence, a_{eff}^2 in equation (3) gives the angle-averaged contributions of thermal, turbulent, and magnetic support.

Myers et al. (1991) have used statistical arguments to deduce that many cloud cores have prolate, rather than oblate, shapes. They note that this result poses difficulty for the above view that cloud cores represent quasi-equilibrium structures in which quasi-static (poloidal) magnetic fields play an important part in the support against self-gravitation (see also Bonnell and Bastien 1991). Tomisaka (1991) points out that equilibrium clouds with prolate shapes become possible if they possess toroidal magnetic fields of substantial strength. Even without toroidal fields, oblate shapes represent a natural theoretical expectation only if we model the effects of molecular-cloud turbulence as an *isotropic* pressure. If "turbulent" support arises, instead, from the propagation and dissipation of nonlinear Alfven waves in an inhomogeneous medium (see Arons and Max 1975, as well as the discussion on pp. 36-37 of Shu, Adams, and Lizano 1987), then the largest amount of momentum transfer might well occur *parallel* to the direction of the mean field $\mathbf{B}_0$, rather than perpendicular to it. For example, if ambipolar diffusion causes $\mathbf{B}_0$ to become nearly straight and uniform, as occurs in the detailed calculations before collapse ensues, then the mean field exerts little stress on average, and the anisotropic forces associated with the fluctuating component $\delta \mathbf{B}$ in total $\mathbf{B} = \mathbf{B}_0 + \delta \mathbf{B}$ may actually cause cloud cores to assume prolate instead of oblate shapes. This tendency for condensations to stretch out along the direction of the mean field would only be enhanced by tidal forces if cloud cores tend to form as individual links on a magnetic "sausage" (cf. Elmegreen 1985 and Lizano and Shu 1987). Bastien et al.'s (1991) study of the fragmentation of elongated cylindrical clouds into such individual links differs from that of Lizano and Shu (1989) in that the former authors ignore magnetic effects and they assume that the core-formation process occurs by dynamical collapse rather than by quasi-static contraction.

5. Protostar Formation from Collapsing Cloud Cores

Apart from the delicate question of the sense of elongation of the shape function $Q(\theta)$, the prediction (3) with regard to the *radial* variation of ρ agrees reasonably well with observations of isolated cores and Bok globules (see, e.g., Zhou et al. 1990, who comment particularly that the *outer* velocity fields are more representative of quasi-static contraction than dynamical collapse). The solution as written represents an asymptotic result; actual configurations become gravitationally unstable in their central regions before a singular density cusp develops there. However, because the phase of quasi-static contraction produces very high central concentrations in the numerical models, the resulting dynamical behavior would probably closely resemble the inside-out collapse known analytically for the (exact) singular isothermal sphere without rotation (Shu 1977). Such a collapse solution builds up a central protostar at a rate,

$$\dot{M}_{\text{infall}} = 0.975 \, a_{\text{eff}}^3 / G, \tag{5}$$

that remains constant in time as long as the external reservoir of molecular cloud gas (with a r^{-2} density distribution) continues to last. For L1551, where $a_{\text{eff}} \approx 0.35$ km s^{-1}, $\dot{M}_{\text{infall}} = 1 \times 10^{-5}$ $M_\odot$ yr^{-1}.

A rate of infall of this magnitude helps to resolve a controversy that arose in the 1970s between Hayashi and Larson concerning the starting sizes (and luminosities) of pre-main-sequence stars. Since the epochal work of Eddington (1930), astronomers have known that – independent virtually of the stage of evoution, main-sequence, pre-main-sequence, or post-main-sequence – radiative stars satisfy a mass-luminosity relationship. In other words, as long as stars retain radiative interiors, their luminosity depends virtually only on their mass, and they evolve (at constant mass M_*) horizontally across the Hertzsprung-Russell diagram (at constant luminosity L_*). In particular, the contraction of pre-main-sequence stars

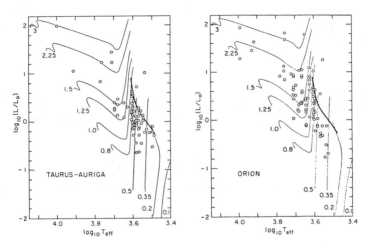

Figure 5. Hertzsprung-Russell diagrams from Cohen and Kuhi (1979) showing theoretical pre-main-sequence tracks and T Tauri stars in the Taurus-Auriga and Orion cloud complexes. The numbers label stellar masses in $M_\odot$, and the heavy solid curve is Stahler's theoretical "birthline." (From Stahler 1983.)

toward the main-sequence has this behavior on the so-called "radiative Henyey tracks" (Henyey, LeLevier, and Levee 1955). (In actual practice, the luminosity of low-mass stars with radiative interiors depends on the inverse square-root of the radius R_* in addition to the 5.5 power of the mass M_*, so that the corresponding Henyey tracks have a slight slope in the H-R diagram; see Fig. 5.)

The important work of Hayashi, Hoshi, and Sugimoto (1962) showed, however, that horizontal evolution across the Hertzsprung-Russell diagram could never carry a quasistatic star below a certain limiting effective temperature T_H, dependent (weakly) on stellar mass M_* and luminosity L_*. Because gas opacities (due mainly to H^- and molecules) of cool stellar photospheres have a limited capacity to trap infrared photons, there exists a "forbidden zone" (dependent on M_* and L_* for a given envelope composition) to which all quasistatic stars must remain to the left. When post-main-sequence stars expanding to the right in the H-R diagram encounter this temperature barrier, their effective temperatures $T_{\rm eff}$ are constrained to equal the limiting value T_H, and the stars ascend the "red-giant branch," with the excess luminosity above the (horizontal) radiative value carried by the appearance of envelope convection. Conversely, when pre-main-sequence stars contract from larger to smaller sizes, they first descend (almost vertically at constant $T_{\rm eff}$) down convective Hayashi tracks until the luminosity falls to a level that can be transported radiatively. The star than crosses the H-R diagram (almost horizontally at constant L_*) along a radiative Henyey track to join the main-sequence, where its radiative losses are offset by hydrogen nuclear reactions.

The question then naturally arises: In the collapse of an interstellar cloud (fragment), where does a star actually make its first appearance as an optically visible (quasistatic) object on the entire possible length of a Hayashi-Henyey pre-main-sequence track? A very conservative estimate proceeds as follows. The existence of a minimum temperature for gas photospheres implies that stars cannot begin their quasistatic life arbitrarily high up on a Hayashi track. A fully convective star, approximated as a polytrope of index 3/2, has a self-gravitational energy content equal to

$$W = -\frac{6}{7}\frac{GM_*^2}{R_*}. \tag{6}$$

In quasistatic contraction, the virial theorem states that half of the gravitational energy released goes into radiation; the other half goes to raising the internal thermal energy content $U = -W/2$. Setting the rate of radiative loss equal to L_*, we get

$$L_* = -\frac{1}{2}\frac{dW}{dt} = -\frac{3}{7}\frac{GM_*^2}{R_*^2}\frac{dR_*}{dt}. \tag{7}$$

With $L_* = 4\pi R_*^2 \sigma T_{\text{eff}}^4$, we obtain the characteristic (Kelvin-Helmholtz) contraction timescale as

$$t_{\text{KH}} \equiv -\frac{R_*}{dR_*/dt} = \frac{3GM_*^2/7R_*}{4\pi R_*^2 \sigma T_{\text{eff}}^4}. \tag{8}$$

For fixed M_* and T_{eff}, t_{KH} becomes very short for large values of R_*. On the other hand, the orbit time at the surface of the star,

$$t_{\text{dyn}} \equiv \frac{R_*}{(GM_*/R_*)^{1/2}}, \tag{9}$$

increases with increasing R_*. For large R_*, therefore, there comes a point when a star with an ordinary gas photosphere loses heat faster than it can contract quasistatically to make up the losses. With $M_* = 1\ M_\odot$ and $T_{\text{eff}} = T_H \approx 4000$ K, t_{KH} equals t_{dyn} when $R_* \approx 500\ R_\odot$. Beyond this size, the evolution from interstellar conditions to stellar ones must involve dynamical collapse. Gaustad (1963) and others first reached a complementary conclusion by considering the problem from the other end, namely, the radiative losses that would be suffered by a contracting cloud. They concluded that such losses, under optically thin conditions, would keep the interstellar gas nearly isothermal, and, therefore, a phase of dynamical collapse would necessarily precede the quasistatic phase of pre-main-sequence evolution.

Following general arguments propounded by Cameron, Hayashi (1966) set an upper limit for the beginning value of R_* that proved an order of magnitude smaller yet than 500 $R_\odot$. Condensed to its essentials, the argument proceeds as follows. After the completion of the dynamic collapse phase, we have a quasistatic object that satisfies the virial theorem,

$$W + 2U = 0. \tag{10}$$

Independent of the collapse process, an energy budget relates the hydrostatic object to the original gas cloud from which it formed. The original cloud started much colder and more extended than the final object, so we may consider the initial thermal and gravitational energies to be essentially zero. The energy budget then reads

$$W + U + I + \Delta = 0, \tag{11}$$

where I equals the total energy required to break the molecules in the original state into atoms and then into ions and electrons, and where Δ equals the total energy radiated into space. If we invoke equation (10) to eliminate U from equation (11), we obtain

$$-W = 2(I + \Delta),$$

which yields

$$R_* = \frac{3}{7}\frac{GM_*^2}{(I+\Delta)} \qquad (12)$$

when we make use of equation (6).

Hayashi obtained a maximum starting size for the pre-main-sequence phase by assuming that no binding energy at all is radiated away, $\Delta = 0$. If all of the gravitational energy released goes into heat and breaking up molecules and atoms, the starting size would equal

$$R_{\rm H} = \frac{3}{7}\frac{GM_*}{\chi},$$

where $\chi \equiv I/M_*$ is the dissociation and ionization energy per unit mass for a gas of cosmic composition. For $M_* = 1\ M_\odot$ and $\chi = 1.6 \times 10^{13}$ erg g^{-1}, we get $R_{\rm H} \approx 50\ R_\odot$, an order of magnitude smaller than the objects's size, 500 $R_\odot$, below which hydrostatic equilibrium can be maintained. Narita, Nakano, and Hayashi (1970) completed the argument by demonstrating that if interstellar clouds were to collapse *homologously* in a dynamical state, reaching free-fall from interstellar dimensions to solar-system dimensions, the subsequent infall to stellar dimensions (which they followed by direct numerical simulation) would indeed radiate away negligible energy Δ, yielding a stellar size in agreement with the simple estimate $R_{\rm H}$.

Controversy arose because hydrodynamic calculations by Bodenheimer and Sweigart (1968) and others, starting from *interstellar* conditions, demonstrated that the collapse of isothermal clouds initially not far removed from equilibria would proceed not homologously, but would develop instead $1/r^2$ density profiles (see also the discussion of the previous section). After some initial transients, the subsequent evolution would then correspond to *mass accretion by a growing protostar* (Larson 1969a). In particular, Larson (1969b) showed that the protostar during the main accretion phase would be virtually invisible at optical wavelengths, but would manifest itself as a bright infrared source. If the timescale to accumulate the star, once a hydrostatic protostar has formed at the center, measures more on the order of $\sim 10^5$ yr, rather than the ~ 10 yr assumed by Narita et al., then Larson obtained substantial radiative losses Δ, and starting sizes for the pre-main-sequence evolution of $1\ M_\odot$ stars of about 2 $R_\odot$. Such stars begin their lives as optically visible objects on the radiative portions of the contraction tracks, missing the wholly convective stages altogether, in disagreement with the observational evidence for T Tauri stars (Cohen and Kuhi 1979).

Numerous subsequent numerical calculations (e.g., Appenzeller and Tscharnuter 1975), as well as the analysis of the route to molecular-cloud core-collapse (see §4), support Larson's position that the formation of individual stars occurs by a gradual process of accretion, with mass-infall rates of $\dot{M}_{\rm infall} \sim 10^{-5}\ M_\odot$ yr^{-1} (or somewhat smaller) typical for sunlike stars. Nevertheless, the subject remained controversial, with Westbrook and Tarter (1975) criticizing Larson's numerical methods, and Hayashi (1970), his treatment of the jump conditions for radiative shocks.

The debate was effectively settled by two developments. (See, however, Tscharnuter's chapter for an account of the additional issues that arose regarding the stability of the position of the accretion shock, and the possibility that the "first hydrostatic core" might suffer large oscillations before the protostar settles into the more quiescent "main-accretion" phase.) An improved numerical treatment by Winkler and Newman (1980a,b) obtained differences of detail, but agreed with Larson in his central conclusions. A complete reformulation of the entire problem, including a more rigorous treatment of the radiative shock-jump conditions and allowance for deuterium burning plus the onset of convection, by Stahler, Shu, and Taam (1980, hereafter SST) also recovered most of Larson's results, but, more importantly, it removed the discrepancy in comparison with observations: a 1 $M_\odot$ star, accumulated at a constant accretion rate of $1 \times 10^{-5} M_\odot$, starts its pre-main-sequence life with a radius of $\sim 5\ R_\odot$, and therefore spends considerable time descending a convective Hayashi track.

SST commented that results for lower-mass cases could be obtained from their 1 $M_\odot$ calculation simply by terminating the infall (with $M_{\text{infall}} = 1 \times 10^{-5}$ $M_\odot$ yr^{-1}) earlier. When Stahler (1983) carried out this task and plotted up the results on the H-R diagram, he obtained his well-known "stellar birthline" (see Fig. 5 again). Stahler's "birthline" fitted well the upper envelope of the data points from four very different stellar groupings in the Cohen and Kuhi (1979) study, prompting Stahler to claim this fact as evidence for a certain "universality" for (a) quasi-spherical accretion, and (b) the infall rate $M_{\text{infall}} = 1 \times 10^{-5}$ $M_\odot$ yr^{-1}. This point of view was challenged by Mercer-Smith, Cameron, and Epstein (1984), who pointed out that the buildup of the central star was more likely to take place realistically by accretion from a disk rather than by direct infall (see also §§6 and 7). Shu and Terebey (1984) offered a resolution of this budding controversy by noting that deuterium burning in a low-mass protostar effectively thermostats the mass-radius relation to yield a "universal birthline" that depends little on the details of the actual accretion process (either the rate or the geometry) – an argument substantiated, within limits, by later detailed studies (Stahler 1988).

6. Infrared Appearance of Rotating Protostellar Objects

An important test of any theory of star formation is to compare its predictions concerning the emergent radiation expected theoretically for each of the stages of evolution. Before we can carry out this task productively for the protostellar stages (*b* and *c* in Fig. 1), we need to modify the self-similar solution (§4) for the collapse of an isothermal cloud core to take into account the presence of rotation. Because ambipolar diffusion occurs relatively slowly, one might imagine that magnetic braking has time to torque down the cloud core sufficiently during the quasi-static condensation stage to allow even the direct formation of single stars, without subsequent angular momentum difficulties (see, e.g., Mouschovias 1978). Realistic estimates yield a somewhat different picture. Before the onset of dynamical collapse, magnetic braking can enforce more-or-less rigid rotation of the core at the angular velocity Ω of its surroundings Mestel 1965, 1985; Mouschovias and Paleologou 1981). Once dynamical collapse starts, however, the infall velocities quickly become super-magnetosonic, and any initial angular momentum possessed by a fluid element carries into the interior. In the inside-out scenario presented above, dynamical collapse is initiated while much of the mass still has an *extended configuration*. For example, to contain 1 $M_\odot$, the wave of falling must typically engulf material out to $\sim 10^{17}$ cm. Even if such a core were perfectly magnetically coupled to its envelope before collapse, it would typically rotate at too large an initial angular speed Ω to allow the formation of just a single star with a dimension of $\sim 10^{11.5}$ cm. The simplest solution to the core's residual angular momentum problem is to form a star *plus a disk*.

Probably as a consequence of cloud magnetic braking, Arquilla and Goldsmith (1986) and Fuller and Myers (1987) find empirically that rotation does not usually play a dynamically important role on scales of a molecular cloud core and larger. This fact suggests the possibility of a perturbational analysis for the dynamical collapse problem (Terebey, Shu, and Cassen 1984), with core rotation at an initially uniform rate Ω treatable as a small correction, in the outer parts, to the spherical self-similar solution for the collapse of a singular isothermal configuration. Large departures from sphericity occur only at radii comparable to or smaller than a centrifugal radius,

$$R_C \equiv \frac{G^3 M^3 \Omega^2}{16 a_{\text{eff}}^8} \tag{13}$$

where $M \equiv \dot{M}_{\text{infall}} t$ equals the total mass that has fallen in at time t. Typical combinations of a_{eff}, Ω, and M yield values for $R_C \sim 10^{15}$ cm.

For $R_* \sim 3 \times 10^{11}$ cm $\ll R_C \sim 10^{15}$ cm, the bulk of the freely-falling matter (on parabolic streamlines because of the nonzero values of the specific angular momentum) does not strike the star directly, but forms a disk of size $\sim R_C$ that swirls around the protostar. Radiative transfer calculations for the emergent spectral energy distribution in an infall model of this type give quite good fits of the data for IRS 5 L1551, if we choose $a_{\text{eff}} = 0.35$ km s^{-1}, $\Omega = 1 \times 10^{-13}$ rad s^{-1}, and $M = 1$ $M_\odot$ (Fig. 6).

Fermi is supposed to have remarked that with three free parameters, he could fit an elephant. Figure 6 gives an empirical proof of this claim. Indeed, one can readily discern a tail at millimeter frequencies, attached to a broad back at far- and mid-infrared frequencies, with a 10 micron silicate absorption feature separating the shoulder from the head and a 3.1 μm water-ice feature defining a tusk. The trunk of the elephant emerges from the droop to optical frequencies, although the models do not reproduce the slight raising of the trunk seen in the data points because they do not properly account for the presence of scattered near-infrared and optical light. By Fermi's standards, then, our failure to catch the trumpet call of the elephant, without the need to introduce additional free parameters, prevents us from claiming a complete victory.

However, we note that the three independent parameters that go into the fits, a_{eff}, Ω, and M, represent an irreducible set from an *a priori* theory, which have fundamental dynamical implications apart from the spectral energy fits. For example, the value $a_{\text{eff}} = 0.35$ km s^{-1} well describes the properties of the core density distribution around L1551 (cf. eq. [3] and the review of Evans 1991). The good agreement of the sizes of the spatial maps at millimeter and submillimeter wavelengths (Walker, Adams, and Lada 1990) and at infrared wavelengths (Butner et al. 1991) with the predictions of the radiative-transfer calculations of our model also bodes well for the values of a_{eff} and M, derived in the spectral energy fits on the basis of providing the correct total optical depth to the central star and the absolute luminosity scale (from protostar theory; 5). Finally, the numerical value

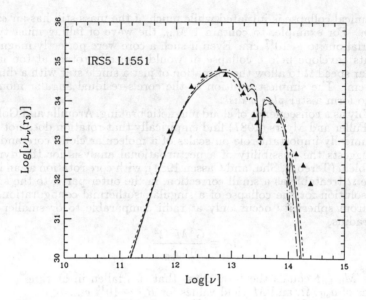

Figure 6. Spectral energy distribution for the bipolar outflow source L1551. The fit for the data points represented by the solid curve results from choosing $a_{\text{eff}} = 0.35$ km s^{-1}, $\Omega = 1 \times 10^{-13}$ s^{-1}, and $M = 0.975\, a_{\text{eff}}^3 t/G = 1\, M_\odot$. (From ALS.)

of Ω, chosen to fit the depth of the silicate absorption feature, yields a predicted size for the circumstellar disk, $R_C = 42$ AU, which compares well with the limits 45 ± 20 AU deduced for this source by Keene and Masson (1990) from radio interferometric measurements of the thermal emission from dust grains in the putative disk. More recently, Goodman et al. (1991, see also Menten and Walmsley 1985) report the detection of a velocity gradient ~ 4 km s^{-1} pc^{-1}, localized to the core region of L1551, consistent in magnitude with the model value $\Omega = 1 \times 10^{-13}$ s^{-1}, although the result may have been contaminated by the velocity shear introduced by the outflow.

A similar modeling of the spectral energy distribution of HH 7-11 suggests that this source also possesses infall at a rate $\dot{M}_{\text{infall}} \gtrsim 1 \times 10^{-5}\, M_\odot$ yr^{-1} (F. C. Adams, private communication). Our ability to fit the spectral energy distribution of famous *outflow* sources like L1551 and HH 7-11 with pure *inflow* models bears directly on the fundamental paradox of §3. Many of the observed embedded systems (those in stage c of Fig. 1) must possess *both* outflow and inflow, with the blowing off of a small polar cap in deeply embedded sources making little difference for the problem of infrared reprocessing in the rest of the infalling envelope. The challenge to observers is to find spectroscopic evidence, in the low-luminosity sources, for the (rotating) collapse along the equatorial directions in an anlogous manner that this task has been accomplished for high-luminosity protostars (see, e.g., Ho and Haschick 1986, Keto et al. 1987, Welch et al. 1987, Rudolph et al. 1990). In any case, here then lies a possible answer to the question: How can a star (SVS 13, the driving source in HH 7-11) form if it loses mass (through a neutral wind emerging from the poles) at the enormous rate $\dot{M}_w = 3 \times 10^{-6}\ M_\odot\ \mathrm{yr}^{-1}$? Clearly, a net gain can still take place if the star is being "force-fed" (through the equatorial regions) at a rate $\dot{M}_{\mathrm{infall}}$ which is $\gtrsim 3$ times larger yet! This empirical finding answers the "how" of our original question, but it does not address the question "why" a newborn star should be losing mass. A clue to the resolution of the latter isssue lies in an examination of the very different way in which a star, in contrast to a disk, may deal with its angular momentum legacy.

7. Protostar Formation by Disk Accretion

Notice that no mass scale that we can identify with stars emerges naturally in the theoretical picture of infall summarized in the previous section. To obtain a stellar mass $M_* \approx \dot{M}_{\mathrm{infall}} t_{\mathrm{infall}}$ from an object (a giant molecular cloud typically), whose characteristic mass scale much exceeds anything that we normally associate with stars, requires us to choose a small total duration t_{infall} (say, 10^5 yr) over which infall occurs. This duration might be set, for example, by the time required for an incipient outflow from the star to completely reverse the inflow. In spherical simulations of protostar formation and evolution (SST; Palla and Stahler 1990), the only event of significance to take place on such a timescale concerns the onset of deuterium burning and the establishment of an outer convection zone at the bottom of which the deuterium burns at a rate equal to that which accretion brings in a fresh supply. In low-mass protostars ($< 2\ M_\odot$, i.e., precursors to T Tauri stars), deuterium burning occurs only slightly off-center; in intermediate-mass stars (2 - 8 $M_\odot$, i.e., precursors to Herbig Ae and Be stars), it occurs in a thin shell. In either case, the important feature is that the process induces an outer convection zone – a condition, when combined with rapid stellar rotation, that Shu and Terebey (1984) speculated would be conducive to dynamo action and to the appearance of strong magnetic fields on the surface of the star.

Calculations by Picklum and Shu (in preparation) demonstrate that accretion through a disk rather than by spherical infall does not modify the above conclusions appreciably, except to lower by a significant factor the emergent photon luminosity for a given rate of mass accretion. The latter effect may alleviate the discrepancies in time scales inferred for embedded protostars and revealed pre-main-sequence stars noted by Kenyon et al. (1990; see also Hartmann et al. 1991).

For star formation via the collapse of a cloud core modeled initially as a uniformly rotating singular isothermal sphere, the fraction of mass that suffers direct infall onto the star compared to that which first lands in the disk equals $1.29\ (R_*/R_C)^{1/3}$ (Adams and Shu 1986). This expression yields a small number (several percent) for stars with radii $\sim$ a few $R_\odot$ and disks with radii $\sim$ 10 - 100 AU. The small cross-sectional area of stars in comparison with their disks leads us to expect generically that most of the mass from a collapsing molecular cloud core, which eventually ends up inside the star, must first make its way through the disk (an original point of view argued first by Cameron 1962; see also Mercer-Smith, Cameron, and Epstein 1984).

The process of disk accretion is not well understood in any astrophysical system. The most frequently invoked mechanism involves the inward transport of mass and the outward transport of angular momentum by the friction associated with some form of anomalous viscosity in a differentially rotating disk (e.g.,

Lynden-Bell and Pringle 1974). In the case of the nebular disks that surround young stellar objects, Lin and Papaloizou (1985; see also Ruden and Lin 1986) identified thermal convection in the disk as a plausible source of the turbulent viscosity, and this may well hold for the optically thick dusty disks believed to surround all classical T Tauri stars. More recently, Balbus and Hawley (1991, see also Hawley and Balbus 1991) have rediscovered a powerful magnetohydrodynamic instability (Chandrasekhar 1960, 1965; Fricke 1969) that afflicts weakly-magnetized differentially-rotating systems in which the angular velocity of rotation decreases outwards (as occurs in all known astrophysical disks). These developments hold great promise for providing a physical foundation for the central assumption of standard accretion-disk theory that the viscosity has an anomalous magnitude much higher than molecular values (for a review, see Pringle 1981).

However, in the most extreme cases (the so-called "flat spectrum sources," see §10 and the chapter by Lada) the far-infrared radiation (relative to the near-infrared values) exceeds by a factor approaching 10^3 what one would have naively predicted by the steady-state viscous model (Adams, Lada, and Shu 1988). The observed disks also have masses, as estimated from their submillimeter and millimeter emission, that compare favorably with those contained in the central stars (Adams, Emerson, and Fuller 1990; Beckwith et al. 1990; Keene and Masson 1990).

The above considerations suggest an interesting scenario for protostellar disk accretion. The overall time scale for viscous transport of mass and angular momentum scales roughly as

$$t_{\text{vis}} \sim \alpha^{-1}(R/H)^2 t_{\text{rot}}, \tag{14}$$

where (R/H) equals the aspect ratio of the disk (radius to vertical scale height), t_{rot} is the rotation period at the outer edge of the disk, and α is a dimensionless parameter in the so-called "alpha" prescription for the anomalous viscosity. For application to nebular disks, typically, $R/H \gtrsim 10$, $t_{\text{rot}} \sim 10^3$ yr, and $\alpha < 1$ (e.g., $\alpha \sim 10^{-2}$ in the convection models of Ruden and Lin [1986]); thus, $t_{\text{vis}} > 10^5$ yr. The statistics of disks around classical T Tauri stars sets a lifetime of $\sim 10^6$ - 10^7 yr (see Strom et al. 1989). This represents only a lower limit for t_{vis} since the disappearance of near-infrared radiation only implies the disappearance of small dust particles surrounding the central star, and not necessarily the viscous transport of all of the material, gas and dust, from the entire disk.

Suppose rotating infall piles matter into the disk faster than viscous accretion can transfer it to the central star. (Models for the main accretion phase of low-mass protostars suggest that $M/\dot{M} \sim 10^5$ yr.) The disk would then build up mass relative to the star until the disk becomes comparably massive. At this point, strong gravitational instabilities (nonaxisymmetric density waves) may develop in the disk that result, in the nonlinear regime, in the inward transport of mass and the outward transport of angular momentum. The energy dissipation associated with this "wave dredging" differs, however, from the viscous mechanism, and we might expect to see a different resultant radial distribution of temperature in the disk and a different emergent spectrum (see §10). In particular, for the transport to occur *globally* (as indicated by the observations of flat-spectrum sources), from the star-disk interface to the outer edge of the disk $\sim 10^4$ times larger, we can restrict our search of unstable modes in nearly Keplerian disks to one-sided disturbances (see Fig. 7), in which circular streamlines distort to elliptical ones with foci at the star (Adams, Ruden, and Shu 1989; herafter ARS).

Unless such instabilities run away catastrophically to form a companion star

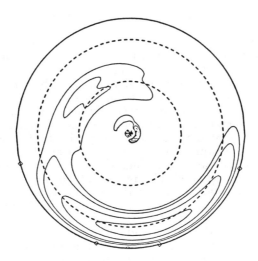

Equidensity Contours

Figure 7. Contour plot of the lowest-order, growing, $m = 1$, mode in a star/disk system where the disk's mass equals the star's mass, but the disk's radius exceeds the star's radius by a factor 10^4. (From ARS.)

(a possible origin for binary systems), we may speculate that they self-regulate the amount of mass in the disk during the infall phases (the second and third panels of Fig. 1) so that it never consists of a large fraction of the total. Thus, we expect that some appreciable fraction f_1 of the total infall rate onto the system (mostly onto the disk), $\dot{M}_{\rm infall}$, must make its way through the disk at an accretion rate $\dot{M}_{\rm acc}$, eventually to be deposited onto the star,

$$\dot{M}_{\rm acc} = f_1 \dot{M}_{\rm infall}. \quad (15)$$

Since the detailed calculations of ARS demonstrate that disks with temperature profiles consistent with the observational requirements suffer strong one-armed spiral instability (see Fig. 7) if they have masses equal, say, to one-half or one-fourth of the total, a reasonable estimate for f_1 ranges from 1/2 to 3/4. We emphasize that this estimate for f_1 represents a *time-averaged* value; the instantaneous rate at which mass from the disk actually empties onto the star may suffer large fluctuations about an average value. In particular, the arguments of §10 suggest that disk accretion induced by global gravitational instabilities may have intrinsic tendencies to occur in a nonsteady manner.

8. Stellar Winds and Bipolar Flows: Theory

The matter entering the protostar from the disk at the rate $\dot{M}_{\rm acc}$ carries a relatively large specific angular momentum. For example, compared to the specific angular momentum of disk matter circling in orbit just outside of a star's equator, the angular momentum per unit mass of a uniformly rotating polytrope of index 1.5 (a good model for a fully convective low-mass protostar) is a small number $b = 0.136$, even if the star were to rotate at the brink of rupture (James 1964). Thus, the disk feeds material to the star that contains, per gram, about $b^{-1} \approx 7$ times more angular momentum than can be absorbed dynamically by the star. Hence, once the star begins to accrete matter from a centrifugally supported disk, it can increase its mass M_*, at most, only by an additional small fraction b before it reaches breakup. In practice, the star may be able to accept even less matter from the disk because even the matter accumulated by the star through direct infall carries nonzero values of specific angular momentum (see, e.g., Durisen et al. 1989).

Since the mass that can be accumulated by a star through direct infall from a realistically rotating molecular cloud core amounts to only several percent of a solar mass, and since the additional amount that can be accreted from a disk amounts to a small fraction of this tiny value, how do stars of sunlike masses and larger ever form?

Paczynski (1991) and Popham and Narayan (1991) calculated star plus disk models where the star, spun to (slightly faster than) breakup, can viscously transport angular momentum to an adjoining disk, and thereby continue to accrete an indefinite amount of mass. This model presumes the existence of a sufficiently efficient viscous coupling between star and disk as to transport away the excess specific angular momentum (above the fraction b) brought in by disk accretion, an assumption that may not hold if the mechanism of disk accretion is gravitational rather than viscous. In any case, their mechanism of continuous outward angular-momentum transport cannot work for T Tauri stars, which are observed to rotate at speeds significantly less than break-up. Furthermore, the almost ubiquitous presence of energetic bipolar flows among deeply embedded sources suggests that quiescient viscous torques do not provide the primary source of relief for the angular momentum difficulty of protostars in their main accretion phase.

Such objects evidently find a more spectacular way of shedding the excess angular momentum brought in by disk accretion. SLRN proposed that the protostar could fling off the excess in a powerful wind once the protostar gets spun to breakup by the disk, if the star possesses sufficiently strong surface magnetic fields (see also Hartmann and MacGregor 1982).

Schematically, the process works as follows. Like other rotating stars with outer convection zones, a protostar probably has a network of open and closed magnetic field lines that protrude from its surface (Fig. 8). An ordinary (ionized) stellar wind blows out along the open field lines; this O-wind may have a higher intensity than that characteristic for a normal star of the same spectral type, luminosity class, and rotation rate, because of the enhancement of stellar dynamo action associated with circulation currents induced by the disk through Ekman pumping (see §9).

If open field lines circulate into the equatorial regions of the protostar, which spins at breakup by assumption, the O-wind may intensify even more via the X-celerator mechanism (SLRN) and become an extraordinary wind. The magneto-centrifugal acceleration in the X-wind takes place much as described in the section on the Blandford-Payne mechanism for disk winds, except that the geometry here no longer selects against field lines (equivalent to streamlines for a conducting gas) that emerge in the "downhill" directions of the effective (corotating) potential. To

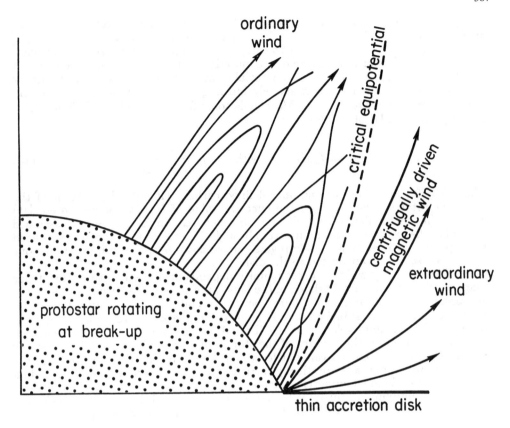

Figure 8. Schematic model for heavy mass-loss driven magnetocentrifugally from the equator of a protostar forced to rotate at breakup because of accretion through a Keplerian disk. All streamlines to the right of the critical equipotential (marked as a dashed curve) correspond to possible "downhill" paths for electrically conducting gas attached to magnetic lines of force that rotate rigidly with the star. The large slip velocities present in the interfaces between the X-wind and the O-wind and between the X-wind and the disk may lead to turbulent mixing layers that have observable radiative signatures. (Adapted from SLRN.)

carry away the requisite mass loss rate $\dot{M}_w$, the sonic transition of the X-wind must occur typically in the photosphere of the star, where the material, being cool, has a relatively low ionization fraction. Thus, the X-wind is predominantly a *neutral* or *lightly ionized* wind (Natta et al. 1987; Ruden, Glassgold, and Shu 1990).

Consider the quasi-steady state wherein spinup by disk accretion at a rate $\dot{M}_{acc}$ is balanced by spin-down via a magnetized stellar wind at a rate $\dot{M}_w$ in such a way that at each stage of the process the star continues to rotate exactly at breakup. Suppose the specific angular momentum carried away in the X-wind, measured as an average over all mass-carrying streamlines, equals some multiple $\bar{J}$ of the specific angular momentum of the material in circular orbit at the equator of the star. If we further assume the equatorial radius to be proportional to the stellar mass, as roughly true for a star actively burning deuterium at its center, we find that the requisite mass loss rate $\dot{M}_w$ equals some fraction f_2 of the disk accretion rate $\dot{M}_{acc}$ (cf. eq. [1] of SLRN):

$$\dot{M}_w = f_2 \dot{M}_{acc} \qquad \text{with} \qquad f_2 = \frac{1-2b}{\bar{J}-2b}, \tag{16}$$

where $b = 0.136$ is the pure number defined earlier. (If only the equator of the protostar rotates near breakup, the effective value of b that appears in eq. [16] could be [much] smaller.) To derive the above expression for f_2, we have assumed that all of the excess angular momentum brought in by accretion above that needed to keep the star exactly at breakup gets carried away by the wind, i.e., that none of it gets viscously transported to the disk.

The extent by which $\bar{J}$ exceeds unity depends on the strength of the stellar magnetic field in the neighborhood of the X-point of the equipotential. For escaping streamlines to have a finite terminal velocity v_w at infinity, the (rotating) magnetic field lines must exert sufficient torque as to make $\bar{J}$ exceed 3/2. For example, in order for the terminal velocity of the X-wind to have a measured value equal to the breakup speed at the equator of the star (typically ~ 150 km s^{-1} for low-mass protostars), $\bar{J}$ must have a value at least equal to 2. In fact, because some of the angular momentum is asymptotically carried away by magnetic stresses rather than all by the fluid, detailed numerical calculations suggest that the actual needed value of $\bar{J} \approx 3.6$, implying a fiducial value for $f_2 \approx 0.2$ Combining equations (15) and (16), we now obtain a relationship between inflow and outflow rates,

$$\dot{M}_w = f \dot{M}_{infall}, \tag{17}$$

with $f \equiv f_1 f_2 \approx 0.1 - 0.2$ typically, in very rough agreement with the observations.

On the other hand, if the magnetic field lines emerging from the neighborhood of the X-point of the effective potential had infinite strength, X-wind gas would be flung to infinity in the entire sector from the equator to the critical equipotential surface (the dotted curve in Fig. 8). Since this equipotential surface (the last rigid "wire" to which an attached bead could still slide "downhill") bends up to turn asymptotically parallel to the rotation axis of the star (at a radius = $\sqrt{3} R_e$), the strong X-wind could act to focus and collimate an otherwise isotropic and more mild O-wind into an (ionized) optical jet (one on each side of the equator).

For stellar magnetic fields of finite strength, the situation becomes more complicated. The method of matched asymptotic expansions provides a tractable approach to the resultant magnetohydrodynamic problem if we adopt the assumption of axial symmetry (Shu, Lizano, Ruden, and Najita, work in progress). Figure 9 gives a sample solution for a case with a dimensionless stellar field strength that produces Alfven crossing at a distance equal to one additional stellar radius from the equator of the star. In this calculation, we have used upper and lower boundary conditions that artificially constrain the limiting streamlines, $\psi = 0$ and $\psi = 1$, to

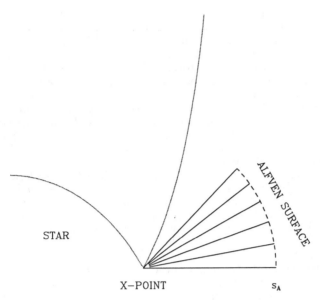

Figure 9. Spreading of streamlines for an X-celerator driven wind in the meridional plane. The calculation is carried out in a frame that corotates with the star. The bottom-most streamline has been constrained to flow horizontally across the surface of a perfectly flat disk; the uppermost streamline has also been taken to be straight for sake of simplicity. In a more realistic model, the slip interfaces between the X-wind and the O-wind and between the X-wind and the disk surface are likely to be turbulent. A dashed curve marks the Alfven surface, where the magnetic stresses no longer dominate over the gas inertia. The dotted curve marks the location of the critical equipotential surface that, together with its counterpart in the lower half of the diagram (not drawn), crosses the stellar surface at an X-point. (From Shu, Lizano, Ruden, and Najita 1991.)

be perfectly straight (see below). Interior to the Alfven surface defined by $\mathcal{A} = 0$, where $\mathcal{A}$ is an Alfven discriminant, the stream function ψ satisfies an elliptic partial differential equation:

$$\nabla \cdot (\mathcal{A} \nabla \psi) = \mathcal{Q}. \tag{18}$$

This equation has the formal structure of steady-state heat conduction; here, however, we are concerned with the spreading of streamlines, and not the spreading of heat. The source term $\mathcal{Q}$ responsible for streamline spreading in the meridional plane equals a sum of three terms that are each proportional to the derivatives (across streamlines) of quantities that are conserved along streamlines: Bernoulli's constant $H(\psi)$, the ratio of magnetic and mass fluxes $\beta(\psi)$, and the specific angular momentum carried in both matter and field $J(\psi)$. Smooth passage through the sonic transition near the X-point of the effective potential yields $H(\psi)$; for the part of the problem much beyond the immediate neighborhood of the X-point of the critical equipotential, $H(\psi)$ goes to 0 in the limit of a cold flow. The requirement that the flow accelerates smoothly through the Alfven surface, $\mathcal{A} = 0$, places a constraint on the normal derivative of ψ, $\nabla \mathcal{A} \cdot \nabla \psi = \mathcal{Q}$, that determines the spatial location and shape of the Alfven surface and the functional form of $J(\psi)$ if we are given $\beta(\psi)$. In other words, the geometry of the problem is entirely determined if we know the strength of the stellar magnetic field and the way that mass is loaded onto the open field lines at the equatorial belt of the protostar.

For the dimensional parameters that approximately apply to HH 7-11 or L1551, we find that a top terminal speed of ~ 150 km s^{-1} (corresponding to $\bar{J} \approx 3.6$) can be achieved when the equator of the protostar has open photospheric magnetic fields of the strength ~ 4 kilogauss. Such fields appear reasonable if we extrapolate from the product of field-strength and filling-factor, 1 - 2 kilogauss, inferred for weak T Tauri stars by Basri and Marcy (1991) from the Zeeman broadening of spectral lines with high Landé-g factors.

It may be informative to elucidate why we believe the X-celerator mechanism to work for a star, but not for a disk (§3). To carry the observed mass-loss rate, the density of the wind at the sonic surface must be large, close to photospheric values. Thus, the sound speed is low, and the sonic transition can occur only near X-points where the effective gravity vanishes, since in a frame which corotates with the footpoint of the magnetic field, the magnetic stresses cannot help the gas make a sonic transition (see eq. [3] of SLRN). On the other hand, if the gas is to continue to accelerate to higher speeds by magneto-centrifugal flinging after having made the sonic transition, the Lorentz force must be able to overcome both inertia and gravity. Since the densities are very high at the sonic surface, a smooth transition from a gas-pressure driven flow to a magnetohydrodynamically driven flow requires either the magnetic field $\mathbf{B}$, or its curl, $\nabla \times \mathbf{B}$, to be large, or both (cf. eq. [2]). The geometry of a thin disk requires $\nabla \times \mathbf{B}$ to be large, but to keep the gas sub-Alfvenic in the post-sonic-transition region, $\mathbf{B}$ also has to be large. This combination proves fatal, we believe, to disk-wind models in the protostellar context (but see the chapter by Pudritz for a different point of view).

An X-celerator driven wind from a protostar (or star-disk interface) fares better, because we require only $\mathbf{B}$ to be large, and we can relatively easily believe that magnetic fields *rooted in the deep convection zones of a rapidly-rotating star* could achieve the requisite values. High-mass stars that lack outer convection zones have greater theoretical difficulty generating surface magnetic fields, although the instability mechanism discussed by Balbus and Hawley (1991, see also Fricke 1969 for specific application to the case of stars) offers a promising mechanism for such objects to generate moderate-to-small magnetic fields provided only that they suffer differential rotation of dynamical significance for the structure of the star (with Ω decreasing radially outward). Indeed, X-celerator driven winds in such a situation may offer an explanation why the terminal velocities of winds from high-mass protostars have such low values (100 - 200 km s^{-1}) in comparison with the high values (1000 - 2000 km s^{-1}) that characterize (probably radiation-driven) winds from main-sequence and evolved stars of early spectral type (see, e.g., Chiosi and Maeder 1986).

In Figure 10, we plot magnetic field patterns and fluid trajectories of the flow in the equatorial plane of our X-celerator model. In a frame that rotates with the same angular speed as the star, streamlines and field lines coincide (dotted curves). In a stationary frame, however, the trajectory of an element of electrically conducting gas takes the form indicated by the dashed curve. The fluid element starts by making a sonic crossing close to the photosphere of the star. The acceleration to sonic speeds (7 or 8 km s^{-1} typically) can occur purely by the outward push of

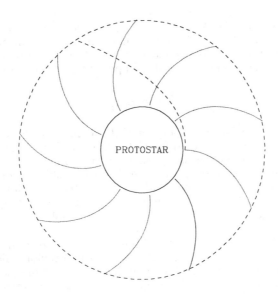

Figure 10. Field lines (dotted curves) and trajectory of a fluid element (dashed curve) for an X-celerator driven wind in the equatorial plane of an inertial frame of reference. The protostar rotates in a counter-clockwise sense, so that open field lines form spirals that trail in the sense of rotation. The dashed circle marks the location of the Alfven surface in a calculation where we have explicitly assumed axial symmetry (From Najita, Shu, Lizano, and Ruden 1991.)

photospheric gas pressure in a small equatorial belt where the effective gravity nearly vanishes because we have assumed that the protostar rotates at breakup speeds. As the open magnetic line of force to which the gas is tied rotates in a counter-clockwise direction with the star, the excess rotation (above local Keplerian values) enforced by the field accelerates the gas along the field line (with negligible slip for the typical degrees of ionization computed by RGS), until the gas reaches the Alfven velocity at the position of the dashed circle. At this point, if the gas speed exceeds the local escape velocity by a significant margin, the Alfven speed will also typically equal a healthy fraction (say, $\sim 90\%$) of the final terminal speed. Because the magnetic field does not have infinite strength, it cannot keep the gas corotating indefinitely with the star, and the pattern of field lines (in either the rotating or inertial frame) form trailing spirals. The field lines and streamlines form a similar lagging pattern at other latitudes, except that they climb in the vertical direction (toward the rotational poles) as the gas flows away from the protostar.

As already mentioned, the boundary conditions in Figure 9 on the uppermost and bottom-most streamlines have been taken to be perfectly straight so as not to bias the results for the interior. When we do this, the intermediate streamlines have a tendency also to remain straight, with the gas at higher latitudes having larger densities and slower terminal speeds (for a magnetic field distribution that starts off approximately uniform at the sonic surface). We are currently trying to compute the amount of bending which takes place when the shape of the upper streamline is left more realistically as a free boundary (pressure balance between an X-wind and an O-wind). When such a calculation becomes available, we will be able to estimate the degree of collimation of the X-wind; in particular, we will be able to compute the angle dependence $P(\theta)$ of the rate of momentum injection (needed for any a priori calculation of the shapes of the swept-up shells of molecular gas):

$$\frac{\dot{M}_w v_w}{4\pi} P(\theta). \tag{19}$$

Even without detailed knowledge of $P(\theta)$, a simple order of magnitude estimate using the theory developed so far yields the velocity of the swept-up bipolar lobes of molecular gas roughly as the geometrical mean between the terminal speed of the wind v_w and the effective sound speed of the ambient molecular cloud core a_{eff} (Shu, Ruden, Lada, and Lizano 1991). For v_w measuring hundreds of km s^{-1}, and a_{eff} ranging from a fraction of a km s^{-1} (low-mass cores) to greater than 1 km s^{-1} (high-mass cores), we would then predict typical lobe speeds $\sim (v_w a_{\text{eff}})^{1/2} \sim$ 10 km s^{-1}, in rough agreement with the observations.

For well-collimated flows, $P(\theta)$ will have values larger than unity for a small range of angles θ near the two poles, 0 and π. Our previous discussion leads us to suspect that younger sources, which may have stronger magnetic fields, possess more highly collimated outflows. We speculate that as the sources age, their rotation speeds progressively fall below the critical rate (see §9), their magnetic fields weaken, and their outflows fan out more in polar angle and sweep out an increasingly greater solid angle of the overlying envelope of gas and dust. In this fashion may the protostars in stage c of Figure 1 become optically revealed as the T Tauri stars of stage d.

9. Revealed T Tauri Stars

The ability to study T Tauri stars at optical and near-infrared wavelengths has yielded tremendous observational dividends. Foremost in importance among the discoveries of the past decade has been the realization that many of the photometric and spectroscopic peculiarities (variability, strong emission lines, infrared and ultraviolet excesses, strong outflows) long known to characterize these systems (cf. the reviews of Herbig 1962, Kuhi 1978, Cohen 1984, Imhoff and Appenzeller 1987, Bertout 1987, Appenzeller and Mundt 1991) may owe their explanation to the presence of circumstellar disks. Several independent lines of evidence lead to the conclusion that circumstellar disks surround many young stellar objects, e.g., the large polarizations often seen in these sources (Elsasser and Staude 1978, Bastien and Menard 1988, Bastien et al. 1989), or the asymmetry of forbidden O I line profiles (Appenzeller et al. 1985, Edwards et al. 1987), or the motions of SiO masers (Plambeck, Wright, and Carlstrom 1990). Here, however, we shall concentrate on the nature and implications of the infrared and ultraviolet excesses in T Tauri stars.

We begin with the ultraviolet excess (Kuhi 1974, Herbig and Goodrich 1986), which has been plausibly linked by Hartmann and Kenyon (1987), and Bertout, Basri, and Bouvier (1988) to the action of a boundary layer between the star and the disk. For a conventional boundary layer to arise (gas heated to high temperatures by the frictional rubbing of a rapidly-rotating disk against a slowly-rotating star), a disk must abut the star.

Cabrit et al. (1989) find an interesting correlation between the strength of the near-infrared excess (a measure of the amount of disk just beyond the boundary layer) and the strength of H-α emission (a measure of T Tauri wind-power). Systems inferred to be missing inner disks (radii equal to a few stellar radii) exhibit relatively little wind-power (weak T Tauri stars), whereas systems with appreciable inner disks have relatively strong winds (classical T Tauri stars). Since millimeter and submillimeter investigations reveal no strong biases with respect to the question of whether weak and classical T Tauri stars possess outer disks (radii equal to thousands of stellar radii), Edwards et al. (1991) interpret the finding to imply that pre-main-sequence and protostellar winds are driven by disk accretion (one that extends virtually right up to the surface of the star). At first sight, this interpretation seems to bode well for the X-celerator theory, which pinpoints the interface between star and disk as the source of the wind. In particular, theories that use self-similar solutions for disk-winds (Blandford and Payne 1982, Konigl 1989) have a difficult time explaining why missing just a small portion (the innermost part) of a disk should completely change the nature of the solution. However, a deeper probing of the T Tauri results shows that they also pose a severe problem, potentially fatal, for X-celerator models.

For the X-celerator model of §8 to work, we need to posit that the equator of the star, to which the open magnetic field lines are tied, rotates at breakup. The very fact that T Tauri stars possess boundary layers (if this is the correct interpretation for the ultraviolet excess) demonstrates, however, that their equatorial regions are not rotating at breakup. Even more damaging, direct spectroscopic investigation shows that most T Tauri stars rotate at speeds about an order of magnitude *slower* than breakup (Vogel and Kuhi 1981, Bouvier et al. 1986, Hartmann et al. 1987; see the review of Bouvier 1991). Yet, Edwards et al. (1991) deduce that T Tauri stars also satisfy an inflow-outflow relationship of the form (16), $\dot{M}_w = f_2 \dot{M}_{\rm acc}$, with f_2 numerically not very different from the value that we have deduced from X-celerator theory ($f_2 \approx 0.2$).

How can we resolve the discrepancy? Galli and Shu (1991) propose that a partial solution may be found by asking how an accretion disk actually tries to spin up a slowly-rotating star. Their answer involves, not the inward diffusion of vorticity as in the models of Paczynski (1991) and Popham and Narayan (1991), but *Ekman pumping*. The mathematics for Ekman pumping is fairly involved; however, the basic physics is simple, and easily explained by analogy to spinup in a cup of tea.

Suppose we place a cup of tea on a turntable spinning at angular speed Ω, and we ask how long it takes for the liquid, initially at rest, to spin up to the same angular speed as the boundaries of the cup (Fig. 11). If spinup occurred by the diffusion of vorticity from the boundary layer to the interior, the theoretical answer, in order of magnitude, would be given by the familiar formula:

$$t_{\rm diff} \sim L^2/\nu, \qquad (20)$$

where L represents the typical dimension of the cup, and ν gives the kinematic viscosity of tea. Putting in characteristic numbers, we would deduce $t_{\rm diff} \sim$ tens of minutes. In fact, the actual spinup time empirically measures more in the neighborhood of tens of seconds, and is theoretically given as the geometric mean between

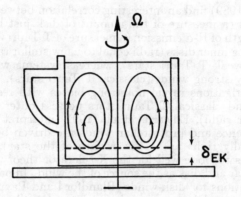

Figure 11. Ekman pumping in a teacup (schematic). A thin Ekman layer δ_{Ek} forms above the bounding surface perpendicular to the axis of rotation, into which fluid from the interior circulates.

the viscous diffusive timescale (20) and the overturn timescale Ω^{-1}:

$$t_{\text{spinup}} \sim (t_{\text{diff}}/\Omega)^{1/2}. \tag{21}$$

The formula (21) represents essentially the time that it takes tea to circulate from the interior of the cup to the boundaries, where it can quickly match its rotational speed to that of the cup. In other words, Ekman pumping sets up a secondary circulation that brings the tea to the boundary layer rather than waiting for the effects of the accelerating agent to diffuse to the interior. Because the bottom of the cup speeds up the tea in contact with it relative to the tea in the interior, the pressure gradient needed to balance the centrifugal force of rotation differs in the two regions. The difference in pressure distribution outside and inside the boundary layer then pumps the fluid from the former to the latter in such a way as to set up the secondary circulation depicted in Figure 11.

We believe an analogous situation to hold for the case of T Tauri stars, except that gravity, in addition to pressure and inertial terms, enters the equation for force balance. Schematically, the presence of an abutting disk tries to spin up the equatorial regions of a star faster than its poles. The excess of centrifugal support in the equatorial regions necessitates less vertical pressure support against gravity there than elsewhere. As a consequence, the unbalanced horizontal pressure gradients will push gas from the surface layers of the rest of the star toward the equator. Continuity requires those streamlines converging onto the equator of the star that do not blow outwards in a wind to resubmerge into the interior of the star, establishing a circulation pattern that looks as depicted in Figure 12.

Figure 12 represents a detailed calculation from Galli's (1990) Ph.D. thesis of Ekman pumping in a polytrope of index 1.5 (to represent a fully convective star). The calculation, however, contains only one of two possible effects that can lead to a secondary circulation: the *barotropic* response to the spinup induced by the frictional torque of an adjacent disk highly concentrated toward the equatorial

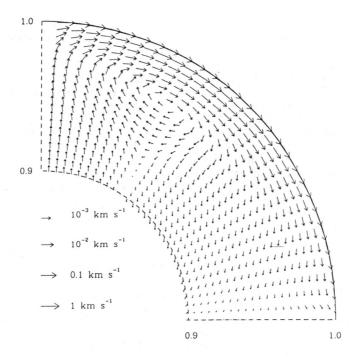

Figure 12. Ekman pumping in a T Tauri star modeled via the spinup of a (uniformly rotating) polytrope of index 1.5 with turbulent viscosity estimated by convective mixing-length theory. A secondary circulation pattern sets up in the meridional plane when the star is subjected to a shear stress at its surface that is highly concentrated toward the equatorial plane. The velocity scale corresponds to a model in which $M_* = 0.8\ M_\odot$, $R_* = 3\ R_\odot$, $\Omega_* R_* = 20$ km s^{-1}, and $\nu = 10^{14}$ cm^2 s^{-1}. The radial scale from $0.9\ R_*$ to $1.0\ R_*$ has been expanded for clarity of display. (From Galli 1990.)

regions. The rotational evolution, with $d\Omega/dt > 0$, is assumed to proceed keeping the zeroth-order rotation spatially uniform, which leads to perturbations in the angular speed of the stellar interior that are stratified across cylinders. If the true surfaces of constant Ω in T Tauri stars correspond – as they do in the case of the convection zone of the Sun (Dziembowski et al. 1989) – not to cylinders, but to radial cones, a second effect can arise because the surfaces of constant pressure do not then correspond to surfaces of constant density. In this situation, the *baroclinic* generation and maintenance of meridional mass flow may compete with or even dominate over the barotropic part. In other words, the meridional flows induced by a quasi-steady balance between spinup through a disk and spin-down via a wind may make the application of Fig. 12 to actual T Tauri stars somewhat oversimplified.

Nevertheless, Figure 12 shows a robust qualitiative feature of great importance to our current discussion: the circulation of matter in the photospheric and subphotospheric layers toward the equator. If such layers carry stellar magnetic field lines along with them, then we can easily visualize how open field lines of kilogauss strength might be brought from the rest of the star that rotates slowly (as required to satisfy the spectroscopic observations of T Tauri stars) to an equatorial band that rotates quite rapidly (by definition, at "breakup" when we enter the disk proper). Conceivably, the Doppler imaging (Vogt 1981) of classical T Tauri stars could test whether circulation currents of this geometry cause starspots to migrate toward the equator on timescales (radius of star divided by photospheric circulation speed) of weeks to months. If so, the resulting ejection of matter in a powerful magnetocentrifugal wind along open field lines might then qualitatively proceed, more-or-less, as we described in §8 for an X-celerator driven flow. Since it takes on the order of a day for circulation-driven, inhomogeneous, magnetic structures to migrate through the sonic surface of a wind originating from an equatorial belt equal to a few percent of the radius of the T Tauri star, we can now understand why optical spectral lines believed to be wind diagnostics (in contrast to general photospheric diagnostics) should show profiles that vary wildly on that sort of timescale (Basri, private communication).

The above speculation, that T Tauri winds originate basically in a highly magnetized and rapidly rotating boundary layer, amounts essentially to a compromise solution between a stellar-driven wind and a disk-driven wind. In this compromise, we make use of the known ability of a star to hold onto very strong magnetic fields, and we take advantage of the natural tendency for disks to possess very rapid rotation. More work, however, needs to be done on this problem before we can claim that we can safely import to this complex situation the main features of the solutions for more simple models.

We should note, however, that *closed* field lines may play a competing role to open ones. If stellar circulation currents force closed field lines of sufficient strength to thread through the disk (a process made easier at a late stage when the disk accretion rate onto the star drops to small values), and if the star becomes magnetically coupled largely to those parts of the disk that rotate fairly slowly, then we might have another explanation why T Tauri stars have projected rotational speeds $v \sin i \lesssim 15$ km s^{-1} (see Konigl 1991). For this braking mechanism to work, however, the same magnetic torques associated with fields threading the gaseous disk need to empty out much of the rapidly-rotating inner portions of the disk – which would otherwise act to spin up the star, rather than spin it down. Weak T Tauri stars may represent (evolved) systems missing such inner disks, and it is interesting to note that they do not possess the powerful (X-celerator driven) winds that classical T Tauri stars do.

10. The Disks Inferred for T Tauri Stars

We turn now to the dynamics of the disk proper, and the origin of the infrared excesses of T Tauri stars. Since the pioneering work of Mendoza (1966, 1968), astronomers have known that T Tauri stars emit significantly more infrared radiation than other stars of its spectral type (typically K subgiants). From the start, thermal emission from circumstellar dust has been suspected as the culprit; however, the geometry of the dust distribution remained obscure until Lynden-Bell and Pringle (1974) suggested that it might lie within a viscous accretion disk. In particular, Lynden-Bell and Pringle pointed out that an optically thick disk, vertically thin

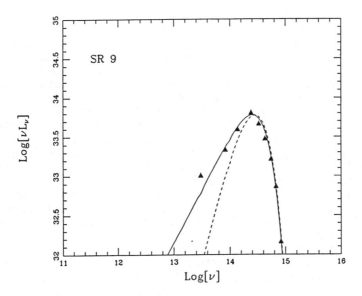

Figure 13. A steep-spectrum T Tauri star, SR 9. The dashed curve shows the expected spectrum for a star alone; the solid, for a star plus a spatially flat, but radially extended, disk that locally reprocesses the starlight that falls on it. (From ALS.)

but spatially extended in distance ϖ from the rotation axis over several orders of magnitude, and possessing a power-law distribution of temperature,

$$T_D \propto \varpi^{-p}, \quad (22)$$

would exhibit a power-law infrared energy distribution,

$$\nu F_\nu \propto \nu^n \quad \text{where} \quad n \equiv 4 - \frac{2}{p}. \quad (23)$$

Viscous accretion-disk theory for a system in quasi-steady-state predicts $p = 3/4$, i.e., $n = 4/3$.

Rucinski (1985) examined the observational evidence and concluded that, while the infrared excesses of T Tauri stars did well approximate a power-law distribution, the mean value of n is appreciably smaller than 4/3 (see also Rydgren and Zak 1987). Adams, Lada, and Shu (1987, herafter ALS) noted that the presence of considerable infrared emission without a correspondingly large visual extinction of the central star implies that the circumstellar distribution of dust must exist in orbit in virtually a single plane about the star. In such a situation (Fig. 13), a *non-accreting*, optically thick, flat, and radially extended disk would intercept 25% of the starlight and reprocess it to near- to far-infrared wavelengths, with disk-temperature and spectral-energy distributions also very nearly satisfying equations

(22) and (23), except that the coefficient of proportionality would be exactly known. Unshielded large dust particles at a distance r from a star would, of course, have a temperature distribution that satisfies $T_D \propto r^{-1/2}$. However, astronomers believe that young nebular disks are completely thick at optical wavelengths even vertically through the disk, so that starlight cannot shine directly on dust particles by propagating through the midplane, but must come at oblique angles from near the limbs of a star of size R_*. This effect introduces an extra geometrical factor of essentially R_*/r into the relation between absorbed starlight ($\propto [R_*/r]r^{-2}$) and emitted infrared radiation ($\propto T_D^4$), which makes the reprocessing temperature law $T_D \propto r^{-3/4}$ rather than $r^{-1/2}$. To remind ourselves of this important difference between the reprocessing in an opaque flat disk with faces perpendicular to the z direction and in an airless planetary surface with projected area perpendicular to r, we write the former law as $T_D \propto \varpi^{-3/4}$.

ALS also demonstrated that the few cases of T Tauri systems that did show the canonical value $n = 4/3$ could indeed be explained in terms of *passive* or *reprocessing* disks; whereas, those disks that have the most extreme infrared excesses and, therefore, exhibit the most quantitative evidence for a nonstellar contribution to the intrinsic luminosity, correspond, not to $n = 4/3$, but to $n = 0$ (Adams, Lada, and Shu 1988; see Fig. 14). Flat-spectrum sources with $n = 0$ require $p = 1/2$, i.e., $T_D \propto \varpi^{-1/2}$. A comprehensive survey by Beckwith et al. (1990) fitted temperature power-laws to the infrared excesses of classical T Tauri stars, and found that the derived values for the exponent p range from the extremes $p = 3/4$ (steep-spectrum sources) to $p = 1/2$ (flat-spectrum sources), with more sources having the latter value than the former.

Kenyon and Hartmann (1987) proposed that *geometric flaring* in purely reprocessing disks could flatten out the spectral energy distributions of the spatially flat models that predict $n = 4/3$, and thereby remove much of the discrepancy between theory and observation. A greater interception and scattering of stellar photons would also better account for the relatively large fractional polarization detected for some T Tauri stars (Bastien et al. 1989). In contrast, Hartmann and Kenyon (1988) note that the source FU Orionis, which may be undergoing enhanced accretion through a disk (for a dissenting view, see Herbig 1989), *does* show the canonical value $n = 4/3$ (see also ALS), and they conclude that nebular disks may transfer much, if not most, of their mass onto the central stars through such recurrent episodic outbursts.

There exists a difficulty with explaining the most extreme flat-spectrum T Tauri stars (which includes some of the most famous examples in the class: T Tauri itself, HL Tau, DG Tau, etc.) as passive, geometrically-flared disks. To reprocess the starlight into the observed amounts of infrared excess would require unrealistic ratios of the dust photospheric height H to disk radius ϖ. Appendix B of Ruden and Pollack (1991; see also Kusaka et al. 1970) gives the temperature of a flared disk illuminated by a central star of radius R_* and effective temperature T_* (at a radius $\varpi \gg R_*$) as

$$T_D(\varpi) = T_* \left[\frac{2}{3\pi} \left(\frac{R_*}{\varpi}\right)^3 + \frac{1}{2} \left(\frac{R_*}{\varpi}\right)^2 \left(\frac{H}{\varpi}\right) \left(\frac{d\ln H}{d\ln \varpi} - 1\right) \right]^{1/4} \qquad (24)$$

Notice that the canonical law $T_D \propto \varpi^{-3/4}$ dominates at small radii ϖ (where H

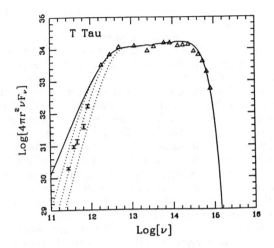

Figure 14. A flat-spectrum T Tauri star, T Tau itself. The curves show models (from Adams, Lada, and Shu 1988) for a star plus a spatially flat disk, with the latter heated by stellar photons as well as possessing an intrinsic source of luminosity with a temperature distribution, $T_D \propto \varpi^{-p}$, where $p = 0.515$. The disk extends from the stellar surface to a radius $R_D = 120$ AU, the minimum value needed to fit the extent of the flat portion of the infrared data points (triangles). The actual size also cannot be much larger before suffering truncation by the companion. (The system corresponds to a binary, but the model is insensitive to which star has the flat-spectrum disk, or even if both stars have disks.) The disk model in the solid curve has infinite mass and is optically thick at all frequencies; the dotted curves that turn over progressively toward higher frequencies because of declining optical depth have disk masses, respectively, of 1, 0.1, and 0.01 $M_\odot$. Measurements at submillimeter and millimeter frequencies (data points with error bars) can establish the disk mass, provided the dust opacities in the disk correspond with those of the model. (From Adams, Emerson, and Fuller 1990.)

$\ll R_*$), while shallower temperature gradients can arise at larger radii if H varies as a power of ϖ with an exponent s greater than unity (a geometrically flared disk). If the dust does not settle to the midplane but remains suspended in the gas (an inconsistent assumption if reprocessing disks are not turbulent), vertical hydrostatic equilibrium requires $H \propto \Omega^{-1} T_D^{1/2}$, so that $s = 9/7$ and $p = 3/7$ in the outer parts of a self-consistent, Keplerian, reprocessing disk. Such a disk typically requires $H/\varpi \gtrsim 1/3$ at large ϖ to produce an appreciable departure from the canonical case (Kenyon and Hartmann 1987). A circumstellar layer of dust with such an aspect ratio would conflict, however, with the observed fact that unobscured T Tauri stars are an order of magnitude more numerous than heavily extincted sources (in the Taurus molecular cloud). Moreover, for sources not seen *through* the disk, since $n = +4/3$ for $p = 3/4$ but $n = -2/3$ for $p = 3/7$, the spectrum would *fall* from the near- to mid-infrared but then *rise* toward the far-infrared. Reprocessing of starlight in either flared disks or warped disks cannot generally reproduce the *single* power law-index, $n = 0$ from the near-infrared to the far-infrared, observed for the most extreme flat-spectrum sources.

Because neither viscous accretion nor passive reprocessing can adequately explain the single exponent $n = 0$, Adams, Lada, and Shu (1988) speculated that the flat-spectrum T Tauri stars must have *active* disks where the processes of transport of mass (inward) and angular momentum (outward) involve *nonlocal* mechanisms. ARS worked out one such mechanism: the generation of one-armed spiral density waves ($m = 1$ disturbances) by eccentric gravitational instabilities that involve the displacement of both star and disk from the center of mass of the system. Shu, Tremaine, Adams, and Ruden (1990, hereafter STAR) gave an analytic demonstration, under a restrictive set of circumstances, that such modal eccentric instabilities would arise whenever the disk mass as a fraction of the whole (star plus disk) exceeds $3/4\pi \approx 0.24$. They also pointed out that the temperature law $T_D \propto \varpi^{-1/2}$ plays a special role in the linear theory of $m = 1$ disturbances in that this distribution keeps the fractional amplitude of the wave (ratio of disturbance to unperturbed surface densities) constant throughout much of the disk. Since the fractional amplitude in a full theory provides a measure of the nonlinearity of the wave, STAR speculated that nonlinear propagation and dissipation of excited $m = 1$ waves would provide the feedback that accounts for the temperature law $T_D \propto \varpi^{-1/2}$ in the first place. E. Ostriker and Shu (work in progress) are pursuing the correct nonlinear derivation of this result, and they are also investigating the possibility that stellar companions in binary systems may provide an alternative source for the excitation of $m = 1$ density waves.

We find it informative to comment on the basic reason why either low-freqeuncy unstable normal modes or fairly distant companions can provide, in principle, a redistribution of mass and angular-momentum of a different qualitative nature than viscous accretion. A disturbance which rotates at a uniform angular speed Ω_p (see §11) contains excess energy E_{wave} and angular momentum J_{wave} above the unperturbed values. These are related to each other by the formula (cf. Lynden-Bell and Kalnajs 1972),

$$E_{\text{wave}} = \Omega_p J_{\text{wave}}. \tag{25}$$

The same proportionality applies *locally* to the wave energy densities $\mathcal{E}_{\text{wave}}$ and angular-momentum densities $\mathcal{J}_{\text{wave}}$ (per unit area),

$$\mathcal{E}_{\text{wave}} = \Omega_p \mathcal{J}_{\text{wave}}, \tag{26}$$

if the spiral waves are tightly wrapped (Lin and Shu 1964, Toomre 1969, Shu 1970), as they inevitably are for disturbances in most of a radially extended nebular disk (cf. ARS and STAR). Moreover, $\mathcal{E}_{\text{wave}}$ and $\mathcal{J}_{\text{wave}}$ are both positive or both negative depending on whether the wave rotates faster or slower than the matter; i.e., $\mathcal{E}_{\text{wave}}$ and $\mathcal{J}_{\text{wave}} > 0$ outside corotation, $\Omega_p > \Omega$, whereas $\mathcal{E}_{\text{wave}}$ and $\mathcal{J}_{\text{wave}} < 0$ inside corotation, $\Omega_p < \Omega$, where $\Omega(\varpi)$ represents the local angular speed of the disk.

On the other hand, in order for a gas element to remain in *circular* orbits when some external torque changes its angular momentum at a rate $\dot{J}$, its energy must change at a rate

$$\dot{E}_{\text{circ}} = \Omega \dot{J}. \tag{27}$$

When spiral density waves inside (outside) corotation dissipate they give a negative (positive) amount of angular momentum to the local gas, causing that gas to move to a radially smaller (larger) orbit. However, the energy lost (gained) by the gas element (a factor of Ω_p times the angular momentum lost [gained]) in the process of damping the wave will generally be inappropriate for the gas to remain in a circular orbit with the modified angular momentum. The excess orbital energy will be converted to heat at a rate

$$\dot{H} = (\Omega_p - \Omega)\dot{J}, \qquad (28)$$

when the orbit of the gas element circularizes by hydrodynamic dissipative mechanisms.

Note that since the gas loses (gains) angular momentum inside (outside) corotation, the rate of conversion of energy into heat, $\dot{H}$, is everywhere positive. Note also that deep inside the disk where $\Omega \gg \Omega_p$, the rate of heat gain is very large per unit rate of angular momentum transferred. In other words, although a slowly rotating disturbance in the outer part of the disk (either a normal mode or a companion star) cannot directly supply much energy to the inner disk (only a factor Ω_p times $\dot{J}$), it potentially comprises a huge reservoir for dumping the angular momentum of the system. Removing angular momentum from the part of the disk inside of corotation (by direct near-resonant coupling in $m = 1$ density waves) results in accretion. When the basic orbits cannot get very noncircular (because tightly wrapped waves cannot yield highly eccentric orbits without producing either radiative damping or streamline crossing and dissipative shocks), the heat released by this nonviscous form of accretion will provide, we believe, the luminosity excess that one sees in the flat-spectrum T Tauri stars.

In the inner portions of the disk, where $\Omega \gg \Omega_p$, but still much beyond the star-disk boundary layer, we note that equation (28) for the disk heat-input differs only by a factor of 3 from the corresponding formula given by viscous accretion-disk theory. (A much bigger difference exists beyond the corotation circle, where $\Omega_p \gg \Omega$, but the physical dimensions of realistic disks may not extend to regions where the wave energy much dominates over the accretion energy.) The ability of wave transport to produce flat-spectrum sources may then depend critically on the fact that $m = 1$ wave-dissipation yields a nonuniform rate of local disk accretion $\dot{M}_{acc}$, with $\dot{M}_{acc} \propto \varpi$ needed to produce a flat-spectrum source.

It might be argued that viscous-accretion disk models could also accomodate flat-spectrum sources if we allow ourselves the luxury of assuming $\dot{M}_{acc} \propto \varpi$. Hartmann and Kenyon (1988) rightly dismissed this possibility since, independent of the details for the specification of the viscosity, the diffusive nature of the viscous process (if stable) inevitably yields a long-term evolution that satisfies the approximate condition, $\dot{M}_{acc} =$ a spatial constant, in the relevant portions of the disk (cf. Ruden and Lin 1986 or Ruden and Pollack 1991). The same criticism does not apply to accretion disks driven by global instabilities. Indeed, the dynamics of quasi-steady wave generation, propagation, and dissipation does *not* generally correspond to the condition of steady-state accretion. The implication that the (self-gravitating) disks in flat-spectrum T Tauri stars have much larger accretion rates in their outermost parts (say, $\sim 10^{-5}$ $M_\odot$ yr^{-1}) than their innermost parts (say, 10^{-8} - 10^{-7} $M_\odot$ yr^{-1}) may have important observational consequences. In particular, the tendency to pile up matter at a bottle-neck at the smallest radii leads, for a temperature distribution fixed by wave-dissipation independent of the surface density distribution (see STAR), to a local lowering of Toomre's (1964) Q parameter. If Q falls sufficiently, violent $m > 1$ gravitational instabilities may make an appearance (see, e.g., Papaloizou and Savonije 1991; Tohline and Hachisu 1990). Such modes act over a much more restricted radial range than $m = 1$ modes, but they may induce a sudden release of the piled-up material onto the central star. Two very different gravitational modes of mass transport may therefore act: $m = 1$ waves, driven by distant orbiting companions or mild instabilities characteristic of the outer disk; and $m > 1$ instabilities with high growth rates characteristic of the inner disk. The relative ineffectiveness of the first mechanism for inducing mass transport at small ϖ leads to large amounts of matter caught in the "throat;" the clearing of this material by the second mechanism may then yield the sporadic "coughs" that Hartmann et al. (1991) identify as FU Orionis outbursts. Much more work would have to be performed on this scenario, however, before we could characterize it as more than an interesting speculation.

11. Binary Stars and Planetary Systems

In their studies, ARS and STAR suggested that the existence of binary systems among the T Tauri stars may itself represent a manifestation of the prior action of $m = 1$ gravitational instabilities (see also the chapter by Pringle). Recent numerical simulations using smooth-particle-hydrodynamics codes (Lattanzio and Monaghan 1991; Bastien et al. 1991; Adams and Benz, private communication) indicate that the formation of companion objects by nonaxisymmetric disturbances (in particular, $m = 1$) in massive nebular disks may indeed be a viable physical process. Nevertheless, because the actual accumulation of mass occurs, at least in some systems, via *gradual* infall of a single molecular cloud core (cf. §§4 and 7), we should not directly apply the results of simulations of fully-accumulated stars plus disks to the actual binary-formation process. We believe it much more likely that *both* stars start off small (perhaps as the result of a runaway $m = 1$ gravitational instability), that the two protostars separately acquire inner accretion disks, that they may also jointly acquire a circumbinary disk, and that buildup of the entire system occurs substantially while the system lies deeply embedded within a common infalling envelope. Haro 6-10 and IRAS 16293-2422 may represent examples of such systems (see Zinnecker 1989; Blake and Carlstrom, private communication).

An interesting aspect of the realization that binary-star formation may occur in the presence of circumstellar, or even circumbinary, disks concerns the eccentricity of the resulting orbits. Duquennoy and Mayor (1990) and, especially, Mathieu et al. (1989) have emphasized the important observational point that most binaries are probably born with eccentric orbits, and that the shorter-period ones only later get circularized by stellar tidal processes. This state of affairs should be contrasted with that applicable to planetary systems, where current belief holds that most planets are born with nearly circular orbits. How do we reconcile this difference if we believe that companion stars and planets are both born from disks?

A qualitative answer to this question may come from the work of Goldreich and Tremaine (1980, hereafter GT) on the interactions of satellites and disks (see also Ward 1986). For satellite orbits of small eccentricity e and inclination i, a Fourier expansion of the potential associated with the satellite's gravity introduces a series of disturbances that have wave frequency ω given by (cf., e.g., Shu 1984)

$$\omega = m\Omega_s \pm n\kappa_s \pm p\mu_s, \tag{29}$$

where m, n, and p are positive integers, with κ_s and μ_s being the epicylic and vertical oscillation frequencies, respectively, when we expand the satellite's eccentric and inclined orbit about the circular orbit with angular rotation speed Ω_s. The integer m yields the angular dependence $\propto e^{-im\varphi}$ of the disturbance potential, whereas n and p represent the number of powers of e and $\sin i$ that enter the coefficient of the specific term in the series expansion. We refer to the terms corresponding to $n = p = 0$, which can arise even if the satellite orbit is circular, as being of *lowest order*; higher-order terms depend on the orbit having nonzero eccentricity e or inclination i.

Since the time dependence of the disturbance potential $\propto e^{i\omega t}$ and the angular dependence $\propto e^{-im\varphi}$, the pattern speed of the disturbance equals

$$\Omega_p = \omega/m. \tag{30}$$

In a steady state (including dissipation), the satellite potential excites a density response in the disk that remains stationary in a frame that rotates at the pattern speed Ω_p. This response achieves its highest values at various resonance locations. *Corotation resonances* occur at radii ϖ where

$$\Omega(\varpi) = \Omega_p; \tag{31}$$

Lindblad resonances occur where

$$\Omega(\varpi) \pm \frac{1}{m}\kappa(\varpi) = \Omega_p; \tag{32}$$

and *vertical resonances* occur where

$$\Omega(\varpi) \pm \frac{1}{m}\mu(\varpi) = \Omega_p; \tag{33}$$

with $\Omega(\varpi)$, $\kappa(\varpi)$, and $\mu(\varpi)$ representing, respectively, the circular, epicyclic, and vertical oscillation frequencies of the local gas. *Inner* resonances occur where the minus sign is chosen in equations (32) and (33); *outer* resonances, where the plus sign is chosen. For a nonaxisymmetric disturbance with given pattern speed Ω_p, we generally encounter in succession an inner resonance, a corotation circle, and an outer resonance, as we move radially outwards through the disk. For a Keplerian disk, all three frequencies are equal, $\Omega = \kappa = \mu$, and special significance attaches to $m = 1$ disturbances because the entire disk can be in near-inner Lindblad or vertical resonance with a distant companion (where $\Omega_p \approx 0$).

Consider now the gravitational back-reaction of the density response in the disk on the exciting satellite. The gravitational potential associated with the density response, being nonaxisymmetric and time-dependent in an inertial frame (but time-independent in a frame that rotates at Ω_p), conserves neither the orbital angular momentum J_s nor the orbital energy E_s of the satellite; however, the combination $E_s - \Omega_p J_s =$ Jacobi's constant *is* conserved. As a consequence, the back-reaction from the disk pumps energy and angular momentum into the orbit of the satellite in the ratio Ω_p, $\dot{E}_s = \Omega_p \dot{J}_s$. For the lowest-order resonances (corotation or Lindblad), where $n = p = 0$, equations (29) and (30) states that $\Omega_p = \Omega_s$; consequently, these lowest-order resonances contribute energy and angular momentum changes to the satellite in exactly the ratio as to keep circular orbits circular. Eccentricity changes in the satellite's orbit arise only from the contributions of the higher-order resonances. In particular, GT show that higher-order Lindblad resonance (inner or outer), contribute to eccentricity *growth*, whereas the associated corotation resonances contribute to eccentricity *damping*.

In an untruncated disk of uniform surface density, GT demonstrate that the sum of the effects of all resonances (where the back-reaction on the satellite is calculated as equal and opposite to the *linear* response of the disk to the satellite) has a slight dominance of corotation resonances over Lindblad resonances, so that the net effect amounts to eccentricity damping (see also Ward 1986 for estimates of the effects of disk surface density gradients). This presumably represents the approximate state of affairs for planets embedded in nebular disks, provided they do not open up very large gaps (Lin and Papaloizou 1986a,b). For a massive companion such as another star formed out of the disk around a primary, however, not only would the companion be born with an initial eccentricity (if the formation mechanism involves the $m = 1$ eccentric instability discussed earlier), but wide gaps opened up on either side of the secondary star may lead to the disappearance of the strongest corotation resonances. The remaining (lower members of high-order) Lindblad resonances may then dominate the disk-satellite interactions, and the eccentricity of the companion star's orbit would be amplified in time. (See the numerical simulations of Artymowycz et al. [1991] concerning the effects of a circumbinary disk. See also the case of GW Ori, which appears to be a counter-example of a binary system with both a nearly *circular* orbit and one or more massive disks [Mathieu, Adams, and Latham 1991].) The final eccentricity acquired remains uncertain, however, to the extent that the action of the Lindblad resonances may themselves open up gaps – as they are known to do in the case of the Saturnian ring-moonlet system (see the discussion of Cuzzi et al. [1985]). Also uncertain remains the question of how much higher an efficiency for sweeping up the matter of a disk by a companion star would the inducement of large eccentricities allow.

In summary, the orbital characteristics of newly-formed planets and companion stars might qualitatively differ from one another because the two types of objects originate by completely different processes. On the other hand, even if they both descend from disks, we see that good mechanistic reasons exist why binary stars might acquire appreciably eccentric orbits, whereas planets usually do not. The difference in outcomes – binary stars or planets – might then depend on differences in initial conditions, namely, the total amount of angular momentum J_{tot} contained in the collapsing system. If J_{tot} is small, a substantial central mass could accumulate by direct infall before disk formation occurs. The stabilizing influence of a massive center of attraction may then prevent any runaway gravitational instabilties in the disk, leaving only the possibility of planetary condensation by chemical or physical means. If J_{tot} is large, the ratio of disk to star mass may reach order unity before the central object has grown sufficiently to prevent the gravitational fragmentation of the disk to form another body on an eccentric orbit. In this case, continued infall builds up a binary-star system.

This work was supported in part by a NSF grant and in part under the auspices of a special NASA astrophysics theory program which funds a joint Center for Star Formation Studies at NASA-Ames Research Center, U. C. Berkeley, and U. C. Santa Cruz.

12. References

Adams, F. C., Emerson, J. P., and Fuller, G. A. 1990, *Ap. J.*, **357**, 606.
Adams, F. C., Lada, C. J., and Shu, F. H. 1987, *Ap. J.*, **312**, 788. (ALS)
Adams, F. C., Lada, C. J., and Shu, F. H. 1988, *Ap. J.*, **326**, 865.
Adams, F. C., Ruden, S. P., and Shu, F. H. 1989, *Ap. J.*, **347**, 959. (ARS)
Adams, F. C., and Shu, F. H. 1986, *Ap. J.*, **308**, 836.
Ambartsumian, V. A. 1980, *Ann. Rev. Astr. Ap.*, **18**, 1.
Appenzeller, I., Jankovics, I., and Oestreicher, R. 1985, *Astr. Ap.*, **141**, 108.
Appenzeller, I., and Mundt, R. 1991, *Astr. and Ap.*, in press.
Appenzeller, I., and Tscharnuter, W. 1975, *Astr. Ap.*, **30**, 423.
Arons, J., and Max, C. 1975, *Ap. J. (Letters)*, **196**, L77.
Arquilla, R., and Goldsmith, P. F. 1986, *Ap. J.*, **303**, 356.
Artymowicz, A., Clarke, C. J., Lubow, S. H., and Pringle, J. E. 1991, *Ap. J. (Letters)*, **370**, L35.
Bachiller, R., and Cernicharo, J. 1991, *Astr. Ap.*, in press.
Balbus, S., and Hawley, J. 1991, *Ap. J.*, in press.
Bally, J. 1987, *Irish Astr. J.*, **17**, 270.
Bally, J., and Lada, C. J. 1983, *Ap. J.*, **265**, 824.
Basri, G., and Marcy, G. 1991, preprint.
Bastien, P., Arcoragi, J-P., Benz, W., Bonnell, I., and Martel, H. 1991, *Ap. J.*, submitted.
Bastien, P., and Menard, F. 1988, *Ap. J.*, **326**, 334.
Bastien, P., Carmelle, R., and Nadeau, R. 1989, *Ap. J.*, **339**, 1089.
Batrla, W., and Menten, K. M. 1985, *Ap. J. (Letters)*, **298**, L19.
Beckwith, S., Sargent, A. I., Chini, R., and Gusten, R. 1990, *A. J.*, **99**, 924.
Bertout, C. 1987, in *Circumstellar Matter*, ed. I. Appenzeller and C. Jordan (Dordrecht: Reidel), p. 45.
Bertout, C. 1989, *Ann. Rev. Astr. Ap.*, **27**, 351.
Bertout, C., Basri, G., and Bouvier, J. 1988, *Ap. J.*, **330**, 350.
Bieging, J. H., Cohen, M., and Schwartz, P. R. 1984, *Ap. J.*, **282**, 699.
Blandford, R. D. 1976, *M.N.R.A.S.*, **176**, 465.
Blandford, R. D., and Payne, D. G. 1982, *M.N.R.A.S.*, **199**, 883.
Blitz, L. 1987, in *Physical Processes in Interstellar Clouds*, ed. G. Morfill and M. Scholer (Dordrecht: Reidel), p. 35.
Blitz, L. 1990, in *The Evolution of the Interstellar Medium*, ed. L. Blitz (San Francisco: Astronomical Society of the Pacific), p. 273.
Blitz, L., and Shu, F. H. 1980, *Ap. J.*, **238**, 148.
Bodenheimer, P., and Sweigart, A. 1968, *Ap. J.*, **152**, 515.
Bonnell, I., and Bastien, P. 1991, *Ap. J.*, in press
Bouvier, J. 1991, in *Angular Momentum Evolution of Young Stars*, NATO ARW, ed. S. Catalano and J. R. Stauffer, in press
Bouvier, J., Bertout, C., Benz, W., and Mayor, M. 1986, *Astr. Ap.*, **165**, 110.
Butner, H. M., Evans, N. J., Lester, D. F., Levreault, R. M., and Strom, S. E. 1991, *Ap. J.*, in press.
Cabrit, S., Edwards, S., Strom, S. E., and Strom, K. M. 1989, *Ap. J.*, **354**, 687.
Cameron, A. G. W. 1962, *Icarus*, **1**, 13.
Chandrasekhar, S. 1960, *Proc. Nat. Acad. Sci.*, **46**, 253.
Chandrasekhar, S. 1965, *Hydrodynamic and Hydromagnetic Stability* (Oxford: Clarendon Press), pp. 384-389.

Chiosi, C., and Maeder, A. 1986, *Ann. Rev. Astr. Ap.*, **24**, 329.
Cohen, M. 1984, *Phys. Rep.*, **116(4)**, 173.
Cohen, M., and Kuhi, L. 1979, *Ap. J. Suppl.*, **41**, 743.
Cudworth, K. M., and Herbig, G. H. 1979, *A. J.*, **84**, 548.
Cuzzi, J. N., Lissauer, J. J., Esposito, L. W., Holberg, J. B., Marouf, E. A., Tyler, G. L., and Boichot, A. 1984, in *Planetary Rings*, ed. R. Greenberg and A. Brahic (Tucson: University of Arizona Press), p. 73.
DeCampli, W. 1981, *Ap. J.*, **244**, 124.
Duquennoy, A., and Mayor, M. 1990, in *Proc. XI European Astr. Meeting of the IAU*, ed. M. Vasquez (Cambridge Univ. Press), in press.
Durisen, R. H., Yang, S., Cassen, P., and Stahler, S. W. 1989, *Ap. J.*, **345**, 959.
Dziembowski, W. A., Goode, P. R., and Libbrecht, K. G. 1989, *Ap. J. (Letters)*, **337**, L53.
Eddington, A. S. 1930, *The Internal Constitution of the Stars* (Cambridge University Press).
Edwards, S., Cabrit, S., Strom, S. E., Ingeborg, H., Strom, K. M., and Andersen, E. 1987, *Ap. J.*, **321**, 473.
Edwards, S., Mundt, R., and Ray, T. 1991, in *Protostars and Planets III*, ed. E. Levy and J. Lunine (Tucson: University of Arizona Press), submitted.
Elmegreen, B. G. 1979, *Ap. J.*, **232**, 729.
Elmegreen, B. G. 1985, in *Protostars and Planets II*, ed. D. C. Black and M. S. Matthews (Tucson: University of Arizona Press), p. 33.
Elmegreen, B. G., and Clemens, C. 1985, *Ap. J.*, **294**, 523.
Elmegreen, B. G., and Lada, C. J. 1977, *Ap. J.*, **214**, 31.
Elsasser, H., and Staude, H. J. 1978, *Astr. Ap.*, **70**, L3.
Evans, N. J. 1991, in *Frontiers of Stellar Evolution*, ed. D. L. Lambert (San Francisco: Astronomical Society of the Pacific), in press.
Fricke, K. 1969, *Astr. Ap.*, **1**, 388.
Fukui, Y. 1989, in *Low Mass Star Formation and Pre-Main Sequence Objects*, ed. B. Reipurth, ESO Conf. and Workshop Proc. No. 33, p. 95.
Fuller, G. A., and Myers, P. C. 1987, in *Physical Processes in Interstellar Clouds*, ed. M. Scholer (Dordrecht: Reidel), p. 137.
Galli, D. 1990, Ph.D. Thesis, University of Florence.
Galli, D., and Shu, F. H. 1991, in preparation.
Gaustad, J. E. 1963, *Ap. J.*, **138**, 1050.
Giovanardi, C., Lizano, S., Natta, A., Palla, F., Evans, N. J., and Heiles, C. E. 1991, in preparation.
Goldreich, P., and Tremaine, S. 1980, *Ap. J.*, **241**, 425. (GT)
Goodman, A. A., Benson, P. J., Fuller, G. A., and Myers, P. C. 1991, in preparation.

Gusten, R., and Mezger, P. G. 1982, *Vistas Astron.*, **26**, 159.
Hartmann, L., and Kenyon, S. J. 1987, *Ap. J.*, **312**, 243.
Hartmann, L., and Kenyon, S. J. 1988, in *Formation and Evolution of Low Mass Stars*, ed. A. K. Dupree and M. T. V. T. Lago (Dordrecht: Kluwer Academic Pub.), p. 163.
Hartmann, L., Kenyon, S. J., and Hartigan, P. 1991, in *Protostars and Planets III*, ed. E. Levy and J. Lunine (Tucson: University of Arizona Press), submitted.
Hartmann, L., Jones, B. F., Stauffer, J. R., and Kenyon, S. J. (1991), *A. J.*, in press.
Hartmann, L., and MacGregor, K. 1982, *Ap. J.*, **259**, 180.

Hartmann, L., Soderblom, D. R., and Stauffer, J. R. 1987, *Ap. J.*, **93**, 907.
Hawley, J., and Balbus, S. 1991, *Ap. J.*, in press.
Hayashi, C. 1966, *Ann. Rev. Astr. Ap.*, **4**, 171.
Hayashi, C. 1970, *Mem. Soc. R. Sci. Liege*, **19**, 127.
Hayashi, C., Hoshi, R., and Sugimoto, D. 1962, *Prog. Theor. Phys. Suppl.*, **22**, 1.
Henriksen, R. N., and Rayburn, D. R. 1971, *M.N.R.A.S.*, **152**, 323.
Henyey, L. G., LeLevier, R., and Levee, R. D. 1955, *P. A. S. P.*, **67**, 154.
Herbig, G. H. 1962, *Adv. Astron. Astrophys.*, **4**, 337.
Herbig, G. H. 1989, in *Low Mass Star Formation and Pre-Main Sequence Objects*, ed. B. Reipurth, ESO Conf. and Workshop Proc. No. 33, p. 233.
Herbig, G. H., and Goodrich, R. W. 1986, *Ap. J.*, **309**, 294.
Ho, P. C. T., and Haschick, A. D. 1986, *Ap. J.*, **305**, 714.
Imhoff, C. L., and Appenzeller, I. 1987, in *Scientific Accomplishments of the IAU*, ed. Y. Kondo (Dordrecht: Reidel), p. 295.
James, R. A. 1964, *Ap. J.*, **140**, 552.
Kaifu, N., Suzuki, S., Hasegawa, T., Morimoto, M., Inatani, J., et al. 1984, *Astr. Ap.*, **134**, 7.
Keene, J., and Masson, C. 1990, *Ap. J.*, **355**, 635.
Kenyon, S. J., and Hartmann, L. 1987, *Ap. J.*, **323**, 714.
Kenyon, S. J., Hartmann, L., Strom, K. M., and Strom, S. E. 1990, *A. J.*, **99**, 869.
Keto, E., Ho, P. T. P., and Haschick, A. D. 1987, *Ap. J.*, **318**, 712.
Konigl, A. 1989, *Ap. J.*, **342**, 208.
Konigl, A. 1991, *Ap. J. (Letters)*, **370**, L39.
Koo, B. C. 1989, *Ap. J.*, **318**, 331.
Knapp, G. R., Kuiper, T. B., Knapp, S. L., and Brown, R. L. 1976, *Ap. J.*, **206**, 443.
Kuhi, L. 1974, *Astr. Ap. Suppl.*, **15**, 47.
Kuhi, L. 1978, in *Protostars and Planets*, ed. T. Gehrels (Tucson: University of Arizona Press), p. 708.
Kusaka, T., Nakano, T., and Hayashi, C. 1970, *Prog. Theor. Phys.*, **44**, 1580.
Lada, C. J. 1985, *Ann. Rev. Astr. Ap.*, **23**, 267.
Lada, C. J., Margulis, M., and Dearborn, D. 1984, *Ap. J.*, **285**, 141.
Lada, C. J., and Shu, F. H. 1990, *Science*, **248**, 564.
Larson, R. B. 1969a, *M.N.R.A.S.*, **145**, 271.
Larson, R. B. 1969b, *M.N.R.A.S.*, **145**, 297.
Larson, R. B. 1986, *M.N.R.A.S.*, **218**, 409.
Lattanzio, J., and Monaghan, J. 1991, preprint.
Levreault, R. M. 1985, Ph.D. Thesis, University of Texas, Austin.
Lin, C. C., and Shu, F. H. 1964, *Ap. J.*, **140**, 646
Lin, D. N. C., and Papaloizou, J. 1985, in *Protostars and Planets II*, ed. D. C. Black and M. S. Matthews (Tucson: University of Arizona Press), p. 981.
Lin, D. N. C., and Papaloizou, J. 1986a, *Ap. J.*, **307**, 395.
Lin, D. N. C., and Papaloizou, J. 1986b, *Ap. J.*, **309**, 846.
Lizano, S., Heiles, C., Rodriguez, L. F., Koo, B., Shu, F. H., Hasegawa, T., Hayashi, S., and Mirabel, I. F. 1988, *Ap. J.*, **328**, 763.
Lizano, S., and Shu, F. H. 1989, *Ap. J.*, **342**, 834.
Lovelace, R. V. E. 1976, *Nature*, **262**, 649.
Lovelace, R. V. E., Berk, H. L., and Contopoulos, J. 1991, preprint.
Lynden-Bell, D., and Kalnajs, A. 1972, *M.N.R.A.S.*, **157**, 1.
Lynden-Bell, D., and Pringle, J. E. 1974, *M.N.R.A.S.*, **168**, 603.

Lynds, B. T. 1962, *Ap. J. Suppl.*, **7**, 1.
Margulis, M., and Snell, R. L. 1989, *Ap. J.*, **343**, 779.
Masson, C., Mundy, L., and Keene, J. 1990, *Ap. J. (Letters)*, **357**, L25.
Mathieu, R. D., Adams, F. C., and Latham, D. W. 1991, *A. J.*, in press.
Mathieu, R. D., Walter, F. M., and Myers, P. C. 1989, *A. J.*, **98**, 987.
Matsumoto, R., Horiuchi, T., Shibata, K., and Hanawa, T. 1988, *Publ. Astr. Soc. Japan*, **40**, 171.
Mendoza, E. E. 1966, *Ap. J.*, **143**, 1010.
Mendoza, E. E. 1968, *Ap. J.*, **151**, 977.
Menten, K. M., and Walmsley, C. M. 1985, *Astr. Ap.*, **146**, 367.
Mercer-Smith, J. A., Cameron, A. G. W., and Epstein, R. I. 1984, *Ap. J.*, **287**, 445.
Mestel, L. 1965, *Q.J.R.A.S.*, **6**, 161.
Mestel, L. 1985, in *Protostars and Planets II*, ed. D. C. Black and M. S. Matthews (Tucson: University of Arizona Press), p. 320.
Mestel, L., and Spitzer, L. 1956, *M.N.R.A.S.*, **116**, 503.
Mezger, P. G., and Smith, L. F. 1977, in *Star Formation, IAU Symp. No. 75*, ed. T de Jong and A. Maeder (Dorecht: Reidel), p. 133.
Moriarty-Schieven, G. H., and Snell, R. L. 1988, *Ap. J.*, **332**, 364.
Moriarty-Schieven, G. H., and Wannier, P. G. 1991, preprint.
Mouschovias, T. Ch. 1976, *Ap. J.*, **207**, 141.
Mouschovias, T. Ch. 1978, in *Protostars and Planets*, ed. T. Gehrels (Tucson: University of Arizona Press), p. 209.
Mouschovias, T. Ch., and Paleologou, E. V. 1981, *Ap. J.*, **246**, 48.
Mouschovias, T. Ch., and Spitzer, L. 1976, *Ap. J.*, **210**, 326.
Mundt, R. 1985, in *Protostars and Planets II*, ed. D. C. Black and M. S. Matthews (Tucson: University of Arizona Press), p. 414.
Mundt, R., and Fried, J. W. 1983, *Ap. J. (Letters)*, L83.
Myers, P. C., Fuller, G., Goodman, A. A., and Benson, P. J. 1991, *Ap. J.*, in press.
Myers, P., and Benson, P. J. 1983, *Ap. J.*, **266**, 309.
Myers, P. C., and Goodman, A. A. 1988a, *Ap. J.*, **326**, L27.
Myers, P. C., and Goodman, A. A. 1988b, *Ap. J.* **329**, 392.
Najita, J., Shu, F. H., Lizano, S., and Ruden, S. 1991, in preparation.
Nakano, T. 1979, *Publ. Astron. Soc. Japan*, **31**, 697.
Narita, S., Nakano, T., and Hayashi, C. 1970, *Prog. Theor. Phys.*, **43**, 942.
Natta, A., Giovanardi, C., Palla, F., and Evans, N. J. 1987, *Ap. J.*, **332**, 921.
Paczynski, B. 1991, *Ap. J.*, **370**, 597.
Palla, F., and Stahler, S. 1990, *Ap. J. (Letters)*, **360**, L47.
Papaloizou, J., and Savonije, G. J. 1991, *M.N.R.A.S.*, in press.
Parker, E. 1966, *Ap. J.*, **145**, 811.
Plambeck, R. L., Wright, M. C. H., and Carlstrom, J. E. 1990, *Ap. J. (Letters)*, **348**, L65.
Popham, R., and Narayan, R. 1991, *Ap. J.*, **370**, 604.
Pringle, J. E. 1981, *Ann. Rev. Astr. Ap.*, **19**, 137.
Pringle, J. E. 1989, *M.N.R.A.S.*, **236**, 107.
Pudritz, R. E., and Norman, C. A. 1983, *Ap. J.*, **274**, 677.
Pudritz, R. E., and Norman, C. A. 1986, *Ap. J.*, **301**, 571.
Reipurth, B. 1989, in *Low Mass Star Formation and Pre-Main Sequence Objects*, ESO Conf. and Workshop. Proc. No. 33, p. 247.
Rodriguez, L. F. 1990, in *The Evolution of the Interstellar Medium*, ed. L. Blitz

(San Francisco: Astronomical Society of the Pacific), p. 183.
Rucinski, S. M. 1985, *A. J.*, **90**, 2321.
Ruden, S. P., Glassgold, A., and Shu, F. H. 1990, *Ap. J.*, **361**, 546.
Ruden, S. P., and Lin, D. N. C. 1986, *Ap. J.*, **308**, 883.
Ruden, S. P., and Pollack, J. B. 1991, preprint.
Rudolph, A., Welch, W. J., Palmer, P., and Dubrulle, B. 1990, *Ap. J.*, **363**, 528.
Ruiz, A., Alonso, J. L., and Mirabel, F. 1991, preprint.
Rydgren, A. E., and Zak, D. S. 1987, in P.A.S.P., **99**, 141.
Scalo, J. 1986, *Fund. Cosmic Phys.*, **11**, 1.
Schwartz, R. D. 1983, *Ann. Rev. Astr. Ap.*, **21**, 209.
Shu, F. H. 1970, *Ap. J.*, **160**, 99.
Shu, F. H. 1977, *Ap. J.*, **214**, 488.
Shu, F. H. 1983, *Ap. J.*, **273**, 202.
Shu, F. H. 1984, in *Planetary Rings*, ed. R. Greenberg and A. Brahic (Tucson: University of Arizona Press), p. 513.
Shu, F. H. 1987, in *Star Formation in Galaxies*, ed. C. J. L. Persson, NASA Conf. Pub. 2466, 743.
Shu, F. H. 1991, in *Frontiers of Stellar Evolution*, ed. D. L. Lambert (San Francisco, Astronomical Society of the Pacific), in press.
Shu, F. H., Adams, F. C., and Lizano, S. 1987, *Ann. Rev. Astr. Ap.*, **25**, 23.
Shu, F. H., Lizano, S., and Adams, F. C. 1987, in *Star Forming Regions, IAU Symp. No. 115*, ed. J. Jugaku and M. Peimbert (Dordrecht: Reidel), p. 417.
Shu, F. H., Lizano, S., Ruden, S. P., and Najita, J. 1988, *Ap. J.(Letters)*, **328**, L19. (SLRN)
Shu, F. H., Lizano, S., Ruden, S., and Najita, J. 1991, in preparation.
Shu, F. H., Najita, J., Galli, D., Ostriker, E., and Lizano, S. 1991, in *Protostars and Planets III*, ed. E. Levy and J. Lunine (Tucson: University of Arizona Press), submitted.
Shu, F. H., Ruden, S. P., Lada, C. J., and Lizano, S. 1991, *Ap. J. (Letters)*, **370**, L31.
Shu, F. H., and Terebey, S. 1984, in *Cool Stars, Stellar Systems, and the Sun*, ed. S. Baliunas and L. Hartmann (Berlin: Spinger-Verlag), p. 78.
Shu, F. H., Tremaine, S., Adams, F. C., and Ruden, S. P. 1990 *Ap. J.*, **358**, 495. (STAR)
Snell, R. L. 1987, in *Star-Forming Regions, IAU Symp. No. 115*, ed. M. Peimbert and J. Jugaku (Dordrecht:Reidel), 213.
Snell, R. L., Loren, R. B., and Plambeck, R. L. 1980, *Ap. J. (Letters)*, **239**, L17.
Snell, R. L., and Schloerb, F. B. 1985, *Ap. J.*, **295**, 490.
Stahler, S. W. 1983, *Ap. J.*, **274**, 822.
Stahler, S. W. 1988, *Ap. J.*, **332**, 529.
Stahler, S. W., Shu, F. H., and Taam, R. E. 1980, *Ap. J.*, **241**, 637. (SST)
Stella, L., and Rosner, R. 1984, *Ap. J.*, **277**, 312.
Stocke, J. T., Hartigan, P. M., Strom, S. E., Strom, K. M., Anderson, E. R., Hartmann, L., and Kenyon, S. J. 1988, *Ap. J. Suppl.*, **68**, 229.
Strom, S. E., Grasdalen, G. L., and Strom, K. M. 1974, *Ap. J.*, **191**, 111.
Strom, K. M., Strom, S. E., Edwards, S., Cabrit, S., and Skrutskie, M. 1989, *A. J.*, **97**, 1451.
Strom, K. M., Strom, S. E., and Vrba, F. J. 1976, *A. J.*, **81**, 320.
Terebey, S., Shu, F. H., and Cassen, P. 1984, *Ap. J.*, **286**, 529.

Tohline, J. E., and Hachisu, I. 1990, *Ap. J.*, **361**, 394.
Tomisaka, K., Ikeuchi, S., and Nakamura, T. 1988, *Ap. J.*, **326**, 208.
Tomisaka, K., Ikeuchi, S., and Nakamura, T. 1989, *Ap. J.*, **341**, 220.
Tomisaka, K. 1991, preprint.
Toomre, A. 1964, *Ap. J.*, **139**, 1217.
Toomre, A. 1969, *Ap. J.*, **158**, 899.
Uchida, Y. and Shibata, K. 1985, *Pub. Astr. Soc. Japan*, **37**, 515.
Uchida, Y., Kaifu, N., Shibata, K., Hayashi, S. S., and Hasegawa, T. 1987, in *Star-Forming Regions, IAU Symp. No. 115*, ed. M. Peimbert and J. Jugaku (Dordrecht:Reidel), p. 287.
Vogel, S. S., and Kuhi, L. V. 1981, *Ap. J.*. **245**, 960.
Vogt, S. 1981, *Ap. J.*, **250**, 327.
Walker, C., Adams, F. C., and Lada, C. J. 1990, *Ap. J.*, **349**, 515.
Walter, F. M., and Boyd, W. T. 1991, *Ap. J.*, **370**, 318.
Ward, W. 1986, *Icarus*, **67**, 164.
Welch, W. J., Dreher, J. W., Jackson, J. M., Terebey, S., and Vogel, S. N. 1987, *Science*, **238**, 1550.
Welch, W. J., Vogel, S. N., Plambeck, R. L., Wright, M. C. H., and Bieging, J. H. 1985, *Science*, **228**, 1329.
Westbrook, C. K., and Tarter, C. B. 1975, *Ap. J.*, **200**, 48.
Winkler, K. H., and Newman, M. J. 1980a, *Ap. J.*, **236**, 201.
Winkler, K. H., and Newman, M. J. 1980b, *Ap. J.*, **238**, 311.
Zhou, S., Evans, N. J., Butner, H. M., Kutner, M. L., Leung, C. M., and Mundy, L. G. 1990, *Ap. J.*, **363**, 168.
Zinnecker, H. 1989, in *Low Mass Star Formation and Pre-Main Sequence Objects*, ed. B. Reipurth, ESO Conf. and Workshop Proc. No. 33, 233.

NUMERICAL STUDIES OF CLOUD COLLAPSE

W. M. TSCHARNUTER
Institut für Theoretische Astrophysik der Universität
Im Neuenheimer Feld 561
W-6900 Heidelberg 1
Germany

ABSTRACT. A review is given of the various attempts that have been made in the past two decades to simulate the stellar formation process by hydrodynamical calculations including self-gravitation and radiation. As a general basis, the classical results on spherically symmetric collapse are outlined and the key rôle of the large disparity between the three characteristic timescales pertinent to protostellar collapse, i.e., the (dynamical) free-fall time, the (thermal) Kelvin-Helmholtz time, and the accretion time is emphasized for designing efficient numerical codes. After a brief discussion of the physical ingredients and newly developed numerical tools, recent results on the stability of protostellar accretion flows and cores with spherical (1-D) and axial (2-D) symmetry are presented. Turbulent friction in the 2-D case, parametrized by the 'α-viscosity', gravitational torques according to the development of non-axisymmetric (spiral-like or bar-like) structures in the full 3-D models are shown to be efficient mechanisms of angular momentum transport.

1. Introduction

Observations show that stars form predominantly in the cold and dense cores of interstellar molecular clouds. Typical numbers for the density and temperature derived from observational data lead to the conclusion that almost all cloud cores ought to be gravitationally unstable, unless the magnetic field permeating the interstellar medium is strong enough to stabilize molecular clouds as a whole. Measured field strengths (for a useful compilation, see Fiebig and Güsten 1989) as well as theoretical investigations (e.g., Shu et al. 1987) make it highly probable that the typical magnetic flux is large enough to compensate the clouds' (self-) gravity. However, above a certain density, ambipolar diffusion becomes efficient and weakens the stabilizing effect of magnetic forces, particularly in the densest regions. At which number densities decoupling of the magnetic field from the gas actually occurs is somewhat controversial among the various authors. Critical values range from 10^4–$10^6 cm^{-3}$ (Mouschovias et al. 1985) to 10^9–$10^{11} cm^{-3}$ (Nakano 1984, 1985). But eventually dynamical collapse will start off and the star formation process is initiated (see chapters by T. Mouschovias and F. Shu, respectively). Further empirical support is given to this scenario in a recent paper by Schulz et al. (1991), in which dense clumps observed in NGC 2024 are interpreted to be protostellar condensations in the very early, isothermal stage.

Thus collapse is the ultimate fate of (clumps in) cloud cores, but there remains the disturbing fact that no protostars in the so-called 'main accretion' phase which, according to theory, should last about $10^5 yr$ have been detected as yet. By contrast, mass outflow from young stellar objects (YSO) is observed, either in bipolar nebulae or in jets, rather than overall mass inflow. It is believed that once again magnetic fields play the dominant rôle in linking accretion onto the YSO from the ambient disk-like nebula with the collimated mass outflow in the direction of the rotation axis of the system (see chapter by R. Pudritz). In this way magnetic fields, presumably re-generated by dynamo processes in the YSO and/or the disk, are likely to determine the 'late' stages of stellar formation.

The big challenge for theory is to model the transition from the undifferentiated isothermal initial state of clumpy protostellar fragments to the rather differentiated structure of the (pre-) T Tau state, where the central star has already separated out from the ambient disk. After all, there is a gap in the range of 12–15 orders of magnitude in the density between the two states which are accessible to observations at present. Gravitational collapse is the only process that is known to be able to fill this gap in a natural way. The dynamics of the collapse is most probably not affected by magnetic forces, because otherwise dynamical contraction is not able to commence. However, it cannot be ruled out that, subsequent to the overall collapse, magnetic torques become important on timescales longer than the (local) free-fall time and become, in addition to 'turbulent' friction and gravitational torques, the main source of 'viscosity'. This is not only needed for maintaining the overall mass inflow by enhanced redistribution of angular momentum but also for spinning down the central parts of the collapsed fragment considerably, thus forming the (pressure supported) central star proper together with a small transition zone, the so-called boundary layer (BL), as the link to the (centrifugally supported) disk-like ambient nebula. At the same time, nevertheless, accretion onto the star is still continuing, which tries to spin up the star again, unless most of the surplus angular momentum advected inward is lost by — presumably magnetically driven — mass outflow. In this context it is interesting to note that, even if the core were rotating critically, continuous mass accretion driven by outward directed viscous angular momentum transport *through* the disk would be possible (Popham and Narayan 1990).

Because of the complexity inherent to the processes briefly outlined above, it is not surprising that the numerical simulations do not cover all details necessary for comparison with observational findings. In particular, magnetic fields have not at all been taken into consideration for modelling the collapse phase. As long as no particular topology of the magnetic field has to be taken into consideration in order to channel outflow or to power jets, a crude, zeroth order approximation would be to subsume magnetic torques as 'magnetic' viscosity under the α-ansatz for 'turbulent' viscosity. This implies that α can then take any value (even greater than 1!) and has to be fitted to observations. This idea has been followed in axisymmetric (2-D) models (see subsection 6.2). 3-D model calculations have shown that rapid and efficient redistribution of angular momentum is made most effectively by gravitational torques if genuine three-dimensional, spiral-like or bar-like, structures develop (for a most recent review with emphasis on the formation of the solar nebula see, e.g., Boss 1987; Tscharnuter and Boss 1991, and references cited therein).

In this chapter we restrict ourselves to the discussion of protostellar collapse models based on the numerical solution of the full system of structure equations. Collapse models that make use of asymptotic similarity solutions (cf. Shu 1977; Stahler et al. 1980a,b, 1981) are not discussed here. Section 2 deals with the classical results for spherically symmetric collapse and general relations independent of symmetry constraints. In sections 3, 4, and 5 the basic physics, mathematics, and numerical tools, respectively, are discussed. As an illustration, some typical model sequences are presented in section 6.

2. Classical Collapse Models

In 1969 a breakthrough in our understanding of stellar formation came with R. B. Larson's famous paper on the collapse of a (non-magnetic, non-rotating) spherically symmetric gas cloud of total mass $M = 1 M_\odot$ on the verge of being Jeans-unstable. For typical temperatures $T \approx 10K$ in the dark cores of interstellar molecular clouds the critical mean density $\bar{\rho}$ for the onset of gravitational instability is about $10^{-19} g\, cm^{-3}$. In general, the so-called 'Jeans-mass' M_J for an isothermal self-gravitating gas sphere consisting mainly of molecular hydrogen and helium (according to the cosmic abundances) is given by

$$\frac{M_J}{1 M_\odot} \approx \left(\frac{\bar{\rho}}{10^{-19} g\, cm^{-3}}\right)^{-1/2} \left(\frac{T}{10K}\right)^{3/2}. \tag{2.1}$$

The corresponding Jeans-radius R_J then is

$$R_J = \left(\frac{3 M_J}{4 \pi \bar{\rho}}\right)^{1/3}. \tag{2.2}$$

If $M > M_J$ or, equivalently, if the cloud radius $R > R_J$, given the temperature T and mean density $\bar{\rho}$, no hydrostatic equilibrium is possible and the cloud starts to collapse.

Despite the fact that spherically symmetric (1-D) collapse flows are very particular solutions of the general three-dimensional problem, they exhibit, nevertheless, characteristic properties and features which seem to be inherent to any self-gravitating flow and are widely independent of the symmetry assumed. This is why 1-D models for protostellar collapse and star formation, which reflect just a very crude and highly idealized approximation to reality, have been taken as a useful guideline for more realistic and also much more complex model calculations with less stringent geometry constraints. In addition, the assumption of spherical symmetry allows a rather complete description of the stellar formation process, although there are still 'dark corners' in the building of the numerical models to be explored (cf. sect. 6 below).

One major difficulty to model the collapse and the ensuing accretion phase in a consistent way relates to the fact that the various timescales pertinent to the flow (dynamical, thermal, mass accretion) differ by several orders of magnitude. Expressed in mathematical terms, the system of the (non-linear) structure equations is a *stiff* one, and implicit numerical methods as well as iteration procedures which are in general extremely expensive in terms of computer time have to be applied.

Larson (1969) was the first to solve this rather difficult technical problem. However, he had to *assume* that the accreting protostellar core is in *strict* hydrostatic equilibrium. Thus the dynamical timescale of the core, which is orders of magnitude shorter than the natural evolutionary accretion timescale, is eliminated and cannot cause trouble anymore for the simulation. Large time steps in the numerical model sequences allow for a rather efficient calculation of the main accretion phase. In the following subsection 2.1. the essential results of Larson's early calculations are compiled.

2.1. GENERAL FEATURES OF COLLAPSE FLOWS

The 'canonical' initial values for protostellar collapse models correspond to an isothermal, homogeneous gas cloud of total mass M, density ρ_i, and temperature T_i, on the verge of being Jeans-unstable according to eq. (2.1), e.g., $M = 1 M_\odot$, $\rho_i = 10^{-19} g\, cm^{-3}$, $T_i = 10K$, and the cloud radius $R \approx 0.1 pc$. The collapse timescale, which is of the order of the free-fall time t_{ff} of a pressureless dust cloud, is proportional to $\rho_i^{-1/2}$ (cf. subsec. 2.2. below) and thus strongly dependent on the initial density. Fortunately, the qualitative patterns of the flow that develop during the collapse do not depend sensitively on the boundary conditions, unless highly artificial prescriptions are chosen. Usually two cases are considered:

a) the collapse takes place within a constant (spherical) volume, i.e., $R = const.$,
b) the outer boundary R of the cloud moves according to the assumption that the external pressure $P_{ex} = const.$,

and in both cases the equivalent temperature of the external radiation field T_{ex} impinging on the cloud is assumed to be a known quantity. Five characteristic phases can be discerned in the evolution of the collapse:

1. As long as the central density $\rho_c < 10^{-13} g\, cm^{-3}$ the collapse is isothermal ($T \approx T_i \approx T_{ex}$), since the compressional energy is easily radiated away in the far infrared (FIR) via the microscopic interstellar dust particles suspended in the gas flow, and proceeds in an extremely non-homologous fashion. Hence, the density increases most rapidly in the central part of the collapsing fragment, while the outer layers are 'left behind', developing a density distribution $\rho \propto r^{-2}$ (r is the radial distance from the center). After only a little longer than one free-fall time ($\approx 10^5 yr$), the structure of a collapsing, asymptotically singular isothermal gas sphere is established.
2. When ρ_c rises above $10^{-13} g\, cm^{-3}$ the innermost layers become optically thick for IR-radiation, the temperature rises, and pressure forces are able to halt the collapse. A quasi-hydrostatic core — the first core — forms. Its linear dimension is only a few astronomical units (as compared to about $0.1 pc$ for the original cloud radius), within which just a few percent of a solar mass are contained.
3. When the central temperature T_c rises above $2 \cdot 10^3 K$ hydrogen molecules begin to dissociate. This causes the first adiabatic exponent, defined as $\Gamma_1 = (\partial \log P / \partial \log \rho)_S$, i.e., as the logarithmic derivative of the pressure by the density at constant entropy S, to drop below the critical value $\Gamma_1^{crit} = 4/3$ indicating dynamical instability for spherically symmetric hydrostatic equilibria. Again, the pressure support can no longer sustain the increasing gravitational pull, in particular near the center, a second collapse — now almost adiabatic — commences. The bounce occurs at stellar densities $\rho_c \approx 10^{-2} g\, cm^{-3}$,

when hydrogen is already substantially ionized so that $\Gamma_1 > 4/3$ again. The end result is a tiny star-like core — the second (stellar) core — which quickly grows to 0.01–$0.02 M_\odot$ in less than $10^2 yr$.
4. Beginning of the main accretion phase. If the stellar core remains vibrationally and dynamically stable, which is anticipated if the core is assumed to be strictly hydrostatic, the evolutionary timescale t_{acc} is now dependent on the mean density of the collapsing cloud as a whole and is of the order of the initial free-fall time, rather independent of the boundary conditions mentioned above (fixed volume or fixed external pressure). The protostar's luminosity is produced by the accretion shock, where the kinetic energy of the infalling material is thermalized and finally transformed into radiation. The density distribution in the almost free-falling envelope changes from the r^{-2}-law to $\rho \propto r^{-3/2}$.
5. Depending on the mass available for accretion onto the protostellar core, there are two possible evolutionary tracks toward the main sequence:
   a) Massive protostars ($M > 3 M_\odot$) arrive at the main sequence, while accretion is still going on ('cocoon' stars),
   b) For low-mass protostars ($M \leq 3 M_\odot$) accretion becomes inefficient before thermonuclear hydrogen burning ignites. The 'naked' stellar 'core' starts to contract toward the main sequence in a Kelvin-Helmholtz time t_{KH} and its luminosity is generated by the slow quasi-hydrostatic contraction according to the virial theorem: one half of the gravitational energy released is used for heating the pre-main-sequence star, the other half is radiated away.

For our canonical $1 M_\odot$-case the dynamical evolution leads to a relatively compact, low-luminosity object at the bottom of the Hayashi track.

These general results obtained by Larson have been confirmed by various authors (Appenzeller and Tscharnuter, 1974, 1975; Tscharnuter and Winkler, 1979; Winkler and Newman, 1980a,b; Balluch, 1988). Much effort has gone into trying to get rid of the key assumption that protostellar cores be strictly hydrostatic. Surprisingly enough, it turned out that there is a major difficulty in carrying out a clean hydrodynamical calculation which would cover the transition stage between the formation of the stellar core embryo and the beginning of the main accretion phase. As a matter of fact, such clean model sequences which satisfy all demands on numerical accuracy and reliability do not yet exist. The reason for this shortcoming is discussed in section 6.

2.2. TIMESCALES

As pointed out in the previous subsection, the specific difficulties of modelling protostellar collapse arise from the huge disparity of the timescales inherent to the problem. There are also large variations of the physical quantities both in space and time. As an example, the density varies over more than 20 orders of magnitude, and the temperature by about a factor of 10^6. Three timescales are relevant:

1. The (mean) free-fall time t_{ff} is defined as the period of time during which a non-rotating, non-magnetic pressureless dust cloud of (mean) density $\bar{\rho}$ collapses

completely,

$$t_{ff} = \sqrt{\frac{3\pi}{32\,G\,\rho}} \qquad (2.3a)$$

where G is the gravitational constant. t_{ff} is also the appropriate unit of time measuring the duration of the collapse of gas clouds and the period of stellar oscillations. Note that t_{ff} depends only on the density ρ. For a hydrostatic self-gravitating star or protostellar core, t_{ff} can also be interpreted as the time a sound wave takes to travel through the core's interior. If R_{core} is the core radius and $c_s = (\frac{\mathcal{R}}{\overline{\mu}}\overline{T})^{1/2}$ the sound velocity corresponding to the core's mean temperature $\overline{T}$ and mean molecular weight $\overline{\mu}$ ($\mathcal{R}$ is the universal gas constant), the free-fall time is of the order of the crossing time of a sound wave,

$$t_{ff} \approx \frac{R_{core}}{c_s}. \qquad (2.3b)$$

The link between these two definitions of t_{ff} is again the virial theorem. If M_{core} denotes the mass of the core we find

$$c_s^2 = \frac{\mathcal{R}}{\overline{\mu}}\overline{T} \approx G\,\frac{M_{core}}{R_{core}}. \qquad (2.4)$$

2. The thermal or Kelvin-Helmholtz timescale t_{KH} of the core is defined by

$$t_{KH} = \frac{|W|}{L} \approx 2.4 \times 10^{-6} \kappa \frac{M_{core}^3}{R_{core}^5\,\overline{T}^4} \quad seconds. \qquad (2.5)$$

W is the gravitational energy, L the luminosity, and κ the opacity, i.e., the Rosseland mean of the absorption coefficient. t_{KH} is of the same order as the time a heat wave takes to propagate through the (optically thick) core.

3. If $\dot{M}$ denotes the mass accretion rate per unit time the accretion timescale is defined by

$$t_{acc} = \frac{M_{core}}{\dot{M}}. \qquad (2.6)$$

There are characteristic relations between these three timescales. For protostellar cores the inequalities

$$t_{ff} \ll t_{KH}, \quad t_{ff} \ll t_{acc} \qquad (2.7)$$

are always valid. As a consequence of the fact that the free-fall time is much — orders of magnitude! — smaller than the thermal and the accretion time, protostellar cores may be considered to be in the state of almost perfect hydrostatic equilibrium. However, this assumption, though simplifying the calculations considerably, becomes doubtful if oscillations of the core with periods of the order of t_{ff} are to be excited. In fact, crude estimates first made by Wuchterl (1989, 1990) led him to the conjecture that newly formed protostellar core 'embryos' are altogether vibrationally unstable due to a strong κ-mechanism. This has been confirmed by Balluch (1989) with a global linear stability analysis which is being published in a series of forthcoming papers (1991a,b,c). He was able to show that protostellar accretion flows are intrinsically unstable. However, with the hydrodynamical codes available to date we are not yet able to answer the crucial question of how large the amplitudes of the core oscillations will actually grow in the non-linear regime.

There is no clear-cut relation between the two remaining timescales, t_{KH} and t_{acc}. The first case, $t_{KH} < t_{acc}$, expresses the fact that accretion has become inefficient and the protostellar core evolves toward the main sequence on a *thermal* time scale, similar to a classical pre-main-sequence star ($t_{acc} \to \infty$). The luminosity is generated by the release of gravitational energy due to the *slow*, quasi-hydrostatic contraction. The other possibility, $t_{KH} > t_{acc}$, characterizes rapid accretion which is typical for the main accretion phase in spherical models. The stellar material cannot adjust thermally and evolves adiabatically. The luminosity of the protostar is generated at the accretion shock by the sudden dissipation of the kinetic energy of the infalling material into heat which, in turn, is transformed into radiation.

3. Basic Physics

In this section a brief overview is given about the basic physical processes which have to be taken into account in order to establish realistic models of protostellar collapse. The raw material in the cold cores of molecular clouds is a mixture of hydrogen in molecular form, helium, and heavier elements according to cosmic (Population I) abundances (70% hydrogen, 27% helium, 3% heavier elements by weight). About 1% of the total mass is in the form of sub-micron dust particles. Since in the deep interior of stars all atoms are practically fully ionized, it is clear that every possible dissociation and ionization state is reached during star formation. Thus the minimum requirement for a realistic approch is to take at least those species into account which pertain to the most abundant elements, i.e., H_2, H, H^+ (molecular, atomic, and ionized hydrogen, respectively), and He, He^+, He^{++} (neutral, first, and second ionization state of helium, respectively). The heavier elements and the dust give just a minor contribution as far as thermodynamics is concerned. However, they are important sources of opacity. Dust, for example, is the main absorber (and emitter) of IR-radiation in the low temperature domain, where it can survive; at higher temperatures ($\geq T_{sub}$, the maximum 'sublimation' temperature above which no dust exists, $T_{sub} \approx 1700\,K$) even the most refractory grains sublimate and the opacity drops by several orders of magnitude.

3.1. EQUATION OF STATE

The rich structure of the realistic equation of state (EOS) — as compared to the EOS of a perfect gas — is easiest to visualize by plotting derivatives of the various state variables. As an example, figure 3.1 shows the variation of the first adiabatic exponent $\Gamma_1 = (\partial \log P / \partial \log \rho)_S$ as a function of density ρ and specific internal energy e.

Γ_1 is an important quantity indicating the dynamical stability behavior of protostellar cores (see subsection 2.1). To improve the accuracy of the numerical solution for the energy balance, the internal energy e is taken as the primary quantity, whereas the temperature T is considered as a derived variable depending on e

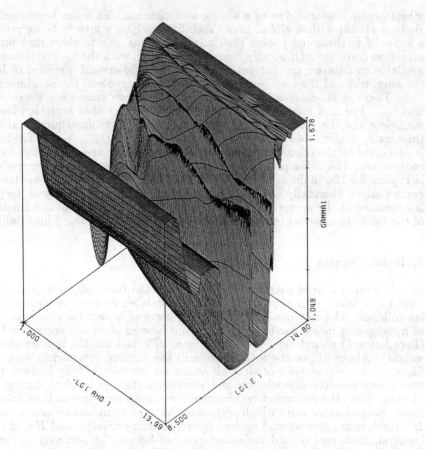

Figure 3.1. Adiabatic exponent Γ_1 as a function of log specific internal energy $[erg\, g^{-1}]$ and log density $[g\, cm^{-3}]$. The corresponding temperature range extends from $10\,K$ to about $2 \times 10^6\,K$. For very low internal energy (low temperature) the material behaves like a perfect monatomic gas ($\Gamma_1 = 5/3$). With increasing temperature, Γ_1 starts to drop when first H_2 rotational degrees of freedom are excited (first small 'valley'). Vibrational excitation and subsequent dissociation bring Γ_1 down to almost 1 (first deep U-shaped valley). After completion of H_2-dissociation Γ_1 rises steeply but ionisation commences and Γ_1 drops again (second U-shaped valley). The remaining two smaller valleys relate to the first and second ionization of helium. On the plateau at the high end of the internal energy, again $\Gamma_1 = 5/3$ for a completely ionized perfect gas plasma. Pressure ionisation and effects of (partial) degeneracy of the electrons are included. (From Wuchterl 1990).

and on the density ρ. Likewise, the pressure P is a given function of e and ρ. Tables of $P = P(\rho, e)$ and $T = T(\rho, e)$ have been constructed by Wuchterl (1990) and two-dimensional rational splines are used for interpolation. In overlapping regions of the (ρ, e)-plane there is good agreement with the numbers given by Däppen *et al.* (1988).

3.2. OPACITY

To solve the complete radiative transfer problem frequency dependent absorption and scattering coefficients, κ_ν^{abs} and κ_ν^{sca}, respectively, for the protostellar material are needed. Up to now only crude approximations of the energy transport by radiation have been adopted in order to make the complicated interplay between radiative, thermal, and dynamical processes in protostellar flows tractable at all. This is why for collapse calculations certain mean frequency-integrated quantities are used so that the transfer problem is reduced to the so-called 'grey' case. For example, if the condition of local thermodynamic equilibrium (LTE) is valid, the specific intensity I_ν equals locally, by definition, the black-body radiation obeying Planck's law $B_\nu(T)$ for the (local) temperature T. The approriate average value κ_R or simply κ of the 'extinction' coefficient $\kappa_\nu = \kappa_\nu^{abs} + \kappa_\nu^{sca}$ is then the so-called *Rosseland* mean or the 'opacity'. It is defined in such a way that the total radiative energy flux over all frequencies is conserved when changing from the frequency-dependent to the grey case, which yields

$$\frac{1}{\kappa_R} = \int_0^\infty \frac{1}{\kappa_\nu} \frac{\partial B_\nu}{\partial T} d\nu \bigg/ \int_0^\infty \frac{\partial B_\nu}{\partial T} d\nu \ . \tag{3.1}$$

In figure 3.2 the surface of κ_R is displayed over the log density – log temperature-plane.

For assumptions that are less stringent than LTE further average values must be introduced, e.g., the 'Planck-mean' of κ_ν^{abs} or averages over the actual local radiation field (for more details see, e.g., Tscharnuter 1985; Mihalas 1986; Winkler and Norman 1986). A self-consistent solution can in principle be obtained by the following iteration process: temperature distribution $\to$ frequency-dependent absorption and scattering coefficients $\xrightarrow{transport\ equ.}$ frequency-dependent radiation field $\xrightarrow{frequ.-averaging}$ grey coefficients $\xrightarrow{energy\ equ.}$ temperature distribution $\to\ \ldots$

For practical purposes, however, mostly the grey case with κ_R taken as the absorption coefficient has been considered. Successful steps forward to a more consistent solution of the transfer problem in protostellar envelopes have been done by Yorke (1979, 1985, 1986, 1988) under the *assumption* that there is a central source of luminosity. In a recent paper by Bodenheimer *et al.* (1990) the spectral appearance of an axisymmetric (2-D) model of the collapsing solar nebula is calculated with correct averaging but without applying the iteration procedure, briefly sketched above, for complete consistency.

3.3. TURBULENT VISCOSITY

One of the major uncertainties of accretion theory is how to generate the amount of viscosity needed in order to explain timescales of variability, density distribution, etc. of accreting systems which are well documented by observations. A good example are so-called cataclysmic variables (CV), which may justly be referred to as a 'standard' laboratory for testing accretion theory. Concerning CVs, for instance, torques that are responsible for transporting angular momentum outward so that mass is able to flow inward within an accretion disk, interpreted as *viscous* torques, have to be much stronger than those related to natural, molecular viscosity,

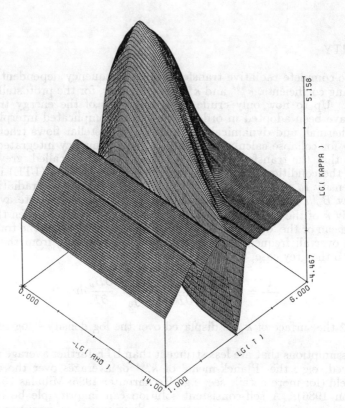

Figure 3.2. Rosseland mean κ of the mass absorption coefficient. $\log \kappa$ in $[cm^2 g^{-1}]$ is displayed as a function of log temperature in $[K]$ and log density in $[g\, cm^{-3}]$. At low temperatures dust is the main absorber of IR-radiation. Up to about $150\,K$ the dust particles are ice-coated, sublimation of ice and volatile material causes an intermediate drop of the opacity. At the sublimation temperature $T_{sub} \approx 1700\,K$ even the most refractory particles evaporate. Molecules remain as very inefficient absorbers so that the opacity drops by several orders of magnitude (opacity gap). For still higher temperatures $\geq 2000-2500\,K$, H^- becomes a major opacity source. κ rises steeply. The first ridge on the 'mountain' is due to H and first He ionisation, the second, smaller ridge is due to the second He ionization. For very high temperatures $\geq 10^5 K$, depending on the density, electron scattering dominates, i.e., $\kappa = const$. (From Wuchterl 1990).

in order to account for the empirical findings. As a rule, the Reynolds numbers in astrophysical flows are in general extremely large. This leads to the conjecture that *turbulence* should be ubiquitous in space. Empirically, this seems to be indeed true. However, the problem is that no basic theory of turbulence is available yet. It is not possible to derive transport coefficients, e.g., 'turbulent' viscosity, from first principles. Shakura (1972) made the fruitful suggestion of parameterizing the coefficient of turbulent 'eddy' viscosity η_t in the form ('α ansatz', see also Shakura and Sunyaev 1973):

$$\eta_t = \frac{1}{3}\alpha \cdot \rho \cdot \ell \cdot c_s \tag{3.2}$$

where ρ and c_s are the density and sound velocity, respectively; ℓ is the typical length scale of the largest eddies and α is a free parameter measuring the typical turbulent velocity in units of the (local) sound speed. Because supersonic turbulence is hard to maintain — hydrodynamical shocks would dissipate the kinetic energy of the eddies very quickly — α ought to be smaller than unity. Interestingly enough, theoretical models for CVs always require $\alpha = 0.1\ldots 1$. Of course, if magnetic torques are taken into consideration, $\alpha > 1$ is possible, but α should not exceed the ratio between the Alfvén velocity v_A and the speed of sound, thus $\alpha \leq v_A/c_s$, because otherwise magneto-hydrodynamical shocks will cause strong damping.

The second 'free' parameter is the typical ('correlation') length ℓ. For thin disks the vertical scale height is a good estimate, since no eddy can be larger than the vertical extension of the disk, provided that we can assume turbulence to be essentially isotropic. The condition of dealing with a thin, non-gravitating, Keplerian disk is fulfilled in CVs but not necessarily in protostellar systems. Nevertheless, with regard to the crude approximations that have to be made anyway, a plausible estimate for ℓ is

$$\ell = \frac{c_s}{\Omega} \tag{3.3}$$

where r is the distance from the axis of rotation and Ω the angular velocity in the equatorial plane. Expression (3.3) is a direct consequence of the (assumed) hydrostatic vertical structure of the disk and Keplerian rotation with $r\Omega > c_s$ so that the vertical scale hight $h \approx \ell < r$ (for a technical modification, see Tscharnuter 1987b). Actual calculations have been made with $\alpha = 0.1$ (see subsection 6.2).

4. Structure Equations

The equations which govern the evolution of protostellar collapse express first and foremost the conservation laws of (non-relativistic) physics, i.e., there is 1 equation for mass conservation, 1 for the energy balance, and there are 3 equations for momentum conservation. Further equations relate to the (scalar) gravitational potential describing Newtonian self-gravitation, to the energy balance of radiation, and radiative energy transport. This system of 10 partial differential equations (PDE) is augmented by additional equations defining the coordinate system to be adjusted to the spatial variation of the state variables (equations for the adaptive grid after discretising the system of PDE). The constitutive relations are of algebraic form, e.g., the equation of state and the temperature (or entropy) as a function of the density and internal energy (see section 3). Disregarding the equations for the transformation of the coordinate system changing with time, e.g., co-moving or Lagrangian coordinates as a special case, we have 12 physical equations for the following 12 quantities: ρ (density), $\mathbf{u}$ (velocity vector, 3 independent components), e (specific internal energy), Φ (gravitational potential), J ($\propto$ radiation energy density), $\mathbf{H}$ ($\propto$ radiative flux vector, 3 independent components), P (gas pressure) and T (temperature).

For the sake of simplicity, we write down the structure equations in a fixed, Eulerian coordinate system by making use of compact operator and vector notation. q_t means the partial time derivative of the quantity q; ∇, $\nabla\cdot$, $\nabla\!:$, are the operators of the gradient, the divergence, and the vector divergence, respectively, and $\mathbf{u} \otimes \mathbf{v}$ denotes the tensor product of the vectors $\mathbf{u}$ and $\mathbf{v}$. The system of equations then reads:

1. Continuity equation (mass conservation):

$$\rho_t + \nabla \cdot (\rho \mathbf{u}) = 0. \tag{4.1}$$

2. Equation of motion (momentum conservation):

$$(\rho \mathbf{u})_t + \nabla : (\rho \mathbf{u} \otimes \mathbf{u}) + \rho \nabla \Phi + \nabla P + \frac{4\pi \kappa \rho}{c} \mathbf{H} = \mathbf{w_Q}. \tag{4.2}$$

Here $\mathbf{w_Q}$ is short for the (viscous) forces and c is the speed of light. Equ. (4.2) contains on the left hand side the inertia terms (first and second one), gravitation, gas pressure, and radiation pressure (terms 3–5), κ is the Rosseland mean of the absorption coefficient. We have restricted ourselves to the grey radiative transport (see subsection 3.2). The viscous force $\mathbf{w_Q} = \nabla : \mathbf{Q}$, where $\mathbf{Q}$ is the viscous (Reynolds) stress tensor. The structure of $\mathbf{Q}$ is discussed in more detail, by using tensor notation, in subsection 5.2 below.

3. Energy equation (1st law of thermodynamics):

$$(\rho e)_t + \nabla \cdot (\rho e \mathbf{u}) + P \nabla \cdot \mathbf{u} = 4\pi \rho \kappa \left(J - \frac{\sigma}{\pi} T^4 \right) + \rho (\epsilon_Q + \epsilon_{nuc}). \tag{4.3}$$

ϵ_Q, ϵ_{nuc} is the viscous and thermonuclear energy generation, respectively. σ denotes the Stefan-Boltzmann constant.

4. Poisson equation (self-gravity):

$$\nabla \cdot (\nabla \Phi) - 4\pi G \rho = 0 \tag{4.4}$$

5. Radiation energy balance:

$$\frac{1}{c} \left(J_t + \frac{1}{3} \nabla \cdot (J \mathbf{u}) \right) + \nabla \cdot \mathbf{H} + \rho \kappa \left(J - \frac{\sigma}{\pi} T^4 \right) = 0 \tag{4.5}$$

This equation for the radiation energy density $E_{rad} = (4\pi/c)J$ is obtained by using the diffusion or Eddington approximation for the radiative flux $\mathbf{F}_{rad}$:

$$\mathbf{F}_{rad} = 4\pi \mathbf{H} = -\frac{4\pi}{3\kappa \rho} \nabla J \tag{4.6}$$

One can do better with spherical symmetry by introducing the so-called Eddington or anisotropy factor (Yorke 1979, 1985). Finally, for particularly chosen coordinates (cartesian, spherical polar, ...) the various components and terms of the equations (4.1)–(4.6) have to be worked out explicitly; modifications have to be made if moving coordinates are adopted (see, e.g., Tscharnuter and Winkler 1979; Tscharnuter 1985, 1987b).

For completion of the above set of structure equations, boundary conditions have to be added. Boundary conditions are essential ingredients of physical models, as they connect the interior solution, which is to be determined, with the 'known' exterior, or can express certain regularity conditions for the state variables for 'singular' points or lines, e.g., at the center or on the axis of rotation. For example, the gravitational potential and the net flux of radiation are determined by conditions at the outer boundary. Loss and gain of mass (cf. Wuchterl 1991a,b) and/or the transfer rate of angular momentum across the boundary can be modelled by choosing appropriate boundary conditions. Coupled with a particular set of initial data, these choices determine the resulting evolutionary sequence.

5. Numerical Tools

To become numerically tractable, the structure equations (4.1)–(4.6) have to be discretized on some suitably chosen coordinate system. For protostellar collapse models one commonly uses spherical polar coordinates (r, θ, ϕ), where r is the distance from the center, θ and ϕ is the polar and azimuthal angle, respectively, or cylindrical coordinates (s, ϕ, z), where s is the distance from the rotational axis, ϕ the azimuthal angle, and z the vertical coordinate ($z = 0$ is the equatorial plane). In the sequel, we will, for the sake of definiteness, only deal with spherical polar coordinates. Since a complete description of all technical details is certainly beyond the scope of this chapter, we will just briefly outline the basic ideas. Details can be found in Boss (1979, 1984), Tscharnuter and Winkler (1979), Winkler and Norman (1986), Norman and Winkler (1986), Tscharnuter (1987b).

5.1. FINITE DIFFERENCES

In order to keep the global conservation properties of the structure equations as complete as possible for the numerical calculations, we must choose a transformation of equ. (4.1)–(4.5). These are the balance equations proper for mass, momentum, and the various forms of energy (internal and radiation) that have to be considered — equ. (4.6) is a pure *transport* equation defining *fluxes* — in order to arrive at the so-called *conservation form* of the discretized equations. For this purpose we first write down the differential equations in integral form. As an example, the continuity equ. (4.1), i.e., the property of mass conservation, is then expressed as

$$\frac{d}{dt}\left[\int_{V(t)} \rho\, d\tau\right] + \int_{\partial V} \rho\,(\mathbf{u}_{rel} \cdot d\mathbf{S}) = 0. \qquad (5.1)$$

In words: the change of the total mass contained in a closed volume $V(t)$ (which may vary with time, e.g., by using co-moving, Lagrangian coordinates) equals the net mass gain or loss through the (oriented) surface ∂V of the volume (mounted with the same orientation) according to the mass flux $\rho \mathbf{u}_{rel}$ relative to the (moving) coordinate system; $d\tau$ and $d\mathbf{S}$ are the (oriented) volume and surface elements, respectively. For Eulerian coordinates which are fixed in space $\mathbf{u}_{rel} = \mathbf{u}$, the volume $V(t) \equiv V$ does not depend on time. Hence, d/dt and $\int$ can be interchanged in (5.1), and applying the Gaussian integral theorem to the surface integral yields (changing notation from $d/dt \to \partial/\partial t$) we end up with

$$\int_V \left[\frac{\partial \rho}{\partial t} + \nabla \cdot (\rho \mathbf{u}) \right] d\tau = 0. \tag{5.2}$$

Since the volume V is arbitrary the integrand must vanish identically. Besides notation, we are thus exactly back to equ. (4.1). Similar procedures can be applied to the remaining equations.

As an illustration of the simplest, 1-D case, we now discretize equ. (5.1) on an r-grid as

$$R = r_1 > r_2 > \ldots r_{m-1} > r_m = 0. \tag{5.3}$$

m is the total number of radial gridpoints. Scalars are attached to the *interior* of the $m-1$ cells (r_j, r_{j+1}), $j = 1, 2, \ldots, m-1$, the radial component of vectors, e.g., the radial velocity is attached to the m cell *boundaries* r_j, $j = 1, 2, \ldots, m$. This is called a staggered numerical grid (see fig. 5.1).

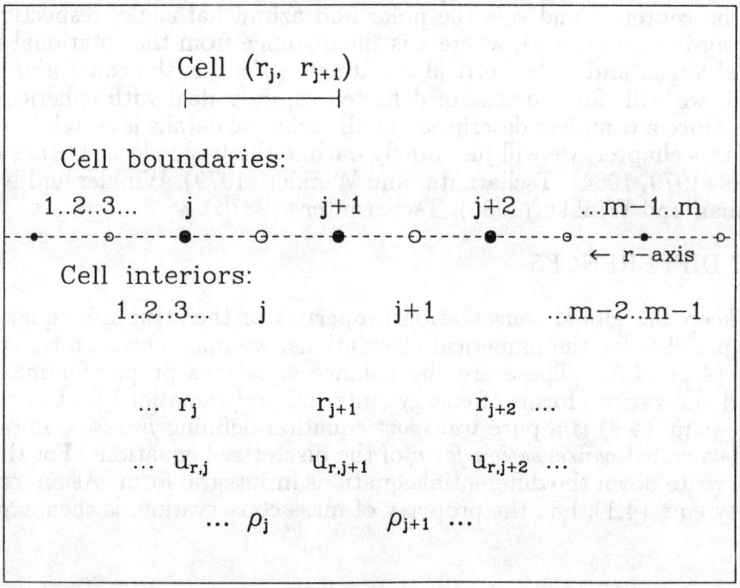

```
            Cell (rⱼ, rⱼ₊₁)
            ├─────────────┤

    Cell boundaries:
    1..2..3...    j        j+1        j+2       ...m-1..m
    --•-----◦-----•-----◦-----•-----◦-----•-----◦-----•---
                                                      ← r-axis
    Cell interiors:
    1..2..3...    j            j+1           ...m-2..m-1

         ... rⱼ           rⱼ₊₁           rⱼ₊₂ ...

         ... uᵣ,ⱼ         uᵣ,ⱼ₊₁         uᵣ,ⱼ₊₂ ...

              ... ρⱼ              ρⱼ₊₁ ...
```

Figure 5.1. Staggered numerical grid. Scalars are attached to the interior of the cells, vectors are defined at the boundaries.

If δt denotes the discrete time step and $\Delta Vol_j = (r_j^3 - r_{j+1}^3)/3$ proportional to the volume of the spherical shell between r_j and r_{j+1}, we have approximately for the 'volume' term

$$\frac{1}{4\pi} \cdot \frac{d}{dt} \left[\int_{V(t)} \rho \, d\tau \right]_j \approx \frac{\delta(\rho_j \Delta Vol_j)}{\delta t} \tag{5.4}$$

where $\delta q = q^{(n+1)} - q^{(n)}$ is short for the difference of the quantity q between the 'new' time level $(n+1)$ and the 'old' one (n).

The second or 'surface' term in equ. (5.1) is often referred to as the 'advection' term. Generally, advection terms deserve particular attention. As a matter of fact, the numerical accuracy attainable for a given number m of gridpoints depends decisively on how they are taken into account. For reasons of numerical stability, gradients of physical quantities to be advected should be taken one-sided so that only cells are involved from which the flow is coming from ('upwind'-differencing). This is physically quite plausible, since in case of supersonic gas velocities no information can be received, at a fixed point, from the downstream regions. For collapse flows which develop strong shocks the formulation of the advection terms is of crucial importance. A first-order approach is the so-called *donor-cell differencing*. It is defined by

$$\frac{1}{4\pi} \cdot \left[\int_{\partial V} \rho \left(\mathbf{u}_{rel} \cdot d\mathbf{S} \right) \right]_j \approx r_j^2 \left(u_{rel} \right)_j \rho_j^{\pm} - r_{j+1}^2 \left(u_{rel} \right)_{j+1} \rho_{j+1}^{\pm} \qquad (5.5)$$

where $\rho^{\pm}$ is to be taken according to the flow direction:

$$\rho_j^{\pm} = \begin{cases} \rho_j & \text{if } (u_{rel})_j \geq 0; \\ \rho_{j+1} & \text{otherwise.} \end{cases} \qquad (5.6)$$

Donor-cell differencing is only first order in space. Therefore, it is not very accurate, though easy to handle, if it is adopted for static grids. However, no severe accuracy problems arise for adaptive grids, which can become arbitrarily tight locally, if steep gradients were to form. For higher order 'monotonic' advection schemes, see Winkler and Norman (1986).

As a last example we consider the pressure term in the equation of motion:

$$\frac{1}{4\pi} \cdot \left[\int_V \frac{\partial P}{\partial r} \, d\tau \right]_j \approx r_j^2 \cdot (P_{j-1} - P_j), \qquad (5.7)$$

i.e., the pressure force is evaluated numerically as surface area × pressure difference between the two cells adjacent to the surface under consideration.

Depending on which time level the purely spatial terms in equ. (4.1)–(4.6) are evaluated, we have to distinguish between two basically different numerical integration methods:
1. The time differences of the physical quantities are determined by evaluating the remaining terms at time level (n), at which all data needed are known. The updating of the state variables for the advanced time level $(n+1)$ is then a straightforward extrapolation. The respective methods are called *explicit* finite difference methods with *forward* time differences.
2. The time differences of the physical quantities are determined by evaluating the remaining terms at time level $(n+1)$, at which the exact values for the physical variables are not known a priori. Since the discretized algebraic equations are non-linear, an iterative process has to be applied in order to achieve consistency. The respective methods are called *implicit* finite difference methods with *backward* time differences. Modifications can be made in considering a time level in between (n) and $(n+1)$; we are then dealing with *centered* or 'second order' time differences.

Explicit methods suffer from the great disadvantage that there is an intimate relation between spatial zoning and the maximum permissible time step. This is the well known Courant-Friedrichs-Lewy (CFL-) condition for warranting numerical stability of explicit numerical schemes (Courant et al. 1928). The CFL-condition reads:

$$\delta t < \min_{grid} \frac{\Delta r}{c_s + |u_{rel}|}, \qquad (5.8)$$

i.e., the time step becomes shorter with increasing spatial resolution and with higher (local) sound and relative velocities. For protostellar collapse flows the time step has to be orders of magnitude larger than the CFL-limit in order to be able to cover the accumulation phase. So, *explicit* methods for calculating hydrodynamical models of stellar formation are very inefficient, unless one is interested in variations on the short dynamical free-fall timescale $(t_{ff})_{core}$.

By contrast, *implicit* schemes are unconditionally CFL-stable. The disadvantage here is that one needs robust and efficient iteration algorithms, e.g., the well known Newton-Raphson-Henyey method, but to achieve convergence is sometimes a matter of good luck. There are huge demands on computer power with respect to superfast CPUs as well as to storage requirements. That is why all 3-D and most 2-D calculations have been carried out with explicit hydrocodes, except for the part pertaining to radiative transfer (cf. Boss 1987). However, a major step forward in modelling star formation will become possible, only when powerful implicit numerical algorithms for 2-D and, in particular, for 3-D flows have been developed (cf. Tscharnuter and Boss 1991).

5.2. ARTIFICIAL VISCOSITY

The development and propagation of shock fronts which appear as discontinuities in the hydrodynamical approximation are a common feature of self-gravitating flows. Difference schemes as discussed above are not able to treat shocks, since finite differences are good approximations only if continuity and differentiability conditions (existence of smooth gradients and continuous second derivatives) for the physical variables are fulfilled. This technical difficulty can be surmounted efficiently by broadening shock fronts over a certain prescribed length scale ℓ_{av} which is achieved by introducing additional viscous stresses $\mathbf{Q}_{av}$. With slab-symmetry $\mathbf{Q}_{av}$ reduces to a scalar q_{av} ('viscous pressure' = −'viscous stress'). The most commonly used form for q_{av} given by v. Neumann und Richtmyer (1950) is

$$q_{av} = \begin{cases} \ell_{av}^2 \rho \left(\frac{\partial u}{\partial x}\right)^2 & \text{if } \frac{\partial u}{\partial x} < 0; \\ 0 & \text{otherwise.} \end{cases} \qquad (5.9)$$

u is the (only non-trivial) velocity component in the x-direction. This formulation is often called 'artificial' or 'pseudo' viscosity. ℓ_{av} expresses the length scale over which a shock is to be broadened (if an adaptive grid is used), or is of the order of the (fixed) local width of the grid. It turned out that the v. Neumann-Richtmyer prescription for artificial viscosity becomes quite inefficient for smoothing shocks in collapse and accretion flows, e.g., with spherical symmetry ($x \equiv r$), because oscillations of the protostellar core about the equilibrium are artificially excited. For axisymmetric models things are even more sensitive to a 'correct' formulation of the artificial viscous stresses. It has become evident that the following properties must hold:

1. Shock fronts have to be broadened according to the local mesh size of the numerical grid (fixed grid) or according to a prescribed length scale (adaptive grid).
2. Regions in the flow which are expanding have to be free of artificial viscosity.
3. The transition of a free supersonic flow into a quasi-hydrostatic region must not introduce perturbations of the equilibrium by the discrete structure of the finite differences.
4. It is a necessary condition that a spherically symmetric, homologous contraction, i.e., $u \propto r$, must not be affected by artificial viscosity.

Now it is not very difficult to choose the appropriate generalization of equ. (5.9). The clue to that is a description of the artificial viscosity term in a coordinate-invariant way, i.e., a tensor formulation. This has been done by Tscharnuter and Winkler (1979). If we define $\overline{(\nabla \mathbf{u})}$ to be short for the symmetrized tensor of the gradient of the velocity field and if $\mathbf{I}$ is the unity tensor, the stress tensor of artificial viscosity respecting all demands listed in items 1.–4. above reads

$$\mathbf{Q}_{av} = \begin{cases} -\ell_{av}^2 \rho \left(\nabla \cdot \mathbf{u} \right) \left[\overline{(\nabla \mathbf{u})} - \tfrac{1}{3} (\nabla \cdot \mathbf{u}) \mathbf{I} \right] & \text{if } (\nabla \cdot \mathbf{u}) < 0; \\ 0 & \text{otherwise.} \end{cases} \quad (5.10)$$

Hence the coefficient of artificial viscosity η_{av} is given by

$$\eta_{av} = \begin{cases} -\ell_{av}^2 \cdot (\nabla \cdot \mathbf{u}) & \text{if } (\nabla \cdot \mathbf{u}) < 0; \\ 0 & \text{otherwise.} \end{cases} \quad (5.11)$$

The total viscosity coefficient η, consisting of the 'turbulent' viscosity η_t according to equ. (3.2) and 'artificial' viscosity η_{av} according to equ. (5.11), now reads $\eta = \eta_t + \eta_{av}$. Thus the total viscous stress tensor is

$$\mathbf{Q} = \eta \left[\overline{(\nabla \mathbf{u})} - \frac{1}{3} (\nabla \cdot \mathbf{u}) \mathbf{I} \right], \quad (5.12)$$

and the corresponding viscous forces, as used in equ. (4.2), are given by applying the vector-divergence $\mathbf{w_Q} = \nabla : \mathbf{Q}$. The viscous energy dissipation rate ϵ_Q can also be easily derived. Details for an actual evaluation in spherical polar coordinates are found in Tscharnuter and Winkler (1979).

5.3. ADAPTIVE GRID

For 1-dimensional problems Dorfi and Drury (1987) solved the problem of how to distribute a given amount of zones over a given domain of integration in such a way that structures which develop within the flow (shocks, hydrostatic parts, ...) are 'recognized' by the numerical scheme and resolved in a most economic way. The distribution of gridpoints depends on the distribution of the physical quantities and is defined not until the iteration for one implicit timestep has converged. Unfortunately, no straightforward extension to higher dimensional problems is known.

The basic idea for attacking this somewhat tricky problem is to distribute the gridpoints equidistantly along a composite graph constructed by the physical quantities and solve for the 'projected' distribution of the r-coordinate. This gives rise to the definition of the so-called *resolution function* Ψ_j at mesh point j, $2 \leq j \leq m-2$ on our r-scale (5.3).

$$\Psi_j = \sqrt{1 + \sum_q g_q \left[\frac{q_j - q_{j+1}}{q_{norm}} \cdot n_j)\right]^2} \qquad (5.13)$$

where g_q are weight factors of the order of unity and

$$n_j = \frac{r_{norm}}{r_j - r_{j+1}} \qquad (5.14)$$

is the *gridpoint concentration*; r_{norm}, q_{norm} are suitable normalization factors for the radii and the various physical parameters q, respectively. Two more parameters, α_g and τ_g, govern the 'rigidity' of grid with regard to changes in space and time, respectively. After 'spatial smoothing' $\tilde{n}_j = n_j - \alpha_g (\alpha_g + 1)(n_{j-1} - 2n_j + n_{j+1})$ and time 'retardation' $\hat{n}_j = \tilde{n}_j + \frac{\tau_g}{\delta t} \delta \tilde{n}_j$ the *grid equation* reads:

$$\frac{\hat{n}_{j-1}}{\Psi_{j-1}} = \frac{\hat{n}_j}{\Psi_j}, \quad j = 3, 4, \ldots, m-3. \qquad (5.15)$$

Equ (5.15) is an additional, implicit, non-linear equation for the distribution of the interior meshpoints of the grid r_j and has to be solved iteratively along with the physical structure equations for the time level $(n+1)$. For further details and test examples, see Tscharnuter (1987b) and the the original paper by Dorfi and Drury (1987).

6. Recent Results

In the following three subsections we give selected examples of numerical simulations pertaining to collapse, or more generally, self-gravitating flows with various degrees of symmetry. New results and starting points of future developments are discussed for the non-rotating 1-D collapse, which turned out to be more complex than thought before. Consequences of angular momentum transport in rotating protostars are emphasized and illustrated on the basis of 2-D and 3-D models. Because of space limitation, no attempts have been made to achieve completeness in discussing all details of existing calculations.

6.1. SPHERICALLY SYMMETRIC (1-D) MODELS

The remarkable revival of investigating the 'simple' spherically symmetric collapse came with a new generation of numerical tools (see section 5) and a greatly improved treatment of the material functions. The starting point was the surprising finding that the second, stellar core embryo seems to be vibrationally unstable. Moreover, marginal dynamical instability due to a low mass-average $\bar{\Gamma}_1$ around 4/3 in the core induces oscillations of large amplitudes (Tscharnuter 1987a, 1989, 1990). In addition, instabilities of the accretion shock in the late stages of the main accretion phase (cf. section 2) occurred in model calculations carried out by Balluch (1988).

An indication of the reality of core oscillations excited by κ-mechanisms, acting in the first, optically thick (dust feature of κ) as well as in the second, stellar core (κ-features due to H^- and ionization of atomic hydrogen), was found by Wuchterl (1989, 1990).

A complete, linear stability analysis of a strictly stationary 'model' accretion flow and the comparison with the solution of the time-dependent structure equations (see section 4) has been made by Balluch (1989, 1991a,b,c). The stationary solution is obtained by integrating the reduced system of the time-independent, ordinary differential equations, given the mass infall rate $\dot{M}$ together with appropriate outer and inner boundary conditions (e.g., to simulate the outer layers of the quasi-hydrostatic core where the density becomes large and the velocity very low), for the supersonic and subsonic part separately; the link between these two domains is the accretion shock, treated as a mathematical discontinuity whose location is determined by applying the well-known Rankine-Hugoniot jump conditions (see, e.g., Courant and Friedrichs 1948), which relate the values of the physical variables upstream with those downstream from the shock. Balluch (1991a) was able to show that almost complete agreement with the full time-dependent solution is established if both artificial tensor viscosity and the adaptive grid, as described in the previous section, is used. The resolution of the flow in the vicinity of the artificially broadened shock $\Delta r/r \simeq 10^{-11}$; the temperature jump at the shock, which is most sensitive to local errors, is less than 1.5% off the exact Rankine-Hugoniot value, otherwise the solutions are practically identical.

The linear stability analysis of the strictly stationary solution yields the following results (Balluch 1991b):
1. For a perfect gas and constant opacity there is an oscillatory unstable location of the accretion shock due to critical cooling, provided $\dot{M} \geq 10^{-6} M_\odot \, yr^{-1}$.
2. In the 'real' case with more realistic material functions, the κ-mechanism can drive an additional *vibrational* instability in the outer layers of the core. This is valid for the whole range of core masses $0.01 \leq M_{core} \leq 1 M_\odot$ and mass flow rates $10^{-3} \geq \dot{M} \geq 10^{-7}$, rather independent of the outer boundary conditions.

Similar instabilities are found for the first, optically thick cores, before becoming dynamically unstable due to dissociation of molecular hydrogen. In this case the κ-mechanism works on the basis of the dust features in the opacity. The instabilities 1.+2. above should first become visible in the velocity field and the radiative flux within the 'settling' zone downstream from the shock, where the mass elements find eventually their hydrostatic equilibrium; sooner or later the location of the shock front will also start to oscillate.

Fig. 6.1 illustrates the instability of the accretion shock during the final stages of accretion, as was found earlier by Balluch (1988) in a collapse calculation which started out with an initial density 10^4 times enhanced with respect to the critical Jeans density. The collapse timescale t_{ff} is about $10^3 yr$.

Since the evolutionary timescale t_{ff} is short, the end product of the accretion is a much more voluminous and luminous pre-main sequence object than is found in the classical calculations, e.g., by Larson (1969). This is because the optical depth of the collapsing fragment is higher from the beginning and the mass elements have not enough time to radiate the compressional energy away.

In the third paper Balluch (1991c) refers to the most accurate collapse calculation ever made and found very good agreement with all predictions of the linear stability analysis, as far as the first, optically thick core is concerned. The artificial

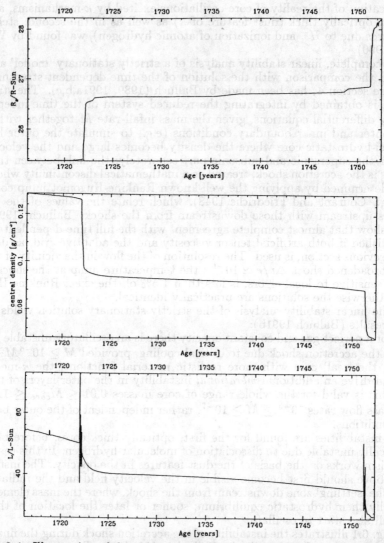

Figure 6.1. Illustration of the accretion shock instability. The three panels display (beginning from above): the location of the shock in units of the solar radius, the central density of the core $[g\ cm^{-3}]$, and the shock luminosity in units of the solar luminosity, respectively, as a function of time. (From Balluch 1988).

viscosity is chosen to be of the order of molecular viscosity. The core oscillations grow to very large amplitude ('hiccups' of the first core). After eight eruptions the dynamical ($\Gamma_1 < 4/3$) instability suddenly triggers the second collapse.

The main problem of following the second collapse numerically with sufficient accuracy relates to the conservation of the total energy. It is not difficult to estimate that the binding energy (total internal energy + gravitational energy) of the second, stellar core is very low, since the embryonic core forms out of an almost perfect adiabatic collapse of the first core and the length scales are reduced by about two orders of magnitude. Radiative losses are negligible in this phase, but it cannot be ruled out that small numerical errors may shift the binding energy to positive values. If, in addition, the mean $\overline{\Gamma}_1 \approx 4/3$, the amplitudes of the oscillations could be a simple consequence of a purely numerical excitation mechanism. To be on the safe side computationally, the accuracy of the total energy conservation must be guaranteed to less than 1% of the binding energy of the first core. This has not yet been achieved by available computer codes.

6.2. AXIALLY SYMMETRIC (2-D) MODELS

Early work on the isothermal collapse of a rotating protostellar cloud strongly suggested the dynamical formation of a self-gravitating ring (instead of a centrally condensed configuration) which is very likely to break up into two or more pieces. These sub-fragments are able to contract dynamically, form rings that again fragment by transforming spin angular momentum of the parent fragment into orbital angular momentum of the respective sub-fragments. In that way both the fragmentation and the angular momentum problem can be solved within one coherent process (Bodenheimer 1978). However, Norman et al. (1980) showed how sensitive the results are with regard to the local conservation of angular momentum. He suggested that rings will form, if angular momentum is gradually transported inward due to a slight local redistribution of angular momentum by the advection terms. In using a higher order scheme, he found an asymptotic evolution toward a 'singular', i.e., centrally condensed flat, configuration for the canonical homogeneous, rigidly rotating initial model. Thus, rings are not expected to develop if angular momentum is systematically transported outward by frictional processes. As a matter of fact, if one uses the α-prescription for turbulent viscosity (see subsect. 3.3) ring structures will never appear. Instead, a single central condensation will form. This is why 2-D collapse models for rotating protostars including turbulent friction are so attractive for simulating the evolution of the (pre-)solar nebula.

As an illustration of how efficiently the α-viscosity works, fig. 6.2 shows the evolution of the core center in the (logarithmic) central density – central temperature ($\log \rho_c$– $\log T_c$) diagram for a simulation of solar nebula formation (high initial mean density, low specific angular momentum; see Tscharnuter 1987b)
A fully implicit, pseudo-spectral 2-D hydrocode with finite differences in the r-direction and an expansion into Legendre polynomials for the θ-direction, i.e., the polar angle of a spherical polar coordinate system (r, θ, ϕ), has been used. Legendre polynomials (or general spherical harmonics) are eigenfunctions of the angular part of the Laplace operator $\nabla \cdot \nabla$. For this reason the Poisson equ. (4.4) is particularly simple to treat numerically.

The initial free-fall time t_{ff} is only slightly longer than $700\,yr$, the initial mean angular velocity is $2.3 \cdot 10^{-11} s^{-1}$. Capital letters A–I mark interesting stages in the evolution. Between A and B the Kelvin-Helmholtz time $t_{KH} \ll t_{ff}$, therefore the fragment cools down because the cloud is transparent to IR-radiation. Shortly after B the central parts of the collapsing fragment becomes optically thick $t_{KH} \gg t_{ff}$ and the evolution proceeds adiabatically. At point C molecular hydrogen dissociates. The second collapse is able to start off, because a substantial amount of angular momentum is removed from the central parts of the quasi-hydrostatic

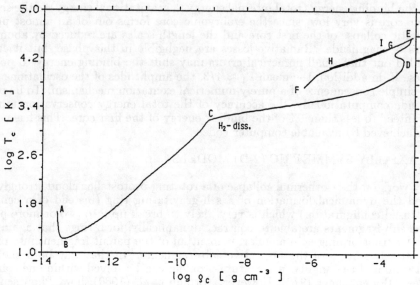

Figure 6.2. Central density ($\log \rho_c$) – central temperature ($\log T_c$) diagram for a 2-D model of solar nebula formation. (From Tscharnuter 1987b).

first core. D marks the stage where angular momentum effects again halt the collapse. Between D and E the excess angular momentum is transported outward and the innermost, pressure-supported core region behaves eventually like a spherically symmetric stellar object.

Oscillations are excited (E–F–G–H–I), which might be related to the recent findings for the 1-D models (see previous subsection), although there is no hope for a rigorous stability analysis for rotating axisymmetric cores.

Fig. 6.3a-d displays the transition from D to E in the respective contour plots of meridional cross sections and the distribution of the radial grid points. The onset of the quasi-adiabatic core oscillations is the reason why one cannot continue the calculations further into the accretion phase on a viscous timescale $t_{visc} \gg (t_{ff})_{core}$.

6.3. 3-D MODELS

In an extensive numerical study of 3-D protostellar collapse, Boss (1989) found that an initially nearly axisymmetric cloud will become significantly non-axisymmetric during dynamic contraction. He claims that this non-axisymmetry results from the non-linear coupling to the supersonic infall velocities and rotational instability of the nebula, which need not be unstable with respect to axisymmetric perturbations according to the Toomre (1964) criterion. Thus spiral-like or bar-like structures are very likely to develop, and gravitational torques become an efficient mechanism for redistributing angular momentum in the cloud. Although the accuracy of 3-D calculations is moderate — in comparison with the 1-D simulations referred to in subsect. 6.1 — a qualitative estimate of the strength of the gravitational torques can be made. The appropriate quantity which indicates rotational instability is the

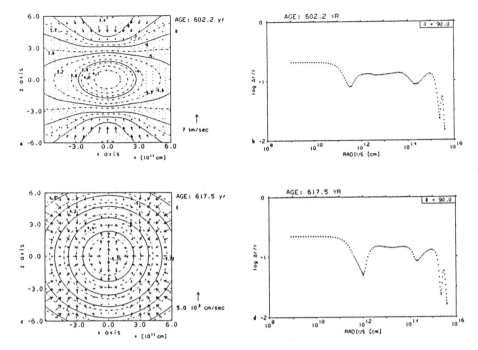

Figure 6.3. Visualization of the transport of angular momentum out of the innermost layers of the stellar core (panels a and c on the left hand side) and relative widths ($\Delta r/r$) of the radial grid zones (panels b and c on the right hand side) at labels D and E, respectively. Full lines are equi-density contours, dash-dotted and dotted lines refer to the temperature and angular velocity, respectively. The numbers given on the plot are logarithms of density (right upper side), temperature (left upper side), and angular velocity (right lower side). Arrows indicate the (projected) velocity field, the single arrow on the right hand side gives the scale. (From Tscharnuter 1987b).

ratio β of rotational energy E_{rot} and gravitational energy W, thus

$$\beta = \frac{E_{rot}}{|W|}. \tag{6.1}$$

Due to local conservation of angular momentum even an initially very low β will become large in the central regions where a rapidly rotating flattened 'core' forms. Eventually $\beta_{core} > \beta_{crit} \approx 0.27$, which is the critical number for the onset of dynamical instability (Ostriker and Bodenheimer 1973). It was speculated that such cores would undergo dynamical fission.
Surprisingly, Durison and Tohline (1985) were able to show that for $\beta_{core} > 0.27$ a bar-mode instability develops and grows into the non-linear regime. This leads to ejection of core material (assumed to obey a polytropic law with index 3/2) and angular momentum in two trailing spiral arms. The interesting result is that there is no tendency to form one or more discrete bodies from the ejected material. A rather smooth ring-like structure forms around the remnant with $\beta < 0.2$. Fig. 6.4

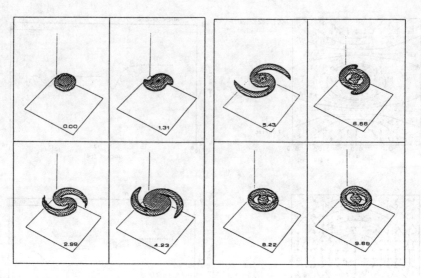

Figure 6.4. Evolution of a rapidly rotating polytrope of index 3/2. Numbers indicate the time in units of the initial rotation period of the center. No binary or multiple stellar system is formed. (From Durison and Tohline 1985).

shows this 'fission' process in a time sequence for which the changing surface of fixed density is displayed.

7. Conclusions

Despite the fact that we have made remarkable progress in our understanding of protostellar collapse flows during the past two decades, we are not yet able to simulate the joint evolution of the accumulating central object and the ambient disk-like nebula in a rigorous and consistent way. Explicit numerical methods suffer from the stringent limitation given by the CFL-condition (5.8) for the maximum admissible timestep, which is determined by the 'hot' star-like protostellar core. This is the reason why only the formation of the disk can be studied in some detail with explicit 2-D and 3-D codes. The most important result so far obtained is that gravitational torques are very efficient in redistributing angular momentum, if non-axisymmetric (bar-like or spiral-like) structures develop in self-gravitating disks.

In order to improve spatial resolution and accuracy of the 3-D models, it would be desirable to develop a suitable technique for handling higher dimensional self-adaptive numerical grids along with a robust implicit 3-D hydrocode.

In the axisymmetric models angular momentum transport due to turbulent friction should be investigated in more detail. Since there is no consistent theory of deriving the transport coefficients (viscosity, heat conduction, electrical resistivity, ...) from first principles, the heuristic α-viscosity is to be treated as a free parameter that has to be adjusted to empirical — astrophysical and/or cosmochemical

— data.

The κ-mechanism working in protostellar cores and the influence of Γ_1 on the stability behavior should be investigated in more detail by carrying out numerical calculations of maximum accuracy.

Magnetic fields presumably do not play a decisive rôle during the dynamical collapse. However, in the ensuing long-lasting accretion phase dynamo processes might be able to regenerate magnetic flux, and magnetic forces should be taken into account in the models. But for the moment being, it is my opinion that "the solution of the nonmagnetic collapse and accretion problem should have a firm basis consisting of reliable and sufficiently accurate results, before still more complexity is introduced in the models." (Tscharnuter and Boss 1991).

Acknowledgements. I am indebted to Drs. W. J. Duschl and J. Kirk for reading parts of the manuscript and for many suggestions that greatly helped to improve the readability of the text.

8. References

Appenzeller, I., and Tscharnuter W.M. 1974, *Astron. Astrophys.*, **30**, 423
Appenzeller, I., and Tscharnuter W.M. 1975, *Astron. Astrophys.*, **40**, 397
Balluch, M. 1988, *Astron. Astrophys.*, **200**, 58
Balluch, M. 1989, Ph.D. thesis, Universität Heidelberg, FRG
Balluch, M. 1991a, *Astron. Astrophys.*, in press
Balluch, M. 1991b, *Astron. Astrophys.*, in press
Balluch, M. 1991c, *Astron. Astrophys.*, in press
Bodenheimer, P. 1978, *Astrophys. J.*, **224**, 488
Bodenheimer, P., Yorke, H.W., Różyczka, M., Tohline, J.E. 1990, *Astrophys. J.*, **355**, 651
Boss, A.P. 1979, Ph.D. thesis, University of California, Santa Barbara
Boss, A.P. 1984, *Astrophys. J.*, **277**, 768
Boss, A.P. 1987, in *Proceedings from the Summer School on Interstellar Processes*, D. Hollenbach and H. Tronson (eds.), Reidel, Dordrecht, 321
Boss, A.P. 1989, *Astrophys. J.*, **345**, 554
Boss, A.P., and Yorke, H.W. 1990, *Astrophys. J.*, **353**, 236
Courant, R., and Friedrichs, K.O. 1948, in *Supersonic Flow and Shock Waves*, H. Bohr, R. Courant, and J. J. Stoker (eds.), Interscience Publishers, 116ff
Courant, R., Friedrichs, K.O., and Lewy, H. 1928, *Math. Ann.*, **100**, 32
Däppen, W., Mihalas, D., Hummer, D.G., and Weibel Mihalas, B. 1988, *Astrophys. J.*, **332**, 261
Durisen, R.H., and Tohline, J.E. 1985, in *Protostars and Planets II*, D. C. Black and M. S. Matthews (eds.), University of Arizona Press, 534
Dorfi, E.A., and Drury, L.O'C. 1987, *J. Comput. Phys.*, **69**, 175
Fiebig, D., and Güsten, R. 1989, *Astron. Astrophys.*, **214**, 333
Larson, R.B. 1969, *Monthly Notic. Roy. Astron. Soc.*, **145**, 271
Mihalas D. 1986, in *Astrophysical Radiation Hydrodynamics*, K.-H. A. Winkler and M. L. Norman (eds.), D. Reidel, Dordrecht, 45
Mouschovias, T.Ch., Paleologou, E.V., and Fiedler, R.A. 1985, *Astrophys. J.*, **291**, 772

Nakano, T. 1984, *Fund. Cosmic Phys.*, **9**, 139
Nakano, T. 1985, *Pub. Astr. Soc. Japan*, **37**, 69
Norman, M.L., Wilson, J.R., and Barton, R.T. 1980, *Astrophys. J.*, **239**, 968
Norman, M.L., and Winkler, K.-H.A. 1986, in *Astrophysical Radiation Hydrodynamics*, K.-H. A. Winkler and M. L. Norman (eds.), D. Reidel, Dordrecht, 187
Ostriker, J.P., and Bodenheimer, P. 1973, *Astrophys. J.*, **180**, 171
Popham, R., Narayan, R. 1990, *Preprints of the Steward Observatory*, **No. 952**, submitted to *Astrophys. J.*
Schulz, A., Güsten, R., Zylka, R., and Serabyn, E. 1991, MPI für Radiostronomie, Bonn, Preprint Series No. 426, to appear in *Astron. Astrophys.*
Shakura, N.I. 1972, *Astron. Zhurn.*, **49.5**, 921
Shakura, N.I., Sunyaev, R.A. 1973, *Astron. Astrophys.*, **24**, 337
Shu, F.H. 1977, *Astrophys. J.*, **214**, 488
Stahler, S.W., Shu, F.H., and Taam, R.E. 1980a, *Astrophys. J.*, **241**, 637
Stahler, S.W., Shu, F.H., and Taam, R.E. 1980b, *Astrophys. J.*, **242**, 226
Stahler, S.W., Shu, F.H., and Taam, R.E. 1981, *Astrophys. J.*, **248**, 727
Toomre, A. 1964, *strophys. J.*, **139**, 1217
Tscharnuter, W.M. 1985, in *Birth and Infancy of Stars*, Les Houches, Session XLI, R. Lucas, A. Omont, and R. Stora (eds.), North-Holland, 601
Tscharnuter, W.M. 1987a, in *Physical Processes in Comets, Stars, and Active Galaxies*, Proceedings, W. Hillebrandt, E. Meyer-Hofmeister, and H. C. Thomas (eds.), Springer Verlag, 96
Tscharnuter, W.M. 1987b, *Astron. Astrophys.*, **188**, 55
Tscharnuter, W.M. 1989, in *Low Mass Star Formation and Pre-Main Sequence Objects*, ESO Workshop, B. Reipurth (ed.), 75
Tscharnuter, W.M. 1990, in *Physical Processes in Fragmentation and Star Formation*, R. Capuzzo-Dolcetta, C. Chiosi, and A. Di Fazio (eds.), Kluwer, 293
Tscharnuter W.M., and Boss, A.P. 1991, in *Protostars and Planets III*, E. H. Levy, J. I. Lunine, and M. S. Matthews (eds.), The University of Arizona Press, Tucson, in press
Tscharnuter W.M., and Winkler, K-H.A. 1979, *Computer Phys. Communications*, **80**, 171
Winkler, K-H.A., and Newman, M.J. 1980a, *Astrophys. J.*, **236**, 201
Von Neumann, J., Richtmyer, R.D. 1950, *J. Appl. Phys.*, **21**, 232
Winkler, K-H.A., and Newman, M.J. 1980b, *Astrophys. J.*, **238**, 311
Winkler, K-H.A., and Norman, M.L. 1986, in *Astrophysical Radiation Hydrodynamics*, K.-H. A. Winkler and M. L. Norman (eds.), D. Reidel, Dordrecht, 71
Wuchterl, G. 1989, Ph.D. thesis, Universität Wien, Austria
Wuchterl, G. 1990, *Astron. Astrophys.*, **238**, 83
Wuchterl, G. 1991a, *Icarus*, **91**, in press
Wuchterl, G. 1991b, *Icarus*, **91**, in press
Yorke, H.W. 1979, *Astron. Astrophys.*, **80**, 308
Yorke, H.W. 1985, in *Birth and Infancy of Stars*, Les Houches, Session XLI, R. Lucas, A. Omont, and R. Stora (eds.), North-Holland, 645
Yorke, H.W. 1986, in *Astrophysical Radiation Hydrodynamics*, K.-H. A. Winkler and M. L. Norman (eds.), D. Reidel, Dordrecht, 141
Yorke, H.W. 1988, in *Proc. 18th Advanced Course of the Swiss Society of Astrophys. and Astronomy, Radiation in Moving Gaseous Media*, Y. Chmielewsky, and T. Lanz (eds.), Geneva Observatory Sauverny-Versoix, 193

BINARY STAR FORMATION

J. E. PRINGLE*
Space Telescope Science Institute[†]
3700 San Martin Drive
Baltimore, MD 21218
USA

ABSTRACT. Any theory of star formation which fails to take cognizance of the fact that most stars are in binary systems is seriously incomplete. Here we review the various mechanisms that have been proposed for the formation of binary stars and argue that there is still much work to be done.

1. Introduction

We now have what one might call a 'cartoon picture' of how stars form (Shu, Adams and Lizano, 1987; Lada and Shu, 1990). We start with the dense core of a molecular cloud, in hydrostatic equilibrium and supported by thermal and, at least in part, magnetic pressure. Then, thanks to the magic of ambipolar diffusion, the magnetic support is gradually withdrawn and the core collapses. A slight rotation of the initial core implies collapse to a centrifugally supported disc. This evolves, allowing angular momentum transport outwards and the formation of a central stellar object. From the disc and/or the star come magnetic winds and bipolar flows, and in the disc perhaps form planets, asteroids, comets and the like. At the end of the day, when all the fuss has died down, we end up with what for astronomical purposes we call a star. Unfortunately, it is a single star.

And yet it has long been recognized that many stars are not single stars, but occur rather as members of binary systems (Michell 1767, 1784; Herschel 1802). Indeed if one adds together the fractions of stars which are binaries found in the various binary period ranges (Halbwachs 1983, 1986; Abt 1983; Garmany, Conti and Massey 1980) one is driven to the conclusion that most stars are members of binary star systems. Moreover many of these are members of higher order multiple systems (Poveda 1988; Herczeg 1988; Abt 1983; Mazeh 1990).

Since this paper is intended as a discussion of the theoretical aspects of (binary) star formation, I do not intend to provide a review of the relevant observations. However one does need to be aware of the kind of predictions which a good theory needs to be able to make and the kind of observational constraint to which it

---

* Affiliated with the Astrophysics Division, Space Sciences Department of ESA.
[†] Operated by AURA under contract with NASA.

is likely to be subject. As an example it is useful to consider the paper by Duquennoy and Mayor (1990) in which they discuss the results of 12 years of radial velocity observations of a group of G-type stars in the solar neighborhood. Even with such a formidable mass of data it is clear that considerable effort must be devoted to the circumvention of selection effects before the final numbers are presented. The three main aspects of the observations of interest to us here are the distributions of period (or separation), of eccentricity and of mass ratio. Bearing these in mind, some general comments are in order. Since the work of Kuiper (1935a,b; 1955) there has been general agreement that there is no strongly preferred length scale in binary separations, and the distribution seems to be moderately uniform at least in the range $0.1-10^4$ pc. The eccentricity distribution shows a wide spread of values covering almost the entire permissible range of 0 to 1 with good evidence that at shorter periods (separation $\lesssim 1$ au) there is a dearth of high eccentricity systems and that at very short periods ($P \lesssim 10^d$) all orbits are circular, presumably due to tidal effects. The distribution of mass ratios has been much more controversial, and a brief perusal of the literature will show that the observed mass ratio distribution has varied wildly from year to year and from author to author. The reason for this is a strong dependence on selection effects and therefore systematic errors, which render the results to be highly subjective. Duquennoy and Mayor find for their stellar sample that the mass ratio distribution is consistent with random pairings from an initial mass function. This in essence should be the null hypothesis against which observations need to be compared, but here again care needs to be taken in describing which is the relevant initial mass function and how the sampling is to be made (Tout 1990).

For convenience I have split the various binary star formation mechanisms under four headings, although it will be evident that there is a certain degree of interrelation between them. Similarly it will be evident that many of the ideas currently under consideration are not really new but are reformulations of previous ideas which have been around for some time. I have not attempted to trace these ideas back in detail to their historical origins, but I have given some of the earlier references so that the interested reader may undertake his/her own historical analysis. Some historical discussion may be found in Chapter 11 of Tassoul (1978), though unfortunately he does not give details of all the references quoted. Two other recent reviews of the formation of binary and multiple star systems are given by Boss (1991) and by Bodenheimer *et al.* (1991).

2. Fission

One way of forming a binary star is to take a single star and split it into two. And an obvious agent to use to achieve such a split is centrifugal force. The idea behind this stems from consideration of the equilibrium shape of a uniformly rotating, liquid body such as the Earth (see the discussions by Lamb (1932), Chapter XII and Jeans (1929), Chapter VIII) and the realization that as the angular velocity for such a body is increased the stable equilibrium shape shifts from being an oblate (MacLaurin) spheroid, to an ellipsoid (Jacobi), to a pear-shaped figure and thence (speculatively) to a fissional split into two drops. Jeans (1929, Chapter IX) however noted a discrepancy between these ideas, and the idea enshrined in Laplace's nebula hypothesis which is that a body which is forced to rotate up to 'break-up' speeds would tend to shed excess angular momentum through losing rings of material at the equator. He suggested that the determining factor as to the outcome might be the compressibility of the matter out of which the rotating and self-gravitating body is comprised, so that incompressible bodies would undergo fission and compressible (centrally condensed) bodies would not. He suggested that the dividing point between the two regimes is an adiabatic index of $\gamma = 2.2$ (where pressure p and density ρ are related by $p \propto \rho^\gamma$) or equivalently $n \equiv 1/(\gamma-1) = 0.83$.

Why should a star spin up to 'break-up' speed? All stars are losing energy, and if the only energy available to a star is its own self-gravitational energy then as it forms and evolves it must contract. Such contraction coupled with angular momentum conservation leads inevitably to increasing angular velocity. In more modern terms this is equivalent to assuming that a newborn star appears with some rotation in the Hertzsprung–Russell diagram at large radius at the top of the convective (Hayashi) track and from there evolves rapidly towards smaller radius through a series of hydrostatic equilibria down the Hayashi track and then more slowly across the diagram on the radiative (Henyey) track to the ignition of hydrogen burning and the main sequence (see, for example, Novotny (1973), Chapter 7). More recently it has become generally realised that stars appear on the H–R diagram much closer to the bottom of the convective track (*e.g.*, Stahler 1983) so that the rapid contraction phase does not take place. Rather, prior to its initial appearance on the H–R diagram, the star forms by means of accretion from a circumstellar accretion disc (*e.g.*, Mercer-Smith, Cameron and Epstein 1982). During this phase, rapid spin up of the central object is possible since the material accreting onto the central star arrives with Keplerian angular velocity. Thus the possibility of spinning a star up to 'break-up' still exists in the modern context and the possibility of fission needs to be addressed.

Fission was strongly advocated by Jeans (1929, Chapter XI) as the formation mechanism for close binary stars (periods $P \lesssim 55d$)—that is the spectroscopic as opposed to the visual binaries. Indeed with the fission during contraction picture in mind he went on to write: "After a binary system has been formed by fission, each of its two components may undergo a further shrinkage, ... and these conditions may produce fission in the components, thus generating triple or multiple systems." [This idea foreshadows the concept of continued fragmentation—see Section 5]. Shortly afterwards Kuiper (1935b) concluded on observational grounds that "the fission theory must be completely abandoned." He elaborated his arguments in a later paper (Kuiper 1955) and his main reasons were a) that the close and wide binary populations are the same (*i.e.*, the subdivision into visual and spectroscopic is a selection effect) so that only one formation mechanism is required, and b) that Jean's had already argued that compressible objects such as stars cannot undergo fission. At about the same time Lyttleton (1953) gave technical arguments that fission could not occur and also pointed out that even if it did occur the process could only produce a binary system with a large mass ratio ($q \lesssim 0.1$). This is because if the split to form a binary occurs on a dynamical timescale then to conserve mass and angular momentum not only the size but also the mass distribution must be approximately conserved. Since the original rapidly spinning single star is centrally condensed, this means that the final binary must be also—in other words it must have a large mass ratio (Pringle 1989).

That could have been the end of the matter, but apparently fission is too attractive a proposition to abandon that easily and the matter was firmly reopened by Ostriker (1970) who pointed out the shakiness of Lyttleton's technical arguments that fission could not occur and (forgetting the mass ratio problem) concluded that "although fission remains unproven, there are now no strong theoretical arguments against the process, and there is considerable observational support for its existence." In view of the inability of the analytic approach to predict accurately the outcome of instabilities present in rapidly rotating bodies as the instability progresses to the non-linear regime (*i.e.*, fission), it was clear that a numerical calculation of the fission process was urgently required. An attempt was made by Lucy (1977) using his newly developed smoothed particle hydrodynamics (SPH). He followed the evolution of a rapidly rotating $\gamma = 5/3$ polytrope and found that it developed an $m = 2$ (bar) instability and then transferred angular momentum outwards via gravitational torques, shedding a small swarm of high angular momentum debris from the equator. Although the swarm did not coagulate into a single body he was able to confirm Lyttleton's argument that if fission occurs it gives rise to large mass ratios and to conclude that belief in fission had been strengthened. The experiment was repeated by Gingold and Monahan (1978) using marginally better resolution and following the calculation for longer. They ran two cases—one with $\gamma = 3$ ($n = 0.5$) and one with $\gamma = 5/3$ ($n = 1.5$), on either side of Jeans' critical value. For the $\gamma = 3$ case they found a split into two more or less equal parts, whereas for the $\gamma = 5/3$ case they more or less duplicated Lucy's result except that they followed the evolution of the swarm of debris for long enough that it coagulated to form a secondary body (of low mass). Thus while the idea of fission was not totally defunct, it was already clear that it was not the general panacea for forming close binaries that had been hoped (see for example the review by Lucy, 1981).

Even so, some doubts still remained about the fission process because of the crude numerical revolution available in the early calculations and because of lingering doubts about the soundness of SPH as opposed to grid based numerical methods. For this reason a thorough numerical investigation was undertaken by Durisen *et al.* (1986) who followed the evolution of a rapidly rotating $\gamma = 5/3$ polytrope using two different grid based codes (due to Boss and to Tohline) and an SPH code (due to Gingold and Monahan). The results were unequivocal and the authors concluded: "Binary fission is not the direct outcome of dynamical bar mode instability in $n = 3/2(\gamma = 5/3)$ polytropes. Instead mass and angular momentum are ejected through the formation of two material spiral arms. Evolutions followed through to an end state show a stable central nonaxisymmetric remnant surrounded by a ring or disk of debris."

In conclusion, then, it appears that even if one sets a stellar-type polytrope rotating sufficiently fast that it is dynamically unstable, it does not split into two bodies. Rather, the instability uses the nonaxisymmetry (gravitational torques) to redistribute angular momentum until stability is achieved. As a corollary, it is evident that since the instability and subsequent angular momentum redistribution occur on a dynamical timescale such a rapidly spinning configuration cannot be achieved either by slow gravitational construction or by gradual accretion from a centrifugally supported disc. Fission is not a process which is relevant to the formation of binary stars.

3. Capture

By the capture process of binary star formation we mean the forming of a binary star from two previously single stars which formed independently and gravitationally unbound from one another. Of course two stars interacting gravitationally as point masses conserve net energy in that if they are intially unbound they remain so. Thus one has to consider processes involving either more than two stars or the dissipation of energy. The capture process is recognized as being of significance in the formation of low mass X-ray binaries within globular clusters (*e.g.*, van den Heuvel 1983). Here, an overabundance of such binary systems compared to the field is attributed to the capture process being enhanced by the high stellar densities ($\sim 10^5$ stars pc^{-3}) and relatively low velocity dispersion (~ 10 km s^{-1}) (Clark 1975). Hills (1976) pointed out that the three-body process is too inefficient to be worth considering even here. The most likely process at present considered is the one proposed by Fabian, Pringle and Rees (1975) who suggested that the excess orbital energy in a two body encounter can be taken up in stellar non-radial oscillation excited by tides. However as a general mechanism for binary star formation this is a non-starter simply because such a close encounter is required to produce capture that the formation rate cannot differ much from the star-star collision rate which is minimal even in a globular cluster (Hills and Day 1976) and because the binaries so formed are very close in the sense that the periastron distances do not greatly exceed the stellar radii.

Nevertheless it must remembered that at the premainsequence/protostellar stage stars are larger and have matter surrounding them so that slightly different conditions pertain. Silk (1978) pointed out that the "extended accretion envelopes around protostars provide a medium in which viscous drag is important" and argued that such an envelope can in principal give rise to binary formation through capture. More recently, and more specifically, Larson (1990) drew attention to the growing body of observational evidence for the presence of circumstellar discs around low mass stars with masses $\sim 10^{-2} - 10^{-1} M_\odot$ and radii ≥ 100 a.u. (Cohen *et al.* 1989, Strom *et al.* 1989, Beckwith *et al.* 1990). He suggested that such discs could lead to significant binary star formation rates by enabling the disipation of orbital energy. He took a stellar density of 10^4 stars pc^{-3} and velocity dispersion of 1.5kms^{-1}as being representative of the core of the Trapezium cluster and estimated a capture rate per star of 0.4 Myr^{-1}. He made clear, however, that binaries formed would tend to be typically quite wide and eccentric. Clarke and Pringle (1991a) have undertaken a more detailed investigation of the disc capture process. They argued that the stellar density taken by Larson seems rather high compared to average regions in which stars form. More importantly they found that while collisions of low relative velocity between a star and a circumstellar disc can indeed produce a bound system, it is more probable that before such a low velocity encounter can take place, one or more high velocity encounters will have occurred. The outcome of a high velocity encounter is to disrupt the disc and to prevent any subsequent capture process taking place. They conclude that for this reason alone, the binary formation rates due to disc-capture must be revised downwards by an order of magnitude or so.

In conclusion it would seem that the formation of binary systems by the capture of a stellar passer-by through the dissipation of orbital energy by means of a circumstellar disc around a newly forming single star remains a possibility, particularly for regions of very high stellar density (comparable to the cores of globular clusters) and for wide (that is, loosely bound) binaries. However in the binary formation process as a whole, it seems unlikely that disc-capture plays a major role.

4. Independent condensations/separate nuclei

Jeans (1929) took the view that long period binaries were "systems formed out of independent condensations in the parent nebulae", whereas Kuiper (1955) insisted that "a single mechanism *i.e.*, independent condensation, must be held responsible for the origin of *all* binaries." In a sense this mechanism can be (and has been) criticised as being merely a mechanism for passing the buck. In other words, given that single stars form this "mechanism" at its simplest level is just an assumption that they form in the right places—in particular that they form close enough together to give the observed parameter distributions for binary systems. What is required, then, is to try to fold this particular idea in with the understanding of the single star formation process which we have arrived at so far.

At a basic level one can ask: how close together can stars form? The answer to this must be of order the Jeans' length. In terms of the densest regions found in molecular clouds (for example the cloud cores detected in NH_3 by Myers and Benson, 1983) which are thought to be the sites of recent or imminent star formation the relevant Jeans' length is of order 10^4 a.u. But if we form two stars independently at a separation of $\sim 10^4$ a.u. we simply end up with a binary system with a period of $\sim 10^6 y$. Alternatively one can assume that a small cluster of N independent stars is formed with interstellar distances $\sim 10^4$ a.u. The $N = 3$ case is discussed in detail by Valtonen and Mikkola (1990). Such a triple star system is well known to be unstable and the outcome, which occurs typically within ~ 20 initial crossing times, is a single star plus a binary. The binary is highly eccentric and has a final separation of order a few times less than the initial separation. Thus assuming $N = 3$ has the effect of producing a somewhat more tightly bound binary at the expense of generating a single star for each binary formed. This trend continues for larger values of N. The cases $N = 4$ and $N = 5$ are considered by Harrington (1974, 1975) and the cases $N = 10$ and $N = 25$ by van Albada (1968a, b). In all cases the final state is one (or sometimes two) binary sytems of compactness increasing with N, but with high eccentricity ($e \sim 0.8 - 0.9$) and with a resulting large number of single stars. Thus allowing stars to form independently, but bound to each other, does not yield either the observed distribution of binary separations or the observed frequency of binary stars.

The real situation, however, is likely to be more complicated than implied so far. For example if stars form in close proximity to one another—in effect as separate nuclei in a collapsing cloud core—then the dynamical collapse timescales for accretion onto the separate nuclei are comparable to the orbital timescales of the nuclei about one another. A numerical investigation of such a process was given by Larson (1978). He took an early version of an SPH code using only 150 particles and followed the evolution of a gravitationally unstable cloud. He found that decreasing the pressure in the initial cloud had the effect of allowing the cloud to break up into more fragments. In effect the number of 'stars' formed corresponded roughly to the number of Jeans' masses present in the initial configuration. The numerical approach used by Larson gives a crude representation of hydrodynamics. Nevertheless one should still ask why this collapse calculation gave rise to binary (and multiple) systems whereas the grid based methods failed (see section 5). One reason is that he was able to follow the evolution for many dynamical timescales, whereas grid based codes at present cannot. Another is that because of the small number of particles used, his initial configuration was rather lumpy and contained in effect seed nuclei.

It must be said, however, that the initial configuration considered here of a cloud containing seed nuclei and many Jeans' masses does not sit easily with the current conventional idea that gravitational collapse occurs through a series of hydrostatic equilibria in which magnetic support gradually seeps away (Shu, Adams and Lizano 1987). Indeed for this picture to be valid it is necessary to put aside the conventional picture and to assume that the initial collapse is an altogether more dynamical process. For example Pringle (1989) suggests that collapse might be instigated by collision of dense fragments within a cloud and that this could lead to an initial configuration similar to that envisaged above. Such ideas are amenable to test by bringing modern computer power to bear on the problem but such calculations have yet to be carried out. Nevertheless we may still ask whether a dynamically induced collapse with a number of seed nuclei both accreting and mutually interacting is capable of giving rise to the observed range of binary parameters—in particular the range of binary separations ($e.g.$, $0.1 - 10^4$ a.u. Duqennoy and Mayor 1990)—given that the initial separation of such seed nuclei must be order the Jeans' length, $i.e.$, $\sim 10^4$ a.u.

Pringle (1989) showed that if the rotation of the cloud as a whole is comparable to the rotation of the matter accreting onto the seed nuclei then the sizes of the centrifugally supported discs formed round each nucleus were comparable to the periastion distances of the mutual orbits of the nuclei and argued that therefore interaction between the discs was inevitable. Clarke and Pringle (1991b) considered the process in more detail, though still in a highly simplified form. They took as an initial configuration two or three stars, surrounded by massive accretion discs and followed the subsequent evolution. They allowed the disc masses and radii to evolve with time and followed the orbital dynamics, taking account of the dissipative effects of star-disc encounters. For the two body case they found that the eccentricity and separation of the final binary depended simply on the initial rotation rate of the cloud, and that in this case the range of binary separations would need to be due simply to a range in cloud rotation rates. In contrast for the three body case they found that for a single cloud rotation rate, small variations in the initial stellar velocities coupled with the chaotic nature of three body dynamics (Valtonen and Mikkola 1991) gave rise to a broad spread of final binary separations in line with those observed. They further noted that the presence of dissipation gave rise to the frequent occurrence of stable hierarchical triple systems, and thus decreased the number of single stars so formed.

In conclusion, the idea of independent condensations or seed nuclei is an attractive one for the formation of binary and multiple systems when put in the context of our current understanding of the star formation process. The picture implies, however, realisation that the onset of gravitational collapse is usually a more complicated, and in particular dynamic, process than has been assumed hitherto. Subsequently, although the formation of each individual star is likely to follow lines similar to those drawn out in the standard picture with discs, winds, outflows, jets, etc., the interactions between the various stars as they form renders the whole much more complicated. There is much more work to be done in this area before it can be accepted as the primary scenario for binary star formation, but it nevertheless remains a promising avenue for further research.

5. Fragmentation/continued fragmentation

In a seminal paper Larson (1969) computed the gravitational collapse of a spherically symmetric non-rotating cloud of gas of uniform temperature and density ρ. He found that the collapse was non-homologous with the central regions collapsing first and the density distrubution with radius, r, becoming of the form $\rho \propto r^{-2}$ on of order a free fall time or two. At that point a dense core developed, but most of the matter was still at large radius. Subsequently the matter accreted onto the dense core at a more or less steady rate (cf. Shu 1977). Larson (1972) then undertook a similar computation in cylindrical symmetry starting with a rotating cloud. He found that as matter fell in conserving angular momentum, centrifugal forces hindered collapse in the direction perpendicular to the rotation axis and led to the formation of a centrifugally supported ring. Larson speculated that such a ring must be gravitationally unstable and would be likely to split up to form a binary or multiple sytem. Note that this scenario is analogous to the fission picture envisaged by Jeans (1929) and others except that in this case the support against gravity is provided solely by centrifugal force and not by pressure. For this reason many of the objections raised in Section 2 do not apply.

This led to a debate as to whether one did (Black and Bodenheimer 1976) or did not (Tscharnuter 1975) get a ring as the outcome of such a collapse, and it became evident that whether one did or did not depended critically on the initial conditions as well as on the details of the numerical method employed (Norman, Wilson and Barton 1980, Boss and Haber 1982). Independent of the outcome of such deliberations it was clear that collapse to a centrifugally supported disc or ring was a possibility and that investigation of the fission of such a body could be of relevance to binary star formation. For example Cook and Harlow (1978) and Norman and Wilson (1978) computed numerically the evolution of a dynamically unstable ring and showed that a small $m = 2$ or $m = 3$ perturbation grew rapidly and led to fission into two or three bodies, respectively. More recently Adams, Ruden, and Shu (1989) showed analytically in the linear regime that the $m = 1$ mode applied to a massive disc around a central star led to dynamical instability and speculated that this might lead to the formation of a binary companion to the original star.

This general approach was criticised by Bodenheimer, Tohline and Black (1980) on the grounds that setting up a manifestly unstable configuration and showing that it breaks apart, begged the question of how it got into the dynamically unstable state in the first place (cf Section 2). They argued that technically one should apply any perturbation before the collapse was initiated and then follow the collapse process as a whole. Difficulties now emerge since any initial imposed perturbation tends to be smoothed out both by any pressure forces present in the initial configuration and also by the differential rotation brought about by local angular momentum conservation in a non-homologous collapse. Thus for initial states with large rotation (so that centrifugal forces do not permit much collapse to occur) and with large enough perturbations (relative to pressure) plausible binary systems (that is self gravitating nuclei containing a large fraction of the initial cloud mass) were found to occur. On the other hand while initial states with little rotation do sometimes give the impression that seed nuclei have formed at small radius (corresponding to centrifugal support), so little of the original mass is involved in these, that it is difficult to make the case that we are seeing the beginnings of a binary star (see also Boss 1986). As mentioned earlier (Section 4) grid based codes cannot at present follow the evolution for much more than a freefall time, at which stage (Larson 1969) most of the initial material has yet to fall in.

These difficulties notwithstanding there has also developed the idea of not just simple fragmentation during the first collapse but continued or hierarchical fragmentation. This is the exact analogy of Jeans' concept of hierarchical fission (Section 2). The idea was taken up by Heintz (1969) who assumed the "Darwin-Jeans" model of fission operating on a largely homogeneous cloud and envisaged a "cascade of successive decays." Bodenheimer (1978) details the fission-cascade in the hierarchical fragmentation picture (see also Boss 1988). The idea is that the initial collapse occurs to form a ring or disc which then fragments in the manner outlined above. Each fragment is then assumed to undergo a further collapse until it too is centrifugally limited, whereupon it too fragments, and so on. Boss (1988) envisages two or three hierarchies and that in this way multiple star systems and also close binaries can be formed. After the presentation of these ideas by Bodenheimer (1981), Tayler voiced the opinion that if only a few percent of the mass goes into fragments at each fragmentation stage then hierarchical fragmentation seems a very inefficient way of forming stars.

The major criticism of the fragmentation picture as currently envisaged, which is inherent in the comments made by Bodenheimer, Tohline and Black (1980), is one of timescales (Pringle, 1989). If one envisages the collapse of a cloud of initial radius R_o, down to a centrifugally supported radius of βR_o ($\beta < 1$), then the dynamical timescale of the disc or ring formed at βR_o is less than the timescale on which the disc is formed by a factor $\beta^{3/2}$. To form a 1 a.u. binary from an initial cloud of radius 10^4 a.u., this ratio is $\sim 10^{-6}$. This means that for disc-fission to proceed in such circumstances one requires that the dynamical instability which gives rise to the fission be suppressed for $\sim 10^6$ dynamical timescales and only be allowed to act when the material has all had a chance to a fall in. Similar criticisms apply to the ordinary fission process as well (see Section 2).

That said, it is evident that there is a degree of overlap between these ideas and those elaborated in Section 4. Discounting hierarchies, the main difference between the two processes is in the degree of inhomogeneity envisaged in the initial configuration. Those involved in the fragmentation calculations have felt obliged to mimimize inhomogeneity of the initial state and this, coupled with the restriction of the numerical approach to grid based methods has prevented pursuit of the evolution for more than an initial dynamical time. Indeed the picture outlined in Section 4 can be regarded as a kind of 'prompt initial fragmentation' in which the initial conditions are similar to those usually considered in fragmentation calculations but much more inhomogeneous and lacking in symmetry. It seems likely that a Lagrangian numerical method such as SPH could be a more fruitful way of tackling such a problem.

6. Concluding Remarks

We have reviewed the various mechanisms so far proposed for the formation of binary stars. As will be evident to the reader, all the proposed mechanisms have problems—some more than others. It is also clear, however, that there are a number of avenues open which are conducive to further useful research. As more results come in it will gradually become apparent which areas of the current theories of single star formation can be retained and, if necessary, adapted, and which must be abandoned. Until then, our theories of star formation are seriously incomplete. Only when we have discovered why most of the stars around us are members of binary and multiple systems will we be able to claim that some progress has been made.

4. References

Adams, F. C., Ruden, S. P., and Shu, F. H. 1989, *Ap. J.*, **347**, 959.
Abt, H. A. 1983, *Ann. Rev. A. A.*, **21**, 343.
Beckwith, S. V. W., Sargent, A. I., Chini, R. S., and Guŝten, R. 1990, *A. J.*, **99**, 924.
Black, D. C., and Bodenheimer, P. 1976, *Ap. J.*, **206**, 138.
Bodenheimer, P. 1978, *Ap. J.*, **224**, 488
Bodenheimer, P. 1981, in *Proc. IAU Symp 93*, "Fundamental Problems in the Theory of Stellar Evolution", D. Sugimoto, D. Q. Lamb and D. Schramm (eds.), Reidel, Dordrecht, 5.
Bodenheimer, P., Ruzmaikina, T., and Mathieu, R. D. 1991, in *Protostars and Planets III*, in press.
Bodenheimer, P., Tohline, J. E., and Black, D. C. 1980, *Ap. J.*, **242**, 209.
Boss, A. P. 1986, *Ap. J. Suppl.*, **62**, 519.
Boss, A. P. 1988, *Comm. in Ap.*, **XII**, 169.
Boss, A. P., and Haber, J. G. 1982, *Ap. J.*, **225**, 240.
Boss, A. P. 1991, in "Interacting Binary Stars," J. Sahade, G. McCluskey and Y. Kondo (eds.), Kluwer, Dordrecht.
Clark, G. W. 1975, *Ap. J. (Letters)*, **199**, L143.
Clarke, C. J., and Pringle, J. E. 1991a, *M.N.R.A.S.*, in press.
Clarke, C. J., and Pringle, J. E. 1991b, *M.N.R.A.S.*, in press.
Cohen, M., Emerson, J. P., and Beichman, C. A. 1989, *Ap. J.*, **339**, 455.
Cook, T. L., and Harlow, F. H. 1988, *Ap. J.*, **224**, 1005.
Duquennoy, A., and Mayor M. 1990, in *Proc. XI European Astr. Meeting of IAU*, M. Vasquez (ed.), Cambridge University Press.
Durisen, R. H., Gingold, R. A., Tohline, J. E., and Boss, A. P. 1986 *Ap. J.*, **305**, 281
Fabian, A. C., Pringle, J. E., and Rees, M. J. 1975, *M.N.R.A.S.*, **172**, 15p.
Garmany, C. D., Conti, P. S., and Massey, P. 1980, *Ap. J.*, **242**, 1063.
Gingold, R. A., and Monahan, J. J. 1978, *M.N.R.A.S.*, **184**, 481.
Halbwachs, J. L. 1980, *Astr. Ap.*, **128**, 399.
Halbwachs, J. L. 1986, *Astr. Ap.*, **168**, 161.
Harrington, R. S. 1974, *Cel. Mech.*, **9**, 465.
Harrington, R. S. 1975, *A. J.*, **60**, 1081.

Heintz, W. D. 1969, *Journal RAS Can.*, **63**, 275.
Herchel, W. 1802, *Phil. Trans. Roy. Soc.*, xciii, 340.
Herczeg, T. J. 1988, *Ap. Sp. Sci.*, **142**, 89.
Hills, J. G. 1976, *M.N.R.A.S.*, **175**, 1p.
Hills, J. G., and Day, C. A. 1976, *Ap. Lett.*, **17**, 87.
Jeans, J. H. 1929, "Astronomy and Cosmogony", 2nd ed., Cambridge University Press.
Kuiper, G. P. 1935a, *Pub. A.S.P.*, **47**, 15.
Kuiper, G. P. 1935b, *Pub. A.S.P.*, **47**, 121.
Kuiper, G. P. 1955c, *Pub. A.S.P.*, **67**, 387.
Lada, C. J., and Shu, F. H. 1990, *Science*, **248**, 564.
Lamb, H. 1932, "Hydrodynamics", 6th ed., Cambridge University Press.
Larson, R. B. 1969, *M.N.R.A.S.*, **145**, 211.
Larson, R. B. 1972, *M.N.R.A.S.*, **156**, 437.
Larson, R. B. 1978, *M.N.R.A.S.*, **184**, 69.
Larson, R. B. 1990, in "Physical Processes in Fragmentation and Star Formation", R. Capuzzo-Dolatta, C. Chiosi, and A. Di Fazio (eds.), Kluwer, Dordrecht, 389.
Lucy, L. B. 1977, *A. J.*, **82**, 1013.
Lucy, L. B. 1981, in *Proc. IAU Symp 93*, "Fundamental Problems in the Theory of Stellar Evolution.", D. Sugimoto, D. Q. Lamb and D. Schramm (eds.), Reidel, Dordrecht, 73.
Lyttleton, R. A. 1953, in "The Stability of Rotating Liquid Masses", Chapter X, Cambridge University Press.
Mazeh, T. 1990, *A. J.*, **99**, 675.
Mercer-Smith, A., Cameron, A. G. W., and Epstein, R. I. 1982, *Ap. J.*, **279**, 363.
Michell, J. 1767, *Phil. Trans. Roy. Soc.*, lxxii, 97.
Michell, J. 1784, *Phil. Trans. Roy. Soc.*, lxxiv, 56.
Myers, P. C., and Benson, P. J. 1983, *Ap. J.*, **266**, 309.
Norman, M. L., and Wilson, J. R. 1978, *Ap. J.*, **224**, 497.
Norman, M. L., Wilson, J. R., and Barton, R. T. 1980, *Ap. J.*, **239**, 968.
Novotny, E. 1973, "Introduction to Stellar Atmospheres and Interiors", Oxford University Press.
Ostriker, J. P. 1970, in "Stellar Rotation", A. Shettebak (ed.), Gordon and Breach, New York, 147.
Poveda, A. 1988, *Ap. Sp. Sci.*, **142**, 67.
Pringle, J. E. 1989, *M.N.R.A.S.*, **239**, 361.
Shu, F. H. 1977, *Ap. J.*, **214**, 488.
Shu, F. H., Adams, F. C., and Lizano, S. 1987 *Ann. Rev. A. A.*, **25**, 23.
Silk, J. 1978, in "Protostars and Planets", T. Gehrels (ed.), Univ. of Arizona Press, 172.
Stahler, S. W. 1983, *Ap. J.*, **274**, 822.
Strom, K. M., Strom, S. E., Edwards, S., Cabrit, S., and Strutskie, M. F. 1989, *A. J.*, **97**, 1451.
Tassoul, J-L. 1978, "Theory of Rotating Stars", Princeton University Press.
Tout, C. A. 1991. *M.N.R.A.S.*, in press.
Tscharnuter, W. 1975, *Astr. Ap.*, **39**, 207.
Valtonen, M., and Mikkola, S. 1991, *Ann. Rev. A. A.*, in press.
van Albada, T. S. 1968a, *Bull. Astr. Inst. Neth.*, **19**, 479.
van Albada, T. S. 1968b, *Bull. Astr. Inst. Neth.*, **20**, 57.

van den Heuvel, E. P. J. 1983, in "Accretion Driven Stellar X-ray Sources", E. P. H. van de Heuvel and W. H. G. Levin (feds.), s Cambridge University Press, 303.

Lee Hartmann and Jim Pringle

SINGLE-STAGE FRAGMENTATION AND A MODERN THEORY OF STAR FORMATION

TELEMACHOS CH. MOUSCHOVIAS
University of Illinois at Urbana-Champaign
Departments of Physics and Astronomy
1002 West Green street
Urbana, IL 61801, U. S. A.

ABSTRACT. A basic theory of single-stage fragmentation (or core formation) in magnetic clouds is presented. The *decay* of hydromagnetic waves due to ambipolar diffusion, not the generation of perturbations, is identified as the cause of such fragmentation. The condition for its onset selects a *range* of length scales and densities in typical present-day molecular clouds. Protostellar masses 0.1 - 10 $M_\odot$ are selected, with a preference toward the smaller masses. Many independent fragments can form in magnetically supported clouds. The role of three critical natural length scales (Alfvén, thermal, and magnetic) is explained.

The relation $B_c = const \times (\rho_c T_c)^{1/2}$ between the magnetic field strength and the gas density and temperature in a contracting cloud's central flux tubes is reexamined, accounting for the effect of hydromagnetic waves and fragmentation. Key results from recent numerical calculations of the *self-initiated* collapse (due to ambipolar diffusion) of axisymmetric clouds are summarized. They support the analytical conclusions and, in addition, determine the structure, sizes, masses, and magnetic fluxes of protostellar cores (up to densities 3×10^9 cm^{-3}). A modern theory of star formation is described.

1. INTRODUCTION

In Paper I in this volume (p. 61) we presented a quantitative description of the underlying physics, identified the most basic processes, and obtained key results relevant to the formulation of a theory of star formation. In this paper we focus on theoretical *results* and their applications, and we synthesize the collective knowledge acquired from them into a modern theory of star formation. The description is physical, not formal, but accurate.

We first summarize in § 2.1 some of the key discoveries responsible for our current understanding of the role of magnetic fields in star formation. Then, after we present some new concepts and new analytical and numerical results on fragmentation (or core formation) and its (observable) consequences in §§ 2.2 - 2.5, we propose in § 3 a modern theory of star formation. Observations of the density structure, velocity field, mass, and magnetic field of protostellar fragments with densities 10^4 - 10^9 cm^{-3} are needed for *testing* the predictions of the theory. The differential mass-to-flux ratio $dm(\Phi_B)/d\Phi_B$ in clouds and fragments (or cores) is the most needed observational *input* to the theory. At present, as explained in Paper I, it is treated as a free (but not arbitrary) function in the theory, within the observationally allowed limits and theoretical constraints from stability considerations.

Diverse phenomena taking place in massive clouds and relating to star formation have been understood on the basis of a simple theoretical *result* (not assumption); namely, that even relatively weak magnetic fields are capable of supporting virtually all observed molecular clouds against gravitational collapse (Mouschovias 1976a, b; hereafter TM76a, TM76b, etc.; Mouschovias and Spitzer 1976; TM77a, TM78; see, also, the review TM87a). This is not to suggest that the interstellar magnetic field is the only agent supporting molecular clouds. After all, magnetic forces act only perpendicular to the field lines. Star formation, however, cannot take place by one-dimensional (along field lines) motion (Spitzer 1968), and, unless the mass-to-flux ratio exceeds a critical value, a frozen-in magnetic field prevents collapse even if the external pressure were to be infinite; a large external pressure would only transform the cloud into a thin sheet with its plane perpendicular to the field lines. We therefore begin our discussion with a brief description of the consequences of magnetic support of massive clouds and the historical context in which they were obtained.

2. BASIC BUILDING BLOCKS OF A THEORY OF STAR FORMATION

2.1. Consequences of Magnetic Support of Clouds

1. While in the 1970s the primary focus of theoretical research efforts on star formation was the question of what triggers the collapse of molecular clouds (e.g., see 1977, IAU Symp. No. 75, *Star Formation*), we have argued from a purely theoretical point of view, at the same time that Zuckerman and Palmer (1974) were arguing from an observational point of view, that the question must really be: what *prevents* the molecular clouds from collapsing and copiously forming stars? We showed that magnetic fields, in fact much weaker than the previously predicted $B \propto \rho^{2/3}$ (Mestel 1966), *can* support very massive clouds (TM75b, TM76a, TM76b, Mouschovias and Spitzer 1976, and reviews Spitzer 1978, TM78).

2. The same calculations mentioned above showed that magnetically supercritical *clouds* (as opposed to fragments within clouds) should, if they exist at all, be very rare in nature. The magnetic tension force was found capable of supporting extended envelopes and halting the contraction of a cloud at progressively smaller radii as long as the magnetic field is frozen in the matter. We therefore proposed that *ambipolar diffusion* is responsible for permitting contraction in the core and only in the core of self-gravitating clouds (TM76b). Without it, the formation of stars would be a rare phenomenon, even if one allows for occasional external triggers to set clouds into collapse. The observed inefficiency of the star formation process was also attributed to the demonstrated effective magnetic support of envelopes.

3. Contrary to the then accepted dogma, that ambipolar diffusion is not important below $n_n \sim 10^9$ cm^{-3} (Nakano and Tademaru 1972), we demonstrated that it in fact sets in rapidly enough ($\tau_{AD} \lesssim 10^7$ yr) in the cores of molecular clouds and leads to star formation through a *single-stage* (as opposed to hierarchical) *fragmentation* (TM77a, TM77b; see, also, review TM81) --this process has since been renamed "core formation." The angular momentum problem of star formation is resolved by magnetic braking, and the entire range of periods (and, therefore, angular momenta) of binary stars from 10 hr to 100 yr is explained naturally. The induced contraction was *quasistatic* by construction, in accordance with the original proposal by Spitzer (1968). It was also discovered and emphasized that, contrary to earlier notions that ambipolar diffusion leads to the loss of flux by a cloud as a whole (Mestel and Spitzer 1956; Nakano and Tademaru 1972; Nakano 1976, 1977), the essence of ambipolar diffusion is a *redistribution* of flux in the *central* flux tubes of a cloud, not a loss of flux by the whole cloud (TM78, TM79b)[1] --see footnote on next page.

4. It was also recognized and emphasized from the outset that the *differential* mass-to-flux ratio $(dm/d\Phi_B)$ is a crucial quantity for a cloud's evolution (TM76a, TM76b, TM78). The same total mass of a cloud distributed differently in its various flux tubes makes the

difference between a reclusive life for the cloud (in equilibrium) and an eventful, exciting existence, with rapid collapse and star formation in its core (e.g., see TM78, p. 218).

5. Not only does ambipolar diffusion permit star formation in cloud cores (TM76b) but it also *initiates* the collapse of cores (and only of cores) of otherwise quiescent clouds (TM79b). The ambipolar diffusion time scale was found to be typically 3 - 4 orders of magnitude smaller in the core than in the envelope ($\sim 10^5$ yr *versus* $\gtrsim 10^8$ yr) --see summary in Paper I, § 2.3.3.

6. For most clouds, which are magnetically supported, the free-fall time scale is an irrelevant time scale. The ambipolar diffusion time scale (more precisely, the flux-loss time scale in the core) *is* the evolutionary time scale (TM77a, TM79b).

7. The magnetic braking time scale is short enough to set H I clouds in nearly synchronous galactocentric orbits and to keep the angular velocity of molecular clouds near galactic-rotation levels up to densities $\sim 10^3$ cm^{-3} (TM77a, TM78). Before those papers, the latest work on the subject had concluded that "The efficiency of [magnetic] braking is never so great as to keep the cloud even roughly corotating with the background" (Gillis, Mestel, and Paris 1974).

8. The time scale ($\tau_\perp$) for magnetic braking of a perpendicular rotator is much shorter than that of an aligned rotator ($\tau_\parallel$) of similar physical parameters (TM77a, Mouschovias and Paleologou 1979, 1980; hereafter, MP79, MP80, etc.). Thus most clouds would tend to become aligned rotators in time. However, fragments within a perpendicular rotator can exhibit the phenomenon of *retrograde rotation* with respect to the sense of their revolution

---

[1]Although Nakano (1976, 1977) refers to his arbitrarily selected one-, ten-, and hundred-solar-mass objects as "fragments," in his formal description the surrounding medium is assumed to have a negligible density and a negligible magnetic field; i.e., these objects are clouds in their own right. Moreover, each such object is envisioned as losing flux *as a whole*; no redistribution of mass within its flux tubes is considered. Also, in these papers it was *not* realized that ambipolar diffusion itself is responsible for the onset of collapse and for the selection of the fragment mass. The Parker (1966) instability is explicitly named as responsible for fragmentation(!), and the masses of these objects are put in arbitrarily by the author. What these papers did recognize and demonstrate is that, if some agent leads to a significant density (and magnetic field) enhancement (> 5) compared to that of the background, then ambipolar diffusion may further enhance this density perturbation by driving the flux out. The fragmentation talked about, but not calculated, in these papers is *hierarchical*, not the single-stage fragmentation (or core formation) which we proposed, as is also made abundantly clear in the review by Nakano (1981, p. 65), who names the process "repeated fragmentation" and identifies it with the hierarchical fragmentation inferred earlier by Mestel (1965). Statements in the literature to the contrary simply have no basis in fact.

We cannot overemphasize in this context that the paper TM77a, contrary to the statements by Nakano (1990, § 6), was not concerned with the magnetic flux problem of star formation. What that paper demonstrated is that ambipolar diffusion occurs rapidly enough at densities $\simeq 10^4 - 10^6$ cm^{-3} to crucially affect the evolution of molecular clouds and lead to star formation. Nowhere did it claim that the magnetic flux problem is resolved at such low densities. The distinction between the onset of ambipolar diffusion at a significant rate to affect a cloud's evolution and the resolution of the magnetic flux problem is of the essence. The fine paper by Nakano (1979) still fails to recognize the importance of ambipolar diffusion at typical molecular cloud densities, for it starts the (quasistatic) evolution of the model cloud at a density 10^7 cm^{-3}.

in the cloud. It has been observed in at least two independent studies (Clark and Johnson 1981; Young et al. 1981). This feature may be preserved and show up as retrograde rotation in stellar and planetary systems, as a direct and natural consequence of magnetic braking. The expression for the magnetic braking time scale ($\tau_\perp$ or $\tau_\parallel$) can be recovered from a simple physical argument: *it is the time required by the torsional Alfvén waves to propagate far enough into the external medium to affect a moment of inertia equal to that of the cloud.*

9. Detailed collapse calculations determined that the contraction time scale is typically ten times longer than the free-fall time (TM83b). This result and those from the magnetic braking and ambipolar diffusion calculations were combined and applied to ~ 1 $M_\odot$ cores to explain the observed (Vogel and Kuhi 1981; Wolff et al. 1982) rotational velocities of stars of spectral type O5 to F5.

10. Alfvénic or superAlfvénic collapse traps the torsional Alfvén waves in a cloud, thereby marking the onset of angular momentum conservation (TM78). TM87a shows that conservation of angular momentum due to such contraction (as opposed to $\tau_\parallel \simeq \tau_{AD}$) does not change TM77's conclusion/explanation of binary periods. (Nakano 1990 ignores TM87a and claims this result as new.)

11. Magnetically linked fragments in molecular clouds tend to trap rotational kinetic energy (and mean-square angular momentum) among them, with the consequence that the angular momentum problem is resolved and recreated, perhaps several times, before it is irreversibly resolved (Mouschovias and Morton 1985a, b; hereafter MM85a, MM85b, etc.). Fragments can spin near or even above breakup as a result of that trapping and sharing of rotational kinetic energy.

12. Magnetic braking remains effective even in the presence of ambipolar diffusion for typical (subAlfvénically contracting) molecular cloud cores; it can resolve the angular momentum problem even for single stars and planetary systems (MP86).

13. The observed supersonic (but not superAlfvénic) spectral line widths in molecular clouds can be the result of long-wavelength hydromagnetic waves, which are indistinguishable from large-scale oscillations of a cloud and decay only on the relatively long ambipolar diffusion time scale in cloud envelopes (TM75b; see, also, acknowledgement in Arons and Max 1975; TM87a).

14. As a result of ambipolar diffusion, narrowing and eventual thermalization of line widths should be exhibited by observations of molecules sampling progressively higher densities (and smaller length scales) --see TM87a. This effect has been observed by Baudry et al. (1981) in TMC 2, in which the measured line widths for ^{13}CO, $H^{12}CO^+$, $H^{13}CO^+$, and $C^{18}O$ are 1.1, 1.0, 0.75, and 0.6 km s^{-1}, respectively.

2.2. Critical Length Scales and Protostellar Masses

Three natural *critical* length scales play a crucial role in the formation and evolution of fragments (or cores) in self-gravitating, magnetic interstellar clouds. We refer to them as the *Alfvén length scale* λ_A, the *thermal length scale* $\lambda_{T,cr}$, and the *magnetic length scale* $\lambda_{M,cr}$ [see extensive discussion in TM87a, § 2.2.6(b); TM90b, § 4; TM91, § 4]. They are given by

$$\lambda_A = \pi v_A \tau_{ni}, \qquad \lambda_{T,cr} = 1.4 C \tau_{ff}, \qquad \lambda_{M,cr} = 0.62 v_A \tau_{ff}, \qquad (1a, b, c)$$

where the quantity v_A is the Alfvén speed in the neutrals, C the isothermal speed of sound, τ_{ni} the neutral-ion collision time, and τ_{ff} the free-fall time. In brief, their physical meaning is as follows.

Alfvén waves with wavelengths smaller than λ_A cannot propagate in the neutrals; they decay because of ambipolar diffusion. The self-gravity of a spherical, isothermal cloud (or fragment) becomes stronger than the opposing thermal-pressure forces if the radius is greater than $\lambda_{T,cr}$, which is identical with the Bonnor-Ebert (1955 - 1957) critical radius. In a qualitative sense, the meaning of $\lambda_{T,cr}$ analogous to that of λ_A is that sound waves with wavelengths longer than $\lambda_{T,cr}$ are damped by self-gravity and, if thermal pressure is the only opposing force, collapse ensues. Similarly, a spherical or oblate cloud (or fragment) whose polar radius (along the field lines) exceeds $\lambda_{M,cr}$ cannot be magnetically supported. The length scale $\lambda_{M,cr}$ is completely equivalent to the Mouschovias and Spitzer (1976) critical mass-to-flux ratio $(M/\Phi_B)_{crit} = (63G)^{-1/2}$ (see Paper I, eq. [22a]). (Recall that the critical mass-to-flux ratio in the *central* flux tube can be up to 50% greater than this value; see Paper I, eq. [22b] and associated discussion.) Qualitatively, in terms of the wave picture, hydromagnetic waves in a cold cloud of wavelength longer than $\lambda_{M,cr}$ contain enough mass-to-flux ratio within one wavelength along field lines and any radius perpendicular to the field lines to become gravitationally unstable.

At typical molecular cloud parameters, the three length scales are given by

$$\lambda_A = 0.29 \left(\frac{B}{30~\mu G}\right) \left(\frac{1}{10^{1-2k}}\right) \left(\frac{10^3~\text{cm}^{-3}}{n_n}\right)^{0.5+k} \left(\frac{3 \times 10^{-3}~\text{cm}^{-3}}{K}\right) \text{pc}, \quad (2a)$$

$$\lambda_{T,cr} = 0.29 \left(\frac{T}{10~\text{K}}\right)^{1/2} \left(\frac{10^3~\text{cm}^{-3}}{n_n}\right)^{1/2} \text{pc}, \quad (2b)$$

$$\lambda_{M,cr} = 0.91 \left(\frac{B}{30~\mu G}\right) \left(\frac{10^3~\text{cm}^{-3}}{n_n}\right) \text{pc}, \quad (2c)$$

where we have used the following expressions for the Alfvén speed, the neutral-ion collision time, the sound speed, and the free-fall time:

$$v_A = \frac{B}{(4\pi\rho_n)^{1/2}} = 1.4 \left(\frac{B}{30~\mu G}\right) \left(\frac{10^3~\text{cm}^{-3}}{n_n}\right)^{1/2} \text{km s}^{-1}, \quad (3a)$$

$$\tau_{ni} = 6.7 \times 10^{3+2k} \left(\frac{3 \times 10^{-3}~\text{cm}^{-3}}{K}\right) \left(\frac{10^3~\text{cm}^{-3}}{n_n}\right)^{k} \text{yr}, \quad (3b)$$

$$C = \left(\frac{k_B T}{\mu m_H}\right)^{1/2} = 0.19 \left(\frac{T}{10~\text{K}}\right)^{1/2} \text{km s}^{-1}, \quad (3c)$$

$$\tau_{ff} = \left(\frac{3\pi}{32 G \rho_n}\right)^{1/2} = 1.1 \times 10^6 \left(\frac{10^3~\text{cm}^{-3}}{n_n}\right)^{1/2} \text{yr}, \quad (3d)$$

as well as the relation between the ion and neutral densities

$$n_i = 3 \times 10^{-3} \left(\frac{K}{3 \times 10^{-3}~\text{cm}^{-3}}\right) \left(\frac{n_n}{10^5~\text{cm}^{-3}}\right)^{k} \text{cm}^{-3}. \quad (3e)$$

The algebraic expression for τ_{ni} is given by the second part of equation (12i) of Paper I. As in Paper I, a collision cross section appropriate for HCO^+-H_2 collisions has been used, and equation (3e) is valid in the approximate density range $10^3 \leq n_n \leq 10^9$ cm^{-3}. The quantity μ

in equation (3c) is the mean mass per particle in units of the atomic-hydrogen mass, m_H. The "canonical" values of k and K are $1/2$ and 3×10^{-3} cm^{-3}, respectively. In the deep interiors of gravitationally contracting molecular clouds, the magnetic field strength B is not independent of the gas density n_n and temperature T; it varies as $B \propto (n_n T)^{1/2}$, often written as $B_c \propto \rho_c^\kappa$, $\kappa = 1/3 - 1/2$, and with the value $\kappa = 1/2$ having a special meaning (TM76b; see, also, Paper I, §§ 5.2.2 and 5.2.3; and § 2.3 below).

In the context of cosmic-ray propagation and scattering off magnetic inhomogeneities in the diffuse intercloud medium, Kulsrud and Pearce (1969) found that small-amplitude Alfvén waves of wavelength smaller than λ_A cannot propagate in the predominantly neutral intercloud gas. It is only relatively recently, however, that the role of the length scale λ_A in the initiation of fragmentation (or core formation) and the selection of protostellar masses $\sim$ 1 $M_\odot$ in otherwise quiescent molecular clouds ($n_n \sim 10^4$ cm^{-3}) has been recognized (TM87a; see, also, TM91). If hydromagnetic waves provide partial support against gravity, at least along field lines, ambipolar diffusion removes this support over length scales smaller than λ_A. Thus, masses $\simeq 1 - 0.1$ $M_\odot$ are selected naturally by the decay of Alfvén waves in molecular cloud interiors in the respective density range $10^4 - 10^6$ cm^{-3}, characteristic of quiescent molecular clouds. The near equality of $\lambda_{T,cr}$ and λ_A (see eqs. [2a] and [2b]) implies that gravitational forces can in fact induce contraction against thermal pressure over such length scales, while ($\lambda_{M,cr}$ being greater than $\lambda_{T,cr}$ and λ_A) the envelopes of molecular clouds remain magnetically supported. Many such fragments can form essentially independently inside a molecular cloud. This, we have suggested, is the origin of the plethora of fragments (or cores) observed in molecular clouds (e.g., see review by Myers 1985).

Since in the density range $10^3 - 10^4$ cm^{-3}, at which ambipolar diffusion sets in at a rate rapid enough to begin to affect a cloud's evolution ($\tau_{AD} \simeq 6 - 2 \times 10^6$ yr, respectively), we have that $\lambda_{T,cr} \simeq \lambda_A$ ($\simeq 0.1 - 0.3$ pc $< \lambda_{M,cr}$), the "initial" core mass (~ 1 $M_\odot$) cannot be prevented by magnetic or thermal-pressure forces from eventually collapsing and forming a protostar, despite the fact that $\lambda_{T,cr}$ and λ_A (and the mass inside these length scales) evaluated at the *instantaneous* density of the core decrease as $n_n^{-1/2}$ (for $k = 1/2 = \kappa$) upon contraction (see eqs. [2b] and [2a]). In other words, a runaway compact core of mass significantly less than the original core mass is expected to form, but the original core mass represents a reservoir of matter (with a spatially rapidly decreasing density, and infall speeds comparable to the sound speed) which will eventually accrete onto the compact core and thus determine the protostellar mass. *Because of effective magnetic support, matter significantly farther than the original $\lambda_A \simeq \lambda_{T,cr}$ from the core's center can accrete only extremely slowly, and, therefore, it is not available for increasing the protostellar mass by any appreciable amount. At the later stages of contraction, accretion may be further limited by the presence of angular momentum. Moreover, at a distance greater than $\simeq 2\lambda_{T,cr}$ from the center of a forming core, gravitationally driven ambipolar diffusion can lead to the formation of another, independent core much more rapidly than it would take for this material to be accreted by the first core. Each core can form either a binary or a single star, depending on its leftover angular momentum. We have suggested that this is the origin of stellar clusters in otherwise quiescent molecular clouds.*

The absolute and relative variations of the three critical length scales with density are

$$\lambda_A \propto \frac{1}{\rho_n^{0.5+k-\kappa}}, \qquad \lambda_{T,cr} \propto \left(\frac{T}{\rho_n}\right)^{1/2}, \qquad \lambda_{M,cr} \propto \frac{1}{\rho_n^{1-\kappa}}, \qquad (4a, b, c)$$

$$\frac{\lambda_{T,cr}}{\lambda_A} \propto \frac{1}{\rho_n^{\kappa-k}}, \qquad \frac{\lambda_{M,cr}}{\lambda_A} \propto \frac{1}{\rho_n^{0.5-k}}, \qquad \frac{\lambda_{T,cr}}{\lambda_{M,cr}} \propto \rho_n^{0.5-\kappa}, \qquad (5a, b, c)$$

where we have used the relation $B \propto \rho^\kappa$. It is clear from the relations (5a) - (5c) that the only way in which the initial relative magnitudes of the three length scales ($\lambda_{M,cr} > \lambda_{T,cr} \simeq \lambda_A$) will be preserved during the formation and contraction of a cloud's core is for the condition $k = \kappa = 1/2$ to be satisfied at all times. Detailed (axisymmetric) collapse calculations in the presence of ambipolar diffusion have shown that $\kappa = 0.4 - 0.5$ for all observationally allowed values of the free parameters of the problem, with the value $\kappa = 0.47$ being the most typical (Fiedler and Mouschovias 1991a, b; Morton and Mouschovias 1991a, b; TM90b, TM91). Elmegreen's (1979) work indicates that $k \simeq 0.5$. A similar value was obtained by Nakano (1979). Falgarone and Perault (1987) argue for $k = 0.4$.

At relatively low densities ($\lesssim 10^3$ cm^{-3}), as a self-gravitating cloud's central flux tubes contract, force balance along field lines is established relatively rapidly and lateral contraction proceeds only as rapidly as magnetic forces allow (TM76b). This is so for both isothermal and nonisothermal contraction. It is so even if hydromagnetic waves and subsonic turbulence provide partial support against gravity along field lines (see Paper I, § 5.2.3). As reviewed in Paper I, such contraction, with the magnetic field frozen in the matter, establishes the relation

$$B_c/B_0 = (\rho_c/\rho_0)^\kappa, \qquad \kappa = 1/2 \qquad (6a, b)$$

in the *central* part of the cloud. (For the variation of κ with position inside a cloud, see TM78.) Since gravity balances the thermal-pressure forces along field lines and the magnetic forces perpendicular to the field lines in a critical central flux tube, it is intuitively clear that in this part of a cloud the thermal, magnetic, and gravitational pressures will be comparable in magnitude; i.e., $P_c \sim \pi G \sigma_{m,c} \sim B_c^2/8\pi$, where $\sigma_{m,c}$ is the column density of the central flux tube. All the critical states determined by TM76b have a ratio of magnetic and thermal pressures,

$$\alpha = \frac{B^2}{8\pi\rho_n C^2} = 26.0 \left(\frac{B}{30\ \mu G}\right)^2 \left[\frac{10^3\ \text{cm}^{-3}}{n_n}\right] \left(\frac{10\ \text{K}}{T}\right), \qquad (7)$$

evaluated *at the center*, always in the range 0.5 - 0.7. However, as is seen from equation (7), in which all quantities have been normalized to typical values of quiescent molecular cloud *envelopes*, magnetic forces dominate the thermal-pressure forces in the envelopes. This *bimodal opposition to gravity* (by magnetic forces in the envelope and thermal-pressure forces in the core), exacerbated by ambipolar diffusion, has important implications for star formation, not the least consequential of which is the introduction of a break in the slope of the density profile logρ-log$r$; i.e., *a characteristic length is introduced in the problem which plays a key role in determining protostellar masses* (TM91; see, also, § 2.5 below). In §§ 2.3 and 2.4 we explain the basic physics of the proposed theory, which predicts a B_c-ρ_c relation even in the presence of ambipolar diffusion, accounts for fragmentation in magnetic clouds (including its effect on the B_c-ρ_c relation), predicts a range of densities at which fragmentation should occur, and obtains naturally $\sim 1\ M_\odot$ as a typical protostellar mass.

2.3. The B_c - ρ_c Relation Revisited

In Paper I, § 5.2.3, we summarized the physical origin and validity of the relation $B_c = $const $\times (\rho_c T_c)^{1/2}$ between the magnetic field strength, the gas density, and the (effective) temperature in the central part of a contracting cloud, but we did not discuss the proportionality constant. We now examine this relation more closely by using simple analytical means. In what follows, as in TM91 and in Paper I, one may replace the isothermal sound speed C (or the temperature T_c) with an effective sound speed C_{eff} (or an effective temperature $T_{c,\text{eff}}$) in order to account for possible hydromagnetic-wave or subsonic-turbulence support along field lines. Such replacement must be made in applications to molecular clouds not exhibiting significant flattening along field lines.

We have seen that balance of forces along field lines implies the relation

$$C^2 \simeq 2\pi G \rho_c Z^2 \qquad (8)$$

(Paper I, eq. [117a]), where Z is the polar radius (or half-thickness) of the cloud. Since $\sigma_{m,c}$ and Z are related by $\sigma_{m,c} \equiv M_c/\pi R^2 = 2Z\rho_c$, where R is the equatorial radius of the central flux tube under consideration, we may eliminate Z from equation (8) in favor of $\sigma_{m,c}$ and ρ_c to find that

$$\rho_c \simeq \frac{\pi G \sigma_{m,c}^2}{2C^2}. \qquad (9)$$

Equation (9) gives that density at which balance of forces along field lines (in the central flux tube) is established when the column density is $\sigma_{m,c}$ and the temperature T_c.

However, a magnetically critical central flux tube has a column density determined by the central magnetic field strength, and given by

$$\sigma_{m,c} = 1.5\,(63G)^{-1/2} B_c \qquad (10)$$

(Paper I, eqs. [22b] and [22a]). Hence, $\sigma_{m,c}$ may be eliminated between equations (9) and (10) to find a relation between B_c and the product $\rho_c T_c$; namely,

$$\alpha_c \equiv \frac{B_c^2}{8\pi \rho_c C^2} \simeq 0.7, \qquad (11a)$$

or, equivalently,

$$B_c \simeq 50 \left[\frac{n_c}{10^4\,\text{cm}^{-3}} \frac{T_{c,\text{eff}}}{10^2\,\text{K}} \right]^{1/2} \mu\text{G}, \qquad (11b)$$

where n_c is the density of neutrals in the central flux tube under consideration, and the effective temperature $T_{c,\text{eff}}$ is defined in terms of C^2 and the mean-square wave speed u^2 by the relation

$$n_c k_B T_{c,\text{eff}} = \rho_c (C^2 + u^2), \qquad (12)$$

assuming that the wave pressure can be written as $P_w = \rho_c u^2$. Note that equation (11a) is in good agreement with the values of α_c found by TM76b, as well as those found recently by Tomisaka, Ikeuchi, and Nakamura (1988), characterizing states on the verge of collapse against thermal and magnetic forces. The effective temperature has been normalized to 100 K in equation (11b) because the Alfvén speed in typical molecular clouds is expected to be $\gtrsim 1$ km s^{-1} (see eq. [3a]); hence, equation (11a) implies that $C_{\text{eff}} \simeq 1$ km s^{-1}. The *actual* temperature of typical H I clouds, which are presumably the progenitors of molecular clouds, is $\sim 10^2$ K and that of the interiors of quiescent molecular clouds ~ 10 K. This would seem to imply that isothermality breaks down during the formation of molecular clouds. However, observed line widths in molecular clouds are often ~ 1 km s^{-1}. One may therefore conclude that approximate "isothermality," in the sense $T_{c,\text{eff}} \simeq \text{const}$, is maintained during this phase of cloud evolution. For this reason, the relation $B_c \propto \rho_c^{1/2}$ remains valid in the inner part of a gravitationally contracting cloud with a frozen-in magnetic field. In what follows, we examine how single-stage fragmentation (or core formation) sets in, how it progresses, how it selects ~ 1 $M_\odot$ as a typical protostellar mass, and how it affects the above B_c-ρ_c relation.

2.4. Fragmentation and its Effect on the B_c - ρ_c Relation: A Simple Theory

We consider the breakup of a cloud's critical central flux tube, in which equation (11) is obeyed, into N fragments of equal mass (and, therefore, equal column density). (Breakup of subcritical flux tubes is studied below.) The number N and the cause of fragmentation are left arbitrary for the moment. This fragmentation process has reduced the mass-to-flux ratio of each fragment to $N^{-1}(dm/d\Phi_B)_{c,\text{crit}}$, where $(dm/d\Phi_B)_{c,\text{crit}}$ is the critical mass-to-flux ratio of the parent flux tube and is given by equations (22b) and (22a) of Paper I. Since each fragment is now characterized by a subcritical mass-to-flux ratio, it will wait for gravitationally and quasistatically driven ambipolar diffusion to increase its mass-to-flux ratio back to the critical value, at which stage rapid collapse may ensue if the fragment is also thermally supercritical. *Problem: At which central density $\rho_{c,\text{fr,cr}}$ will a fragment reestablish force balance along field lines and acquire a critical mass-to-flux ratio (because of ambipolar diffusion perpendicular to the field lines)?*

If quasistatic contraction due to ambipolar diffusion progresses in a (subcritical) fragment with the field lines almost "held in place" (in reality, the field strength is expected to increase somewhat; see TM78), i.e., if $B_c \simeq const$, then the fragment's column density must increase by a factor N in order to reach the critical state. Actually, detailed numerical studies in an axisymmetric geometry have shown that, typically, the field strength increases by a factor ≤ 2 during this phase of contraction (Fiedler and Mouschovias 1991a, b; Morton and Mouschovias 1991a, b; see, also, TM90b, Fig. 2c). Therefore, the column density must increase by the factor $2N$ before the fragment becomes magnetically critical (see eq. [10]).

Since the density at which balance of forces along field lines is (re)established is given by equation (9), with $\sigma_{m,c}$ on the right-hand side now being the critical column density, it follows that $\rho_{c,\text{fr,cr}}$ will be greater than the density at which the breakup along field lines began (call it $\rho_{c,\text{fr,0}}$) by the factor $(2N)^2$ --assuming that $T_{c,\text{eff}} = const$ in the process. Altogether, then, if we know the density $\rho_{c,\text{fr,0}}$ at which fragmentation takes place and the resulting number of fragments (or cores) in the central flux tube of a cloud, we can calculate the density at which both force balance along field lines will be (re)established *and* the mass-to-flux ratio of the fragment will become critical. In fact, we can also calculate how long this process will take since the contraction time is essentially the "initial" flux-loss time scale at the density $\rho_{c,\text{fr,0}}$, namely, $\tau_{\Phi,c0} = 0.5\tau_{\text{AD},c0} = 0.3(\tau_{\text{ff}}^2/\tau_{\text{ni}})_{c0}$ (Paper I, § 2.3.1).

For the typical case of three fragments along a flux tube, as originally argued by MM85b (p. 206 and Fig. 2), we find that the density will increase by the factor $(2N)^2 = 36$ before equation (11b) is reestablished for the fragment. The number $N \simeq 3$ is also suggested by the relative magnitude of the critical length scales, discussed in § 2.2, under typical conditions in quiescent molecular clouds. Equations (2b) and (2c) show that $\lambda_{M,\text{cr}} \simeq 3 \lambda_{T,\text{cr}}$, suggesting that a magnetically critical flux tube may break up into three fragments by a Jeans-like instability along field lines. Actually, a more precise value of $\lambda_{M,\text{cr}}$ for the *central* critical flux tube of a cloud is up to 50% greater than that given by equation (2c), which was obtained from equation (22a), rather than equation (22b), of Paper I. Thus it may be the case that $\lambda_{M,\text{cr}} \simeq 5 \lambda_{T,\text{cr}}$. Three fragments still seem a reasonable number, allowing for less than 100% efficiency of the fragmentation process and/or length scales somewhat greater than $\lambda_{T,\text{cr}}$ to grow. In these estimates, we have used the actual rather than the effective temperature of the central flux tube because, as we have already discussed in § 2.2 and as seen from equations (2a) and (2b), ambipolar diffusion deprives a fragment of hydromagnetic-wave support over length scales λ_A ($\simeq \lambda_{T,\text{cr}}$).

The above considerations make it clear why we have disfavored hierarchical fragmentation. Since magnetically supercritical clouds (as opposed to cores) are extremely rare, if they exist at all, *each stage* of hierarchical fragmentation would inevitably involve magnetically subcritical fragments, whose evolution toward critical conditions would altogether be too slow; other fragments, formed in the single-stage fragmentation (or core formation) we have envisioned (TM77a, TM77b; also, TM81), can form stars much more rapidly in the neighborhood. The stars themselves would tend to disrupt hierarchical fragmentation either directly by disrupting the fragments or indirectly by raising the temperature of the gas and thus providing better thermal support and, therefore, slowing ambipolar diffusion down. (For the effect of thermal pressure on τ_{AD}, see eq. [14].)

We have also ignored rotation in developing this basic theory of fragmentation. As reviewed in Paper I, § 3.3, magnetic braking, even in the presence of ambipolar diffusion, remains effective enough to keep the centrifugal forces at manageable levels. Besides, we are not attempting to build a complete theory of fragmentation here. We are only interested in identifying the underlying basic physics of the process, especially the role of magnetic fields, and in obtaining approximate results in a simple but reliable way. Some time in the future these considerations will probably seem simplistic. However, since virtually nothing is known at present about the role of magnetic fields in fragmentation, searching for the most elementary and most basic physical principles seems to us to be the way to proceed.

A question of at least observational importance remains unanswered: *At which density does breakup along field lines occur*? Actually, phrased in this fashion, this question can lead theoretical investigations astray, for it presupposes that the density is both the proper and the only parameter determining the onset of fragmentation. Of more fundamental theoretical importance is the question *what conditions(s) must be satisfied in order for fragmentation to take place*? We have suggested that the decay of hydromagnetic waves due to ambipolar diffusion over length scales smaller than λ_A (see eq. [2a]) marks the onset of fragmentation in *magnetically supported* (hence, *subcritical*) molecular clouds, provided that $\lambda_{T,cr}$ is comparable to λ_A, so that gravitational contraction can take place [TM87a, § 2.2.6(b)]. In other words, the searches for answers to the question of where the perturbations that initiate fragmentation come from have, in our view, been completely misplaced --much like the searches for triggers for the collapse of molecular clouds. *What we believe initiates fragmentation is the **decay**, due to ambipolar diffusion, of unavoidably present hydromagnetic disturbances in molecular clouds.* Since a region of size $\simeq \lambda_A$ cannot become gravitationally unstable if the thermal length scale $\lambda_{T,cr}$ is significantly greater than λ_A (but still smaller than $\lambda_{M,cr}$ in a magnetically supported cloud), and since a (magnetically subcritical) region of size $\simeq \lambda_{T,cr} \gg \lambda_A$ can be supported by wave pressure and/or grow only extremely slowly because of ambipolar diffusion, it seems that *the condition for self-initiated fragmentation (or core formation) is $\lambda_{T,cr} \simeq \lambda_A$*. The condition $\lambda_A \simeq \lambda_{T,cr}$ is fulfilled not at a unique density but in a range of densities, approximately 10^3 - 10^6 cm^{-3}. The corresponding masses are approximately 10 - 0.1 $M_\odot$ (TM87a). Thus, protostellar masses of the correct order are selected quite naturally by this single-stage fragmentation mechanism. In present-day molecular clouds, masses ≤ 1 $M_\odot$ are favored by this process.

In summary, ambipolar diffusion not only initiates the collapse of the central flux tubes of otherwise magnetically supported molecular clouds, but it also initiates the breakup of these flux tubes along their lengths into fragments (typically $\simeq 3$), each with typical mass $\sim 1 M_\odot$. It is the *removal* of hydromagnetic perturbations, not their generation, that initiates fragmentation. Several independent flux tubes can suffer this breakup.

Figure 1 exhibits the effect of ambipolar diffusion and consequent single-stage fragmentation (or core formation) on the exponent κ in the relation $B_c \propto \rho_c^\kappa$. An effective temperature $T_{c,eff} \simeq 270$ K was assumed (relatively arbitrarily and for illustrative purposes only) for the diffuse stages of contraction, up to the density $n_{c,fr,0} \simeq 3 \times 10^3$ cm^{-3}, at which fragmentation of the magnetically critical central flux tube sets in at a significant rate due to the damping of hydromagnetic waves by ambipolar diffusion. The *dashed line* shows the B_c-ρ_c relation (slope $\kappa = 1/2$) in the absence of fragmentation (i.e., with the cloud having a single massive critical central part or "core") and for the same effective temperature as in the more diffuse stages of contraction. (A decrease of $T_{c,eff}$ upon contraction would have the same effect as a decreasing κ on this diagram.) Three fragments (or cores) were assumed to form at the density 3×10^3 cm^{-3} for the reasons explained above. In accordance with the discussion at the beginning of this section, quasistatic contraction due to ambipolar diffusion is responsible for the nearly horizontal part of the *solid curve* (B_c increases by only a factor of 2 while the density increases by a factor of 36). At a density $n_{c,fr,cr} \simeq 10^5$ cm^{-3}, a critical mass-to-flux ratio is reestablished in each of the three fragments, and further contraction proceeds with $\kappa = 1/2$. More precisely, detailed numerical modeling shows that the exponent at this late stage of contraction of a fragment is $\kappa = 0.47$ (see § 2.5). Once critical conditions are established in each fragment, we assume that further contraction takes place at a constant $T_{c,eff} \simeq 27$ K. An interested reader may use equation (11b) with different $T_{c,eff}$ from the values used above and a different number of fragments, and then plot a family of pairs of curves like the pair shown in Figure 1. Fragmentation in subcritical clouds delays the state $\kappa = 1/2$ until after critical conditions are established by ambipolar diffusion.

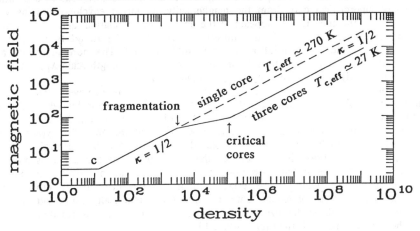

Figure 1. *Central magnetic field* (in μG) *as a function of central density* (in cm^{-3}) *in a contracting, fragmenting model cloud.* Up to a density $n_{c0} \simeq 14$ cm^{-3} (point c), gravity is too weak to induce compression perpendicular to the field lines, and the field B_c is essentially constant (= 3 μG). Balance of forces along field lines *and* a critical central mass-to-flux ratio are established at n_{c0}. Subsequently, $B_c \propto n_c{}^\kappa$, with $\kappa = 1/2$. Damping of waves by ambipolar diffusion induces fragmentation of the central flux tube into three "cores" at a density $n_{c,fr,0} = 3 \times 10^3$ cm^{-3} (point marked "*fragmentation*"). Each fragment (or core) evolves along the *solid curve* until ambipolar diffusion increases its mass-to-flux ratio to the critical value for collapse, at a density $n_{c,fr,cr} \simeq 10^5$ cm^{-3} (point labeled "*critical cores*"). Beyond $n_{c,fr,cr}$, $\kappa = 1/2$ is reestablished in each fragment. The *dashed curve* would represent the evolution of the *critical* central flux tube in the absence of fragmentation.

2.5. Numerical Modeling of Axisymmetric Collapse due to Ambipolar Diffusion

Recent, accurate and extensive numerical work follows the evolution of model clouds to an enhancement of the central density by a factor 10^6, from 3×10^3 to 3×10^9 cm^{-3} (Morton and Mouschovias 1991a, b; Fiedler and Mouschovias 1991a, b; see, also, TM91). We consider axisymmetric model clouds which are initially in exact equilibrium states, with gravity being balanced by thermal-pressure forces along field lines and by magnetic plus thermal-pressure forces perpendicular to field lines. If the magnetic field remained frozen in the matter, the clouds would exist in these states indefinitely. Ambipolar diffusion induces quasistatic contraction in the deep interiors, as discussed analytically in Paper I, § 2.3. The mass-to-flux ratio increases in the central flux tubes until a magnetically critical (and, for these $T = 10$ K models, a thermally supercritical) state is reached. Beyond this stage, the contraction of the now supercritical core accelerates, and its magnetic flux (for K and k in eq. [3e] near their "canonical" values of 3×10^{-3} cm^{-3} and $1/2$, respectively) remains nearly trapped inside, while the envelope is very well supported by magnetic forces. The quasistatic contraction phase lasts however long is necessary to increase the central mass-to-flux ratio to its critical value; thus this phase is longer the more subcritical a cloud is. Once a supercritical core forms, it runs away in density (i.e., it collapses rapidly) from the magnetically supported envelope and, since thermal-pressure forces eventually dominate the magnetic forces in this region, a power-law density profile *tends* to be established, as found by nonmagnetic calculations (e.g., see Larson 1969; Shu 1977). However, the equatorial radius separating the magnetically supercritical core from the magnetically supported envelope remains virtually unchanged from the time it first appears to the end of the calculation, and so does the mass inside and the mass accretion (better, infall) rate across it. We found typical protostellar masses $1 - 2$ $M_\odot$, and typical infall rates $\lesssim 5 \times 10^{-7}$ $M_\odot$ yr^{-1}. The more subcritical a cloud is initially, the more important the role of the critical thermal length scale $\lambda_{T,cr}$ (appropriate to a flattened cloud) is in determining the protostellar mass.

As explained in § 2.4, a critical mass-to-flux ratio is achieved after a density enhancement which is essentially equal to the square of the factor by which the central mass-to-flux ratio is initially below critical. At radii beyond the rapidly evolving critical core (the contraction time scale of the core is $< 10^3$ yr at the end of a typical run), the evolution is controlled by very slow ambipolar diffusion. As a critical central mass-to-flux ratio is approached, a break in the slope of the logρ-log$r$ profile appears. It is more severe the more subcritical the initial mass-to-flux ratio is. As time progresses, the power-law density profile which tends to be established inside the magnetically (and thermally) supercritical region does *not* extend outward, into the magnetically supported envelope. It extends *inward* instead, as the size (and mass) of the innermost, uniform-density, compact part of the core shrinks with time (cf. TM91, Fig. 3).

Figures 2a - 2c show the density in the equatorial plane, normalized to the reference density $n_{c0} = 3 \times 10^3$ cm^{-3}, as a function of radius at seven different times for three different models characterized by initial central mass-to-flux ratios subcritical by a factor 2/3, 1/3, and 1/10, respectively. The unit of length is 0.0457 pc in all figures. The initial central *equilibrium* density and magnetic field are $n_c(t = 0) = 4.35$, 3.24, and 3.02×10^3 cm^{-3}, and $B_c(t = 0) = 18.7$, 32.8, and 102.2 μG. (One may think of the three clouds as having the same mass, 191 $M_\odot$, ion and neutral densities, etc., but a magnetic field in the reference state $B_{c0} = 15.3$, 31.4, and 101.8 μG, respectively.) The times are chosen so as to have a central density enhancement by a factor of 10 from one time to the next, beginning with the initial state at $t = 0$. These times are $t = 0$, 6.0988, 7.0457, 7.2041, 7.2377, 7.2460, and 7.2482×10^6 yr for Figure 2a; $t = 0$, 13.726, 15.380, 15.608, 15.650, 15.660, and 15.662×10^6 yr for Figure 2b; and $t = 0$, 16.593, 21.502, 22.176, 22.254, 22.268, and

22.271 × 10^6 yr for Figure 2c. The *open circle* on each curve is the instantaneous half-thickness of the cloud along the (z-)axis of symmetry, which is along the magnetic field. The *star* is the critical thermal length scale $\lambda_{T,cr}$ appropriate to a thin disk. The *filled ellipse*, present only after a magnetically critical mass-to-flux ratio is achieved in the core, denotes the instantaneous location (equatorial radius) of the critical mass-to-flux ratio.

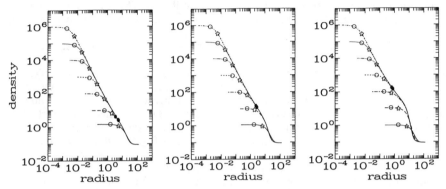

Figure 2a (*left*), **2b** (*center*), and **2c** (*right*). *The density as a function of radius in the equatorial plane of three model clouds at seven different times.* The initial central mass-to-flux ratio is subcritical by the factor 2/3, 1/3, and 1/10, respectively. See text for density normalization, unit of length, time for each curve, meaning of *open circles*, *stars*, and *filled ellipses*.

It is clear from these figures that the smaller the initial mass-to-flux ratio is than its critical value, the longer it takes to establish a critical core, the smaller the core's size (and mass), which is virtually unchanged during the evolution, and the more pronounced the break in the slope of the logρ-log$r$ profile, i.e., the sharper the separation into a runaway supercritical core and a magnetically supported envelope. Once a magnetically critical core is established, the compact, uniform-density, nearly spherical central part of it has size determined by the instantaneous value of the critical thermal length scale, $\lambda_{T,cr}$.

The infall speed has a maximum just beyond $\lambda_{T,cr}$; at the end of each run, it is in the range 0.7 - 0.5C for the above three models, respectively. It decreases rapidly to negligible values in the subcritical envelope, near the cloud boundary --unless one imposes initial conditions that imply steep magnetic field gradients and/or unrealistically small ion density in the envelope. The logB_c-logρ_c curves for these three models are shown in Figure 3; recall the discussion of § 2.4. A parameter study is given by Morton and Mouschovias 1991b.

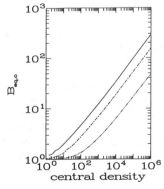

Figure 3. *Central equatorial magnetic field as a function of central density* for the three models of Figures 2a (*solid curve*), 2b (*dash-dot*), and 2c (*dash-3 dot*). $B_{eq,c}$ and $n_{eq,c}$ are normalized to the initial values in the reference state (B_{c0} = 15.3, 31.4, and 101.8 μG, respectively; and n_{c0} = 3 × 10^3 cm^{-3} for all three models). $B_{eq,c}$ increases by only a factor $\simeq$ 2 during the subcritical, quasistatic contraction phase. A slope κ = 0.47 is established asymptotically, after a critical mass-to-flux ratio is achieved.

3. A MODERN THEORY OF STAR FORMATION

Remarkable agreement has been found between theory and observations from scales of ~ 1 kpc to scales of $\sim 10^{-3}$ pc (Heiles 1987; TM81, TM87a). The quantitative calculations described above and in Paper I suggest a very specific and novel sequence of events leading to star formation, earlier versions of which were described in TM77a, TM78, and TM87a. In what follows, the term *collapse* means indefinite contraction, but not necessarily free fall.

3.1. Cloud Formation and Early Evolution: Magnetic Braking and Onset of Collapse

We have seen that the most likely mechanism responsible for the formation of interstellar clouds and cloud complexes is the Parker (1966) instability triggered in the galactic disk by spiral density waves or by any other large-scale disturbance (TM74, TM75a; Mouschovias, Shu, and Woodward 1974; Shu 1974; Blitz and Shu 1980). The significance of the cloud formation mechanism stems mainly from the fact that it determines the distribution of forces (such as gravitational, magnetic, thermal, and centrifugal) within a cloud. Soon after a cloud's formation, thermal-pressure and centrifugal forces lose memory of initial conditions. The differential mass-to-flux ratio $[dm(\Phi_B)/d\Phi_B]$ determines the relative magnitude of gravitational and magnetic forces, and affects crucially the subsequent evolution of a cloud (TM76a, TM76b). One can obtain reasonable theoretical estimates for this function, but no observational determination exists as yet. Observations such as those by Bregman *et al.* (1983) and by Schwarz *et al.* (1986) constitute an important first step toward the determination of $dm/d\Phi_B$. Similar and more extensive observations for dense clouds are needed.

For a cloud to collapse, two conditions must be satisfied. Its mass-to-flux ratio *and* the external pressure must exceed certain critical values (see Paper I, eqs. [129a] and [129b]). Since, as we have emphasized repeatedly, the collapse of a cloud *as a whole* should be a rare phenomenon, it is more appropriate to think of critical conditions being achieved in a cloud's central flux tube(s), while the bulk of the cloud has a subcritical mass-to-flux ratio (i.e., it remains magnetically supported).

A cloud which has not contracted sufficiently to become self-gravitating could either disperse once the cause of its formation is removed or exist in pressure equilibrium with the intercloud medium. In the latter case, it will await an external disturbance (such as a supernova shock wave or an ionization front) to either disperse or implode it --or implode part of it and disperse the rest. Theoretical considerations, as seen above and in Paper I, and the observational fact that velocity fields characteristic of gravitational collapse are not observed suggest that very few, if any, clouds acquire a supercritical mass-to-flux ratio from the outset and begin to collapse as a whole. Instead, *clouds reach first a relatively quiescent state* (near hydrostatic equilibrium with magnetic, thermal-pressure, and possibly hydromagnetic-wave-pressure forces balancing gravity), with possible leftover stable oscillations about the equilibrium state (TM75b). Centrifugal forces remain small at this stage due to magnetic braking, which operates over a time scale $\tau_\parallel$ or $\tau_\perp$, depending on whether the angular momentum vector J is parallel or perpendicular to the magnetic field, respectively (see summary in Paper I, § 3.3.4). Since $\tau_\perp \ll \tau_\parallel$, *most clouds and fragments tend to become aligned rotators in time.*

Molecules form over time scales shorter than evolutionary time scales once the density becomes sufficiently large. Collapse may receive a boost or even be initiated by the decrease in sound speed due to the larger molecular weight and the possible decrease in temperature that may accompany the formation of molecules and/or the decay of subsonic turbulence. The electrical conductivity is so large that the magnetic field remains frozen in the matter (Spitzer 1962) and lengthens the collapse time scale (even after the onset of ambipolar diffusion in the core; see below) to at least $10\tau_{ff}$ (TM83b).

3.2. Consequences of Ambipolar Diffusion: Fragmentation and Selection of Protostellar Masses $\sim 1\ M_\odot$

Ambipolar diffusion is unavoidable in the core of a self-gravitating, magnetically supported cloud. The essence of ambipolar diffusion is to permit self-gravity to redistribute mass in the central flux tubes of a cloud and thereby lead to a critical mass-to-flux ratio and *self-initiated* collapse (TM77a, TM78, TM79b). Moreover, it is directly responsible for removing wave-pressure support over length scales smaller than the Alfvén length scale λ_A; since in typical molecular clouds the critical thermal length scale $\lambda_{T,cr}$ is comparable to λ_A in the density *range* (not at a unique density) $10^3 - 10^6$ cm^{-3}, ambipolar diffusion is directly responsible for single-stage fragmentation (or core formation) with fragment masses in the corresponding range $10 - 0.1\ M_\odot$ (TM87a, TM89, TM91; see, also, § 2.4 above). (Conditions in present-day molecular clouds favor masses $\lesssim 1\ M_\odot$.) We are thus suggesting that it is the *removal* of perturbations, not their appearance, which is the root cause of fragmentation of the central flux tubes of a typical molecular cloud into "cores" with masses $\sim 1\ M_\odot$, and that *the condition for the onset of fragmentation at a significant rate is* $\lambda_A \simeq \lambda_{T,cr}$. Eventually, dense cores achieve a critical mass-to-flux ratio. At that stage, cool cores are usually, but not always, thermally supercritical as well, and rapid collapse ensues.

If ambipolar diffusion in cloud cores can prevent the magnetic field from increasing much above a few mG up to densities 10^{11-12} cm^{-3}, then the bulk of the (or even the entire) magnetic flux problem of star formation will be resolved. This is so because the ratio of the flux of a protostellar fragment and the flux of a magnetic star (of the same mass) is

$$\frac{\Phi_{B,c}}{\Phi_{B,*}} = \frac{B_c \pi R_c^2}{B_* \pi R_*^2} = \frac{B_c}{B_*}\left(\frac{\rho_*}{\rho_c}\right)^{2/3} \simeq \left[\frac{B_c/(10^{-3}\ G)}{B_*/(10^4\ G)}\right]\left(\frac{10^{12}\ \text{cm}^{-3}}{n_c}\right)^{2/3}, \qquad (13)$$

where we have taken $\rho_* \simeq 0.3$ g cm^{-3} as a representative mean density of magnetic A stars. *The distinction between the onset of rapid ambipolar diffusion and the actual resolution of the magnetic flux problem is of the essence.* Approximate analytical calculations (TM77a, TM78, TM79b) as well as axisymmetric collapse calculations (Fiedler and Mouschovias 1991a, b; Morton and Mouschovias 1991a, b; see summary in TM90b, TM91, and § 2.5 above) have shown that rapid ambipolar diffusion can indeed set in by the stage $n_c \sim 10^6$ cm^{-3} in cloud cores and can crucially affect a core's evolution. However, the issue of whether ambipolar diffusion can resolve the magnetic flux problem is still not completely settled.

Up to the stage that ambipolar diffusion increases the mass-to-flux ratio of the core above the critical value, the contraction remains *quasistatic* and the time scale τ_{AD} for ambipolar diffusion (in an axisymmetric geometry) is given by (see TM89, eq. [20]):

$$\tau_{AD} = \frac{8}{1.4\pi^2}\frac{\tau_{ff}^2}{\tau_{ni}}\left(1 - 8\tau_{ff}^2/\pi^2\tau_s^2\right)^{-1}, \qquad \text{for } 8\tau_{ff}^2/\pi^2\tau_s^2 < 1, \qquad (14)$$

where the free-fall time τ_{ff} and the neutral-ion collision time τ_{ni} are given by equations (3d) and (3b), respectively, and τ_s is the sound-crossing time across the radius of the core. For *magnetically supported* clouds the term in brackets in equation (14) is essentially equal to unity and the expression for the ambipolar diffusion time scale in the form $\tau_{AD} = 0.5\tau_{ff}^2/\tau_{ni}$, valid for quasistatic contraction, is a property of the general equations governing ambipolar diffusion and does not depend on geometry (see summary in Paper I, § 2.3.1). In the cores of magnetically supported clouds, equation (14) can be put in the useful form

$$\tau_{AD} = 2 \times 10^6\ (x/10^{-7})\quad \text{yr}, \qquad (15)$$

where $x \equiv n_i/n_{H2}$. The characteristic time τ_Φ for the reduction of the magnetic flux of a core is actually smaller than τ_{AD} by exactly a factor of 2.

After ambipolar diffusion has increased the mass-to-flux ratio of a quasistatically contracting core of a typical molecular cloud above the critical value given by equation (22b) of Paper I, accelerated contraction may ensue (if thermal pressure cannot support the core), in which case equation (14) no longer describes the rate at which ambipolar diffusion progresses. One must use instead equation (23) of TM89 (see Paper I, eq. [25]), which shows that it becomes more difficult for a core to continue to lose magnetic flux. [This is partly the reason for which, for typical molecular cloud cores, the exponent κ in the relation $B_c \propto \rho_c^\kappa$ becomes nearly equal to 1/2 even in the presence of ambipolar diffusion *after* critical conditions are (re)established in a core; see Figures 1 and 3 and associated discussion.] In Paper I we examined how much magnetic flux is expected to be lost and whether the magnetic flux problem of star formation can be resolved by ambipolar diffusion at $n_c < 10^{12}$ cm^{-3}. The case is still too close to call at the present time.

In summary, *the quasistatic, subAlfvénic, self-initiated collapse of cloud cores*, which may turn dynamic only during the late stages of contraction, after thermally *and* magnetically supercritical conditions are established because of ambipolar diffusion, is the most common case (TM78, App. B). The decay of hydromagnetic waves due to ambipolar diffusion is directly responsible for single-stage fragmentation (or core formation) in molecular cloud interiors, and leads to a natural selection of protostellar masses $\lesssim 1$ M$_\odot$ (TM87a; see § 2.4 above). The *subAlfvénic* nature of the contraction is crucial to the above sequence of events, for it allows torsional Alfvén waves to escape the contracting cores and thereby transport angular momentum away and keep centrifugal forces relatively weak (magnetic braking); see TM78 (p. 220), TM87a, §§ 3.5.1 and 3.6. Alfvénic or superAlfvénic collapse (e.g., due to implosion) leads to a very different sequence of events and, most likely, to a different IMF; i.e., more high-mass stars (TM87a, § 3.6).

3.2.1. Disk Formation and Bipolar Outflows.

That a contracting core should form a disk with its axis of symmetry aligned with the common direction of the magnetic field and the angular momentum vectors is an inevitable consequence of the inherent anisotropy of magnetic and centrifugal forces. However, it is only recently that theoretical calculations have *predicted*, as opposed to assume, what the velocity field and the distribution of mass and magnetic flux in the disk are, and what fraction of that mass should eventually find its way into a protostar (see summary in TM90b, TM91, and § 2.5 above). To the best of our knowledge, there is as yet no theory that can predict, or just explain, the commonly observed bipolar outflows in regions of low-mass star formation, although ideas abound; promising new ideas and results are discussed in the chapter by Shu.

3.2.2. Formation of Binary Stars and of Intermediate- and Low-Mass Stars in Cloud Cores.

For subAlfvénically contracting cores, it was demonstrated early on that magnetic braking followed by ambipolar diffusion can explain the entire range of periods of binary stars from 10 hr to 100 yr (TM77). Moreover, a single maximum in the distribution of the number of binaries as a function of (orbital) period was predicted, in disagreement with earlier but in agreement with more recent observations (see review by Abt 1983). It has also been shown that, in subAlfvénically contracting cores, magnetic braking can be effective even while ambipolar diffusion is in progress, so that sufficient angular momentum is lost to the envelope for Sun-Jupiter "binaries" and single stars to form without interference from centrifugal forces (see MP86; or TM87b, § 4; or Paper I, § 4).

In the case of magnetically connected fragments, trapping of torsional Alfvén waves sets each one of them in a series of high-spin and low-spin states (MM85a, MM85b). Thus the angular momentum problem for each fragment may be resolved, recreated, and then resolved again, perhaps several times, before the angular momentum of the system of fragments is reduced sufficiently for stars to form.

Since low-mass fragments owe their existence to ambipolar diffusion, they reach the dynamical stage of contraction and protostar formation before the more massive, outlying fragments, which may have formed while the magnetic field is nearly frozen in the matter. However, once a massive star has formed, it evolves relatively rapidly and either the expansion of the resulting H II region or the eventual supernova explosion changes the evolution of other, especially massive, fragments substantially. In particular, a supernova may induce rapid collapse of nearby fragments with masses $\sim 10^1 - 10^3$ $M_\odot$, while it may disperse the least massive ones, and may lead to more efficient star formation.

3.3. Low- versus High-Mass Star Formation

Although collapse at subAlfvénic speeds in cloud cores is the most common case (TM78, Appendix B) and leads to the formation of low- and intermediate-mass stars and close binaries (see TM87a, fig. on p. 486), Alfvénic or superAlfvénic collapse of a cloud or fragment (e.g., due to implosion by a shock) leads to a very different sequence of events and, most likely, to a different IMF; i.e., more high-mass stars and wide binary systems (TM87a). One idea is that massive star formation is the result of the collapse of clouds with supercritical masses, $M > M_{crit}$ (e.g., see Mestel 1985; Lizano and Shu 1987). This idea is qualitatively different from our ideas based on the externally triggered superAlfvénic collapse of a cloud (or fragment) of mass $M \lesssim M_{crit}$ in a cloud complex (see TM89, § 1.4). The difference and its consequences are summarized as follows.

First, supercritical cloud masses are rarely, if ever, observed. Second, the contraction of a cloud with $M > M_{crit}$ leads to a high degree of flattening and to a rapid increase of the magnetic tension force (see TM78). It therefore remains to be demonstrated whether this kind of contraction can lead to a preferential formation of massive stars. On the contrary, during the superAlfvénic implosion of a clump with mass $M \lesssim M_{crit}$ in a cloud complex, magnetic tension cannot stop or significantly slow down the collapse and, in addition, the torsional Alfvén waves remain trapped within the clump. Contraction with angular momentum nearly conserved implies that centrifugal forces will become progressively more important. Depending, however, on the precise density at which rapid collapse begins, they may not increase sufficiently to prevent the formation of very wide binaries, including rapidly rotating, massive single stars as members. For example, if angular momentum begins to be conserved above a density 10^4 cm^{-3}, the angular momentum left over in a fragment (due to earlier efficient magnetic braking) is exactly what is required to form a relatively wide (visual) binary system, with a period $\tau_b = 56$ yr. This argument does not depend on mass. Binaries forming according to this scenario will have periods determined by the density $n_{J,cr}$, above which angular momentum is conserved:

$$\tau_b = 56 \left[\frac{10^8 \text{ yr}}{\tau_{J,cr}} \right]^3 \left[\frac{10^4 \text{ cm}^{-3}}{n_{J,cr}} \right]^2 \text{ yr}, \qquad (16)$$

where $\tau_{J,cr}$ is the period of rotation of the parent fragment at the onset of conservation of angular momentum (see TM77a, eq. [5]). *Note that conservation of angular momentum during rapid collapse is due to trapping of the torsional Alfvén waves inside the cloud, not due to effective (fast) ambipolar diffusion* (TM87a, § 3.6). [Nakano 1991, eq. (7), normalizes $n_{J,cr}$ to 10^5 cm^{-3} and τ_b to 3 yr, converts the rotation period $\tau_{J,cr}$ to an angular velocity, and then presents both equation (16) and the trapping of the torsional Alfvén waves as new!] This has significant implications for massive star formation.

If Alfvénic or superAlfvénic implosion begins at $n < 10^3$ cm^{-3}, centrifugal forces will lead to a quasi-equilibrium configuration by balancing gravity perpendicular to the axis of rotation. Further contraction takes place only as rapidly as magnetic braking can remove angular momentum to the envelope. Since the fragment still has most of its magnetic flux trapped inside at this stage and since it was *magnetically supported before implosion* (a key difference from the supercritical-mass idea of Mestel 1985 and of Lizano and Shu 1987), *magnetic forces will also contribute significantly to this quasi-equilibrium, and both magnetic braking and ambipolar diffusion are expected to occur simultaneously, over comparable time scales*. Although this complex configuration has not been studied in any satisfactory fashion, it seems likely that stars which will form in such fragments will be characterized by both rapid rotation *and* strong magnetic fields. Since these are signatures of massive stars, TM87a suggested, at least tentatively, that this is how such stars form.

Although calculations do exist, they have been pursued only up to densities 3×10^9 cm^{-3} and, therefore, they do not presently allow definitive conclusions on the issue of whether a protostar (low-mass or massive) becomes opaque while possessing a relatively weak magnetic field which will be amplified by a dynamo process, or whether it does so while having too much magnetic flux which will be dissipated during this or subsequent stages of contraction. We believe, nevertheless, that it is during the relatively late, opaque stages of protostar formation that its magnetic field may detach from the background. This view is based on calculations (see TM76a, TM76b, TM78) which show that "pinching forces" (Mestel 1966) and conditions conducive to magnetic detachment do not arise during the earlier, relatively diffuse stages of fragment contraction and protostar formation. Once magnetic detachment takes place, further loss of angular momentum becomes very inefficient (compared to earlier stages) and occurs through a magnetic wind.

4. SUMMARY

Many phenomena have been predicted or explained as a direct consequence of the single theoretical *result*, not assumption, that molecular clouds are primarily magnetically supported. Effective magnetic braking and self-initiated formation and collapse of cloud cores in otherwise quiescent molecular clouds are two of the most important consequences. We identify the *decay* of hydromagnetic waves due to ambipolar diffusion, not the generation of perturbations, as the root cause of single-stage (as opposed to hierarchical) fragmentation (or core formation); it selects naturally a *range* of masses $0.1 - 10$ M$_\odot$ that cannot be prevented from collapsing. Ambipolar diffusion first permits quasistatic contraction of these masses, and eventually leads to magnetically and thermally supercritical fragments and rapid collapse. Analytical arguments and numerical modeling show that accretion (better, infall) across the virtually stationary boundary separating the supercritical core from the magnetically supported (subcritical) envelope is controlled by ambipolar diffusion and is very slow (typically $< 5 \times 10^{-7}$ M$_\odot$ yr^{-1}). A power-law density profile tends to be established, but only inside the supercritical region. As time progresses, the power-law does *not* spread outward into the magnetically supported envelope; it extends, instead, *inward* because of the shrinking of the size (and mass) of the compact, rapidly evolving ($\tau_{contr} < 10^3$ yr at $n_c \sim 10^9$ cm^{-3}), uniform-density, innermost part of the supercritical region.

The Alfvén (λ_A), thermal ($\lambda_{T,cr}$), and magnetic ($\lambda_{M,cr}$) critical length scales play an important role in the fragmentation of magnetic flux tubes and the selection of protostellar masses. The condition for fragmentation is $\lambda_A \simeq \lambda_{T,cr}$ ($< \lambda_{M,cr}$), which is satisfied in the density range $10^3 - 10^6$ cm^{-3} in present-day molecular clouds. Fragmentation *along* a magnetic flux tube decreases the mass-to-flux ratio and thereby tends to introduce a "plateau" in the $\log B_c$ - $\log \rho_c$ relation between the magnetic field strength and the gas density, as each (subcritical) fragment contracts quasistatically (because of ambipolar diffusion) toward establishing a critical mass-to-flux ratio. Once a critical state is established, $B_c \propto \rho_c^{1/2}$ [assuming that $T_{c,\text{eff}} = const$; otherwise, $B_c \propto (\rho_c T_{c,\text{eff}})^{1/2}$]. Compact cores ($\lesssim 0.1$ M$_\odot$, $n_c \sim 10^9$ cm^{-3}), milliGauss fields, and accretion disks are some of the observable predictions of the calculations. We have synthesized the theoretical results into a modern theory of star formation which accounts naturally for the formation of close binaries, wide binaries, intermediate- and low-mass stars, Sun-Jupiter "binaries," and massive stars. Although improvements will undoubtedly have to be made, the fact that so many of the theoretical predictions have been confirmed by observations inspires confidence that at least the basic physical processes have been correctly identified and that the theoretical investigations are on the right track.

Acknowledgements. This work was supported in part by the National Science Foundation.

Abt, H. A. 1983, *ARA&A*, **21**, 343
Arons, J., and Max, C. E. 1975, *ApJ*, **196**, L77
Baudry, A., Cernicharo, J., Perault, M., de la Noe, J., and Despois, D. 1981, *A&A*, **194**, 101
Blitz, L., and Shu, F. H. 1980, *ApJ.*, **238**, 148
Bonnor, W. B. 1956, *MNRAS*, **116**, 351
Bregman, J. D., Troland, T. H., Forster, J. R., Schwarz, U. J., Goss, W. M., and Heiles, C. 1983, *A&A*, **118**, 157
Clark, F. O., and Johnson, D. R. 1981, *ApJ*, **247**, 104
Ebert, R. 1955, *Zs. Ap.*, **37**, 217
———. 1957, *Zs. Ap.*, **42**, 263
Elmegreen, B. G. 1979, *ApJ*, **232**, 729
Falgarone, E., and Pérault, M. 1987, in *Physical Processes in Interstellar Clouds*, ed. G. E. Morfill and M. Scholer (Dordrecht: Reidel), 59
Fiedler, R. A., and Mouschovias, T. Ch. 1991a, *ApJ, to be submitted*
———. 1991b, *ApJ, to be submitted*
Gillis, J., Mestel, L., and Paris, R. B. 1974, *Ap. & Space Sci.*, **27**, 167
Goldsmith, P. F., and Arquilla, R. 1985, in *Protostars & Planets II*, ed. D. C. Black, and M. S. Matthews (Tucson: Univ. of Arizona Press), 137
Heiles, C. 1987, in *Physical Processes in Interstellar Clouds*, ed. G. E. Morfill and M. Scholer (Dordrecht: Reidel), 429
Kulsrud, R., and Pearce, W. 1969, *ApJ*, **156**, 445
Larson, R. B., 1969, *MNRAS*, **145**, 271
Lizano, S., and Shu, F. H. 1987, in *Physical Processes in Interstellar Clouds*, ed. G. E. Morfill and M. Scholer (Dordrecht: Reidel), 173
———. 1989, *ApJ*, **342**, 834
Mestel, L. 1965, *QJRAS*, **6**, 265
———. 1966, *MNRAS*, **133**, 265
———. 1985, in *Protostars and Planets II*, ed. D. C. Black, and M. S. Matthews (Tucson: Univ. of Arizona Press), 81
Mestel, L., and Spitzer, L., Jr. 1956, *MNRAS*, **116**, 503
Morton, S. A., and Mouschovias, T. Ch. 1991a, *ApJ, in preparation*
———. 1991b, *ApJ, in prepartion*
Mouschovias, T. Ch. 1974, *ApJ*, **192**, 37
———. 1975b, *Ph.D. Thesis*, University of California at Berkeley
———. 1976a, *ApJ*, **206**, 753
———. 1976b, *ApJ*, **207**, 141

———. 1977a, *ApJ*, **211**, 147
———. 1977b, in *Star Formation*, ed. T. de Jong and A. Maeder (Dordrecht: Reidel), 276
———. 1978, in *Protostars and Planets*, ed. T. Gehrels (Tucson: U. of Arizona Press), 209
———. 1979b, *ApJ*, **228**, 475
———. 1981, in *Fundamental Problems in the Theory of Stellar Evolution*, ed. D. Sugimoto, D. Q. Lamb, and D. N. Schramm (Dordrecht: Reidel), 27
———. 1983b, in *Solar and Stellar Magnetic Fields: Origins and Coronal Effects*, ed. J. O. Stenflo (Dordrecht: Reidel), 479
———. 1987a, in *Physical Processes in Interstellar Clouds*, ed. G. E. Morfill and M. Scholer (Dordrecht: Reidel), 453
———. 1987b, in *Physical Processes in Interstellar Clouds*, ed. G. E. Morfill and M. Scholer (Dordrecht: Reidel), 491
———. 1989, in *The Physics and Chemistry of Interstellar Molecular Clouds*, ed. G. Winnewisser and J. T. Armstrong (Berlin: Springer-Verlag), 297
———. 1990b, in *Physical Processes in Fragmentation and Star Formation*, ed. R. Capuzzo-Dolcetta, C. Chiosi, and A. di Fazio (Dordrecht: Kluwer), 117
———. 1991, *ApJ*, May 20
Mouschovias, T. Ch., and Morton, S. A. 1985a, *ApJ*, **298**, 190
———. 1985b, *ApJ*, **298**, 205
Mouschovias, T. Ch., and Paleologou, E. V. 1979, *ApJ*, **230**, 204
———. 1980, *ApJ*, **237**, 877
———. 1986, *ApJ*, **308**, 781
Mouschovias, T. Ch., Paleologou, E. V., and Fiedler, R. A. 1985, *ApJ*, **291**, 772
Mouschovias, T. Ch., Shu, F. H., and Woodward, R. 1974, *A&A*, **33**, 73
Mouschovias, T. Ch., and Spitzer, L., Jr. 1976, *ApJ*, **210**, 326
Myers, P. C. 1985, in *Protostars & Planets II*, ed. D. C. Black and M. S. Matthews (Tucson: University of Arizona Press), 81
Nakano, T. 1976, *PASJ*, **28**, 355
———. 1977, *PASJ*, **29**, 197
———. 1979, *PASJ*, **31**, 697
———. 1981, in *Progress in Theoretical Physics*, No. 70, 54
———. 1990, *MNRAS*, **242**, 535
———. 1991, in *Fragmentation of Molecular Clouds and Star Formation*, IAU Symp. No. 147, in press
Nakano, T., and Tademaru, T. 1972, *ApJ*, **173**, 87
Parker, E. N. 1966, *ApJ*, **145**, 811
Schwarz, U. J., Troland, T. H., Albinson, J. S., Bregman, J. D., Goss, W. M., and Heiles, C. 1986, *ApJ*, **301**, 320
Shu, F. H. 1974, *A&A*, **33**, 55
———. 1977, *ApJ*, **214**, 488
Spitzer, L., Jr. 1962, *Physics of Fully Ionized Gases*, 2nd ed. (New York: Interscience)
———. 1968, in *Stars and Stellar Systems*, Vol. 7, *Nebulae and Interstellar Matter*, ed. B. Middlehurst and L. H. Aller (Chicago: Univ. of Chicago Press), 1
———. 1978, *Physical Processes in the Interstellar Medium* (New York: Wiley-Interscience)
Tomisaka, K., Ikeuchi, S., and Nakamura, T. 1988, *ApJ.*, **335**, 239
Vogel, S. N., and Kuhi, L. V. 1981, *ApJ*, **245**, 960
Wolff, S. C., Edwards, S., and Preston, G. W. 1982, *ApJ*, **252**, 322
Young, J. S., Langer, W. D., Goldsmith, F., and Wilson, R. W. 1981, *ApJ*, **251**, L81
Zuckerman, B., and Palmer, 1974, *ARA&A*, **12**, 279

III. PHYSICS OF EARLY STELLAR EVOLUTION AND STELLAR WINDS

R. Pudritz, A. Blaauw, J. Bally, G. Fuller

MOLECULAR OUTFLOWS: OBSERVED PROPERTIES

JOHN BALLY
AT&T Bell Laboratories
HOH-L245, Holmdel, NJ 07733

ADAIR P. LANE
Department of Astronomy
Boston University
725 Commonwealth Ave.
Boston, MA 02215

ABSTRACT. Molecular outflows from young stellar objects represent one of the earliest manifestations of the birth of a young star. Recent observations have found flows with CO velocities in excess of 150 km s^{-1}, discrete molecular "bullets", and a high degree of collimation. In this article we review the general properties of molecular outflows, highlight some recent observations which illustrate certain characteristics of the outflow phenomenon, discuss some of the statistical properties of the nearly 200 outflows currently known, and discuss models of the CO emission region.

1. Introduction

In 1976, it was realized that broad (over 100 km s^{-1} wide) line wings seen towards the core of the Orion molecular cloud in ^{12}CO are produced by the release of energy from a deeply embedded object (Zuckerman, Kuiper, and Kuiper 1976; Kwan and Scoville 1976). The source of the outflow turned out to be the luminous young stellar object IRc2. Since this seminal discovery, nearly 200 molecular outflows have been found by their broad mm-wavelength ^{12}CO emission lines (Bally and Lada 1983; Lada 1985; Snell et al. 1988; Fukui 1989; Wilking, Blackwell, and Mundy 1990; McCutcheon et al. 1990). Since star formation is the result of the gravitational collapse of a cloud core, it came as a surprise that one of the most striking manifestations of the birth of a young star is the production of highly supersonic mass outflow.

It is now generally accepted that young stellar objects (YSO's) of all masses produce outflows during their birth. During this phase, which lasts for at least 10^3 to over 10^5 years, a fast supersonic wind, usually collimated into oppositely directed lobes, inflates a pair of cavities in the surrounding molecular gas. The gas displaced from the cavities is swept into shells or partial shells and is detectable as the bipolar CO outflow. The most recent observations indicate that bipolar outflows may represent the earliest detectable phase of a young star's life. Several young stellar objects, such as the driving source of the NGC 2264 G flow (Margulis et al. 1990) and the ρ-Oph A flow (Andre et al. 1990a,b), were first identified by their CO outflows. These YSO's remain undetected at wavelengths as long as the IRAS 100 μm band.

The source of the ρ-Oph A flow (VLA 1623 - Figure 4) has only been seen by its sub-mm wavelength (850 μm and 1 mm) and 6 cm radio continuum emission.

In this review, we will discuss the basic observational characteristics of molecular outflows, the relationship to other tracers of outflow activity, the limitations imposed on our understanding due to selection effects, and models of the nature of the CO emission region.

2. Molecular Outflow Characteristics

Our present understanding of the outflow phenomenon and its relationship to star formation is limited to a large extent by selection effects. The first molecular outflows were discovered by accident while mapping the structure of star forming molecular clouds in the CO lines. By 1980, serendipity had led to the identification of about a half dozen outflows, mostly associated with high luminosity YSO's (Orion A, AFGL 961, Cepheus A, GL 490, and L1551 were the first examples). Most of the early discoveries have CO line wings which are well over 30 km s^{-1} in extent, arise from a confined region around an embedded infrared source, and exhibit large ^{12}CO to ^{13}CO line ratios which demonstrates that the high velocity gas has lower optical depth and column density than the molecular cloud core harboring the YSO. The case for outflow in these sources is strong since the maximum ^{12}CO velocities are much greater than the escape velocity from the cloud core, which can be estimated from the cloud mass and size.

During the 1980's several groups surveyed known infrared sources and other categories of young objects such as T Tauri stars, Herbig-Haro objects, and dark cloud cores in ^{12}CO, searching for broad emission lines (Torrelles et al. 1983; Bally and Lada 1983; Edwards and Snell 1982, 1983, 1984; Levreault 1988; see Lada 1985 and Fukui 1989 for more references). This work led to the discovery of dozens of outflows, some with very low velocity amplitudes which were associated with low luminosity YSO's. This work demonstrated the following general characteristics of molecular outflows:

(1) The majority of molecular outflows exhibit bipolar structure, with redshifted gas located on one side of the source and blueshifted gas located on the other. The degree of collimation (ratio of the observed major axis to the minor axis length) ranges from about 1 (poorly collimated) to over 10 (a highly collimated jet-like object). Most outflows are poorly collimated. Although this may result from projection effects for some flows (flow oriented mostly along the line-of-sight), statistical analysis (Lada 1985) indicates that on average, intrinsic collimation is poor. Figures 1 through 8 show several examples of bipolar outflow structure. Some outflows exhibit more complex morphology, including multiple outflow lobes (see for example IRAS 1629-2422 - Walker et al. 1988) and flow re-direction with alternating redshifted and blueshifted lobes (Cepheus A - Bally and Lane 1991; see Figure 5). These complexities may be produced by variations in the structure of the environment with which the outflow is interacting, by precession of the underlying wind source combined with time variability, or by the existence of multiple outflow sources within the same region.

(2) YSO's of all masses and luminosities produce outflows. Outflows have been identified surrounding sources with luminosities as low as 0.2 L$_\odot$ and greater than 10^6 L$_\odot$.

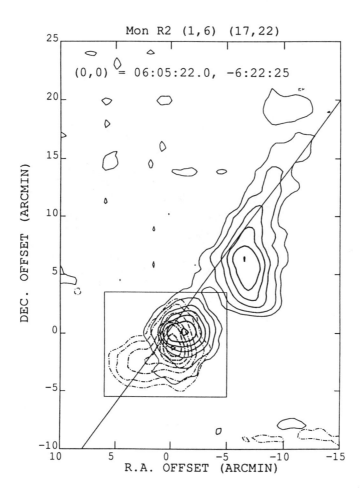

Figure 1. The Mon R2 outflow, one of the largest in the sky, as seen at 100" resolution with the Bell Laboratories 7-m antenna (Wolf, Lada, and Bally 1990) in the J=1-0 ^{12}CO transition. Located at a distance of about 730 pc, the flow has a length of about 4 pc, and is produced by a luminous cluster of stars having a total luminosity of about 4×10^4 $L_\odot$. The total mass of outflowing gas is estimated to be nearly 200 $M_\odot$ making this one of the most massive flows known. The rectangular box indicates the area shown in Figure 2. The diagonal line shows the orientation of the "slit spectrum" shown in Figure 3. Solid contours show the blueshifted emission between V_{LSR} = 1 and 6 km s^{-1}. The segmented contours indicate the redshifted lobe of the outflow between V_{LSR} = 17 to 22 km s^{-1}. Contour levers for both lobes are set at 1,2,4,6,8,10,12 K km s^{-1}.

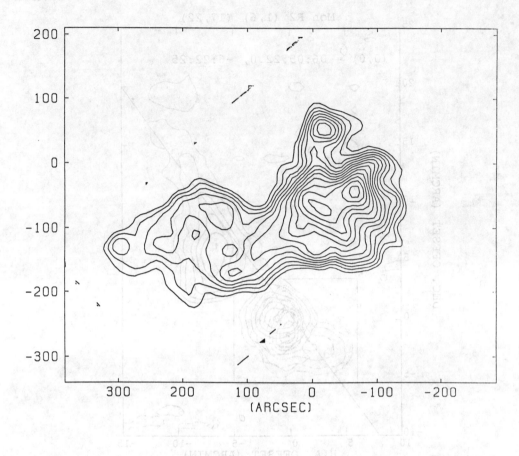

Figure 2. A high angular resolution map of the redshifted lobe of the Mon R2 flow as mapped with the 15" beam of the Nobeyama 45-meter telescope (Bally, Takano, and Hayashi - in preparation). Despite the high angular resolution, shell structure is not evident in the structure of this flow which resembles a meandering river. The velocity range is V_{LSR} = 17 to 22 km s^{-1} with contours every 0.5 K km s^{-1}, starting at 2 K km s^{-1}.

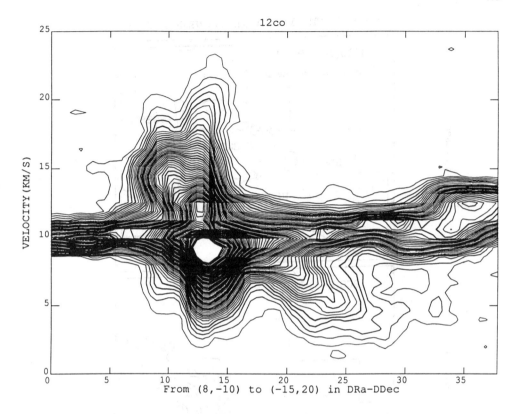

Figure 3. A spatial-velocity diagram or a "slit spectrum" showing the structure of the velocity field along the major axis of the Mon R2 outflow (along the line indicated in Figure 1) in the J=1-0 line of ^{12}CO. Contour levels are displayed at every 0.3 K from 0.3 to 21 K. This figure illustrates the apparent increases of the radial velocity with increasing distance from the central source, located 13' from the left edge, in this case one or more members of the Mon R2 infrared cluster. There are several distinct kinematic components in this outflow including a spatially confined high velocity component centered about 13' from the left edge of the map, and a spatially extended component with a smaller total velocity extent.

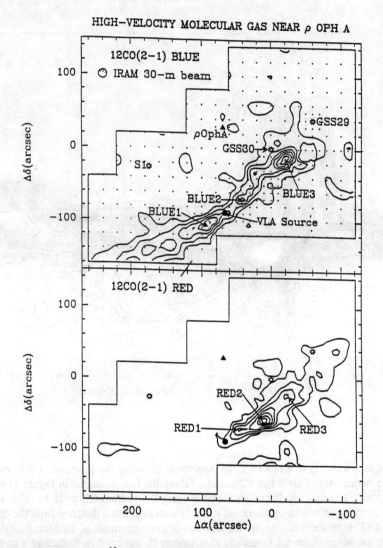

Figure 4. Maps of J=2-1 ^{12}CO emission from the highly collimated outflow ρ-Oph A (taken from Andre et al. 1990). The upper panel shows the blueshifted portion of the outflow integrated from V_{LSR} = -10 to 0 km s^{-1} while the lower panel shows the redshifted gas between 8 and 18 km s^{-1}. The outflow has a total mass of about 0.02 $M_\odot$. The contour interval is 3 K km s^{-1}. The near infrared sources GSS 29, 30 and S1 are shown by stars. Andre et al. (1990) argue that the source of the outflow is the faint 6 cm radio source designated VLA 1623. Although detected at a wavelength of 1 mm with the 13" beam of the IRAM 30 telescope, the source was not detected at any infrared wavelength below 350μm by either the IRAS satellite or from the ground. This source must have a luminosity under 10 $L_\odot$. This flow demonstrates that bipolar molecular flows can be one of the earliest manifestations of the birth of a young star in a dense molecular cloud core. The high degree of collimation and the presence of both red and blueshifted gas towards the north-west suggests that this flow lies mostly along the plane of the sky.

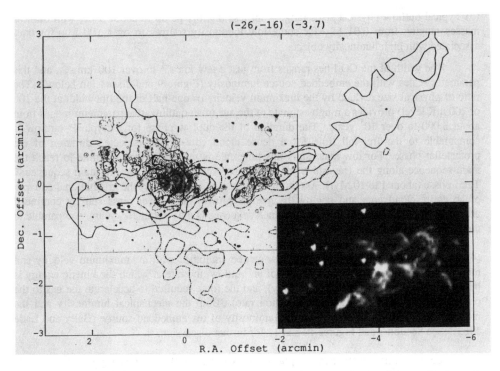

Figure 5. A 15" resolution map of the high velocity CO J = 1-0 emission in Cepheus A superimposed on a map showing the near-infrared emission observed through a 1 percent bandpass filter centered on the 2.122 μm line of H_2. In the CO map, solid contours show the blueshifted emission (V_{LSR} = -26 to -16 km s^{-1}) while dashed contours show the redshifted gas (V_{LSR} = -3 to 7 km s^{-1}). Contour levels are displayed at 3, 6, and 12 Kkm s^{-1}. The map is based on a 1,900 point map obtained with the NRO 45 meter telescope (Bally, Hayashi, and Hayashi - unpublished). The near-IR contour map (Bally and Lane 1991) shows the shock-excited molecular hydrogen emission and near-infrared reflection nebulosity in this region. The emission around (-1.5,0), near the right side of the near-infrared image, coincides with the Herbig-Haro object GGD 37 and is pure H_2 line emission. The inset photograph shows a blow-up of the structure of the H_2 filaments associated with GGD 37 which is located near the start of the blue shifted lobe extending to the west from (-1.5,0). The arc of near-IR emission near (1.7,0.6) also consists of pure H_2 line emission. This feature lies along the axis-of-symmetry of the two ridges of blueshifted CO emission extending to the east and northeast and the near-IR reflection nebulosity emerging from the vicinity of (0,0). Although the eastern-portion of this outflow is shell-like, the western part looks like a curved ridge with a flow reversal (from redshifted to blueshifted gas) near the position of the Herbig-Haro object GGD 37.

(3) Typical outflow physical sizes range from less that 0.1 pc up to about 5 pc, with outflow masses ranging from 0.01 $M_\odot$ for some low luminosity sources to over 100 $M_\odot$ for ones associated with high luminosity objects.

The width of the CO lines ranges from just a few km s^{-1} to over 100 km s^{-1}, and this parameter scales with the embedded source luminosity (Figure 9 and discussion below). The ratio of physical size divided by the maximum velocity (= one-half of the linewidth at the 100 or 200 mK level) provides a rough estimate of the age of an outflow; this parameter ranges from about 3,000 to over 10^5 years. The duration of the outflow phase for a "typical" outflow is comparable to the free-fall time for a dense cloud core or the expected duration of the protostellar phase. For low mass stars this timescale is much less than the time to reach the main-sequence along the Hayashi track or the observed lifetime of the pre-main sequence T Tauri phase (about 1 to 10 Myr). Assuming that massive stars form by the gravitational collapse of an entire cloud core and not by the gradual growth of a lower mass star by continued accretion, the time required to reach the main sequence is only about 10^5 years, comparable to the duration of the outflow phase.

(4) The physical dimensions of an outflow can be combined with the maximum velocity and total mass to estimate the kinetic energy of an outflow, the rate at which the kinetic energy is being supplied (the mechanical luminosity), and the force required to accelerate the gas to the observed velocity (the momentum injection rate). Both the mechanical luminosity and the momentum injection rate scale with the luminosity of the embedded source (Bally and Lada 1983). See Section 3 and Figures 10 and 11 for further discussion.

(5) The outflow emission tends to be optically thick in the ^{12}CO lines with a typical ^{12}CO/^{13}CO intensity ratio of about 20. The CO optical depth tends to decrease with increasing velocity offset from the line center. Some old outflows such as Mon R2 (Figures 1 - 3) exhibit very optically thick CO line wings which are bright in ^{13}CO. Young outflows, such as Orion A, tend to have optically thinner ^{12}CO wings which are hard to detect in ^{13}CO.

(6) The low intensity of the high velocity gas is best explained in terms of a small beam filling factor, which can range from a maximum value of 1 to under 0.01. The filling factor tends to decrease with increasing velocity, suggesting that most of the mass is moving slowly and only a small fraction of the gas is moving rapidly.

(7) The outflow phase occurs very early in the life of a young star and may coincide with the main accretion phase of the protostellar object. The most successful early surveys were those which targeted highly embedded infrared sources with no visible stellar counterparts. Visible low mass pre-main sequence objects such as T Tauri stars which are not highly embedded are rarely associated with molecular outflows. Recently, several outflows have been discovered which have central sources invisible at wavelengths below 350µm (NGC 2264 G - Margulis et al. 1990; ρ-Oph A, shown in Figure 4; and L 1448 - Bachiller et al. 1990).

(8) Many molecular outflows exhibit other manifestations of outflow activity in the radio, infrared, and optical wavelength regions. Shock-excited gas in the form of Herbig-Haro objects and optical jets is sometimes found to lie inside the boundaries of the molecular outflows (see

Reipurth - this volume). When the molecular flow is located in a dense environment, it is frequently a source of 2.12μm H_2 line radiation, indicating a post-shock temperature of at least 2000 K (Lane 1989). We will discuss these points in greater detail in section 3.3. Many embedded YSO's are surrounded by H_2O masers (Comoretto et al. 1990). Some moderate luminosity sources, too cool to emit Lyman continuum radiation which can ionize hydrogen, exhibit weak thermal radio continuum emission in the cm-wavelength regime. In a few cases, such as L1551 IRS 5, this plasma is collimated into a well-defined radio jet, coincident with an optical jet (Snell et al. 1985). In some cases, such as the radio triplet in the Serpens outflow (Rodriguez et al. 1989) and the radio source associated with the Herbig-Haro object GGD 37 in Cepheus A (Hughes and Moriarty-Schieven 1990) the radio emission exhibits large (several hundred km s^{-1}) proper motions. At least one radio source associated with an outflow has a non-thermal origin (V571 Ori - Yusef-Zadeh et al. 1990).

(9) High luminosity sources are frequently associated with ultra-compact HII regions and extended regions of photodissociated gas. The DR 21 outflow, one of the most massive known, contains a bright HII region, powerful masers, extensive photodissociation regions detected by their far-infrared line emission, and strong molecular hydrogen "jets" seen in the 2 μm lines (see Lane et al. 1990; Garden et al. 1991). The Mon R2 outflow shown in Figures 1 through 3 emerges from the vicinity of a highly embedded cluster of infrared sources, some of which are associated with H_2O masers and compact HII regions.

The short evolutionary timescales for massive stars and their tendency to cluster suggests that such massive flows may be formed by multiple energy injection events from different YSOs in the same cluster. A variety of mechanisms not available for low luminosity objects may contribute to the inflation of the molecular outflow lobes including ionized stellar winds, the expansion of HII regions, and ionization-driven ablation flows (see Bally and Scoville 1982 for an early model of an ionization- driven ablation flow from a disk). When the mass flux from a region exceeds a critical value, ionization fronts can become trapped near the source and the accelerated gas recombines to form a neutral wind.

(10) Most molecular outflow sources which have been searched in high density tracers such as CS or ammonia are found to be embedded in dense, compact molecular cores.

(11) Some outflows (NGC 2071, L1551, HH 7-11, and DR 21) show compact high velocity regions of 21 cm atomic hydrogen line emission (Bally and Stark 1983; Lizano et al. 1988; also see Figure 8). This component may be produced by a neutral atomic wind which may be the primary source of energy and momentum for the molecular component. The winds may be time-variable (Reipurth 1989b; Bally and Lane 1991).

(12) The total energy and momentum injection rate into the surrounding cloud is very large. Since outflows are embedded within molecular clouds, this energy and momentum is efficiently coupled to the cloud. Therefore, molecular outflows may make a major contribution to the support of molecular clouds by the generation of internal motion and turbulence which tends to oppose gravitational collapse. Outflows may play a crucial role in the self-regulation of the star formation rate within a molecular cloud (see Bally and Lada 1983; Lada 1985; and Fukui 1989 for discussions).

3. Recent Developments

During the middle and late 1980's, many outflows have been mapped in great detail and with high sensitivity with single-dish beam sizes as small as 11". For the first time, the morphological complexity of the bipolar outflow phenomenon became apparent. Observations with optical CCDs and near-infrared array cameras have begun to explore the detailed relationships between the CO outflows and other outflow tracers. Additional surveys have increased the number of known outflows to nearly 200 (e.g., Wouterloot, Henkel, and Walmsley 1989; Snell, Dickman, and Huang 1990; Wilking, Blackwell, Mundy 1990; McCutcheon et al. 1991). These and other recent studies have to a large extent confirmed the previously noted trends, and added some important new developments. Many of the recent outflow discoveries are associated with low luminosity sources. Both high and low luminosity sources have been found to exhibit extremely high velocity (EHV) CO emission.

3.1 EHV CO OUTFLOWS

Six outflows are now known to exhibit faint CO components with linewidths ranging from 100 to over 300 km s^{-1}. However, most outflows have not been searched with sufficient sensitivity to detect these faint high velocity features. While three of the EHV CO flows are associated with luminous sources (NGC 2071, CRL490, and S140 - L$\approx$10^3 to 10^4 L$_\odot$; Margulis and Snell 1989), three others are driven by low luminosity sources including SVS-13 in the HH 7-11 outflow (90 L$_\odot$; Koo 1990; Masson, Mundy, and Keene 1990; Bachiller and Cernicharo 1991), the U-star in L1448 (L < 11 L$_\odot$; Bachiller et al. 1990), and the flow in L723 (Margulis and Snell 1989).

Recent observations show that the EHV gas in HH 7-11 in CO (J=2-1, and J=3-2) is confined to two knots of emission extending over a 60" region (see Figures 6 and 7) and located in the vicinity of the shocks traced by H$_2$ and the stationary HCO$^+$ clumps observed by Rudolph and Welch (1988). Unlike the classical CO line wings, the EHV CO is seen as high velocity CO cloudlets ("bullets") with a velocity of about 100 km s^{-1} or greater with respect to the molecular cloud and with linewidths of about 20 km s^{-1}. The spatial extent of the EHV CO is somewhat smaller than the lower velocity "classical" CO flow as mapped by Liseau, Sandell, and Knee (1988) and Bachiller and Cernicharo (1991) who find that the extent of the CO flow is about 30" by 120". The velocity of the EHV CO is comparable to that of the optical HH objects in this region. The total mass in the EHV CO component is estimated to be about 0.03M$_\odot$, about 1% of the total outflow mass. Although the momentum in the EHV component is about an order of magnitude less than that in the classical flow (2.2 vs. 11 M$_\odot$ km s^{-1}), the mechanical power is about two orders of magnitude larger (450 vs. 3.3 L$_\odot$). The EHV CO may form in a fast neutral wind emerging from SVS 13 or it may represent gas swept from the surrounding cloud and accelerated to the observed velocities by momentum transfer from an even faster wind. In the former model, the CO may form by ion-molecule reactions within a mostly atomic wind, in which H is atomic but some of the carbon is converted to molecules (see Figure 8 and discussion in the chapters by Shu and by Natta and Giovanelli in this volume). CO may have a lower relative abundance with respect to hydrogen than is typical of

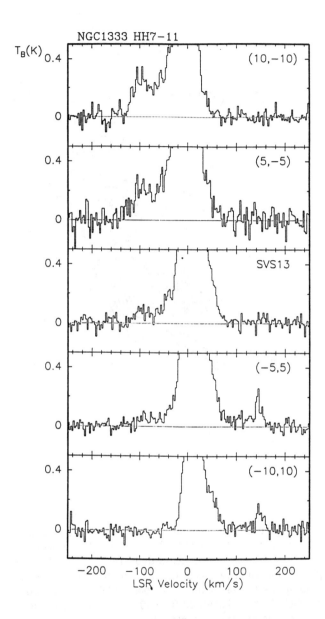

Figure 6. Profiles of J = 2-1 ^{12}CO emission showing the extremely high velocity (EHV) CO emission and molecular bullets in the NGC 1333 HH 7-11 outflow. Taken from Bachiller and Cernicharo (1991). The full CO velocity extent of this flow is greater than 300 km s^{-1}; individual bullets have radial velocities in excess of 150 km s^{-1} with respect to the molecular cloud.

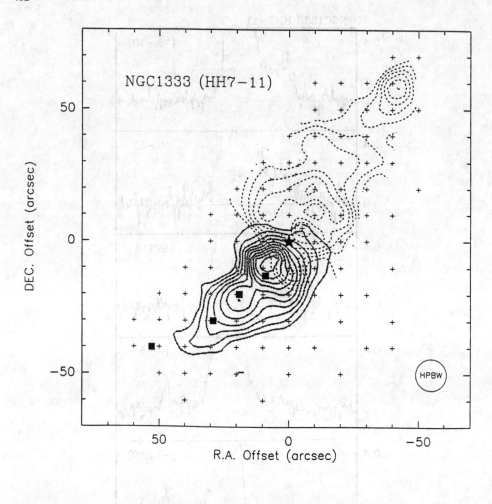

Figure 7. CO J = 2-1 maps of the 2 $M_\odot$ bipolar outflow emerging from the 60 to 150 $L_\odot$ source SVS 13 (marked by a star) in the NGC 1333 cloud (taken from Bachiller and Cernicharo 1991). The blueshifted lobe is closely associated with the Herbig-Haro objects HH 7-11 which exhibit large proper motions mostly along the axis of the flow (indicated by the dark rectangles). The map shows the blueshifted CO lobe (solid) integrated from -150 to -20 km s^{-1}, while the redshifted lobe is integrated from 30 to 170 km s^{-1}. The first contour is at 6 K km s^{-1}; subsequent contours are spaced at intervals of 3 K km s^{-1}.

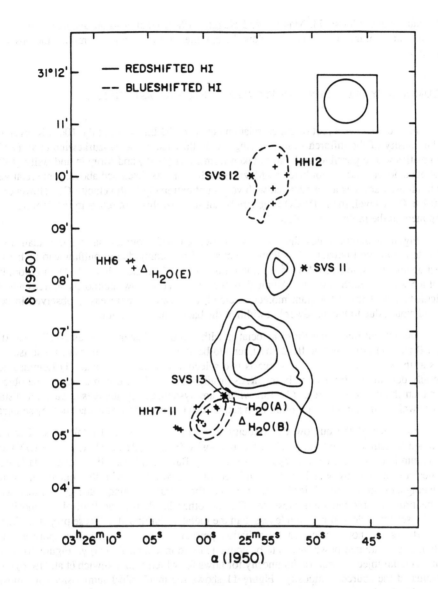

Figure 8. The NGC 1333 HH 7-11 flow is one of the few which has been detected and imaged in the 21 cm line of atomic hydrogen. This figure, taken from Rodriguez et al. (1990) shows contours of redshifted (solid; V_{LSR} = 18.3 to 38.9 km s^{-1}) and blueshifted (dashed; V_{LSR} = -22.9 to -2.3 km s^{-1}) HI emission observed with the VLA. Contours show levels at 3,4,5, and 6 mJy/beam. The beam is shown by the circle in the inset box. Extremely high velocity HI has also been detected from this flow using the Arecibo 300 m dish (Lizano et al. 1988), but this component is too faint to image with the VLA.

molecular clouds (Glassgold, Mamon, and Huggins 1989) leading to an underestimate of the mass and momentum in the EHV CO flow. Instabilities in the flow may produce the observed "bullets".

3.2 LUMINOSITY DEPENDENCE OF FLOW PROPERTIES AND STATISTICS

Figure 9 shows a plot of the correlation between CO linewidth at the 100 mK level and the luminosity of the infrared source driving the outflow using the Mozurkewich et al. (1986) luminosities where possible, and values taken from Lada (1985) and Morgan and Bally (1991) for about a dozen sources not listed by Mozurkewich et al. As discussed above, some relatively low luminosity sources are associated with very faint extremely high velocity CO. However, at the 0.1 to 0.2K level, the EHV CO is generally not seen, so this plot refers to the "normal" CO component of the molecular outflow.

Figure 9 demonstrates that outflows associated with sources having less than about $10\,L_\odot$ tend to have linewidths less than 10 km s^{-1}. Broadening due to outflow can be hard to detect against the standard GMC ^{12}CO line profile, which in regions like L1641 in Orion, has about a 6 km s^{-1} width at 0.1 K. Although outflows formed by low luminosity sources may be difficult to detect toward a giant molecular cloud, they can be more easily observed in cold isolated clouds due to the narrower linewidth of the background emission.

Distant outflows, and those associated with infrared clusters such as NGC 2024 (the Orion B jet) and Orion A (the IRc2 flow), tend to be assigned higher luminosities than isolated flows at the same distance. This suggests two selection effects in the data: (1) Luminosities, generally determined from large beam IRAS or airplane based measurements, include objects other than the outflow source. (2) Linewidths are systematically underestimated for distant sources which suffer more beam-dilution in single-dish CO measurements than nearby sources.

Estimates of the outflow CO velocity V, spatial extent R, and total mass M can be combined to estimate the total mechanical luminosity ($= MV^2/2\tau_{dyn}$ where $\tau_{dyn} = R/V$) and momentum injection rate ($= MV/\tau_{dyn}$). As shown by Bally and Lada (1983) and Lada (1985), the mechanical luminosity and momentum injection rate correlate with the source luminosity. On average, the momentum injection rate is two or three orders of magnitude greater than what can be provided by radiation pressure. On the other hand, the mechanical luminosity is typically several orders of magnitude less than the radiative luminosity. These properties imply that outflows are not accelerated by radiation pressure alone. Nevertheless, only a small fraction of the radiated power needs to be converted into mechanical energy. Figure 10 shows the momentum injection rate vs. luminosity for flows for which Mozurkewich et al. (1986) have re-evaluated the source luminosity. Figure 11 shows the mechanical luminosity / luminosity relationship for the same objects.

As is discussed above and shown in Figure 9, there is evidence that source confusion and clustering leads to a systematic overestimate of the radiative source luminosities, especially at the high luminosity end. Therefore the true ratio of mechanical luminosity to actual source luminosity may be larger than that given by Bally and Lada (1983), Lada (1985), or even by Mozurkewich et al. (1986) by about an order of magnitude or or more. As can be seen in

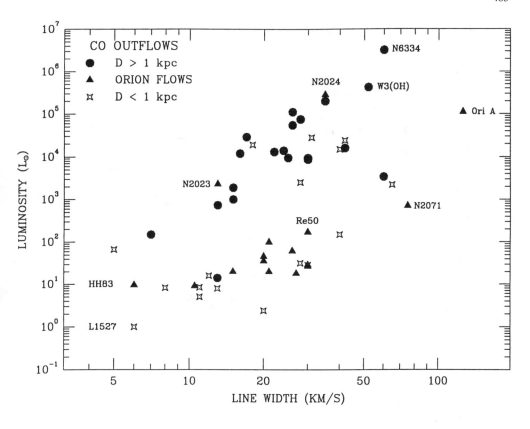

Figure 9. A plot showing the linewidth / luminosity relationship for outflows with well-determined luminosities. The luminosities were taken from Mozurkewich et al. (1986) while the linewidth comes from a variety of references. This plot highlights two selection effects: (1) for very distant sources, the linewidth may be systematically underestimated due to severe beam dilution effects which are especially large for the highest velocity components in outflows. (2) Clustering may lead to systematic overestimates of the luminosity of outflow sources, since in many cases, such as NGC 2023 or Orion A, many individual stars lie within a few arc-minutes of the actual outflow source. On this plot, isolated and nearby outflows lie systematically below outflows located in crowded and/or distant regions.

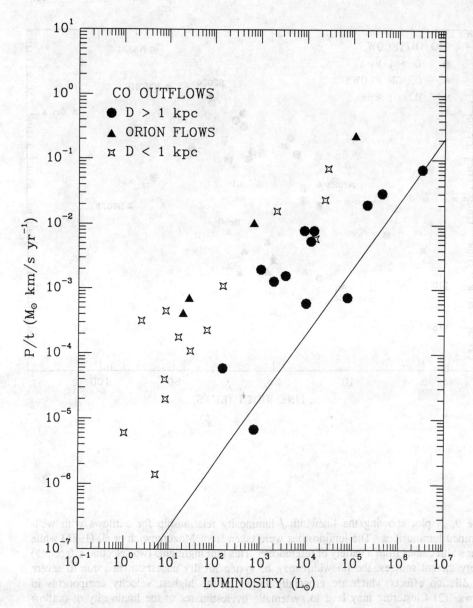

Figure 10. The relationship between the momentum injection rate and central source luminosity, adapted from Mozurkewich et al. (1986). The diagonal line shows the momentum injection rate of the radiation field. Most sources lie above this line by several orders of magnitude indicating that radiation pressure can not drive molecular outflows.

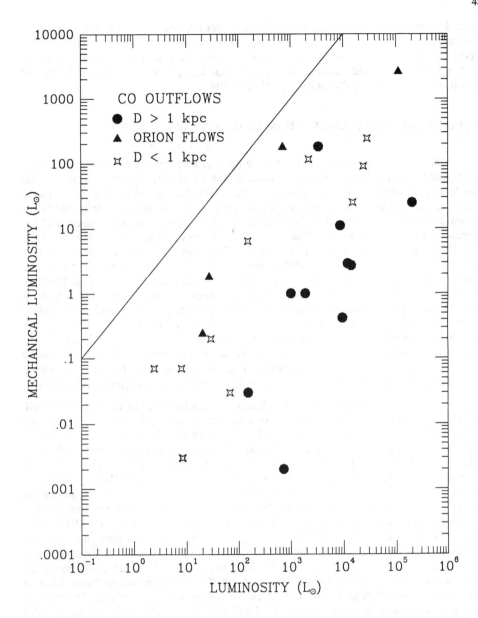

Figure 11. The relationship between the mechanical luminosity of an outflow (taken from Lada 1985) and the source luminosity taken from Mozurkewich et al. (1986). On the diagonal line, the mechanical luminosity equals the radiative luminosity.

Figure 10, the correlation between momentum injection rate and source luminosity is biased towards greater radiative luminosity by the inclusion of distant (> 1 kpc) objects. This bias also effects the CO mechanical luminosity / luminosity relationship in a similar way. From the Orion and nearer isolated sources, we estimate that $L_{mech} \approx 0.006 L_{IR}^{1.25}$ $L_\odot$ and that the momentum generation rate, $P/\tau \approx 5 \times 10^{-6} L_{IR}^{0.9}$ $M_\odot$ yr^{-1} kms^{-1}

3.3 OPTICAL AND NEAR-IR OBSERVATIONS OF MOLECULAR OUTFLOWS

While much work remains to be done in order to investigate the relationships among optical, infrared, and millimeter-wave line emission from the vicinity of young stars undergoing mass loss, it is already clear that CO flows, Herbig-Haro objects, and shock-excited near-IR lines of molecular hydrogen are highly complimentary tracers of outflow activity. While the mm-wave CO lines trace the bulk of the (cold) accelerated gas, the average gas velocity of this component is low (1 to 20 km s^{-1}). The near-infrared H_2 lines tend to have linewidths of order 20 to 100 km s^{-1}. The optical lines seen in Herbig-Haro objects exhibit the highest velocities, with values ranging from less than 100 to over 600 km s^{-1}. The shock-excited optical and near-IR line emission, as well as mid- and far-infrared cooling line radiation, such as that from [OI], [CII], OH, and CO, can be used to study gas which has been heated to thousands of degrees Kelvin by the dynamical interaction between high velocity outflows and ambient cloud material. Since the cooling time for the hot component of molecular outflows is short, typically 1 to 100 years, the infrared and optical lines trace the present location of shocks and energy dissipation in the flow. In contrast, mm-wave emission lines such as CO are collisionally excited at the 10 to 50 K temperature typical of molecular clouds. Therefore, these lines trace the total amount of mass and momentum in an outflow averaged over the lifetime of the flow.

A catalog of 184 Herbig-Haro (HH) objects has been compiled by von Hippel, Burnell, and Williams (1988), and about 45 new HH objects have recently been reported (Reipurth and Graham 1988; Reipurth 1989a,b; Ogura 1990; and Ogura and Walsh 1991; see review by Reipurth in this volume). The majority of HH objects occur in groups, with each group probably associated with a specific outflow source. About 230 individual HH objects are catalogued which can be assigned to about 90 individual outflows. Lists of HH objects and CO outflows are necessarily incomplete, making statistical comparisons of these two manifestations of mass loss difficult. Most of the recently discovered HH objects have not yet been searched for CO outflows, and conversely, most outflows have not been searched for HH objects. Nonetheless, when the catalogs of HH objects reported in the references above are compared with lists of CO outflows in Lada (1985) and Fukui (1989), the overlap of these tracers is already striking. Out of 40 CO outflows that have been searched optically, 26 contain HH objects within their boundaries. On the other hand, out of 31 HH object complexes which have been observed with high sensitivity in CO, 26 have associated CO outflows. The frequent occurrence of HH objects among the sample of well-studied CO outflow regions suggests that HH objects are an important feature of outflow activity.

The nearest CO outflows tend to be the ones associated with HH objects. Good examples include L1551-IRS5 and HL/XZ Tau, the HH 7-11 complex, and the NGC 1999 region in Orion. In most cases (although not all), the visible HH objects are located within the blue (approaching) lobe of the CO outflow which usually suffers less extinction than the red (receding) lobe.

Although almost all HH objects have been found in regions of low mass star formation, this probably reflects a bias since low mass star-forming regions are more numerous, are found closer to the Sun, and tend to be located in smaller cloud cores suffering less extinction than high mass star-forming regions. Indeed, a number of HH objects (5 to 10% of the known sources) are associated with high luminosity sources such as the Orion/IRc2 outflow (M42-HH objects, Axon and Taylor 1984) and Cepheus A (GGD37 and HH-NE, Lenzen 1988).

Nearly 20 of the HH objects associated with low luminosity YSO's in the Orion region occur in portions of the L1641 and L1630 clouds which have been well-mapped in ^{12}CO. However, only a few of these HH objects have been detected as CO outflows. This may be due to the previously mentioned selection effect resulting from the difficulty in distinguishing low-velocity accelerated CO against the turbulent background emission of a typical molecular cloud. Since the CO linewidth of outflows correlates with the luminosity of the driving star (see Fig. 9), fainter sources having lower velocity CO flows are harder to distinguish against the background emission. The ubiquity of optical outflows and HH objects is dramatically illustrated by the recent work of Reipurth (1989a) who shows that at least 11 distinct optical flows from young stellar objects are present in a 30'x30' field in the northern part of the L1641 region of Orion. Only three CO outflows have been detected so far in this region (HH 1-2 and HH 35, Leverault 1988; HH 83, Bally, Castets, and Duvert 1991 - in preparation).

Many molecular outflows have associated H_2 emission at near-infrared wavelengths, usually detected in the most prominent line, the v=1-0 S(1) line at 2.122 μm. In most cases, the near-IR H_2 lines are excited by shocks propagating into relatively dense molecular gas (n > $10^4 cm^{-3}$). About 60 regions have been searched for H_2 emission: out of 42 detections, 32 are CO outflows and 29 are HH objects, with 19 being both HH objects and CO outflows. Although molecular hydrogen emission is found in the same vicinity as the low excitation optical line emission which defines HH objects, the H_2 emission region frequently exhibits different morphology (Lane and Bally 1986, Harvey et al. 1986, Lane 1989). The structural differences can be understood as a consequence of the shock temperature and density structure and the mechanism of line excitation (Lane and Bally 1991, in preparation).

The most luminous sources of H_2 emission tend to be associated with energetic, luminous molecular outflows such as Orion/IRc2, NGC 2071, DR21, and Cepheus A (see Lane 1989, and references therein). The close association between HH objects near low luminosity outflows and shock-excited molecular hydrogen emission gives us some confidence that the excitation conditions in the high luminosity outflow sources are similar in nature to those found in the HH objects. Therefore, the 2 μm lines of H_2 can be used to infer the presence of extensive shock-excited gas in these sources which is similar in nature to the HH objects. In highly obscured sources, it may be possible to use the near-infrared lines to identify knots of H_2 emission as invisible, near-IR `Herbig-Haro objects'.

Progress in understanding of the relationship between Herbig-Haro objects, near infrared H_2 emission, and molecular outflows requires systematic surveys of large samples of flows in the optical, near-infrared, and mm-wavelength regions.

4. Outflow Models

A key step in understanding outflows is the de-projection of the observed 3-dimensional phase space (spatial-spatial-velocity) data into an actual model of the flow with 3 spatial and 3 velocity dimensions. Since the direct problem is too complex to solve, the most common approach has been to construct a 3-dimensional flow model, and to project this onto the plane of the sky and the radial velocity direction. Sets of models can be compared with real data in order to constrain the distribution and velocity field of the observed CO in real flows (Cabrit and Bertout 1986; Meyers-Rice and Lada 1991).

These investigations show that the orientation of the flow axis with respect to the line-of-sight plays a major role in the appearance of the flow. Figure 12, taken from Cabrit and Bertout (1986), illustrates this point. For flow parallel to the outflow walls, the inclination angle alone can cause a flow the look highly collimated or poorly collimated. Meyers-Rice and Lada (1991) show that outflow statistics are consistent with flow vectors parallel to the outflow axis or cavity walls. In this case, the highest velocity flows are expected, on average, to show poor collimation while the lower velocity flows should show higher degrees of collimation since they flow mostly orthogonal to the line-of-sight. On the other hand, if outflows expand primarily orthogonal to the flow axis, as might be expected for a pair of inflating bubbles, then there ought to be little systematic variation of the flow velocity with elongation. Statistics show that if a vector component orthogonal to the flow axis exists, it must be small compared to the parallel component. Bipolar outflows are more jet-like than bubble-like in their kinematic behavior. For parallel flow vectors, the estimated dynamical lifetime of the outflow depends on the inclination angle. For the same object, the lifetime estimate will be short if the flow is viewed end-on and long if it lies close to the plane of the sky. Therefore, the derived mechanical parameters are very sensitive to the inclination of the outflow and to the orientation of the flow vector with respect to the flow axis.

The radial velocity and proper motion of any associated Herbig-Haro objects can be used to constrain the orientation of a molecular flow. Such measurements are available for HH objects associated with several outflows, the best examples being HH 1-2 (orientation mostly in the plane of the sky), RNO 43 (in the plane of the sky), L1551 IRS 5 (inclined about 45°), and HH 7-11 (inclined about 45°). More measurements of this type are needed to study the intrinsic collimation angles, flow lifetimes, and systematic biases in the determination of the mechanical parameters of outflows.

Models can be used to analyze spatial-velocity diagrams ("slit spectra") such as that shown in Figure 3. Figure 13 (taken from Meyers-Rice and Lada 1991) demonstrates that the near side of the receding shell and the far side of the approaching shell produce a low velocity gradient ridge in spatial-velocity diagrams taken along the projected flow axis. Due to its greater inclination with respect to the line-of-sight, the far side of the receding lobe and the near side of the approaching lobe exhibit a larger velocity gradient and subtend a shorter projected angular extent. The maximum velocity separation between the near and far walls of the flow occurs on the flow axis, and the two components merge into a single feature in slit-spectra taken tangent to the projected edge of a flow. In some flows, such as Mon R2, the velocity data provides evidence for shell structure when this structure is not evident in the spatial data.

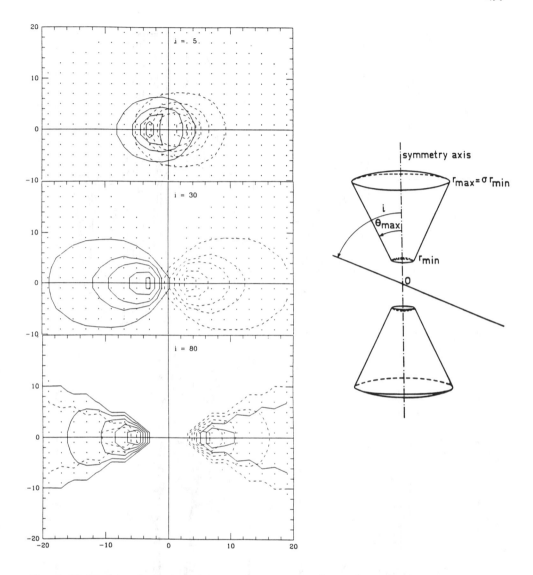

Figure 12. Models of a conical bipolar outflow with a 30° degree opening angle as seen from the source. The source is assumed to have a thin shell of CO emission with flow vectors parallel to the flow surface. This model is viewed from three different angles (5°, 30°, and 80°). A nearly end-on flow (top left) shows poor collimation but a large radial velocity extent. A flow whose axis lies close to the plane of the sky appears collimated and has a smaller radial velocity extent. In this case, both sides of the outflow show both redshifted and blueshifted gas. Solid contours indicate blueshifted gas while dashed contours show redshifted gas.

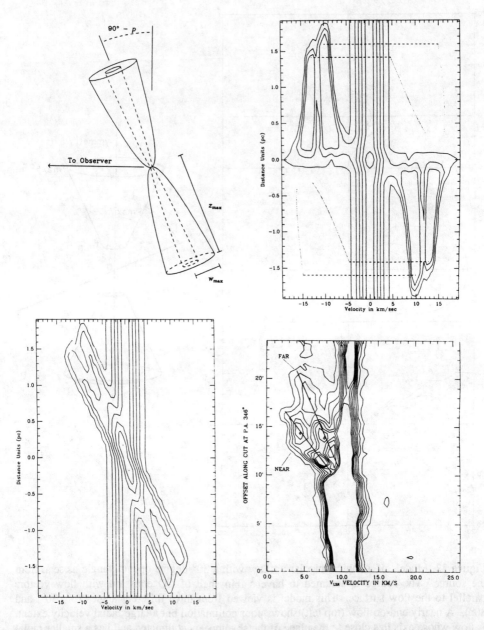

Figure 13. Diagrams taken from Meyers-Rice and Lada (1991) which illustrate the spatial-kinematic behavior of a thin-shell conical bipolar outflow model in which the flow vector field is tangential to the flow walls. *Upper left:* The model. *Upper right:* Model with constant flow amplitude, $|V| = 23.0$ km s^{-1}. *Lower left:* Model with linear acceleration (Hubble flow) in which $|V| = 23.0(R/R_{max})$. *Lower right:* The blue-shifted lobe of the Mon R2 flow, which shows the kinematic signature of shell structure with a Hubble law acceleration.

The thickness of the CO emitting layer in an outflow can range from a thin shell in which most of the gas is confined to the outflow cavity walls, to thick shells, or even flows which occupy the entire outflow volume, like water flowing down a pipe. The low observed CO intensities indicate that most of the gas at any particular velocity is confined to a small fraction of the projected outflow area. In some large angular diameter outflows, such as L1551 IRS 5, the CO emission is produced in a shell with a thickness ($\Delta R / R$) of about 0.1. Other outflows do not show clear signatures of shell structure (see for example Cepheus A - Bally and Lane 1991).

Many outflows exhibit better collimation with increasing radial velocity as well as a systematic increase in radial velocity along the outflow axis. As seen in Figure 13, in some flows, the radial velocity increases (or decreases) linearly with increasing distance from the source, resulting in a self-similar "Hubble" flow. As discussed by Shu et al. (1991), these properties can be understood in terms of a combination of projection effects and the acceleration of the swept up layer along the walls of a curved cavity where a wind interacts with a medium with an isothermal ($\rho \propto r^{-2}$) density distribution modified by rotational flattening.

If a large portion of the CO emission is confined to the walls of the outflow cavity, what fills the cavity itself? Observations of the optically thin ^{13}CO species have shown that the outflow cavities are mostly devoid of stationary molecular gas. On the other hand, towards the nearest and least obscured outflows, such as the blue lobe of the L1551 IRS 5 flow, Herbig-Haro objects and an optical jet can be seen, suggesting that cavity is filled with tenuous high velocity gas, which "lights-up" when it passes through shocks, which excite optical emission lines. Direct evidence for a fast atomic wind filling the outflow lobes has been found for NGC 2071 (Bally and Stark 1983), HH 7-11 and L1551 (Lizano et al. 1988), and DR 21 (Russell, Padman, and Bally 1991 - in preparation) in the form of extremely high velocity 21 cm HI line wings localized to the outflow region.

Discussion of theoretical models of molecular outflows is beyond the scope of this review and are discussed elsewhere in this book (see chapters by Shu and by and Pudritz, Pelletier, and Gomez de Castro).

References

Andre, Ph., Martin-Pintado, J. Despois, D., and Montemerle, T. 1990a *Astron. Astrophys.*, **236**, 180.
Andre, Ph., Montemerle, T., Feigelson, E.D., and Steppe, H. 1990b *Astron. Astrophys.*, **240**, 321.
Axon, D.J., and Taylor, K. 1984 *M.N.R.A.S.*, **207**, 241.
Bachiller, R., and Cernicharo, J. 1991 *Astron. Astrophys.*, in press.
Bachiller, R., Cernicharo, J., Martin-Pintado, J., Tafalla, M., and Lazareff, B. 1990 *Astron. Astrophys.*, **231**, 174.
Bally, J., and Lada, C.J. 1983 *Ap.J.*, **265**, 824.
Bally, J., and Lane, A.P. 1991 in *Astrophysics with Infrared Arrays*, eds. K. Pilakowski and R. Elston (Astron. Soc. of the Pacific Conference Series), in press.
Bally, J., and Scoville, N.Z. 1982 *Ap.J.*, **255**, 497.

Bally, J., and Stark, A.A. 1983 *Ap.J.(Letters)*, **266**, L61.
Cabrit, S., and Bertout C. 1986 *Ap.J.*,**307**, 313.
Comoretto, G., et al. 1990 *Astron. Astrophys. Suppl.*, **84**, 179.
Edwards, S., and Snell, R.L. 1982 *Ap.J.*,**261**,151.
Edwards, S., and Snell, R.L. 1983 *Ap.J.*,**270**,605.
Edwards, S., and Snell, R.L. 1984 *Ap.J.*,**281**,237.
Fukui, Y. 1989 *Low Mass Star Formation and Pre-Main Sequence Objects*, ed. Bo Reipurth, (ESO, Garching), p. 95.
Garden, R.P., Geballe, T.R., Gatley, I., and Nadeau, D. 1991 *Ap.J.*, **366**, 474.
Glassgold, A.E., Mamon, G.A., and Huggins, P.J. 1989 *Ap.J.(Letters)*, **336**, L29.
Harvey, P.M., Joy, M., Lester, D.F., and Wilking, B.A. 1986 *Ap.J.*, **301**, 346.
Hughes, V.A., and Moriarty-Schieven, G. 1991 *Ap.J.*, **360**, 215.
Koo, B.C. 1990 *Ap.J.*, **337**, 318.
Kwan, J., and Scoville, N. Z. 1976 *Ap.J.(Letters)*, **210**, L39.
Lada, C.J. 1985 *Ann. Rev. Astron. Astrophys.*, **23**, 267.
Lane, A.P. 1989 *Low Mass Star Formation and Pre-Main Sequence Objects*, ed. Bo Reipurth, (ESO, Garching), p. 331.
Lane, A.P., and Bally, J. 1986 *Ap.J.*, **310**, 820.
Lane, A.P., Haas, M.R., Hollenbach, D.J., and Erickson, E.F. 1990 *Ap.J.*, **361**, 132.
Lenzen, R. 1988 *Astron. Astrophys.*, **190**, 269.
Levreault, R.M. 1988 *Ap.J.Suppl.*, **67**, 283.
Liseau, R., Sandell, G. and Knee, L.B.G. 1988 *Astron. Astrophys.*, **192**, 153.
Lizano, S. Heiles, C. Rodriguez, L.F., Koo, B.C., Shu, F.H., Hasegawa, T., Hayashi, S.S., and Mirabel, I.F. 1988 *Ap.J.*, **328**, 763.
Margulis, M. and Snell, R.L. 1989 *Ap.J.*, **343**, 779.
Margulis, M., Lada, C.J., Hasegawa, T., Hayashi, S.S., Hayashi, M., Kaifu, N., Gatley, I., Greene, T.P., and Young, E.T. 1990 *Ap.J.*, **325**, 615.
Masson, C.R., Mundy, L.G., and Keene, J. 1990 *Ap.J.(Letters)*, **357**, L25.
McCutcheon, W.H., Dewdney, P.E., Purton, C.R., Sato, T. 1991 *A.J.* (in press).
Meyers-Rice, and Lada, C.J. 1991 *Ap.J.*,**368**, 445.
Morgan, J.A., and Bally, J. 1991 *Ap.J.*, in press.
Mozurkewich, D., Schwartz, P.R., and Smith, H.A. 1986 *Ap.J.*, **311**, 371.
Ogura, K. 1990, *Publ. Astron. Soc. Pac.*, **102**, 1336.
Ogura, K., and Walsh, J. R. 1991, *A.J.*, **101**, 185.
Reipurth, B. 1989a *Astron. Astrophys.*, **220**, 249.
Reipurth, B. 1989b, in *Low Mass Star Formation and Pre-Main Sequence Objects*, ed. B. Reipurth (ESO, Garching), p. 247.
Reipurth, B. and Graham, J.A. 1988 *Astron. Astrophys.*,**202**, 219.
Rodriguez, L. F., Curiel, S., Moran, J. M., Mirabel, I. F., Roth, M., and Garay, G. 1989, *Ap. J.*, **346**, L85.
Rodriguez, L.F., Lizano, S., Canto, J., Escalante, V., and Mirabel, I.F. 1990 *Ap.J.*, **365**, 261.
Rudolph, A. and Welch, W. 1988 *Ap.J.(Letters)*, **326**, L31.
Shu, F. H., Ruden, S. P., Lada, C. J., and Lizano, S. 1991 *Ap.J.(Letters)*, **370**, L31.
Snell, R.L., Bally, J., Strom, S.E., and Strom, K.M. 1985, *Ap.J.*, **290**, 587.

Snell, R.L., Dickman, R.L., and Huang, Y.-L. 1990, *Ap. J.*, **352**, 139.
Snell, R.L., Huang, Y.L., Dickman, R.L., and Claussen, M.J. 1988 *Ap.J.*, **325**, 853.
Torrelles, J.M., Rodriguez, L.F., Canto, J., Carral, P., and Marcaide, J. 1983 *Ap.J.*, 274, 214.
von Hippel, T. Burnell, S.J.B., and Williams, P.M. 1988 *Astr. Astrophys. Suppl.*, **74**, 431.
Walker, C.K., Lada, C.J., Young, E.T., and Margulis, M. 1988 *Ap.J.*, **332**, 335.
Wilking, B. A., Blackwell, J.H., and Mundy, L.G. 1990 *A.J.*, **100**, 758.
Wolf, G.A., Lada, C.J., and Bally, J. 1990 *A.J.*, **100**, 1892.
Wouterloot, J. G. A., Henkel, C., and Walmsley, C. M. 1989, *Astron. Astrophys.*, **215**, 131.
Yusef-Zadeh, F., Cornwell, T. J., Reipurth, B., and Roth, M. 1990, *Ap.J.(Letters)*, **348**, L61.
Zuckerman, B., Kuiper, T.B.H., and Kuiper, E.N.R. 1976 *Ap.J.(Letters)*, **209**, L137.

S. Guilloteau and John Bally at special session on outflows.

Bo Reipurth, Hans Zinnecker, and Charles Lada

HERBIG-HARO OBJECTS

BO REIPURTH
European Southern Observatory
Casilla 19001
Santiago 19
Chile

ABSTRACT. This article reviews our current understanding of the Herbig-Haro phenomenon. HH objects are shocks occurring in supersonic flows from very young stars. The history of HH research is outlined, and observations at various wavelengths summarized. Particular attention is paid to the highly collimated Herbig-Haro jets, whose morphology, kinematics and physical properties are discussed. Shock models for the jet structure and the working surface of jets are reviewed, and two regions with HH jets are examined in detail. The energy sources of HH flows are, as a class, among the youngest stars known, and it is argued that HH objects are related to disk accretion events in their driving sources.

1. Introduction

Herbig-Haro objects have for a long time been recognized as important phenomena in star formation regions, but it is only recently that we are beginning to glimpse how they are tied into the global star formation process and to decipher the information they carry about events in the earliest stages of the life of a star. It is a rapidly developing subject, as witnessed by the number and evolving content of recent reviews (e.g. Schwartz 1983a, Dyson 1987, Mundt 1988, Reipurth 1989a, Böhm 1990, Raga 1991). In the present overview, I will stress the historical development of the subject as well as the latest results, and list extensive references to the literature, thus hopefully providing a tool useful for students of star formation.

In the course of objective prism surveys of dark clouds, Herbig (1948, 1950, 1951, 1952) and Haro (1952, 1953) discovered a number of small objects, which display characteristic emission line spectra of hydrogen and forbidden low excitation lines of mainly [OI], [NII] and [SII], with at most very weak continua. On direct photographic plates, Herbig and Haro found these objects to be tiny, almost semi-stellar nebulae, with a tendency to be found in groups, often aligned in strings of objects. Ambartsumian (1954) subsequently coined the term Herbig-Haro objects for these tiny nebulae. From their location in dark clouds often containing nebulous stars or the newly recognized T Tauri stars (cf. Joy 1942, Ambartsumian 1947), early researchers concluded that the HH objects were somehow related to the process of star birth.

In order to better separate what constitute individual peculiarities and what are general characteristics, surveys have been made to find and study new HH objects. In the early years, Herbig and Haro found about 40 objects, which are listed in the catalogue of Herbig (1974). Subsequently more objects have been found by Strom et al. (1974), Schwartz (1977a), Adams et al. (1979), Mundt and Fried (1983), Reipurth (1985a), Malin et al. (1987), Mundt et al. (1988), Reipurth and Graham (1988), Heyer and Graham (1989), and Reipurth and Olberg (1991), either by objective prism surveys, or from surveys of direct Schmidt plates supplemented with additional spectra or interference filter CCD images. Attempts to find HH objects on direct plates purely on morphological grounds have not been very successful, the surveys of Gyulbudaghian et al. (1978) and Reipurth (1981) contain only a few bona fide HH objects. The HH objects known up to 1987 are listed in a catalogue by von Hippel et al. (1988), which is by now rather incomplete. To the above mentioned references must be added a number of papers each containing one or a few new HH objects. In total, well over 150 HH objects are known today.

2. The Early Years: Observations and Models

The co-existence of Herbig-Haro objects and T Tauri stars in heavily obscured regions, and particularly the direct associations found in the case of for example T Tauri and HL Tauri (Herbig 1969a), strongly suggested to early workers an organic connection between the two types of objects. For a while, they considered the possibility that HH objects could be some visible manifestation preliminary to star formation, but the appearance and disappearance of knots in HH 1 and 2 reported by Herbig (1969b, 1973) and the failure to detect infrared stars inside HH knots (Mendoza 1969) made this view untenable.

The first infrared source associated with an HH object was discovered near HH 100 by Strom et al. (1974a). The fact that the source is displaced from the HH objects led Strom et al. (1974b) to propose a model in which HH objects are seen as small reflection nebulae, illuminated through cavities in the surrounding cloud by circumstellar emission from nearby embedded stars. Detection of significant polarization in some knots of HH 24 by Strom et al. (1974c), with an origin in an embedded infrared source, supported this picture. However, a detailed study of HH 24 by Schmidt and Miller (1979) demonstrated conclusively that the emission-line knots are unpolarized and thus formed *in situ*, but are intermixed with patches of polarized reflection nebulae from an embedded star.

With the gradual realization that powerful winds emanate from young stars, other models became possible. Already in the late fifties, Osterbrock (1958) suggested that the line emission in HH objects might be due to energetic outflows from young stars. Detailed studies of the spectra of HH objects associated with T Tauri and their resemblance to the emission of supernova remnants led Schwartz (1975) to suggest that HH objects are radiative shocks resulting from the interaction of a supersonic stellar wind with the surrounding medium. This concept was expounded in greater detail by Schwartz (1978) in the shocked cloudlet model. In this view, a stellar wind from an embedded young star with a mass-loss rate of 10^{-5} to 10^{-6} $M_\odot yr^{-1}$ impinges upon small ambient cloudlets, forming bow shocks, which face towards the driving source. A large number of observational characteristics of HH objects find a natural explanation within this scenario, including their low excitation spectra, radial velocities, luminosities and variational time scales.

Another approach was taken by Norman and Silk (1979) in their interstellar bullet model. They considered the break-up of a circumstellar shell caused by instabilities when a supersonic wind first interacts with its surroundings. Rayleigh-Taylor spikes form and are swept up by the wind. Initially extremely dense and confined by ram pressure, the spikes provide the sites of H_2O masers. These interstellar bullets are accelerated, expand and are expelled to distances of ~ 1 pc before themselves being broken up by instabilities. At this later stage they are identified with HH objects.

A third model, the wind cavity model, was proposed by Cantó (1980) and Cantó and Rodríguez (1980), who considered the stationary flow pattern around a young star which blows off a strong isotropic stellar wind, while located at the edge of a dense cloud. The static configuration attained in an environment with a pressure gradient developing into a constant–pressure medium is an ovoid-shaped cavity. Two shocks occur, the first where the wind impinges against the oblique walls of the cavity and is refracted towards its tip. Here, a second shock arises, as the flow converges and shocks against itself. This last shock, occurring at the tip of the ovoid shaped cavity, is identified with an HH object in this model. Barral and Cantó (1981) extended the model to operate in the middle of a gaseous disk.

The one observational fact which was not well established at the time when the above models were conceived was the ubiquity of high proper motions of HH objects. Cudworth and Herbig (1979) demonstrated that HH 28 and HH 29 have tangential motions of about 145 km s^{-1} in directions away from the embedded source L1551 IRS 5, a proper motion already noticed by Luyten (1971). Shortly thereafter, Herbig and Jones (1981) established the supersonic cross motions of the prototype objects HH 1 and HH 2, in which individual knots were shown to have tangential velocities in the range 100 to 350 km s^{-1}, with HH 1 and HH 2 moving in opposite directions, away from a deeply embedded infrared source now accepted as the driving source of the complex (Pravdo et al. 1985, Rodríguez et al. 1985, Roth et al. 1989). Fig. 1a shows a CCD image of HH 1 and 2, and fig. 1b shows the kinematics of the objects, the location of the energy source and also of the Cohen-Schwartz star, which was previously thought to be the driving source. Subsequently, it has been shown that large proper motions are common properties of HH objects (e.g. Jones and Herbig 1982, Herbig and Jones 1983, Jones et al. 1984, Schwartz et al. 1984, Jones and Walker 1985).

The large proper motions of HH objects have been explained with varying success by the three models outlined above. In the shocked cloudlet model it was not clear, and still is not, if a wind can accelerate a cloudlet to the velocities observed without destroying it. Besides the problem of stability of a wind-accelerated cloudlet, it can be argued that interstellar clouds do not behave like tennis-balls: by the time the backside of a cloudlet realizes that the front is affected by a wind, a shock has transversed the whole cloud and may in the process have destroyed it. The preliminary numerical results of Różyczka and Tenorio-Tagle (1985) indicate that cloudlets indeed are likely to be destroyed when affected by a powerful wind. Much work remains to be done on this subject. Still, there are a number of HH objects which appear to be best understood in the context of the shocked cloudlet model, so although it clearly fails for more collimated HH flows (sect.5), the model may well, properly adapted, be relevant in certain cases. The interstellar bullet model obviously has no problem in explaining the observed large space motions of HH objects, but is largely abandoned today for other reasons. In particular, the scenario of a Rayleigh-Taylor unstable circumstellar shell from which fragments

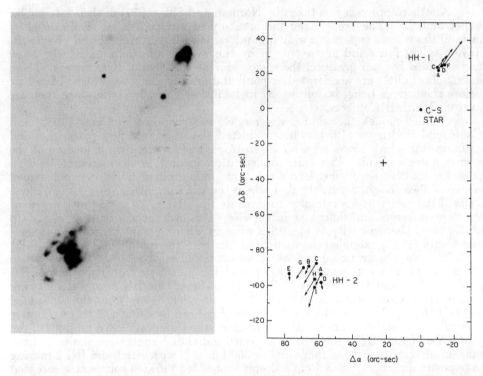

Figure 1. (a) An Hα CCD image of HH 1 and 2. (b) Proper motions of HH 1 and 2, from Herbig and Jones (1981). The cross shows the location of the VLA source driving the HH complex.

are ejected as H_2O masers is not readily reconciled with the general lack of masers around HH energy sources (e.g. Haschick et al. 1982), and present ideas of episodic formations of HH objects in collimated flows. Finally, the stellar wind cavity model produced in its original form only stationary shocks, and so was unable to produce any proper motions, unless the possible existence of instabilities or other time-dependent effects in the flow is invoked (Cantó 1985). The model remains interesting in a modified form, in which it works as an engine to form jets (Cantó et al. 1988, Tenorio-Tagle et al. 1988).

Such was the overall state of affairs, when in the early eighties it was realized that Herbig-Haro flows may take the form of highly collimated supersonic jets. This discovery has had a fundamental impact on our understanding of the Herbig-Haro phenomenon, and is dealt with in more detail in sect.5. More details on the early years of HH research can be found in the review paper by Schwartz (1983a) and in a special issue of Revista Mexicana (vol.7) in honor of Guillermo Haro.

3. Herbig-Haro Objects at Different Wavelengths

3.1   Optical Spectra

The brightest known HH object, HH 1, has been the subject of the most detailed spectroscopic studies up to now.   Its spectrum was first described by Herbig (1951), and subsequent spectrophotometric studies (e.g. Böhm 1956, Osterbrock 1958, Haro and Minkowski 1960, Böhm et al. 1973) further established its peculiar nature.   In an HH object, permitted and forbidden emission from neutral atoms (e.g. [OI], OI, [CI], [NI]) and from ions of low excitation energy (e.g. CaII, [CaII], [FeII], MgII, [SII]) are much stronger than in common photo-ionized nebulae.   However, line-ratios can vary greatly from one HH object to another, and HH objects are often classified according to their level of "excitation" (e.g. Böhm 1983).   HH 1 is the prototype of a high excitation object: for example, its [SII]6717/Hα ratio is very low ($\sim$0.2), its [OIII]5007/Hβ ratio is high ($\sim$0.6), its [NI]5198/5200 to Hβ ratio is low ($\sim$0.2), and its [CI]9848/Hα ratio is very low ($\sim$0.02) (Brugel et al. 1981a, Solf et al. 1988).   HH 7, on the other hand, is a typical low excitation object, with a very high [SII]6717/Hα ratio ($\sim$2.8), a low [OIII]5007/Hα ratio ($\sim$0.1), a very high [NI]5198/5200 to Hβ ratio ($\sim$1.8) and a very high [CI]9848/Hα ratio of $\sim$1.3 (Böhm et al. 1980, 1983).   Most HH objects have line-ratios which lie within these values.   Fig.2 illustrates a variety of HH spectra in the region around Hα.

In addition to their distinct spectral features, it was found early on that HH objects are characterized by very considerable radial velocities of the order of a hundred km s^{-1} (e.g. Schwartz and Dopita 1980).   Cantó (1981) showed that 75% of HH objects exhibit negative radial velocities, while the rest have positive velocities.   This discrepancy arises because the molecular clouds associated with HH objects tend to obscure the objects moving away from us.

For a derivation of line-ratios, and a meaningful interpretation of them, HH spectra must be de-reddened.   This correction can be significant, because HH objects mostly are located in regions of rather high extinction.   The best method to correctly determine the reddening of an HH object is the one of Miller (1968). In this, the sum of the auroral lines of [SII] at $\lambda\lambda$ 4068/4076 are compared with the sum of the transauroral lines of [SII] at $\lambda\lambda$ 10318/10337.   These lines appear in very different parts of the spectrum, but originate from the same upper energy level.   Their theoretical ratio is therefore fully determined by the ratios of their transition probabilities and their frequencies; but the observed ratio is additionally affected by reddening.   Adopting an extinction curve then leads directly to an E(B-V) for the observed object.   While easy in principle, in practice it is not trivial to get accurate, good signal-to-noise observations of the [SII] lines beyond 1 μm. Another similar method is to compare Balmer and Paschen lines which come from the same upper level; for a careful discussion of these reddening methods see Solf et al. (1988).   For spectra that do not have sufficient spectral coverage to include both sets of lines, more indirect methods exist (e.g. Dopita 1978a).

In addition to emission lines, HH objects have very weak continua (Herbig 1951, Böhm 1956, Böhm et al. 1974, Brugel et al. 1981b).   Although in a few HH objects one observes red continua due to reflected light from the stellar energy source, the majority of continua are blue, and indeed keep rising into the ultraviolet (see sect 3.3).   These blue continua were interpreted as the result of a collisionally enhanced two photon decay mechanism in neutral atomic hydrogen by Dopita et

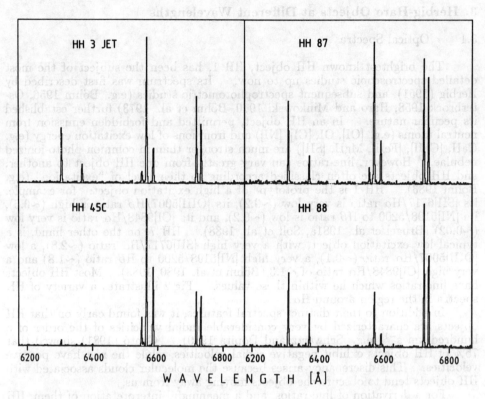

Figure 2. Low resolution spectra of four HH objects, showing the large variation of emission line strengths. The lines seen are [OI] 6300, [OI] 6363, [NII] 6548, Hα, [NII] 6584, [SII] 6717 and [SII] 6731.

al. (1982a) and Brugel et al. (1982). Recent high signal-to-noise data for HH 1 corroborate this suggestion (Solf et al. 1988).

An important step towards interpreting the spectra of HH objects came with the realization that their line spectra could originate in the cooling regions behind shock waves (Schwartz 1975,1978). This was borne out in more detailed comparisons between observed line ratios and plane-parallel shock wave models (Dopita 1978a, Raymond 1979). The best fits to the observed data came for shocks with velocities of typically 70 - 100 km s^{-1} moving into a rather low density, almost neutral medium. However, none of the models could reconcile the simultaneous presence of low excitation lines of [OI] and [SII] with highly ionized species like CIV. The introduction of bow shock models has resolved many of the inconsistencies inherent in the plane-parallel shock geometry. A bow shock produces a range of shock velocities, because only the component of the flow velocity perpendicular to the bow shock is thermalized. A single bow shock can therefore produce high excitation near its apex and simultaneously much lower excitation lines from the oblique shocks along its wings (e.g. Hartmann and Raymond 1984, Choe et al. 1985, Raga and Böhm 1985,1986, Hartigan et al. 1987, Raymond et al. 1988).

The relevance of bow shock models for HH objects is supported not only by the line ratios they produce, but also by the line shapes they predict. The kinematics of the flow in the recombination region behind the shock can be accurately predicted, and by integrating the emission over an appropriate volume (e.g. corresponding to what is seen through a spectrograph slit) a direct comparison with observations can be made. The results can be presented either as line profiles (e.g. Hartigan et al. 1987, Raymond et al. 1988) or also including the spatial dimension as position-velocity diagrams (e.g. Choe et al. 1985, Raga and Böhm 1985,1986, Hartigan et al. 1990).

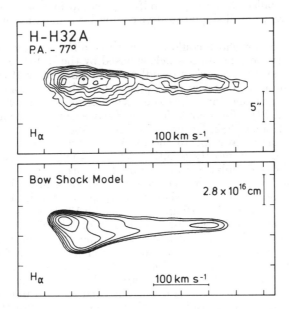

Figure 3. An observed Hα long slit spectrum of HH 32A and the predicted position-velocity diagram from a bow shock model. From Solf et al. (1986).

Fig. 3 shows a comparison of the observed long-slit spectrum of the Hα line in HH 32A with a position-velocity diagram calculated to reproduce the observed features, from a detailed study of the HH 32 complex by Solf et al. (1986). On the spatial (vertical) axis, the source is located towards the bottom, and on the velocity (horizontal) axis lower (more negative) velocities are towards left, higher (more positive) towards right. The immediately striking feature of the observed line profile is its very large width, of more than 350 km s^{-1}. This can readily be reproduced by a bow shock because it, as distinct from a plane-parallel shock, contains a large range of velocities projected along the line of sight, in addition to the thermal broadening of the emission from the hot post-shock gas. The observed emission clearly has two peaks, which are rather well reproduced in the model; they arise from regions near the bow apex and from the bow wings, respectively (see also Raga et al.1986). Finally, the line has a distinct triangular shape. This asymmetry is related to the angle under which the bow shock is observed: for a bow shock moving in the plane of the sky the line is completely symmetric, but for bow shocks moving at increasingly small angles to the line of sight the asymmetry becomes more and more pronounced (Raga and Böhm 1986, Hartigan et al. 1987). The angle of the flow is a parameter to which the models are very sensitive. The parameters which Solf et al. (1986) found best match the observation in fig.3 are: bow shock velocity 300 km s^{-1} at the stagnation point, and an angle to the line of sight of 20°, with the object moving away from the observer. Hartigan et al. (1987) present simple analytic formulae that can be used to estimate the shock velocity and bow shock orientation from a single high resolution spectrum of a low excitation line.

Although the qualitative agreements between observations and models are good, it must be recalled that many simplifying assumptions are made for the calculations, and that very detailed quantitative comparisons must await more physically realistic models.

In principle, a bow shock could arise in three different ways in an HH object. Either it could curve around an obstacle shocked by the stellar wind from a young star, thus facing the star ("shocked cloudlet model", Schwartz 1978). Or it could form around a clump of material as it plows through the ambient medium after being ejected from a young star or its close environment ("interstellar bullet model", Norman and Silk 1979). Or it could form similarly around the head of a collimated jet ("jet model", Königl 1982, Dyson 1984, Mundt 1985). In the latter two cases the bow shock is facing away from the driving source. As discussed in sect.5, observational evidence is mostly pointing towards the jet model. Nonetheless, almost all bow shock calculations up to now have been made for flows around a rigid sphere, in order to simplify calculations. More realistic bow shock models have begun to appear, with bow shocks occurring at the head of nonadiabatic collimated jets (e.g. Raga 1988, Blondin et al. 1989,1990), see sect. 5.1e.

In recent years a large number of studies of HH objects, based on high-resolution spectra of a few selected lines, have appeared. In addition to papers already mentioned, these include the works of Böhm and Solf (1985), Hartigan et al. (1986a,b), Bührke et al. (1986), Solf (1987), Solf and Böhm (1987), Noriega-Crespo et al. (1989), Reipurth and Heathcote (1990). Ideally, spectroscopic observations should provide both the large spectral range needed to yield numerous line-ratios, and the high spectral and spatial resolution necessary to map the kinematics of the flow in each emission line. Usually these requirements are, for instrumental reasons, mutually exclusive, but lately such data have started to appear (e.g. Solf et al. 1988, Böhm and Solf 1990, Solf and Böhm 1991).

For further details on the optical emission line spectra of HH objects, see the reviews of Böhm (1990) and Raga (1991).

3.2 Infrared Spectra

The near-infrared spectral region contains a number of emission lines of molecular hydrogen, which have proven to be important diagnostics for the molecular component of the gas in HH objects. The vibrational and rotational transitions of the H_2 molecule can be excited either by shocks (via collisional excitation in gas heated by the shocks to a few thousand degrees) or by absorption of ultraviolet radiation. The line strengths produced by the collisional and the fluorescent processes are significantly different, and so it is easy to distinguish between these mechanisms. The strongest and most commonly studied line is the 2.12 μm v=1-0 S(1) line of H_2. This line was first detected in HH objects around T Tauri by Beckwith et al. (1978). Elias (1980) detected the line in a number of southern HH objects, and from a study of line-ratios concluded that the molecular gas was dense ($\sim 10^4$ cm^{-3}) and heated by shock waves. Subsequently, a large number of HH objects were searched for H$_2$ emission, and many were detected and mapped with single-beam techniques (Brown et al. 1983, Simon and Joyce 1983, Zealey et al. 1984, 1986, Lane and Bally 1986, Lightfoot and Glencross 1986, Sandell et al. 1987, Schwartz et al. 1987, Zealey et al.1989). It is now clear that molecular hydrogen can contribute sometimes significantly to postshock cooling, and that shock-codes should incorporate both atomic and molecular processes. A considerable fraction of the HH objects detected have low-excitation optical spectra formed in low velocity or oblique shocks.

H$_2$ line profiles have been studied in high resolution infrared spectra obtained by Doyon and Nadeau (1988), Zinnecker et al.(1989) and Brand et al. (1989). The width of the S(1) line is typically only half of the width of the Hα line in the same objects, and the observed maximum velocities of the H$_2$ lines are also lower. Zinnecker et al. (1989) favor two mechanisms for the generation of the H$_2$ emission, either shock heating of external molecular gas in the wings of a bow shock or shock heating of molecular gas entrained at the boundary of the flow.

The similarities and differences between distributions of optical and molecular shock regions may provide essential clues to the flow structure in HH objects. Following the emergence of infrared array imagers, it is now possible to study the H$_2$ emission distribution at resolutions comparable to optical images. Such studies have been performed by Schwartz et al. (1988), Hartigan et al. (1989), Garden et al. (1990), Wilking et al. (1990) and Lane et al. (1991). The technique and new results are discussed in a recent review by Lane (1989). The overall distribution of H$_2$, which traces low velocity shocks around 10 to 40 km s^{-1}, is the same as the optical emission, caused by faster shocks, typically with velocities of a hundred to several hundred km s^{-1}. But there are subtle and important differences. A particularly intriguing case is the Cepheus A region (e.g. Hartigan et al. 1986, Lenzen 1988), which has been mapped at arcsecond resolution (Lane 1989, Lane et al. 1991). The GGD 37 HH object has an overall bow shock morphology, with complex bow shaped substructures and extremely broad optical emission lines indicating shock velocities up to almost 500 km s^{-1}. The same morphology is seen in molecular hydrogen, but the arcs are softer and less protruding. It appears that the H$_2$ emission occurs mainly in the wings of the bow shocks, where the shocks are oblique and thus much weaker (Lane et al. 1991), consistent with the much narrower linewidths of 25-45 km s^{-1} found by Doyon and Nadeau (1988).

Little work has been done on HH objects at far-infrared wavelengths. The [OI] line at 63μm is often the dominant coolant in shocks for postshock temperatures between 10^3 and 10^4K (e.g. Hollenbach 1985). Detections of this line in HH objects have been made by Cohen et al.(1988). Observations of this line are a potential tool for detecting HH objects, which are moving through embedded windblown cavities.

3.3 Ultraviolet Spectra

It came as a surprise when Ortolani and D'Odorico (1980) discovered a strong continuum and emission lines in the ultraviolet spectrum of HH 1, given its relative optical faintness and not insignificant reddening. Subsequently, more than a dozen studies of ultraviolet spectra of HH objects have confirmed the importance of uv radiation from these objects. A detailed review of this subject has recently appeared (Brugel 1989), so only a brief summary is given here.

The uv spectra of high-excitation objects are dominated by emission lines of CIV 1548/1551 and CIII] 1907/1909; low-excitation objects, on the other hand, almost exclusively show fluorescent lines of the Lyman bands of the H_2 molecule (Schwartz 1983b). The uv continuum rises towards shorter wavelengths, particularly rapidly between 1700 Å and 1620 Å, with a peak around 1575 Å. This continuum is likely to be a continuation of the observed blue continuum, and is also believed to be due to a collisionally enhanced two photon decay mechanism in neutral atomic hydrogen (Brugel et al. 1982), but an additional component due to fluorescent H_2 appears to play a role at the shortest wavelengths (Böhm et al. 1987). Mundt and Witt (1983) cautioned that the uv continuum of HH 1 and 2 could be seriously contaminated by extended reflected light, but Böhm et al.(1987) and Lee et al. (1988) showed such a possible component to be minor. They also measured the spatial extent of various HH objects in uv and optical lines, and compared with expectations from shock models.

Interpretations of uv observations are extremely sensitive to the reddening corrections applied. Although the amount of extinction can be reasonably determined using the Miller (1968) method, the shape of the extinction curve may differ from region to region. Böhm-Vitense et al. (1982) argued that flatter than normal extinction curves (like the θ Ori curve) may be appropriate.

All the brightest HH objects have now been observed in the uv (Brown et al. 1981, Böhm et al, 1981,1984, Brugel et al. 1985, Schwartz et al. 1985, Cameron and Liseau 1990, and abovementioned references) and the practical limit of the IUE satellite has been reached. Further progress is expected from the Hubble Space Telescope.

3.4 Radio Observations

The study of Herbig-Haro objects at radio wavelengths is still in its infancy. HH 1 and 2, which are among the brightest HH objects known, were the first HH objects detected in radio continuum (Pravdo et al. 1985). In a more detailed study both these objects were detected in 20, 6 and 2 cm continuum observations at the VLA (Rodríguez et al. 1990). Both HH objects have flat spectra, consistent with optically thin free-free emission. The central energy source was also detected and was found to be elongated along the axis defined by the HH objects, but unresolved at an angle perpendicular to this. An upper limit of 0.1" for the minor axis suggests that collimation is present on scales as small as 50 AU. This behaviour is very similar to what was found for the L1551 IRS 5 source (e.g. Cohen et al. 1982). Rodríguez et al. (1990) conclude that the morphology of the central source and its flux and angular size dependence with frequency are suggestive of a bipolar confined jet, presumably the embedded counterpart to the optically visible jet (Strom et al. 1985).

A second pair of HH objects, HH 80 and 81, were detected at 6 cm by Rodríguez and Reipurth (1989), with flux densities of $\sim$1 mJy, similar to HH 1 and 2. But HH 80/81 are more than 3 times as distant as HH 1/2, and therefore intrinsically about 10 times as luminous in the radio regime. Optically HH 80/81 are the intrinsically most luminous HH objects known (Reipurth and Graham 1988). Also in this case the suspected driving source, IRAS 18162-2048, was detected as a very elongated object pointing to the HH objects, with a major axis of $\sim$1 pc. This large size compared to the size of the HH 1/2 and L1551 sources ($\sim$0.01 pc) is probably related to its much larger luminosity (about 20000 L$_\odot$ versus less than 100 L$_\odot$); HH 80/81 appear to be formed by a massive young B-star. Another region of HH objects most probably driven by a more massive star is Cepheus A (e.g. Hartigan and Lada 1985, Lenzen 1988). Here a group of seven HH objects have been detected in radio emission by Hughes and Moriarty-Schieven (1990).

At present it is not possible to distinguish between a photo-ionized and a shock-ionized thermal radio source from its continuum alone (Curiel et al. 1987). The distinction could possibly be made by observing radio recombination lines, but they are still beyond the sensitivity of presently available detectors.

It is an interesting possibility that radio continuum observations could be a tool to detect deeply embedded, optically invisible HH objects. Perhaps the unusual triple radio source in Serpens might be such a case. Here the central object is the IRAS source 18273+0113, a 300 L$_\odot$ embedded star. Two lobes are symmetrically placed about 6 arcseconds on either side of the source.

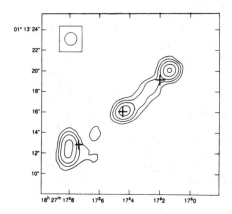

Figure 4. A 6 cm VLA map of the triple radio source in Serpens, made in March 1989. The crosses show the positions of the features in June 1978. The high proper motion is obvious. From Rodríguez et al. (1989).

Fig. 4 is from a study by Rodríguez et al. (1989) and shows a 6 cm VLA map from March 1989. The crosses show the positions of the same objects in June 1978 and reveal that the lobes have a large cross motion of 300±100 km s^{-1}. The kinematic age of the fastmoving lobes is merely about 50 years. The spectral index of the central source is $\alpha = 0.1 \pm 0.1$ (where α is given by $S_\nu \propto \nu^\alpha$), but both lobes have indices $\alpha < -0.6$, suggesting that the emission is nonthermal. Rodríguez et al. (1990) note that synchrotron radiation generated by relativistic electrons in the environment of pre-main sequence stars is not easily explained, given the absence of large gravitational potentials. Instead they speculate that an acceleration mechanism involving shocks could work. Cruzius-Wätzel (1990) has shown that the particles responsible for the non-thermal emission indeed can be accelerated in the bow shocks of bullets or jets moving with super-Alfvénic velocity.

Another object, probably similar to the Serpens triple source, is located in Orion. A compact radio source, coincident with a faint 2 μm source, shows significant linearly polarized emission at 6 cm and a non-thermal spectrum of

$\alpha \sim -1$, and an extended radio tail or jet stretches almost 3 arcminutes (0.4 pc) from this source (Yusef-Zadeh et al. 1990). In this case optical emission is associated with the object; the so-called "Orion Streamers" which consist of long, narrow filaments with dimensions of about 10×300 arcseconds and with an HH-like emission spectrum (Reipurth and Sandell 1985) are displaced about 10 arcseconds from the radio lobe. Because of their unusual morphology the Orion Streamers can hardly be called Herbig-Haro objects in the classical sense, and the case illustrates some of the more rare phenomena which may occur in star forming regions. The study of extended synchrotron emission regions around young stars could provide important new insights into shocks moving through environments with high gas density and high magnetic field strength.

4. A Picture Gallery of Herbig-Haro Objects

On photographic plates, HH objects appear as groups of small semi-stellar fuzzy nebulae (e.g. Herbig 1974). Because some of the most prominent emission lines in HH objects are emitted between 6300 Å and 6800 Å, modern red-sensitive CCD detectors have caused a revolution in the detailed study of their morphology and spectra. Deep CCD images reveal that HH objects consist of myriads of knots, bows, wisps and streamers of luminous gas, testifying to the very complex dynamical processes which shape them. Broadband CCD images are not ideal for imaging of HH objects, because they usually blend together a number of emission lines. Narrowband interference filter images, which isolate the contribution of individual lines are powerful means to trace the distribution of shocks at various levels of excitation. Hα filters (which in most cases are really Hα/[NII] filters, because they often include the [NII] 6548/6584 lines) are used to trace gas of relatively high excitation, [SII] 6717/6731 filters display the distribution of gas of low excitation, and [OIII] 5007 filters are good for showing regions of the highest excitation. Ideally, to such filters should be added narrowband continuum filters at wavelengths close to the emission line filters for a proper subtraction of possible stellar reflection components.

If a small nebula located towards a nearby molecular cloud shows up clearly in a [SII] filter, but not in a neighboring continuum filter or a broadband filter in the far red like a Gunn z filter (which mostly transmit stellar continuum light), then it is almost certainly an HH object. Note that this is not always true if the emission filter is an Hα filter, partly because Hα emission can be a very strong component of light reflected from a young embedded star, partly because all sorts of small scale ionized structures show up at the surfaces of molecular clouds. Some care should therefore be excercized when identifying new HH objects, and a spectrum remains the best way of proving the HH nature of a candidate object.

Numerous papers have appeared in recent years containing CCD images of HH objects. They include the works of Mundt et al. (1984), Hartigan and Lada (1985), Strom et al. (1986), Ray (1987), Reipurth and Graham (1988), Raga and Mateo (1988a), and Reipurth (1989b). Other papers have focused specifically on the highly collimated subset of HH objects, the Herbig-Haro jets; these will be further discussed in sect.5.

Here a gallery of CCD images of HH objects is presented, giving examples of 5 regions containing HH objects. Fig. 5 shows a [SII] image of HH 75, a complex of small and large knots in the southern constellation Vela, stretching over a full

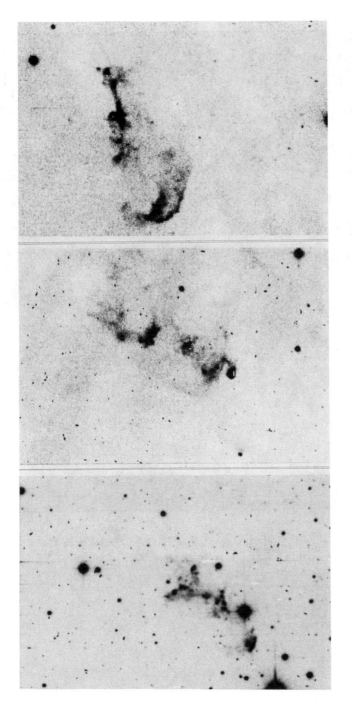

Figure 5. A [SII] CCD image of HH 75 in Vela. The field is 2.5 × 4 arcminutes.

Figure 6. An Hα CCD image of HH 84 in Orion. The field is 2.5 × 4 arcminutes.

Figure 7. An Hα CCD image of HH 91 in Orion. The field is 2 × 3.2 arcminutes.

Figure 8. An Hα + [SII] CCD image of the RNO 43 complex in the λ Orionis molecular shell. The field is 7.5 × 7.5 arcminutes.

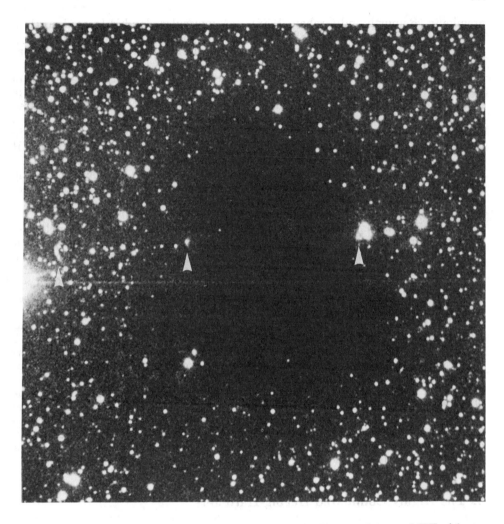

Figure 9. An Hα CCD image of the Bok globule B335 and several HH objects. The axis of the HH objects, which are marked, pass through an embedded far-infrared source in the center of the globule.

arcminute (Reipurth and Graham 1988). A molecular cloud is located north of the complex, terminating where the HH objects appear. Presumably the energy source is hidden somewhere along the long axis of the complex, inside the molecular cloud north-west of the object.

Fig. 6 shows an Hα image of HH 84, a fine group of HH knots located in Orion, in the L1641 cloud (Reipurth 1989b). Its long axis is parallel to that of many other HH flows in the region, e.g. HH 1/2 and HH 34, a direction shared by polarization vectors, suggesting it is defined by the large scale magnetic field in the region. The total extent of the bright knot groups is 100 arcseconds, which in Orion corresponds to a projected dimension of 0.23 pc, a value typical for many HH flows. It is noteworthy that despite the lumpy structure of the complex, virtually all groups of knots are interconnected by low level emission. This highly winding chain of HH knots might be an initially more collimated flow in which instabilities have set up large wiggles (see sect.5 for further details). The energy source has not been located but is likely to be found in a cloud north-west of the object.

Fig. 7 shows an Hα image of the HH 91 complex in the Orion B molecular cloud. Here the source is embedded at the center of the image, from which it ejects material to the east and the west. The region shown is only the central part of a large complex, with several more knots and bow shocks stretching over 9 arcminutes, corresponding to 1.2 pc in projection (Gredel and Reipurth 1991).

Fig. 8 shows a sum of Hα and [SII] images of RNO43, located in the large molecular shell surrounding λ Orionis. RNO43, which was discovered by Cohen (1980), is the bright object in the upper left corner of the figure. Proper motions show that it is moving away with more than 300 km/sec from an embedded IRAS source in the middle of the image (Jones et al. 1984). A large fragmented counterbow shock has been discovered on the opposite side of the source, making it a major bipolar HH flow of more than 1.1 pc extent (Reipurth and Bally 1991).

Fig. 9 shows an Hα image of the Bok globule B335. At the center of the globule there is a deeply embedded far-infrared source (Keene et al. 1983). Several HH knots have been found on a line through the source (Vrba et al. 1986, Reipurth and Heathcote 1991b), which coincides with the axis of an extended bipolar molecular outflow (e.g. Cabrit et al. 1988). The easternmost of these HH objects is seen to be a fine little bow shock. The source is likely to be a very young star.

5. The Highly Collimated Herbig-Haro Jets

5.1 Jets and Bow Shocks

5.1a *Morphological Properties of Jets*

A particularly intriguing subset of Herbig-Haro objects is formed by the highly collimated HH jets. The first jet recognized as such was the HH 46/47 jet (Dopita et al. 1982b), which remains one of the finest cases known (see sect. 5.2). Mundt and Fried (1983) and Mundt et al. (1983, 1984) did CCD imaging of young stars and known HH objects and identified various HH jets.

The prototypical case is the HH 34 jet, discovered by Reipurth (1985b), which illustrates all of the properties of its class. Fig. 10 shows a CCD image of the HH 34 complex. The strikingly collimated jet emerges from a faint but optically visible emission line star. Over about the first 10 arcseconds the jet is very faint as it flows from the small cloud core in which the source was born. It suddenly brightens and the main body of the jet consists of a chain of distinct knots out to a distance of 26 arcseconds from the source. 54 arcseconds from the source there is a faint bow shaped knot and finally, about 100 arcseconds from the source, corresponding to 0.22 pc in projection, the large knotty bow shock HH 34 is located. The whole complex is lying close to the plane of the sky, with the jet and HH 34 in the approaching lobe (Reipurth et al. 1986). On the opposite side of the source, symmetric with HH 34, there is a faint counter bow shock, HH 34N, discovered by Bührke et al. (1988). No counter jet is found in the receding lobe.

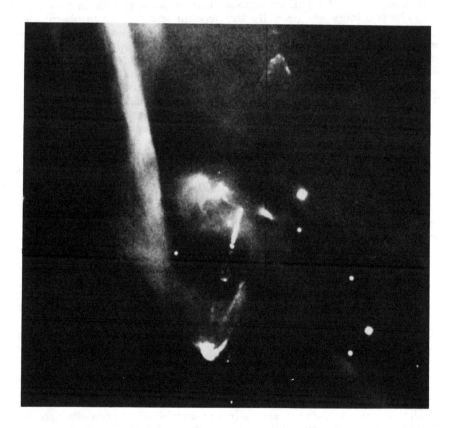

Figure 10. A [SII] CCD image of the HH 34 bipolar jet-complex in Orion. The faint star at the base of the jet is the driving source.

The opening angle of jets are in general small, of the order of a few degrees. Length-to-width ratios vary considerably, from fairly poorly collimated jets with a ratio of 5 or so to extremely high ratios exceeding 100 (e.g. HH 111). The widths of jets are mostly poorly determined, but are in a few cases known well; the HH 34 jet for example has a width of 0.7 arcseconds or 300 AU (Raga and Mateo 1988b, Bührke et al. 1988, Raga et al 1991) and a width of about 0.8 arcseconds was found for HH 111 (Reipurth et al. 1991).

All jets have a knotty sub-structure, with typical semi-regular spacings of one to a few arcseconds. These knots do not always trace a perfectly straight line, it is common that jets show considerable wiggling (e.g. Strom et al. 1983, Raga and Mateo 1987, Reipurth and Olberg 1991). The knots provide a means to measure transverse motions, when present, of jets; proper motions have been measured quite accurately for several jets by now: the L1551 IRS 5 jet (Neckel and Staude 1987, Campbell et al. 1988), HH 34 (Reipurth 1989a, Heathcote and Reipurth 1991) and several jets in the HL Tau region (Mundt et al. 1990). Typical transverse motions are around 200 - 300 km s^{-1}.

In many cases HH jets are bipolar, tracing an approaching and a receding lobe. The two lobes are generally very asymmetric. Uneven obscuration may play a role, but cases appear to exist where the asymmetries are real. This could either be because the jet-flow is intrinsically asymmetric, but could also result from differences in the ambient medium into which the jets flow (e.g. Mundt et al. 1987). Typical dimensions of a lobe in a jet complex are around 0.2 - 0.25 pc, but a few very large flows are known, the current record is, by a large margin, the bipolar jet from Z CMa with a total extent of 3.6 pc (Poetzel et al. 1989).

5.1b *Kinematics, Physical Conditions and Entrainment of Ambient Material*

Long slit spectra along a jet axis give information on the variation of radial velocity, line width, excitation and electron density in the flow. Such studies have now been made for many jets, see for example Solf (1987), Meaburn and Dyson (1987), Mundt et al. (1987), Bührke et al. (1988), Reipurth (1989b,c), Mundt et al. (1990), Hartigan et al. (1990), Heathcote and Reipurth (1991, see also Reipurth 1989a). Typical radial velocities are high, around 100 - 200 km s^{-1}. Jets show velocity variations, which are often complex but rarely major. When an ordered motion is found it is mostly a decrease in the radial velocity; in one case, HH 83, a major increase in radial velocity is measured as one moves away from the source (Reipurth 1989b).

Electron densities as derived from the [SII] 6717/6731 ratio vary considerably, with values ranging typically from a few hundred to a few thousand electrons per cm^3. When changing systematically along the jet, the electron density more often decreases, which in the simplest interpretation suggests diverging flows (e.g. Mundt et al. 1987). It is not so easy to determine the actual gas density of jets, because they are likely to have rather low ionization fractions, so that the gas number density could be considerably larger than the observed electron density. In the model calculations of Raga et al. (1990), the hydrogen ionization fraction is found to be at most 10%.

The excitation of jets is generally low, and jets are normally prominent in CCD images through a [SII] filter, and much weaker in Hα images. The low excitation character of HH jets and their very high space motions would appear contradictory, but can be reconciled if the shocks in jets are very oblique (see e.g. Raga 1989). Jet models consisting of a train of crossing shock cells have been explored in recent years (see sect. 5.1d).

The velocity dispersions in HH jets are, with some exceptions, generally low or moderately low. In HH 34, for example, the widths of the [SII] 6717/6731 lines are less than 30 km s^{-1} (Bührke et al. 1988). In several jets, one sees a splitting of the lines, e.g. in HH 24 (Solf 1987), in HH 46/47 (Meaburn and Dyson 1987), in HH 34 (Bührke et al. 1988, Reipurth 1989a, Heathcote and Reipurth

1991) and in several jets in the HL Tau region (Mundt et al. 1990). In HH 24, Solf (1987) found two velocity components close to the source, a high velocity component displaying a gentle velocity decrease away from the source, and a weaker lower velocity component with a steep acceleration. Solf (1987) interprets this in terms of a jet model with an inner cylindrical core region and a co-axial envelope. The inner high velocity core is the jet itself and the envelope is ambient material dragged along with the jet; their mutual interaction will slightly slow down the jet and accelerate the entrained material.

A first step towards a more complete hydrodynamical model describing the turbulent mixing layer formed at the outer boundary of a jet has been developed by Cantó and Raga (1991).

5.1c *Jets and Magnetic Fields*

The role which magnetic fields play in HH flows is unknown, although various theoretical jet models invoke their presence (see e.g. Uchida 1989 and sect. 5.1f). The more indirect question of the orientation of the local magnetic field in regions with HH flows has been addressed by various authors. If interstellar clouds preferentially contract along the local magnetic field lines, circumstellar disks might be aligned perpendicular to the field, from which one could possibly expect an alignment of outflows with the field lines. Such an approximate relation was found for molecular outflows by Cohen et al. (1985). Heyer et al. (1987) find the HH 7-11 flow oriented parallel to the polarization vectors of background stars, but the main axis of the nearby HH 12 is offset by 60° from the magnetic field lines. They also find that HH 33/40 lies parallel to the polarization vectors, and Reipurth (1989b) showed that this direction is shared by almost all the flows in this part of the L1641 cloud. The well defined line of HH objects in B335 lie within about 20° of the magnetic field direction determined by Vrba et al. (1986). Another group of HH jets which share a common direction lies in the HL Tau region, but here the local magnetic field is also considerably offset from the preferred flow direction (Mundt et al. 1988, 1990). These authors also consider how a jet can bend as it flows through a non-parallel magnetic field.

More detailed work on the collapse of interstellar clouds dominated either by magnetic forces or by gravitational forces (e.g. Lizano and Shu 1989) can perhaps eventually explain the alignment or misalignment of HH flows with the ambient magnetic field.

5.1d *Models of Jet Structure*

A wealth of theoretical work has been done on extragalactic jets, but these are generally considered adiabatic, and for stellar jets the radiative energy losses can be important and can affect the behaviour of the flow. Ideally, one would like to have time-dependent, non-adiabatic calculations with high spatial and temporal resolution, but this is still a computationally prohibitive task. First steps in this direction have, however, been made (e.g. Tenorio-Tagle et al. 1988, Raga 1988, Blondin et al. 1990).

High spatial resolution can be achieved if one relinquishes the time-dependence, and, as long as one considers regions away from the working surface, much information on jet structure can be obtained from stationary solutions with a realistic treatment of the relevant atomic processes. Such steady, non-adiabatic jet models have been studied by, among others, Falle et al. (1987), Cantó et al. (1989) and Raga et al. (1990b,1991b). They find that, given their assumptions (see the original papers for details), a chain of steady crossing-shock patterns is developed. A jet which initially has a pressure several times that of the homogeneous environment will expand freely in a Mach cone, and consequently the jet pressure drops below that of the ambient medium (see fig.11).

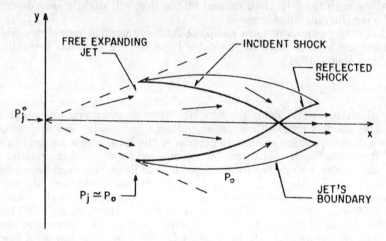

Figure 11. The first crossing-shock within an underexpanded jet. The jet expands freely until its pressure falls below the pressure of the ambient medium P_0. It is then reconfined by an incident conical shock and turned along the axis through a reflected conical shock. From Cantó et al. (1989).

The resulting underpressured flow is then recollimated by a curved "incident shock". Where the incident shock converges on the jet axis, it sets up a "reflected shock". The jet material, which was refocused after the incident shock, will again be approximately parallel to the symmetry axis after passing through the reflected shock. Since it is now again overpressured compared to its environment it will once more expand in a Mach cone, and the whole pattern is repeated.

An advantage of such steady jet calculations is that it is possible to extensively explore parameter space, and see the effect of varying individual parameters. One can also predict the jet appearance in various emission lines and calculate theoretical long-slit spectra (Raga et al. 1990b,1991b). Comparisons with observations, however, reveal that the models still are not very realistic. If the knots observed in HH jets are identified with the predicted crossing-shock cells, one finds a calculated length-to-width ratio of $\sim 1.4 \times M_j$ (where M_j is the Mach number of the jet, typically around 20) but in HH 34 the observed length-to-width ratio is an order of magnitude smaller (Raga and Mateo 1988b, Bührke et al. 1988). Furthermore, observed knots in jets show large proper motions, which clearly cannot be reproduced in these stationary models. Perhaps the observed knots are due to other knot formation mechanisms, like Kelvin-Helmholtz instabilities (e.g. Ray et al. 1988) or time-variable sources (e.g. Kim and Raga 1991). Finally, the models predict narrow line widths, in qualitative agreement with observations, but they cannot reproduce the changes in radial velocity observed along many jets. Stationary jet models are therefore useful first steps, but more detailed comparisons with observations must await high-resolution time-dependent models. Further details are given in the review by Raga (1989).

5.1e The Working Surface of a Jet

As in any two fluids that collide supersonically, two shocks occur when a supersonic jet rams through the ambient medium: a shock in which material from the environment is accelerated, called the bow shock, and a shock in which the jet material is decelerated, called the Mach disk or the jet shock. The whole double shock structure is often called the working surface. The shocks are separated by a contact discontinuity.

Hartigan (1989) has studied the relative brightness of the bow shock and the Mach disk, and explored the conditions under which either of the shocks will be dominant. He finds that, unless one of the shocks is so strong that it does not cool radiatively, then the Mach disk and the apex of the bow shock should have comparable surface brightnesses, if the jet and environmental densities do not differ by more than one or two orders of magnitude. He also concludes that if a jet is much denser than the environment then the bow shock will be brighter, and conversely, a jet significantly less dense than the ambient material will produce a brighter Mach disk.

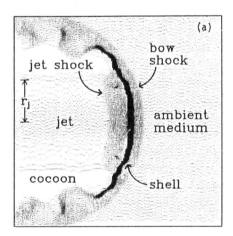
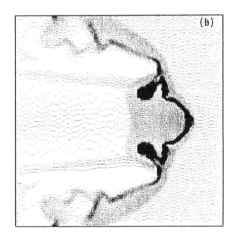

Figure 12. Density maps of the working surface of a non-adiabatic jet after (a) 800 yr, and (b) 1200 yr, from Blondin et al. (1989). The flow is clearly dynamically unstable.

For a "heavy" jet the bow shock will thus dominate the emission and in this case the jet is often in models replaced by a rigid non-emitting obstacle. Although this clearly is an approximation it allows for relatively simple calculations. The detailed predictions of line-ratios and spectral line shapes and the successful comparisons with observations have already been discussed in sect. 3.1. Raga (1986) has also calculated the expected appearance in various interference filters for bow shocks of varying velocities, and find a reasonable qualitative agreement with observations (see also Reipurth et al. 1991). These models, however, failed to reproduce the sub-structure of knots or condensations observed in most bow shocks. In HH 1 and 2, for example, one observes a multitude of knots. One possibility could be that each knot represents an individual bow shock (Hartigan et al. 1987). An alternative view has been suggested by Raga and Böhm (1987) and Raga et al. (1988) who carried out time-dependent bow shock calculations. They find that thermal instabilities in the flow produce condensations, morphologically similar to the observed knots, and they additionally find that the expected kinematics is qualitatively similar to the observed proper motions.

Recently, time-dependent calculations have been made of a working surface where a real non-adiabatic gaseous jet is placed instead of the rigid obstacle (Raga 1988, Blondin et al. 1989, 1990). Fig.12 shows density maps of the working surface of a jet after 800 yr and 1200 yr from Blondin et al. (1989). They find that a strongly cooling jet develops a shell of cold, high-density gas between the bow shock and the jet shock. As the figure illustrates, such a shell is dynamically unstable, and may fragment into separate clumps.

5.1f *Origin and Collimation of Jets*

It is not yet known how HH jets are formed, but a wealth of theoretical scenarios have been suggested, and the topic is rapidly developing. The supersonic material in jets may originate as winds from a disk, or from the boundary layer between a disk and a star, or from the star itself. These various wind-models are discussed in other chapters of this book. Some of the winds arise with a certain measure of collimation, others depend on subsequent collimation mechanisms. Here brief reference will be made only to some studies of nozzles specifically aimed at explaining HH jets. The wind origin is in the following assumed to be stellar, but properly scaled and with minor modifications such nozzles will operate with other types of winds.

One possible mechanism for the formation of a collimated outflow was proposed by Cantó (1980), see also sect.2. He considered the effect of an isotropic stellar wind with a stratified disk-like environment, and studied the properties of the two ensuing elongated cavities. A fundamental assumption here is that the flow is highly non-adiabatic. Because of the very high radiative losses, the cavities are lined with narrow layers of cold gas which flow towards the apex of the cavities, where they collide and shock. Königl (1982) studied the flow under the opposite assumption that the flow is adiabatic, and showed that this leads to the formation of two oppositely directed de Laval nozzles. These nozzles will then collimate the originally isotropic stellar wind after it has gone through a shock.

It is interesting to compare these two models which correspond to opposite limits of high and low cooling efficiencies, but the comparison is not straightforward because they employ different analytic approximations to solve the hydrodynamic equations. Raga and Cantó (1989) therefore extended Königl's calculations of de Laval nozzle flows to the non-adiabatic case, and studied a range of intermediate cases with moderate cooling. In all cases a more or less collimated outflow is produced, but the narrower and better collimated jets are formed when the radiative energy losses are larger and for wind velocities higher than a few hundred km per sec. Kim and Raga (1991) has studied the flow behaviour through such non-adiabatic de Laval nozzles with time-dependence, assuming an intrinsically variable energy source. Velocity and density variations of the wind can cause large velocity variations in the jet.

A potentially serious shortcoming of these models is that magnetic fields are ignored. Such fields could play an important role in collimating a wind (e.g. Pudritz and Norman 1986, Kwan and Tademaru 1988, Uchida 1989, Gómez de Castro and Pudritz 1991), but such mechanisms are dealt with in detail elsewhere in this book.

A comparison of jet formation mechanisms with observations is likely to be difficult, because of the high extinction and small scales involved. High resolution radio continuum observations may be important (e.g Cohen et al. 1982) and the high resolution spectral imaging technique which Solf (1989) has employed for T Tauri stars appears very promising.

5.2 The HH 46/47 Jet Complex

The HH 46/47 jet complex, discovered by Schwartz (1977b), emanates from a Bok globule in the Gum Nebula (Bok 1978) containing an embedded infrared source (Emerson et al. 1984, Cohen et al. 1984b, Graham and Heyer 1989). It is a bipolar HH system, with an approaching lobe containing a fine jet ending in a bright bow shock (47A) and further out a fainter, more extended bow shock (47D), and a receding lobe with a faint counter-jet and another bow shock (47C) (Dopita et al. 1982b, Graham and Elias 1983). Proper motion studies show the two lobes moving away from the source (Schwartz et al. 1984). At the base of the jet there is a large reflection nebula which is strongly variable (Graham 1987) and highly polarized (Scarrott and Warren-Smith 1988). Spectra of the reflection nebula show that the underlying star is a T Tauri star with a rich emission line spectrum (Dopita 1978b, Graham and Heyer 1989). The detailed kinematics of the complex has been discussed by Meaburn and Dyson (1987) and Hartigan et al. (1990).

Recently a series of very high-resolution interference filter CCD images were acquired of HH 46/47 (Reipurth and Heathcote 1991a). Fig. 13 shows a [SII] 6717/6731 image taken in 0.7 arcsecond seeing, and Fig. 14 the same field through an $H\alpha$/[NII] filter in 0.6 arcsecond seeing. In the [SII] image the body of the jet is clearly seen to consist of numerous well defined knots, which meander and wiggle out from the source, located behind an obscuring ridge between the jet and counter-jet. At larger distances from the source the jet becomes fainter, wider and splits into two strands before it merges with the bright bow shock 47A, where the flow rams into the ambient medium. Further out is the bow shock 47D. In the $H\alpha$ image the appearance of the jet complex is very different. Here the trunk of the jet is now seen as a chain of smoky wisps and puffs, the counter-jet has disappeared, and the outer bow shock 47D is now much more pronounced. These differences are illustrated in fig. 15, where the two images have been subtracted, so that $H\alpha$ strong (high excitation) regions are black, and [SII] strong (low excitation) regions are white. Stellar continuum is effectively subtracted this way, but since the star has strong $H\alpha$ emission the reflection nebula shows up as black. Detailed study of the difference image is hampered by the presence of the bright rims and the reflection nebula. Nonetheless, from the least confused regions one can see that the knots in the jet consist of a bright [SII] core with a coating of $H\alpha$ emission on the side facing the flow direction. Where the jet bends, the $H\alpha$ photocenter is always displaced towards the outside of the bend, relative to the [SII] photocenter. In the more pronounced cases the $H\alpha$ emission takes the form of small streamers, typically swept backward at an angle of about 45° to the axis of the jet. The [SII] bright material is likely to represent shocks in the jet, while the $H\alpha$ bright wisps

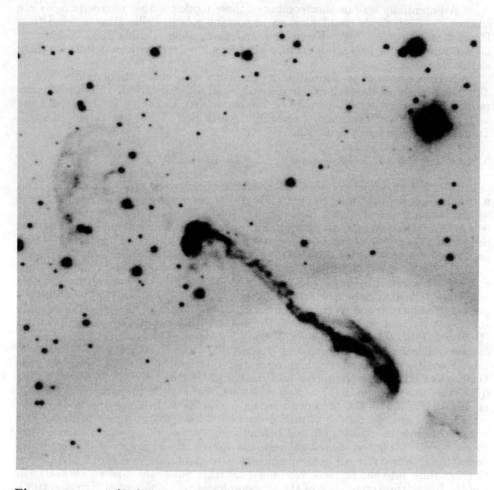

Figure 13. A [SII] high resolution CCD image of the HH 46/47 jet complex, obtained at the ESO New Technology Telescope in a seeing of 0.7 arcsecond. The jet emanates from an infrared source embedded in the Bok globule. The field is slightly more than 2 × 2 arcminutes.

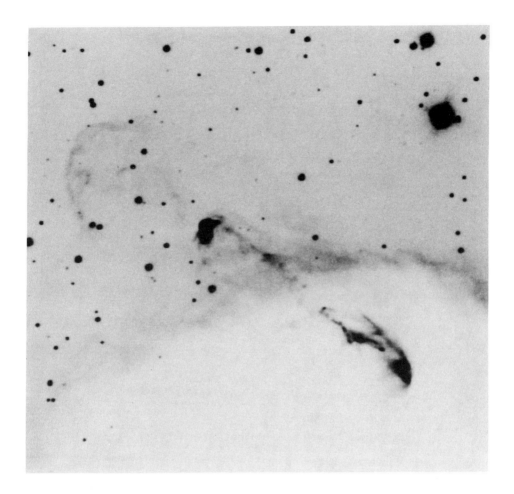

Figure 14. An Hα high resolution CCD image of HH 46/47 from the ESO NTT, in 0.6 arcsecond seeing. In Hα, the body of the jet consists of wispy shocks surrounding the well defined knots seen in [SII].

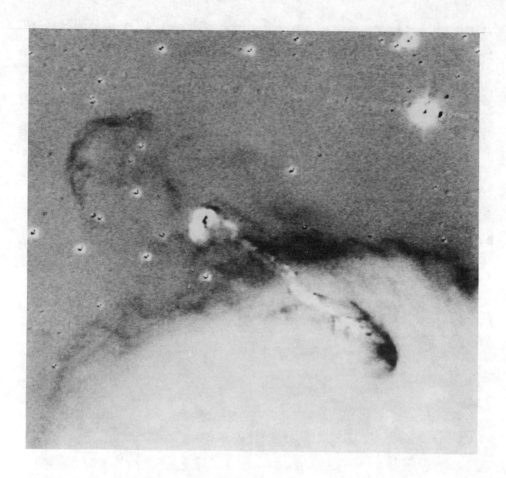

Figure 15. An Hα – [SII] difference image, from the images shown in figs. 13 and 14. Hα strong regions are black, and [SII] strong regions are white. This difference image illustrates the excitation distribution along the HH flow.

and streamers are the shocks driven into the ambient medium due to the passage of the jet.

In the bright head of the jet, 47A, one sees two separate regions: a low excitation region where the flow faces the ambient medium and behind it a high excitation region where the jet feeds into the head. These regions are likely to be the two shocks of the working surface: the bow shock where ambient material shocks, and the Mach disk where jet material shocks. In this case the bow shock is weaker than the Mach disk, because it flows in the wake of a previous working surface, 47D, which has accelerated and compressed the gas through which it passed and into which 47A is now moving (for details, see Reipurth and Heathcote 1991a).

For the outer bow shock, 47D, one sees that it is predominantly of higher excitation, but that also a faint inner [SII] strong shock is visible, suggesting the presence of another Mach disk.

The Herbig-Haro flow provides a record of the recent history of activity of the energy source, which suggests that the source has undergone episodic events of more energetic mass loss. Such events are possibly linked to periods of more active accretion from a circumstellar disk as manifested in the FUor events (see sect. 6). Raga et al. (1990a) modelled a jet produced by a time-variable source and compared with data for HH 46/47; the evidence suggests a recent outflow episode, extending from a thousand years ago to the present, and a previous outflow which started approximately 5000 to 6000 years ago and ended 1500 to 2500 years ago.

5.3   The HH 111 Jet Complex

Perhaps one of the finest collimated HH jets is the HH 111 jet in Orion. It emerges from a dense cloud core in the L1617 cloud, which harbors a 25 $L_\odot$ infrared source (Reipurth 1989c). Fig. 16 shows the whole complex in a large field CCD image. The energy source, a class I source in the classification scheme of Lada (1987), is located at the center of the cloud core, which is outlined by faint Hα emission caused by the OB stars in the Ori Ia association. Light from the source illuminates a conical reflection nebula through which the jet emerges. The flow terminates in a bow shock 142 arcseconds from the source, corresponding to 0.32 pc in projection. On the opposite side of the source, at a separation of 200 and 226 arsec, respectively, two other bow shocks are found. The total extent of the whole complex is more than 0.8 pc in projection. Spectroscopic observations show moderate radial velocities, with the western lobe containing the jet approaching us, and the eastern lobe receding. Proper motions have been determined for the flow, they are large, of the order of 300 km s^{-1} and directed away from the source (Reipurth et al. 1991). Fig. 17 shows the intensity profile of the jet in two Hα images, from 1987 and 1990; the shift due to the proper motion is obvious. From the proper motion and radial velocities one can derive an inclination of the jet to the plane of the sky of only about 10°.

The jet is very bright in [SII], suggesting highly oblique shocks. From the [SII] lines the electron density can be determined, and it is found to decrease steeply with distance from the source. This is often the case for HH jets (Mundt et al. 1987), and in the simplest interpretation suggests a constant mass flow through a slightly widening channel. But the electron density derived from the [SII] lines refer to the recombination region in the post-shock gas. Although a simple relation exists between pre-shock and post-shock density as a function of shock velocity (Dopita 1978a), a jet consists of many shocks, and so the density may become a

524

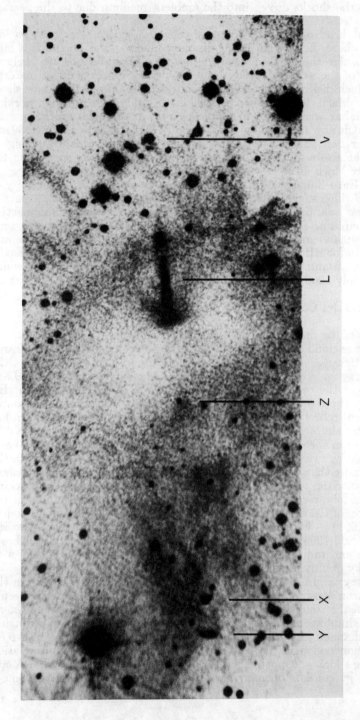

Figure 16. A red broadband CCD mosaic of the HH 111 jet complex, from the ESO 3.6m telescope. The jet emerges from an IRAS source embedded in a cloud core outlined by faint bright rims. Several bow shocks (V, L, X, Y) can be seen on either side of the cloud core. A bright knot (Z) in a very faint counter jet is also visible.

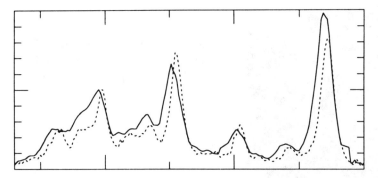

Figure 17. Intensity profile of the HH 111 jet in Hα, observed in 1987 (full line) and in 1990 (dotted line). The driving source is towards the left. The proper motion of individual knots is clearly visible.

complex function of the flow geometry.

An Hα - [SII] difference image of the approaching lobe (fig.18) shows an interesting distribution of excitation conditions. The outer bow shock is seen to consist of an exterior high excitation arc, curving around an interior low-excitation region. This is interpreted as the division of the whole working surface into a bow shock proper, where the ambient gas shocks, and a Mach disk where the jet material shocks (Reipurth et al. 1991). The body of the jet is of low excitation, but begins and ends with a region of high excitation. This region is seen in a contour plot in fig.19. The original well sampled CCD image of seeing 0.6″ has here been deconvolved to an effective seeing of 0.4″. The jet is well resolved and has a width of about 0.8″. It is seen to consist of numerous brighter and fainter knots with a slight wiggling. The top of the jet is clearly extended into another high excitation bow shock very similar to the outer bow shock. From their distances to the source and their space motions, and taking the simplest case of constant motion, one finds that the outflow triggering the outer bow shock was initiated less than a thousand years ago, and only a few hundred years ago for the inner bow shock.

At the base of the flow, the first bright knot is resolved into two side by side knots with a separation of 1.1 arcseconds, corresponding to about 500 AU. No similar feature has been seen in other jets, which makes its interpretation more difficult. It appears significant that the excitation of the double knot is distinct from the rest of the jet, and that the double knot is found just where the jet abruptly emanates from the surface of the cloud. A possible explanation is that the source has had a short period of enhanced mass loss, producing a higher density region travelling away from the source. If this denser region intersects oblique shocks it might form a ring shaped shock, which in projection would be seen as a double knot. The oblique shocks could be those of the "cavity model" of Cantó (1980) or perhaps internal crossing shocks in the jet flow (e.g. Cantó et al. 1989). If this interpretation is correct, one can predict that the double knot should narrow or widen in the next few years, depending on whether the oblique shock is converging or diverging (Reipurth et al. 1991).

A careful analysis with astrometric accuracy of the positions of the knots in

Figure 18. An Hα − [SII] difference image of the blue lobe in the HH 111 jet complex. Hα strong regions are black and [SII] strong regions are white.

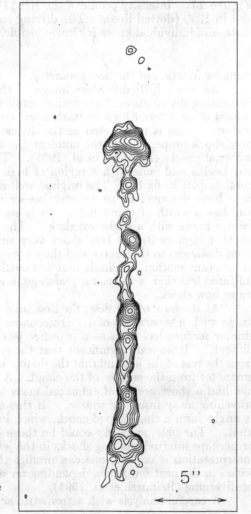

Figure 19. A contour plot of the HH 111 jet, from an Hα CCD image deconvolved to an effective seeing of 0.4 arcsecond. The bow shock where the jet terminates is evident, as is the double knot at the base of the jet.

the Hα and [SII] images shows that almost all knots in the body of the jet have a constant displacement of $0.4'' - 0.5''$, so that the Hα emission is further away from the source than the [SII] emission. This is not what one would expect from a jet with crossing shocks, in fact in the crossing shock cells studied in the steady jet models of Raga et al. (1990b) the higher excitation lines are always formed closer to the source. The observed excitation distribution is more reminiscent of the "internal working surfaces" formed as a result of time variations of the source and studied by Raga et al. (1990a).

The high resolution images of HH 111 demonstrate that by reaching sub-arcsecond resolution and resolving the structures in the jet, it is possible to make comparisons with theory that have a direct impact on our choice of jet model.

5.4 Herbig-Haro Jets and Molecular Outflows

Molecular outflows are abundant in star forming clouds and have an important impact on their surroundings (e.g. Lada 1988, Fukui 1989). Numerous young stellar objects are known which drive both Herbig-Haro objects and molecular outflows. But it is also common to find HH flows with no evidence for molecular outflows, and vice versa. A statistical study of the relationship between these two outflow phenomena is hampered by prohibitive selection effects: molecular outflows are likely to be most common where much of the molecular cloud around a newborn star still remains, but the high obscuration is likely to render HH objects invisible. Conversely, the most extended HH flows lie in the plane of the sky, which make for small projections of the molecular flow velocity vectors onto the line of sight. Moreover, few unbiased surveys for molecular outflows have been made, and most searches are heavily weighted towards high-luminosity embedded infrared sources, only few of which are known to produce HH objects.

Some possible evidence for a direct link between the two phenomena has been given by Fridlund et al. (1984,1989). They find that the L1551 molecular outflow has a sudden velocity shift at the position of HH 29 located in its blue lobe: a blueshifted ^{12}CO wing increases in velocity going from the source to the HH object, but disappears abruptly at the position of the HH object.

Snell et al. (1985) determined the momentum and mass loss rates of the L1551 molecular flow and argued that if only a small fraction of the material of the HH jet is ionized, then this jet may have enough momentum to drive the molecular outflow. However, Snell et al. considered the jet a simple homogeneous cylindrical flow, an assumption which must be reviewed in the light of more advanced jet models. Indeed, Mundt et al. (1987) and Ray et al. (1990) have argued, based on an interpretation of the knots in jets as crossing shocks, that the mass loss and momentum rates are one or two orders of magnitude too small to be responsible for molecular outflows. The small mass loss and momentum rates of these authors arise mainly because they apply a correction for shock compression in the jet flow. But Raga (1991a,b) suggests, based on an analysis of recent jet models (Falle et al. 1987, Cantó et al. 1989, Raga et al. 1990b) that this correction is incorrectly applied, and finds mass loss and momentum rates on average a factor 300 larger than those given by Mundt et al. (1987). For details of the arguments, see Raga (1991b). A central point, as stressed by Raga, is that any determination of mass loss and momentum rates is heavily model dependent.

With the revised mass loss and momentum rates for HH jets, they could, at least in principle, be responsible for the molecular outflows. But it does not necessarily follow that they actually *are* driving these molecular outflows. It could very well be that both HH jets and molecular outflows are different manifestations of the wind emerging from the central source. For a discussion of winds and their relationship to molecular outflows see Natta (1989). It is particularly interesting that Lizano et al. (1988) found a massive, high-velocity neutral wind emanating from the source of the HH 7-11 flow. It is conceivable that the HH jets are only the central, partly ionized core of a more extended neutral wind.

6. Herbig-Haro Energy Sources and Disk Accretion Events

6.1 Surveys and Source Properties

Some of the first identified associations between HH objects and T Tauri stars included T Tau, HL Tau, R Mon and AS 353A (Herbig 1968, 1969a, 1974). Following the discovery of an embedded infrared source next to HH 100 by Strom et al. (1974a), surveys around HH objects at near-infrared wavelengths were carried out, and a number of likely energy sources were identified (Cohen and Schwartz 1980, 1983, Reipurth and Wamsteker 1983). Far-infrared observations with airborne instruments followed (Fridlund et al. 1980, Cohen et al. 1984a,b, 1985). Early suggestions that HH objects themselves were detected at far-infrared wavelengths have not stood the test of time. Finally, IRAS offered a sensitive and unbiased survey so that a majority of HH objects now have a reasonably plausible energy source identified (Cohen and Schwartz 1987, Cohen 1990). The main problem of IRAS was its poor spatial resolution, and when an HH object has no identified source candidate it is most often because it lies in a confused region. The high resolution and insensitivity to extinction make radio continuum observations a potentially interesting survey method, but with present sensitivities only a few sources have been found in this way (see sect. 3.4).

The availability of near-infrared arrays makes surveys much easier than with previous single-beam raster scan techniques. Fig. 20 shows the energy source of HH 83, IRAS 05311-0631, which is embedded in the L1641 molecular cloud, at J (1.2 μm), H (1.6 μm) and K (2.2 μm), from Moneti and Reipurth (1991). The very red and heavily obscured source, which is invisible at optical wavelengths, gradually stands out at increasing wavelengths compared to the foreground star. Such images can also give information on the environment of embedded sources seen in reflected light (e.g. Graham and Heyer 1989, Roth et al. 1989).

The luminosities of HH sources range from very low, e.g. 0.15 $L_\odot$ for Th 28 (Krautter et al. 1984, Krautter 1986, Graham and Heyer 1988) or 0.9$L_\odot$ for DG Tau (Jones and Cohen 1986) or 0.26 $L_\odot$ for HH 55 IRS (Heyer and Graham 1990) to very high, exceeding 10^4 $L_\odot$, as in the Ceph A region and the region around IRc2 in Orion (Axon and Taylor 1984). Recently, HH objects have also been discovered around intermediate luminosity Herbig Ae/Be stars like LkHα 234 (Ray et al. 1990). It thus appears that young stars of all luminosities can produce HH objects. The large majority of sources, however, are low luminosity stars with luminosities in the range 1 to 100 $L_\odot$ (Cohen and Schwartz 1987).

Spectra of HH energy sources can be obtained for those few that are visible, or by observing reflection nebulae near the embedded sources (e.g Cohen et al. 1986). When identifiable, the spectra are usually those of low-mass late-type T Tauri stars, which are almost invariably of the rich emission line advanced type (class 4 and 5

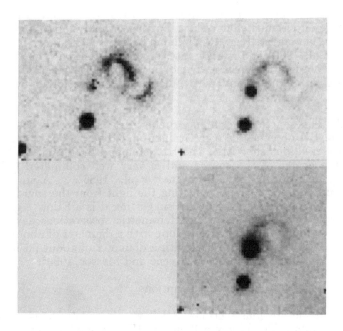

Figure 20. The region of HH 83 observed at three near-infrared wavelengths, J, H and K (clock-wise from upper left). The source gradually becomes more and more visible. A reflection nebula is associated with the source. From Moneti and Reipurth (1991).

of Herbig 1962). A few sources have also been identified as FU Orionis objects, like HH 57 IRS (Graham and Frogel 1985, Reipurth 1985c), L1551 IRS 5 (Mundt et al. 1985, Carr et al. 1987, Stocke et al. 1988) and Z CMa (Hartmann et al. 1989).

6.2 Three well studied Energy Sources

AS 353A This is one of the brightest of the optically visible HH driving sources, and it has been studied extensively with low and high resolution spectroscopy (e.g. Herbig and Jones 1983, Mundt et al. 1983, Hartigan et al. 1986, Böhm and Raga 1987, Eislöffel et al. 1990). The star has a broad Hα emission line with a blueshifted absorption component at -200 km s^{-1}. Also the Sodium doublet shows highly blueshifted absorption cores. The star has a rich emission line T Tau type spectrum, from which high densities in the radiating circumstellar material is inferred, but no stellar absorption lines. It has only moderate reddening, which allows a strong blue continuum to be detected. The M_v of AS 353A is that of a subgiant, and the star definetely lies above the main sequence.

HL Tau Cohen and Kuhi (1979) find this rich emission line star to be heavily veiled, with some indication of an underlying late K spectrum. Its infrared excess together with the presence of very strong ice and silicate absorptions at 3.1μm and 10 µm, respectively, suggested to Cohen (1983) that HL Tau is seen through an almost edge-on circumstellar disk. Further support for this view came from polarization measurements (e.g. Gledhill and Scarrott 1989), near-infrared imaging (e.g. Monin et al. 1989) and speckle interferometry (Beckwith et al. 1984). ^{13}CO interferometer observations (Sargent and Beckwith 1987) revealed a 0.1 $M_\odot$ disk about 100 AU in diameter and perpendicular to the jet found by Mundt et al. (1988).

L1551 IRS 5 This source drives a major molecular outflow (e.g. Snell et al. 1985), and a chain of HH objects (see e.g. Graham and Heyer 1990). VLA centimeter continuum observations show that the HH jet found by Mundt and Fried (1983) extends to the center of the source and is very well collimated (Bieging et al. 1984, Bieging and Cohen 1985, Rodríguez et al. 1986). A small optical and infrared nebula outlines a cavity reflecting the light from the source (e.g. Moneti et al. 1988), which has spectral features suggesting it is a FUor (e.g. Mundt et al. 1985). Millimeter and centimeter interferometric observations have revealed two dust structures around the star, an envelope with a diameter of about 4000 AU and a mass of 2 – 3 $M_\odot$, and a dense disk with a diameter of about 90 AU and a mass of about 0.6 $M_\odot$ (Sargent et al. 1988, Keene and Masson 1990).

6.3   HH objects and FU Orionis Eruptions

As evidenced in the previous sections, HH flows often show signs of being formed in several outbursts. The highly collimated HH jets provide a convenient glimpse of the recent time-history of activity of the underlying source, modulated by the interaction with the ambient medium. The presence of several bow shocks in HH 111 led Reipurth (1989c) to suggest that the energy source had gone through several FU Orionis like eruptions.

FUor events have been studied extensively by Herbig (1966,1977,1989), see also the reviews by Reipurth (1990) and Hartmann et al. (1991) and elsewhere in this book. They have been modelled as disk accretion events by Hartmann and Kenyon (1985,1987,1990 and references therein), who find that the general FUor characteristics can be explained by the accretion of material through a circumstellar disk at rates of the order of 10^{-4} $M_\odot$ yr^{-1}. FUors possess massive, very high velocity winds (Bastian and Mundt 1985, Croswell et al. 1987), which are likely to have a major effect on the ambient medium. An attractive possibility is that FUors are responsible for the HH flows (Dopita 1978b, Reipurth 1985c).

Three FUors out of the less than ten known cases have confirmed HH objects associated. No other class of young stars have anywhere near such a high percentage of association with HH objects. Yet most HH energy sources, when observable, show T Tauri characteristics. This is, however, not surprising when one takes into account the time scales of the phenomena involved. HH flows have typical dynamic time scales of several thousand years. From the limited information available on FUors it appears that they decay on timescales of a hundred to a few hundred years. When observing an HH energy source it is therefore an order of magnitude less likely to be a FUor. A key observation is here that HH energy sources which are T Tauri stars almost inevitably are the active rich emission line objects of Herbig's class 4 or 5. This suggests that a FUor event leaves the star in an elevated state of activity for an extended period. In an HH jet like HH 34 we see the jet extend all the way to the visible source, so the star must still be in the process of creating the jet. The necessary energy is either derived from the tail-end of the accretion event or is gradually being released from the star after it was spun up in the FUor event and stored rotational energy as a fly-wheel. For further details, see Reipurth (1989a).

Other types of observations suggest that there is a direct relation between accretion and mass outflow. Particularly noteworthy is the study by Cabrit et al. (1990), who find a clear relation between Hα and forbidden line emission, which are wind diagnostics, with the excess infrared luminosity arising from circumstellar material.

FUor events are rare, and it has been argued that they are not sufficiently frequent to be responsible for HH flows (Mundt et al. 1990). But it must be kept in mind that FUors are defined by optical characteristics, and we are not yet able to identify embedded stars in FUor stages from infrared signatures alone. In other words, present FUor statistics sample stars which have shed much of their circumstellar material and have become visible. It is entirely possible that accretion events are more frequent when stars are newborn and still shrouded in placental material. Highly luminous infrared sources are not so common, but it is even conceivable that such early eruptions, while more frequent, are also weaker (e.g. L1551 IRS 5 has a luminosity of only 36 $L_\odot$), but well channeled because of the large amounts of circumstellar material. Conversely, the FUor eruptions we observe in visible stars may, despite their violence, have lost some or all of their means to collimate their outflows, and consequently they are less likely to produce HH flows. Clearly, more observations are needed to put such speculations on a firmer footing.

As a class, HH energy sources are among the youngest stars known, and studies of them and their associated HH flows hold the promise of providing crucial insights into the very earliest stellar evolutionary phases.

Acknowledgements

I am grateful to Steve Heathcote and Alex Raga for many discussions and their critical readings of the manuscript.

References

Adams, M.T., Strom, K.M., Strom, S.E.: 1979, *ApJ* **230**, L183
Ambartsumian, V.A.: 1947, *Stellar Evolution and Astrophysics* (Yerevan, Acad. Sci. Armenian SSR.).
Ambartsumian, V.A.: 1954, Comm. Byurakan Obs. No.13
Axon, D.J., Allen, D.A., Bailey, J., Hough, J.H., Ward, M.J., Jameson, R.F.: 1982, *MNRAS* **200**, 239
Axon, D.J., Taylor, K.: 1984, *MNRAS* **207**, 241
Barral, J.F., Cantó, J.: 1981, *Rev. Mex. Astron. Astrofis.* **5**, 101.
Bastian, U., Mundt, R.: 1985, *A&A* **144**, 57
Beckwith, S., Gatley, I., Matthews, K., Neugebauer, G.: 1978, *ApJ* **223**, L41.
Beckwith, S., Zuckerman, B., Skrutskie, M.F., Dyck, H.M.: 1984, *ApJ* **287**, 793
Bieging, J.H., Cohen, M., Schwartz, P.R.: 1984, *ApJ* **282**, 699
Bieging, J.H., Cohen, M.: 1985, *ApJ* **289**, L5
Blondin, J.M., Königl, A., Fryxell, B.A.: 1989, *ApJ* **337**, L37
Blondin, J.M., Fryxell, B.A., Königl, A.: 1990, *ApJ* **360**, 370
Böhm, K.H.: 1956, *ApJ* **123**, 379
Böhm, K.H.: 1983, *Rev. Mex. Astron. Astrofis.* **7**, 55
Böhm, K.H.: 1990, in *Structure and Dynamics of the Interstellar Medium*, eds. G. Tenorio-Tagle, M. Moles, J. Melnick, Springer, p.282
Böhm, K.H., Perry, J.F., Schwartz, R.D.: 1973, *ApJ* **179**, 149.
Böhm, K.H., Schwartz, R.D., Siegmund, W.A.: 1974, *ApJ* **193**, 353
Böhm, K.H., Brugel, E.W., Mannery, E.: 1980, *ApJ* **235**, L137
Böhm, K.H., Böhm-Vitense, E., Brugel, E.W.: 1981, *ApJ* **245**, L113
Böhm, K.H., Brugel, E.W., Olmsted, E.: 1983, *A&A* **125**, 23
Böhm, K.H., Böhm-Vitense, E.: 1984, *ApJ* **277**, 216
Böhm, K.H., Solf, J.: 1985, *ApJ* **294**, 533
Böhm, K.H., Bührke, Th., Raga, A.C., Brugel, E.W., Witt, A.N., Mundt, R.: 1987, *ApJ* **316**, 349
Böhm, K.H., Raga, A.C.: 1987, *PASP* **99**, 265
Böhm, K.H., Solf, J.: 1990, *ApJ* **348**, 297
Böhm-Vitense, E., Böhm, K.H., Cardelli, J.A., Nemec, J.M.: 1982, *ApJ* **262**, 224
Bok, B.J.: 1978, *PASP* **90**, 489
Brand, P.W.J.L., Toner, M.P., Geballe, T.R., Webster, A.S.: 1989, *MNRAS* **237**, 1009
Brown, A., Jordan, C., Millar, T.J., Gondhalekar, P., Wilson, R.: 1981, *Nature* **290**, 34
Brown, A., Millar, T.J., Williams, P.M., Zealey, W.J.: 1983, *MNRAS* **203**, 785
Brugel, E.W.: 1989, in ESO-Workshop on *Low Mass Star Formation and Pre-Main Sequence Objects*, ed. Bo Reipurth, p.311
Brugel, E.W., Böhm, K.H., Mannery, E.: 1981a, *ApJ Suppl.* **47**, 117
Brugel, E.W., Böhm, K.H., Mannery, E.: 1981b, *ApJ* **243**, 874
Brugel, E.W., Shull, J.M., Seab, C.G.: 1982, *ApJ* **262**, L35
Brugel, E.W., Böhm, K.H., Shull, J.M., Böhm-Vitense, E.: 1985, *ApJ* **292**, L75
Bührke, T., Brugel, E.W., Mundt, R.: 1986, *A&A* **163**, 83
Bührke, T., Mundt, R., Ray, T.P.: 1988, *A&A* **200**, 99
Cabrit, S., Goldsmith, P.F., Snell, R.L.: 1988, *ApJ* **334**, 196
Cabrit, S., Edwards, S., Strom, S.E., Strom, K.M.: 1990, *ApJ* **354**, 687
Cameron, M., Liseau, R.: 1990, *A&A* **240**, 409

Campbell, B., Persson, S.E., Strom, S.E., Grasdalen, G.L.: 1988, *Astron. J.* **95**, 1173
Cantó, J.: 1980, *A&A* **86**, 327
Cantó, J.: 1981, in *Investigating the Universe*, ed. F.D. Kahn, Dordrecht: Reidel, p.95
Cantó, J.: 1985, in *Nearby Molecular Clouds*, ed. G. Serra, Lecture Notes in Physics vol. 237, Springer-Verlag, p.181
Cantó, J, Rodríguez, L.F.: 1980, *ApJ* **239**, 982
Cantó, J., Tenorio-Tagle, G., Różyczka, M.: 1988, *A&A* **192**, 287
Cantó, J., Raga, A.C., Binette, L.: 1989, *Rev. Mex. Astron. Astrofis.* **17**, 65
Cantó, J., Raga, A.C.: 1991, in press
Carr, J.S., Harvey, P.M., Lester, D.F.: 1987, *ApJ* **321**, L71
Choe, S.U., Böhm, K.H., Solf, J.: 1985, *ApJ* **288**, 338
Cohen, M.: 1980, *Astron. J.* **85**, 29
Cohen, M.: 1983, *ApJ* **270**, L69
Cohen, M.: 1990, *ApJ* **354**, 701 (erratum: *ApJ* **362**, 758)
Cohen, M., Kuhi, L.V.: 1979, *ApJ Suppl.* **41**, 743
Cohen, M., Schwartz, R.D.: 1980, *MNRAS* **191**, 165
Cohen, M., Bieging, J.H., Schwartz, P.R.: 1982, *ApJ* **253**, 707
Cohen, M., Schwartz, R.D.: 1983, *ApJ* **265**, 877
Cohen, M., Harvey, P.M., Schwartz, R.D., Wilking, B.A.: 1984a, *ApJ* **278**, 671
Cohen, M., Schwartz, R.D., Harvey, P.M., Wilking, B.A.: 1984b, *ApJ* **281**, 250
Cohen, M., Harvey, P.M., Schwartz, R.D.: 1985, *ApJ* **296**, 633
Cohen, M., Dopita, M.A., Schwartz, R.D.: 1986, *ApJ* **307**, L21
Cohen, M., Schwartz, R.D.: 1987, *ApJ* **316**, 311
Cohen, M., Hollenbach, D.J., Haas, M.R., Erickson, E.F.: 1988, *ApJ* **329**, 863
Cohen, R.J., Rowland, P.R., Blair, M.M.: 1985, *MNRAS* **210**, 425
Croswell, K., Hartmann, L., Avrett, E.H.: 1987, *ApJ* **312**, 227
Crusius-Wätzel, A.R.: 1990, *ApJ* **361**, L49
Cudworth, K.M., Herbig, G.H.: 1979, *Astron. J.* **84**, 548
Curiel, S., Cantó, J., Rodríguez, L.F.: 1987, *Rev. Mex. Astron. Astrofis.* **14**, 595
Dopita, M.A.: 1978a, *ApJ Suppl.* **37**, 117
Dopita, M.A.: 1978b, *A&A* **63**, 237
Dopita, M.A., Binette, L., Schwartz, R.D.: 1982a, *ApJ* **261**, 183
Dopita, M.A., Schwartz, R.D., Evans, I.: 1982b, *ApJ* **263**, L73
Doyon, R., Nadeau, D.: 1988, *ApJ* **334**, 883
Dyson, J.E.: 1984, RAL Conf. Proc., *Gas in the Interstellar Medium*, ed. P. Gondhalekar, p.107
Dyson, J.E.: 1987, in IAU Symp. 122 *Circumstellar Matter*, I. Appenzeller and C. Jordan (eds.), Reidel, Dordrecht, p159
Eislöffel, J., Solf, J., Böhm, K.H.: 1990, *A&A* **237**, 369
Elias, J.H.: 1980, *ApJ* **241**, 728
Emerson, J.P., Harris, S., Jennings, R.E., Beichman, C.A., Baud, B., Beintema, D.A., Marsden, P.L., Wesselius, P.R.: 1984, *ApJ* **278**, L49
Falle, S.A.E.G., Innes, D., Wilson, M.J.: 1987, *MNRAS* **225**, 741
Fridlund, C.V.M., Nordh, H.L., Van Duinen, R.J., Aalders, J.W.G., Sargent, A.I.: 1980, *A&A* **91**, L1
Fridlund, C.V.M., Sandqvist, Aa., Nordh, H.L., Olofsson, G.: 1984, *A&A* **137**, L17
Fridlund, C.V.M., Sandqvist, Aa., Nordh, H.L., Olofsson, G.: 1989, *A&A* **213**, 310
Fukui, Y.: 1989, in *Low Mass Star Formation and Pre-Main Sequence Objects*,

ed. Bo Reipurth, p.95
Garden, R.P., Russell, A.P.G., Burton, M.G.: 1990, *ApJ* **354**, 232
Gledhill, T.M., Scarrott, S.M.: 1989, *MNRAS* **236**, 139
Gómez de Castro, A.I., Pudritz, R.E.: 1991, in press
Graham, J.A.: 1987, *PASP* **99**, 1174
Graham, J.A., Elias, J.H.: 1983, *ApJ* **272**, 615
Graham, J.A., Frogel, J.A.: 1985, *ApJ* **289**, 331
Graham, J.A., Heyer, M.H.: 1988, *PASP* **100**, 1529
Graham, J.A., Heyer, M.H.: 1989, *PASP* **101**, 573
Graham, J.A.: Heyer, M.H.: 1990, *PASP* **102**, 972
Gredel, R., Reipurth, B.: 1991, in preparation
Gyulbudaghian, A.L., Glushkov, Yu.I., Denisyuk, E.K.: 1978, *ApJ* **224**, L137
Haro, G.: 1952, *ApJ* **115**, 572
Haro, G.: 1953, *ApJ* **117**, 73
Haro, G., Minkowski, R.: 1960, *Astron. J.* **65**, 490
Hartigan, P.: 1989, *ApJ* **339**, 987
Hartigan, P., Lada, C.J.: 1985, *ApJ Suppl.* **59**, 383
Hartigan, P., Mundt, R., Stocke, J.: 1986a, *Astron. J.* **91** 1357
Hartigan, P., Lada., C.J., Stocke, J., Tapia, S.: 1986b, *Astron. J.* **92**, 1155
Hartigan, P., Graham, J.A.: 1987, *Astron. J.* **93**, 913
Hartigan, P., Raymond, J., Hartmann, L.: 1987, *ApJ* **316**, 323
Hartigan, P., Curiel, S., Raymond, J.: 1989, *ApJ* **347**, L31
Hartigan, P., Raymond, J., Meaburn, J.: 1990, *ApJ* **362**, 624
Hartmann, L., Raymond, J.: 1984, *ApJ* **276**, 560
Hartmann, L., Kenyon, S.: 1985, *ApJ* **299**, 462
Hartmann, L., Kenyon, S.: 1987, *ApJ* **312**, 243
Hartmann, L., Kenyon, S.: 1990, *ApJ* **349**, 190
Hartmann, L., Kenyon, S.J., Hewett, R., Edwards, S., Strom, K.M., Strom, S.E., Stauffer, J.R.: 1989, *ApJ* **338**, 1001
Hartmann, L., Kenyon, S., Hartigan, P.: 1991, in *Protostars and Planets III*, in press
Haschick, A.D., Moran, J.M., Rodríguez, L.F., Ho, P.T.P.: 1982, *ApJ* **265**, 281
Heathcote, S., Reipurth, B.: 1991, in preparation
Herbig, G.H.: 1948, Ph.D. Thesis, University of California
Herbig, G.H.: 1950, *ApJ* **111**, 11
Herbig, G.H.: 1951, *ApJ* **113**, 697
Herbig, G.H.: 1952, *J.R. Astron. Soc. Can.* **46**, 222
Herbig, G.H.: 1962, *Advances Astron. Astrophys.* **1**, 47
Herbig, G.H.: 1966, *Vistas in Astronomy* **8**, 109
Herbig, G.H.: 1968, *ApJ* **152**, 439
Herbig, G.H.: 1969a, *Mem. Soc. Roy. Sci. Liege*, 5th series *19*, 13
Herbig, G.H.: 1969b, in IAU Coll, *Non-Periodic Phenomena in Variable Stars*, ed. L. Detre, Reidel, Dordrecht, p.75
Herbig, G.H.: 1973, *IBVS*, no.832
Herbig, G.H.: 1974, Lick Obs. Bull. No.658
Herbig, G.H.: 1977, *ApJ* **217**, 693
Herbig, G.H.: 1989, in ESO Workshop on *Low Mass Star Formation and Pre-Main Sequence Objects*, ed. Bo Reipurth, p.233
Herbig, G.H., Jones, B.F.: 1981, *Astron. J.* **86**, 1232
Herbig, G.H., Jones, B.F.: 1983, *Astron. J.* **88**, 1040

Heyer, M.H., Strom, S.E., Strom, K.M.: 1987, *Astron. J.* **94**, 1653
Heyer, M.H., Graham, J.A.: 1989, *PASP* **101**, 816
Heyer, M.H., Graham, J.A.: 1990, *PASP* **102**, 117
Hollenback, D.J.: 1985, *Icarus* **61**, 36
Hughes, V.A., Moriarty-Schieven, G.: 1990, *ApJ* **360**, 215
Jones, B.F., Herbig, G.H.: 1982, *Astron. J.* **87**, 1223
Jones, B.F., Cohen, M., Sirk, M., Jarrett, R.: 1984, *Astron. J.* **89**, 1404
Jones, B.F., Walker, M.F.: 1985, *Astron. J.* **90**, 1320
Jones, B.F., Cohen, M.: 1986, *ApJ* **311**, L23
Joy, A.H.: 1942, *PASP* **54**, 15
Keene, J., Davidson, J.A., Harper, D.A., Hildebrand, R.H., Jaffe, D.T., Loewenstein, R.F., Low, F.J., Pernic, R.: 1983, *ApJ* **274**, L43
Keene, J., Masson, C.R.: 1990, *ApJ* **355**, 635
Kim, S.H., Raga, A.C.: 1991, in press
Königl, A.: 1982, *ApJ* **261**, 115
Krautter, J.: 1986, *A&A* **161**, 195
Krautter, J., Reipurth, B., Eichendorf, W.: 1984, *A&A* **133**, 169
Kwan, J., Tademaru, E.: 1988, *ApJ* **332**, L41
Lada, C.J.: 1987, in IAU Symp. no. 115 *Star Forming Regions*, eds. M. Peimbert, J. Jugaku (Reidel), p.1
Lada, C.J.: 1988, in *Galactic and Extragalactic Star Formation*, eds. R.E. Pudritz, M. Fich (Kluwer), p.5
Lane, A.P.: 1989, in ESO-Workshop on *Low Mass Star Formation and Pre-Main Sequence Objects*, ed. Bo Reipurth, 331
Lane, A.P., Bally, J.: 1986, *ApJ* **310**, 820
Lane, A.P., Bally, J., Hartigan, P.: 1991, in preparation
Lee, M.G., Böhm, K.H., Temple, S.D., Raga, A.C., Mateo, M.L., Brugel, E.W., Mundt, R.: 1988, *Astron. J.* **96**, 1690
Lenzen, R.: 1988, *A&A* **190**, 269
Lightfoot, J.F., Glencross, W.M.: 1986, *MNRAS* **221**, 993
Lizano, S., Heiles, C., Rodríguez, L.F., Koo, B.C., Shu, F.H., Hasegawa, T., Hayashi, S., Mirabel, I.F.: 1988, *ApJ* **328**, 763
Lizano, S., Shu, F.H.: 1989, *ApJ* **342**, 834
Luyten, W.J.: 1971, *The Hyades*, Univ. of Minnesota Press, Minneapolis
Malin, D.F., Ogura, K., Walsh, J.R.: 1987, *MNRAS* **227**, 361
Meaburn, J., Dyson, J.: 1987, *MNRAS* **225**, 863
Mendoza, E.E.: 1969, Mém. Soc. Roy, Sci. Liège, Ser.5, 19, 305
Miller, J.S.: 1968, *ApJ* **154**, L57
Moneti, A., Forrest, W.J., Pipher, J.L., Woodward, C.E.: 1988, *ApJ* **327**, 870
Moneti, A., Reipurth, B.: 1991, in preparation
Monin, J.-L., Pudritz, R.E., Rouan, D., Lacombe, F.: 1989, *A&A* **215**, L1
Mundt, R.: 1985, in *Protostars and Planets II*, eds. J. Black and M. Matthews (Tucson, Univ. of Arizona Press), p. 414
Mundt, R.: 1988, in *Formation and Evolution of Low Mass Stars*, NATO ASI vol.241, eds. A.K. Dupree, M.T.V.T. Lago, Kluwer Academic, p.257
Mundt, R., Fried, J.W.: 1983, *ApJ* **274**, L83
Mundt, R., Stocke, J., Stockman, S.: 1983, *ApJ* **265**, L71
Mundt, R., Witt, A.N.: 1983, *ApJ* **270**, L59
Mundt, R., Bührke, T., Fried, J.W., Neckel, T., Sarcander, M., Stocke, J.: 1984, *A&A* **140**, 17

Mundt, R., Stocke, J., Strom, S.E., Strom, K.M., Andersson, E.R.: 1985, *ApJ* **297**, L41
Mundt, R., Brugel, E.W., Bührke, T.: 1987, *ApJ* **319**, 275
Mundt, R., Ray, T., Bührke, T.: 1988, *ApJ* **333**, L69
Mundt, R., Ray, T.P., Bührke, T., Raga, A.C., Solf, J.: 1990, *A&A* **232**, 37
Natta, A.: 1989, in ESO-Workshop on *Low Mass Star Formation and Pre-Main Sequence Objects*, ed. Bo Reipurth, p.365
Neckel, T., Staude, H.J.: 1987, *ApJ* **322**, L27
Noriega-Crespo, A. Böhm, K.H., Raga, A.C.: 1989, *Astron. J.* **98**, 1388
Norman, C., Silk, J.: 1979, *ApJ* **228**, 197
Ortolani, S., D'Odorico, S.: 1980, *A&A* **83**, L8
Osterbrock, D.E.: 1958, *PASP* **70**, 399
Poetzel, R., Mundt, R., Ray, T.P.: 1989, *A&A* **224**, L13
Pravdo, S.H., Rodríguez, L.F., Curiel, S., Cantó, H., Torreles, J.M., Becker, R.H., Sellgren, K.M.: 1985, *ApJ* **293**, L35
Pudritz, R.E., Norman, C.A.: 1986, *ApJ* **301**, 571
Raga, A.C.: 1986, *Astron. J.* **92**, 637
Raga, A.C.: 1988, *ApJ* **335**, 820
Raga, A.C.: 1989, in ESO Workshop on *Low Mass Star Formation and Pre-Main Sequence Objects*, ed. Bo Reipurth, p.281
Raga, A.C.: 1991a, in proceedings of the Third Haystack Conference: *Skylines*, ed. A.D. Haschick, in press
Raga, A.C.: 1991b, in press
Raga, A.C., Böhm, K.H.: 1985, *ApJ Suppl.* **58**, 201
Raga, A.C., Böhm, K.H.: 1986, *ApJ* **308**, 829
Raga, A.C., Böhm, K.H., Solf, J.: 1986, *Astron. J.* **92**, 119
Raga, A.C., Böhm, K.H.: 1987, *ApJ* **323**, 193
Raga, A.C., Mateo, M.: 1987, *Astron. J.* **94**, 684
Raga, A.C., Mateo, M., Böhm, K.H., Solf, J.: 1988, *Astron. J.* **95**, 1783
Raga, A.C., Mateo, M.: 1988a, *Rev. Mex. Astron. Astrofis.* **16**, 13
Raga, A.C., Mateo, M.: 1988b, *Astron. J.* **95**, 543
Raga, A.C., Cantó, J.: 1989, *ApJ* **344**, 404
Raga, A.C., Cantó, J., Binette, L., Calvet, N.: 1990a, *ApJ* **364**, 601
Raga, A.C., Binette, L., Cantó, J.: 1990b, *ApJ* **360**, 612
Raga, A.C., Mundt, R., Ray, T.P.: 1991a, in press
Raga, A.C., Biro, S., Cantó, J., Binette, L.: 1991b, *Rev. Mex. Astron. Astrofis.*, in press
Ray, T.P.: 1987, *A&A* **171**, 145
Ray, T.P., Bührke, T., Mundt, R.: 1988, in *Formation and Evolution of Low Mass Stars*, eds. A.K. Dupree, M.T.V.T. Lago (Kluwer), p.281
Ray, T.P., Poetzel, R., Mundt, R.: 1990, in *Molecular Clouds*, eds. R.A. James, T.J. Millar (Cambridge Univ Press), in press
Ray, T.P., Poetzel, R., Solf, J., Mundt, R.: 1990, *ApJ* **357**, L45
Raymond, J.C.: 1979, *ApJ Suppl.* **39**, 1
Raymond, J.C., Hartigan, P., Hartmann, L.: 1988, *ApJ* **326**, 323
Reipurth, B.: 1981, *A&A Suppl.* **44**, 379
Reipurth, B.: 1985a, *A&A Suppl.* **61**, 319
Reipurth, B.: 1985b, in ESO – IRAM – Onsala Workshop on *(Sub)-Millimeter Astronomy*, eds. P. Shaver, K. Kjär, p.459
Reipurth, B.: 1985c, *A&A* **143**, 435

Reipurth, B.: 1989a, in ESO Workshop on *Low Mass Star Formation and Pre-Main Sequence Objects*, ed. Bo Reipurth, p.247
Reipurth, B.: 1989b, *A&A* **220**, 249
Reipurth, B.: 1989c, *Nature* **340**, 42
Reipurth, B.: 1990, in *Flare Stars in Star Clusters, Associations and the Solar Vicinity*, eds. L.V. Mirzoyan et al., IAU Symposium No.137, p.229
Reipurth, B., Wamsteker, W.: 1983, *A&A* **119**, 14
Reipurth, B., Sandell, G.: 1985, *A&A* **150**, 307
Reipurth, B., Bally, J., Graham, J.A., Lane, A.P., Zealey, W.J.: 1986, *A&A* **164**, 51
Reipurth, B., Graham, J.A.: 1988, *A&A* **202**, 219
Reipurth, B., Heathcote, S.: 1990, *A&A* **229**, 527
Reipurth, B., Olberg, M.: 1991, *A&A*, in press
Reipurth, B., Raga, A.C., Heathcote, S.: 1991, in press
Reipurth, B., Heathcote, S.: 1991a,b, in preparation
Reipurth, B., Bally, J.: 1991, in press
Rodríguez, L.F., Roth, M., Tapia, M.: 1985, *MNRAS* **214**, 9p
Rodríguez, L.F., Cantó, J., Torrelles, J.M., Ho, P.T.P.: 1986, *ApJ* **301**, L25
Rodríguez, L.F., Curiel, S., Moran, J.M., Mirabel, I.F., Roth, M., Garay, G.: 1989, *ApJ* **346**, L85
Rodríguez, L.F., Reipurth, B.: 1989, *Rev. Mex. Astron. Astrofis.* **17**, 59
Rodríguez, L.F., Ho, P.T.P., Torrelles, J.M., Curiel, S., Cantó, J.: 1990, *ApJ* **352**, 645
Roth, M., Tapia, M., Rubio, M., Rodríguez, L.F.: 1989, *A&A* **222**, 211
Różyczka, M., Tenorio-Tagle, G.: 1985, *Acta Astron.* **35**, 213
Sandell, G., Zealey, W.J., Williams, P.M., Taylor, K.N.R., Storey, J.V.: 1987, *A&A* **182**, 237
Sargent, A.I., Beckwith, S.: 1987, *ApJ* **323**, 294
Sargent, A.I., Beckwith, S., Keene, J., Masson, C.: 1988, *ApJ* **333**, 936
Scarrott, S.M., Warren-Smith, R.F.: 1988, *MNRAS* **232**, 725
Schmidt, G.D., Miller, J.S.: 1979, *ApJ* **234**, L191
Schwartz, R.D.: 1975, *ApJ* **195**, 631
Schwartz, R.D.: 1977a, *ApJ Suppl.* **35**, 161
Schwartz, R.D.: 1977b, *ApJ* **212**, L25
Schwartz, R.D.: 1978, *ApJ* 223, 884
Schwartz, R.D.: 1983a, *ARA&A* **21**, 209
Schwartz, R.D.: 1983b, *ApJ* **268**, L37
Schwartz, R.D., Dopita, M.A.: 1980, *ApJ* **236**, 543
Schwartz, R.D., Jones, B.F., Sirk, M.: 1984, *Astron. J.* **89**, 1735
Schwartz, R.D., Dopita, M.A., Cohen, M.: 1985, *Astron. J.* **90**, 1820
Schwartz, R.D., Cohen, M., Williams, P.M.: 1987, *ApJ* **322**, 403
Schwartz, R.D., Williams, P.M., Cohen, M., Jennings, D.G.: 1988, *ApJ* **334**, L99
Simon, T., Joyce, R.R.: 1983, *ApJ* **265**, 864
Solf, J.: 1987, *A&A* **184**, 322
Solf, J.: 1989, in ESO Workshop on *Low Mass Star Formation and Pre-Main Sequence Objects*, ed. Bo Reipurth, p.399
Solf, J., Böhm, K.H., Raga, A.C.: 1986, *ApJ* **305**, 795
Solf, J., Böhm, K.H.: 1987, *Astron. J.* **93**, 1172
Solf, J., Böhm, K.H., Raga, A.C.: 1988, *ApJ* **334**, 229
Solf, J., Böhm, K.H.: 1991, in press

Snell, R.L., Bally, J., Strom, S.E., Strom, K.M.: 1985, *ApJ* **290**, 587
Stocke, J.T., Hartigan, P.M., Strom, S.E., Strom, K.M., Andersson, E.R., Hartmann, L.W., Kenyon, S.J.: 1988, *ApJ Suppl.* **68**, 229
Strom, K.M., Strom, K.E., Grasdalen, G.L.: 1974a, *ApJ* **187**, 83
Strom, S.E., Grasdalen, G.L., Strom, K.M.: 1974b, *ApJ* **191**, 111
Strom, K.M., Strom, S.E., Kinman, T.D.: 1974c, *ApJ* **191**, L93
Strom, K.M, Strom, S.E., Stocke, J.: 1983, *ApJ* **271**, L23
Strom, S.E., Strom, K.M., Grasdalen, G.L., Sellgren, K., Wolff, S., Morgan, J., Stocke, J., Mundt, R.: 1985, *Astron. J.* **90**, 2281
Strom, K.M., Strom, S.E., Wolff, S.C., Morgan, J., Wenz, M.: 1986, *ApJ Suppl.* **62**, 39
Tenorio-Tagle, G., Cantó, J., Różyczka, M.: 1988, *A&A* **202**, 256
Uchida, Y.: 1989, in ESO Workshop on *Low Mass Star Formation and Pre-Main Sequence Objects*, ed. Bo Reipurth, p. 141
Von Hippel, T., Burnell, S.J., Williams, P.M.: 1988, *A&A Suppl.* **62**, 39
Vrba, F.J., Luginbuhl, C.B., Strom, S.E., Strom, K.M., Heyer, M.H.: 1986, *Astron. J.* **92**, 633
Wilking, B.A., Schwartz, R.D., Mundy, L.G., Schultz, A.S.B.: 1990, *Astron. J.* **99**, 344
Yusef-Zadeh, F., Cornwell, T.J., Reipurth, B., Roth, M.: 1990, *ApJ* **348**, L61
Zealey, W.J., Williams, P.M., Sandell, G.: 1984, *A&A* **140**, L31
Zealey, W.J., Williams, P.M., Taylor, K.N.R., Storey, J.W.V., Sandell G.: 1986, *A&A* **158**, L9
Zealey, W.J., Mundt, R., Ray, T.P., Sandell, G., Geballe, T., Taylor, K.N.R., Williams, P.M., Zinnecker, H.: 1989, *Proc. Astron. Soc. Aust.* **8**, 62
Zinnecker, H., Mundt, R., Geballe, T.R., Zealey, W.J.: 1989, *ApJ* **342**, 337

THE PHYSICS OF DISK WINDS

Ralph E. Pudritz
CITA, University of Toronto, Toronto, Ont. M5S 1A1

Guy Pelletier
Groupe d'Astrophysique, Observatoire de Grenoble, Grenoble.

Ana I. Gomez de Castro
Dept. of Physics, McMaster University, Hamilton, Ont. L8S 4M1

1. INTRODUCTION

The last decade of research on star formation has established two important facts, namely that most if not all young stellar objects (henceforth YSO's) are associated with both outflows and disks. The obvious question is whether or not these two phenomena are physically related. The outflow data clearly show that radiation cannot be the driving agent for the outflows, so that hydromagnetic drives that employ rotation and strong magnetic fields are favoured. Since both the central YSO and surrounding accretion disk are natural reservoirs in this regard, the theoretical and observational debate has increasingly turned to discriminating between these two possibilities. If outflows are stellar, then interaction between the wind and a surrounding flaring disk may redirect it into bipolar outflow. If outflows are disk winds however, the connection between outflows and disks is more profound. In the latter case, the outflow is intrinsically bipolar and disk accretion rates and wind mass loss rates are fundamentally connected since the disk angular momentum is carried off by the wind. The mechanical energy of the outflow derives from the gravitational binding energy released by the accretion flow of material in the disk. Moreover, evolution of the disk, a widely held notion, would imply that outflows must also evolve with time.

In this paper, we review the physics of disk winds and show that a hydromagnetic disk wind may be the simplest, self-consistent theory available for understanding a multitude of observed outflow properties. We first briefly outline several of the key observational developments that constrain any theory of outflows from YSOs.

2. OBSERVATIONAL CONSTRAINTS

2.1 Correlation between winds, stars, and disks

Outflow phenomena occur on a multitude of scales around young stellar ob-

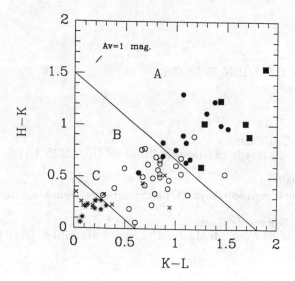

Figure 1. Infrared colours of YSOs with emission lines. Continuum stars and strong [OI] emitters, open circles; normal TTS and faint [OI] emitters, solid circles; LkCa II stars and X-ray sources, solid triangles and squares.

jects as many of the authors in this volume discuss at length. On the largest scales, the molecular outflows (fractions of a pc) have long been known to require something other than a radiative drive. The reason is that the observed CO outflows have several orders of magnitude more momentum than could be provided by the central radiation field of the star (Bally and Lada 1983, Lada 1985, Cabrit and Bertout 1989). The most recent analysis of the observations by Cabrit and Bertout (1989) finds that

$$\frac{F_w c}{L_{bol}} = 400 \left(\frac{L_{bol}}{L_\odot}\right)^{-0.2}$$

$$\frac{L_w}{L_{bol}} = 4 \times 10^{-3}$$

$$v_w = 6 km s^{-1} \frac{L_{bol}}{L_\odot} \qquad (2.1)$$

where the observed thrust in the CO flow is $F_w \equiv \dot{M}_w v_w$ and the mechanical luminosity is $L_w = (1/2)\dot{M}_w v_w^2$. The observations then motivate theoretical models that rely upon pressure gradients or hydromagnetic drives.

Hydromagnetic drives use either magnetic pressure, or centrifugal force in order to accelerate the outflow. Some attempt has been made to find evidence for rotation in the CO outflow as suggested in Pudritz and Norman (1986). To date, there is no completely convincing evidence of rotating molecular outflow, but some authors have argued that it exists (eg. Uchida et al 1987).

The work on forbidden line emission from T-Tauri stars over the last few years has opened up a whole new way of testing theories of star formation and the physics of outflow. The observed correlation between the strength of optical forbidden line emission (eg [OI], [NII], etc) and infrared excess suggests that there is a connection between outflow and disks (Edwards et al 1987, Edwards et al 1989, Cabrit et al 1990). The line emission does not correlate very well with the properties of the central star which suggests that the disk may play a primary role in the outflow mechanism.

Recently Gomez de Castro and Pudritz (1991) have plotted the data on emission line strengths directly in the infrared colour-colour diagram using published infrared data of Rydgren et al (1976), Cohen and Kuhi (1979), Elias (1978), Herbig et al (1986) and Walter et al(1988). Figure 1 shows the colours of strong [OI] (λ6300) emitting stars; faint [OI] emitting stars, and normal TTS; and finally the LkCaII stars and X-ray sources. The figure strikingly confirms the results of earlier workers that strongly embedded sources appear to have intense line emission and optical jets. As one becomes less embedded, these features fade out. Note that the likely range in colour for standard accretion disk models is $K - L \simeq 0.4 - 0.9$ (eg. Bertout et al 1988) which have outflows characterized by weak [OI] line emission according to our diagram.

It has long been known that observed T-Tauri stars do not rotate very quickly (eg Bertout et al 1988, Bouvier 1990). The mean projected rotation speed is a small fraction of the break up speed of a T-Tauri star, being only $< vsini >= 25$ km s^{-1} while $v_{break-up} \simeq 300$ km s^{-1} (Bouvier 1990). This is a very important constraint since hydromagnetic stellar winds require stellar rotation rates to be near break-up if the observed high speed winds are to be produced. The accretion disk material however always does rotate near break-up so that high speed wind can be produced from the inner parts of an accretion disk. The mystery facing all models is how to keep the central T-Tauri star from spinning up due to accretion of gas from the surrounding disk. We (GdeC/P) have looked for a correlation of the rotation speeds of T-Tauri stars with infrared excess and our results are plotted in Figure 2. It is clear that no such correlation exists so that the mechanism for producing forbidden line emission regions is unlikely to be a stellar hydromagnetic wind.

2.2 Observed Disk Properties; Mass and Temperature

Submillimetre observations of young stellar objects constrain the mass of surrounding dusty disks (Beckwith et al 1989, Chini, 1989). The main uncertainties arise from the largely unknown grain properties which could give errors as large as a factor of 10 in the dust emissivity. Observations of 87 pre-main sequence stars in Taurus-Auriga have detected emission in 37 (42%) of the sources. The instrumental sensitivity would allow any disk of mass $M_d > 10^{-4} M_\odot$ to be detected. Disk masses are surprisingly high (by the standard discussion of even five years ago) lying in the range $0.01 \leq (M_d/M_\odot) \leq 1.0$ with a mean value of $< M_d/M_\odot > \simeq 0.02$. The mean disk temperature at 1 AU is 100 K.

The disk temperature as a function of disk radius is inferred from submillimetre spectra with the result that $T(r) = T_o(r_o/r)^{0.5}$ (eg Chini 1989). This is surprising since both standard viscous accretion disk models, or the reprocessing of stellar radiation predict $r^{-3/4}$ laws. While it is possible to carefully arrange a flaring disk in order to reproduce the result, it is equally likely that something more fundamental is going on in these systems. For example, if magnetic fields thread

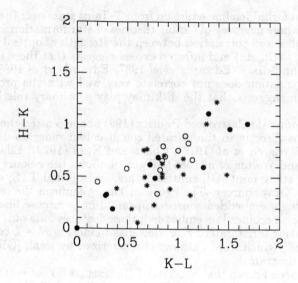

Figure 2. Infrared colours of PMS sources in Taurus with mearsured *vsini*. Asterisks are slow rotators, $vsini < 15$ km s^{-1}; open circles are intermediated rotators, $15 < vsini < 30$ km s^{-1}; filled circles $vsini > 20$ km s^{-1} (values taken from Hartmann and Stauffer, 1989 and Hartmann et al 1986).

through protostellar disks then they can heat disks by the process of ambipolar diffusion. In this mechanism, the magnetic force on ions in a partially ionized gas pushes them with respect to the neutrals. The resulting ion-neutral collisions heat the gas. Thus, at every disk radius, a portion of the gravitational binding energy in a magnetized disk can be carried off as the mechanical energy of the wind, and the remainder is dissipated in the disk as heat. Since the ratio of these two energy losses varies from point to point in the disk in a manner that depends upon the details of the ionization mechanism, one does not expect an $r^{-3/4}$ law in general (He Nan and Pudritz, in preparation).

Having briefly sketched some of the observational reasons for taking disk wind theory seriously, we turn to examine some of the basic physics of this process.

3. 2-D DISK WINDS: BASIC PHYSICS

Winds that originate from the surfaces or coronae of accretion disks are inefficient in carrying off angular momentum if they are not magnetized. Consider a particle at radius r_o near the disk surface. If a purely hydrodynamic wind is at work, then each wind particle only carries off its own angular momentum per unit mass $\Omega(r_o)r_o^2$ where $\Omega(r_o)$ is the local, possibly Keplerian rotation frequency of the disk. In order for a purely hydrodynamic wind to carry off disk angular momentum, the wind mass loss rate $\dot{M}_w(r_o)$ must be comparable to accretion rate $\dot{M}_a$ through the disk. This is of little interest for star formation however, since little growth of the central star would occur.

The presence of a strong magnetic field however, drastically alters the physics of the problem. If the field is strong enough, then it enforces the corotation of the particle with the rotor. Under these circumstances, the angular momentum carried by the particle increases by the square of its lever arm $(r_l/r_o)^2$ compared to particles in the disk. We discuss how this lever arm is calculated later in this section. Thus, hydromagnetic disk winds are highly efficient in the sense that wind mass loss rates can be a tiny fraction of the accretion rates and still effectively transport away disk angular momentum. It is interesting that all systems with measured mass outflow rates and disk accretion rates, such as TTS, and cataclysmic variables (eg Cannizzo and Pudritz 1988) have mass loss rates that are between 0.1 and 0.01 of the mass accretion rate. Similar relations between these rates do not exist in models that feature angular momentum transport in disks by viscous or wave torques since radial and not bipolar angular momentum flux is invoked. We will therefore only concern ourselves with the action of MHD disk wind models in the rest of this review.

The geometry of magnetic fields within protostellar accretion disks is likely to be a consequence of how disks formed within strongly magnetized molecular clouds. Since molecular clouds have apparently well organized, dynamically significant magnetic fields (see eg Moneti et al 1984) collapse along field lines may be how disks form as much early work has always supposed. The size and mass of protostellar disks depends upon how much angular momentum was stripped out of the gas by hydromagnetic waves prior to collapse (see eg Tscharnuter's chapter). If clouds did not possess strong magnetic fields, then it would be possible to generate them within disks by dynamo action (eg Pudritz 1981). The important difference between dynamo generated and threading magnetic field geometry is that the favoured mode generated by dynamo action in a disk is quadropolar (closed). Some form of coronal activity needs to be invoked to open up such structures in order that outflow could be set up. We follow the standard assumption that disks are threaded by largely open magnetic field lines along which the disk originally collapsed. Hopefully this will eventually be accessible to direct observational scrutiny.

3.1 Centrifugally Driven Winds

Figure 3 is a cartoon of the expected magnetic field structure in and near a protostellar accretion disk. The vertical structure of the disk is determined by the condition that gas and magnetic pressure are in hydrostatic balance with the gravitational field of the central star and also the disk if it is sufficiently massive. A magnetic field line threading the disk at a radius r_o emerges from it and into the disk corona and beyond. If the disk has formed by Jeans collapse in the ordered field then the disk has contracted significantly. This pulls in the magnetic field with respect to its ultimate anchor point in the farther reaches of the cloud creating an hourglass-shaped magnetic field structure. The magnetic field in the disk corona is nearly force free, that is magnetic field dominates gas pressure (Blandford and Payne 1982, Pelletier and Pudritz 1991, henceforth BP and PP respectively). The field lines behave dynamically somewhat like rigid wires. Now picture a parcel of gas attached to this field line. If the gas is sufficiently cold, then the only forces acting upon this bead are the gravity $\mathbf{F_g}$ of the central star and the centrifugal force $\mathbf{F_c}$ due to the Keplerian rotation of the particle. Decomposing these into components parallel and perpendicular to the rigid magnetic wire, we see that the centrifugal force presses the particle against the wire at the same time as it accelerates the parcel outwards along it. This outward acceleration is retarded by

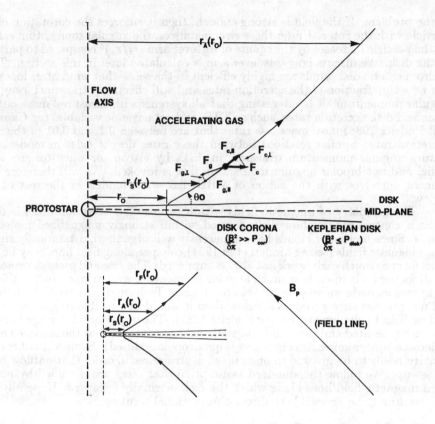

Figure 3. Structure of a protostellar accretion disk, its threading magnetic field, and the associated hydromagnetic disk wind

the relevant component of the gravitational force of the central star (if disks are self gravitating, this of course also contributes). At some point far enough above the disk surface, $F_{c,ll} = F_{g,ll}$. Below this point, the gas is nearly in hydrostatic equilibrium in a disk corona. Beyond this point, an outflow that is centrifugally driven commences. Further along the field line we find two important points in the flow, known as the Alfvén point $r_A(r_o)$ and the fast magnetosonic point $r_F(r_o)$. These are the approximate points at which the gas ceases to accelerate, and where collimation of the flow sets in, respectively.

A few words about the role played by the magnetic field in the drive mechanism are in order. The field is crucial because it enforces the corotation of the accelerating gas with the disk. However, the driving force is centrifugal. It is clear that driving requires the magnetic field to open up at some angle θ from vertical.

In fact, a simple calculation of the bead on a wire (Henricksen and Rayburn 1971) problem discussed above shows that the field line must be bent by at least 30° with respect to the vertical if cold outflow is to occur (BP). We note that if the gas is warm, then this critical angle for the initiation of outflow can be somewhat smaller (PP). Gas in the centrifugal drive can continue to accelerate as long as the wire is effectively rigid. In order to make this precise, we restrict ourselves only to the case of axisymmetric flow (ie independent of ϕ in cylindrical co-ordinates). It is convenient to separate out the toroidal component of all vector under these circumstances, calling everything else the *poloidal* component. Thus the magnetic field is $\mathbf{B} = \mathbf{B}_\phi + \mathbf{B_p}$, and the gas velocity is $\mathbf{v} = \mathbf{v}_\phi + \mathbf{v_p}$. The field is no longer able to effectively enforce co-rotation beyond the point where the energy density in the flow is comparable to the poloidal field energy density; $\rho v_p^2 = B_p^2/4\pi$. The radius that the magnetic field line has at this special point is the Alfvén radius $r_A(r_o)$, which is a function of the position of the footpoint r_o of the field line. Another way of expressing this physics is that when the poloidal flow speed reaches the Alfvén speed, acceleration is nearly over. Another signal speed in a magnetized gas is the fast magnetosonic (henceforth FM) speed, which is the Alfvén speed in the total magnetic field.

It is useful to assign Mach numbers to the ratio between the poloidal speed of the gas, and the Alfvén or FM speeds. These are the Alfvénic and FM Mach numbers;

$$m \equiv \frac{(4\pi\rho)^{1/2} v_p}{B_p}$$

$$n \equiv \frac{(4\pi\rho)^{1/2} v_p}{(B_p^2 + B_\phi^2)^{1/2}}. \tag{3.1}$$

Flows typically become strongly super-Alfvénic $m >> 1$ beyond the Alfvén surface defined by $m = 1$. The FM point is achieved when $n = 1$ and under some circumstances can become super FM. Note that always $n < m$.

There is one final signal speed in a magnetized gas that is the analogue of the sound speed, namely, the slow magnetosonic speed. The point along a field line at which this speed is achieved is the slow point $r_s(r_o)$. For cold MHD flow (pressure unimportant for accelerating the gas) this point occurs very near the disk. Thus, for cold MHD flows which are to be expected for disk winds, the ordering of the various *critical points* in the wind (points at which the flow achieves each of the three types of signal speed) is

$$r_o \lesssim r_s(r_o) << r_A(r_o) << r_F(r_o)$$

Moreover, one finds that gravity, while important for determining gas properties at the slow point, is unimportant at the Alfvénic and FM points.

3.2 Conservation Laws: Flow Along Field Lines

The most important and useful insights into complicated physics problems arise from the use of conservation laws. It is fortunate that if one assumes that flows are steady (time independent) and axisymmetric, that very powerful relations between the magnetic field and velocity of the gas can be derived. If we could ignore forces perpendicular to the magnetic field lines, these conservation laws would suffice to solve for the flow. This 1-D theory is the basis of several of the classic analysis of MHD winds that have appeared in the early literature of the subject.

One begins with Maxwell and Faraday who tell us that a magnetic field can be induced by the curl of the electric field. In a moving conductor, $\mathbf{E} = -\mathbf{v} \times \mathbf{B}/c$ so that in steady state the velocity field and the magnetic field are linked;

$$\nabla \times (\mathbf{v} \times \mathbf{B}) = 0 \tag{3.2}$$

The remaining easy relations are that mass and magnetic flux are conserved,

$$\nabla \cdot (\rho \mathbf{v}) = 0 \tag{3.3}$$

$$\nabla \cdot (\mathbf{B}) = 0 \tag{3.4}$$

Assuming henceforth that flows are axisymmetric, the conservation of magnetic flux shows that the poloidal magnetic field can be derived from a single scalar potential which is just the poloidal magnetic flux;

$$\mathbf{B_p} = \frac{1}{r} \nabla a \times \hat{\phi} \tag{3.5}$$

These ideal MHD equations lead to 2 basic relations linking velocity and magnetic field. As first shown by Chandrasekhar (1956), Mestel (1961), and Schatzman (1962), these three results show that the poloidal velocity and magnetic field are parallel to one another, ie

$$\rho \mathbf{v_p} = \frac{\kappa(a)}{4\pi} \mathbf{B_p} \tag{3.6}$$

where $\kappa(a)$ is as yet an unknown function of the flux. The result shows that field and streamlines are contained in surfaces $a = const$.

The second feature that drops out of the induction equation is a relation between the toroidal magnetic and velocity fields.

$$B_\phi \mathbf{v_p} = [\Omega - \Omega_K(a)] r \mathbf{B_p} \tag{3.7}$$

Here, Ω is the angular velocity of the gas at any point in the flow, while $\Omega_K(a)$ is that of the rotor and as such must be a function of a. This result shows that we expect the toroidal speed to be small at the rotor, but it may increase as one moves out along a magnetic surface. Another relation amongst the toroidal fields can be derived from the angular momentum equation (toroidal component of the steady state equation of motion), namely, that the total specific angular momentum of the gas is comprised of a material and magnetic contibutions, and that their total is conserved along each magnetic surface $a = const$,

$$l(a) = \Omega r^2 - \frac{rB_\phi}{\kappa} = const \equiv \Omega_K r_A(a)^2 \tag{3.8}$$

We may algebraically solve equations (3.7) and (3.8) for the toroidal field components. As has been known for a long time, these solutions are singular at the Alfvén surface unless we put a condition on the unknown function,

$$\kappa(a) = [4\pi \rho_A(a)]^{1/2} \tag{3.9}$$

where $\rho_A(a)$ is the density of the gas at the Alfvén surface. This exercise has taught us that by ensuring the regularity of solutions through critical points, we can constrain the flow by determining unknown functions. Thus, the conservation laws allow us to solve for the angular velocity of the flowing gas as well as its toroidal field;

$$\Omega = \Omega_K[1-g]$$
$$B_\phi = -\kappa \frac{\rho}{\rho_A}\Omega_K r g \qquad (3.10)$$

where g is a monotonically varying function that increases from the rotor to infinity and $0 < g < 1$ (see PP). These results show that as one moves farther from the axis along a field line, the angular velocity of the gas decreases while the toroidal magnetic field increases. Note however that while the material angular velocity decreases, the actual angular momentum per unit of mass of the gas *increases* since angular momentum is extracted from the disk by the MHD torque. The flow angular momentum equation (3.8) shows that angular momentum in the gas and field are juggled between one another in such a way that the total angular momentum on that field line is precisely constant.

The final conservation law is of course, energy conservation along a surface of $a = const$, ie the Bernoulli equation

$$E(a) = (1/2)(v_p^2 + \Omega^2 r^2) + \Phi + h + \Omega_K(\Omega_K r_A^2 - \Omega r^2) \qquad (3.11)$$

Note that because we were able to solve for the magnetic fields in terms of the velocity field, there is no explicit appearance of the magnetic field in the equation although the last term is the magnetic contribution.

We now turn to reviewing known properties of flows as found using assumptions about their underlying geometry.

3.3 1-D Solutions

In a 1-D flow problem, the various dynamical variables are functions of the the single variable s; the scalar distance along the streamline or flux surface. As an example, if the magnetic geometry is known then we can solve the flow using the equations above. The basic equations are the conservation laws derived in the previous subsection plus the condition that the flow passes through the 3 critical points r_s, r_A and r_F. For cold MHD flows, pressure is unimportant for driving the gas, and the condition on r_s is trivially satisfied. In order to solve the equation for the velocity of the gas, one must input conditions at the boundary r_o; $B(r_o)$, $\Omega_K(r_o)$, and $\rho_K(r_o)$. Figure 4 shows the solutions of a flow that is warm, ie , for which pressure is also an important constituent, taken from the pioneering work of Weber and Davis (1967). Their paper, entitled the *Angular Momentum of the Solar Wind*, examines the flow of gas in a radial magnetic field geometry. They found that there are two branches of solutions after the flow has gone through all the critical points. One of these is supersonic and super Alfvénic , while the other (magnetic analogue of Parker's "breeze" solution) is subsonic.

A fundamental result of 1-D analysis of MHD winds was obtained by Michel (1969), who found that the terminal speed along a flow line (essentially the Alfvèn point) was simply related to the rotation speed of the rotor, and the lever arm;

$$v_\infty = (2/3)^{1/2} v_{rot}(r_A/r_o)$$

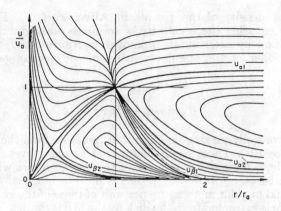

Figure 4. The Weber and Davis solution (1967) for hot MHD wind in a radial stellar magnetic field (1-D). Note that two branches are possible beyond the Alfvén point; one is super-Alfvénic, and the other is subsonic.

By combining the two sides of this result, one has a parameter, the so-called *Michel parameter*, which can be considered a constant on any particular field line. Pudritz and Norman (1986) made use of this in their analysis of disk winds. In fact, particular solutions of 2-D wind theory known as self-similar solutions, also have this property as we shall see.

3.4 2-D Solutions

The difficulty in solving a 2-D, MHD flow is that forces are transmitted at point to point in directions normal to any given magnetic surface. As we've seen, the accelerating gas pushes normal to such surfaces. A more important effect is that the magnetic pressure in the outflow is one of the major decollimating agents. Thus, even in a stellar wind problem, it cannot be assumed that the field lines will remain strictly radial. The flow problem is 3-D in general since we have to self consistently solve for the magnetic structure and the flow along it. If axisymmetry is assumed, then it is best to find flux surfaces $a = const$ where the scalar potential is a function of the cylindrical co-ordinates (r, z).

How then are flows collimated in general? Near to the disk, but not within it, the magnetic field is essentially force free. This situation arises because the gas in a disk corona is so dilute that neither gas pressure gradients nor gravity are comparable to magnetic forces. Collimation of the flow in this regime is a struggle between magnetic pressure and centrifugal force of the outflowing gas with the constraining magnetic tension. *Within* the disk, the magnetic field must participate in hydrostatic balance and the configuration is not force-free. Over a pressure scale height the direction of the magnetic field will change and can become vertical at the midplane. This must be so since for hydrostatic balance to be possible in the disk, its magnetic pressure cannot exceed disk gas pressure, otherwise the disk field becomes buoyant. Far from the disk, the magnetic field becomes largely toroidal. The magnetic pressure is overcome by the magnetic hoop stress applied through the toroidal field component.

The additional equation that must be solved for the 2-D problem is the component of the equations of motion that is perpendicular to the magnetic flux surface at each surface. In general, this is a technically difficult problem. The approach taken by most authors has been to look for self-similar solutions, that is, axisymmetric solutions that have no intrinsic scale (eg. Chan and Henricksen 1980 for the theory of radio jets, BP for disk winds, and Konigl 1989 for disk winds from protostellar disks). This assumption has the advantage of mathematical simplicity which can act as a guide for more general solutions. It must be noted however, that there is no good physical reason to think that disk winds are self-similar. Just because quantities have power law behaviour with radius on the disk, it does not follow that the flow need be self-similar. PP have uncovered a rich class of solutions which do not rely on this assumption. The most daunting prospect is to face the full 2-D problem. The force balance perpendicular to magnetic flux surfaces is governed by a form of the Grad-Shafranov equation, known to plasma physicists interested in magnetic confinement in Tokomaks.

Recently some authors have questioned the ability of disk winds to operate in protostars (eg. Lada and Shu 1990). One of the issues is whether or not the magnetic field lines that emerge from a disk bend sufficiently away from the flow axis that a centrifugal wind can really get started (simple arguments in BP show the angle between field and the normal to the disk must be greater than $30°$). Since the magnetic field that threads the disk must be perpendicular to the disk midplane, these authors argue that a disk corona does not provide enough pressure to push out the field lines to opening angles large enough for the outflow to get started. While the argument is correct for a disk corona, it ignores the fact that *the magnetic pressure within the disk must be less than or equal to the disk gas pressure*. The field in the disk therefore is significantly deflected from the vertical because it is plays an intimate role in establishing vertical hydrostatic balance of the disk. Another point is that the BP constraint is only for a cold flow. A flow can be initiated from a sufficiently warm corona even if the field lines are bent less than $30°$ from the vertical (PP). A second issue that has often been raised about disk wind models (Pudritz and Norman, 1986) is their seeming reliance on a sufficiently massive rotor. Current observations seem to find smaller disk masses. In this regard, we emphasize that the basic mechanism of disk winds is the same whether one has massive, or massless (Keplerian) disks, namely, gravitational binding energy released by the accreting gas is converted to mechanical energy of the flow. If the underlying disk is not a massive rotor but a Keplerian accretion disk, then the energy that is carried by the outflow reflects the accretion rate through the disk and the depth of the gravitational potential well created by the central star. Since Pudritz and Norman (1986) investigated the case of massive disks, we accordingly discuss only Keplerian disks.

3.5 Blandford and Payne Similarity Solutions

Given the central role that these solutions have had in the subject of disk winds, we emphasize a few essential points. A self-similar model means that the

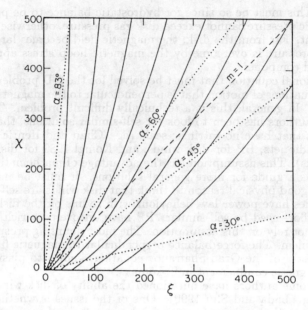

Figure 5a. Shape of the field lines and Alfvén surface ($m = 1$) for self similar wind in the BP theory (reproduced from BP with permission).

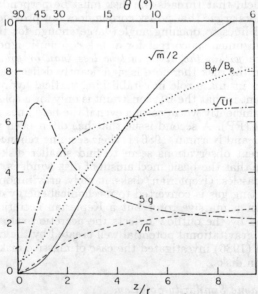

Figure 5b. Various dynamical quantities in the BP theory along a single magnetic surface $a = const$ in the flow, for standard parameters. Note $g \equiv v_\phi/v_K(r_o)$.

physics of the wind does not change as one moves out across the disk or flow. This immediatly tell us that the ratio of the terminal speed of the flow to the rotor speed must scale with the lever arm, and nothing more. Thus the Alfvén surface must be

$$\frac{v_A(r_o)}{v_K(r_o)} \propto \frac{r_A(r_o)}{r_o} = const \qquad (3.12)$$

ie., the Alfvén surface is a cone. In this solution, the outflow has the highest terminal flow on the axis because it originates from the most quickly rotating part of the accretion disk. The distribution of magnetic flux on the disk must be very special in order to have a self-similar result, $B(r_o) \propto r_o^{-5/4}$ so that the magnetic flux distribution on the disk is $a(r_o) \propto r_o^{3/4}$. Any other power laws would not give self-similar solutions. The field lines are reproduced in Figure (5a) from BP.

The BP solutions theory has two important control parameters that completely fix the solution:
1. dimensionless magnetic flux parameter which measures how much magnetic flux threads the disk,

$$\kappa_{BP} \equiv (1 + \zeta_o'^2)^{1/2} \frac{\kappa(a) v_K(r_o)}{B(r_o)} = const \qquad (3.13a)$$

where $\theta_o \equiv cotan^{-1}\zeta_o'$ is the angle of the poloidal field with the disk as it leaves the disk surface; 2. dimensionless angular momentum parameter

$$\lambda_{BP} = \frac{l(r_o)}{l_K(r_o)} = const \qquad (3.13b)$$

BP take as standard parameter values for extragalactic radio jets $\kappa_{BP} \simeq 0.03$ and $\lambda_{BP} = 30$ with a jet opening angle $\theta_o = 6°$. For these values the lever arm $r_A/r_o = 5$ so that the terminal speed along each field line is 5 times the Kepler rotation speed of the field line at the point that it threads through the accretion disk. Figure 4 in BP is reproduced in Figure (5b), and it illustrates the points made in our general remarks. The Alfvénic Mach number m achieved in these cold flows with the standard parameters is 400, while the terminal FM Mach number $n_\infty \simeq 6$. In addition the field becomes tightly wrapped with $B_\phi/B_p \simeq 8$. A final important point is that the rotation speed of the accelerating gas never achieves high values with asymptotic values being only $v_\phi/v_K \simeq 0.4$. Thus the accelerated gas rotates less quickly than it did at its source on the disk.

3.6 Magnetic Collimation of Outflows

Optical jets are highly collimated outflows and no theory can be complete without a mechanism that explains this. Much of the physics of jet collimation has already been discussed in the context of extragalactic radio jets (eg. see review Rees 1984). Here we only note that magnetized jets have an intrinsic collimation mechanism, namely, the pinching stress due to strong toroidal fields in the jet. Far away from the source of the flow, we have seen that MHD winds have magnetic field that is predominantly toroidal. The magnetic force $\propto \mathbf{J} \times \mathbf{B}$ is directed *radially inwards* since $\mathbf{J} \propto \nabla \times \mathbf{B}$ (Maxwell) is predominantly along the axis of the jet under these circumstances. Thus, independent of any external agent such as the gas pressure of the surrounding core or molecular cloud through which a jet is propagating, flow collimation is possible.

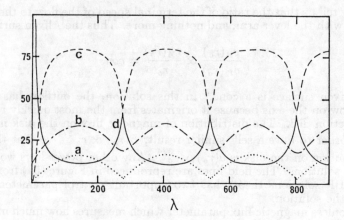

Figure 6a. Oscillations in the radius of a self-similar jet solution in the presence of external pressure (from Chan and Henricksen 1980).

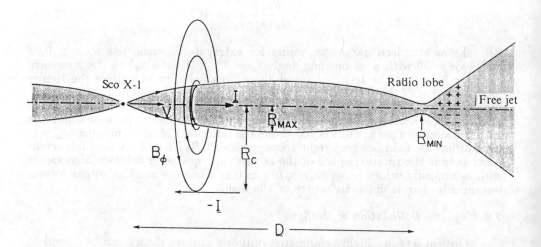

Figure 6b. Recollimation of self similar MHD jets illustrating the presence of a magnetic focal point (from Achterberg, Blandford, and Payne 1983, reproduced with permission)

Self similar solutions to MHD flow equations have shown that collimation is a generic property of MHD flows. Chan and Henricksen (1980) and BP both found that flow collimates to a cylinder, and can even "overdo" it and recollimate towards the axis. In the former paper, oscillations in disk radius depend upon the mismatch between the jet and the external pressure and the solutions are reproduced in Figure 6a. BP found that independent of the external pressure, flows with n higher than some critical FM number n_T recollimate towards the axis while for $1 \leq n \leq n_T$, cylindrical collimation is achieved.

Recollimation of flows is an extremely important property because shocks can develop at these magnetic "focal points". The speed at which a flow collimates itself perpendicular to the flow axis is the Alfvén speed in the toroidal field. The focal distance of the flow is then easily estimated (Achterberg, Blandford, and Goldreich 1983) by taking the ratio of the flow speed v_p to the transverse collimation speed. If D is the distance from the disk to the focal point, and r_{max} is the maximum radius of the jet ($r_{max} \geq r_F$), then

$$\frac{D}{r_{max}} \simeq 2 \cdot \frac{(4\pi\rho)^{1/2} v_p}{B_\phi} \leq 2n \qquad (3.14)$$

A cartoon of this physics is shown in Figure 6b which is reproduced from Achterberg et al. We return to these important scalings in the last section of this review.

3.7 Historical Interlude

Much of the content of the previous discussion of the MHD equations was developed in the 50's, 60's and 70's by theorists interested in MHD winds from magnetized stars. It is interesting to chart how these ideas appeared in relation to other important advances in astronomy over the last decades. In 1958 Parker wrote the seminal paper on the theory of hydrodynamic, 1-D winds which he applied to the sun. Within a few years, the great potential for hydromagnetic winds to strongly brake rotating stars was realized by Mestel (1961) and Schatzman (1962) and the first quantitative treatment of 1-D MHD winds was made by Weber and Davis (1967). Note that just prior to this work, quasars and pulsars were discovered. The first serious attempt at a 2-D stellar hydromagnetic wind theory was made by Pneumann and Kopp (1971) and Okamoto (1974, 1975) although Mestel (1968) had made a start on it. Pulsars had by now captured the attention of theorists and Michel (1969) extended the Weber and Davis 1-D wind ideas to relativistic wind torques on fast rotating neutron stars. Radio jets and quasars were extensively discussed in the theoretical literature, stimulated by the pioneering work of Rees (see his 1984 review) on the nature of extragalactic radio jets. Blandford and Znajek (1977) extended the theory of hydromagnetic winds from rotating stars to the ultimate in exotica; namely rotating black holes. In 1982, BP, realizing the importance of disks in the flow and collimation problem, proposed that extragalactic jets could be formed and collimated if they originated from disks around black holes. The magnetic fields in such quasar disks could be generated by dynamo action within them (Pudritz 1981).

The history of bipolar outflows from young stellar objects mirrors the forgoing history in some interesting ways. Beginning with the seminal discovery of Snell, Plambeck, and Loren (1980) that bipolar flows exist, the field rapidly developed two competing ideas. Hartmann and MacGregor (1982) argued that the outflow

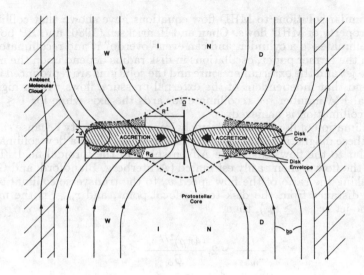

Figure 7a. A centrifugally driven disk wind drives accretion flow through the underlying accretion disk since it carries off disk angular momentum (from Pudritz 1985).

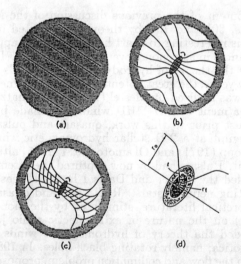

Figure 7b. Torsional Alfvén waves generated in a radially collapsing disk drives off a low speed bipolar outflow (Uchida and Shibata 1985, reproduced with permission).

could originate from a fast rotating magnetized protostar whose MHD wind was redirected into bipolar flow by a surrounding disk. Draine (1983) argued that the field lines threading a rapidly rotating protostar would twist up and that the resulting flux of torsional Alfvén waves would push a bubble into the surrounding molecular gas. Konigl (1982) invoked a pure stellar wind focussed by the de Laval mechanism. The problem with hydromagnetic stellar wind models for YSO's is that the observed stars at least, rotate far too slowly. Pudritz and Norman (1983) therefore proposed that the winds originated from the surrounding disks. This was meant to be a prediction for the necessity of disks around YSO's since at the time, there was no or very little evidence for disks. In a series of papers (Pudritz 1985, 1988; Pudritz and Norman 1986) it was established that the extraction of angular momentum from the surrounding disk by the wind drives an accretion flow that can result in the formation of the central star. The theory linked outflow and accretion in a fundamental way (see Figure 7a). This is the aspect of disk winds that has no counterpoint in stars. Further developments in disk wind theory were made by Sakurai (1985), and Lovelace et al (1987). Konigl (1989) extended the BP theory by invoking self-similar accretion disks as the source of a BP wind. Pelletier and Pudritz 1991 have developed a truly 2-D theory of MHD disk winds which have the BP theory as a special case. Some work on the idea of hydromagnetic winds from protostars is also being pursued (Shu et al 1988).

A different variant of disk wind theory has been advocated by Shibata and Uchida over the last 5 years. If one assumes the disk is not in steady state but is collapsing radially, then these authors showed that it is possible to generate a very strong toroidal field. In their models, these authors begin with an open uniform magnetic field that threads through the disk. The disk is then allowed to collapse radially inwards. Since the disk is rotating, the threading field is wound up in the plane of the disk. The only way the field can relax in this scheme is that a torsional (Alfvén) wave propagates away from the disk along the field lines. The effect is somewhat like a coiled spring that pushes gas away in the vertical directions (Figure 7b). The collimation of the flow is actually intrinsic to the initial conditions since the original field is taken to be straight and perpendicular to the disk surface. The ejection velocity these authors find is $\simeq 1 - 2V_K$ where V_K is the Kepler velocity at the inner edge of the initial disk (Shibata and Uchida 1986). This should be compared with the result (3.12) for centrifugally driven winds where the lever arm $r_A/r_o \geq 5$ for flows studied by BP and Pudritz and Norman (1986). As this theory is numerical in nature, it is still difficult to say whether the flow studies by Uchida and Shibata settles down into centrifugally driven wind once equilibrium of the disk is achieved. The appearance of additional sophisticated MHD numerical calculations (eg Stone 1990), will allow some of these questions to be studied exhaustively.

Having reviewed the basic issues of MHD disk winds, we turn to examine the crucial physics of angular momentum extraction from accretion disks.

4. ANGULAR MOMENTUM EXTRACTION FROM DISKS

We have examined the angular momentum equation of the wind, but since the wind originates from the disk surface, we must in principle examine the disk angular momentum equation as well. This new boundary condition gives a very powerful, general character to the theory since it will turn out that the details of mass loss from the disk are immaterial for the problem. The disk angular momentum equation

is
$$\nabla \cdot (\rho \mathbf{v_p} \Omega r^2 - \frac{rB_\phi \mathbf{B_p}}{4\pi}) = \nabla \cdot (\tau_{\text{visc}}) \tag{4.1}$$

It is simple to show from this result that the magnetic torque in any reasonable magnetized disk dominates viscous torque. If one scales the viscous torque as Shakura and Sunyaev did (1973) then $\tau_{visc} \simeq \alpha r p$ where p is the gas pressure in the disk and $\alpha \leq 1$ is the famous parameter. It follows that

$$\frac{MHD \ Wind \ Torque}{Viscous \ Torque} = \frac{B_z^2}{4\pi p} \frac{r_A}{\alpha H} \tag{4.2}$$

where H is the disk half thickness. Thus if the threading field is roughly in equipartition with the gas pressure in the disk, then the wind torque is overwhelmingly stronger than viscous torque. The reason is simply that the former has a lever arm of r_A while the latter has a lever arm of only αH! Thus disk winds likely play the major role in fixing disk structure through the steady state angular momentum equation. They would also predict different density and temperature profiles throughout the disks in principle. This would not be inconsistent with the observations (see section 2.)

By vertically averaging equation (4.1) and ignoring the contribution of viscous torque one finds a much more physical form of the equation;

$$\dot{M}_a \frac{d(\Omega_K r_o^2)}{dr_o} = 2\frac{d\dot{M}_w}{dr_o}(l(a) - \Omega_K r_o^2) \tag{4.3}$$

(see Pudritz 1985, PP). In steady state, $\dot{M}_a$ is a constant, and for a Keplerian rotation curve for the disk, we see that the wind mass loss rate of the disk from every radius is *fixed* by the disk accretion rate and the lever arm of the flow. *The details of the mass loss mechanism are unimportant in steady state!*

The important fact that the details of the mass loss mechanism do not matter has a precedent if we consider the problem of steady state viscous disks. The viscous torque in steady state disks scales as the product of the disk surface density and the disk viscosity $\nu \Sigma$. In order that the torque carry off the disk angular momentum in a Keplerian disk therefore, the disk angular momentum equation constrains Σ if the viscosity parameter is given. For wind torques, knowledge of the lever arm fixes the disk wind mass loss rate in just the same way. Thus, the boundary condition provided by steady state transport in the disks allows the disk wind mass loss rate to be found at every disk radius. The incremental mass loss from an annulus of width dr_o at a radius r_o in the disk is found by a slight rearrangement of equation (4.3), ie;

$$d\dot{M}_w = -\frac{\dot{M}_a r_e^2}{6\Omega_K^{7/3} r_A^2} d\Omega_K \tag{4.4}$$

In order of magnitude, the results are that $\dot{M}_a \simeq (r_A/r_o)^2 \dot{M}_w$ so for a lever arm of 10 the wind mass loss rate from a disk is merely 10^{-2} of the accretion rate. We have heard several speakers suggest that accretion rates in young stellar objects lie in the range $10^{-6} - 10^{-7} M_\odot yr^{-1}$. These accretion rates can be provided by disk mass loss rates of only $10^{-8} - 10^{-9} M_\odot yr^{-1}$. Thus, star formation can certainly proceed by wind driven, accretion flow through disks.

We turn now to examine the physics of disk winds when the artificial constraints of self similarity are removed.

5. BEYOND SELF SIMILAR SOLUTIONS

In this section we highlight the basic approach and results of Pelletier and Pudritz (1991) on 2-D disk winds. The requirements are that
  1. energy conservation (Bernoulli-equation) and force balance between neighbouring magnetic surfaces (Grad-Shafranov equation) be maintained at every point in the flow,
  2. the solutions flow smoothly through the three critical surfaces, r_s, r_A, and r_F,
  3. steady state, axisymmetric solutions are sought, and
  4. the disk angular momentum equation is satisfied, and the disk is assumed to be Keplerian (it is trivial to use any rotation curve we wish).

5.1 Control Parameters

As usual, one casts the equations in dimensionless form; a procedure that allows one to isolate the basic controlling parameters in the problem. Writing the poloidal potential as $a(r,z) = a_e \psi(r,z)$ where a_e is the flux through the outermost radius r_e of the disk ($v_{K,e}$ is the Keplerian velocity of material at this radius), and B_e as the magnetic field strength of the disk field at r_e, a fundamental control parameter measuring the magnetic flux through the disk is

$$\epsilon \equiv \frac{\dot{M}_a v_{K,e}}{3 r_e^2 B_e^2} \quad (5.1)$$

For a solar mass star and an assumed disk radius $r_e = 100 AU$, magnetic field $B_e = 10^{-3}$ G, and accretion rate of 3×10^{-7} M$_\odot$ yr^{-1}, the value of this parameter is $\epsilon = 0.84$. We have isolated a second crucial quantity related to the Michel parameter found in 1-D. On making the Grad-Shafranov equation dimensionless, one encounters the parameter

$$\lambda \equiv \frac{\Omega_K^2 r_A^2}{V_A^2} \mid \nabla r_A \mid_A^2 \quad (5.2)$$

which is basically the Michel parameter (see section 3.3) for full 2-D flow. In this expression, V_A is the particle speed at the Alfvèn surface, and the gradient of r_A is to be evaluated at the Alfvén surface. Since the Bernoulli equation tells us that the terminal speed of the wind along any flux surface is $v_{p,\infty} \simeq 2^{1/2} \Omega_K r_A$, then one anticipates that λ will be nearly constant everywhere in the flow. In fact, we assume henceforth (as in PP), that $\lambda = const$. We note that the self-similar theory of BP has this property, but we will find that a much broader class of solutions does as well.

There is still one unknown function in the problem, namely, the distribution of magnetic flux across the underlying accretion disk. In general, one should solve disk structure equations and then match onto the disk wind solution in order to self consistently determine this. It is convenient to suppose that this is a power law in radius, ie

$$\psi = \omega^{-1/\alpha} \propto r_o^{3/2\alpha} \quad (5.3)$$

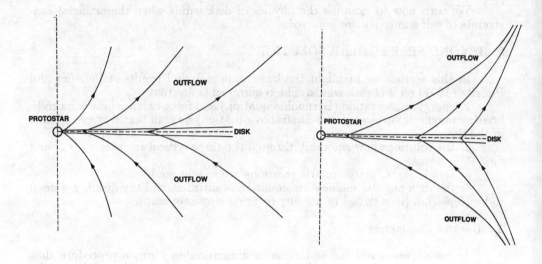

Figure 8. Magnetic field and Alfvén surface geometries for non-self similar models: lotus ($\alpha < 2$), and cylindrical sheath ($\alpha > 2$) respectively.

where the dimensionless angular velocity is $\omega \equiv \Omega/\Omega_{K,e}$ and where we note that the self-similar model is for $\alpha = 2$. For aficionados, we note that we can readily translate our parameters ϵ and λ into their BP counterparts (using $\alpha = 2$) to find

$$\lambda_{BP} = (3/8)^2 \frac{\lambda}{\epsilon^2}$$

$$\frac{\kappa_{BP}}{(1+\zeta_o'^2)^{1/2}} = (8/3)^3 \frac{\epsilon^3}{\lambda} \quad (5.4)$$

5.2 Alfvén Surface

We look for solutions to the equations assuming that $\lambda = const$. As we see from equation (5.2) and (5.4), this implies a differential equation for the Alfvén surface. By solving this, we have a solution for the Alfvén surface, ie

$$\frac{r_A(r_o)}{r_o} = f_A(\alpha) \frac{\lambda^{1/2}}{\epsilon} \psi^{2-\alpha} \propto r_o^{(3/\alpha)-(3/2)} \quad (5.5)$$

where f_A is a function of the index α and is of order unity. Notice that for $\alpha = 2$, one finds the BP result that $r_A/r_o = const$. The important generalization here is that r_A is not in general a conical surface. Its exact shape depends upon the distribution of magnetic flux on the accretion disk. A second important feature of these results is that there are two general classes of solution separated by the self-similar model. For (i) $\alpha < 2$, the Alfvén surface grows more rapidly than r_o, so the magnetic field lines must open out more as one moves out in the disk. This lotus-like opening of the field is in distinct contrast to the second case (ii) $\alpha > 2$, wherein the Alfvén surface diverges more slowly than r_o. In that case, the field lines must collect together in something like a cylindrical sheath. We illustrate these two classes of non self-similar solutions in Figure 8.

A third important aspect of this solution is the scaling of the Alfvén surface with the control parameters of the problem. From equations (5.1) and (5.5) we see that $r_A/r_o \propto (B_e^2/\dot{M}_a)$. Thus, if the disk field weakens, the Alfvén surface is strongly affected and moves closer to the disk thereby reducing the lever arm of the flow. A reduction in the disk accretion rate on the other hand drives up the lever arm. Since both the accretion rate and disk field are expected to decrease with time, it appears that the weakening of the field may play a major role in the evolution of wind.

5.3 Terminal Speed of the Flow

The terminal speed along any field line in the flow is $2^{1/2}\Omega_K(a)r_A(a)$ which we have already noted derives from the Bernoulli equation. Having just found the Alfvén surface, this then becomes

$$v_{p,\infty} = 2^{1/2} v_{K,e} f_A \frac{\lambda^{1/2}}{\epsilon} \psi^{2-(4\alpha/3)} \propto r_o^{(3/\alpha)-2} \qquad (5.6)$$

For self similar flows, we check that the terminal speed for flow originating at a position r_o on the disk scales as the Keplerian speed. However, a wide variety of solutions is possible. Note that if $\alpha < 3/2$ then the flow is slowest at the axis and higher terminal speeds are achieved by material leaving the outer reaches of the disk. This seems rather unphysical. For $\alpha > 3/2$, the highest terminal speeds are achieved by material leaving the inner parts of the disk. These solutions confirm the crucial fact about disk winds is that a wide range of terminal speeds of material in the wind is achieved depending upon where in the disk the flow started. As an example, for self similar flow from a Keplerian disk, material accelerated from 0.1 AU on the disk acheives terminal speeds 160 times larger than gas that is accelerated from 100 AU if the lever arm is 5. Clearly a huge range in velocities is sampled by looking at various parts of the flow.

We note that the terminal speed scales as $v_{p,\infty} \propto B_e^2/\dot{M}_a$ so that a drop in the disk magnetic field with time implies a decrease in the disk wind speed. This has important consequences for understanding the observational correlations discussed in section 2 of this review, and we return to this point in the next section.

5.4 Wind Mass Loss Rate

Solving equation (4.4) using the solution for r_A gives

$$\frac{\dot{M}_w}{\dot{M}_a} = f_{\dot{M}_w} \frac{\epsilon^2}{\lambda} \chi(\psi)$$

$$\chi(\psi) \equiv [\psi_i^{2\alpha-4} - \psi^{2\alpha-4}] \quad (\alpha \neq 2)$$

$$\chi(\psi) \equiv ln(\psi/\psi_i) \quad (\alpha = 2) \tag{5.7}$$

where ψ_i is the dimensionless magnetic flux function at the inner edge of the disk. The result for $\alpha = 2$ is just that found by BP so that once again we find that their theory is a special case of the more general 2-D equations with $\lambda = const$. This confirms the small wind mass loss rate needed to drive a much larger accretion flow. Since $\dot{M}_w/\dot{M}_a \propto B_e^{-2}$ we see that weakening the disk magnetic field actually increases the disk wind mass loss rate relative to the accretion rate. Under these circumstances, the wind becomes less efficient in extracting angular momentum so more gas is required for the job.

5.5 Mechanical Luminosity of the Wind

Adding up the total mechanical luminosity of the flow from both sides of the disk we find

$$L_w = \frac{1}{2}\frac{G\dot{M}_a M_*}{r_i}[1 - (r_i/r_o)] \tag{5.8}$$

This result shows that with no dissipation of the field in the disk, the flow can extract up to half the total gravitational binding energy that can be released in the accretion process in the form of mechanical energy of the wind. This is the explanation of the correlation suggested in equation (2.1) (see also Pudritz and Norman 1986). A hydromagnetic wind is an excellent extractor of gravitational binding energy! The result says that $L_w/L_{bol} \simeq const$ if the bolometric luminosity is largely due to accretion flow.

5.6 Wind Thrust

The flux of momentum in the bipolar wind is

$$F_w = f_{F_w}(\alpha)\frac{\epsilon}{\lambda^{1/2}}\dot{M}_a v_{K,e}[\psi_i^{(2\alpha/3)-2} - \psi^{(2\alpha/3)-2}] \tag{5.9}$$

Thus, the thrust is largest in the centre of the flow and decreases outwards. A weakening of the field implies that the wind thrust increases since the mass outflow rate increases.

Many of the other variables in the flow have analytic solutions and can be found in PP. The crucial issue of flow collimation, found by BP, generalizes nicely to the full 2-D problem. We find that flows with $n > n_T$ recollimate towards the axis, as BP found for self similar models. This motivates us to look more deeply into the consequences of flow recollimation, to which we now turn.

6. THE ORIGIN AND EVOLUTION OF OPTICAL JETS

We return, finally, to consider again the observational issues raised in section 2 of this review. Since the line emission correlates best with the presence of a disk, Gomez de Castro and Pudritz (1990) have proposed that optical emission lines and jets originate at magnetic focal points in hydromagnetic disk winds. This is a very different model than that of Hartmann and Raymond (1988) who found reasonable

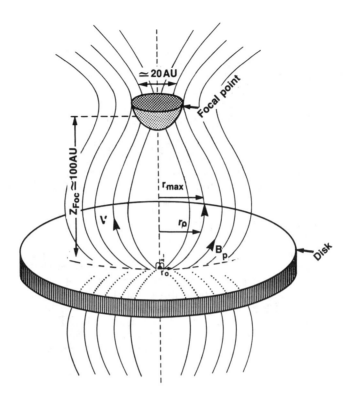

Figure 9. Recollimation of high speed disk wind produces a shock at the magnetic focal points. Line emission and optical jets may originate at these points (from Gomez de Castro and Pudritz 1990).

fits to the line profiles by assuming that a stellar wind shocks a surrounding, significantly flared accretion disk. It is not clear that such a theory would predict the striking correlation between line intensity and infrared excess discussed in section 2.

We have seen that focal points only occur if the flow achieves high enough speed, $n \geq n_T \simeq [(3\lambda^{1/2}/2^{1/2})-3]^{1/2} \simeq 5.0$. Under these conditions, the focal point stands above the disk at a distance of $D/r_{max} \simeq 10$ so that it forms at roughly 100 AU above the disk surface. Since the source of the outflow is the accretion disk, this theory predicts that the evolution of the flow is intrinsically coupled to the evolution of the disk. Figure 9 shows a cartoon of the basic picture.

In general, forbidden lines provide extremely useful information about the wind structure on small scales around the star. The three main points can be summarized as follows:

1) The profiles show blueshifted but not redshifted components. This indicates that the receding part of the wind is hidden by a disk (Jankovics et al., 1983). Bipolar winds are the best suited to reproduce the profiles (Edwards et al.,

1987).

2) Very low excitation species as [O I] and [S II] are observed in high velocity (200 km/s) wind. In purely hydrodynamic wind models, the only way to reproduce the line intensities is by oblique shocks.

3) In many YSOs, most of the [O I] and [S II] line emission is split in two components centered at 0 and ∼-150 km/s. However not all the species have these two components. Higher excitation lines ([N II]) only show blueshifted component while low excitation lines (like [O I]$_{5577}$) are observed at 0 velocity (Edwards et al., 1989).

In the converging geometry of the "magnetic focus" a wide variety of shock obliquity is indeed possible. Using the MHD shock jump conditions, one can show that the post shock magnetic field must be significantly amplified and rotated away from the flow axis compared with the pre shock field. We find that for our standard accretion rates that post shock, jet gas moves at 200 km s^{-1}, has a density of $\simeq 10^6$ cm^{-3}, and must be largely neutral (an electron fraction of less than 0.1). While detailed line profile calculations for such oblique MHD shocks remain to be completed, the prospects are encouraging.

Finally, we come to the issue of evolution of the optical jets and line emission. We have already noted that the MHD flow equations predict that the outflow speeds and mass loss rates are very sensitive to the magnetic field threading the disk. We have proposed that since ambipolar diffusion of the disk magnetic field must occur, that even a factor of 2 reduction in its strength over the life of the disk may be sufficient to reduce the outflow speed below the critical FM number n_T. In that case, the flow no longer recollimates, and the line emission, if it occurs at all, will be very faint. Thus, disk evolution and jet evolution are inextricably connected.

7. CONCLUSIONS

We have reviewed key observational constraints on the physics of jets and disk winds and have been guided by them to the view that hydromagnetic disk winds may be the simplest way of accounting for a host of complicated observational trends. The merits of disk winds are readily summarized:

1. flows are intrinsically bipolar on all scales down to the star,
2. the mechanical luminosity of outflows is the gravitational binding energy released during accretion,
3. flows transport disk angular momentum so that a profound relation between the disk accretion rate and wind mass loss rate must be established,
4. disk winds naturally collimate to cylindrical flow and at high enough speed, recollimate into magnetic focal points,
5. jets likely originate at shocks that form at magnetic focal points,
6. the evolution of jets and flows is intrinsically coupled to the evolution of accretion disks,
7. there is no dynamical reason why up to one half of the mass of a young star could not have been put in place by wind driven accretion flow through disks.

The really exciting aspect of our subject is that most of our theories are open to test by current or planned instruments!

We wish to thank the organizers, Nick Kylafis and Charlie Lada, for putting together a really interesting and stimulating meeting. The financial assistance given to REP and AIGdeC are greatly appreciated. AIGdeC thanks the Spanish

Ministry of Education for financial support through project DGICYT PB87-0167. This research was supported by a NATO collaborative grant, Universite Joseph Fournier and the Observatoire de Grenoble, CITA through the award of a Reinhardt Fellowship to REP, and NSERC of Canada.

REFERENCES

Achterberg, A., Blandford, R.D., and Goldreich, P., 1983; *Nature*, **304**, 607
Appenzeller, I., and Mundt, R., 1989; *Astron. Ap. Rev.*, **3**, 56
Bally, J., and Lada, C.J., 1983; *Ap. J.*, **265**, 824
Beckwith, S.V.W., Sargent, A.I., Koresko, C.D., and Weintraub, D.A., 1989; *Ap. J.*, **343**, 393
Blandford, R.D., and Znajek, R.L., 1977; *M.N.R.A.S.*, **179**, 433
Blandford, R.D., and Payne, D.R., 1982; *M.N.R.A.S.*, **199**, 883
Bertout, C., Basri, G., and Bouvier, J., 1988; *Ap. J.*, **330**, 350
Bouvier, J., 1990; preprint
Cabrit, S., and Bertout, C., 1990; preprint
Cabrit, S., Edwards, S., Strom, S.E., and Strom, K.M., 1990; *Ap. J.*, **354**, 687
Cannizzo, J.K., and Pudritz, R.E., 1988; *Ap. J.*, **327**, 840
Chan, K.L., and Henriksen, R.N., 1980; *Ap. J.*, **241**, 534
Chandrasekhar, S., 1956; *Ap. J.*, **124**, 232
Chini, R., 1989; ESO Workshop on *Low Mass Star Formation and Pre-Main Sequence Objects*, ed. B. Reipurth, p. 173
Cohen, M., and Kuhi, L.V., 1979; *Ap. J. Suppl.*, **41**, 743
Cohen, M., Emerson, J.P., and Beichman, C.A., 1989; *Ap. J.*, **339**, 455
Draine, B.T., 1983; *Ap. J.*, **270**, 519.
Edwards, S., Cabrit, S., Strom, S.E., Heyer, I., Strom, K.M., and Anderson, E., 1987; *Ap. J.*, **321**, 473
Edwards, S., Cabrit, S., Ghandour, L.O., Strom, S.E., 1989; ESO Workshop on *Low Mass Star Formation and Pre-Main Sequence Objects*, ed. B. Reipurth, p.385
Elias, J.H., 1978; *Ap. J.*, **224**, 857
Gomez de Castro, A.I., and Pudritz, R.E., 1990; submitted to Ap. J.
Hartmann, L., and McGregor, K.B., 1982; *Ap. J.*, **259**, 180
Hartmann, L., Hewett, R., Stahler, S., and Mathieu, R.D., 1986; *Ap. J.*, **309**, 275
Hartmann, L., and Stauffer, J.R., 1989; *Astron. J.*, **97**, 873
Hartmann, L., and Raymond, J.C., 1988; *Ap. J.*, **337**, 903
Henriksen, R.N., Rayburn, D.R., 1971; *M.N.R.A.S.*, **152**, 323
Herbig, G.H., Vrba, F.J., and Rydgren, A.E., 1986; *Astron. J.*, **91**, 575
Jankovics, S., Appenzeller, I., and Krautter, J., 1983; *P.A.S.P.*, **95**, 883
Konigl, A., 1982; *Ap. J.*, **261**, 115
Konigl, A., 1989; *Ap. J.*, **342**, 208
Lada, C.J., 1985; *Ann. Rev. Astron. Astrophys.*, **23**, 267
Lada, C.J., and Shu, F.H, 1990, *Science*, **248**, 564
Lovelace, R.V.E., Wang, J.C.L., and Sulkanen, M.E., 1987; *Ap.J.*, **315**, 504
Mestel, L., 1961; *M.N.R.A.S.*, **122**, 473
Mestel, L., 1968; *M.N.R.A.S.*, **138**, 359
Mestel, L., and Paris, R.B., 1979; *M.N.R.A.S.*, **187**, 337

Michel, F.C., 1969; *Ap. J.,*, **158**, 727
Moneti, A., Pipher, J.L., Helfer, H.L., McMillan, R.S., and Perry, M.L., 1984;
    Ap. J., **282**, 508
Okamoto, 1975; *M.N.R.A.S.*, **173**, 357
Parker, I.N., 1958; *Ap. J.*, **128**, 664
Pelletier, G., and Pudritz, R.E., 1991; submitted to Ap. J.
Pneumann, J.W., and Kopp, R.A., 1971; *Solar Phys.*, **18**, 258
Pudritz, R.E., 1981; *M.N.R.A.S.*, **195**, 897
Pudritz, R.E., 1985; *Ap. J.*, **293**, 216
Pudritz, R.E., 1988; NATO ASI on *Galactic and Extragalactic Star Formation*,
    eds. R.E. Pudritz and M. Fich, p. 135
Pudritz, R.E., and Norman, C.A., 1983; *Ap. J.*, **274**, 677
Pudritz, R.E., and Norman, C. A., 1986; *Ap. J.*, **301**, 571
Rees, M.J., 1984; *Ann. Rev. Astron. Ap.*, **22**, 471
Rydgren, A.E., Strom, S.E., and Strom, K.M., 1976; *Ap. J. Suppl.*, **30**, 307
Sakurai, T., 1985; *Astr. Ap.*, **152**, 121
Schatzman, E., 1962; *Astr. Ap.*, **25**, 18
Shakura, N.I., and Sunyaev, R.A., 1973; *Astr. Ap.*, **24**, 337
Shibata, K., and Uchida, Y., 1986; *Publ. Astron. Soc. Japan*, **38**, 631.
Shu, F.H., Lizano, S., Ruden, S.P., Najita, J., 1988; *Ap. J. Lett*, **328**, L19
Snell, R.L., Loren, R.B., and Plambeck, R.L., 1980; *Ap. J. Lett*, **239**, L7
Stone, J., 1990; Ph.D. Thesis, University of Urbana, Illinois
Uchida, Y., Kaifu, N., Shibata, K., Hayashi, S.-S.,
    Hasegawa, T., and Hamatake, H., 1987; *P.A.S.J.*, **39**, 907
Uchida, Y., and Shibata, K., 1984; *Pub Astr. Soc. Japan*, **36**, 105
Walter, F.M., Brown, A., Mathieu, R.D., Myers, P.C., and Vrba, F.J., 1988;
    Astron. J., **96**, 297
Weber, E.J., and Davis, L. Jr., 1967; *Ap. J.*, **148**, 217

IONIZED WINDS FROM YOUNG STELLAR OBJECTS

Nino Panagia*
Space Telescope Science Institute, Baltimore, USA.

Abstract — The observations and the theory of the radio and infrared emission from ionized winds of young stellar objects are reviewed and discussed.

1. Introduction

Young stellar objects (YSO) or, equivalently, pre-main-sequence (PMS) stars, are most easily observed in the infrared and in the radio because in these spectral domains the heavy extinction associated with a star forming region is only a minor problem, if at all, so that the whole region can be studied thoroughly. In this way one is able: i) to search for recently formed stars and do statistical studies on the rate of star formation, ii) to determine their luminosity, hence, to study luminosity functions and initial mass functions down to low masses, iii) to study their spectra and, thus, to determine the prevailing conditions at and near the surface of a newly born star and its relations with the surrounding environment. This third point is the main interest of my review. In fact, I shall limit myself to considering the observations concerning the processes of outflows from, and accretion onto, YSOs and the theory necessary to interpret them. Moreover, I shall consider only the case of *ionized* winds; the physics and phenomenology of neutral winds from YSOs is the subject of Antonella Natta's review (this volume). Section 2 discusses the radiative processes relevant in stellar outflows. The main observational results are presented in Section 3. A discussion of the statistical properties of stellar winds from YSOs is given in section 4.

2. Emission from Extended Outflows: A Theoretical Overview

In this section we shall review the problem of radiation transfer applied to the case of stellar outflows. Quite of number of authors have contributed to this field, both for the formation of the continuum (see among others, Panagia and Felli

---

* Affiliated to the Astrophysics Division, Space Sciences Department of ESA; on leave from University of Catania.

1975, Wright and Barlow 1975, Olnon 1975, Marsh 1975, Chiuderi and Torricelli Ciamponi 1978, Felli and Panagia 1981, Schmid-Burgk 1982, Castor and Simon 1983, Lamers and Waters 1984, Reynolds 1986, Felli and Panagia 1990) and the lines (Krolik and Smith 1981, Simon et al. 1983, Rodriguez and Cantó 1983, Felli et al. 1985a, Hamann and Simon 1986, Panagia et al. 1990). The most important points and results are outlined and discussed in the following subsections.

2.1. Assumptions - Basic Transfer Problem

In order to make the transfer problem easily tractable the following assumptions are usually made (e.g., Panagia and Felli 1975):

1. The wind is spherically symmetric with a steady mass loss, $\dot{M}$.
2. The wind is fully ionized. This condition, combined with the continuity equation gives the behaviour of the electron density n_e with the distance, r, from the center of the star

$$n_e = \frac{\dot{M}}{4\pi r^2 v(r) \mu_e m_H} \quad (1)$$

where $v(r)$ is the wind velocity, μ_e is the mean atomic weight per electron

$$\mu_e = \frac{\sum_i X_i m_i}{\sum_i X_i Z_i} \quad (2)$$

with X_i, m_i and Z_i being the abundance by number, the mass and the charge of the i-th ion, and m_H is mass of a hydrogen atom.

3. The wind velocity increases with radius like a power law

$$v = v_o (r/r_o)^\gamma \quad (3)$$

where v_o and r_o are the initial velocity and radius, respectively, until it reaches its terminal value v_∞ at a radius r_1

$$v = v_\infty \qquad r \geq r_1 = r_o (v_\infty/v_o)^{1/\gamma} \quad (4)$$

4. The wind is isothermal at the same temperature as the stellar photosphere, i.e.,

$$T_{wind} = T_{star} \quad (5)$$

5. The wind material is in quasi-LTE conditions, in the sense that the Kirchoff law is valid

$$j_\nu = \kappa_\nu B_\nu \quad (6)$$

where j_ν and κ_ν are the emissivity and the absorption coefficient, respectively. With these assumptions the radiative transfer equation along any line of sight can be written as

$$\frac{dI}{ds} = -\kappa I + j \quad (7)$$

where s is the current coordinate along the line of sight (see Figure 1 which illustrates the geometry of the problem). Its solution is

$$I = I_o e^{-\tau} + B(1 - e^{-\tau}) \quad (8)$$

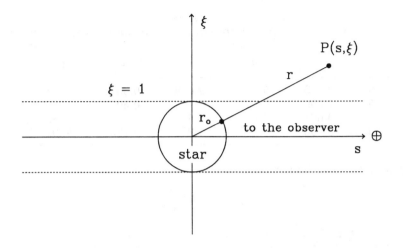

Fig. 1 – The geometry of the star+wind system.

where I_o is the background intensity and τ is the optical depth as integrated along the line of sight, i.e.

$$\tau = 2 \int_o^\infty \kappa ds \qquad (9)$$

With our assumptions it is possible to express the total optical along any line of sight (assuming the opacity to be proportional to n_e^2) as

$$\tau = \tau_o \xi^{-(3+2\gamma)} \qquad (10)$$

where τ_o represents the optical depth for impact parameter equal to unity, i.e., for the line of sight tangent to the stellar disk.

The radiation that we receive from the system star+wind is the intensity integrated over all lines of sight parallel to line joining us to center of the star. It is immediately clear that for $\xi < 1$, i.e., for lines of sight which project on the stellar disk,

$$\begin{aligned} I_o &= B \\ I &= Be^{-\tau} + B(1-e^{-\tau}) = B \end{aligned} \quad \xi < 1 \qquad (11)$$

In other words the presence of a wind in front of the star does not affect the radiation we receive from the stellar disk because the amount of radiation which is absorbed is perfectly compensated by the emission of the wind.

For impact parameters greater than the stellar radius there is no background radiation so that we have simply

$$I = B(1-e^{-\tau}) \qquad \xi \geq 1 \qquad (12)$$

The total emission received from the star+wind system will then be given by

$$F_\nu = \int I_\nu d\Omega = \frac{1}{4\pi d^2} 4\pi R^2 \pi B_\nu \left[1 + 2\int_1^\infty (1-e^{-\tau})\xi d\xi\right] \qquad (13)$$

Table 1
Coefficients of the continuous emission from the wind

γ	a	b	c	d	$J(\gamma)$	α
0.0	2.000	0.125	2.679	0.667	0.785	0.600
0.5	1.000	0.167	1.772	0.500	0.667	0.950
1.0	0.667	0.188	1.489	0.400	0.589	1.160
2.0	0.400	0.208	1.276	0.286	0.491	1.400

Equation (12) can be rewritten as

$$F_\nu = \frac{4\pi R^2}{4\pi d^2} \pi B_\nu [1 + \epsilon] \tag{14}$$

where R is the photospheric radius and d is the distance; ϵ represents the excess of radiation due to the presence of the wind and is given by

$$\epsilon = 2 \int_1^\infty (1 - e^{-\tau}) \xi d\xi \tag{15}$$

Equations (14) and (15) are the basic transfer equations which govern both for the continuum and the emission line problems. Analytical solutions to eq. (15) can be obtained for two cases, which eventually turn out to cover all relevant possibilities (Panagia and Felli 1990):

  i. The quasi-optically thin case, which is valid for moderate optical depths ($\tau_o \lesssim 3$). In this case by expanding equation (15) to second order terms and after some manipulation one obtains

$$\epsilon = \frac{a\tau_o}{1 + b\tau_o} \tag{16}$$

where the coefficients a and b are simple functions of the velocity slope γ:

$$a = \frac{2}{1 + 2\gamma} \qquad b = \frac{1 + 2\gamma}{8 + 8\gamma} \tag{17}$$

Values of a and b for some values of γ are given in Table 1. Since in general $a \gg b$ and $b \ll 1$ the radiation excess ϵ increases essentially linearly with the optical depth and only for relatively high values of τ (say $> 1/2$) the rise departs appreciably from linearity.

  ii. The optically thick case: here the optical depth is large ($\tau_o > 3$) and therefore one can write

$$1 + \epsilon = 1 + 2 \int_1^\infty \left\{1 - \exp[-\tau_o \xi^{-(3+2\gamma)}]\right\} \xi d\xi \simeq \int_o^\infty \{1 - \exp[-\tau_o \xi^{-(3+2\gamma)}]\} \xi d\xi \tag{18}$$

This equation correspond to a gamma function, so that we obtain:

$$1 + \epsilon = \Gamma\left(1 - \frac{2}{2\gamma + 3}\right) \tau_o^{\frac{2}{2\gamma+3}} = c\tau_o^d \tag{19}$$

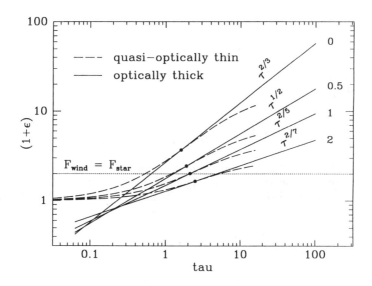

Fig. 2 – The quantity $(1+\epsilon)$ as a function of the optical depth: curves for both the quasi-optically thin (dashed lines) and the optically thick (solid lines) cases are shown for some values of γ. The points where the two approximations meet are marked as well as the line of equal flux contributed by the wind as emitted by the star.

Table 1 reports values of the coefficients $c = \Gamma[1 - 2/(2\gamma + 3)]$ and $d = 2/(2\gamma + 3)$ for some values of γ. It is apparent that in this case the increase of the radiation excess with the optical depth is systematically slower than linear.

If we plot the quantity $(1+\epsilon)$ as computed with eq. (16) and (19), respectively, we see (Fig. 2) that the two curves join quite smoothly with each other for optical depths around 2. Detailed numerical calculations show that where the two curves intersect each other the analytical formulae give the correct value of ϵ to within better than 1.7%. Therefore, one can safely use eq. (16) for lower values of the optical depth and eq. (19) for higher values, thus covering the whole possible range with two simple formulae. Finally, it is worth noting that the excess radiation becomes equal to the original stellar emission for values of the optical depth of the order of unity: in particular 0.53, 2.09 and 4.82 for γ values of 0, 1 and 2, respectively (cf fig. 2).

2.2. The continuum

The main emission mechanism in the infrared and in the radio is the free-free (or bremsstrahlung) radiation (f-f). The free bound transitions (i.e. recombinations of free electrons to bound levels), which dominate the optical emission, are of modest importance in the near infrared (they provide 20% of the continuum at 2.2 μm) and become completely negligible at longer wavelengths. Therefore, for the sake of simplicity in the following discussion we shall consider only the f-f process. The f-f opacity per unit length, which we express as the product of the electron and ionic densities times a function k_ν, is given by

$$\kappa_\nu = n_e n_i k_\nu = 3.69 \times 10^8 n_e n_i \left[1 - e^{-\frac{h\nu}{kT}}\right] g_{ff} T^{-1/2} \nu^3 \qquad (20)$$

At radio frequencies k_ν can be approximated quite accurately as (Mezger and Henderson, 1967)

$$k_\nu = 8.44 \times 10^{-28} \left[\frac{\nu}{10 \text{ GHz}}\right]^{-2.1} \left[\frac{T}{10^4 K}\right]^{-1.35} \qquad (21)$$

Although this formula is accurate only in the radio domain, for qualitative discussions it can also be used advantageously at infrared wavelengths as well. Inserting eq. (20) into (10) the optical depth at impact parameter equal to unity is evaluated to be

$$\tau_o = 2J(\gamma) n_o^2 R k_\nu \propto \frac{\dot{M}^2}{R^3 v_o^2} \nu^{-2.1} T^{-1.35} \qquad (22)$$

with $J(\gamma)$ defined as

$$J(\gamma) = \frac{\sqrt{\pi}}{2} \frac{\Gamma(\gamma + \frac{3}{2})}{\Gamma(\gamma + 2)} \qquad (23)$$

Values of $J(\gamma)$ for some values of γ are given in Table 1. It is straightforward now to estimate the emitted flux by using equations (14), (16) and (19). In the quasi-optically thin case the flux is virtually proportional to τ_o and therefore the main functional dependences of the excess flux are

$$F_{wind} = \frac{4\pi R^2}{4\pi d^2} \pi B_\nu \epsilon \propto \frac{R^2}{d^2} B_\nu \tau_o \propto \frac{\dot{M}^2}{v_o^2} R^{-1} T^{-1.35} \nu^{-2.1} B_\nu \qquad (24)$$

On the other hand, in the optically thick case the total flux is

$$F_{total} = \frac{4\pi R^2}{4\pi d^2} \pi B_\nu (1 + \epsilon) \propto \frac{R^2}{d^2} B_\nu \tau_o^{\frac{2}{2\gamma + 3}} \qquad (25)$$

which, expanding the dependences of τ_o upon the various parameters, becomes:

$$F_\nu \propto d^{-2} R^{\frac{4\gamma}{2\gamma+3}} T^{\frac{2\gamma+0.3}{2\gamma+3}} \nu^{\frac{4\gamma+1.8}{2\gamma+3}} \left(\frac{\dot{M}}{v_o}\right)^{\frac{4}{2\gamma+3}} \qquad (26)$$

The first important point is the frequency dependence which correspond to a power law whose spectral index α ($F_\nu \propto \nu^\alpha$) is given by

$$\alpha = \frac{4\gamma + 1.8}{2\gamma + 3} \qquad (27)$$

i.e., always lower than the black body slope of 2 (see Table 1). This is because at lower frequency the opacity is higher (cf. eq. (21)) and, therefore, the radius at which the optical depth is of the order of unity becomes larger and the effective emitting surface increases. In other words, the emission correspond to that of a black body (as expected for a well behaved, optically thick, thermal source) whose size increases with wavelength, so that the overall decrease of flux with wavelength is slower than it is for a Planck curve. In fact, we can define an effective radius to be that of a black body at the wind temperature which emits the same flux as the wind. Therefore, from eq. (14) we obtain

$$R_{eff} \simeq \sqrt{1+\epsilon}\, R \qquad (28)$$

Therefore, for large τ_o the effective radius becomes a simple function of the frequency:

$$R_{eff} \propto \nu^{-\frac{4.2}{2\gamma+3}} \qquad (29)$$

Such an effective radius can be considerably larger than the photospheric radius. For example, in the radio the size of the emitting radius can be several thousands of stellar radii for typical values of the mass loss. Therefore, since the accelerating force becomes weak at large radii, the velocity in the radio emitting region can be expected to be essentially constant. In the constant velocity case ($\gamma = 0$) one recovers the "classical" formulae of the radio emission from winds (Panagia and Felli 1975)

$$F_\nu = 5.12 \left[\frac{\nu}{10\text{ GHz}}\right]^{0.6} \left[\frac{T}{10^4 K}\right]^{0.1} \left[\frac{\dot{M}}{10^{-6}\frac{M_\odot}{\text{yr}}}\right]^{4/3} \left[\frac{v_\infty}{100\text{ km s}^{-1}}\right]^{-4/3} \times$$

$$\left[\frac{d}{1\text{ kpc}}\right]^{-2} \text{ m Jy} \qquad (30)$$

$$R_\nu = 6.23 \times 10^{14} \left[\frac{\nu}{10\text{ GHz}}\right]^{-0.7} \left[\frac{T}{10^4\ K}\right]^{-0.45}$$

$$\left[\frac{\dot{M}}{10^{-6}\frac{M_\odot}{\text{yr}}}\right]^{2/3} \left[\frac{v_\infty}{100\text{ km s}^{-1}}\right]^{-2/3} \text{ cm} \qquad (31)$$

$$\theta = 0.042 \left[\frac{\nu}{10\text{ GHz}}\right]^{-0.7} \left[\frac{T}{10^4\ K}\right]^{-0.45} \left[\frac{\dot{M}}{10^{-6}\frac{M_\odot}{\text{yr}}}\right]^{2/3}$$

$$\left[\frac{v_\infty}{100\text{ km s}^{-1}}\right]^{-2/3} \left[\frac{d}{1\text{ kpc}}\right]^{-1} \text{ arcsec} \qquad (32)$$

In this case a mass losing star is expected to be detected as having a characteristic $\nu^{0.6}$ radio spectrum and a source size of few hundredths of an arcsec. We note that the theory is quite simple and the radio flux depends in a simple manner on $\dot{M}$. Therefore, radio measurements can provide reliable determinations of the mass loss rate provided that the terminal velocity can be estimated independently. Even so, to do it right one has to make sure that the radio spectrum is really that of a wind, i.e., that the spectral index is ~ 0.6 indeed. As discussed by Felli and Panagia (1981) this requires making radio measurements at three widely spaced frequencies.

2.3. The emission lines

The problem of line radiation transfer is much more complicate because it involves bound levels which may not be in any simple equilibrium and because by its very nature the line opacity is a rapidly varying function of the frequency. However, the problem can be simplified by adding the following assumptions to those already listed in section 2.1:

• The expansion velocity of the wind is much higher than the sound speed of the gas which is in the range 10-30 km s^{-1}. This condition is generally well verified because the observed expansion velocities are not less than 100 km s^{-1}.

• The line opacity is much larger than that of the continuum, so that the two transfer problems can be decoupled: this is certainly true in the IR but is not verified at radio wavelengths.

With these assumptions the line radiative transfer is reduced to a local problem in that a photon can interact only with matter at the place of "perfect resonance" where due to systematic motions an atom is perfectly "tuned" for absorbing "that" photon. This is the so-called Sobolev approximation from the name of the russian astronomer who introduced it (Sobolev 1957). Following Castor (1970) the line optical depth can be expressed as

$$\tau(\nu,\xi) = \frac{\pi e^2}{mc} gf \frac{\frac{N_l}{g_l} - \frac{N_u}{g_u}}{\nu_o \frac{v(\rho)}{c}} R \frac{\rho}{1 + \frac{s^2}{\rho^2}(\gamma - 1)} \tag{33}$$

where N_l, N_u and g_l, g_u are the population and the statistical weight of the lower and upper levels of the transitions. The quantity ρ is the radial distance in the plane (s, ξ), i.e., $\rho^2 = s^2 + \xi^2$ and like s and ξ is expressed in units of stellar radii. In the case of hydrogen lines, which is the case we will be dealing with, the population of the excited levels is proportional to the square of the ionized matter density and, therefore, the optical depth can be written as

$$\tau(\nu,\xi) = \tau_o \frac{\rho^{-3(\gamma+1)}}{1 + \frac{s^2}{\rho^2}(\gamma - 1)} \tag{34}$$

where

$$\tau_o = \frac{\pi e^2}{mc} gf \frac{\frac{N_l}{g_l} - \frac{N_u}{g_u}}{\nu_o v_o/c} R \propto \frac{\dot{M}^2}{R^3 v_o^3} \tag{35}$$

represents the optical depth at $\xi = 1$ and at the center of the line, $\nu = \nu_o$. Note that all dependences on the frequency are "hidden" in the relationship between the projected velocity, hence the observed frequency, and the geometric coordinates s and ρ. As an illustration, let us consider the case of $v \propto r^{1/2}$. Figure 3 shows the surfaces of constant projected velocity (as seen by an observer located at "infinite distance" on the right hand side of the figure) as well as some surfaces of constant optical depth. Note that at any value of ρ the optical depth (cf. equation 34) can vary only between τ_o (for $\xi = \rho$ and $s = 0$) and τ_o/γ (for $\xi = 0$ and $s = \rho$) and, therefore, the surfaces of constant optical depth are almost spherical.

As we have done for the continuum we can distinguish two extreme cases, i.e. optically thin and optically thick cases. However, since the transfer problem is

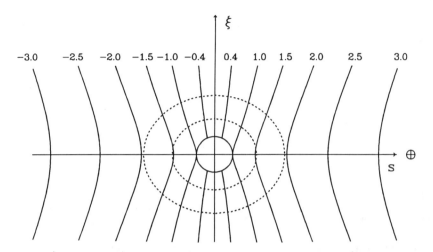

Fig. 3 – The projected surfaces of equal velocity (in units of the initial velocity) for the case of $v \propto r^{1/2}$. The loci of constant optical depth are also shown (dashed lines) for τ values of 10 (inner curve) and 1 (outer curve).

rather more complicated than it is for the continuum, here we will limit ourselves to discuss the simplest concepts and to present the most straightforward results.

In the optically thin case the total (i.e. integrated over the line profile) the line intensity is proportional to the volume emission measure of the wind (and, therefore, to the line optical depth)

$$I_{tot} \propto \int n_e n_i 4\pi r^2 dr \propto \frac{\dot{M}^2}{v_o^2 R} \qquad (36)$$

and the line width is proportional to the initial velocity through a constant which depends exclusively on the index γ. As long the terminal velocity is much higher than the initial velocity, the half-power-full-width (HPFW) of a line can be expressed as (Panagia et al. 1987):

$$\mathrm{HPFW} = 2^{\frac{4\gamma+1}{3\gamma+1}}(v_o/c)\nu_o \qquad (37)$$

In the optically thick case there is no simple general solution. Two cases have been considered by Simon et al. (1983), i.e., $\gamma = 0$ and $\gamma = 1$, who solved the line radiative transfer problem numerically and expressed the results in formulae whose functional dependence was derived from simplified theory and the coefficient was determined from a fit to the numerical results. Krolik and Smith (1981) have approached this problem in a general but very simplified manner which we follow here to illustrate the main characteristics of the emission lines formed in an optically thick wind. In such an approach, the size of emitting region is taken to be equal to the radius at which the optical depth becomes unity, i.e.

$$\xi_{eff} \simeq \rho(\tau = 1, s = 0) \simeq \tau_o^{\frac{1}{3(\gamma+1)}} \qquad (38)$$

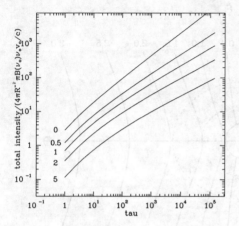 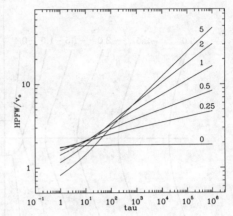

Fig. 4 – The total intensity of emission lines for several values of γ.

Fig. 5 – The half-power-full-width of emission lines for several values of γ.

Thus, the intensity at the peak of the line will essentially be that of the black body with such a size, i.e.

$$I_{peak} \simeq 4\pi R^2 \pi B(\nu_o) \xi_{eff}^2 \simeq 4\pi R^2 \pi B(\nu_o) \tau_o^{\frac{2}{3(\gamma+1)}} \tag{39}$$

Similarly, the effective width of the line will correspond to the maximum velocity at which the optical depth is still equal to 1. This occurs at $\xi = 0$ (cf. eq. (34)) at a value of $\rho = \rho(\tau=1, \xi=0)$, so that the frequency displacement at which the line is still strong will be

$$\Delta\nu = \frac{v_{max}}{c}\nu_o = \frac{v_o}{c}\left(\frac{\tau_o}{\gamma}\right)^{\frac{\gamma}{3(\gamma+1)}} \nu_o \tag{40}$$

This formula essentially provides an estimate of the HPFW, whose functional dependence on the stellar and the wind parameters is

$$\text{HPFW} \simeq 2\Delta\nu \propto \dot{M}^{\frac{2\gamma}{3(1+\gamma)}} R^{-\frac{\gamma}{1+\gamma}} \nu_o^{\frac{1}{1+\gamma}} \tag{41}$$

Hence, the total line intensity can approximately be estimated as

$$I_{tot} \simeq I_{peak} \times \text{HPFW} = 4\pi R^2 \pi B(\nu_o) \tau_o^{\frac{2+\gamma}{3(\gamma+1)}} \nu_o \frac{v_o}{c} \tag{42}$$

It is interesting to note that this formula recovers the functional dependences derived by Simon et al. (1983) and extends their validity to any value of γ. However, due to the crudeness of the approach these formulae are accurate only to within a factor of two as compared with numerical results. More detailed calculations have been made by Panagia, Oliva and Felli (1990) both using refined analytical approximations and performing numerical calculations for a large variety of cases. Figures 4 and 5 present the results of the numerical calculations of the total intensity and the HPFW for emission lines formed in a wind where the optical depth is proportional to the square of the density, as appropriate for hydrogen lines. The analytical formulae they derive for the cases of high optical depths agree with the numerical results to better than 30% in general and better than 10% in most cases:

$$I_{peak} = 4\pi R^2 \pi B(\nu_o) \Gamma\left(1 - \frac{2}{3(\gamma+1)}\right) \tau_o^{\frac{2}{3(\gamma+1)}} \qquad (43)$$

$$I_{tot} = 4\pi R^2 \pi B(\nu_o) \frac{4}{2+\gamma} \left[\Gamma\left(1 - \frac{2}{3(\gamma+1)}\right)\right]^{1+\frac{\gamma}{2}} \tau_o^{\frac{2+\gamma}{3(\gamma+1)}} \frac{v_o}{c} \nu_o \qquad (44)$$

It also should be noted that at high optical depths, the higher the value of γ the wider the lines become both in the HPFW and the extension of wings: this is a natural consequence of the faster increase of the velocity with radius for higher values of γ.

2.4. Extension to more general cases

In this section we will briefly consider the effects of releasing some of the basic assumptions we have made in sections 2.1 and 2.3.

2.4.1. Incomplete ionization - We can identify two main possibilities:
i) The wind is ionized bounded, i.e. the wind is fully ionized up to a certain radius and becomes mostly neutral beyond that: this the the analog to a Strömgren sphere for the case of an outflow. This problem has been studied in detail by Felli and Panagia (1981) and Simon et al. (1983) who have shown that if such a situation occurs at all, the boundary is likely to be very close to the base of the flow. As a consequence, the IR emission may not be strongly affected while the radio flux can drastically be reduced by orders of magnitude. The maximum rate of mass loss for which a wind can be fully ionized by a star with a Lyman continuum photon flux of N_L is (Felli and Panagia, 1981)

$$\dot{M}_c = 5.75 \times 10^{-6} (1+2\gamma)^{1/2} \left[\frac{N_L}{10^{49}\,\text{s}^{-1}}\right]^{1/2} \left[\frac{v_o}{100\,\text{km s}^{-1}}\right] \left[\frac{R}{10\,R_\odot}\right]^{1/2} \left[\frac{T}{10^4\,K}\right]^{0.4} \frac{M_\odot}{\text{yr}} \qquad (45)$$

Correspondingly the maximum radio flux density which a wind ionized by such a star can emit is

$$F_{max}(wind) = 52.7 (1+2\gamma)^{2/3} \left[\frac{\nu}{10\,\text{GHz}}\right]^{0.6} \left[\frac{N_L}{10^{49}\,\text{s}^{-1}}\right]^{2/3} \left[\frac{R}{10\,R_\odot}\right]^{2/3} \left[\frac{v_\infty}{v_o}\right]^{-4/3} \left[\frac{d}{1\,\text{kpc}}\right]^{-2} \text{mJy} \qquad (46)$$

It is clear then that a wind can never be a very strong source by emitting thermal radiation, i.e. its flux density is unlikely to exceed a few tens of mJy.
ii) The gas is partially ionized everywhere throughout the wind. Natta et al. (1988) have shown that this situation may occur in the case of low luminosity protostars or young stars. In this case one can still apply the same kind of theory

which has been outlined in the previous sections but now one should allow for the incomplete ionization in writing the dependence of the electron and ion density on the radius. In other words, one has to include the extra dependence on the ionization fraction in addition to those already present in the continuity equation. In the simplest case of a constant ionization fraction, all the equations discussed so far still hold with the only change that the mass loss rate should be replaced by the the product of the mass loss rate times the ionization fraction itself. The net effect is that in order to account for any given observed wind emission the required mass loss rate will be 1/(ionization fraction) times higher than in the case of complete ionization.

2.4.2 The outflow is not spherically symmetric: this case has been considered in quite some detail by Schmid-Burgk (1982) and Reynolds (1986). As for the slope of the continuum, the lack of spherical symmetry may affect it appreciably only if the fraction of solid angle filled by the flow varies systematically with the radius. For example, Reynolds has shown that in the case of collimated jets with cross section varying like $w(r) = w_o(r/r_o)^\eta$ and a density within the jet $n_e(r) = n_o(r/r_o)^{-\beta}$ the radio continuum spectral index is given by

$$\alpha(\beta,\eta) = 2 - 2.1\frac{1+\eta}{2\beta - \eta} \tag{47}$$

As for the absolute value of the flux it is found that, irrespective of the details of the source geometry, the formulae derived for a spherically symmetric wind are still valid, to within a factor of two, for more complex configurations as long as any ratio of the structural length scales in the source does not exceeds a value of about 10. In other words, only highly anisotropic structures can give results appreciably deviant from the predictions of the spherically symmetric models. Among these alarming cases one can consider a wind which consists solely of a number of narrow jets: in this case the radiating efficiency can be higher than in a spherical wind by as much as an order of magnitude and, therefore, the mass loss rate derived with spherical wind formulae may be strongly overestimated. For example, Reynolds (1986) has shown that a narrow conical jet with opening angle θ_o and seen with an inclination i can produce the same radio flux with a much smaller mass loss rate given by

$$\dot{M}(jet) = \dot{M}(\text{spherical wind}) \times 0.20\theta_o^{3/4}(\sin i)^{-1/4} \tag{48}$$

2.4.3 Time variability - One of the basic assumptions is that the mass loss process is steady with time. On the other, variability is an observed fact and the more so in the case of young stellar objects. A discussion of the main effects has been presented by Felli et al. (1985b). Summarizing it briefly here, one can distinguish two cases: i) variation of the mass loss rate, and ii) variation of the ionization. In the former case one expects to detect short term variability in the optical and IR but not in the radio because even if the $\dot{M}$ variation is instantaneous the transit time of the front is

$$t_{transit} \simeq R_{eff}/v = 19.3\left[\frac{R_{eff}}{10\ R_\odot}\right]\left[\frac{v}{100\ \text{km s}^{-1}}\right]^{-1}\ \text{hours} \tag{49}$$

and, therefore, it is of the order of a few hours in the near IR but it can be of the order of weeks or months in the radio. Whether or not one sees any variation at all in the radio depends on the time scale of the intrinsic variation: if it occurs over a time much shorter than a few weeks, then the variations will completely be averaged out in the radio. In the case in which the ionization varies, the situation is different. Again assuming that the ionizing agent undergoes a sudden variation, the time scale of the phenomena is set by the recombination time which varies like a higher power of the radius:

$$t_{rec} = \frac{1}{n_e \beta_2} = 7.4 \left[\frac{\dot{M}}{10^{-6} \, M_\odot \, \text{yr}^{-1}} \right]^{-1} \left[\frac{R_{eff}}{10 \, R_\odot} \right]^2 \left[\frac{v}{100 \, \text{km s}^{-1}} \right] \left[\frac{\mu_e}{1.2} \right] \left[\frac{T}{10^4 \, K} \right]^{0.8} \text{seconds} \quad (50)$$

Therefore, one expects a much more prompt reaction in the optical and IR, which originate in the inner and denser layers, than in the radio. These points stress the importance of making simultaneous measurements at widely different frequencies in order to understand the nature and the origin of the variability.

2.4.4 Accretion - All throughout this theoretical section we have assumed that the emitting gas is flowing away from a young star. But the reverse situation, *i.e.* mass infall or accretion, may occur since we are dealing with just forming objects. And its occurrence is actually detected in some instances by the observation of inverse P Cyg profiles of optical lines, i.e. lines with a red absorption and a blue emission, which are the unambiguous sign of gas falling onto the stellar surface. In these case, most of the theory remains valid apart for an obvious change in the velocity sign. In fact, one cannot distinguish the velocity sign, whether the mass flows in or out of a star, from observations in the continuum because its flux is solely determined by the density radial behaviour which is invariant with respect to a sign inversion as long as the modulus variation of the velocity stays the same. On the other hand, in the case of accretion one does not expect a lower velocity closer to the star because gravity will unavoidably accelerate the matter as it falls onto the star. As a consequence, the expected velocity slope is negative ($\gamma < 0$) whereas in the case of a wind both the theory and the observations suggest an outwardly acceleration, i.e. $\gamma > 0$. Of special interest is the case of accretion in free fall conditions which entails $v \propto r^{1/2}$. It is easy to verify that with the resulting density distribution $n \propto r^{-3/2}$ the volume emission measure

$$\int n_e n_i dV \propto n_o^2 r_o^3 \ln(r/r_o) \quad (51)$$

would diverge with increasing radius. Therefore, either the distance at which the inflow originates is suitably small or the gas will unavoidably be ionization bounded, i.e. neutral in the outer parts of the flow, for any accretion rate. For this reason the continuous spectrum will remain essentially flat even at those frequencies at which the flow is optically thick until the effective emitting radius will become comparable to the outer boundary. Felli *et al.* (1982) give an approximate expression for the radio emission. After some manipulation it can be written as

$$F_\nu \text{ (accretion)} = 8.18 \left[\frac{\nu}{10 \text{ GHz}}\right]^{-0.1} \left[\frac{T}{10^4 \text{ K}}\right]^{-0.35}$$
$$\left[\frac{\dot{M}_{accr}}{10^{-7} M_\odot \text{ yr}^{-1}}\right] \left[\frac{M}{M_\odot}\right]^{-1} \left[\frac{d}{1 \text{ kpc}}\right]^{-2} \ln\left(\frac{r_\infty}{r_c}\right) \text{ mJy} \quad (52)$$

where $\dot{M}_{accr}$ is the accretion rate, M is the stellar mass, r_c is the radius at which the radial optical depth is equal to $3/4$:

$$r_c = 3.66 \times 10^{14} \left[\frac{\nu}{10 \text{ GHz}}\right]^{-1.05} \left[\frac{T}{10^4 \text{ K}}\right]^{-0.68}$$
$$\left[\frac{\dot{M}_{accr}}{10^{-7} M_\odot \text{ yr}^{-1}}\right] \left[\frac{M}{M_\odot}\right]^{-1} \left[\frac{\mu_e}{1.2}\right]^{-1} \text{ cm} \quad (53)$$

r_∞ is the outer boundary, generally corresponding to the accretion radius

$$r_\infty \simeq \left(\frac{v_{escape}}{v_{turb}}\right)^2 r_o = 2.65 \times 10^{16} \left[\frac{M}{M_\odot}\right] \left[\frac{v_{turb}}{1 \text{ km s}^{-1}}\right] \text{ cm} \quad (54)$$

v_{turb} being the turbulent velocity of the cloud where the accretion originates. The spectrum described by equation (51) is essentially flat, corresponding to spectral indexes between -0.1 to $+0.1$. This is possible because, although the central parts can be quite optically thick, still a considerable fraction of the flow remains optically thin. Equation (51) is valid as long as $r_c \ll r_\infty$ or, conversely, as long the frequency is higher than

$$\nu_c = 169 \left[\frac{T}{10^4}\right]^{-0.64} \left[\frac{\dot{M}_{accr}}{10^{-7} M_\odot \text{ yr}^{-1}}\right]^{0.95} \left[\frac{M}{M_\odot}\right]^{-1.43}$$
$$\left[\frac{v_{turb}}{1 \text{ km s}^{-1}}\right]^{1.90} \left[\frac{\mu_e}{1.2}\right]^{-0.95} \text{ MHz} \quad (55)$$

At lower frequencies the spectrum will tend to the ordinary Rayleigh-Jeans tail of a black body, i.e. $F_\nu \propto \nu^2$. It is clear then that the spectrum will remain essentially flat down to frequencies well below a gigahertz and then decline like ν^2 at lower frequencies. Therefore, judging from the radio spectrum any accreting star could be mistaken for a compact HII region, except that the widths of the emission lines would be drastically different, i.e. several hundreds of km s^{-1} for an accretion flow as compared with the few tens of km s^{-1} expected for an HII region.

2.5. Non-thermal emission

The main non-thermal process which is relevant for young stars is synchrotron radiation. It is produced by electrons spiraling in a magnetic field and is regarded as a non-thermal process because it is an efficient process only with energetic electrons and these are abundant enough in non-thermal situations. This aspect makes much harder to uniquely define the parameters of a synchrotron emitting source or even to define the most probable situations, essentially because there are too many free parameters and too many unknowns. Therefore, here we will not discuss this emitting mechanism in detail but rather we shall limit ourselves to describing its main properties. More detailed summaries and discussions on the non-thermal emission processes in stars can be found in Dulk (1985), White (1985) and André (1987). The main characteristics of synchrotron radiation are i) a typical power law spectrum which, unlike to thermal spectra, increases with wavelength (until it starts becoming self-absorbed) ii) high polarization and iii) high brightness. This last property is due to the fact that very high energy particles are involved and, therefore, the limiting brightness temperature can be accordingly very high. Because of the same reason, a non-thermally emitting region can be much more compact than a thermal one. Both the small size, which implies a short transit time, and the high energy density, which implies a short energy loss time, concur in making the typical time scale of temporal variations of the non-thermal radio emission quite short. With the parameters appropriate for a star, such timescales range from hours to few days: this is orders of magnitude faster than it is possible in a thermal region. Therefore, short term variability can be taken as a fourth distinctive property of non-thermal emitting sources.

3. Radio and Infrared Observations of Young Stars

3.1. Ionized winds from YSOs?

In the previous sections we have discussed what type of emission we can expect from ionized winds. Now the question arises: *do YSOs have ionized winds?*. Two simple arguments demonstrate that indeed the winds from a considerable number of YSOs are ionized:

1) A well defined correlation between the radio continuuum flux at 6 cm and the Brγ line intensity has been found by Carr (1988) for a sample of YSOs. The radio flux is an order of magnitude fainter than the value expected for an HII region (*i.e.* an optically thin gas) with the same Brγ intensity. On the other hand, the observed fluxes are perfectly bracketed between the curves predicted for fully ionized, accelerated winds with γ values in the range 0.5-5.

2) As discussed in the following sections, for quite a number of YSOs the radio spectrum has a spectral index 0.6, i.e. the one predicted for a fully ionized wind (*cf.* Section 2).

We conclude that ionized winds are a common feature of YSOs. On the other hand, it is also clear that there are quite a number of YSOs with partially ionized or almost neutral winds (*cf.* Natta, this volume). It is possible that the ionization state of a YSO wind be a function of its evolutionary stage and/or of the initial conditions when the star was formed. For the moment, however, I feel that there is not enough statistics to clarify this point. In what follows I shall discuss the IR and radio observations of YSOs under the implicit assumption that their winds are *fully* ionized.

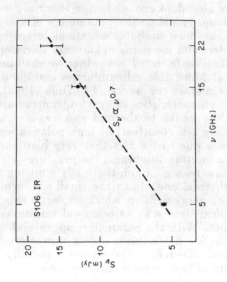

Fig. 7 - The radio spectrum of S 106-IRS4 (from Bally et al. 1983).

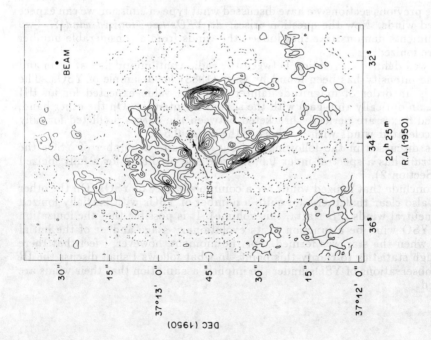

Fig. 6 - The VLA radio map of the S 106 region at 15 GHz (Bally et al. 1983).

3.2. The "well behaved" case of S 106-IRS4

One of the best studied young stars is S 106-IRS4. As its name suggests, it is a star associated with the HII region S 106 and has been discovered in the IR as a point like source (Sibille et al. 1975; Allen and Penston, 1975). In fact it cannot be seen in the optical because of about 21 magnitudes of visual extinction in front of it. The whole region has been the subject of a number of studies in the IR and in the radio (considering only papers relevant to the present discussion, see e.g.: Simon and Fischer 1982, Bally et al. 1983, Felli et al. 1984, 1985c and references therein). The radio observations show S 106-IRS4 to be a point like source (Bally et al. 1983; cf Fig. 6) still unresolved at 22 GHz (diameter $\leq 0.15''$, Felli et al. 1985c) and with a spectral index 0.7 ± 0.1 (cf Fig. 7). In the infrared, shortward of about $10\mu m$, it displays a clear excess relative to the stellar emission (Felli et al. 1984) and hydrogen emission lines (Brα, and Brγ; Simon and Fischer 1982, Felli et al. 1985c) whose HPFW is around 150 km s^{-1} and whose intensity ratios differ markedly from those of lines formed in HII regions. All of these aspects fit perfectly the picture of a star undergoing conspicuous mass loss. Combining all the available information Felli et al. (1984) deduced the following parameters for the star and the wind, for an assumed distance of 600 pc:

Star	Effective Temperature, T_{eff}		28000 K
	Radius, R		12 $R_\odot$
	Luminosity, L		$> 2 \times 10^4$ $L_\odot (\sim 10^5\ L_\odot)$
	Lyman Continuum Photon Flux, N_L		1.7×10^{48} photons s^{-1}
	Inferred Stellar Mass		$> 15 M_\odot$
	Matching Spectral Type		O9.5 V - B0 III
Wind	Mass Loss Rate, $\dot{M}$		$2.6 \times 10^{-6} M_\odot$ yr^{-1}
	Velocity Index, γ		0–0.25
	Initial Velocity, v_o		40–80 km s^{-1}
	Terminal Velocity, v_∞		~ 300 km s^{-1}

Thus, S 106-IRS4 looks like the perfect "theorist's dream", i.e. the object whose observations fit all theoretical predictions. We shall see in the following sections that reality is not always so nice nor so simple. Even in this case the results are not "boring": what at first glance seems just a replica of an ordinary early type star, after a deeper exam reveals a number of peculiar properties:

i) The ratio $\dot{M}/L$ is higher than 3×10^{-11} $M_\odot L_\odot^{-1}$ yr^{-1}: This is already an order of magnitude higher than it is measured in "normal" main sequence (MS) and supergiant stars (Tanzi et al. 1981) and is comparable to the values of WR stars ($[\dot{M}/L] = 4 \times 10^{-11} M_\odot L_\odot^{-1}$ yr^{-1}; Barlow et al. 1981). ii) The momentum carried by the wind exceeds that available in the radiation field: $(\dot{M}v_\infty)/(L/c) \sim 2$. This is intermediate between ordinary OB stars, that have ratios lower than 0.5, and WR stars for which that ratio is higher than 10. iii) The terminal velocity, $v_\infty = 300$ km s^{-1}, is much lower than that measured in MS and SG stars of comparable luminosity ($v_\infty \simeq 1500$ km s^{-1}, see e.g. Panagia and Macchetto, 1982). These aspects suggest that S 106-IRS4 represents an especially "active", short-lived, pre-main-sequence phase of the evolution of a moderately massive star. We will see in section 4 that most PMS stars seem to pass through such a phase before reaching the MS. Another interesting aspect is the marked absence of radio emission in a strip about $5''$ wide passing through the star in the SE-NW direction (cf Fig. 6). It can only be due to lack of ionizing photons in that direction because identical gaps are seen also in the IR, thus excluding the hypothesis of dust absorption occurring in a small disk/ring around the star. The most "natural" explanation is that of a genuine anisotropy of the stellar UV radiation, which is minimum on the equatorial plane and maximum toward the poles (Felli *et al.* 1984). The orientation of this "void" region is perpendicular to the direction of the large scale, bipolar flow and is essentially coplanar with the disk-like molecular region which projects in front of the star. Therefore, we see that one and the same geometric symmetry exists from a very small scale, i.e. the stellar radius $\sim 10^{12}$ cm, up to a very large scale, i.e. the size of the molecular disk of about 3×10^{18} cm. The high degree of "organization" of very different scales suggests that all of these aspects are intimately connected with each other and have a common origin.

3.3. General results of radio observations

Essentially all possibilities considered theoretically in the previous sections are observed in different objects. André (1987) in his review of the radio emission from young stars has defined four broad classes which are based on observed characteristics and can place the various objects into physically meaningful categories:

1) Steady sources, unresolved, steep spectrum with positive spectral index ($\alpha > 0.5$): thermal wind. Examples of this class are, in addition to the obvious S 106-IRS4, the BN object (Moran *et al.* 1983), Lk Hα 101 (Brown *et al.* 1976) Lk Hα 234 (Wilking *et al.* 1986), GL 490 (Campbell *et al.* 1986), CRL 961 (Snell and Bally 1986) and component #2 of Cep A (Hughes, 1985). All of these sources can be explained in terms of spherically symmetric, fully ionized winds with either constant or slightly increasing velocity.

2) Steady source, resolved, flat spectrum: thermal jet or ultracompact HII region. Examples: G129.58-0.04, NGC 2071, S 68-FIRS1, L1551-IRS5 (Snell and Bally 1986). The radio spectrum of this type of sources does not follow a simple power law, the spectral index being steeper at lower frequencies. An alternative to the HII region model is that such a spectrum is due to accretion.

3) Steady source, unresolved, flat spectrum: thermal accretion flow or non-thermal emission? Typical examples are S1, VS 14 and WL 5 in the ζ Oph cloud (André *et al.* 1987). The stable emission observed for these objects cannot be readily explained as due to flare activity but, on the other hand, being unresolved with the VLA their surface brightness seems to be too high for a ordinary thermal source.

4) Variable source, unresolved: non-thermal stellar flares. Examples are: DoAr 21 (= ROX 8), ROX 31 (Feigelson and Montmerle 1985); V410 Tau, HP Tau/G2 and HP Tau/G3 (Cohen and Bieging 1986). The strong variability on short time scales identifies these sources as definitely non-thermal. André, Montmerle and Feigelson (1987) note that their fluxes are comparable to those of post-main sequence RS CVn systems and suggest that they may be associated with the most luminous X-ray selected pre-main-sequence stars.

As remarked by André (1987) the objects in classes 1) and 2), i.e. those which do have stellar winds, show some definite evidence of large scale outflows, e.g. optical jets and/or association with HH objects for T Tau stars, bipolar flows for more powerful IR stars, so that they can easily fit with the same picture as drawn for S 106-IRS4 (cf. section 3.1).

3.4. Radio Surveys

If now we have an idea of what individual stars may do, still it isn't clear how many of them do that and for how long, i.e. we know about the phenomena that may occur but we cannot place them in the evolutionary history of a star. A step forward in that direction can be taken by making complete surveys in regions of star formation. In such a way, not only can one determine the properties of individual sources which were detected but also consider the "non-detections" to perform statistics which can tell us about the duration of the various phases in different classes of objects. Among the studies of this sort done in recent years I have selected three that are particularly relevant for our interests.

1) Bieging, Cohen and Schwartz (1984) — They made VLA observations at 4 frequencies (1.5, 4.9, 15 and 22.5 GHz) of all T Tau stars with $\log L/L_\odot > 0.2$ associated with the Taurus-Auriga dark cloud (46 objects). Their results can be summarized as follows:
- 12 stars out of the 46 (26%) were detected at one frequency at least.
- For 8 of them the spectral index was determined:
    3 had a steep spectrum ($\alpha > 1$) which could be interpreted as thermal emission; however, for one of them, V410 Tau, the fast variability indicates a non-thermal origin of the emission.
    2 had a spectrum consistent with a stellar wind ($\alpha \sim 0.6$)
    2 had a flat spectrum but were unresolved: they could be explained as either non-thermal or accretion emission.
    1 had a negative spectral index (HP Tau/G2, $\alpha = -0.35$) which speaks clearly of non-thermal radiation.
  Therefore, in about half of the cases the emission is of thermal origin and, among these, in about half it originates from stellar winds.
- The missed detection of 34 of the surveyed stars implies an upper limit to their mass loss rate of $\dot{M} \sim 2 \times 10^{-8} M_\odot \text{ yr}^{-1}$ [v/200 km s^{-1}].
- All of the 5 stars associated with HH objects do emit in the radio.

2) Snell and Bally (1986) — They surveyed (complete VLA mapping at 1.4, 5 and 15 GHz, to limiting fluxes of about 0.3 mJy at 5 GHz) 12 regions containing molecular outflows. The main results are:
- Sources were detected in 8 out of the 12 surveyed regions.
- A total of 43 sources were detected. However, based on a statistical argument of source counts, it is concluded that about half of them are likely to be extragalactic.
- 15 of the detected sources were identified with IR sources. In particular, all IR sources associated with the outflows having a bolometric luminosity higher than 100 $L_\odot$ were detected (9 objects). In total, the IR sources detected in the radio amount to about 2/3 of all genuine sources contained in these outflow regions.
- For 7 sources the spectral index was determined:
    3 had a spectral index and a size fully consistent with a stellar wind interpretation: CRL 961, Lk Hα 234 in addition to our friend S 106-IRS4.
    4 had a flat index: two (L1551 IRS5 and NGC 2071 IRS 1-3) are clearly extended and, therefore, are likely to be compact HII regions, whereas all possibilities, except a stellar wind hypothesis, remain possible for the other two.

- For all sources, if the radio emission is formed by a steady, fully ionized wind, the implied mass loss parameters fail by more than an order of magnitude to explain the energetics of the associated molecular outflows.

3) André, Montmerle and Feigelson (1987) — They made a VLA survey of the ρ Oph dark cloud, covering an area of 4 square degrees at 1.4 GHz (limiting fluxes 1-3 mJy) and observing the central part also at 5 GHz with a limiting flux of 1 mJy. Over the whole region, they detected 93 objects of which they judged that only 13 were truly stellar objects. Grouping the detected objects according to a variety of properties, they find:

- 7, i.e. ∼ 60% of the sources, can be identified with near or far IR sources. However, among the three classes defined by Lada (1986; see section 3.5) only class III objects are detected: in particular, the detection rates are 0/5 for class I, 0/21 for class II and 4/6 for class III. Three more IR sources detected in this survey could not be classified within the Lada's scheme.
- 10, i.e. 77% of the detected stellar sources, have a spectral index either flat or increasing with frequency. One of them, DoAr 21 is known to be non-thermal because of its variability. Most of the others are likely to be thermal sources.
- 5 are associated with X-ray stellar sources. Since this cloud contains 47 such stars, only about 10% of the X-ray stars emit in the radio.
- None of the 26 visible PMS (emission-line or T Tau) stars associated with the cloud was detected. This places an upper limit to their mass loss rate of $\dot{M} < 7 \times 10^{-8} M_\odot$ yr^{-1} [v/200 km s^{-1}].

3.5. Infrared observations

Infrared measurements are important for the study of young stars for two main reasons: first because the high extinction of star forming clouds, that prevents us from "penetrating" deeply into them at optical wavelengths, becomes less and less important at longer wavelengths. The second reason is that, especially for very young objects, the original circumstellar cocoon is still so thick that the bulk of the stellar energy is actually radiated in the infrared. In fact, most of the high luminosity YSOs can only be detected in the IR, because their short evolutionary time makes it hard for the circumstellar material to dissipate before the star has reached the main sequence. A historically important example is the case of the Becklin and Neugebauer object (BN) which was discovered as an extremely cool object ($T \sim 700K$) appearing in a map of the Orion Nebula at 2.2 μm and was immediately recognized as a possible protostar (Becklin and Neugebauer, 1967).

A systematic study of the energy distributions of young stellar objects has been done by Wilking and Lada (1983) and Lada and Wilking (1984) who made broad-band photometric observations of the embedded population in the core of the ρ Oph dark cloud. From an analysis of the derived energy distributions (λF_λ vs λ) they identified three main morphological classes (see fig. 8):

Class I – Sources with energy distributions broader than a black body and which are rising longward of 2 μm.
Class II – Sources with an energy distribution broader than a black body but which is flat or decreasing longward of 2 μm.
Class III – Sources whose energy distributions can be modeled with reddened black bodies and show no or little excess near infrared emission.

Class I sources were all invisible and deeply embedded in the cloud whereas class II and III sources were associated mostly with visible stars. In particular,

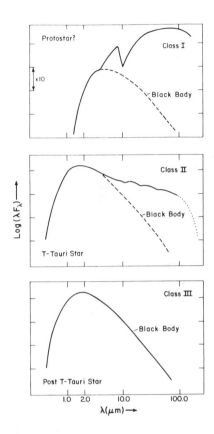

Fig. 8 – Schematic representation of the classification of young stellar objects on the basis of their IR spectra (Lada and Wilking 1984).

nearly all previously known class II objects were classified as T Tau stars. Moreover, the data available in the literature for a number of stars of low and moderate luminosity ($L < 10^3 L_\odot$) fit equally well in this classification scheme. In particular, inspecting the Rydgren et al. (1984) catalog of the infrared spectra for 61 T Tau stars in the Taurus-Auriga complex one finds that among the 41 stars for which the wavelength coverage is complete enough to classify them, 34 falls in class II, 5 in class III and none in class I. Lada (1986) suggests that such an empirical sequence may represent an evolutionary sequence in which class I objects are protostars, just formed and still deeply embedded in their original cloud. The circumstellar envelopes dissipate gradually with time, thus producing class II and III morphologies. The fact that many (but not all) class I objects and many (but not all) T Tau stars are associated with energetic outflows and/or stellar wind activity suggests that these processes represent the agents that dissipate the circumstellar envelopes and drive the evolution from class I to class III. From all of this it is clear that, due to the strong dust emission, in most cases (class I and class II objects) one cannot use IR continuum measurements to safely estimate the properties of the stellar wind. On the other hand, line emission is not affected by the presence of dust, except for the obvious effect of extinction. In addition, lines have the great advantage of carrying information about the velocity of the wind; such an information is completely lost in the continuous radiation. Therefore, IR emission lines are very precious for specifying the detailed properties of the stellar winds. The most intense lines observed from YSOs are generally hydrogen emission lines such as Brα (4.05 μm), Brγ (2.17 μm) or Pfγ (3.74 μm). By using the theory outlined in section 2.3 one can determine both the mass loss rate and the characteristics of the velocity field in the wind. A number of searches have been carried out on YSOs of all luminosities from $\sim 1 L_\odot$ up to $10^6 L_\odot$ (among the most recent studies we recall Thompson 1982, Simon et al. 1983, Thompson et al. 1983, Persson et al. 1984, Smith et al. 1987, Thompson 1987, Evans et al. 1987, Persson et al. 1988). The main result is that strong lines are indeed observed, with a strength much higher than predicted from the measured radio flux if the emitting region is assumed to be optically thin at all relevant wavelengths (the so-called HII region theory). Such a result, surprising at first glance, is the obvious consequence of the wind properties, in particular the high optical depth. In fact, the wind is optically thick longward of few microns and the optical depth in the radio is several orders of magnitude higher than in the IR. Therefore, while both the IR and the radio fluxes are lower than in the optically thin case, the effect is much more conspicuous in the radio where the optical depth is higher. Hence, the ratio IR/radio in a wind can become considerably higher than it is in an HII region.

Another important result obtained from line measurements is the realization (e.g., Thompson 1982) that the ionization in the wind does requires a large number of ionizing photons, in many cases highly in excess to what the star can supply in the form of Lyman continuum photons (i.e., photons capable of ionizing the ground level of hydrogen, $h\nu > 13.6$ eV). This problem will be discussed in some detail in the next section.

4. Statistical Properties of Stellar Winds from Young Stars

Here, we discuss the available information on winds from YSOs to try to understand their nature and their relation to the process of star formation. A more complete account of this analysis can be found in Panagia (1985).

The data were collected from various sources in the literature, namely Cohen et al. (1982), Felli et al. (1982 and 1984), Persson et al. (1984), Simon et al. (1983), Thompson (1982), Thompson et al. (1983). The derived parameters, such as the mass loss and ionization rates, were directly taken from the original articles and may not be fully consistent with each other in that various authors sometimes have used different models to interpret their data. No attempt was made to 'homogeneize' the data and the possible discrepancies arising from the use of different models have been regarded as additional errors which sum up to the observational uncertainties.

The quantities that we can consider to define the properties of YSO winds are:

1. The total radiative luminosity, L_{tot}, which includes the energy observed in the visual and, possibly, in the infrared plus the energy needed to ionize the wind (see point 3).

2. The maximum expansion velocity, v_{max}, as deduced from line profile observations.

3. The wind ionization luminosity, L_{ion}, i.e. the energy required to keep the wind gas steadily ionized. It is computed as the product of the number of ionizations occurring in the wind (as deduced from IR/optical line or IR/radio continuum observations) times the H ionization energy, 13.6 eV.

4. The mass loss rate, $\dot{M}$. It is deduced from either line measurements (mostly

IR lines but sometimes optical, e.g. Hα) or IR/radio continuum data.

5. The momentum flux defined as the product of the mass loss rate times the maximum wind velocity. The actual momentum carried away by the wind can be higher than this estimate, by a factor of ~ 2, because part of the momentum is used to win the gravitational force (Abbott, 1980).

6. The 'mechanical' luminosity, L_{kin}, of the outflowing wind defined as the product of the mass loss rate times $1/2v_{max}^2$. Note that the total kinetic energy carried by the wind away from the star may be higher because part of it has to be used to overcome the gravitational potential and part may have been transformed into heat.

The total radiative luminosity will be used as the independent variable because it is the observational quantity which is essentially determined by the stellar mass and, therefore, best characterizes the properties of the star irrespective of the presence of the wind. Plots of the other quantities as a function of L_{tot} are displayed in Figures 9–13.

The first interesting result is that in all cases the various quantities follow well defined trends as a function of L_{tot}. Clearly, there is some dispersion present in all plots but it is always much smaller than the total variation of the involved quantities and, therefore, there is no doubt that the apparent correlations are real.

The second important aspect is that there is good continuity in the properties of the winds of YSOs over as much as six orders of magnitude in the luminosity from, say, about 1 $L_\odot$ up to more than a million solar luminosities and correspondingly for stellar masses in the range from about 1 $M_\odot$ up to almost 100 $M_\odot$. Such a continuity of properties over a quite broad interval strongly suggests a great uniformity in the processes which induce and govern stellar winds in the PMS phase.

Looking at individual plots, we see (Fig. 9) that the maximum expansion velocity is of few hundred km s^{-1} for *all* stars. Also, it is apparent that v_{max} is either lower or, at most, equal to the escape velocity from the stellar surface ($v_{esc} = 2 GM/R$): this is true whether we compute v_{esc} using the stellar parameters of main sequence (MS) stars or if we adopt larger radii for any given luminosity as appropriate for PMS stars (hereafter we assume that a typical PMS star has an effective temperature a factor of 1.58, or 0.2 dex, lower than, and a luminosity equal to, that of a MS star of equal mass). Moreover, the observed velocities are much lower than those measured for the winds of MS stars, at least for those of high luminosity mass (e.g. Panagia and Macchetto, 1982). This suggests that the acceleration mechanism is different from that operating in MS stars. Furthermore, the fact that the terminal velocity is never higher than the escape velocity at the stellar surface suggests that the acceleration process is rather gradual and must operate up to relatively large distances from the stellar surface.

Looking at Figure 10 we see that L_{ion} is very well correlated with L_{tot} over almost 6 orders of magnitude with a modest dispersion of $\sim \pm 1/2$ dex. The best-fitting power-law relationship is

$$L_{ion} \propto L_{tot}^{1.2 \pm 0.2}$$

Alternatively, the experimental correlation can be expressed as

$$L_{ion}/L_{tot} \simeq 8\%$$

to within a factor of three. In most cases, the ionizing flux is much higher than the stellar Lyman continuum flux but lower than the Balmer continuum flux expected

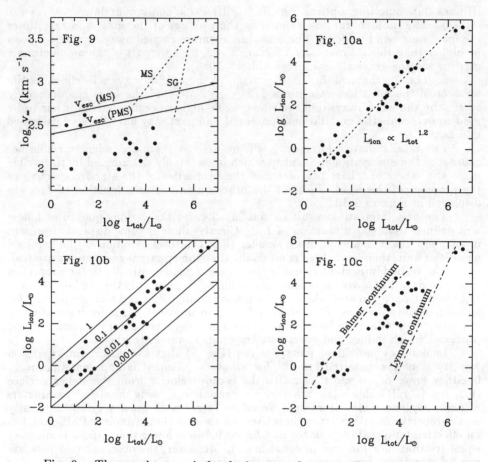

Fig. 9 – The maximum wind velocity as a function of the total luminosity. The terminal velocities for MS star and for supergiant winds are displayed. Also shown are the escape velocities curves for MS and PMS stars.

Fig. 10 – The ionization luminosity as a function of the total luminosity: a) The best-fitting power law is plotted; b) Lines of constant ratio L_{ion}/L_{tot} are shown; c) The curves of the ionizing radiation in the Lyman and the Balmer continua are compared with the data.

on the basis of theoretical stellar atmosphere calculations (cf. Fig. 10c). Therefore, the wind could be ionized by Balmer continuum photons as suggested by Thompson (1982) and Simon *et al.* (1983). This problem has been addressed in detail by Natta *et al.* (1988). Alternatively, there may be a process which is capable to convert into ionization an energy corresponding to about 8% of the total observed stellar luminosity. Such an energy, however, cannot be provided at the expenses of the kinetic energy of the wind because this is generally smaller than 1% of the stellar luminosity (cf. Fig. 13). Among the number of possible processes the most plausible are:

i) The presence of a hot corona at the base of the wind: its temperature should not exceed few hundred thousand, and be higher than several tens of thousand, Kelvin degrees in order to be able to provide the necessary ionizing flux but escape detection both at UV wavelengths (e.g. IUE range 1150-3200 Å) and in the X-ray domain.

ii) Collisional ionization and excitation by thermal processes: this implies that the wind gas temperature be as high as 15000 K in order to make hydrogen 50% ionized. Clearly some heating agent must keep the gas at such a temperature: a possibility is dissipation of Alfven waves which have been proposed to explain the acceleration of the wind (Lago, 1984).

iii) Radiation produced by infall of matter onto the stellar surface: this accretion may represent the terminal phases of the very star formation process. In this case mass accretion and mass loss would coexist with each other: the former would prevail on and near the equatorial plane of the star and would originate from the molecular disk which gave birth to the star (Pudritz, 1985). The latter would dominate in the polar directions and, in addition to the ordinary wind manifestations, would also be responsible for the ionized component of the bipolar flows associated with YSOs (e.g. see the discussion on S 106 by Felli et al. 1984).

As for the mass loss rate and the momentum flux (Figures 11 and 12) we see that again there is a good correlation of both quantities with L_{tot}, of the form

$$\dot{M} \propto L_{tot}^{0.67 \pm 0.04}$$

$$\dot{M}v \propto L_{tot}^{0.68 \pm 0.06}$$

The fact that the dependence of $\dot{M}$ and $\dot{M}v$ on L_{tot} is virtually the same is a direct consequence of the velocity being almost independent of the luminosity. Both the mass loss rate and the momentum flux are considerably higher that what one measures in "ordinary" stars, both MS or supergiant stars (e.g. Panagia and Macchetto, 1982; Felli and Panagia, 1981) while they are comparable, after a suitable scaling, to the values observed in WR stars (e.g. Barlow et al. 1981). This indicates that stellar winds in the PMS phase are much more "active" than they will be in subsequent phases of the evolution. As a corollary, it appears that the mechanism which produces the mass loss is different from the one that operates in MS and supergiant stars but, because of the continuity of observed properties, it is likely to be the same for all YSOs. We also see that the momentum flux $\dot{M}v$ is generally higher than the momentum carried by the radiation, L/c (cf. Fig. 12). Such high values of the momentum flux demand that the wind acceleration be due to either multiple scattering of the stellar radiation or to some non-radiative process (e.g., deposition of Alfven waves: Lago 1984). The former mechanism is able to account for the wind acceleration in early type stars (Panagia and Macchetto, 1982) but in order to work with comparable efficiency for PMS stars would require a high UV luminosity: this condition would also satisfy the ionization requirements but would imply a rather peculiar spectrum in that no trace of a strong UV continuum component has been observed in PMS stars with IUE (e.g., Gahm 1980 and references therein).

Finally, an inspection of Figure 13 reveals that the mechanical luminosity appears to correlate quite well with the radiative luminosity, i.e.

$$L_{kin} \propto L_{tot}^{0.76 \pm 0.11}$$

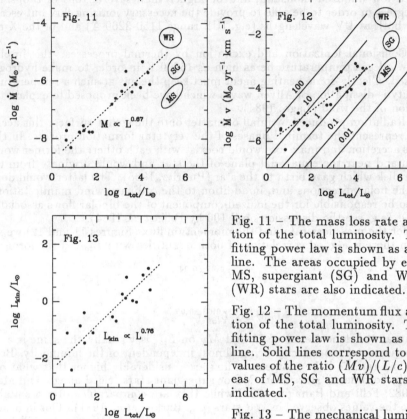

Fig. 11 – The mass loss rate as a function of the total luminosity. The best-fitting power law is shown as a straight line. The areas occupied by early type MS, supergiant (SG) and Wolf-Rayet (WR) stars are also indicated.

Fig. 12 – The momentum flux as a function of the total luminosity. The best-fitting power law is shown as a dashed line. Solid lines correspond to constant values of the ratio $(\dot{M}v)/(L/c)$. The areas of MS, SG and WR stars are also indicated.

Fig. 13 – The mechanical luminosity as a function of the total luminosity. The best-fitting power law is shown as a dashed line.

On the other hand, in all cases L_{kin} corresponds to a minor fraction of the total radiative luminosity

$$L_{kin} \simeq 0.01\, L_{tot}$$

and, therefore, it cannot play any major role in the interplay star/wind/environment.

In summary we can conclude as follows:

1. One and the same mechanism appears to be responsible for the mass loss in all YSOs. The mechanism itself, however, is not identified.

2. The wind ionization requires an agent external to the wind but the ordinary Lyman continuum radiation from the star is not enough for this purpose. Either Balmer continuum ionization or a more exotic processes is required to account for the wind ionization.

3. The acceleration of the wind is likely not to be due to radiation and may be produced by direct deposition of energy into the wind.

Finally it is worth considering the large scale properties of the winds of YSOs and comparing them with the properties of the molecular bipolar flows which appear to be associated with YSOs. Measurements of stellar winds probe regions of modest size, say less than few 10^{14} cm, which corresponds to less than $0.1''$ at the distance of the closest T associations. On the other hand, one can currently observe ordered flows much farther away from the star but which undoubtedly originate from the star itself. This is, for example, the case of DG Tau B which displays an ionized, narrow jet (opening angle less than 5 degrees; Mundt and Fried, 1983) or S106-IRS4 (see section 3.1) These YSOs and bipolar flows or jets appear to be associated with relatively massive molecular disks. They also appear to be aligned with the molecular flows which may extend over an even larger scale and seem to arise from the molecular disks moving approximately perpendicularly to the disk itself (i.e., along the polar axis). Examples of these structures are S 106 (e.g., Felli et al. 1984) or L 1551 (Lada, 1985 and references therein). In all cases, the momentum carried by the molecular flows is more than an order of magnitude higher than that carried by the ionized flow (Lada, 1985). Therefore, although it cannot be the stellar wind to push the molecular flow (because of the momentum problem) nor can the molecular flow directly produce the ionized flow and/or the stellar wind (because the former does not reach regions so close to the central star where the latter is observed to exist) a tight connection between the two phenomena must exist. It is quite possible that they both represent different manifestations of the same general phenomenon governing the formation of stars of any mass.

5. References

Abbott, D. C., 1980, *Ap. J.*, **242**, 1183.
Allen, D. A., and Penston, M. V., 1975, *M.N.R.A.S.*, **172**, 245.
André, Ph., 1987, in "Protostars and Molecular Clouds", eds. T. Montmerle and C. Bertout (Saclay: CEA/Doc).
André, Ph., Montmerle, T., and Feigelson, E. D., 1987, *A. J.*, **93**, 1182.
Bally, J. B., Snell, R. L., and Predmore, R., 1983, *Ap. J.*, **272**, 154.
Barlow, M. J., Smith, L. J. and Willis, A. J., 1981, *M.N.R.A.S.*, **196**, 101.
Becklin, E. E., and Neugebauer, G., 1967, *Ap. J.*, **147**, 799.
Bieging, J. H., Cohen, M., and Schwartz, P. R., 1984, *Ap. J.*, **282**, 699.
Brown, R. L., Broderick, J. J., and Knapp, G. R., 1976, *M.N.R.A.S.*, **175**, 87P.
Campbell, B., Persson, S. E., and McGregor, P. J., 1986, *Ap. J.*, **305**, 336.
Carr, J., 1988, Ph.D. Thesis, The University of Texas at Austin.
Castor, J. I., 1970, *M.N.R.A.S.*, **149**, 111.
Castor, J. I., and Simon, T., 1983, *Ap. J.*, **265**, 304.
Chiuderi, C., and Torricelli Ciamponi, G., 1978, *Astr. Ap.*, **69**, 333.
Cohen, M., and Bieging, J. H., 1986, *A. J.*, **92**, 1396.
Cohen, M., Bieging, J. H. and Schwarz, P. R., 1982, *Ap. J.*, **253**, 707.
Dulk, G. A., 1985, *Ann. Rev. Astron. Astrophys.*, **23**, 169.
Evans, N. J., II, Levreault, R. M., Beckwith, S. and Skrutskie, M., 1987, *Ap. J.*, **320**, 364.
Feigelson, E. D., and Montmerle, T., 1985, *Ap. J. (Letters)*, **289**, L19.
Felli, M. and Panagia, N., 1981, *Astr. Ap.*, **102**, 424.
Felli, M., Gahm, G. F., Harten, R., Liseau, R. and Panagia, N., 1982, *Astr. Ap.*, **107**, 354.
Felli, M., et al. 1984, *Astr. Ap.*, **135**, 261.

Felli, M., Oliva, E., Stanga, R., and Panagia, N., 1985a, in "(Sub)Millimeter Astronomy", eds. P. Shaver and K. Kjaer, p. 503.
Felli, M., Stanga, R., Oliva, E., and Panagia, N., 1985b, *Astr. Ap.*, **151**, 27.
Felli, M., Simon, M., Fischer, J., and Hamann, F., 1985c, *Astr. Ap.*, **145**, 305.
Felli, M., and Panagia, N., 1990, in preparation.
Gahm, G. F., 1980, in "The Universe at Ultraviolet Wavelengths: The First Two Years of IUE," NASA CP-2171, p. 105.
Hamann, F., and Simon, M., 1986, *Ap. J.*, **311**, 909.
Hughes, V. A., 1985, *Ap. J.*, **298**, 830.
Krolik, J. H., and Smith, H. A., 1981, *Ap. J.*, **249**, 628.
Lada, C. J., 1985, *Ann. Rev. Astron. Astrophys.*, **23**, 267.
Lada, C. J., 1986, in "Star Forming Regions", eds. M. Peimbert and J. Jugaku (Dordrecht: Reidel), p. 1.
Lada, C. J., and Wilking, B. A., 1984, *Ap. J.*, **287**, 610.
Lago, M. T. V. T., 1984, *M.N.R.A.S.*, **210**, 323.
Lamers, H. J. G. L. M., and Waters, L. B. F. M., 1984, *Astr. Ap.*, **136**, 37.
Marsh, K. A., 1975, *Ap. J.*, **201**, 190.
Mezger, P. G., and Henderson, A. P., 1967, *Ap. J.*, **147**, 471.
Moran, J. M., Garay, G., Reid, M. J., Genzel, R., Wright, M. C. H. and Plambeck, R. L., 1983, *Ap. J. (Letters)*, **271**, L31.
Mundt, R. and Fried, J. W., 1983, *Ap. J. (Letters)*, **274**, L83.
Natta, A., Giovanardi, C. and Palla, F, 1988, *Ap. J.*, **332**, 921.
Olnon, F., 1975, *Astr. Ap.*, **39**, 217.
Panagia, N., 1985, *Physica Scripta*, **T11**, 71.
Panagia, N. and Felli, M., 1975, *Astr. Ap.*, **39**, 1.
Panagia, N. and Macchetto, F., 1982, *Astr. Ap.*, **106**, 226.
Panagia, N., Oliva, E., and Felli, M., 1990, in preparation.
Persson, S. E., Geballe, T. R., McGregor, P. J., Edwards, S. and Lonsdale C. J., 1984, *Ap. J.*, **286**, 289.
Persson, S. E., McGregor, P. J., and Campbell, B., 1988, *Ap. J.*, **326**, 339.
Pudritz, R. E., 1985, *Ap. J.*, **293**, 216.
Reynolds, S. P., 1986, *Ap. J.*, **304**, 713.
Rodriguez, L. F., and Cantó, J., 1983, *Rev. Mexicana Astron. Astrof.*, **8**, 163.
Rydgren, A. E., Schmelz, J. T., Zak, D. S., and Vrba, F. J., 1984, *Pub. U.S. Naval Obs.*, Second Series, Vol. XXV, Part I.
Schmid-Burgk, J., 1982, *Astr. Ap.*, **108**, 169.
Sibille, F., Bergeat, J., Lunel, M., and Kandel, R., 1975, *Astr. Ap.*, **40**, 441.
Simon, M., Felli, M., Cassar, L., Fischer, J. and Massi, M., 1983, *Ap. J.*, **266**, 623.
Simon, M., and Fischer, J., 1982, *B.A.A.S.*, **14**, 925.
Smith, H. A., Fischer, J., Geballe, T. R., and Schwartz, P. R., 1987, *Ap. J.*, **316**, 265.
Snell, R. L., and Bally, J. B., 1986, *Ap. J.*, **303**, 683.
Sobolev, V. V., 1958, in "Theoretical Astrophysics", ed. V. A. Ambartsumyan (New York: Pergamon Press), p. 482.
Tanzi, E. G., Tarenghi, M., and Panagia, N., 1981, in "Effects of Mass Loss on Stellar Evolution", eds. C. Chiosi and R. Stalio, (Dordrecht: Reidel), p. 51.
Thompson, R. I., 1982, *Ap. J.*, **257**, 171.

Thompson, R. I., 1987, *Ap. J.*, **312**, 784.
Thompson, R. I., Thronson, H. A. and Campbell, B., 1983, *Ap. J.*, **266**, 614.
White, R. L., 1985, *Ap. J.*, **289**, 698.
Wilking, B. A., and Lada, C. J., 1983, *Ap. J.*, **274**, 698.
Wilking, B. A., Mundy, L. G., and Schwartz, R. D., 1986, *Ap. J. (Letters)*, **303**, L61.
Wright, A. E., and Barlow, M. J., 1975, *M.N.R.A.S.*, **170**, 41.

Nino Panagia and Sylvie Cabrit

Antonella Natta

THE PHYSICS OF NEUTRAL WINDS FROM LOW MASS YOUNG STELLAR OBJECTS

A. NATTA and C. GIOVANARDI
Osservatorio di Arcetri
Largo Fermi 5
50125 Firenze
Italia

ABSTRACT. The physical structure of winds ejected by low luminosity, pre–main-sequence stars is investigated, using the results of numerical calculations available in the literature as well as analytical approximations. These latter allows us to derive order-of-magnitude estimates of the most important quantities, such as the ionization fraction. The topics covered in this review include the ionization and excitation structure of hydrogen and of sodium, intensities and profiles of some interesting lines and the gas temperature.

Line profile calculations can be compared to spectroscopic observations of suitable stars to derive wind parameters such as the rate of mass-loss and the gas temperature. An example of a diagnostic of this kind, using various infrared hydrogen recombination lines and the Na D doublet at 5900 Å in T Tauri stars, is briefly discussed. It is found that T Tauri stars showing blue-shifted absorption in the Na I lines have a rate of mass-loss in the range 3×10^{-8}–$3 \times 10^{-7} M_\odot yr^{-1}$, and a gas temperature of 6000–7000 K in the innermost regions of the wind. For those stars which have been searched for CO outflows, the wind $\dot M$ values are in good agreement with those inferred from the molecular outflow observations, confirming the scenario where the outflows are driven by fast stellar winds via a momentum conserving shock.

The temperatures derived from observations are compared to those produced by various heating mechanisms.

1. Introduction

A low mass star, in its path toward the main-sequence, evolves from an accreting core, still deeply embedded in its parental cloud, to an optically visible star in hydrostatic equilibrium, whose mass is basically constant, and which has dispersed most of the surrounding matter. Mass loss is apparently an important phenomenon during most of the protostellar and pre–main-sequence evolution, and probably crucial to the formation process itself (Shu, Adams and Lizano 1987). It is generally believed that mass-loss begins at a very early phase of the evolution, co-existing with the accretion for a significant fraction of time, and that it continues (probably at a decreasing rate) well into the pre–main-sequence phase. However, our understanding of the nature of the mass-loss process has been hampered by the fact that observations at millimetric wavelength (the only possible wavelength for deeply reddened young objects) have revealed aspects of the mass-loss phenomenon different from what was known in optical objects.

Mass outflow in low mass, optically visible pre–main-sequence stars (T Tauri stars, TTS) was first inferred by Herbig 1962 from the existence in some stars of blue-shifted absorption components superimposed on the strongest emission lines, such as H_α, Ca II H and K. The maximum expansion velocities derived from the shift of these absorption components were of the order of 100-200 km s^{-1}, comparable with the escape speed for TTS and with the width of the emission components of the same lines. The first determinations of the mass-loss rate $\dot{M}$ (and the only ones for a long period of years) were obtained by Kuhi 1964, 1966, who found values in the range 0.3 to 5.8x10^{-7} $M_\odot$yr^{-1} for a group of 8 stars, by fitting the observed profiles of some of the brightest lines. The uncertainties of these determinations were however very large (up to two orders of magnitude, according to De Campli 1981); difficulties in identifying an ejection mechanism which could account for the higher values of $\dot{M}$ (De Campli 1981; Hartmann, Edwards and Avrett 1982) have favoured values lower than those originally derived by Kuhi.

In the case of embedded young stellar objects, the evidence for mass outflow has come from molecular observations, mostly of CO emission, which have shown matter moving away from a central core at velocities of 10-20 km s^{-1}, extended over a large area (10^4-10^5 AU), often with bipolar morphologies (see, for a review, Lada 1985 and this book). It is generally believed that the CO outflow is not matter directly ejected by the central core, but rather ambient molecular gas driven by a fast stellar wind (similar to that observed in TTS) through a momentum conserving shock. This model was first proposed by Snell, Loren and Plambeck 1980 for the case of L1551, a remarkable outflow, showing a high degree of collimation, associated with a source of about 30 $L_\odot$ and about 1 $M_\odot$ (Adams, Lada and Shu 1987). If this is the case, then, by equating the momentum content of the outflow to the wind momentum deposition rate integrated over its lifetime (and admitting they have the same lifetime), we must have

$$\dot{M}v_w \simeq \dot{M}_{CO}v_{CO} \qquad (1)$$

where $\dot{M}$ is the wind mass loss rate, v_w and v_{CO} are respectively the wind and the outflow expansion velocities, and $\dot{M}_{CO}$ is a molecular flow rate simply obtained dividing the mass of the high velocity molecular gas by the estimated outflow lifetime. Assuming a value of v_w typical of TTS, since all the rest can be measured, we can use eq.(1) to derive $\dot{M}$. The $\dot{M}$ values derived in this way can be very high. For the case of L1551, Snell and Schloerb 1985 obtain $\dot{M}$=1.5x10^{-6} $M_\odot$yr^{-1}, much larger than any value ever proposed for TTS.

In this context, a particularly interesting class of objects are those TTS which are associated with molecular outflows. Searches for CO ouflows around TTS have been carried out by several authors, for a total of about 50 objects (Edwards and Snell 1982; Kutner et al. 1982; Calvet et al. 1983; Levreault 1988). Of these, only 4 have been positively detected. This small group of objects offers the best chance to test the standard model, and therefore to assess the reliability of the derived $\dot{M}$ for the obscured objects as well.

An important step in this direction was the work of Evans et al. 1987 who measured the luminosity of hydrogen infrared recombination lines for a sample of objects, including the 6 TTS which had been searched for CO outflows. The aim was to compare the observations to the predictions of wind models having the value of $\dot M$ derived from CO data, under the momentum conservation assumption. Near-infrared hydrogen recombination lines are well suited for these studies, due to the reduced extinction compared to the optical lines. For the TTS in the Evans et al. sample (see also Suto, Mizutani, and Maihara 1989), the observed Br_α luminosities appeared consistently lower than the prediction of the wind models. Although other remedies have been suggested, such as energy conserving shocks (Dyson 1984, Kwok and Volk 1985) and accretion disk winds (see Uchida and Shibata 1985, Pudritz and Norman 1986, Pudritz this book), Evans et al. noted that the wind models they used were computed under a set of assumptions: spherical symmetry, LTE, constant temperature, and full ionization. These models had been originally computed by Simon et al. 1983 for studying winds from luminous ($L \geq 100$ $L_\odot$) young stellar objects, where a copious Lyman continuum warrants in fact almost full ionization. It is not improbable that the hypothesis of full ionization could be the one responsible for the observed discrepancy. In fact, if the wind is almost neutral its Br_α luminosity, for the same $\dot M$, will be strongly reduced.

As a consequence of the increasing interest in the mass-loss in low luminosity objects, the problem of the wind diagnostic in TTS has received a new attention. In principle a detailed modeling of the wind should start from a minimum number of assumptions regarding:

i) the photospheric properties, i.e., some steller parameters at the base of the wind such as the radius, density, gas temperature and radiation field.

ii) the acceleration mechanism, i.e., its nature (thermal, acoustical, magnetical, etc.), its spatial simmetry (spherical or localized) and the characteristic parameters for the energy and momentum transfer (such as damping lengths, etc.).

The model should then yield the runs with radius of the geometry and velocity of the flow, of the gas density and temperature, of the excitation and ionization state for the various atoms and finally the intensities and profile shapes of the spectral lines, the continuum radiation, etc., which can be directly compared with the observations. Self-consistent calculations requires solving the coupled equations of dynamics, radiation transfer and thermal balance. Indeed, there have been attempts to build sets of such self-consistent models, mostly within the frame of Alfvenic winds (Hartmann et al. 1982; Holzer, Flåand Leer 1983). More often, in order not to be restricted to any specific mechanism of mass ejection and gas heating, the approach adopted was somewhat different. Some of the physical quantities characterizing the wind (in general the stellar properties and the run with r of the flow velocity v and temperature T_g ,i.e., the quantities directly influenced by the acceleration process) were specified a priori, and used to determine, for different values of the other parameters, such as the mass loss rate $\dot M$, the behaviour of some observables. Examples of this approach can be found in Kuhi 1964, Croswell, Hartmann and Avrett 1987, Natta et al. 1988a, Natta, Giovanardi, and Palla 1988, Natta and Giovanardi 1989, Giovanardi et al. 1990a, Hartmann et al. 1988, Kwan and Alonso-Costa 1988, Alonso-Costa and Kwan 1989. Clearly the hope is to find some diagnostic tool to measure the mass loss rate and the wind temperature of T Tauri stars, with greater confidence than it was possible before, and, if possible, to gain inference on the acceleration mechanism itself.

In this paper, we will discuss in detail how the physical structure of the winds can be computed, under some simplifying assumption, starting with the hydrogen ionization and excitation and recombination line intensity (Section 2). Section 3 presents an example of the ionization and excitation mechanisms for an important trace element, i.e., sodium. In Section 4 we show the results of a diagnostic technique, which makes use of both Na D lines and hydrogen recombination lines and which allows us to derive value of $\dot{M}$ and of the gas temperature T_g in the inner parts of the wind. In Section 5 we examine possible heating mechanisms, with reference to the information on T_g obtained in the previous section.

Most of the following discussion and examples will refer to the inner regions of the winds only, typically the first few stellar radii. This region corresponds to less than 0.01 arcsec at the Taurus cloud distance, and cannot be spatially resolved with the current techniques. All the information is obtained from spectroscopic data, namely from line shapes and intensities. Although this may appear a limitation, the inner region, in fact, gives more direct information on the wind ejection mechanism, because the outer parts are likely to be dominated by secondary processes, resulting from the interaction of the wind with the ambient matter (such as shocks, entrainement of cloud material, etc.).

2. Hydrogen Ionization and Excitation

To model the hydrogen excitation and ionization state in the wind, and then the recombination line intensities and profiles, we will make a number of simplifying assumptions. We will consider winds of pure hydrogen with a constant and known mass loss rate $\dot{M}$. The gas density profile $n_g(r)$ is linked to the velocity profile $v(r)$ through:

$$n_g = \frac{\dot{M}}{\mu 4\pi} \frac{1}{r^2 v}, \qquad (2)$$

where μ is the hydrogen atomic mass. This equation does not imply necessarily spherical symmetry, therefore most of the results will apply to the whole class of flows described by eq. (2).

The velocity profile $v(r)$ is assumed to be known and of the form

$$v(r) = v_0 + v_1(1 - r^{-\alpha}) \qquad (3)$$

Such assumption is not justified by any physical arguments, but has the advantage of reaching the asymptotic value $v_\infty = v_0 + v_1$ at different distances from the star by simply varying the value of the parameter α. To characterize the velocity profile we will use the two quantities v_∞ and $r_{1/2}$, the distance from center in units of the stellar radius $r_\star$ at which $v = v_\infty/2$. Observed line profiles suggest that the acceleration from v_0 to v_∞ occurs very close to the star, within 2-4 stellar radii (Mundt 1984; Beckwith and Natta 1987; Natta and Giovanardi 1989). The value of v_∞ is one of the best determined parameters in TTS winds, from the shift of the blue edge of the absorption component of P-Cygni profiles of bright lines, and it varies between 150 and 300 km s^{-1}. The velocity at the base of the wind, v_0, is not directly constrained by observations. It is reasonable to assume that $v_0 \gtrsim v_{th}$, where v_{th} is the sound speed, which, for temperatures typical of TTS winds, is $\sim$ 10 km s^{-1}; in the following, we have adopted v_0=20 km s^{-1}. It should be kept in mind that wind models like those described here do not account properly for the subsonic region, i.e., a region of thickness of 0.05-0.1 stellar radii.

We will assume that we can apply the large velocity gradient (LVG) approximation and, with it, escape probability methods (Sobolev 1960, Castor 1970) for the line radiation transfer. This implies that the differential flow velocity of the wind (due both to its actual linear acceleration and to the geometrical divergence of the flow) is large compared with the gas random motions (either thermal or turbulent). Within the LVG approximation the line radiation transfer is greatly simplified; in fact we can divide the wind in shells which are radiatively decoupled from each other. This is obtained when the velocity difference between two shells (δv) and the difference between their radii δr satisfy the condition $\delta v/\delta r > v_{random}/r$. From a physical point of view the LVG approximation means that the random motions in the wind are small enough to let the radial flow be the main cause of line broadening. Our results will not strictly apply therefore to very turbulent flows such as those predicted by several Alfvenic models (Hartmann et al. 1982), especially for what concerns the line profiles, but the general picture about the temperature and ionization of the envelope should still hold. Notably, it is the the local nature of the line radiation transfer in the LVG approximation which renders the results largely independent from the flow geometry and therefore applicable to all flows described by eq. (2).

What we will use in practice is a non-LTE model where, at each radius (or for each independent shell), we will write a full set of (steady-state) rate equations. For each m, l level of our model H atom (m is the principal quantum number and l the azimuthal one) the rate equation has the form:

$$n_{ml}\left[\sum_{m'\neq m}\sum_{l'=l\pm 1}\mathcal{R}_{mlm'l'} + \sum_{m'\neq m}\sum_{l'=0}^{m'-1}n_e\mathcal{C}_{mlm'l'} + \sum_{l'=l\pm 1}n_e\mathcal{C}_{mlml'}\right] -$$

$$\left[\sum_{m'\neq m}\sum_{l'=l\pm 1}n_{m'l'}\mathcal{R}_{m'l'ml} + \sum_{m'\neq m}\sum_{l'=0}^{m'-1}n_e n_{m'l'}\mathcal{C}_{m'l'ml} + \sum_{l'=l\pm 1}n_{ml'}n_e\mathcal{C}_{ml'ml}\right] +$$

$$n_{ml}\left(\mathcal{R}_{mlk} + N_e\mathcal{C}_{mlk}\right) - n_e n_k\left(\mathcal{R}_{kml} + n_e\mathcal{C}_{kml}\right) = 0. \quad (4)$$

here n_e is the electron density, n_{ml} is the population of level m,l; R and C refer respectively to radiative and collisional rates, and the index k refers to the continuum. Such a system is non-linear in the unknowns $n_{m,l}$ and n_e, because the radiative rates depend on the local radiation field, which in turn depends on the level populations. Actually, due to the exponential form of the escape probability (see eq. (7)) the non-linearity is rather severe. The model and its properties are discussed in detail by Natta et al. 1988b.

2.1 IONIZATION STATE

Isothermal models are particularly instructive for understanding the processes at work and the general behavior of the wind properties. Some examples are reported in Figure 1, which illustrates the behavior of the ionization degree of the wind ($x_e = n_e/n_g$) as a function of r (note the logarithmic scale). All the models in Figure 1 are for the same central star (here defined by its radius r_* and color temperature $T_\star$) and the same acceleration law but for different wind temperatures and mass loss rates.

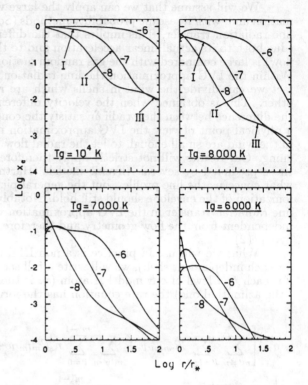

Figure 1. The ionization fraction x_e vs. radius (normalized to the stellar radius $r_\star$). The four panels refer to isothermal winds of different T_g. Each model is identified by the log value of $\dot{M}$. On the two top panels the arrows indicate the regimes where the different H ionization processes discussed in the text (I, II, and III) are dominant. All the models in this figure are for a 20 $L_\odot$ star with $T_\star = 4500$ K, and for an acceleration law with $v_0 = 20$ km s^{-1}, $v_\infty = 320$ km s^{-1} and $r_{1/2} = 1.4 r_\star$ (from Natta et al. 1988b).

Clearly the ionization, even in isothermal winds, varies strongly with r, and even more important is a quite strong function of the wind temperature T_g (although the main ionization mechanism is radiative); the reason for this will become clear when we will discuss in detail the atomic processes at work. In particular we can see that a wind with $\dot{M} = 10^{-6}$ M$_\odot$yr^{-1} is almost fully ionized at $T_g = 8000$ K but neutral at $T_g = 6000$ K. Such a behavior is better illustrated, for a grid of winds with $v(r) = const.$, in Figure 2. Here the maximum ionization fraction in the wind $x_e(max)$ is plotted vs. T_g for three different mass loss rates. Similar results have been obtained by Randich et al. 1990, in calculations referring to expanding uniform slabs.

A coarse idea of the dependence of the ionization state of the (hydrogen) gas on $T_\star$ is given by Figure 3, from Randich 1988. Here, in the $T_g - n_g$ plane, is reported the position of the quasi full ionization point (fractional ionization $x_e = 0.9$) for two different $T_\star$'s; the geometrical radiation dilution factor is fixed at 0.5, i.e., at the stellar surface. At a typical density $n_g = 10^{10}$ cm^{-3} the quasi full ionization is reached at $T_g = 10000$ K for $T_\star = 4500$ K and already at $T_g = 6500$ K for a star at 8000 K.

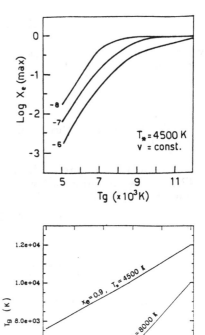

Figure 2. The maximum value of the ionization fraction x_e is plotted as a function of T_g for three different mass-loss rates (each curve is labelled by the corresponding value of Log $\dot{M}$ in $M_\odot yr^{-1}$). In all cases $T_*=4500$ K, $v = const. = 150$ km s^{-1}.

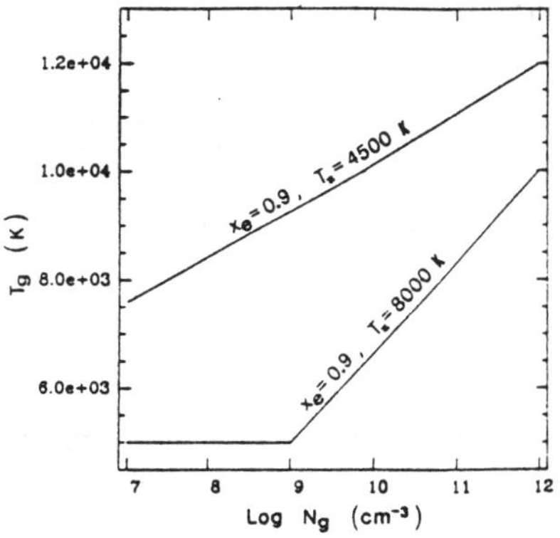

Figure 3. The ionization state for a set of models of homogeneous plane parallel expanding slabs. In the $N_g - T_g$ plane are reported the lines of constant ionization fraction $x_e=0.9$ for two values of T_*. The lines are labelled by the T_* values in K. Regions with $x_e \geq 0.9$ lie above the respective lines. The parameters of this set of models are: column density $N_d=10^{22}$ cm^{-2}, radiation dilution factor $w=0.5$, velocity difference across the slab $\Delta v=300$ km s^{-1} (from Randich et al. 1990).

2.2 DOMINANT IONIZATION PROCESSES

The main result about the physical processes involved in the wind ionization is that the dominant ionization process is photoionization from level $m=2$. The gas is invariably optically thick to the Lyman-continuum photons produced by recombination, and the ionization is primarily by absorption of the Balmer continuum photons emitted by the central star. Thus the electron density n_e is set by the condition:

$$n_2 R_{2k} = n_e^2 \alpha_B \qquad (5)$$

n_2 is the number of H atoms in level $m=2$. R_{2k} is the rate of radiative ionization from level 2, $R_{2k} = w R_{2k}^\star$ where w is the geometrical dilution coefficient and $R_{2k}^\star$ is the rate at the star surface, a function of the spectral type, or T_*, only. α_B is the recombination rate to all levels but level 1.

In turn, the population of level 2 is controlled by level 1, where most of the hydrogen population resides:

$$n_1 n_e C_{12} = n_2 n_e C_{21} + \frac{3}{4} n_2 A_{21} \beta_{12} \qquad (6)$$

where C_{12} and $C21$ are collisional bound-bound rates, A_{21} is the radiative decay rate from level 2 to level 1 and β_{12} is the escape probability for Lyman-α photons:

$$\beta = \frac{1 - e^{-\tau}}{\tau} \simeq \frac{1}{\tau} \qquad (7)$$

We can identify three different regimes for the population of level 2 (crf. Figure 1):

i) (label I in Figure 1). In the inner part of the wind (if $\dot{M} \gtrsim 10^{-8}$ $M_\odot \text{yr}^{-1}$), collisional de-excitation of level 2 dominates over Ly-α decay, and level 1 and 2 are in thermal equilibrium, $(n_2/n_1) = (n_2/n_1)_{\text{LTE}}$. In this region, the ionization fraction ($x_e = n_e/n_g$) can be computed from eq.(5) and (6) taking into account that $n_1 = n_g - n_e$,

$$x_e = \frac{R_{2k}}{2 n_g \alpha_B} \left(\frac{n_2}{n_1}\right)_{\text{LTE}} \left(\sqrt{1 + \frac{4 \alpha_B n_g}{(n_2/n_1)_{\text{LTE}} R_{2k}}} - 1\right) \qquad (8)$$

which for $n_1 \sim n_g$ becomes:

$$x_e = \sqrt{\frac{R_{2k}}{n_g \alpha_B} (n_2/n_1)_{\text{LTE}}} \qquad (9)$$

Since R_{2k} and n_g are both roughly proportional to r^{-2}, we obtain:

$$x_e(r) = const.$$

$$x_e(\dot{M}) \propto \dot{M}^{-1/2}$$

$$x_e(T_g) \propto e^{-\frac{118000}{2 T_g}}$$

ii) (label II in Figure 1). As r increases the density decreases and the Lyman-α decay will dominate over the collisional de-excitation of level 2 for $r \geq r_1$, where r_1 is defined by the condition: $n_e C_{21} \sim A_{21} \beta_{12}$. We can derive an explicit expression for r_1 by making use of some approximations: $n_1 \sim n_g$, $\beta_{12} \sim 1/\tau_{12} \sim (n_1 \sigma_{12} dr/dv)^{-1}$, $dr/dv \sim r/v$. With these approximations and using eq.(8) for x_e it is:

$$r_1 = 8.0 \times 10^{22} \left(\frac{R^*_{2k}}{\alpha_B}\right)^{1/6} C_{21}^{1/3} \left(\frac{n_2}{n_1}\right)_{\text{LTE}}^{1/6} v^{-5/6} \dot{M}^{1/2} \qquad (10)$$

For $r \geq r_1$, the ratio n_2/n_1 is given by:

$$\frac{n_2}{n_1} = \frac{4 C_{12}}{3 A_{21} \beta_{21}} n_e \qquad (11)$$

and the ionization fraction is:

$$x_e = \frac{4C_{12}R_{2k}}{3A_{21}\alpha_B\beta_{21}} \tag{12}$$

where we have assumed again for simplicity that $n_1 \sim n_g$. We can then easily derive:

$$x_e(r) = r^{-3}$$

$$x_e(\dot{M}) \propto \dot{M}$$

$$x_e(T_g) \propto e^{-\frac{118000}{T_g}}$$

iii) (label III in Figure 1). At even larger radii and lower densities, levels $2s$ and $2p$ decouple, because the Ly-α decay of $2p$ occurs at a rate higher than the collisional mixing $n_{2p}C_{ps}$. In this regime, the ionization fraction is given by:

$$x_e = \frac{1}{2}\sqrt{\frac{C_{12}R_{2k}}{3\alpha_B C_{ps}n_g}} \tag{13}$$

And again

$$x_e(r) = const.$$

$$x_e(\dot{M}) \propto \dot{M}^{-1/2}$$

$$x_e(T_g) \propto e^{-\frac{118000}{2T_g}}$$

In all three regimes x_e will also depend on $T_\star$ through the value of $R^\star_{2k}$.

Eq.(9),(12) and (13) show that the ionization fraction reaches a maximum value at a radius where the velocity v reaches its asymptotic value v_∞, remains approximately constant until, for $r \geq r_1$, drops rapidly, approximately as r^{-3}.

The runs, and quantitative behavior, of these analytical approximations for the ionization degree are illustrated in Figure 4 (dashed lines) and compared to numerical model calculations (solid lines) for three different rates of mass loss, $\dot{M} = 10^{-8}$, 10^{-7} and 10^{-6} M$_\odot$ yr^{-1}. T_g, $T_\star$ and the velocity profile are kept fixed. For $\dot{M} = 10^{-8}$, r_1 is less than the star radius and x_e does not reach the maximum value given by eq.(9).

For high values of $\dot{M}$ ($\dot{M} > 5\,10^{-7}$ M$_\odot$yr^{-1}) the contribution of the ionization from excited levels ($n > 2$) is no longer negligible. This effect is also illustrated in Figure 4. While for $\dot{M}=10^{-8}$ and 10^{-7} M$_\odot$yr^{-1} the run of x_e vs. r conforms to the predictions of eq.(9), for $\dot{M}=10^{-6}$ M$_\odot$yr^{-1} eq.(9) underestimates x_e by large factors. In such dense, and relatively cold winds, this is due to the non-negligible optical thickness of the innermost layers to Balmer continuum photons. When this is the case ionization from levels higher than $m=2$ will have an appreciable contribution and x_e will increase accordingly, but only in the very inner layers. The main effect will be instead an overproduction of Balmer continuum photons by radiative recombination which will propagate to the thin layers of the wind in addition to the stellar continuum.

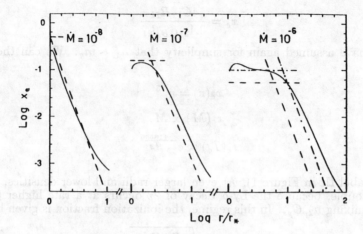

Figure 4. The ionization fraction x_e vs. the normalized radius for three models of isothermal winds. The $\dot{M}$ values are different, as labelled, while for all models T_g=7000 K and $T_\star$=4500 K. The acceleration law is as in Figure 1. The dashed lines are obtained with the analytical approximations discussed in the text, the solid lines show the numerical results, and the dot-dash line shows the analytical results when the emission of the extra number of Balmer photons following radiative recombination in the inner layers is taken into account (from Natta et al. 1988b).

The overall behavior will still be described by eq.(9) but with a different and higher R_{2k}. When this effect is taken into account (dot-dash line in Figure 4) x_e vs. r is again fairly described by the analytical approximation. Similar deviations occur in dense but hot (T_g >8000 K) winds where collisional ionization can effectively increase x_e in the inner shells.

An additional excitation-ionization mechanism has been proposed by Kwan and Alonso-Costa 1988. It consists in using sub-Lyman continuum photons to once ionize the neutral nitrogen, namely from the excited level $2p^3\ ^2D^0$ (ionization potential 12.16 eV). The resulting N II undergoes a charge exchange reaction with an H I atom, producing back an N I and giving rise to an H II ion. Finally the recombination of H II and the subsequent cascade will populate the m =2 level. Also in this case the hydrogen ionization is due to Balmer photons while the direct contribution from charge exchange reactions will be completely negligible. The mentioned authors have computed a full grid of models including this, and other, reactions and found that it becomes effective only for relatively hot stars ($T_\star$ >8500 K).

A last point interesting point about H ionization regards the *ionization freezing* in the wind. We can define a dynamical time $\tau_{dyn} \simeq r/v$ which will increase with r. Due to the gas expansion the recombination time τ_{rec} will also increase with r and, due both to the decrease of x_e and of n_g, much more rapidly than τ_{dyn}. In the inner parts of the wind $\tau_{dyn} > \tau_{rec}$ and actually this fact renders possible the use of time-independent rate equations in a fluid element. As r increases, there is a point in the wind where the two times are equal, the *freezing point*. In the zone of the wind beyond this point the recombination process will no longer be instantaneous respect to the fluid motion and we can think of a freezing of the ionization state irrespectively of the actual local conditions. Depending on the details of the model, especially on $\dot{M}$ and on T_g, the location varies but remains invariably within a few stellar radii, where x_e is still decreasing with r. Therefore the actual ionization of the outer envelope is higher than implied by the time independent calculations we have used here.

2.3 EXCITATION STATE AND LINE INTENSITIES

After having examined the ionization state of the winds, we will now move to study their excitation structure and the consequent H line emission. It is not straightforward to discuss briefly on analytical grounds, as we did for the ionization, the excitation mechanisms of the various levels involved in the formation of the main H recombination lines. Therefore we will briefly summarize the results of the numerical non-LTE modeling of isothermal winds of Natta, Giovanardi, and Palla 1988. A first point regards the amount of departure from LTE of the populations of the atomic levels. (By LTE population we mean the value expected for T_g and for the n_e actually computed at a certain point in the wind, i.e., the convention used in standard recombination theory). As usual the largest departures are observed in the lower levels. For the higher levels (m >5) the values of the departure coefficients b range between 0.8 and 1.5 in most of our models.

A second point regards the optical thickness of the H lines in TTS winds. It is usually expected that the lines of the first series are optically thick, but it is not *a priori* clear what happens to all the rest. Actually it is often assumed, to adopt simpler transfer techniques, that the high series ($m > 2$) are relatively thin. This is not generally the case. The situation for the α lines of the series from $m = 4$ (Pa_α) to $m = 15$ is reported in Figure 5. The optical thickness always exceeds unity and turns out to be always higher than expected in LTE, especially for the most often observed near infrared lines. Notably, for the higher series, even the LTE values predict quite thick lines; this is a consequence of the partial ionization of the wind which allows a consistent population of the excited levels.

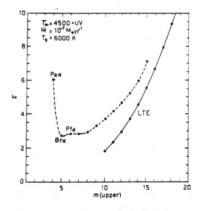

Figure 5. The optical depth of the hydrogen α-lines of the series from 3 to 15 as a function of the upper level quantum number m (dashed line). The wind parameters are T_*=4500 K + UV (see Giovanardi *et al.* 1990 for an exact description of the adopted UV excess), T_g=6000 K, $\dot{M}= 10^{-7} M_\odot yr^{-1}$, v_∞=150 km s^{-1}, $r_{1/2}$=4. The solid line shows the LTE values for the same wind model.

Where are the formation regions of the lines? They are in general very narrow and close to the stellar surface. The situation in a typical wind model is depicted in

Figure 6 for some of the lines. The most extended formation region pertains to H_α, for which the 80% of the luminosity is emitted within 2.2 stellar radii. In the case of Pf_γ and Br_γ, which have virtually identical formation regions, 80% of the total intensity is already emitted within $1.3r_*$ (this means $0.3r_*$ from the photosphere!); this fact places them in a very good position for studies of the inner shells of the wind where most of the wind acceleration takes place.

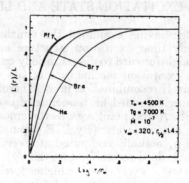

Figure 6. The contribution of the layers between r_* and r to the integrated emerging luminosity of various H lines in a representative isothermal wind. Same acceleration law as in Figure 1 (from Natta et al. 1988b).

The expected luminosities of the α (emission) lines of the various series are presented in Figure 7, from Pa_α (1.88 μm) down to 14α (138 μm); as indicated the luminosities of the higher series can be inferred with good accuracy using LTE values. It is also clear that there is a drop in luminosity of 5 decades going from the near to the far infrared, which renders T Tauri winds rather weak objects for the FIR spectrographs.

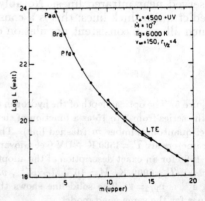

Figure 7. Luminosity of the hydrogen α-lines as a function of the quantum number of the upper level m. Same wind parameters as in Figure 5.

3. Ionization of Trace Elements: the Case of Sodium

Trace elements, such as Ca, Mg, and Na, show strong resonance lines in the optical spectra of TTS, often characterized by classical P Cygni profiles, with blue-shifted absorption extending to velocities comparable to those estimated from the width of the H recombination lines. Such lines traditionally provide important information on the physical conditions of TTS winds, and have been modeled by various authors (Hartmann et al. 1989; Natta and Giovanardi 1990; Giovanardi et al. 1990a; Grinin and Mitskevich 1988,1990a). In this Section we will discuss the ionization and excitation of one of those elements, namely sodium.

From the atomic point of view, neutral sodium has a rather simple configuration, with one active electron in the 3S ground state. It is also easily ionized, the ground state ionization potential is 5.139 eV, while the potential for double ionization climbs securely over 47 eV. The Grotrian diagram for the first levels of Na I is illustrated in Figure 8.

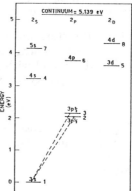

Figure 8. A simplified Grotrian diagram of Na I. The transitions reported refer to the resonance D_2 doublet.

A set of numerical computations has been run for the Na I excitation and ionization structure in isothermal winds (Natta and Giovanardi 1990). Figure 9a illustrates the run vs. r of the sodium ionization fraction for 4 models with different T_g's and the same star, mass loss rate and acceleration law.

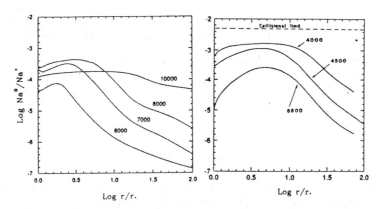

Figure 9. Panel a: the Na ionization fraction vs. the normalized radius for a set of isothermal wind models at different T_g. The curves are labelled by the values of T_g in K, the other wind parameters are : $T_*=4500$ K, $\dot{M}=1\times10^{-8}$ M$_\odot$yr^{-1}, $v_0=3$ km s^{-1}, $v_\infty=150$ km s^{-1}, $r_{1/2}=4$. Panel b: same for different T_* and $T_g=7000$ K. The dashed line near the upper edge indicates the collisional limit (from Natta and Giovanardi 1990).

It is interesting to note that, first, the sodium is invariably heavily ionized, second, the neutral fraction is confined near the star, that is in a region similar to the formation region of the H recombination lines, and, third the overall behavior with T_g is very different from what shown for the hydrogen in the preceeding Section. The variation of the ionization state with T_* is also illustrated in Figure 9b: as the stellar temperature decreases, the fraction of neutral sodium reaches an upper limit corresponding to a collisional ionization regime (dashed line). In analogy to what done for the hydrogen we will now try to isolate the most important ionization processes and their range of influence.

i) Near the star, and for low T_g, radiative ionization from excited levels is of importance. Let us define as $Na(3S)$, $Na(3P)$ the density of Na atoms (cm^{-3}) in level 3S and 3P, and with Na^0, Na^+ the density of neutral and singly ionized sodium. If, for simplicity, we assume that the ionization occurs from level 3P only:

$$Na(3P) R^*_{2k} w = Na^+ n_e \alpha_T \tag{14}$$

where R^*_{2k} is the radiative ionization rate from level 3P at the stellar surface, w is the geometrical dilution factor, n_e the electron density and α_T is the total recombination coefficient. The population of excited levels is set by radiative processes, and therefore:

$$\frac{Na(3P)}{Na(3S)} \propto w \tag{15}$$

then

$$\frac{Na^0}{Na^+} \propto \frac{n_e}{w^2} \tag{16}$$

Eq. (16) shows that in this regime the fraction of neutral sodium increases with r, as seen at small r's in many of the curves of Figure 9. Sodium is more neutral if T_g or $\dot{M}$ are higher, because in both cases the electron density is higher.

ii) Radiative ionization from level 3S becomes important in the outer parts of all winds. In this case :

$$Na(3S) R^*_{1k} w = Na^+ n_e \alpha_T \tag{17}$$

and

$$\frac{Na^0}{Na^+} \propto n_g \frac{x_e}{w} \tag{18}$$

When $v \sim const.$, the ratio Na^0/Na^+ decreases with r as the H ionization fraction x_e, i.e., approximately as r^{-3}. This decrease is seen in all the outer parts of the winds in our grid. The dependence on T_g and $\dot{M}$ is similar to case *i)*.

iii) Collisional ionization from level 3S is important in the hotter winds of our grid. In this regime, the ionization balance can be approximated as:

$$\frac{Na^0}{Na^+} = \frac{\alpha_T}{C_{1k}} \tag{19}$$

which increases sharply when T_g decrease, but does not depend on r or $\dot{M}$. The flat portions of the curves Na^0/Na^+ in Figure 9 are always due to the dominance of collisional ionizations over the other processes. The value given by equation (19) is actually an upper limit to the fraction of neutral sodium in a wind of given T_g. Radiative processes can only cause the fractional abundance of Na^0 to be lower than given by equation (19).

The influence of charge exchange reactions with hydrogen

$$Na^0 + H^+ \leftrightarrow Na^+ + H^0 (m=2) \qquad (20)$$

is not easy to estimate due to the scarcity of the relevant atomic data. From the available low energy cross sections we can infer a rate, for Na^0 in the ground state, of about 4×10^{-11} cm^{-3}s^{-1} (Kushawaha, Burkhardt and Leventhal 1980). It follows that charge exchange is never the dominant ionization agent but can account for about 1/2 of the observed sodium ionization in winds at intermediate temperature ($T_g \sim 8000$ K).

The interplay of the above processes causes the behavior of the sodium ionization seen in Figure 9. As a result, the column density of neutral sodium $N(NaI)$ is not a monotonic function of T_g, the location of the maximum depending on the adopted photospheric radiation field (and in some measure on $\dot{M}$).

4. An Example of Diagnostic of $\dot{M}$ and T_g

In the following we will briefly describe how calculations such as those discussed in the previous sections can be combined with observations of TTS to infer wind parameters, namely the mass-loss rate $\dot{M}$ and the gas temperature T_g. Details of this particular example can be found in Natta and Giovanardi 1989, Giovanardi et al. 1990a.

The diagnostic power of the H recombination lines alone to determine the wind parameters, especially the mass loss rate from a certain object, is not very promising and the reason is illustrated in Figure 10.

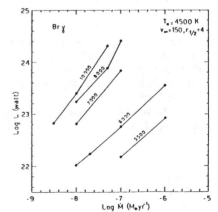

Figure 10. Brγ luminosity as a function of $\dot{M}$ for different values of the gas temperature T_g (reported on the curves). The wind parameters are $T_*=4500$ K, $v_\infty=150$ km s^{-1}, $r_{1/2}=4$.

Here we have plotted the $Br\gamma$ luminosity as a function of the mass loss rate for a grid of isothermal models with different T_g's. It is evident that for the same $\dot M$ we can span 2.5 orders of magnitude in the line luminosity by simply moving T_g from 5500 to 8000 K. That is to say that, even assuming we know precisely the star radiation field, a recombination line luminosity does not depend only on $\dot M$ but it is also a very strong function of the gas temperature. This results from the particular ionization process in these winds, that is photoionization of collisionally populated excited levels. The situation does not improve if, instead of the line luminosity, we choose to use line intensity ratios. These are, first, measured with poor accuracy compared to what is needed, and, second, they are mainly determined by the optical thickness regime of the lines involved and bear the same entangled dependences on the wind parameters (see also Randich et al. 1990). It appears therefore that the determination of the mass loss rate needs some additional information. Clearly, if we find the additional information needed to disentangle the $\dot M - T_g$ connection, we will be able to determine both these quantities in any given object. We also have now an idea of what kind of additional information we need: it must be some observable stemming from a mechanism different from the collisional excitation plus photoionization which provides the H lines emission.

Among the lines commonly observed in T Tauri stars the neutral sodium resonance D_2 doublet (5900 Å) is a very good candidate. More specifically, the depth of the blue-shifted absorption component of the lines is certainly due to neutral sodium in the wind, and it has the important advantage of being an easy measurement, independent of distance and extinction effects. On the theoretical point of view, the absorption depth is strictly linked to the optical depth τ, defined as:

$$\tau = 0.0265 \frac{gf}{\nu_0} \frac{c}{v_\infty} N(NaI) \qquad (21)$$

where gf is the oscillator strength of the line, ν_0 its frequency, c the speed of light, v_∞ the terminal wind velocity and $N(NaI)$ the column density of neutral sodium, whose behaviour with the wind parameters has been described in the previous section. In particular, and very important, τ is not a strong function of T_g (in the range of temperatures we consider), but varies strongly with $\dot M$. Moreover, the contribution to τ of the wind outer layers (outer than about $2r_*$) is in general negligible, so that sodium absorpion and infrared hydrogen lines integrated emission probe the same very inner part of the wind.

We can then build a grid of isothermal models with different T_g and $\dot M$ and compute for each of them the infrared H lines luminosities and the τ of Na I (unfortunately we must build such a grid for each different radiation field and acceleration law). In practice we will map a *theoretical* plane $T_g - Log\dot M$ into an *observable* plane $LogL - \tau$. This procedure is illustrated in Figure 11, for three different H recombination lines, namely Pa_β, Br_γ and Br_α. If, for a certain star we know an H line luminosity and can estimate the Na I τ, we can use these diagrams to infer both T_g and $\dot M$.

The relevant data are presently available only for a few stars and the results of this kind of study are reported in Table 1 (adjusted from Giovanardi et al. 1990a).

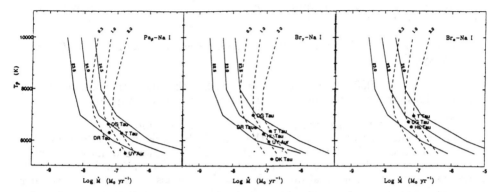

Figure 11. In the plane $\text{Log}\dot{M}\text{-}T_g$, the figure shows as solid lines the location of curves of constant luminosity of the hydrogen lines (Pa_β in Panel a, Br_γ in Panel b, Br_α in Panel c) and as dashed lines the location of curves of equal optical depth in the $3S_{1/2}$-$3P_{3/2}$ Na I line. The solid lines are labelled with the logarithm of the line luminosity in watts, the dashed lines with the value of τ. The wind parameters are $T_* = 4500$ K +UV, $v_\infty = 150$ km s^{-1}, $r_{1/2} = 4$. The dots show some of the stars reported in Table 1 (from Giovanardi et al. 1990).

Table 1

Derived Wind Parameters

	L	ST	v_∞	$\dot{M}$	T_g	$\dot{M}$(CO)
	($L_\odot$)		(km s^{-1})	($M_\odot$yr^{-1})	(K)	($M_\odot$yr^{-1})
RY Tau	17.4	K1	180	≤2(-8)	≥7000	not detected
T Tau	18	K1	159	1.1(-7)	6500	1.5(-7)
DF Tau	5.4	M0.5	140	≤4(-8)	≥6300	-
DG Tau	7.2	C	210	5(-8)	6800	not detected
DK Tau	4.0	K7	200	1.5(-7)	5250	not detected
HL Tau	8.5	C(K7)	200	7(-8)	6400	4(-8)
DL Tau	1.5	C	150	≤2(-8)	≥7000	not detected
DR Tau	5	C	150	5(-8)	6350	not detected
UY Aur	4.3	K7	200	1.6(-7)	5750	-
SU Aur	18	G2	200	≤6(-8)	≥7100	-
RW Aur	5.5	C	260	2(-7)	6250	-
GW Ori	66	G5:	160	4.3(-8)	7800	-
AS 353a	7.2	C	310	2(-7)	7000	1.6(-7)

The results can be summarized by saying that T Tauri stars with Na I blueshifted absorption have :

i) Mass-loss rate $\dot{M}$ in a restricted range between approximately 3×10^{-8} and 3×10^{-7} $M_\odot yr^{-1}$;

ii) low T_g (≤ 7000 K), rather constant within the line formation region (in fact, the values inferred from the three different lines are very similar);

iii) a low ionization fraction ($x_e < 0.1$), i.e. neutral winds.

As we mentioned above, the fact that the TTS winds are mostly neutral is also of importance for the theory of CO outflows. Since the ionized material is only a minor component of the wind, full ionization models for the wind will predict, on the basis of the H recombination line luminosities, mass loss rates much smaller than what is needed to drive the CO outflow. The last column of Table 1 reports the (wind) $\dot{M}$ values derived from the outflow parameters, that is from eq.(1). In the very few cases where we can make a comparison the agreement is excellent and confirms the scenario of the momentum conserving shocks. Although the method above described can only be applied to optical objects, HI 21 cm observations of the obscured central regions of L1551 and HH 7-11 have revealed the presence of fast neutral winds around the central IR sources which are coextensive with the CO lobes. The lobes appear therefore filled with fast (≥ 150 km s^{-1}) neutral material emerging from the central source. Again the observed wind parameters are in excellent agreement with the standard (momentum conserving) wind-outflow scenario (Lizano *et al.* 1988; Giovanardi *et al.* 1990b).

5. The Gas Temperature

Temperatures of 6000-7000 K, as we found in the previous section for the inner regions of TTS winds, cannot be due to radiative processes only, because the central stars are too cold. Mechanical heating, possibly (but not necessarily) caused by the same mechanism which cause the matter ejection, must be responsable for the observed temperatures.

In this section we will review various heating processes, computing for each of them the resulting equilibrium temperature for a wind with parameters typical of the TTS reported in Table 1. As in the derivation of the ionization and excitation structure of the wind , also these temperature calculations will not be the result of self-consistent models. Namely, the runs with r of density and velocity are specified *a priori*, and we will not address the issue of the capability of the ejection mechanism under consideration to produce the assumed velocity gradient and mass-loss rate.

Details on the various processes can be found in Spitzer 1978, Hartmann *et al.* 1982, Kwan and Alsonso-Costa 1988, Ruden *et al.* 1990 and Grinin and Mitskevich 1990b.

5.1 THE HEAT EQUATION

At any given distance r from the star, the gas temperature is given by the heat equation:

$$\frac{3}{2} k n_g v \frac{dT_g}{dr} - k T_g n_g \frac{1}{r^2} \frac{d}{dr}(r^2 v) = \Gamma - \Lambda \tag{22}$$

where n_g is the gas density, v the flow velocity, k the Boltzmann constant.

The first term on the left accounts for the variation of the internal specific energy per unit volume, while the second term on the left represents the work done by the expanding gas. The term on the right is the difference between the gas heating (Γ) and cooling (Λ) rates (in erg s^{-1} cm^{-3}).

In general, Γ and Λ are complicated functions of r, the gas density, velocity, acceleration and temperature. However, once $n_g(r)$ and $v(r)$ are specified, eq.(22) is just an ordinary differential equation, which can be solved by standard methods. At each r, the ionization fraction x_e is computed using the analytical expressions discussed in Section II. The results illustrated in the following refer to a wind model having T_*=4500 K, $\dot{M}$=10^{-7} M$_\odot$yr^{-1}, v_∞=150 km s^{-1}, $r_{1/2}$=4.

5.2 RADIATIVE HEATING AND COOLING

We will consider here heating and cooling due to bound-free transitions of H and H$^-$, to free-free H transitions, Ly$_\alpha$ cooling and cooling due to the H and K lines of Ca II. This last is representative of the cooling due to metal lines, which contribute significantly to the cooling of the solar transition zone.

5.2.1 H$^-$ bound-free transitions.
Heating and cooling rates can be written as:

$$\Gamma(H^-) = n_{H^-} \, w \, \epsilon_A g_A \tag{23}$$

$$\Lambda(H^-) = n_{H_*^-} \left(\epsilon_S g_S + w \, \epsilon_I g_I \right) \tag{24}$$

where w is the geometrical dilution factor, n_{H^-} is the H$^-$ density (in cm^{-3}) and $n_{H_*^-}$ is the LTE H$^-$ density, given by the Saha equation:

$$n_{H_*^-} = 2.89 \times 10^{-22} \, n_e n_H \left(\frac{5040}{T_g}\right)^{1.5} e^{\frac{8749}{T_g}} \tag{25}$$

The quantities ϵ_A, ϵ_S and ϵ_I are the mean energy (in erg cm^{-3} s^{-1}) per ionization, spontaneous recombination and stimulated recombination, respectively:

$$\epsilon_A = 4\pi \frac{\int_{\nu_0}^\infty (h\nu - h\nu_0) \frac{\sigma_\nu}{h\nu} B_\nu(T_*) \, d\nu}{\int_{\nu_0}^\infty \frac{\sigma_\nu}{h\nu} B_\nu(T_*) \, d\nu} \tag{26}$$

$$\epsilon_S = 4\pi \frac{\int_{\nu_0}^\infty (h\nu - h\nu_0) \frac{\sigma_\nu}{h\nu} e^{-h\nu/kT_g} \frac{2h\nu^3}{c^2} \, d\nu}{\int_{\nu_0}^\infty \frac{\sigma_\nu}{h\nu} e^{-h\nu/kT_g} \frac{2h\nu^3}{c^2} \, d\nu} \tag{27}$$

$$\epsilon_I = 4\pi \frac{\int_{\nu_0}^\infty (h\nu - h\nu_0) \frac{\sigma_\nu}{h\nu} e^{-h\nu/kT_g} B_\nu(T_*) \, d\nu}{\int_{\nu_0}^\infty \frac{\sigma_\nu}{h\nu} e^{-h\nu/kT_g} B_\nu(T_*) \, d\nu} \tag{28}$$

and g_A, g_S and g_I are the rates (in units of cm^3 s^{-1}) of the three processes:

$$g_A = 4\pi \int_{\nu_0}^{\infty} \frac{\sigma_\nu}{h\nu} B_\nu(T_*)\, d\nu \tag{29}$$

$$g_S = 4\pi \int_{\nu_0}^{\infty} \frac{\sigma_\nu}{h\nu} e^{-h\nu/kT_g} \frac{2h\nu^3}{c^2}\, d\nu \tag{30}$$

$$g_I = 4\pi \int_{\nu_0}^{\infty} \frac{\sigma_\nu}{h\nu} e^{-h\nu/kT_g} B_\nu(T_*)\, d\nu \tag{31}$$

In eq.(26) to (31), ν_0 is the threshold frequency for dissociating H^- ($\nu_0 = 1.815 \times 10^{14}$ s^{-1}), c is the speed of light, h the Planck constant, σ_ν the dissociation cross section (Wishart 1979). Note that ϵ_A, g_A depend on the radiation field at the stellar surface only, ϵ_S, g_S on the gas temperature only, and ϵ_I, g_I on both.

The abundance of H^-, which appears in eq.(23) is computed by balancing the free-bound formation rate with the destruction rate. This last includes bound-free dissociation and the associative detachment ($H + H^- \leftrightarrow H_2 + e$, with cross section $\gamma = 2 \times 10^{-9}$ cm^3 s^{-1}), which is the most important collisional destruction process. It is convenient to express n_{H^-} as a function of $n_{H_*^-}$ by defining a departure coefficient b given by:

$$b = \frac{g_S + w\, g_I + n_H \gamma}{w\, g_A + n_H \gamma} \tag{32}$$

When associative detachment is negligible, $b \propto w^{-1}$ and becomes very large at large radii. If collisional processes dominate, $b \sim 1$.

Eq.(23) and (24) can then be rewritten as:

$$\Gamma(H^-) = b\, n_{H_*^-}\, w\, \epsilon_A g_A \tag{33}$$

$$\Lambda(H^-) = n_{H_*^-}\, (\epsilon_S g_S + w\, \epsilon_I g_I) \tag{34}$$

In a static layer of gas, the heat equation has a solution given by $\Lambda = \Gamma$. If, as it is often the case, stimulated recombinations and collisional processes can be neglected, the equilibrium condition is satisfied when $g_A = g_S$, or, in the approximation $h\nu_0 \gg kT_*$, when $T_g = T_*$. Because g_A does not depend on r, a static solution yields a constant gas temperature T_g. This results (i.e., $T_g \sim T_*$ at all radii) is typical of situation where bound-free and free-bound processes determine both the thermal equilibrium and the formation equilibrium. The dashed line labelled (1) in Figure 12 shows the equilibrium temperature due to H^- cooling and heating. Note that T_g is somewhat lower than T_* (3650K for $T_* = 4500$K), because $h\nu_0/kT_* \sim 1.9$ only.

5.2.2 H bound-free and free-free.
Both processes are discussed in detail by Spitzer 1978.

5.2.3 Ly-α cooling.
Using the escape probability β_{12} (see eq.(7)), it is:

$$\Lambda_{Ly-\alpha} = h\nu_{Ly-\alpha}\, n_2\, A_{21}\, \beta_{12} \tag{35}$$

with the same meaning of symbols as in Section 2. Making use of the fact that $\beta_{12} \sim 1/\tau_{12}$, it turns out:

$$\Lambda_{Ly-\alpha} = 2.29 \times 10^5 \frac{n_2 g_1}{n_1 g_2} \frac{dv}{dr} \tag{36}$$

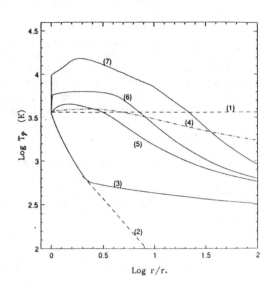

Figure 12. Gas temperature as a function of radius for different heating and cooling processes. Curve (1): H$^-$ heating and cooling for a stationay layer of gas; curve (2): adiabatic expansion cooling; curve (3): adiabatic expansion cooling, radiative heating and cooling; curve (4): friction heating is added to the above processes; curve (5), (6) and (7): magnetic wave heating is added to adiabatic expansion and radiative processes, with $\alpha=10^{-4}$, 10^{-3} and 1, respectively. The wind parameters are: T_*=4500 K, T_g=7000 K, $\dot{M}=10^{-7}$ M$_\odot$yr^{-1}, v_∞=150 km s^{-1}, $r_{1/2}$=4.

5.2.4 *Ca II H and K lines*. The two lines at 3933 Å ($4S_{1/2}$-$4P_{1/2}$, $4S_{1/2}$-$4P_{3/2}$) are optically thick. Assuming that the two P levels are in thermal equilibrium, the total cooling due to the two lines can be written, in analogy to what done for Lyα, as:

$$\Lambda_{CaII} = h\nu\, n(4P)\, A_{21}\, \beta_{12} \tag{37}$$

where A_{21}=1.48x10^8 s^{-1}. If the level 4P can only decay to 4S, via radiative and collisional de-excitation (i.e., neglecting de-excitations via the infrared triplet), and $n(4S) \sim n(CaII) \sim n(Ca)$, we can re-write eq.(37) as:

$$\Lambda_{CaII} \sim 10^{-17}\, n_g\, n_e\, C_{12}\, \left[1 + \frac{n_e C_{21}}{A_{21}\beta_{12}}\right]^{-1} \tag{38}$$

where we have assumed a Ca abundance of 2x10^{-6}. C_{12} and C_{21} are the collisional bound-bound rates, given by:

$$C_{21} \sim 8 \times 10^{-9}\, T_g^{-1/2} \quad (cm^3 s^{-1}) \tag{39}$$

$$C_{12} = C_{21}\, e^{-\frac{36615}{T_g}} \quad (cm^3 s^{-1}) \tag{40}$$

The escape probability, under the same assumptions, is given by:

$$\beta_{12} \sim 2.8 \times 10^{-12}\, n_g\, \frac{dv}{dr} \tag{41}$$

Eq.(38) assumes that the excitation of level 4P is due to electronic collisions. Because the ionization fraction of the winds is generally quite low, it is possible that in fact collisions with hydrogen atoms are more effective in exciting the level 4P, and should be taken into account. Neglecting them, however, does not change much the resulting temperature, because, when the ionization fraction is very low, radiative cooling in total is negligible with respect to the adiabatic cooling (see below).

In a cold radiation field, such as that in TTS, the radiative thermal balance of a static layer of gas is dominated by the H^- formation and dissociation equilibrium. The winds are very cold and neutral, with a small ionization fraction caused by metals of low ionization potential, such as sodium. All the other radiative processes that we have just listed (and in particular metal lines cooling) become important only when other non-radiative heating processes enhance the gas temperature, and hence the gas ionization fraction, to higher values.

5.3 ADIABATIC EXPANSION COOLING

The second term on the left in eq.(22) is often defined as adiabatic expansion cooling, and it is the dominant cooling process in most of the winds (in some cases, radiative cooling may be important in the very inner parts of winds; Hartmann *et al.* 1982 and in the following). If radiative terms can be neglected, eq.(22) can be rewritten as:

$$\frac{3}{2}\frac{dT_g}{dr} = \frac{1}{r^2 v}\frac{d}{dr}(r^2 v) \qquad (42)$$

yielding the well known analytical solution (cfr. dashed line (2) in Figure 12):

$$T_g \propto (r^2 v)^{-2/3} \propto n_g^{2/3} \propto r^{-4/3} \quad (if \quad v = const.) \qquad (43)$$

The curve labelled (2) in Figure 12 is the solution of eq.(22) when only the adiabatic expansion cooling is taken into account, while curve (3) is the solution when also the radiative heating and cooling are included. The wind is very cold ($T_g \ll 3500$ K), and T_g drops sharply with r, according to eq.(43). Notice, however, that at large radii, the heating due to H^- dissociation does contribute significantly to the thermal balance, causing a flattening from the $r^{-4/3}$ law.

5.4 FRICTION HEATING

This process is typical of those acceleration mechanisms which act on the ionized component of the wind only, such as magnetic fields. Then the acceleration of neutral atoms is due to collisions with the protons, which also heat the gas (*ambipolar diffusion*; Spitzer 1978, Ruden *et al.* 1990). The heating rate per unit volume is the product of the drag force between ions and neutrals f times the drift velocity between ions and neutrals v_D:

$$\Gamma_{friction} = f\, v_D \qquad (44)$$

Assuming, as stated at the beginning of this section, that the velocity profile is known, and so is the velocity gradient dv/dr, the drag force f must satisfy the force equation:

$$\rho v \frac{dv}{dr} = -\frac{GM_* \rho}{r^2} + f \qquad (45)$$

where ρ is the mass density of neutrals and M_* is the mass of the central star, as well as momentum conservation:

$$\rho v_D = \frac{f}{n_g \langle x_i \sigma v_D \rangle} \qquad (46)$$

For small values of v_D, the average collision rate between ions and neutrals $\langle x_i \sigma v_D \rangle$ is approximately $3.26 \times 10^{-9} x_{H^+} + 2.4 \times 10^{-9} x_M$ (Draine 1980), where x_M is the fractional abundance of ionized metals ($x_M = 10^{-5}$).

By combining eq.(44), (45) and (46), $\Gamma_{friction}$ can be expressed as:

$$\Gamma_{friction} = \mu_H \frac{\left[\frac{GM_*}{r^2} + v\frac{dv}{dr}\right]^2}{\langle x_i \sigma v_D \rangle} \qquad (47)$$

When the flow velocity becomes constant (i.e., $\frac{dv}{dr} = 0$), $\Gamma_{friction} \propto r^{-4}$. The friction heating is effective only near the star.

The curve labelled (4) in Figure 12 shows the equilibrium temperature for a wind where radiative heating and cooling, adiabatic cooling and friction heating are considered. The wind is warmer compared to curve (3), but it is still much cooler than observed ($T_g \lesssim 4000$ K at all radii).

Notice that, in this formalism, $\Gamma_{friction}$ increases when x_e decreases. However, when the ionized component of the wind is very small, collisional acceleration of the neutral particles by the ionized ones become very unefficient. The drag force f can be much smaller than the value given by eq.(45).

5.5 HEATING DUE TO DISSIPATION OF ALFVEN WAVES

We consider here the dissipation of Alfvén waves as an example of wave heating. Again, as in the previous case, we will assume that $v(r)$ is known. In other words, we will consider the wave energy as a free parameter, not caring if it is enough to produce the required acceleration.

Following Hollweg 1973 we write the heating rate due to Alfvén wave dissipation as:

$$\Gamma_{wave} = \frac{1}{8\pi} \left[-\frac{1}{r^2} \frac{d}{dr} r^2 \left(\langle \delta B \rangle^2 (\frac{3}{2}v + v_a) \right) + \frac{v}{2} \frac{d \langle \delta B \rangle^2}{dr} \right] \qquad (48)$$

where $\langle \delta B \rangle^2 / 8\pi$ is the energy density associated with the wave. By introducing a damping length λ for the wave dissipation, eq.(48) can be written as:

$$\Gamma_{wave} = \frac{\langle \delta B \rangle^2}{8\pi} \frac{(v + v_A)}{\lambda} \qquad (49)$$

v_A is the Alfvén velocity:

$$v_A^2 = \frac{B^2}{4\pi\rho} \qquad (50)$$

and B is the unperturbed radial magnetic field.

The wave energy falls off exponentially with radius:

$$\langle \delta B \rangle^2 = \langle \delta B_0 \rangle^2 \frac{M_A^0}{M_A} \frac{(1+M_A^0)^2}{(1+M_A)^2} e^{-\lambda(r/r_*-1)} \tag{51}$$

M_A is the Mach number ($M_A = v/v_A$); the index 0 refers to the values on the stellar surface.

We consider the two quantities B_0 and $\langle \delta B_0 \rangle^2$ to be free parameters. In the three cases shown in Figure 12 (curves (5), (6) and (7)), we choose the wave flux $\langle \delta B_0 \rangle^2 v_A^0 / 8\pi$ to be a fixed fraction α of the photon flux at the stellar surface:

$$\frac{1}{8\pi} \langle \delta B_0 \rangle^2 v_A^0 = \alpha \, (\sigma T_*^4) \tag{52}$$

In all cases, $\langle \delta B_0 \rangle^2 / B_0^2 = 0.2$, and $\lambda = 1 r_*$. For $T_* = 4500$ K, $\dot{M} = 10^{-7}$ M$_\odot$yr^{-1}, the magnetic field at the stellar surface ranges from $B_0 = 226$ G for $\alpha = 1$ to 10 G for $\alpha = 10^{-4}$.

Figure 12 reports three curves, corresponding to $\alpha = 10^{-4}$ (curve (5)), 10^{-3} (curve (6)) and 1 (curve (7)), respectively. In these models, the gas temperature in the inner regions of the wind looks quite different than in the previous cases, being significantly higher (up to 15000 K in curve (7)) and almost constant. The energy input required to substain these higher temperatures, however, is very high compared to the mechanical wind luminosity ($1/2 \dot{M} v_\infty^2 \sim 7 \times 10^{32}$ erg s^{-1}), and even to the radiative luminosity of the star, as shown in Table 2. The wave heating is balanced by radiative cooling processes (mostly H free-bound and Ca II lines emission), while adiabatic cooling becomes important only at large radii. The maximum ionization fraction varies between 2×10^{-4} ($\alpha = 10^{-4}$), 1.4×10^{-2} ($\alpha = 10^{-3}$) and 1 ($\alpha = 1$).

Table 2

Thermal Content of Alvfen Wave Driven Winds

α	$\dot{E}_{thermal}/\dot{E}_{wind}$	$\dot{E}_{thermal}/L_*$
1(-4)	0.02	1.8(-4)
1(-3)	0.11	1.0(-3)
0.01	1.1	0.01
0.05	5.6	0.05
1	107	1.0

5.6 HEATING DUE TO H$_2$ FORMATION

H$_2$ formation liberates for gas heating a binding energy of 4.5 eV. Its effects have been investigated by Ruden *et al.* 1990 in a paper on the thermal structure of winds from young stellar objects, where radiative processes and friction heating were also included. In the low luminosity cases of their grid of models, the temperature drops very steeply in the inner few stellar radii, due to the adiabatic expansion of the winds; the H_2 formation is somewhat effective only when strong collimation is invoked (i.e., when the gas density at a given distance from the star is higher than in the spherically simmetric case at the same $\dot{M}$). Even in those more favorable cases, T_g is never larger than 3000 K at $r \sim 2r_*$.

5.7 COMPARISON WITH THE OBSERVATIONS

The situation described in Figure 12 can be summarized as follows:

i) The cooling due to adiabatic expansion cause a sharp drop with r of the gas temperature from an initial value which is of the order of T_*. In order to obtain almost isothermal winds with $T_g \sim$6000-7000 K, as required by the observations, it is necessary to introduce an additional heating mechanism which dissipates approximately 10% of the wind mechanical luminosity in the first 1-2 r_* from the star.

ii) Frictional heating, as computed from "reasonable" radial acceleration laws, does not heat the gas to the observed temperatures. Also, chemical processes in the gas, such as H_2 formation, while contributing significantly to the gas thermal balance, do not seem able to provide the required energy input (at least in TTS winds).

iii) Wave heating can effectively warm up the gas to the observed temperatures. In the case of Alfvén driven winds, self-consistent models, where both the dynamics and the heat equations are solved together, have been computed by Hartmann et al. 1982 , Natta et al. 1988b. By looking at those results, it seems difficult to match both the observed acceleration and temperatures. In order to accelerate the wind (of given $\dot{M} \sim 10^{-7} M_\odot \text{yr}^{-1}$) to v_∞ of the order of 100-300 km s^{-1} , values of $\alpha \sim 1$ are required. Then, the corresponding gas temperature would be higher than observed ($T_g \gtrsim 10000$ K), and the wind fully ionized.

6. Conclusions

In this paper, we have addressed the question of how to compute the physical structure of winds ejected by low luminosity young stars. In particular, we have discussed the question of the ionization and excitation of hydrogen and trace metals, and described the various processes which can determine the gas temperature. The results of numerical calculations, available in the literature, have been used as a guideline, but, whenever possible, we have described analytical solutions to the various equations.

In general, the methods and the results quoted here are not self-consistent, in the sense that they do not solve together the equations of dynamics, radiation transfer and thermal balance, but rather assume that some of the quantities characterizing the wind are known *a priori*. This approach has been followed successfully by several authors (see, for example, Natta *et al.* 1988a,b; Natta and Giovanardi 1989; Giovanardi *et al.* 1990a; Hartmann *et al.* 1988; Kwan and Alonso-Costa 1988; Alonso-Costa and Kwan 1989) with the purpose of finding good diagnostic tools for the most relevant wind parameters, such as the mass-loss rate $\dot{M}$ and the gas temperature profile $T_g(r)$.

An example diagnostic, using lines from hydrogen and sodium (from Natta and Giovanardi 1989; Giovanardi et al. 1990a), has been discussed in Section IV. The comparison of the model calculations to the observed spectra of a sample of TTS (chosen to have blue-shifted absorption in the Na I doublet at 5900 Å) suggests that TTS winds are cool but not cold (T_g ~5000-7000 K near the star), and almost isothermal in the inner $1-2r_*$. As a consequence, they are mostly neutral, with maximum ionization fraction $x_e \lesssim 10\%$. This particular sample of TTS has a rate of mass-loss in the range $3 \times 10^{-8} - 3 \times 10^{-7}$ $M_\odot yr^{-1}$. This set of conditions (mass-loss rate and gas temperature) requires that the equivalent of approximately 10% of the mechanical wind luminosity (not much less and not much more!) is dissipated into gas heating very near the star. All the ejection-heating mechanisms we have examined in Section V seem unable to produce the right range of temperatures and/or the right velocity gradient.

Some of the TTS studied in Section V have been searched for molecular outflows. The momentum computed for the TTS winds is similar to the momentum required to drive the outflow under the assumption that this last is ambient matter accelerated by the stellar wind in a momentum conserving shock, as first proposed by Snell et al. 1980 for the case of L1551. Most of the stellar wind momentum is carried by *neutral* matter, which would not be detected by an *ionized* matter diagnostic, such as hydrogen recombination lines or free-free emission.

The authors thank Lee Hartmann, Francesco Palla and Frank Shu for interesting discussions. Rino Bandiera and Marco Salvati contributed very useful comments to Section 5. Partial support was provided by ASI grant 1989 to the Osservatorio di Arcetri.

References

Adams, F.C., Lada, C.J., and Shu, F.H. 1987, *Ap.J.*, **312**, 788.
Alonso-Costa, J.L., and Kwan, J. 1989, *Ap.J.*, **338**, 403.
Beckwith, S., and Natta, A. 1987, *Astr.Ap.*, **181**, 57.
Calvet, N., Cantó ,J., and Rodriguez, L.F. 1983, *Ap.J.*, **268**, 739.
Castor, J.I. 1970, *M.N.R.A.S.*, **149**, 111.
Croswell, K., Hartmann, L., and Avrett, E.M. 1987, *Ap.J.*, **312**, 227.
De Campli, W.M. 1981, *Ap.J.*, **244**, 124.
Draine, B.T. 1980, *Ap.J.*, **241**, 1021.
Dyson, J.E. 1984, *Astr.Space Sci.*, **106**, 181.
Edwards, S., and Snell, R.L. 1982, *Ap.J.*, **261**, 151.
Evans, N.J.II, Levreault, R.M., Beckwith, S., and Skrutskie, M. 1987, *Ap.J.*, **320**, 364.
Giovanardi, C., Gennari, S., Natta, A., and Stanga, R. 1990a, *Ap.J.*, **367**, 173.
Giovanardi, C., Evans, N.J.II, Heiles, C., Lizano, S., Natta, A., and Palla, F. 1990b, in preparation.
Grinin, V.P., and Mitskevich, A.S. 1988, *Izvestiya Krimshoy Astroph.Obs.*, **78**, 28.

Grinin V.P., and Mitskevich, A.S. 1990a, *Astrofisika*, in press.
Grinin V.P., and Mitskevich, A.S. 1990b, *Astrofisika*, in press.
Hartmann, L., Calvet, N., Avrett, E.H., and Loeser, R. 1989, *Ap.J.*, **349**, 168.
Hartmann, L., Edwards, S., and Avrett, E.M. 1982, *Ap.J.*, **261**, 279.
Hartmann, L., and MacGregor, K.B. 1980, *Ap.J.*, **242**, 260.
Herbig, G.H. 1962, *Advances in Astr.Ap.*, **1**, 47.
Hollweg, J.V. 1973, *Ap.J.*, **181**, 547.
Holzer, T.E., Flå, T., and Leer, E. 1983, *Ap.J.*, **275**, 808.
Kuhi, L.V. 1964, *Ap.J.*, **140**, 1409.
Kuhi, L.V. 1966, *Ap.J.*, **143**, 991.
Kushawaha, V.S., Burkhardt, C.E., and Leventhal, J.J. 1980, *Phys.Rev.Lett.*, **45**, 1686.
Kutner, M.L., Leung, C.M., Machnik, D.E., and Mead, K.N. 1982, *Ap.J.(Letters)*, **259**, L35.
Kwan, J., and Alonso-Costa, J.L. 1988, *Ap.J.*, **330**, 870.
Kwok, S., and Volk, K. 1985, *Ap.J.*, **299**, 181.
Lada, C.J. 1985, *Ann.Rev.Astr.Ap.*, **23**, 267.
Levreault, R.M. 1988, *Ap.J.*, **330**, 897.
Lizano, S., Heiles, C., Rodriguez, L.F., Koo, B.-C., Shu, F.H., Hasegawa, T., Hayashi, S., and Mirabel, I.F. 1988, *Ap.J.*, **328**, 763.
Mundt, R. 1984, *Ap.J.*, **280**, 749.
Natta, A.,Giovanardi, C., Palla, F. and Evans, N.J.II 1988a, *Ap.J.*, **327**, 817.
Natta, A.,Giovanardi, C., and Palla, F. 1988b, *Ap.J.*, **332**, 921.
Natta, A., and Giovanardi, C. 1990, *Ap.J.*, **356**.
       646 Pudritz, R.E., and Norman, C.A. 1986, *Ap.J.*, **301**, 571.
Randich, M.S. 1988, *Tesi*, Università di Firenze.
Randich, M.S., Giovanardi, C., Natta, A., and Palla, F. 1990, *Astr.Ap.Suppl.*, in press.
Ruden, S.P., Shu, F.H., and Glassgold, A.E. 1990, *Ap.J.*, **361**, 546.
Shu, F.H., Adams, F.C., and Lizano, S. 1987, *Ann.Rev.Astr.Ap.*, **25**, 23.
Simon, M., Felli, M., Cassar, L., Fischer, J., and Massi, M. 1983, *Ap.J.*, **266**, 623.
Snell, R.L., Loren, R.B., and Plambeck, R.L. 1980, *Ap.J.(Letters)*, **239**, L17.
Snell, R.L., and Schloerb, F.P. 1985, *Ap.J.*, **295**, 490.
Sobolev, V.V. 1960, *Moving Envelopes of Stars* (Cambridge: Harvard University Press).
Spitzer, L. 1978, *Physical Processes in the Interstellar Medium* (Wiley: New York).
Suto, H., Mizutani, K., and Maihara, T. 1989, *M.N.R.A.S.*, **239**, 139.
Uchida, Y., and Shibata, K. 1985, *P.A.S.J.*, **37**, 515.
Wishart, A.W. 1979, *M.N.R.A.S.*, **187**, 59p.

Scott Kenyon receives second place poster prize from Blaauw.

EPISODIC PHENOMENA IN EARLY STELLAR EVOLUTION

L. HARTMANN
Harvard-Smithsonian Center for Astrophysics
60 Garden St., Cambridge, MA, USA

ABSTRACT. The outbursts of FU Orionis objects are apparently caused by a rapid increase in mass accretion through a protostellar disk onto a T Tauri star. Accretion rates during outbursts are very large, $\gtrsim 10^{-4} M_\odot \, yr^{-1}$. Event statistics indicate that many, if not most, low-mass stars undergo multiple Fuor events, ultimately accreting at least 5% to 10% of the total stellar mass from the disk during these brief episodes. Models of Fuor disks indicate that they must be massive and hot during outburst, and may be moderately thick in the vertical direction. Fuor disk accretion produces very energetic winds, capable of driving bipolar molecular flows and Herbig-Haro objects. Fuors may also provide insight into very early stages of evolution, when disk accretion may be particularly rapid.

1. Introduction

Roughly half of all young stars appear to posess detectable circumstellar disks with substantial ($\sim 0.01 - 0.1 M_\odot$) mass (Strom *et al.* 1990; Beckwith *et al.* 1990; Adams, Emerson, and Fuller 1990). Of order 0.1 $M_\odot$ appears to accrete onto the central star during optically-visible phases of evolution (Hartmann and Kenyon 1985, 1990). Disk accretion may be even more important during the earliest phases of star formation, when the central protostar is obscured by its dusty infalling envelope (e.g., Shu, Lizano, and Adams 1987). Thus, it seems quite possible that low-mass stars are mostly built up by disk accretion (Mercer-Smith, Cameron, and Epstein 1984).

Essentially all pre-main sequence stars vary in brightness, and much (though not all) of this variability is now attributed to variations in the mass accretion rate through the circumstellar disk. The most spectacular variations in light are observed in the classical FU Orionis (Fuor) outbursts (Herbig 1966, 1977, 1989), during which the objects brighten by a factor of one hundred or more at optical wavelengths. Scott Kenyon and I have attributed these outbursts to dramatic increases in disk accretion of up to three orders of magnitude on a time scale of a year or so (Hartmann and Kenyon 1985, hereafter HK; see also Lin and Papaloizou 1985).

The Fuor outbursts are important for a variety of reasons. Fuors represent energetic phenomena that appear to be quite common (though short-lived) aspects of early evolution. The amount of mass added to the central star during such outbursts, $\sim 0.05 - 0.1 M_\odot$, is not negligible, and may significantly affect the evolution of the star. The extreme variability has important implications for the physics of protostellar disks, and allows one to place interesting constraints on disk properties. Finally, the discovery of Fuor characteristics in the prototype of heavily-embedded bipolar flow sources, L1551 IRS 5 (Mundt *et al.* 1985; Carr, Harvey, and Lester 1987) suggests that rapid disk accretion might be an important feature of protostellar evolution.

Fuors exhibit a number of properties which permit detailed observational analysis. The high accretion rates during outburst cause the disks to become quite hot, so that Fuor disks can be studied in the optical and near-infrared (and even at ultraviolet wavelengths), where high spectral resolution data can be obtained. Accretion rates are so large during outburst that the disk outshines the star by a large factor, simplifying the analysis of the spectral energy distribution. Because of these properties, we have more detailed information on Fuor disks, i.e. T Tauri disks during rapid accretion, than we have for T Tauri disks during normal slow accretion.

Herbig (1966, 1977, 1989) and Reipurth (1990) have reviewed the Fuor phenomenon with special emphasis on observational properties. We have recently added to this list with a review concentrating on connections to solar system problems (Hartmann, Kenyon, and Hartigan 1991), as well as a discussion from the point of view of disk modelling (Hartmann and Kenyon 1991). In this review I shall try to provide more details than has been possible in briefer discussions.

2. Observed Properties of Fuors

The Fuors are found in star-forming regions and are kinematically associated with the gas therein (Herbig 1966, 1977; Hartmann and Kenyon 1985, 1987a,b; Hartmann *et al.* 1989). All members of the class have reflection nebulae (cf. Goodrich 1987), suggesting that they are recently-formed pre-main sequence objects. The strong Li I absorption seen in these objects further emphasizes their youth (Herbig 1966, 1977).

Table 1 lists the Fuors that have been identified at the time of writing, with dates and timescales of outbursts, estimated accretion rates $\dot{M}$ in units of $M_\odot \, yr^{-1}$, and whether ^{12}CO high-velocity flows and optical jets or HH objects are associated with the Fuor.

Fuors were originally defined as objects which exhibited a large increase in optical brightness, typically 5 magnitudes or more (Figure 1). Since the outburst times can be much shorter than the decay times, it seems reasonable to include objects for which the outburst was not detected, but which have spectroscopic similarities to the known Fuors. No outburst has been detected in Z CMa, although it is highly variable (by at least two magnitudes in the optical), and is very similar to FU Ori in its optical spectroscopic properties. Objects like L 1551 IRS 5 are essentially optically invisible because of heavy extinction, making it difficult to say whether any outburst has occurred from historical photographic plate records. The list of objects in Table 1 is thus not selected on the basis of photometric behavior, but attempts to include pre-main sequence objects thought to exhibit extremely rapid disk accretion.

Fuors have rather distinctive spectroscopic properties. All members of the

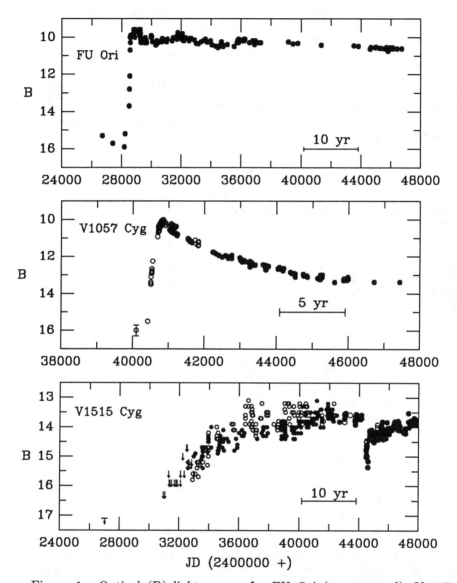

Figure 1. Optical (B) light curves for FU Ori (upper panel), V1057 Cyg (middle panel), and V1515 Cyg (lower panel). The data have been compiled from various sources: V1057 Cyg (KHH; Simon and Joyce 1988; and references therein); V1515 Cyg (Herbig 1977; Landolt 1977; Tsvetkova 1982; Kolotilov and Petrov 1983); FU Ori (Herbig 1977; Kolotilov and Petrov 1985; KHH). Five magnitudes corresponds to a factor of 100 in brightness.

class exhibit an optical spectral type of F or G supergiant. Broad absorption blueshifted by several hundred $km\,s^{-1}$ is observed in the Balmer lines. The Na I resonance lines also show broad blueshifted absorption, sometimes in distinct velocity components or "shells". The emission component in the P Cygni Hα profile is often absent; when present, this emission extends to much smaller velocities redward than the blueshifted absorption. The combination of F or G supergiant spectrum with broad Hα absorption is unknown among T Tauri stars, and is sufficiently remarkable that Mundt et al. (1985) and Stocke et al. (1989) identified the famous bipolar flow source L 1551 IRS 5 as a Fuor based on the low-resolution optical spectrum alone.

Table 1
Fuor Properties

Object	Outburst date	Rise time	Decay time	d(kpc)	$L/L_\odot$	CO flow	Jet/HH
FU Ori	1937	$\sim$ 1 yr	$\sim$ 100 yr	0.5	490	no	no
V1057 Cyg	1970	$\sim$ 1 yr	$\sim$ 10 yr	0.6	350	yes	no
V1515 Cyg	1950's	$\sim$ 20 yr	$\sim$ 30 yr	1.0	200	no	no
V1735 Cyg	$\sim$ 1957-65	< 8 yr	> 20 yr	0.9	>75	yes	no
V346 Nor	>1984	< 5 yr	> 5 yr	0.7	?	yes	yes
L1551 IRS5	?	?	?	0.15	>40	yes	yes
Z CMa	<1860?	?	> 100 yr	1.1	7800	yes	yes
BBW 76	<1930	?	$\sim$ 40 yr	1.7?	?	?	yes

References: Herbig 1977, 1989; Elias 1978; Levreault 1983; Covino et al. 1984; Graham and Frogel 1985; Mundt et al. 1985; Reipurth 1985, 1989b; Cohen and Schwartz 1987; Stocke et al. 1988; Poetzel, Mundt, and Ray 1990; Reipurth et al. 1990; Eisloeffel, Hessman, and Mundt 1990; Ray 1990, private communication; Rodriguez, Hartmann, and Chavira 1990.

The very luminous object Z CMa occasionally exhibits different spectral behavior. Although Z CMa usually exhibits the same type of spectrum as the other Fuors, at maximum light this object exhibits many strong emission lines of Fe II and other low-excitation species, with little evidence for the usual absorption lines (Hessman et al. 1991).

Of particular importance is the single pre-outburst spectrum of a Fuor, V1057 Cyg, shown by Herbig (1977). The spectrum is not of high signal-to-noise, but the general impression is that of a highly-reddened T Tauri star with Balmer, Ca II, Fe I, Fe II emission. At maximum light, the emission lines disappeared in V1057 Cyg, and only Balmer absorption was visible, with a spectral type estimated as $\sim$ early A. For this single Fuor there is no question that the outburst was not the result of dispersal of an absorbing dust screen, but instead involved a qualitative change in the source (Herbig 1977).

At higher spectral resolution, the optical line profiles are complicated and interesting. Absorption "shell" features blueshifted by tens of $km\,s^{-1}$ are seen in the blue spectral region of FU Ori (Herbig 1966), and there is some evidence for unresolved shell features near 5000 Å in the spectra of FU Ori and V1057 Cyg (Hartmann and Kenyon 1985). At wavelengths $\gtrsim$ 6000 Å the line profiles of FU

Ori and V1057 Cyg show two absorption dips more or less symmetrically displaced from line center (Hartmann and Kenyon 1985, 1987a,b). In Z CMa, this line doubling is observed down to ~ 4500 Å (Hartmann et al. 1989). The lines are quite broad, suggesting rapid rotation at velocities of 20 to 120 km s^{-1}, hardly typical of normal supergiants. No evidence has yet been observed for systemic radial velocity variations any larger than a couple of km s^{-1} (Hartmann and Kenyon 1987b).

The infrared emission of Fuors is greatly in excess of that expected from a normal star (Figure 2). Mould et al. (1978) showed that FU Ori and V1057 Cyg exhibited H_2O absorption at 1.6 μm and CO band absorption at 2.2 μm typical of an M giant or supergiant, but inconsistent with the optical spectral type. CO absorption has similarly been observed in Z CMa (Hartmann et al. 1989), L 1551 IRS 5 (Carr, Harvey, and Lester 1987), and V1735 Cyg (Elias 1978). In addition to the cooler spectrum seen at longer wavelengths, the rotational velocity measured in the infrared is substantially smaller than the optical $v \sin i$ (Hartmann and Kenyon 1987a,b; see § 3.2).

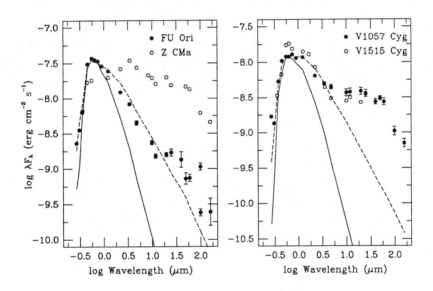

Figure 2. Spectral energy distributions for four Fuors, corrected for extinction. The solid line indicates a single-temperature blackbody; the dashed line shows the energy distribution for the steady accretion disk of Figure 3.

Herbig (1977) originally suggested that Fuor outbursts were repetitive. By counting up the number of known T Tauri stars in the appropriate volume, Herbig estimated that all stars would need to have ~ 10^2 eruptions in their lifetimes to account for the observed frequency of Fuor outbursts. We now recognize that counts of T Tauri stars out to several hundred parsecs are likely to be incomplete. Assuming that the mean star formation rate derived by Miller and Scalo (1979) represents an adequate averages for the solar neighborhood, approximately one star should be formed in a 1 Kpc cylinder centered on the Sun every hundred years. The five Fuors known to have erupted in this volume over the last fifty years imply a repetition rate of ~ 10 per star (Hartmann and Kenyon 1985). This estimate does not include known Fuors which have not apparently erupted in this time period, and it is likely that we have missed some past eruptions, since the frequency of discovery is clearly increasing with time. Thus it is hard to escape the conclusion that Fuor eruptions occur between 10 to 100 times on a typical low-mass star (see also Herbig 1989).

3. Steady Disk Models

The accretion disk model for Fuors (Hartmann and Kenyon 1985; Lin and Papaloizou 1985) initially was motivated by analogy to other astrophysical objects with similar outburst behavior and by the need to explain the observed variation of spectrum with wavelength and the large infrared excess emission (Figure 2). Here I outline some basic properties of accretion disks to demonstrate the essential predictions of a disk model. Although such highly time-dependent objects such as Fuors cannot be fully understood in terms of steady-accretion disk models, the steady-state assumption permits some important results to be obtained without knowing anything about the nature of the viscosity in the disk. The basic equations were derived by Shakura and Sunyaev (1973) and Lynden-Bell and Pringle (1974); a particularly clear exposition can be found in Pringle (1981).

3.1 STEADY DISK SPECTRUM

In the simplest possible situation, the disk is thin, and the particles exhibit essentially Keplerian motion around a central point mass, with a slow, steady, inward drift superposed. In a viscous disk, the shearing motion is dissipated as heat, and this energy released causes the matter to fall in. The energy released in falling from $R + \Delta R$ to R is then just the difference in the energies of the two (nearly-circular) orbits. Since the disk is thin, we expect this energy to be radiated locally. Thus, for a steady infall rate $\dot{M}$, we have an energy balance equation for an annulus of width ΔR at a distance R from the central star,

$$\frac{GM\dot{M}}{2R}\frac{\Delta R}{R} = D(R)\,\Delta R,$$
$$= 2 \times 2\pi R\,\Delta R \sigma T^4. \tag{1}$$

Here it is assumed that the energy dissipated in the annulus, $D(R)\,\Delta R$, is equal to the energy radiated by the surface, supposing that the disk is optically thick and that it radiates from both sides approximately as a blackbody. (The assumption that the disk is optically thick will be justified later.) In general, we should also add in the radiation from the central star absorbed by the disk (Adams and Shu 1986; Adams, Lada, and Shu 1987); however, since the Fuors show such large increases in light, we presume that the *stellar* luminosity is much less than the energy radiated by the disk in outburst, and so the "reprocessing" of the central star's radiation can be neglected.

Using equation (1), one finds the variation of disk surface temperature with radius (Shakura and Sunyaev 1973)

$$T^4 = \frac{GM\dot{M}}{8\pi\sigma R^3}. \tag{2}$$

The exact equation generally used in the literature to model the spectral energy distribution is

$$T^4 = \frac{3GM\dot{M}}{8\pi\sigma R^3}[1 - (R_*/R)^{1/2}], \tag{3}$$

where R_* is the stellar radius. The term in brackets results from the choice of inner boundary condition, namely that the flux of angular momentum at the inner disk radius is $\dot{J} = \dot{M}\Omega_K R_*^2$, where Ω_K is the Keplerian angular velocity at R_* (cf. Pringle 1981). Although equation (3) has the same asymptotic form as (2), at large distances approximately three times as much energy is emitted by the disk than is dissipated locally. This energy comes from the inner regions, which as a consequence have temperatures dropping to zero as R approaches R_*. This result is probably unphysical, but it is not clear what improvements should be made. For most purposes, the differences between (2) and (3) are of little importance, although of course the exact values of mass accretion rate, stellar radius, etc. derived from observations are affected. We use equation (3) in the following discussion.

With this rather modest amount of theory, we are now in position to compare the predictions of the simplest disk model with the observations. Figure 2 shows the spectral energy distribution of a single temperature blackbody (i.e., a star) and the spectrum of a steady disk, compared with the observed spectral energy distributions of four Fuors. The maximum temperature of the disk and its size have been scaled to provide the best match to the data. The simple disk model actually does remarkably good job of matching the infrared excess emission of FU Ori, and represents a reasonable model for the emission of V1057 Cyg and V1515 Cyg shortward of 10 μm. It is worth emphasizing that the infrared excess emission, which extends over a range of almost two orders of magnitude in wavelength, requires a range of temperatures over a *very* large object. For example, with a temperature maximum $\sim$ 7000 K to match the optical spectrum of FU Ori and T $\propto r^{-3/4}$, one infers that the 10 μm emission (dominated by emission at roughly 300 K) comes from a radius $\sim 10^2 R_*$. One very strong point of the disk model is that it is capable of explaining this large excess naturally by positing an optically-thick disk of solar system dimensions. Of course, V1057 Cyg and V1515 Cyg exhibit even more infrared excess emission at long wavelengths, and Z CMa has a much flatter energy distribution than can be explained by this simple disk model. I shall return to these complications briefly at the end of § 6.

Figure 3 provides detailed information for the disk model of Figure 2. The peak surface temperature of the disk, as given by equation (3), occurs at R = 1.36 R_*. Rather than let the disk temperature fall to zero at R_*, as implied by equation (3), which is unphysical, we take the disk temperature to be constant interior to this radius. The fractional contributions of various regions of the disk to the total emission at the specified wavelengths have been calculated assuming that the disk radiates like stars above 3000 K and like blackbodies at lower temperatures. The optical light of the disk model comes from regions $\sim$ 6000 K, within a few stellar radii of the inner disk radius. This maximum temperature has been chosen to

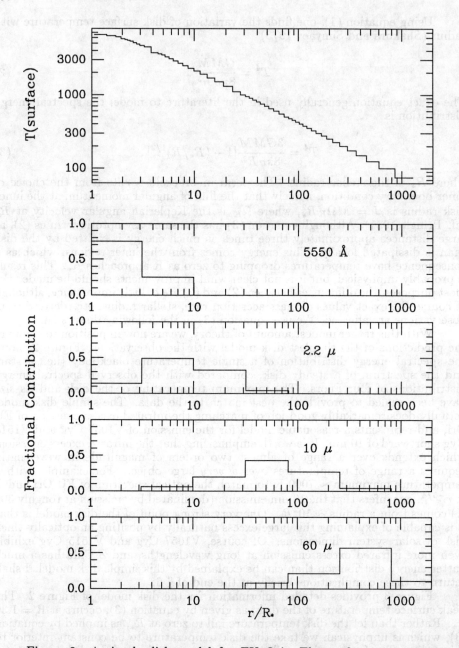

Figure 3. A simple disk model for FU Ori. The surface temperature distribution is given as a function of radius in the uppermost panel, using equation (3). The lower panels show the fractional contributions of different parts of the disk to the total light at four different wavelengths.

match the observed optical spectral type of late F - early G supergiant (Kenyon, Hartmann, and Hewett 1988 = KHH). However, at 2 μm, where the first-overtone $v'' - v' = 2$ CO bands are formed, the light mostly arises from regions with temperatures $\sim 3000K$. Thus, the disk model predicts that the infrared spectrum should look cooler than the optical spectrum. Detailed calculations using the spectra of supergiant stars with the appropriate temperatures shows that the disk model can explain the great strength of the CO bands quantitatively (KHH). (Use of higher-gravity stellar atmospheres results in equivalent widths for the absorption spectra that are too small, both in the optical and in the infrared. The reasons why supergiant stellar atmospheres are appropriate are discussed in the following section.)

3.2 DIFFERENTIAL ROTATION

If the disk is in essentially Keplerian motion, with only a very small inward drift velocity, then the rotational velocities will decrease as $r^{1/2}$. Since the light at long wavelengths is dominated by emission from larger disk radii, one expects the rotational velocity broadening of the spectral lines to decrease with increasing wavelengths. This effect has been observed in FU Ori and in V1057 Cyg (Hartmann and Kenyon 1987a,b). Figure 3 shows that the region of the disk responsible for the 2.2 μm continuum emission is about four times as large in radius than the region responsible for the 0.6 μm emission. Thus, one would expect that the rotational velocity measured in the infrared should be about half of the optical rotational velocity, assuming Keplerian rotation.

Hartmann and Kenyon (1987a,b) measured the rotational broadening in the optical and infrared regions using cross-correlation analysis. The cross-correlation peak is then a measure of the average line profile in the spectral region of interest. Figure 4 shows that the observed rotational broadening at 2.2 μm is $\sim 2/3$ of the rotational broadening observed at 0.6 μm, in reasonable agreement with expectation.

One can make even more detailed tests of the model by synthesizing a disk spectrum. Using the temperature distribution shown in Figure 3, one can produce a detailed disk spectrum by assuming that each annulus radiates like a star of the appropriate effective temperature. The model spectrum can be cross-correlated in exactly the same way as the observations for direct comparison. Figure 4 shows that the model predictions of the ratio of infrared and optical rotational broadening agree quite well with the observations, certainly within uncertainties in the modelling.

The disk model predicts that the variation of rotation with wavelength should be continuous. Welty *et al.* (1990) have also presented some evidence of such a continuous variation of rotational velocity with wavelength in the optical spectra of V1057 Cyg. The effect is subtle ($\sim 10\%$) and uncertain, but consistent with the disk model; the spectrum synthesis indicates that the differential rotation should be small over the limited wavelength baseline available in the optical region.

3.3 LINE PROFILES

The line profiles produced by a simple disk model are qualitatively different than the parabolic shapes produced by a spherical rotating star. In the limit where the rotational velocity is large compared with thermal or turbulent velocities, the regime appropriate to Fuors, the line profiles produced by a rotating, flat, narrow ring as a function of velocity shift from line center ΔV are of the form

$$\phi(\Delta v) = (1 - (\Delta v/v_{max})^2)^{-1/2}, \quad -v_{max} < \Delta v < v_{max}, \tag{4}$$

where v_{max} is the rotational velocity of the ring corrected for its inclination to our line-of-sight. The final line profile observed at a given wavelength is the sum of of profiles from several neighboring annuli with different rotational velocities, and this differential rotation smooths the profile indicated by equation (4), resulting in a rounded "double-peaked" profile.

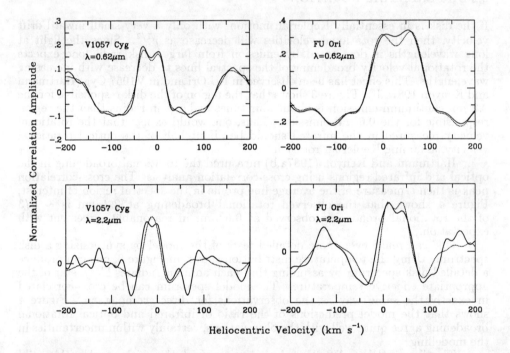

Figure 4. Rotational velocities of FU Ori and V1057 Cyg at two different wavelengths, as indicated by cross-correlation peak widths (solid lines). Slower rotation is observed in the 2.2 μm CO lines than at optical wavelengths. Cross-correlations of disk model spectra (dashed lines) are in good agreement with observations, showing that the observed differential rotation is consistent with Keplerian motion.

The cross-correlations shown in Figure 4 show evidence for double-peaked average line profiles, in reasonable agreement with the model prediction. (The model cross-correlation peak shown in Figure 4 for V1057 Cyg at 2.2 μm is not doubled because the rotational velocity broadening is not large in comparison with the instrumental resolution.) As shown in Figure 5, doubled line profiles are often seen in individual lines in the red spectral region of FU Ori and V1057 Cyg. Moreover, as indicated in Figure 6, disk model spectra synthesized from standard stars as described in the previous section, when convolved with the expected rotational broadening profile (4), provide a remarkably good match for many observations of Z CMa. The agreement between observed and model spectra is not this good shortward of 6000 Å in FU Ori and in V1057 Cyg. We shall explore possible reasons for this in § 6.

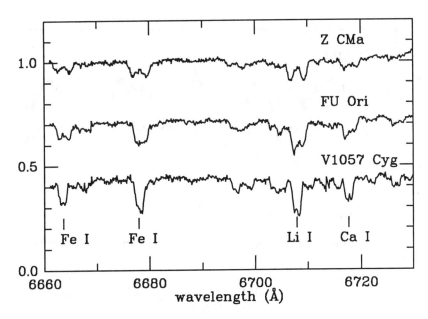

Figure 5. Comparison of optical line profiles in three Fuors. Note the double peaked structure characteristic of a rotating disk (see text), also indicated in the cross-correlation peaks of Figure 4.

3.4 INFERRED PROPERTIES OF THE CENTRAL STARS

One can estimate the inner disk radii and masses of the central object from the observations for comparison with the properties of T Tauri stars. From optical and ultraviolet spectroscopic observations we can estimate T_{max} of the disk; then, from (3), we have

$$T_{max} \simeq 0.488 \left(\frac{3GM_*\dot{M}}{8\pi\sigma R_*^3} \right)^{1/4}. \quad (5)$$

The total flux f received at the Earth from a flat accretion disk of total accretion luminosity L_{acc} with inclination i to the line of sight is

$$f = \frac{L_{acc} \cos i}{2\pi d^2}. \tag{6}$$

We take

$$L_{acc} = \frac{GM_* \dot{M}}{2R_*}. \tag{7}$$

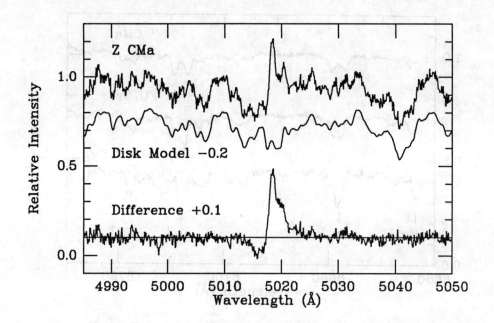

Figure 6. A comparison between a high-resolution optical spectrum of Z CMa and a disk model spectrum. The difference spectrum (lower plot) is essentially featureless, showing very good agreement between model and observation, except for the strong Fe II λ 5018 "P Cygni" line formed in the strong wind of Z CMa. From Welty (1990).

The luminosity is derived by integrating over the observed energy distribution, making an extinction correction from optical photometry based on a disk model and T_{max}, and assuming a distance. Except in the case of Z CMa, the uncertainty in the total luminosity is likely to be a factor of two or less. (There is also some uncertainty in whether the factor of 2 in the denominator of equation (7) should be eliminated to account for reprocessing of boundary layer radiation, as discussed in the following section). From these equations we derive an estimate of the inner disk radius R_*,

$$R_* \simeq 0.12 \left(\frac{L_{acc}}{\sigma T_{max}^4} \right)^{1/2}. \tag{8}$$

Using the luminosities in Table 1 ($= 4\pi d^2 f$), and adopting maximum disk temperatures of 7200 K for FU Ori and V1515 Cyg, and 6580 K for V1057 Cyg, one derives inner disk (stellar) radii for these objects of 4.1 $R_\odot/cos^{1/2} i$, 2.6 $R_\odot/cos^{1/2} i$, and 4.1 $R_\odot/cos^{1/2} i$, respectively. These values are reasonably consistent with the typical radii $R_* \sim 2 - 3 R_\odot$ of T Tauri stars. The estimated inner radius of Z CMa is 10 - 20 $R_\odot$ (Hartmann *et al.* 1989), which is quite a bit larger than the size of a typical T Tauri star, but this result is quite uncertain because the spectral energy distribution is so peculiar (Figure 2).

One can then use the rotational velocity broadening measurements in the optical spectral region to estimate the central mass,

$$M_* sin^2 i \simeq \frac{2R_*(v\,sin\,i)^2}{G} \quad (optical), \qquad (9)$$

where we have assumed that the optical rotation measurements are characteristic of a disk radius $\sim 2R_*$ (see Figure 3). The observed optical rotational velocities of about 65, 45, and 22 km s^{-1} for FU Ori, V1057 Cyg, and V1515 Cyg imply values of $M sin^2 i\, cos^{1/2} i$ of 0.182, 0.087, and 0.013 $M_\odot$, respectively. For $i = 45°$ this implies a mass for FU Ori of 0.43 $M_\odot$. The masses of V1057 Cyg and V1515 Cyg are consistent with stellar masses only if we assume that the objects are seen nearly pole-on; Goodrich (1987) has argued that such is the case, based on the structure of the nebulosity surrounding these objects. For $i = 30°$ the masses of V1057 Cyg and V1515 Cyg would be 0.37 $M_\odot$ and 0.06 $M_\odot$, respectively. The mass of Z CMa is exceedingly uncertain, $\sim 1 - 2 M_\odot$. The estimated accretion rates for most Fuors is $\sim 10^{-4} M_\odot$; for Z CMA, the very uncertain accretion rate is closer to $10^{-3} M_\odot$.

To summarize, the disk model for Fuors naturally explains the observed broad spectral energy distribution and variation of spectral type with wavelength; it *quantitatively* accounts for the variation of rotation with wavelength; and it simply explains much of the peculiar line profile structure ("doubled" lines). The masses and radii of the central objects derived from fitting disk models to observations are reasonably consistent with those of T Tauri stars.

4. Disk Properties

Minimum disk masses for Fuors can be estimated by multiplying the accretion rates inferred above by the duration of the outbursts. The decay time for V1057 Cyg is of order a decade (although the light curve recently has shown a tendency to flatten out; Figure 1), so the minimum mass for this disk is $\sim 10^{-3} M_\odot$, i.e. a "minimum mass" solar nebula. FU Ori has a decay time of 50 years or so, and therefore the amount of mass that will be accreted in the present outburst is $\sim 10^{-2} M_\odot$. For Z CMa, our admittedly crude estimate of the accretion rate is an order of magnitude larger, and although this object has varied irregularly over the last century it has remained relatively bright for most of this time. Thus, the suggested mass in this disk is $\sim 10^{-1} M_\odot$. It should be emphasized that if these objects are repetitive in their outbursts, as seems likely from the event statistics discussed in § II, the *total* disk mass will be $\sim$ ten times larger than these estimates.

To gain further insight into the physical conditions in a typical Fuor disk, it is necessary to consider the implications of the observed time-variability. The radius which includes about half of the optical light arises from a region that is approximately $2 R_*$ (Figure 3), or $\sim 10 R_\odot$. Suppose that the optical outburst is produced by infall of material across this region. Since the rise time for FU Ori and V1057 Cyg is of order one year, the implied infall velocity is $\sim 0.2\,\mathrm{km\,s^{-1}}$. From the luminosity of the outburst, the mass infall rate (falling on to a 1 $M_\odot$ star) must be $\sim 10^{-4} M_\odot\,yr^{-1}$, so with $\dot{M} = 2\pi R \Sigma v_R$ we arrive at an estimate of the surface density in the inner disk of $\Sigma \sim 6\times 10^5\,\mathrm{g\,cm^{-2}}$. Approximately $0.2\,M_\odot$ would be contained in a disk with this surface density within 1 A.U.

Alternatively, as discussed in § 6, the outburst might be produced by the propagation of a thermal instability front. In this case the actual infall velocity would be smaller than $0.2\,\mathrm{km\,s^{-1}}$ (see § 6) and therefore the required surface density would be even higher.

The disk central density and temperature can be estimated by considering the momentum balance equation in the direction perpendicular to the disk,

$$\frac{1}{\rho}\frac{dc_s^2 \rho}{dz} = \frac{GM}{(R^2+z^2)}\frac{z}{(R^2+z^2)^{1/2}}, \qquad (10)$$

where c_s is the sound speed, ρ is the gas density, R is the distance in the disk plane from the center of the star, and z is the distance from the disk midplane. To keep things simple, assume that the disk is roughly isothermal in the vertical direction. Then the solution of eq. (10) is

$$\rho = \rho_0\, exp\left[\frac{GM}{c_s^2}\left(\frac{1}{(R^2+z^2)^{1/2}} - \frac{1}{R}\right)\right], \qquad (11)$$

which for small z reduces to a gaussian distribution

$$\rho \sim \rho_0\, exp\left[\frac{GM}{c_s^2}\left(-\frac{1}{2}\frac{z^2}{R^2}\right)\right] \qquad (12)$$

with a scale height

$$H = \left(\frac{c_s^2 R^3}{GM}\right)^{1/2} = \frac{c_s}{v_\phi}, \qquad (13)$$

where v_ϕ is the Keplerian velocity.

In FU Ori, the inner disk regions must have *surface* temperatures of 7000 K, and the midplane temperature must be larger since the disk is optically thick (see below), so $c_s \gtrsim 10\,\mathrm{km\,s^{-1}}$. Since $v_\phi \sim 10^2\,\mathrm{km\,s^{-1}}$, we have $H \sim 10^{-1} R$. The central density is related to the surface density Σ by

$$\rho_0 = \Sigma/(2\pi)^{1/2} H, \qquad (14)$$

which for the present case is of order $10^{-6}\,g\,cm^{-3}$.

In the relevant temperature and density range, Lin and Papaloizou (1985) have modeled the Rosseland mean opacity (mostly H^-) as $\kappa_R \sim 10^{-36} \rho^{1/3} T^{10}$. Assuming a temperature ≥ 7000 K, the opacity is $\geq 3\,g\,cm^{-2}$, and multiplying by the surface density one finds the Rosseland mean optical depth through the disk $\tau_R \gtrsim 10^6$.

The large optical depth through a Fuor disk justifies the use of standard stellar spectra to model the emergent disk spectrum. In addition to being very opaque, stellar atmospheres also satisfy the approximation of radiative equilibrium to a high degree, and these two properties combine to produce a decreasing temperature proceeding outward and an absorption line spectrum. Although a simple viscous disk has energy deposition at all heights, the volumetric heating rate for a viscosity ν

$$d = (1/2)\nu\rho(Rd\Omega/dR)^2 \qquad (15)$$

depends directly upon the density (Lynden-Bell and Pringle 1974). If one adopts the standard "alpha" viscosity $\nu = \alpha c_s H$, (see § 6), where c_S is the local sound speed and α is a dimensionless parameter that does not vary (too much), the heating rate (15) is mostly sensitive to the local density. For extremely opaque disks like Fuors, the photosphere occurs several scale heights above the disk midplane, where the density is relatively low. Virtually all of the viscous heating in this model occurs in the central scale height of the disk, and the photosphere must transmit an energy flux orders of magnitude greater than the local energy dissipation rate. Thus, radiative equilibrium is a resonable approximation for the disk photosphere, just as for a stellar photosphere.

At sufficiently large heights above the disk midplane, the low-density gas does not radiatively cool very efficiently, and is transparent to the flow of radiation from below. At this point a "chromospheric" temperature rise may occur due to the local viscous dissipation (Shaviv and Wehrse 1986). Indeed, emission is observed in typical chromospheric transitions like the Mg II resonance lines (Ewald, Imhoff, and Giampapa 1986; Kenyon *et al.* 1989).

Finally, we note that the vertical surface gravity of the disk, using $R = 10R_\odot$ and $H/R \sim 10^{-1}$, is $(GM/R^2) \times H/R \sim 10^1 g\, cm\, s^{-2}$. Since supergiant stars have similar surface gravities, we are justified initially in using the spectra of these stars to model the disk spectrum. The absorption line equivalent widths increase with decreasing surface gravity, because the H^- continuum opacity decreases with decreasing pressure, increasing the line-to-continuum opacity ratio. As a practical matter, only supergiants have sufficiently strong absorption lines to match the observations. The spectral modelling assumes that lateral radiative transfer is negligible, which seems reasonable for thin disks, but as Fuor disks may have non-negligible vertical thicknesses (see below), further investigation of this problem is desirable.

5. Boundary Layer

The boundary layer is the region where disk material rotating at Keplerian velocity comes to rest on the surface of the slowly-rotating star (Lynden-Bell and Pringle 1974). Since the kinetic energy of the innermost disk orbit is equal to the energy that must be lost for a particle in the disk to fall in from infinity to R_*, in a steady state the luminosity of the boundary layer is comparable to the accretion luminosity of the disk. The boundary layer emission is thought to be at most $\sim$ one disk scale height in lateral radial distance (Pringle 1989). Thus, in a thin disk model, the boundary layer must be much hotter than the disk to radiate the same amount of energy as the disk, resulting in substantial excess emission at short wavelengths.

FU Ori and Z CMa show no evidence for boundary layer emission in proportion to their accretion luminosity (Kenyon *et al.* 1989), even though the spectra of T Tauri stars indicate significant boundary layer fluxes (Kenyon and Hartmann 1987; Bertout, Basri, and Bouvier 1988; Basri and Bertout 1989; Hartmann and Kenyon 1990a; see chapter by Bertout).

Boundary layer radiation would be suppressed if if the central star is rotating near the breakup velocity. The initial rotational velocity of the central star is unknown for any Fuor. However, most T Tauri stars are very slow rotators, and the amount of material estimated to have fallen onto the central object in known Fuor outbursts is not sufficient to spin the star up to breakup velocity, assuming solid-body rotation.

The absence of observable boundary layer emission from Fuors may be related to finite vertical thicknesses of these disks. Boundary-layer radiation from an optically thick disk should diffuse in the radial direction by a scale height. If the disk is thick, the boundary layer emission would be spread out over a much larger radial distance, resulting in lower temperatures.

To see why Fuor disks may have appreciable vertical thicknesses, consider the possible temperature structure in the disk interior. The trapping of radiation in the optically-thick disk will cause the central regions to become hotter than the surface. We can estimate the magnitude of this effect by considering the temperature distribution of the classical gray atmosphere in LTE and radiative equilibrium:

$$T^4 \simeq 3/4\, T_{eff}^4 (\tau + 2/3) \qquad (16)$$

(Mihalas 1978). The estimates of the previous section suggest disk optical depths $>> 10^4$, which implies that the midplane temperature is at least an order of magnitude larger than the surface temperature. The surface temperature of the inner disk is ~ 6000 K, so this implies that the central disk temperature is $\gtrsim 6 \times 10^4 K$ and $c_s \gtrsim 30\,\mathrm{km\,s^{-1}}$. The Keplerian rotation speed in the inner disk is $\sim 150\,\mathrm{km\,s^{-1}}$, so that $H/R \gtrsim 0.2$! Thus, it may not be strictly accurate to model Fuors as thin disks (cf. Clarke et al. 1990).

Another possibility is that some of the accretion energy normally released in the boundary layer gets deposited in the outer envelope of the central star (Larson 1980). This may be especially plausible since the midplane pressure of the inner disk implied by the estimates of § 4 is more than an order of magnitude larger than the expected photospheric pressure. Since the luminosity of a pre-main sequence star is derived mostly from gravitational contraction, the addition of material to the star at a rate such that $L_{acc} >> L_*$, means that the star must undergo substantial internal readjustment. Detailed calculations of the effect of such rapid accretion onto the stellar surface have not yet been performed. What seems certain is that there are quite good reasons to expect the inner disk emission of Fuors to differ substantially from the emission of T Tauri disks.

6. Outburst Mechanisms

The nature of the mechanism producing the outbursts of Fuors is not understood. However, some general remarks are appropriate.

Fuor disks span a remarkable range of radii - more than a factor of 10^3. Thus, the natural timescales of the disk should vary appreciably from inner to outer edge. In some ways, it is surprising that a steady disk model should explain the spectral energy distribution of FU Ori so well, since it must take a long time for the outer disk to communicate with the inner edge. The outer radius of optically-thick material in the FU Ori disk must be at a distance of ~ 50 A.U. from the central star in order to explain the longest-wavelength emission observed. The fastest timescale is the free-fall time; no accretion can happen faster than this. Unless the outer disk of FU Ori is simply reprocessing radiation from the inner disk (that is, absorbing light from inner regions and reradiating at longer wavelengths), rapid accretion must have been turned on more than the free-fall time (~ 300 years) ago. In turn, this implies that the rapid accretion from large radii began long before the optical outburst, and the Fuor "event" corresponds to the recent propagation of material into the inner regions where the disk radiates at optical wavelengths (cf. § 3).

To make further progress one must make some assumption about the viscosity. Usually the so-called "α" viscosity is employed (Shakura and Sunyaev 1973)

$$\nu = \alpha c_s H, \qquad (17)$$

where α is a dimensionless number. The motivation behind equation (17) is that some type of "turbulence" produces the viscous coupling between adjacent disk annuli needed to transfer angular momentum and allow material to fall in. It is usual to suppose that "turbulent" eddies have a natural scale of the same order as the disk scale height, and have velocities less than or comparable to the sound speed; thus one assumes generally $\alpha \leq 1$. Alternatively, one may suppose that magnetic fields provide viscous effects; this can also be dealt with in terms of the α formalism (Shakura and Sunyaev 1973).

What value of α is appropriate for Fuors? The approximate value can be estimated from the observations if it is assumed that the outburst is due to a thermal instability front (see below) propagating across the inner disk. Lin et al. (1985) estimate that such a front propagates at a speed $\sim \alpha c_s$. Using the estimate made in § 4 for the front speed of $0.2 \, \mathrm{km \, s^{-1}}$, and the estimate of $c_s \sim 30 \, \mathrm{km \, s^{-1}}$ made in the previous section, one arrives at $\alpha \sim 10^{-2}$. Detailed calculations by Clarke et al. (1990) suggest that lower values of $\alpha \sim 10^{-3}$ may be required to make the outburst last long enough.

Adopting the "α" prescription, with the appropriate values needed to explain the time-dependence of the observed light curve, one can calculate the disk structure in detail and examine its stability. One begins with the general steady disk equation relating the disk viscosity, surface density, and accretion rate,

$$\nu \Sigma = \frac{\dot{M}}{3\pi}[1 - (R_*/R)^{1/2}] \qquad (18)$$

(see Pringle 1981 for a derivation), and consider perturbations from the steady state.

It has been recognized for some time that α disks of the appropriate temperatures and densities might exhibit thermal instability. A good discussion of such instabilities for the α disk equations for the simple case where the disk properties can be averaged perpindicular to the disk plane has been given by Faulkner, Lin, and Papaloizou (1983). One can solve equation (18) at a particular radial distance for a variety of disk surface densities. At a particular radius and surface density, there is generally only one value of $\dot{M}$ consistent with the steady α disk equations. However, due to the complicated dependence of gas opacity on temperature in certain regimes, most notably hydrogen ionization, there may be more than one solution to (18) at a given value of Σ. Each solution differs in temperature, and thus differs in the magnitude of the viscosity - and therefore, in the accretion rate.

The general form of such solutions is shown in Figure 7. Different sections of the equilibrium solution have different stability properties. In the region to the lower right of the curve, the disk viscous heating exceeds the radiative cooling; in the region to the upper left of the curve, cooling is greater than heating. Now consider a disk annulus which happens to lie somewhere on the section AB. At a given surface density, if the annulus is perturbed to a higher temperature, this moves the disk upwards to the region where cooling exceeds heating. Therefore, there is a tendency for the disk to cool off and the perturbation is stabilized. On the other hand, a similar temperature perturbation on section BC puts the annulus into the heating > cooling regime, and the annulus heats up still further. Thus the AB and CD sections represent thermally stable equilibrium solutions, but the BC section of the equilibrium curve is thermally unstable. To obtain an outburst, one imagines a disk slowly increasing in surface density, proceeding along the low-$\dot M$ track from A to B. Because BC is thermally unstable, one ends up with a rapid readjustment from B to D, that is, a rapid increase in $\dot M$, producing an outburst in light.

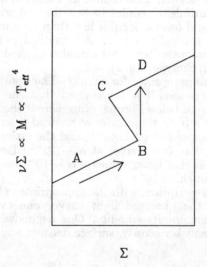

Figure 7. Schematic outline of local thermal instability analysis for α disks, as described in the text. The curves show the relationship of surface density Σ with the viscosity ν times the surface density, which is also proportional to the mass accretion rate and the surface energy flux, all evaulated at a fixed radial distance.

Lin and Papaloizou (1985) tried to use the basic thermal instability model to explain the outbursts of FU Ori. Although Lin and Papaloizou were able to produce increases in brightness of the required magnitude, they could not make the outbursts last longer than a year. This is because there is only a narrow range of parameters at a given radius for which the pre-outburst disk lies near the region BCD. The standard thermal instability cannot be easily triggered at large radii, because the disk temperature is too low to ionize hydrogen. Thus, there is not enough mass in the unstable disk region to continue the outburst. Alternatively, one can say that the outermost radius of the unstable region is not at large enough distances for the material to take a long enough time to fall in.

Clarke *et al.* (1989) suggested that if Fuor disks have a finite thickness, radial transport of energy might change the stability analysis. They presented heuristic calculations which suggested that radial transport might help keep accretion rates high; however, they assumed a steady source of infalling mass, which does not address the question of what keeps the accretion rate high in the outer disk. In a further analysis, Clarke *et al.* (1990) found that they could perturb disks with conditions appropriate to T Tauri stars and produce outbursts of the required amplitude over the necessary timescales of decades. Quite large perturbations of the disk at distances ~ 0.1 A.U. from the central star were required. The radius of the perturbation had to be at least this large to get a long viscous timescale $t_\nu \sim \alpha^{-1}(R/H)^2 \, t(orbit)$ (cf. Pringle 1981) to explain the slow decay in light. Looked at in another way, a mass accretion rate of $10^{-4} M_\odot$ and a decay time of ~ 50 yr implies a total amount of accreted material of $10^{-2} M_\odot$. Adopting the inner disk surface density $\Sigma = 6 \times 10^5 \mathrm{g \, cm^{-2}}$ derived above (cf. § 3), the required mass is contained in a radius $\sim 3 \times 10^{12}$ cm $\sim 0.2 A.U.$

Long-lived accretion episodes are easier to achieve if the outburst or high accretion occurs at large radii. However, the high temperatures necessary for hydrogen ionization instability ($\gtrsim 10^4$K) imply $H/R \gtrsim 1$ at $R > 1$ A.U. (Clarke *et al.* 1990), so this instability is unlikely to be important at large R. It should be emphasized that this analysis assumes a universal, constant value of α. There is no reason why α must be fixed throughout the disk, particularly as the observations imply $\alpha << 1$. At present there is no theory which provides estimates of the mean viscosity which one can adopt confidently, let alone possible variations in α.

Gravitational instabilities are potentially important because Fuor disks are likely to be quite massive. Such instabilities, when sheared by the differential rotation of the disk, produce spiral waves that can transfer angular momentum outward and thus cause accretion (Cassen *et al.* 1981; Larson 1984; Lin and Pringle 1987; Adams *et al.* 1989). For self-gravity to become important, the gravitational field due to the disk must be comparable to the vertical gravity due to the central star, that is, $M_d/M_* \sim H/R$ (Pringle 1981). More precisely, for expected T Tauri disk temperatures Larson (1984) and Pringle (1988) estimated the critical mass for gravitational instabilities to be

$$M(disk) \sim 0.2 \, (R/100 AU)^{1/8} M_*. \qquad (19)$$

If the typical Fuor has a lifetime total of ~ 10 events of 0.01 $M_\odot$ accreted in outbursts, it appears possible that the disk is susceptible to gravitational instabilities.

Because $M_d \sim \Sigma R^2$, and Σ falls off more slowly with radial distance than R^{-2} in most disk models, most of the disk mass is at large R. Thus, gravitational instabilities are most likely to occur in the outer disk, where the timescales of accretion are much too long to explain the short rise times of Fuor outbursts. To explain the Fuor events one could adopt a model combining both types of instabilities. Gravitational instabilities are likely to drive accretion at large radii, and could cause a slow pileup of material in the inner disk until conditions become right for the thermal instability, producing the optical outbursts.

In principle, evidence for this picture could come from the spectral energy distributions. If the accretion rate through the outer disk is larger than that through the inner disk, the spectrum of the pre-Fuor, i.e. the young T Tauri star, should be flatter than the steady disk energy distribution. The "disk" spectra of T Tauri stars are in general substantially flatter than predicted either from steady accretion or from reprocessing of light from the central star by a flat disk (Rucinski 1985; Kenyon and Hartmann 1987; Adams, Lada, and Shu 1987). The possibility of time-dependent accretion producing such energy distributions has not yet been explored.

The perturbation of the T Tauri disk required by Clarke *et al.* (1990) to produce a Fuor outburst might in principle be provided by the close passage of a massive body, such as a binary companion on a highly eccentric orbit. However, repeated passages of binary might disrupt the disk long before multiple outbursts could be achieved (cf. § 2). Perhaps such perturbations are not necessary. If both thermal and gravitational instabilities are involved in Fuor disks, the outbursts might simply be due to the difference in the natural timescales of the outer and inner disk, making it impossible to obtain truly steady accretion from $\sim$ 100 A.U. to $\sim 10^{-2}$ A.U. Gas from the outer disk might pile up in inner regions, where gravitational instabilities are not so effective, until thermal instabilities can be triggered, producing the optical outburst. Much work remains to be done before the plausibility of this scenario is established.

As noted in § 2, all Fuors are surrounded by substantial reflection nebulosities. In addition, as discussed in § 3.1, the spectral energy distributions of V1057 Cyg, V1515 Cyg, Z CMa, L 1551 IRS 5 exhibit large far-infrared excess emission fluxes that are not consistent with a simple accretion disk. This far-infrared emission could conceivably arise in a dust shell, as suggested by Adams, Lada, and Shu (1987) for the small excess apparently present in FU Ori. The Fuor phase may be sufficiently early in the evolution of the central star that infall is still occurring to outer disk radii. This infall could provide the necessary material to keep the outer disk at the edge of gravitational instability, providing a steady inflow of disk material to smaller radii.

7. Mass Loss and Bipolar Flows

Fuors have very powerful winds (Herbig 1977; Bastian and Mundt 1985). The mass loss rate in the best studied case, FU Ori, is $\dot{M}(wind) \sim 10^{-5} M_\odot$ (Croswell, Hartmann, and Avrett 1987; see Figure 8), roughly 10% of the mass accretion rate. Similar efficiencies of ejection to accretion are estimated for the other Fuors. The observed terminal velocities of these winds (300 - 1000 km s^{-1}) are a factor of two to four larger than the estimated escape velocity at the inner edge of the disk, so the kinetic energy in the outflow is at least 10% of the accretion luminosity. This is particularly interesting in view of the results for high-velocity molecular outflow sources, which also suggest $L(flow) \sim 10^{-1} L_*$ (e.g., Lada 1985 and references therein).

Radiation pressure cannot be responsible for Fuor mass ejection, because the surface temperature of the disk is only $\sim$ 6000 K. The low temperature inferred for

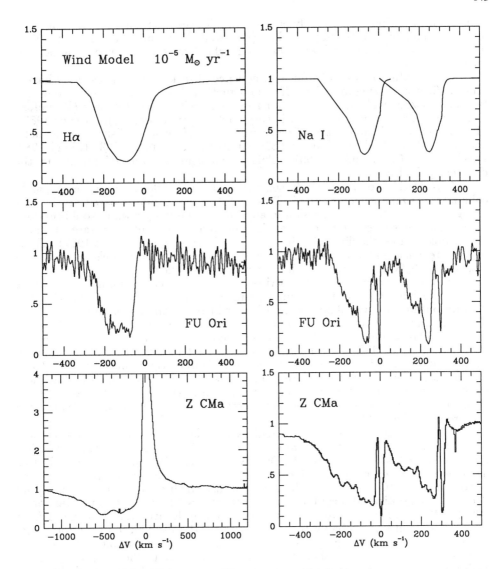

Figure 8. Hα and Na I line profiles calculated by Croswell *et al.* (1987) for a wind model with a mass loss rate of $10^{-5} M_\odot \, yr^{-1}$, compared with observed P Cygni profiles of FU Ori and Z CMa. Red-shifted emission is eliminated only in models that have very collimated flow, assuming that the objects are not being observed pole-on and that the wind does not accelerate extremely rapidly. Note that the Hα absorption of Z CMa extends to very high blueshifts of 1000 km s^{-1}.

material in the wind (also roughly 6000 K) implies that thermal pressure gradients are not important in driving the flow (Croswell, Hartmann, and Avrett 1987). Thus, the most likely wind acceleration mechanism is some type of action by magnetic fields.

Rapid rotation is an obvious consequence of disk accretion and may have a crucial effect in driving mass loss. If the magnetic fields are sufficiently strong, gas tied to rotating field lines can be flung outward with high efficiency (Blandford and Payne 1982), even if the wind is very cold (Hartmann and MacGregor 1982). Material is probably accelerated directly from the surface of the disk along magnetic field lines (Blandford and Payne 1982; Pudritz and Norman 1983; Königl 1989; Lovelace and Contopoulos 1990). This mechanism is described further in the chapter by Pudritz.

Observations of Fuor winds suggest that the outflows are at least partly collimated *as they are accelerated*. This tentative conclusion is based on the lack of high velocity, redshifted emission in any Fuor. An isotropic wind would produce much more redshifted emission than is observed, unless all Fuors are observed nearly pole-on (see the discussion in Croswell, Hartmann, and Avrett 1987). These results are consistent with the magnetically-driven disk wind models (Pudritz and Norman 1983; Königl 1989), which predict collimated flow.

In the previous section I described disk evolution assuming only viscous and gravitational transport of angular momentum are important. In his chapter, Pudritz outlines a quite different model, in which the angular momentum of the accreting material is taken away by the disk wind. The time evolution of such a model could be quite different than that outlined in the previous section. Moreover, in the wind-driven disk model most of the energy released by accretion is taken up by the wind rather than radiated by the disk. Observational mass loss estimates are too uncertain to conclusively test this scenario, but my suspicion is that most of the energy is released dissipatively rather than in the wind. If this is not true, then we need to revise our estimates of accretion rates upwards substantially from the present values $\sim 10^{-4} M_\odot$.

The wind-driven disk model predicts that large amounts of mass are lost from the outer disk. The 2.2 μm v' - v'' 2-0 CO lines show some evidence for blue-shifted absorption in FU Ori (Kenyon and Hartmann 1989). If this asymmetry is interpreted as being due to a wind, the mass loss rate from disk regions at $r \sim 10 R_*$ appears to be substantially smaller than the mass loss rate deduced from optical absorption lines formed at $r \sim 2 R_*$ (Kenyon and Hartmann 1989). (The outer disk mass loss rate computed by Kenyon and Hartmann from the 2μm CO lines is probably too small because the distribution over rotational states was neglected in the optical depth calculation.) The estimates of mass loss rates from infrared studies should improve dramatically in the next few years as high-resolution infrared spectrometers with array detectors become available.

The wind probably must be mechanically heated to keep temperatures sufficiently large to produce Hα absorption (Croswell, Hartmann, and Avrett 1987). Mechanical heating is also suggested by the presence of strong Mg II emission lines in FU Ori and Z CMa (Ewald, Imhoff, and Giampapa 1986; Kenyon *et al.* 1989). MHD waves are an attractive possibility for producing this heating, and the momentum deposited by such waves near the disk might also assist in initiating the mass loss (cf. Blandford and Payne 1982). The chapter by Natta and Giovanardi discusses wind heating in more detail.

The winds of Fuors are particularly interesting because the energetics of some Herbig-Haro (HH) objects and jets seem to require very dense outflows, and Fuor winds are the most massive directly observed outflows among pre-main sequence objects. A causal connection between Fuor winds and HH objects depends upon the (unknown) duty cycle, but even a single "short" burst of mass loss at $\sim 10^{-5} M_\odot$ for 10^2 yr, as in FU Ori, can still produce a mass loss rate of $10^{-7} M_\odot$ averaged over a typical flow dynamical time of 10^4 yr. Reipurth (1989a) and Hartigan, Raymond, and Meaburn (1990) have argued that some HH objects are produced in highly time-variable outflows. Reipurth (1990) has suggested that Fuor eruptions are responsible, noting that Fuors are commonly associated with jets and HH objects (see Table 1). The most famous bipolar flow source, L1551 IRS5, is spectroscopically a member of the Fuor class (Mundt et al. 1985; Carr, Harvey, and Lester 1987; Stocke et al. 1988). The HH57 emission nebula was known before the associated Fuor erupted (Graham and Frogel 1985), and a jet has recently been discovered in Z CMa (Poetzel, Mundt, and Ray 1990). Thus, studies of Fuor winds may provide us with detailed information on the acceleration of gas responsible for many or most of the high-velocity flows in star-forming regions, and produce a clearer insight into the connection between disk accretion and outflow.

My research into the nature of Fuors has been the result of an extremely fruitful collaboration with my friend and colleague, Scott Kenyon. I wish to acknowledge many useful conversations concerning disk models with Doug Lin, Jim Pringle, Cathie Clarke, Robbin Bell, and Nuria Calvet. This research was supported by NASA Grant NAGW-511 and by the Scholarly Studies Program of the Smithsonian Institution.

8. References

Adams, F.C., Emerson, J.P., and Fuller, G.A. 1990. *Astrophys. J.* **357**, 606.
Adams, F.C., Lada, C., and Shu, F.H. 1987. *Astrophys. J.* **312**, 788.
Adams, F.C., and Shu, F.H. 1986. *Astrophys. J.* **308**, 836.
Adams, F.C., Ruden, S.P., and Shu, F.H. 1989, *Astrophys. J.* **347**, 959.
Basri, G., and Bertout, C. 1989, *Astrophys. J.* **341**, 340.
Bastian, U., and Mundt, R. 1985. *Astron. Astrophys.* **144**, 57.
Beckwith, S.V.W., Sargent, A.I., Chini, R.S., and Gusten, R. 1990. *Astron. J.* **99**, 924.
Bertout, C., Basri, G., and Bouvier, J. 1988, *Astrophys. J.* **330**, 350.
Blandford, R.D., and Payne, D.G. 1982. *Mon. Not. Roy. Astron. Soc.* **199**, 883.
Carr, J.S., Harvey, P.M., and Lester, D.F. 1987. *Astrophys. J. Lett.* **321**, L71.
Cassen, P.M., Smith, B.F., Miller, R.H., and Reynolds, R.T. 1981, *Icarus*, **48**, 377.
Clarke, C.J., Lin, D.N.C., and Papaloizou, J.C.B. 1989. *Mon. Not. Roy. Astron. Soc.* **236**, 495.
Clarke, C.J., Lin, D.N.C., and Pringle, J.E 1990. *Mon. Not. Roy. Astron. Soc.* **242**, 439.
Cohen, M., and Schwartz, R.D. 1987, *Astrophys. J.* **316**, 311.
Covino, E., Terranegra, L., Vittone, A.A., Russo, G. 1984. *Astron. J.* **89**, 1868.
Croswell, K., Hartmann, L., and Avrett, E.H. 1987. *Astrophys. J.* **312**, 227.
Eisloeffel, J., Hessman, F.V., and Mundt, R. 1990. *Astron. Ap.*, in press.

Elias, J.H. 1978, *Astrophys. J.* **223**, 859.
Ewald, R., Imhoff, C.L., and Giampapa, M.S. 1986. In *New Insights in Astrophysics*, ed. E.J. Rolfe (ESA SP-263) p. 205.
Faulkner, J., Lin, D.N.C., and Papaloizou, J. 1983. *Mon. Not. Roy. Astron. Soc.* **205**, 359.
Goodrich, R.W. 1987. *Publ. Astron. Soc. Pacific* **99**, 116.
Graham, J.A., and Frogel, J.A. 1985. *Astrophys. J.* **289**, 331.
Hartigan, P., Raymond, J.C., and Meaburn, J. 1990. *Ap. J.*, in press.
Hartmann, L., and Kenyon, S. 1985, *Astrophys. J.* **299**, 462.
Hartmann, L., and Kenyon, S. 1987a, *Astrophys. J.* **312**, 243.
Hartmann, L., and Kenyon, S. 1987b, *Astrophys. J.* **322**, 293.
Hartmann, L. and Kenyon, S.J. 1990, *Astrophys. J.* **349**, 190.
Hartmann, L., and Kenyon, S.J. 1991. In *IAU Colloquium 129, Structure and Emission Properties of Accretion Disks*, ed. S. Collin-Souffrin, in press.
Hartmann, L., Kenyon, S., and Hartigan, P. 1991. In *Protostars and Planets III*, eds. E.H. Levy and J. Lunine (Tucson: University of Arizona Press), in press.
Hartmann, L., Kenyon, S.J., Hewett, R., Edwards, S., Strom, K.M., Strom, S.E., and Stauffer, J.R. 1989. *Astrophys. J.* **338**, 1001.
Hartmann, L., and MacGregor, K.B. 1982. *Astrophys. J.* **259**, 180.
Herbig, G.H. 1966. *Vistas in Astronomy*, **8**, 109.
Herbig, G.H. 1977. *Astrophys. J.* **217**, 693.
Herbig, G.H. 1989. FU Orionis Eruptions. In *ESO Workshop on Low-Mass Star Formation and Pre-Main Sequence Objects*, ed. B. Reipurth (Garching: ESO) p. 233.
Hessman, F.V., Eisloeffel, J., Mundt, R., Hartmann, L., Herbst, W., and Krautter, J. 1991. *Ap. J.*, in press.
Kenyon, S.J., and Hartmann, L. 1987, *Astrophys. J.* **323**, 714.
Kenyon, S.J., and Hartmann, L. 1989. *Astrophys. J.* **342**, 1134.
Kenyon, S.J., Hartmann, L., and Hewett, R. 1988. *Astrophys. J.* **325**, 231.
Kenyon, S.J., Hartmann, L., Imhoff, C.L., and Cassatella, A. 1989. *Astrophys. J.* **344**, 925.
Kolotilov, E.A., and Petrov, P.P. 1983. *Pis'ma Astr. Zh.*, 9171
Kolotilov, E.A., and Petrov, P.P. 1985. *Pis'ma Astr. Zh.*, 11846
Königl, A. 1989. *Astrophys. J.* **342**, 208.
Lada, C.J. 1985, *Ann. Rev. Astron. Ap.* **23**, 267.
Landolt, A.U. 1977. *Publ. Astron. Soc. Pacific* **89**, 704.
Larson, R.B. 1980, *Mon. Not. Roy. Astron. Soc.* **190**, 321.
Larson, R.B. 1984, *Mon. Not. Roy. Astron. Soc.* **206**, 197.
Levreault, R.M. 1983. *Astrophys. J.* **265**, 855.
Lin, D.N.C., and Papaloizou 1985. In *Protostars and Planets II*, eds. D.C. Black and M.C. Matthews (Tucson: University of Arizona Press) p. 981.
Lin, D.N.C., and Pringle, J.E. 1987, *Mon. Not. Roy. Astron. Soc.* **225**, 607.
Lovelace, R.V.E., and Contopoulos, J. 1990. *Ap. J.*, in press.
Lynden-Bell, D., and Pringle, J.E. 1974. *Mon. Not. Roy. Astron. Soc.* **168**, 603.
Mercer-Smith, J.A., Cameron, A.G.W., and Epstein, R.I. 1984, *Astrophys. J.* **279**, 363.
Mihalas, D. 1978. *Stellar Atmospheres*, (San Francisco: Freeman).
Miller, G.E. and Scalo, J.M. 1979. *Astrophys. J. Suppl.* **41**, 513.

Mould, J.R., Hall, D.N.B., Ridgway, S.T., Hintzen, P., and Aaronson, M. 1978. *Astrophys. J. Lett.* **222**, L123.
Mundt, R., Stocke, J., Strom, S.E., Strom, K.M., and Anderson, E.R. 1985. *Astrophys. J. Lett.* **297**, L41.
Poetzel, R., Mundt, R., and Ray, T.P. 1990. *Astron. Astrophys.* **224**, L13.
Pringle, J.E. 1981, *Ann. Rev. Astron. Ap.* **19**, 137.
Pringle, J.E. 1988, in *Formation and Evolution of Low Mass Stars*, ed. A.K. Dupree and M.T.V.T. Lago (Dordrecht: Kluwer), p. 153.
Pringle, J.E. 1989, *Mon. Not. Roy. Astron. Soc.* **236**, 107.
Pudritz, R.E., and Norman, C.A. 1983, *Astrophys. J.* **274**, 677.
Reipurth, B. 1985. In *Proc. ESO-IRAM-Onsala Workshop on Sub Millimeter Astronomy*, eds. P.A. Shaver and K. Kjar, p. 458.
Reipurth, B. 1989a, *Nature*, **340**, 42.
Reipurth, B. 1989b. Observations of Herbig-Haro Objects. In *ESO Workshop on Low-Mass Star Formation and Pre-Main Sequence Objects*, ed. B. Reipurth (ESO: Garching), p. 247.
Reipurth, B. 1990. In *IAU Symposium 137, Flare Stars in Star Clusters, Associations, and the Solar Vicinity*, Byurakan, USSR, in press.
Reipurth, B., Olberg, M., Booth, R. 1990. In preparation.
Rodriguez, L.F., Hartmann, L.W., and Chavira, E. 1990. *Pub. Astr. Soc. Pacific*, in press.
Rucinski, S.M. 1985, *Astron. J.* **90**, 321.
Shakura, N.I., and Sunyaev, R.A. 1973. *Astron. Astrophys.* **24**, 337.
Shaviv, G. and Wehrse, R. 1986. *Astron. Astrophys.* **159**, L5.
Shu, F.H., Adams, F.C., and Lizano, S. 1987. *Ann. Rev. Astron. Ap.* **25**, 23.
Simon, T., and Joyce, R.R. 1988. *Publ. Astron. Soc. Pacific* **100**, 1549.
Stocke, J.T., Hartigan, P.M., Strom, S.E., Strom, K.M., Anderson, E.R., Hartmann, L.W., and Kenyon, S.J. 1988. *Astrophys. J. Suppl.* **68**, 229.
Strom, S.E., Edwards, S., and Skrutskie, M.F. 1990. In *Protostars and Planets III*, eds. E.H. Levy and J. Lunine (Tucson: University of Arizona), in press.
Tsvetkova, K.P. 1982. *I.B.V.S.*, No. 2236.
Welty, A.D. 1990. Ph.D. Thesis, Univ. of Mass.
Welty, A.D., Strom, S.E., Strom, K.M., Hartmann, L.W., Kenyon, S.J., Grasdalen, G.L., and Stauffer, J.R. 1990. *Astrophys. J.* **349**, 328.

Thierry Montmerle, Inge Heyer, and Lee Hartmann

PROPERTIES AND MODELS OF T TAURI STARS

Claude Bertout
Observatoire de Grenoble
Université Joseph Fourier
38041 Grenoble Cedex
France

Gibor Basri
Astronomy Department
University of California
Berkeley, CA 94720
USA

ABSTRACT. This lecture provides (i) a short discussion of observational properties of T Tauri stars as well as a brief reminder of past ideas related to interpreting these observations; (ii) an introduction to the physics of the current model for T Tauri systems, which are now believed to include, besides the star itself, an accretion disk and a boundary layer between star and disk; (iii) a review of results obtained within the framework of this model; and (iv) a discussion of some currently unanswered questions concerning in particular the evolutionary status of T Tauri stars.

1. Historical Background and Key Observations

During the 1940's, A.H. Joy observed the brightest stars of the Taurus-Auriga and Orion dark clouds and found that they have common properties, such as light variability and the presence of emission lines in their optical spectrum. He therefore proposed that they are members of the same class of objects that he then named T Tauri stars (cf. Joy 1945). T Tau itself, the brightest object of the class, had already been studied for a number of years at that time, less because of the star's variability than because of the light variations in Burnham's nebula. Also known as NGC 1555, Burnham's nebula is a bright reflection nebulosity patch located about 30" East of T Tau, and it happened to be the first undubitable case of a variable nebula.

The spectroscopic studies of T Tauri stars were then continued by G.H. Herbig, who proposed classification criteria based solely on spectroscopic diagnostics (Herbig 1962). Indeed, spectroscopic observations of these objects were to become a Lick Observatory's tradition and are still actively pursued today. This systematic work led to publication of several catalogs of T Tauri stars and related objects (e.g. Cohen and Kuhi 1979; Herbig and Bell 1988).

Both Joy and Herbig and, more recently, Finkenzeller and Basri (1987) em-

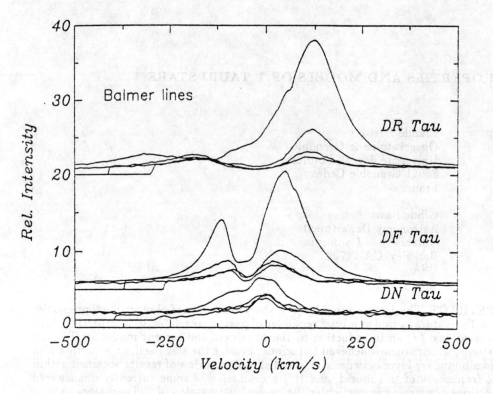

Figure 1. The first four Balmer lines for three representative T Tauri stars. The lines for each star are shown on a common continuum scale. Velocity is measured in the stellar rest frame (from Bertout, Basri, and Cabrit 1990).

phasized the similarities between the metallic emission spectrum of active T Tauri stars and the flash spectrum of the Sun. Herbig (1970) proposed that the emission line spectrum was formed in a deep chromosphere with temperature minimum located at a higher optical depth than in the Sun, and several studies explored this possibility in some detail (e.g. Calvet, Basri, and Kuhi 1984; Herbig and Goodrich 1986). It became clear recently that this class of models requires a magnetic energy flux into the chromosphere that is unreasonably large given their measured rotation rates–which are comparable to those of active late-type stars (Bouvier et al. 1986; Hartmann et al. 1986)–and direct and indirect evidence about surface magnetic field strengths (Basri and Marcy 1990; Calvet and Albarrán 1986; Bouvier 1990; Feigelson, Giampapa, and Vrba 1990).

Understanding the properties of T Tauri stars in fact proved a challenging task (see the reviews by Bertout 1989 and Appenzeller and Mundt 1989). For many years, models were based mainly on spectroscopic data, in particular on the profiles of optical lines, most notably H_α, which has the breadth expected from a fast moving non-stellar region, but remains one of the least understood aspects of the T Tauri phenomenon as well as one of the most fascinating because of the

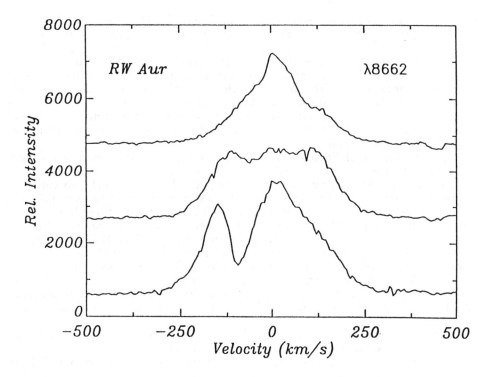

Figure 2. One of the CaII infrared triplet lines for the extreme star RW Aur. The figure shows the appearance of the line on three different nights. The line profiles cover the three main types of line shapes seen in T Tauri stars (from Bertout, Basri, and Cabrit 1990).

variety of line profiles and their temporal changes. First statistics of line shapes were presented by Kuhi (1978) and by Ulrich and Knapp (unpublished catalog). Figure 1 presents typical Balmer line profiles for three bright T Tauri stars. It has recently become possible to spectroscopically monitor emission lines with high resolution and signal-to-noise. A first important result of this monitoring is the fact that the H_α line comes in three basic shapes.
• The most common is broad emission with fairly symmetric far wings extending to 200-400 km/s and a blue-shifted absorption feature near 100 km/s that usually does not go below the continuum. This yields emission peaks with the red usually brighter than the blue.
• Less common is a more or less flat-topped emission feature, possibly with central absorption that can be unshifted or shifted to either side.
• Finally there are fairly symmetric triangular (more sharply peaked than Gaussian) shaped emission lines, often with little or no absorption.
     These shapes can be seen in the other Balmer lines (although for these the absorption goes more often below the continuum) and in other strong emission

lines. The weakest of these seem to have a preference for the triangular shape with little absorption. The large degree of symmetry seen in broad lines argues for a substantial orbital or turbulent component in the velocity broadening. Although the origin of triangular lines is still controversial, such shapes may well arise due to variable turbulent velocities in the line formation region (Basri 1990). In some cases, especially in weaker lines like the Ca II infrared triplet or He I lines, one clearly sees both a low-lying broad component and a narrow, undisplaced emission peak (e.g. Basri 1987). The less active stars tend to show only the narrow component. It is tempting to associate this component with the chromosphere or with filled magnetic loops on the star.

The variability of broad emission lines is puzzling. The same line can have quite a stable appearance in some objects, while in others it undergoes large intensity changes without altering its profile much (e.g., H_α in DF Tau); and yet again in others it changes its shape dramatically. As an example of an extreme case, the same CaII line in the same star (RW Aur) has shown all three basic shapes discussed above (Figure 2), which may mean that they are correspond to different manifestations of a common underlying physical process. The Balmer decrement can be very large, i.e., H_α can be much brighter than the Balmer lines with higher quantum number relative to the local continuum. This probably means that H_α arises from a substantially larger geometrical area than other Balmer lines. These lines can vary in intensity and shape on timescales down to an hour or less (e.g. Mundt and Giampapa 1982; Basri 1990), indicating that they most likely arise in small regions quite near the star.

The spectroscopic properties of T Tauri stars were first thought to be caused by infall of matter onto the star as they were passing through the dark cloud. Matter accretion is actually one of the main recurrent themes in research on T Tauri stars. Apart from the first tentative interpretation mentioned above, which was dismissed after Ambartsumian (1947) recognized that T Tauri stars are young stars that were born within the molecular cloud associated with them, accretion was invoked to explain properties of a subclass of T Tauri stars which display inverse P Cygni absorption components at some of their emission lines (Walker 1972; Wolf, Appenzeller, and Bertout 1977).

The basic picture behind this so-called infall model was the spherical protostellar collapse as computed by Larson (1969) and others, which predicted that newly-born low-mass stars would become visible in the optical range while still surrounded by an extensive infalling envelope. YY Orionis stars were therefore thought to be protostars in this long-lasting accretion phase. Infall velocities observed in YY Orionis stars are consistent with free-fall onto a $\approx 1 M_\odot$ protostellar core. A major advantage of this model was the presence of an energy source external to the star: here the gravitational energy of infalling matter, which could be called upon to explain the young star's activity.

Paradoxally, the second main recurrent theme in interpreting T Tauri line profiles is matter ejection. A few bright T Tauri stars do display P Cygni profiles that can be unambiguously attributed to a stellar wind with mass-loss rate in the range $10^{-8} - 10^{-7} M_\odot \mathrm{yr}^{-1}$ (Kuhi 1964). The physical mechanism that drives such a massive wind from a cool, low-luminosity object is not yet entirely clear. Basic observational and theoretical constraints on wind models are discussed by DeCampli (1981), Hartmann (1986), and Bertout, Basri, and Cabrit (1990).

The complication that appeared when high-resolution spectroscopic observations became possible for the relatively faint YY Orionis objects was the simultaneous presence of both P Cygni and inverse P Cygni absorption components in some lines, most notably CaII K, NaI D, and the high members of the Balmer series (e.g. Bertout et al. 1982; Mundt 1984). Some efforts to interpret these profiles in the framework of a pure infall model (Wagenblast, Bastian, and Bertout 1982) used the fact that it is often difficult to make a distinction between absorption and lack of emission at certain frequencies in the line profile. But high signal-to-noise data soon removed this ambiguity, at least in some cases, and it became evident that infall and outflow were indeed occurring simultaneously in YY Orionis stars, which led to the dismissal of the spherical infall model.

Parallel to these developments, based mainly on optical spectroscopy, research of T Tauri stars benefited from the opening of several windows in the electromagnetic spectrum. Mendoza (1966, 1968) first noticed the near-infrared excess of T Tauri stars, which was subsequently studied systematically (e.g. Rydgren, Strom, and Strom 1976; Cohen and Kuhi 1979). Several models were devised for explaining the near-infrared energy distributions of T Tauri stars. Some were based on the presence of several dust shells, while others explored the possibility that bound-free and free-free transitions were the main contributors to the observed excess; this approach did however require the ratio of selective-to-visual extinction to be rather different from its canonical, average value (cf. Vrba, Strom, and Strom 1976). A problem that appeared in all dust shell models was the apparent discrepancy between the relatively moderate visual extinction of T Tauri stars and the large extinction predicted by the spherical shell models that were able to account for the infrared excesses. This obvious discrepancy was probably partly responsible for attempts to explain the near-IR excess in terms of free-free radiation.

Free-free and bound-free radiation possesses spectral signatures in the ultraviolet and radio ranges. The launch of the IUE ultraviolet satellite allowed one to study the line and continuum spectra in the range 1300–2800Å (see the review by Imhoff and Appenzeller 1987) and accompanying observations of the Balmer jump region can be made from the ground (e.g. Kuhi 1974). A T Tauri star atlas of blue spectra extending down to 3100 Å is being prepared by G. Basri (priv. comm.), and a series of simultaneous spectroscopic and photometric observations ranging from the IUE range to about 5μm will be published shortly (Basri et al. in preparation). These studies show that the UV continuum of T Tauri stars is caused primarily by Balmer continuum emission. Furthermore, they demonstrate that the UV excess is related to the so-called veiling of the optical spectrum first noted by Joy (1945). The main effect of veiling is that it makes the photospheric absorption lines appear shallower than in standard stars with similar spectral types (see Section 4). It was attributed to overlying continuous emission from an ionized envelope by Rydgren, Strom, and Strom (1976), which led to the prediction that the Bremstrahlung flux at radio wavelengths should be strong.

Another motivation for studying the radio continuum emission of T Tauri stars was the hope that the slope of the radio flux, which reflects the velocity field, would allow one to distinguish between large-scale infall and outflow of ionized material (Bertout and Thum 1982). As it turned out, the behavior of T Tauri stars in the radio range is, in most cases, far from spectacular. There are a few exceptions, such as V410 Tauri, a weak-emission line T Tauri star which has been observed once during a strong radio flare (Cohen and Bieging 1986). T Tau itself is a fairly strong radio source (about 10 mJy at 15 GHz), but VLA maps revealed that the main radio source was not the optical star but rather its infrared companion, a mysterious, probably protostellar object that was first discovered by speckle techniques (Dyck *et al.* 1982). Extensive surveys have however demonstrated that most T Tauri stars are not strong radio continuum emitters (Bieging, Cohen, and Schwartz 1984; André *et al.* 1986). These findings led to the conclusions that the ionization degree of T Tauri stars winds must be low and that neither the veiling emission nor the UV emission are caused by an extended, optically thin ionized envelope. The few sources that are detected have relatively flat spectral slopes not readily interpreted in terms of winds or accretion. A non-thermal origin of the radio spectrum was suggested in a few cases (cf. André 1987).

Extensive X-ray emission also signals non-thermal phenomena at the surface of stars, and surveys using the *Einstein* X-ray Observatory uncovered the strong X-ray activity of T Tauri stars, which typically emit one per thousand of their luminosity in the X-ray range (e.g. Montmerle *et al.* 1983; Feigelson *et al.* 1988). The temperature of X-ray emitting regions in T Tauri stars is in the range 10^7K, i.e., hotter than in normal solar-type stars. Searches for coronal emission lines in optical spectra of T Tauri stars were unsuccessful, but the UV emission lines of transition region[1] ions such as CIV or SiIV proved to be much stronger in T Tauri stars than in active dwarfs with comparable X-ray fluxes. Thus, there is probably another region—besides the transition region—with a temperature $\approx 10^5$K in T Tauri atmospheres, the exact nature of which remains mysterious to this day.

Besides these results, X-ray studies of regions of star formation had far-reaching consequences; they uncovered a new population of T Tauri stars, now called weak-emission line T Tauri stars (WTTS), that display X-ray emission but lack the more exotic properties of the classical T Tauri stars (CTTS), although they have similar masses and ages (cf. Walter *et al.* 1988). Current proper motion surveys seem to indicate that there are perhaps as many as 2 to 3 times more WTTS than CTTS. The two subclasses differ widely in their spectral energy distributions. Figure 3, which displays overall spectral energy distributions of representative T Tauri stars, illustrates this point. Data presented there are uncorrected for extinction. Visual extinction values are 0 for the WTTS TAP57, 0.4 mag for the moderate CTTS DN Tau, 1.3 mag for the active CTTS DF Tau, and about 1.5 to 2 mag for the extreme CTTS DR Tau. The WTTS TAP57 has a basically normal K7 spectral type, but the three CTTS display various amounts of infrared and ultraviolet flux excesses, the amounts of which appear correlated. Comparing these data with the Balmer lines of Fig. 1, one notices that excess continuum flux and optical emission-line activity are also correlated.

The mid and far infrared spectral ranges, where CTTS emit appreciable amounts of energy, became widely accessible in 1983 with IRAS. Data obtained by this satellite provided in good part the impetus for the current picture of T Tauri stars by showing that many of them have substantial far infrared emission that cannot be explained by spherical dust shell models (Myers *et al.* 1987) because such shells would lead to a much larger visual extinction than observed; this provides a first indication that dust is distributed in a circumstellar disk rather than in a spherical shell. Rucinski (1985) was first to note that the observed spectral slope in the far-infrared roughly compares to what a disk energy distribution is expected to look like according to simple theoretical models. Further indirect evidence for the presence of disks, or rather of optically thick screens in the equator's plane, is given by the neutral oxygen and sulfur forbidden lines that are observed in the spectra of

---

[1] In solar-type stars, the transition region is intermediary between the chromosphere and the corona.

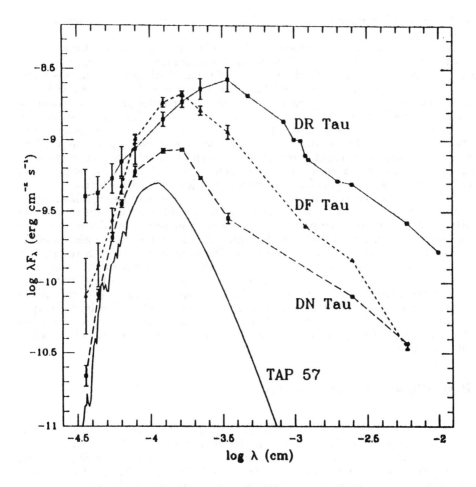

Figure 3. Observed spectral energy distributions from the ultraviolet range (U-filter) to 100μ for three representative CTTS (DN Tau, DF Tau, and DR Tau) and for one WTTS (TAP57), the energy distribution of which has been displaced downwards by 0.3 dex. Error bars indicate the observed variability when available (from Bertout 1989).

stars with prominent signs of outflow. Their profile is often peculiar in that its red part, which traces the part of the wind going away from us, is missing. This was interpreted as an indirect piece of evidence for the presence of an optically thick disk with size about 100 AU by Appenzeller, Oestreicher and Krautter (1984) and Edwards et al. (1987).

At the distances of the closest star-forming regions, such as the Taurus molecular cloud, a solar-system sized disk of 100AU subtends an angle of 0"6. Resolving a disk at near-infrared wavelengths is thus barely possible by using direct imaging with adaptive optics on a 4m telescope, provided the contribution of the stellar object itself to the near-infrared flux is either suppressed (e.g., using coronographic techniques) or independently determined. Such observations should become available soon.

Both infrared speckle interferometric and imaging techniques have already begun to provide direct images of circumstellar structures (e.g., Beckwith, Zuckerman, and Skrutskie 1985; Monin et al. 1990), but distinguishing between disk and other geometries from deconvolution of the data is less than obvious and often involves some faith. Millimeter-wave interferometry currently provides a resolution of a few arcseconds and has proved a valuable tool in mapping flattened cool circumstellar structures around a few objects, most notably HL Tau (e.g., Sargent and Beckwith 1987). For this particular star, millimetric observations reveal a molecular disk with diameter about 4000 AU, i.e., much more extended than the solar nebula. The relationship between the molecular disk and the warmer inner disk "seen" at shorter wavelengths (see the dicussion below) is still unclear.

Several of the more exotic properties of T Tauri stars can be explained by the presence around the star of the circumstellar disk that is the natural outcome of the gravitational collapse of a dense molecular core with non-zero angular momentum. In this picture, both CTTS and WTTS display the characteristic properties of the magnetically active central star, and the specific properties of CTTS result from interaction between the disk and the star. WTTS, on the other hand, have either lost their disks or do not interact with one. In the following, we first discuss the physics of the current "standard model" for a T Tauri system. Although it has many weak points, this model shows remarkable agreement with a large array of observational properties of young stars, as reviewed in Section 3. In Section 4, we then discuss some of the observations that are not currently understood as well as possible solutions.

2. Basic Physics of Accretion Disks

The idealized model of a T Tauri system consists of (i) a central star surrounded by (ii) a geometrically thin, dusty accretion disk that interacts with the star via (iii) a boundary layer. The accretion disk model, originally devised by Shakura and Sunyaev (1973) and Lynden-Bell and Pringle (1974) (hererafter LBP) has been refined and adapted to the case of young stellar objects by several groups (e.g. Adams and Shu 1986; Kenyon and Hartmann 1987; Bertout, Basri, and Bouvier 1988; Adams, Lada, and Shu 1988). It assumes that *local* processes induce a viscous coupling between neighboring disk annuli, thereby transporting angular momentum through the disk. Note that there could be other, more global ways of redistributing angular momentum, e.g., through density waves. These are not considered in this model. Once a physical mechanism for transporting angular momentum is specified, the equations governing the disk structure as well as its evolution in time can be written down.

The basic angular momentum transport mechanism considered by LBP is kinematic viscosity. In a disk where the gas is rotating differentially, any chaotic motions in the gas will give rise to viscous forces (shear viscosity). Gas particles moving along two neighboring streamlines at R and $R + dR$ with angular velocities respectively $\Omega(R)$ and $\Omega(R+dR)$ have different amounts of angular momentum, and chaotic motions lead to angular momentum transport in the sense that a viscous torque $N(R)$ is exerted on the outer streamline by the inner streamline. The rate of working done by the net torque on the ring between R and $R+dR$ is

$$\mathbf{F} \cdot \mathbf{v} = \Omega[N(R+dR) - N(R)] = \Omega \frac{dN}{dR} dR = \left[\frac{d(N\Omega)}{dR} - N\frac{d\Omega}{dR}\right] dR. \quad (1)$$

The first term of the RHS represents global transport of rotational energy by the torques and depends only on the assumed boundary conditions. The second RHS term represent the local rate of transforming mechanical energy into heat, which must be radiated away by the up and down faces of the annulus with area $2 \cdot 2\pi R dR$ if one assumes that the radial radiative flux is zero. The energy dissipation rate per unit disk surface due to viscous torques is therefore

$$D(R) = \frac{N(R)}{4\pi R} \frac{d\Omega}{dR}. \quad (2)$$

Assuming steady-state, angular momentum conservation requires

$$\frac{dN}{dR} = \frac{d\dot{L}}{dR} = \dot{M}\frac{d(R^2\Omega)}{dR}, \quad (3)$$

where $\dot{M}$ represents the constant mass-accretion rate that accompanies angular momentum transport. Thus,

$$N(R) = \dot{M}R^2\Omega + C \quad (4)$$

where C is a constant determined by the inner boundary condition. At large R, C becomes negligible. If one then assumes that the disk is not self-gravitating, then its rotation law is quasi-keplerian and $\Omega = \sqrt{GM_*/R^3}$. The hypothesis of a quasi-keplerian disk also implies that the disk must be geometrically thin. One then finds the following expression for the energy dissipation rate:

$$D(R) = \frac{3GM_*\dot{M}}{8\pi R^3}. \quad (5)$$

Note that no assumption regarding the physical properties of the kinematic viscosity have been made in the above analysis. Nor is the analysis restricted to Keplerian disks. If another rotation law was valid as would be the case, e.g., if the viscous torques were of magnetic rather than kinematic origin, then $D(R)$ would not necessarily be $\propto R^{-3}$.

In order to determine the disk density, one must however assume something about the viscosity since its magnitude determines the angular momentum flow. LBP derive the disk density under the assumption that the kinematic viscosity is constant within the disk. Shakura and Sunyaev (1973) derive another analytical solution based on the same basic assumption with the difference that the viscosity is assumed to be proportional to the local scale height times the local sound speed, with the proportionality constant (the infamous α) being restricted to values smaller or equal to unity. Underlying this ad-hoc formulation are the ideas that turbulent eddies cannot be larger than the disk height and that any supersonic turbulence should rapidly become subsonic because of the formation of internal shocks in the disk. Both formulations are ad-hoc prescriptions that simply parametrize our ignorance of the nature of kinematic viscosity in disks and give qualitatively comparable results for the run of density with radius.

The disk temperature structure, which determines the emitted spectrum, can be computed from the energy dissipation rate by making an assumption about the radiative transfer. As already apparent in the derivation of Eq. 2, LBP hypothesized that viscous energy released in a given disk annulus is radiated away through both faces of that annulus and that the radial radiative flux is zero, which leads to the well-known disk spectrum $\lambda F_\lambda \propto \lambda^{-4/3}$. This is justified in the optically thick, geometrically infinitely thin disks envisioned by Lynden-Bell and Pringle, but the assumption of an infinitely thin disk is clearly a first approximation that should be eliminated in future models for the radiative transfer in disks.

Also important for the disk temperature is the heating by stellar photons. Note that there are two ways in which the central star influences disk properties. Its mass and radius determine the potential well seen by the disk, i.e., the viscous energy dissipation rate, and hence the disk temperature. But the local disk temperature may also depend upon the stellar effective temperature, which determines together with geometrical factors the local rate of heating by the central star. Friedjung (1985), Adams and Shu (1986), Kenyon and Hartmann (1987), and Ruden and Pollack (1991) computed in various approximations the resulting disk temperature at the photospheric level. At large distance R from the star, and assuming that the disk is infinitely thin, one finds that the local rate of heating of a flat disk due to reprocessing of photons originating from a star with radius R_* and effective temperature T_* is

$$F(R) = \frac{2\sigma T_*^4 R_*^3}{3\pi R^3}. \qquad (6)$$

At large distances from the central star, the above assumptions then lead to the following equation for the disk effective temperature $T_D(r)$:

$$\sigma T_D^4(r) = D(R) + F(R) = \frac{3GM_*\dot{M}}{8\pi R^3} + \frac{2\sigma T_*^4 R_*^3}{3\pi R^3} \qquad (7)$$

where the RHS's first term represents the viscous energy dissipation rate and the second term takes into account the reprocessing of stellar photons. Note that both terms are proportional to R^{-3} in flat, infinitely thin Keplerian disks, i.e., the overall emitted spectrum has the same spectral slope as that resulting from purely viscous heating. [2]

Equation 7 describes the temperature far away from the central star. When computing the inner disk structure, one must assume something about the inner boundary of the disk, and more specifically about the way angular momentum is transferred from disk to star. LBP imposed the condition that the star exerts no torque on the inner edge of the disk, which also implies the existence of a boundary

---

[2] We would like to emphasize here that Eqs. 6 and 7 depend crucially on the assumption that the disk is infinitely thin. In the more realistic case of a finite atmospheric structure, many of the incident, grazing stellar photons are absorbed in the outer layers of the atmosphere that they heat up, thus creating a "disk chromosphere." Heating of the disk photospheric layers by reprocessing appears less important in that case (Malbet and Bertout, 1991).

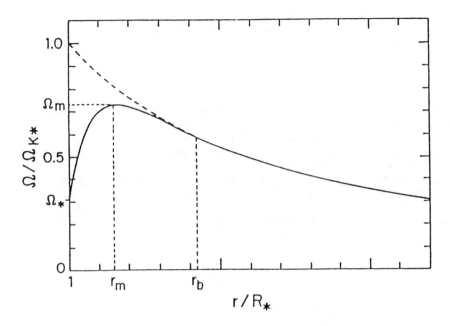

Figure 4. Variation of the angular velocity Ω as a function of stellar radius in the boundary layer. The various symbols are defined in the text (from Regev 1991).

layer (hereafter BL) between the slowly rotating T Tauri stars (typically 20 km/s at the equator) and the inner edge of the Keplerian disk, where matter is circling the star at about 250 km/s. Figure 4 shows the expected angular velocity Ω in the BL. There, r_b is the BL radius in units of stellar radius R_*, r_m is the radius where the derivative of Ω changes sign, i.e., where Ω reaches its maximum Ω_m, Ω_* is the star's equatorial velocity, and Ω_{K*} is the value of the keplerian velocity at the stellar radius.

The LBP inner boundary condition assumes that the extent r_b of the boundary layer is infinitely small and that the angular velocity at the inner disk edge is comparable to the Keplerian velocity at the star, i.e., $\Omega_b = \Omega_m = \Omega_{K*}$. The constant C in Eq. 4 can then be computed from the condition $N(R_*) = 0$. The general expression for the viscous torque is then

$$N(R) = \dot{M}\sqrt{GM_*R_*}\left[1 - \sqrt{\frac{R_*}{R}}\right], \qquad (8)$$

which leads to the well-known expression for the viscous energy dissipation rate:

$$D(R) = \frac{3GM_*\dot{M}}{8\pi R^3}\left[1 - \sqrt{\frac{R_*}{R}}\right]. \qquad (9)$$

Energetic properties of the disk/boundary layer system directly follow from this "standard" inner boundary condition. Integrating Eq. 9 from $r_b \approx R_*$ to ∞, one finds the disk luminosity

$$L_D = \frac{GM_*\dot{M}}{2R_*} = \frac{L_{acc}}{2}. \qquad (10)$$

Since only one half of the total accretion luminosity L_{acc} is dissipated in the disk, one thus concludes that the second half is advected from the disk into the BL; this corresponds to the Keplerian kinetic energy rate at the inner disk radius. Now, however, this energy must be dissipated in a small region at the star's equator instead of over a disk many AU in extent. If the dissipation is radiative, as is usually assumed, this means that the temperature of the radiating region will be far higher. While temperatures in the disk range from 10K far from the star to about 3000K near the star, the boundary layer's temperature will be from 7000 to 12000K (Bertout 1987; Kenyon and Hartmann 1987). The BL thus radiates in the ultraviolet and visible part of the spectrum a luminosity

$$L_{bl} = \frac{L_{acc}}{2}. \qquad (11)$$

As seen below, this is probably the source of the UV excess and optical veiling observed in many CTTS.

It should however be emphasized here that the LBP inner boundary condition maximizes the boundary layer's luminosity. L_{bl} can be reduced in several ways. First, if the star rotates as some angular velocity Ω_*, then the boundary layer luminosity is reduced to

$$L_{bl} = \frac{GM_*\dot{M}}{2R_*} - \frac{\dot{M}R_*^2\Omega_*^2}{2} = \frac{L_{acc}}{2}(1 - \Omega_*^2). \qquad (12)$$

Second, part of the accretion power is used up to spin up the star via the shear at the stellar surface, where the derivative of Ω is non-zero (cf. Regev 1991). Eq. 12 then becomes

$$L_{bl} = \frac{L_{acc}}{2}(1 - \Omega_*)^2. \qquad (13)$$

Third, the boundary layer size need not be infinitely small. The more general case of a finite boundary layer was recently considered in some detail by Duschl and Tscharnuter (1990), who demonstrated that the fraction of accretion luminosity dissipated in the boundary layer is indeed strongly dependent upon the assumed BL size. Finally, some of the accretion energy could be released in non-radiative form, e.g., for driving a wind (cf. Pringle 1989). One should therefore be aware that all comparisons of observed and computed spectral energy distributions done so far assume that half of the accretion luminosity is radiated away in the BL.

The LBP inner boundary condition leads to a singular, isothermal BL. If one then assumes that it is optically thick, its spectrum is a single blackbody with temperature T_{bl} given by

$$T_{bl}^4 = \frac{L_{bl}}{4\pi\sigma R_* r_b}. \qquad (14)$$

Observations to be discussed below indicate that the BL is in fact optically thin at least in some spectral regions. Its temperature, found by solving

$$\int_0^\infty \pi B_\lambda(T_{bl}) \cdot (1 - e^{-\tau_\lambda}) d\lambda = \frac{L_{bl}}{4\pi\sigma R_* r_b}, \qquad (15)$$

is then higher than the value computed by Eq. 14. Notice that the viscosity does not enter in Eq. 14 but that its value is explicitly needed to compute the optical depth τ_λ in Eq. 15. Current models of BL emission use standard LTE stellar atmosphere models to compute T_{bl} iteratively (cf. Basri and Bertout 1989). Improvement over this simple model requires computing the BL's temperature and density structure by self-consistently solving the hydrodynamic equations governing the BL. Such computations are now becoming available (cf. Regev 1991).

3. Comparisons with Observations

Accretion disks and boundary layers in T Tauri systems have been reviewed several times in the last few years (Hartmann and Kenyon 1988; Bertout 1989; Bertout, Basri, Cabrit, 1990). The infrared spectral energy distributions of CTTS are at least as shallow as demanded by the simple flat disk model. Most are actually shallower; Rydgren and Zak (1987) showed that the average slope of CTTS infrared spectra beyond 5μ or so is $\propto \lambda^{-3/4}$ rather than $\lambda^{-4/3}$, which probably means that the simple model discussed above needs refinement. Two basic suggestions have been made to explain the fact that the observed infrared spectra are flatter than the LBP disk model predicts.

The first of these assumes that the disk flares up out of the plane at large radii, thus intercepting (and thermally reemitting) more stellar photons in the outer parts than a thin disk. Some flaring is expected even in classical accretion disks because the local disk height H determined from hydrostatic equilibrium scales as $R^{9/8}$. Kenyon and Hartmann (1987) showed that such a geometry would partially explain the infrared discrepancy if the disk remains opaque over its full height. However, flared-up disks do not really do an adequate job of explaining the truly flat infrared spectra. Also, if disks were generally so flared, one would expect a greater incidence of large extinctions in CTTS. Furthermore, Malbet and Bertout (in preparation) recently computed the vertical structure of T Tauri disks in a self-consistent manner, and found that the optically thick parts of the disk are confined to regions close to the central plane of the disk even if gas and dust remain mixed together for the required length of time, which is another problem. This result is illustrated by Figure 5, which displays the optical thickness distribution in a typical T Tauri disk with mass-accretion rate $1 \cdot 10^{-7} M_\odot/\mathrm{yr}$ as a function of both disk radius and disk height over the mid-plane. The innermost contour corresponds to Rosseland optical thickness $\log\tau = 1.5$, and successive contours differ by 0.5 dex. The heavy-lined contour corresponds to $\tau = 1$, and thus approximately indicates the physical disk size, here about 20 AU in diameter. In order for the far infrared spectrum to steepen up with respect to the usual $\lambda^{-4/3}$ law, the disk should be optically thick all the way out to the outermost contour, which roughly corresponds to $H \approx R^{9/8}$.

The second suggestion, made by Adams, Lada, and Shu (1988), assumes that the temperature distribution in T Tauri disks is flatter than in LBP disks, which

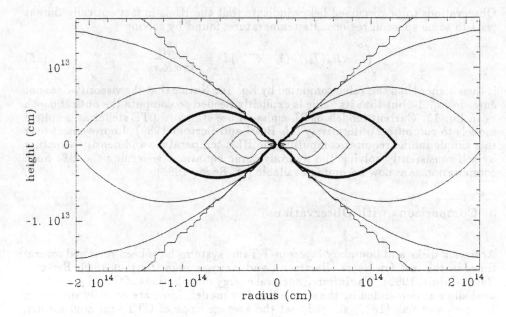

Figure 5. The optical depth distribution of an accretion disk model with parameters typical of a T Tauri disk, as a function of both disk radius R and height H over the mid-plane. Note that the scale is about 10 times larger for H than for R. The heavy-lined contour is for unity Rosseland optical depth, i.e., it roughly shows the observable shape of the disk. Successive contours differ by 0.5 dex (Malbet and Bertout, in preparation).

implies that the disk's temperature distribution does not result from kinematic viscosity. Following this suggestion, it has become common practice to parametrize the temperature law index and to use different values of this parameter to model infrared distributions of young stellar objects, particularly those with flat infrared spectra (e.g., Adams, Emerson, and Fuller 1989, Beckwith *et al.* 1990). Two objections can be made to this procedure, however. Hartmann and Kenyon (1988) showed that in order to get the required temperature gradient one had to assume either massive disks unlike those observed, or unsteady accretion with strong accretion rates in the outer parts of the disk, where it leads to unrealistically small evolutionary timescales. The second problem is that using the temperature law as a free parameter is equivalent to implicitly assuming that the energy dissipation mechanism in the disk differs from star to star, a situation that we find unconfortable. We are more at ease with the idea that the underlying physics of T Tauri disks are the same but that either unforeseen radiative transfer effects in the dusty circumstellar surroundings or the presence of unresolved companions change the appearance of the far-infrared spectrum. That infrared companions can heavily contribute to flattening the overall spectrum has been demonstrated by Leinert and Haas (1989) for Haro 6-10 and by Maihara and Kataza (1991) for T Tau itself.

That many CTTS are close binary systems in fact suggests still another possibility which has not yet been investigated yet, namely that their circumstellar disks might be appreciably warped. This geometry could lead to a flattened infrared spectrum just as flared disks do.

At this point, one must therefore conclude that the far-infrared spectral energy distributions of many T Tauri stars do not fully support the hypothesis that T Tauri disks are really classical accretion disks à la LBP, although there are a few stars, such as DF Tau, which fit the classic model perfectly. A further problem is that the infrared spectrum alone does not allow one to distinguish between passive reprocessing disks and true accretion disks as long as the accretion luminosity is smaller than or comparable to the reprocessed luminosity, i.e., $\dot{M}_{acc} \lesssim 10^{-8} M_\odot/\text{yr}$.

So why do we believe that accretion disks actually surround most CTTS even when the infrared luminosity excess is not decisive? The strongest piece of evidence comes from the blue and ultraviolet spectral ranges. As it turns out, the observations of Balmer continuum emission jumps in CTTS can easily be explained if the boundary layer is optically thin in the Paschen continuum (Basri and Bertout 1989). Even more important, the amount of energy available depends solely on the accretion rate and not on the star's resources, which resolves the mystery of how the photospheric spectrum can be so veiled in some cases, while the disk paradigm explains naturally the observed correlation between the respective amounts of infrared and ultraviolet excesses. Of course, there are still questions regarding the actual geometry and extent of the interface region between disk and star, and to what extent it dissipates energy in non-radiative forms. For example, the maximum velocities present in the boundary layer are similar to those seen in the broad emission line components, suggesting that the broad emission lines are partly formed in the boundary layer and connected regions; this is a topic of current work.

There are fairly few free parameters in the simple models that have been computed so far. Some are associated with the star itself: the stellar effective temperature, mass, and radius. The stellar temperature is known with reasonable certainty from the spectral type, although the translation from type to temperature is untested for these stars with intermediate luminosities. The stellar mass is derived from the comparison of the position of the star in the Hertzsprung–Russell diagram (HRD), with stellar pre-main sequence convective-radiative evolutionary tracks. It is uncertain because the evolutionary tracks themselves are somewhat uncertain (Section 4), and also because the convective tracks are crowded together for the lower mass stars. The effect of disk accretion on the luminosity of the system makes it difficult to precisely estimate the stellar radius. An additional constraint on the radius is provided when both the rotation period (through starspots; cf. Bouvier and Bertout 1989) and the projected rotation velocity (through spectral line broadening) are known. These constrain the radius and inclination jointly.

Another parameter not directly related to the disk is the external extinction to the system. It is difficult to distinguish between circumstellar extinction caused by dust which is near enough to the star to reprocess optical light into the infrared, and true extinction which is due to dust far enough away that the light is fully lost from the beam. Extinctions are estimated from the reddening of the observed light compared to the expected stellar intrinsic spectral energy distribution. Obviously the presence of the boundary layer makes such determinations uncertain, since its intrinsic spectral energy distribution is not known independently.

In addition to those, there are several parameters associated with the disk itself. Of these, the mass and size of the disk are not really a concern, since it is assumed so far that the disk is optically thick, and that its size is sufficiently large not to affect the infrared spectrum. Of course, if one considers the spectrum down into the sub-millimetric part of the spectrum these parameters become important (cf. Beckwith *et al.* 1990). Another parameter in the disk is the value of the α parameter which supplies the viscosity for accretion. Here there is little theoretical guidance. The largest values derived theoretically are of order 10^{-2} from convection (Lin and Papaloizou 1985). Inferred values of α range up to unity or greater in cataclysmic variables (Lin, Williams, Stover 1988). Basri and Bertout (1989) found that setting α to unity was acceptable for most cases, although for lower accretion rates it might easily be of order $10^{-1} - 10^{-2}$. They adopted the philosophy that it should not be used as a free parameter, but fixed arbitrarily at a certain value. For optically thick disks, this leaves the mass accretion rate as the main free parameter for the disk. The other parameter associated with the disk is the inclination of the disk plane to the line of sight, which acts primarily as a scaling parameter on the observed flux, due to foreshortening.

Bertout, Basri, and Bouvier (1988) compared quasi-simultaneous sets of data in the ultraviolet/optical and optical/near-infrared ranges to synthetic spectra computed assuming optically thick disks and boundary layers. They verified that typical T Tauri disks are optically thick over most of their surface so long as $\alpha \leq 1$ and that the spectral energy distribution of typical CTTS can be reproduced from about 0.2 to 10 microns if emission from the isothermal BL is confined to an equatorial region with width comparable to the local disk scale height ($\approx 2\%$ of the stellar radius).

Positive aspects of this simple model are its self-consistency and its small number of free disk parameters (essentially the disk mass-accretion rate and view angle), while a major drawback is the assumption of an optically thick boundary layer. Observed Balmer jumps indicate that the Paschen continuum is at least partially optically thin. Basri and Bertout (1989) therefore computed monochromatic gas opacities in the BL, which is again assumed isothermal. They made the boundary layer width an additional free parameter (rather than α) needed to control the optical depth, and computed emergent spectral energy distributions that they compared to observations of the Balmer and Paschen continuum regions. While the head of the Balmer continuum is optically thick in these models, the Paschen continuum is partially optically thin and the Balmer jump consequently appears in emission. A similar analysis by Kenyon and Hartmann (1990) leads to similar results.

An example of the observations and fit to them for DF Tau appears in Figure 6. Derived main parameters in the model presented are $\dot{M} = 4 \cdot 10^{-7} M_\odot/\text{yr}$, $R_* = 3.7 R_\odot$, $T_{bl} = 6900$ K, $L_* = 2.3$ L$_\odot$, and $L_{acc} = 3$ L$_\odot$. Note that the accretion luminosity is larger than the stellar bolometric luminosity. Line emission from the Balmer lines with high quantum number appears consistent with optically thick line emission from the boundary layer. The emitting area is at most a few percent of the stellar surface area. There is obviously a more extended region of emission which contributes to the flux in the lowest members of the Balmer series, since these are predicted to have very little emission contrast in the simple boundary layer model.

Modelling the spectral energy distributions of T Tauri stars allows one to derive key parameters such as the accretion rate and disk mass. Finding accurate mass-accretion rates from the disk onto the star is important because strong accretion could affect the evolution of the star in the H-R diagram (Kenyon and Hartmann 1990) as well as the evolution of rotation in CTTS. Furthermore, there

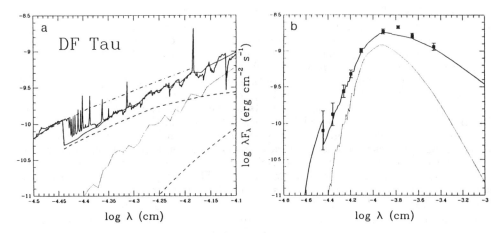

Figure 6. Observations and model fitting for DF Tau. The solid spectrum in (a) is UV Schmidt spectroscopy, and the squares with error bars in (b) are simultaneous ESO photometric data. The solid smooth line in (a) and (b) is the final composite spectrum. The upper dashed line in (a) is the BL contribution, the lower one is the disk contribution. The dash-dotted line is the expected locus of Balmer line peaks, and the dotted line is the stellar photospheric flux. Panel (a) is for the optical range, and Panel (b) shows only the stellar and overall model spectra over the full spectral range (from Basri and Bertout 1989).

is growing evidence of a relationship between mass-accretion and mass-loss (Cohen, Emerson, and Beichman 1989; Cabrit et al. 1989). The efficiency of mass gain/loss conversion is important both for understanding the mass-loss mechanism and for the above issues. By assuming a classical disk model, mass-accretion rates ranging from a few times 10^{-9} to a few times 10^{-7} $M_\odot \text{yr}^{-1}$ can be determined from models of individual stars (Basri and Bertout 1989). Questions remain about the validity of these values: are they highly dependent on the assumed disk model, and are they more or less unique, or can we find several solutions leading to the same spectrum but with very different mass-accretion rates? These questions are discussed in quantitative detail by Bertout and Bouvier (1989) and Bouvier and Bertout (1991).

Bertout and Bouvier (1989) studied the uniqueness problem by constructing maps of the quantity $1/\chi^2$, which measures the goodness of the fit between observed and computed spectral energy distributions, for all parameter couples. They used 6 computational parameters: the stellar radius R_*, the visual extinction in front of the system A_V, the system's view angle i, the accretion rate $\dot{M}_{acc}$, the viscosity parameter α, and the width δ of the emission region associated with the boundary layer. In their analysis of DF Tau[3], for example, the best solutions (with high

---

[3] This analysis is for a set of simultaneous data obtained independently of those shown in Figure 6. Time variations of the mass-accretion rate by a factor of 2 are common in DF Tau.

$1/\chi^2$) span a small range of mass-accretion rates: $\dot{M}_{acc} \approx 1 - 2 \cdot 10^{-7} M_\odot yr^{-1}$. It thus appears that the mass-accretion rate at a given time can be estimated within a factor of two from these models. This result stems from the fact that the mass-accretion rate primarily reflects the relatively well-determined quantity of integrated excess flux in the near infrared.

The two parameters α and δ contain most of the assumed physics for the disk and boundary layer. Given the large range of parameter values that the solution spans, it is reassuring that best fits to the overall spectrum (also including the Balmer jump) were obtained for values $\alpha \approx 1$ and $\delta/R_* \approx 0.02$. This appears physically reasonable and gives us some confidence both in the validity of the underlying disk physics and in the α parametrization. Even more reassuring, the best fits produce continuum veiling compatible with the observed amount (see below).

Similar computations were made for 10 CTTS (using new simultaneous datasets to be published by Bouvier, Basri, and Bertout spanning the ultraviolet to the near-infrared) in the spectral-type range K1–M1. They yield an average mass-accretion rate of $< \dot{M}_{acc} >= (1.4 \pm 1.2) \cdot 10^{-7} M_\odot yr^{-1}$. Because these systems are rather variable, perhaps due to unsteady accretion, observations from the ultraviolet, optical, and near infrared must be gathered at the same time to make confident determinations of system parameters.

Approximate estimates of the mass of CTTS disks can be made by modelling the spectral energy distribution in the sub-millimeter and millimeter range, using the sharp turnover of the spectrum that occurs in that spectral range because the outer disk becomes optically thin. Adams, Emerson, and Fuller (1989) and Beckwith et al. (1990) recently presented such computations, based on the assumption discussed above that the temperature distribution in the disk is a free parameter that can be adjusted to get an overall fit to a given spectral energy distribution. While this ad-hoc assumption introduces some uncertainty in derived disk masses, the main source of uncertainty stems from lack of knowledge of dust opacities in the millimeter range: depending on the assumed opacity law, mass estimates can vary by about one order of magnitude! One thus finds that masses of CTTS disks may range from less than 10^{-2} to perhaps as much as 1 $M_\odot$.

It is clear from the last section that further observational constraints on the parameters of the disk models will be very helpful. One clearly relevant accretion diagnostic is the amount of spectral veiling (actually the amount of excess continuum light) in the optical spectral lines. The strong broad optical emission lines are a second such diagnostic; they should in principle reflect the physical properties of the region where the mass loss originates.

Veiling is defined by

$$r(\lambda) = \frac{f_{obs}(\lambda)}{f_*(\lambda)} - 1,$$

where f_{obs} is the observed flux and f_* the photospheric flux. It can be estimated by comparing the veiled (observed) spectrum with an appropriate spectral standard. One finds that all the same lines are present and in the same ratios, but all the line depths are reduced in the CTTS. One method to derive the veiling is then to add a flat continuum to the standard spectrum until it matches the CTTS spectrum (Figure 7). This was done quantitatively at one wavelength by Hartmann and Kenyon (1990), and over a broad spectral range for one star by Hartigan et al. (1989) and for an extensive sample of stars by Basri and Batalha (1990). Their efforts yield excess light as expected from accretion disk models, with wavelength dependence as predicted by hot boundary layer models. The excess light does not

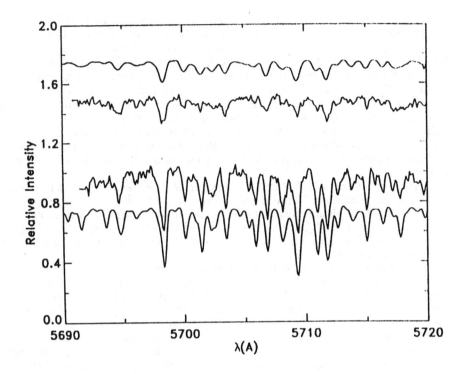

Figure 7. Veiling in the spectra of DF Tau and CI Tau. The upper and lower traces are standard stars spectrograms in the 5690–5720 Å, and the middle traces are observations of the two CTTS DF Tau (upper) and CI Tau (lower) in the same spectral region. The upper standard is M0V; it has been spun up to 22 km/s, the equatorial velocity of DF Tau, and numerically veiled to $r = 2.2$ by addition of a flat continuum. The lower standard is K5V, spun up to 11 km/s and veiled by 0.2. The spectra are all normalized to unity and arbitrarily shifted vertically (from Basri and Batalha 1990).

show features unexpected for the optically thin emission that has been postulated.

A measurement of veiling is useful for several reasons. It allows one to correct the observed spectrum for non-stellar emission independent of extinction corrections. This allows in turn a proper estimation of the extinction from the remaining reddened stellar spectrum, and offers, thereby, a constraint on the radius of the star (assuming distance is known). It also fixes the level of the boundary layer emission. Thus, one should demand that disk models produce the observed veiling in detail, constraining the boundary layer temperature and size along with the extinction, accretion rate, and stellar radius. Obviously veiling by itself does not constrain all these parameters, but inasmuch as it acts as a further constraint on disk models, it may significantly improve the uniqueness of solutions. Since veiling is the result of a particular disk model rather than an input parameter, there is no *a priori* guarantee that models that fit the overall spectral energy distribution best will also give the right veiling value. In their detailed study of DF Tau, Bouvier and Bertout (1991) compared model predictions with empirical average veiling values of 1.6 ± 0.3 at 5000Å and 1.0 ± 0.3 at 6500Å (Basri and Batalha 1990). The acceptable solutions that were found for the disk parameters after the χ^2 minimization procedure mentioned above produced veiling values ranging from 1.55 to 1.9 at 5000Å and from 0.6 to 0.75 at 6500Å, in fair agreement with the observations although they were not obtained simultaneously with the spectral energy distributions. Veiling varies in time by factors which can be as large as 10 in some stars (Basri and Batalha 1990). In order to take full advantage of this additional constraint on the models, it is thus mandatory to measure veiling at the same time as the spectral energy distribution.

Basri and Batalha (1990) and Hartigan et al. (1990) showed that the expected veiling has the relation to infrared excess expected if both are due to accretion and also that the H_α emission flux is closely related to veiling. Cabrit et al. (1990) and Bertout, Basri, and Cabrit (1990) showed that the emission lines in general are correlated with infrared excess, as expected from accretion (and not from a chromospheric origin).

The boundary layer is an obvious candidate for the line emission region (Basri and Bertout 1989), since it is small, fast-moving, turbulent, and energetic. However, inverse P Cygni line profiles seen in YY Orionis stars indicate infall velocities of several hundreds km/s which cannot arise in the boundary layers of classical disk models where accretion velocities are restricted to being subsonic. It is our intuition that strong stellar magnetic loops may interrupt the flow of the disk within a few stellar radii of the surface, and generate the shocks, turbulence, and heating that we have been designating the boundary layer. A modulation of the accretion (emission intensity or veiling) with the stellar rotation period would establish that the emitting region was in the interface between star and disk. There have already been a few reports of "hot spot" modulations (e.g., Bertout, Basri, Bouvier 1988) which could be interpreted as accretion columns onto magnetic stellar regions, where accreting material could reach near free-fall velocities.

It may well be the same thing to say that the lines arise in the boundary layer (now being used in the less restrictive sense of the interface region between star and disk) or that they arise in the base of the wind. In that picture, some material would accrete down through the magnetic loops, while other material would be turned back and flung out in the T Tauri wind. Some relevant theoretical ideas on this still controversial view can be found in Königl (1991).

4. The Evolutionary Status of CTTS and WTTS

One must be careful what is meant by "classical" T Tauri stars and "weak line" T Tauri stars, since there is really a continuum of stars between the two classes. If there is (i) a substantial near infrared excess (and this is best measured from 2 to 5 microns) compared to the flux expected from a star alone, (ii) Balmer continuum emission, and (iii) strong H_α emission (say more than 10Å equivalent width in a late K star), then we conclude from the evidence discussed above that an accretion disk is present, and that we thus have a CTTS. These diagnostics are absent or extremely weak in WTTS. Note that the accretion is the crucial part of this; stars with disks which do not reach their surfaces are WTTS, while a lack of excess emission at any wavelength is required for the star to truly be a "naked" T Tauri star.

The exact relationship between CTTS and WTTS is unclear at present, and work to clarify it should be encouraged since it is one major key for understanding the formation of low-mass stars. A crucial fact is that WTTS are well intermingled with the CTTS both spatially and on the HRD (i.e., temporally). Also, the WTTS have lithium line fluxes similar to the CTTS (Strom et al. 1989b, Basri, Martin, Bertout 1991) after correction for continuum veiling. Lithium is used as age indicator because it is rapidly destroyed by nuclear reactions when the star reaches the main sequence. Histograms of lithium abundances are presented in Figure 8. While both CTTS and WTTS samples are relatively small, there is no significant evidence for differences in lithium abundance distributions. However, a number of stars have lower lithium abundance than expected on the basis of their apparent age and mass, which are both determined from comparing their location in the HRD to theoretical convective-radiative evolutionary tracks. Although these results cast some doubt about the validity of T Tauri stars age calibration, they nevertheless indicate that the two populations have similar age distributions. Furthermore, one also finds that the distributions of rotational speeds in WTTS and CTTS are similar (Walter et al. 1988). These findings are curious because if steady-state accretion proceeds during a sizable time, then disk accretion might contribute a significant fraction of the stellar mass in CTTS, with noticeable consequences in the star's evolution, and should also alter its angular momentum history.

Current estimates based on classical convective-radiative evolutionary tracks indicate that accretion disks last up to a few million years (Strom et al. 1989a), and are a relatively common phenomenon among young solar-type stars, with perhaps less than one-half and more than one-fifth of them appearing currently as CTTS. This is cause for optimism if one had been hoping that planetary systems are a relatively common phenomenon, although it remains to be seen how many of these "solar nebulae" actually give rise to planets.

Two reasons make it difficult at this point to estimate the actual amount of mass that is ultimately accreted by the star even if we were optimistic enough to believe that the average mass-accretion rates derived from our disk models were representative of accretion during most of the CTTS phase. First, it is becoming increasingly clear that either the disk or the boundary layer is also driving the strong wind that characterizes the CTTS phase. Second, disk lifetime estimates depend crucially on the validity of the convective-radiative evolutionary tracks, which may not be relevant for accreting CTTS. The evolution of mass-accreting pre-main sequence stars was recently computed by Stringfellow (1989), who indeed finds that their paths in the HRD bears little resemblance to classical convective-radiative tracks. Furthermore, even if classical tracks are approximately correct, previous determinations of the position of CTTS in the HRD (e.g., Cohen and Kuhi 1979) did not take into account the presence of the disk and must, therefore, be regarded as uncertain by a factor of two or more. These issues have been addressed most directly by Kenyon and Hartmann (1990). For these reasons, age estimates of T Tauri stars based on the HRD appear suspect, as do derived disk lifetimes.

Even though the absolute age of T Tauri stars is doubtful, we assume here the conservative point of view that pre-main sequence stars sharing the same region of the HRD have similar masses and ages, i.e., that young stars do not loop through the HRD during pre-main sequence evolution. The observational evidence reviewed

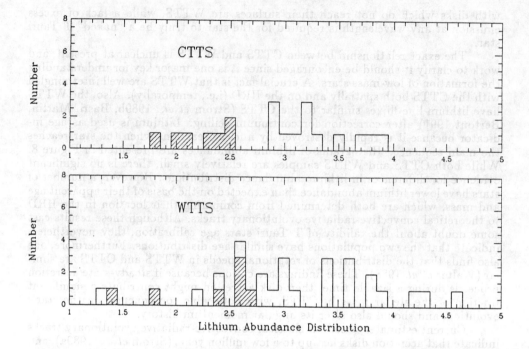

Figure 8. Histograms of the lithium abundance in CTTS and WTTS. It is customary to define lithium abundances log $N(Li)$ on a log $N(H) = 12$ scale. Cross-hatching indicates stars whose derived lithium abundance is markedly deficient compared to the expected value for its mass and age, as determined from its location in the HRD (from Basri, Martin, and Bertout 1991).

above then implies that accretion has little effect on the global stellar properties, perhaps because most of the disk matter is turned into a wind rather than accreted and/or because disk lifetimes are shorter than estimated on the basis of classical convective-radiative tracks. Several different possibilities can then be envisioned for the evolutionary status of CTTS and WTTS.

• WTTS may simply be CTTS which have stopped accreting. Of course, every CTTS must eventually lose its disk since main sequence stars are never observed to have active disks. But if all WTTS were once surrounded by disks with comparable physical properties, WTTS should be older than CTTS on the average, which is not the case. Thus, data demand a range in disk masses and lifetimes that may represent a variety of initial conditions for star formation.

• Alternatively, WTTS may be young stars which never had disks. Some of the WTTS are near the "birthline" for optically visible stars and so may never have been CTTS if one believes in the validity of Stahler's (1983) scenario. The existence of embedded radio sources with blackbody-like spectral energy distributions in the core of the ρ Ophiuchi cloud might also support this possibility (cf. André et al. 1987).

• Another possibility is that many T Tauri stars go through CTTS and WTTS phases several times during early pre-main sequence evolution. In this picture, one could speculate that disk instabilities lead to a large range of recurrent eruptive phenomena from FU Orionis events down to "normal" T Tauri aperiodic variability. Temporal spectroscopic changes seen, e.g., in RY Tau, where the optical spectrum can range from pure absorption to relatively strong emission, could arguably support this hypothesis.

These are probably the three most conservative scenarii for CTTS/WTTS evolution. If it turns out that the evolution of T Tauri stars in the HRD is very different from what we think today, more radical options follow. Obviously, much remains to be done to test all possibilities. Realistic (magnetohydrodynamic) simulations of T Tauri systems' evolution would of course be extremely useful, but because of the complexity of this approach, less ambitious projects such as investigations of the stability and evolution of protostellar disks and boundary layers are welcome first steps. An observational approach to these problems based on systematic spectroscopic and photometric monitoring of some T Tauri stars is also possible, and will offer some clues about the basic physical mechanisms of aperiodic variability in pre-main sequence objects.

5. Summary

Several properties of young low-mass stars can be explained in the framework of a model involving a late-type, active star surrounded by a cool accretion disk and interacting with it through a hot boundary layer. The mechanism leading to angular momentum transport in the disk, and thus to accretion, is unknown at this point. Current simplified quantitative models therefore assume a parametrized viscosity. They are good at reproducing the observed ultraviolet, optical, and near-infrared continuous spectra as well as the spectral veiling. The far-infrared flux is often larger than expected if it were arising in flat accretion disks, and none of the proposed solutions to this problem appears entirely satisfying. While the flux of Balmer lines with high quantum numbers is consistent with an origin in the boundary layer, H_α must originate from a more extended region that must be somehow connected to the accretion process, since veiling and H_α flux are correlated. A correlation between mass accretion and mass outflow is also indicated by the data, but there is no consensus yet about the nature of the mechanism which converts gravitational energy into wind kinetic energy. Young low-mass stars come in two varieties: the CTTS have accretion disks, the WTTS don't. The exact relationship between both subclasses is unclear at the present time.

Acknowledgments. CB is pleased to thank Nick Kylafis and Charlie Lada for their masterly organization of a great NATO-ASI and for financially supporting his participation in the school.

6. References

Adams, F.C., Emerson, J.P., Fuller, G.A. 1989, *Ap. J.* **357**, 606-620.
Adams, F.C., Lada, C.J., Shu, F.H. 1988, *Ap. J.* **326**, 865

Adams, F.C., Shu, F.H. 1986, *Ap. J.* **308**, 836
Ambartsumian, J.A. 1947. *Stellar Evolution and Astrophysics* (Erevan: Acad. Sci. Armenian SSR)
André, P., 1987, *Protostars and Molecular Clouds*, Eds. T. Montmerle, T., C. Bertout (Saclay: CEA/Doc), p. 143.
André, P., Montmerle, T., Feigelson, E.D. 1987, *A. J.* **93**, 1182
Appenzeller, I, Jankovics, I., Oestreicher, R. 1985, *Astron. Astrophys* **141**, 108
Appenzeller, I., Mundt, R. 1989, *Astron. Astroph. Rev.* **1** , 291-334.
Basri, G. 1987, *Fifth Cambridge Workshop on Cool Stars, Stellar Systems, and the Sun*, Eds. Linsky, J.L. and Stencel, R.E., (Berlin: Springer-Verlag) , pp. 411-420.
Basri, G. 1990, *High Resolution Spectroscopy in Astrophysics (Mem. S.A.It.)*, ed. Pallavicini, in press.
Basri, G., Batalha, C. 1990, *Ap. J.* **363**, 654-669.
Basri, G. and Bertout, C. 1989, *Ap. J.* **341**, 340-358.
Basri, G., Marcy, G. 1990, *The Sun and Cool Stars: Activity, Magnetism, Dynamos* (IAU Coll. #130), Touminen (Helsinki), in press.
Basri, G., Martin, E., Bertout, C. 1991. *Astron. Astrophys* , in press.
Beckwith, S.V.W., Sargent, A.I., Chini, R.S., Gušten, R. 1990, *A. J.* **99**, 924-945.
Beckwith, S., Zuckerman, B., Skrutskie, M.F., Dyck, H.M. 1984, *Ap. J.* **287**, 793-800
Bertout, C. 1987, in *Circumstellar Matter*, Eds. I. Appenzeller, I., C. Jordan (Dordrecht: Reidel), p.23
Bertout, C. 1989, *Ann. Rev. Astron. Astrophys.* **27**, 351
Bertout, C., Basri, G., Bouvier, J. 1988, *Ap. J.* **330**, 350
Bertout, C., Basri, G., Cabrit, S. 1990, *The Sun in Time*, Eds. Sonnett, C.P., Giampapa, M.S. (Tucson: University of Arizona Press) , in press
Bertout, C. and Bouvier, J. 1989. it Low Mass Star Formation and Pre-main Sequence Objects, Ed. Reipurth (Munich: ESO Workshop #33), pp. 215-232.
Bertout, C. Carrasco, L., Mundt, R. Wolf, B. 1982, *Astron. Astrophys. Suppl.* **47**, 419
Bertout, C., Thum, C. 1982, *Astron. Astrophys* **107**, 368
Bieging, J.H., Cohen, M., Schwartz, P.R. 1984, *Ap. J.* **282**, 699-708
Bouvier, J. 1990, *A. J.* **99**, 946-964.
Bouvier, J., Bertout, C. 1989, *Astron. Astrophys* **211**, 99
Bouvier, J., Bertout, C. 1991, preprint.
Bouvier, J., Bertout, C., Benz, W., Mayor, M. 1986, *Astron. Astrophys* **165**, 110
Cabrit, S., Edwards, S., Strom, S.E., Strom, K.M. 1990, *Ap. J.* **354**, 687-700.
Calvet, N., Albarrán, J. 1984, *Rev. Mexicana Astron. Astrof.* **9**, 35
Calvet, N., Basri, G., Kuhi, L.V. 1984, *Ap. J.* **277**, 725
Cohen, M., Bieging, J.H. 1986, *A. J.* **92**, 1396
Cohen, M., Emerson, J.P., Beichmann, C.A. 1989, *Ap. J.* **339**, 445-473
Cohen, M., Kuhi, L.V. 1979, *Ap. J. Suppl.* **41**, 743
DeCampli, W.M. 1981, *Ap. J.* **244**, 124
Duschl, W., Tscharnuter, W.M. 1990, *Astron. Astrophys* , in press.
Dyck, H.M. Simon, T., Zuckerman, B. 1982, *Ap. J.* **243**, L89
Edwards, S., Cabrit, S., Strom, S.E., Heyer, I., Strom, K.M., Anderson, E. 1987, *Ap. J.* **321**, 473
Feigelson, E.D., Giampapa, M.S., Vrba, F.J. 1990, *The Sun in Time*, Eds. Sonnett,

C.P., Giampapa, M.S. (Tucson: University of Arizona Press) , in press.
Feigelson, E.D., Jackson, J.M., Mathieu, R.D., Myers, P.C., Walter, F.D. 1987, *A. J.* **94**, 1251
Finkenzeller, U. Basri, G. 1987, *Ap. J.* **318**, 823
Friedjung, M. 1985, *Astron. Astrophys* **146**, 366
Hartigan, P., Hartmann, L., Kenyon, S.J., Hewett, R., Stauffer, J. 1989, *Ap. J. Suppl.* **70**, 899-914.
Hartigan, P., Hartmann, L., Kenyon, S.J., Strom, S.E., Skrutskie, M.E. 1990, *Ap. J.* **354**, L25-L28.
Hartmann, L. 1986. *Fundamentals in Cosmic Physics*, **11**, 279
Hartmann, L., Kenyon, S.J. 1988, in *Formation and Evolution of Low-Mass Stars* Ed. A.K. Dupree (Dordrecht: Reidel), p. 163.
Hartmann, L. and Kenyon, S.J. 1990, *Ap. J.* **349**, 190-196.
Hartmann, L., Hewett, R., Stahler, S., Mathieu, R.D. 1986, *Ap. J.* **309**, 275
Herbig, G.H. 1962, *Advances Astron. Astrophys.*, **1**, 47
Herbig, G.H. 1970, *Mem. Soc. Roy. Sci. Liège* **19**, 13
Herbig, G.H., Bell, K.R. 1988, *Lick Observatory Bulletin No. 1111*
Herbig, G.H., Goodrich, R.W. 1986, *Ap. J.* **309**, 294
Imhoff, C.L., Appenzeller, I. 1987, in *Scientific Accomplishments of the IUE*, Ed. Kondo, Y., (Dordrecht: Reidel) p. 295
Joy, A.H. 1945, *Ap. J.* **102**, 168
Kenyon, S.J., Hartmann, L. 1987, *Ap. J.* **323**, 714
Kenyon, S.J., Hartmann, L. 1990, *Ap. J.* **349**, 197-207.
Königl, A. 1991, *Ap. J. Letters*, submitted.
Kuhi, L.V. 1964, *Ap. J.* **140**, 1409
Kuhi, L.V. 1974, *Astron. Astrophys. Suppl.* **15**, 47
Kuhi, L.V. 1978, *Protostars and Planets* Ed. T. Gehrels (Tucson: University of Arizona Press) p. 708
Larson, R.B. 1969, *Mon. Not. Roy. Astron. Soc.* **145**, 211
Leinert, C., Haas, M. 1989, *Ap. J. Letters* **342**, L39-L42
Lin, D.N.C. and Papaloizou, J. 1985, *Protostars and Planets II*, Eds. Black, D.C., Matthews, M.S. (Tucson: University of Arizona Press) pp. 981-1072.
Lin, D.N.C., Williams, R.E., Stover, R.J. 1988, *Ap. J.* **327**, 234-247.
Lynden-Bell, D., Pringle, J.E. 1974, *Mon. Not. Roy. Astron. Soc.* **168**, 603
Maihara, T., Kataza, H. 1990, *Astron. Astrophys* , in press.
Malbet, F., Bertout, C. 1991, preprint
Mendoza, V.E.E. 1966, *Ap. J.* **143**, 1010
Mendoza, V.E.E. 1968, *Ap. J.* **151**, 977
Monin, J.L., Pudritz, R.E., Lacombe, F., Rouan, D. 1990, *Astron. Astrophys* , in press.
Montmerle, T., Koch-Miramond, L., Falgarone, E., Grindlay, J.E. 1983, *Ap. J.* **269**, 182
Mundt, R. 1984, *Ap. J.* **280**, 749
Mundt, R., Giampapa, M.S. 1982, *Ap. J.* **256**, 156
Myers, P.C., Fuller, G.A., Mathieu, R.D., Beichman, C.A., Benson, P.J., Schild, R.E., Emerson, J.P. 1987, *Ap. J.* **319**, 340.
Pringle, J.E. 1989, *Mon. Not. Roy. Astron. Soc.* **236**, 107
Regev, O. 1991, in *Structure and Emission Properties of Accretion Disks*, IAU Colloq. No. 129, Eds. Bertout, C., Collin-Souffrin, S., Lasota, J.-P., Tran Tranh Van, J. (Paris: Editions Frontières) in press

Rucinski, S.M. 1985, *A. J.* **90**, 2321
Ruden, S.P. , Pollack, J.B. 1991, preprint.
Rydgren, A.E., Strom, S.E., Strom, K.M. 1976, *Ap. J. Suppl.* **30**, 307
Rydgren, A.E., Zak, D.S. 1987, *Publ. Astron. Soc. Pacific* **99**, 141
Sargent, A.I., Beckwith, S. 1987, *Ap. J.* **323**, 294
Shakura, N.I., Sunyaev, R.A. 1973, *Astron. Astrophys* **24**, 337
Stahler, S.W. 1983, *Ap. J.* **274**, 822-829
Stringfellow, G.S. 1989, Evolutionary Scenarios for Low-Mass Stars and Substellar Brown Dwarfs. *Ph.D. Thesis* (Santa Cruz: University of California)
Strom, K.E., Strom, S.E., Edwards, S., Cabrit, S., Skrutskie, M.F. 1989a, *A. J.* **97**, 1451-1470
Strom, K.E., Wilkin, F.P., Strom, S.E., Seaman, R.L. 1989b, *A. J.* **98**, 1444-1450
Wagenblast, R., Bertout, C., Bastian, U. 1982, *Astron. Astrophys* **120**, 6
Walker, M.F. 1972, *Ap. J.* **175**, 89
Walter F.M., Brown, A., Mathieu, R.D., Myers, P.C., Vrba, F.J. 1988, *A. J.* **96**, 297
Wiese, W., Smith, M., Glennon, B. 1966, Atomic Transition Probabilities, NSRDS-NBS 4, Vol. 1
Wolf, B., Appenzeller, I., Bertout, C. 1977, *Astron. Astrophys* **58**, 163

Claude Bertout and Jose Cernicharo

THE X-RAY AND RADIO PROPERTIES OF LOW-MASS PRE-MAIN SEQUENCE STARS

THIERRY MONTMERLE
Service d'Astrophysique, Centre d'Etudes de Saclay
91191 Gif-sur-Yvette Cedex, France

ABSTRACT. X-rays and radio (cm) emission provide a unique opportunity to probe the magnetic fields of pre-main sequence stars. The high luminosities observed in both ranges reveal the ubiquitous evidence for a solar-type magnetic activity, and are mainly explained by the existence of giant magnetic structures (sizes up to ~ 10 stellar radii). However, the strength of the magnetic fields is not significantly larger than in solar active regions. Since low-mass pre-main sequence stars have outer convective layers, like the Sun, it is widely thought that the magnetic fields are produced by the "dynamo mechanism", and there is qualitative evidence to support this idea. The combination of recent results in the radio (mm) range on the cold material, and in the near-IR, reveal broad structural patterns in the circumstellar disks of pre-main sequence stars which may be linked, at least in part, with the presence of stellar magnetic fields.

1. Introduction

The discovery that low-mass pre-main sequence (PMS) stars are copious emitters of X-rays, and the study of their radio emission in the cm (GHz) range, have opened a new window in the knowledge of their outer layers and closeby circumstellar material. As explained in the present review, both kinds of observations allow to have access to general properties of their surface magnetic fields, indirectly in the case of X-rays (§ 2), and now directly in the case of non-thermal radio emission (§ 3). The observations in the X-ray and radio (cm) domains confirm, on a more quantitative ground, and sometimes in a quite spectacular way, the solar-like "magnetic activity" already suspected to exist from optical and UV observations (§ 4). The nature of the activity seen in X-rays and in the radio (cm) ranges is very likely connected with the generation of magnetic fields by the dynamo effect, present in the Sun and in all late-type, cool stars, because of the presence of outer convective layers (§ 5). A paradox arises because there is a strong evidence that the non-thermal radio-emitting stars are very young, while they do not show the near-IR excess thought to be characteristic of young stars. A first answer can be found in the light of recent results in the radio (mm) range, (which probes the cold circumstellar material), on various classes of "young stellar objects" (§ 6). These results suggest that the magnetic fields deduced from the X-rays and the radio (cm) may play an important role in the structure and evolution of the circumstellar material present around low-mass pre-main sequence stars (§ 7).

The reader will find other recent reviews and additional references to the literature, in Montmerle and André 1988 ; Herbig and Bell 1988 ; Bertout 1989 ; Feigelson, Giampapa, and Vrba 1991 ; Montmerle 1990 ; Montmerle, Feigelson, Bouvier, and André 1991 ; and of course in the other papers contained in the present volume.

2. Evidence for magnetic activity in PMS stars : X-rays

2.1. X-RAY EMISSION OF PMS STARS : GENERAL PROPERTIES AND SOLAR-LIKE ACTIVITY

The bulk of the data we now have on the X-ray emission from PMS stars comes from observations using NASA's *Einstein* Observatory. This was the first satellite-borne focusing X-ray telescope, which obtained images in the spectral band ~ 0.4 to ~ 4 keV ("soft" X-rays), with a ≲ 1' resolution, owing to a particular set of special, annular mirrors able to reflect and concentrate X-rays arriving under grazing incidence (for a description, see Giacconi *et al.* 1981). Additional, but more limited, data have been obtained using instruments based on the same principle aboard ESA's EXOSAT satellite (e.g., de Korte *et al.* 1981). Many other X-ray observations will take place in the near future using the NASA/D/UK ROSAT X-ray telescope, equipped with similar mirrors (e.g., Trümper 1984). This satellite has been launched in June 1990, and will provide an all-sky survey (expected to identify several hundred PMS X-ray emitters) and pointed exposures an order of magnitude more sensitive that those obtained with the *Einstein* satellite a decade earlier.

Most of the observed regions were nearby dark clouds belonging to the Gould Belt, well known to undergo active star formation and containing many T Tauri stars : ρ Ophiuchi (Montmerle *et al.* 1983), Taurus-Auriga (Gahm 1981, Feigelson *et al.* 1988, Walter *et al.* 1988), Chameleon (Feigelson and Kriss 1989), and others, all lying at distances ~160 pc. Images within the Orion star-forming complex at ~450 pc distance also revealed dozens of additional X-ray emitting PMS stars (Trapezium region : Ku, Righini-Cohen, and Simon 1982, Caillault and Zoonematkermani 1989 ; L1641 : Strom *et al.* 1990). Some of these regions also contain stars of earlier spectral types which emit X-ray copiously. From these extensive datasets, a variety of PMS stellar X-ray characteristics have been established, which can be summarized as follows.

Intense and variable X-ray emission, between 10^{29} and up to several 10^{31} erg.s^{-1} in the *Einstein* band (i.e., up to 10^6 times the solar X-ray luminosity), is a general property of low-mass PMS stars, and is not confined to any particular subclass. In particular, this emission is present independently of the strength of "classical T Tauri" properties, such as broad emission lines, UV and IR excesses (see, e.g., Bertout and Basri, this volume). For instance, there is no correlation between the X-ray flux and the strength of the Hα emission line. This indicates, in particular, that the X-ray emission is not generally absorbed (or "smothered" as initially suggested by Walter and Kuhi 1981) in dense T Tauri winds, thought to be at least in part responsible for the Hα line emission.

The X-ray spectra from PMS stars are typically centered around 1 keV. When a good PMS X-ray spectrum is available, it is consistent with bremsstrahlung emission from a hot (T ~ 10^7 K) plasma with line-of-sight absorption compatible with optical obscuration (typical A_V = 1-2 ; see § 2.3 for further details on the X-ray emission mechanisms).

The X-ray fluxes are also highly variable, and this variability is currently interpreted in terms of large X-ray flares, which dominate the X-ray emission. By comparison, coronal emission expected on the basis of that prevailing on main sequence stars of the same spectral type seems to be weak, at best barely at the level of sensitivity of *Einstein,* i.e., ≳ 10^{30} erg.s^{-1} at ~ 160 pc. The most extensive data are from the ρ Ophiuchi cloud, for which several dozen X-ray emitting stars have been found in the course of repeated exposures (Montmerle *et al.* 1983, 1984) ; Fig. 1 shows the light curves of representative sources. The observed variations are typically factors of 2-20 in amplitude ; they are statistically similar to, and have the same power-law distribution of peak X-ray emission than solar flares. In addition, there is a trend towards an increase of temperature with increasing luminosity, a feature also encountered in the Sun. These flares thus share many characteristics typical of solar X-ray flares, apart from a scaling factor of up to ~ 10^6 in luminosity, and they are taken as evidence for strong solar-like magnetic activity at the stellar surface.

The X-ray flares from low-mass PMS stars are also an order of magnitude stronger than flares on dMe flare stars (see discussion in Montmerle *et al.* 1983), and are quite similar to flares observed on RS CVn close binary systems (e.g., Agrawal, Rao, and Riegler 1986). However, it is estimated that strong flaring activity takes place only ~ 5 % of the time (Gahm 1988).

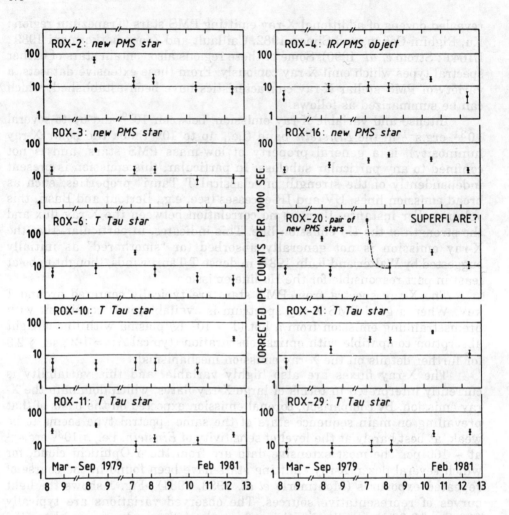

Fig. 1. *Light curves of a selected sample of the ρ Oph X-ray sources, including T Tauri stars known prior to the survey, and stars found subsequently by optical observations to be also T Tauri stars. Their X-ray luminosities are proportional to the number of counts received in the "Imaging Proportional Counter" aboard the Einstein satellite. All sources are seen to vary; source ROX-20 has undergone a major flare, reaching an X-ray luminosity ~ 10^{32} erg s^{-1}. (From Montmerle et al. 1984.)*

2.2. A NEW CLASS OF PMS STARS

A major discovery resulting from the soft X-ray surveys has been the identification of dozens of previously unrecognized PMS stars with "weak T Tauri" properties, i.e., weak or absent emission lines, notably Hα. Initially called "naked" T Tauri stars (Walter 1988), because of the weakness or absence of the Hα line was thought to reflect the absence of circumstellar material, these X-ray discovered stars are now preferably – and more empirically – called "weak-line" T Tauri stars, or "WTTS" for short. The bulk of the formerly cataloged T Tauri stars (e.g., Cohen and Kuhi 1979) are now called "classical" T Tauri stars, or "CTTS" for short. The boundary between the two classes is generally set in reference to the strength of their Hα emission line, which is usually chosen to be 5 to 10 Å in equivalent width, depending on authors. (The idea is that small EWs refer to a purely chromospheric origin for Hα ; see below, § 5.2.)

Estimates of the ratio of WTTS to CTTS populations range from ~ 10:1 for Taurus- Auriga (Walter *et al.* 1988) to ~ 3:1 for L1641 in Orion (Strom *et al.* 1990), to ~ 2:1 in Chamaeleon (Feigelson and Kriss 1989). This last value is consistent with the early results of Montmerle *et al.* (1983) on the ρ Oph cloud. The smaller values are more reliable than the first because the *Einstein* image coverage was very incomplete for the Taurus-Auriga complex. The accuracy of a PMS stellar census in any cloud is limited by several factors : Hα line strengths are time-variable in all PMS stars, permitting stars to cross the WTTS/CTTS boundary ; X-ray data have different sensitivities between and within star-forming regions due to different image exposures, distances and obscurations ; and the large-amplitude intrinsic X-ray variability may cause some stars to disappear in any given X-ray exposure, as vividly shown in the results of Montmerle *et al.* (1983). Despite these limitations, there is strong evidence that the total population of low-mass PMS stars is 3-4 times larger than the population of CTTS (EW[Hα] > 10 Å) alone, and is about 2 times larger than the PMS stellar population known prior to X-ray surveys.

As far as the *Einstein* data can tell, WTTS and CTTS do not differ in their X-ray properties, a fact which reinforces the idea that the mechanism responsible for the X-ray emission must be linked to the star, and not to its circumstellar environment.

2.3. PHYSICAL MECHANISMS OF X-RAY EMISSION

The absorption cross-section for X-rays in the ~ keV range is due to the photoelectric effect on the inner shells of the intervening atoms. It decreases as the third power of the photon energy, and, for absorption by the interstellar gas, it shows a modulation corresponding to the K absorption edges of the elements making up the absorbing material. At 1 keV, for normal abundances, an opacity $\tau = 1$ corresponds to an equivalent gas column density of $N_H \sim 5 \times 10^{21}$ H atom cm^{-2} (see, e.g., Ryter, Cesarsky, and Audouze 1975).

In most astrophysical conditions, there are two main X-ray emission processes : thermal bremsstrahlung (i.e., free-free radiation) from a hot plasma, in general optically thin, and nonthermal emission from highly energetic electrons, either by synchrotron radiation in intense magnetic fields, or by the Compton effect on the ambient radiation field (see, e.g., Rybicki and Lightman 1979). These last two processes are encountered for instance in active galactic nuclei or, for hard X-rays (> 10 keV) in impulsive solar flares (e.g., Bai and Sturrock 1989). But in all well documented cases relating to the emission of PMS stars, the observed spectra show that soft X-rays come from bremsstrahlung emission of an optically thin plasma at $T \sim 1$ keV, typically absorbed at low energies by an interstellar gas column density $N_H \sim 10^{21}$ to 10^{22} cm^{-2}, i.e., corresponding to a visual absorption A_V of $\lesssim$ a few magnitudes within the cloud.

The emissivity η_ν for bremsstrahlung of a plasma having a Maxwellian velocity distribution at temperature T, per unit volume dV, per unit solid angle $d\Omega$ and per unit frequency interval $d\nu$, is given by $\eta_\nu \, d\Omega \, d\nu \, dV = \alpha_\nu \, B_\nu(T) \, d\Omega \, d\nu \, dV$ (Kirchhoff's law), where α_ν is the absorption coefficient, and $B_\nu(T)$ the Planck function. It may be expressed as :

$$\eta_\nu = n_i n_e \left(\frac{16}{3}\right)\left(\frac{2\pi}{3}\right)^{1/2}\left(\frac{e^6 Z^2}{c^3 m_e^2}\right)\left(\frac{m_e}{kT}\right)^{1/2} g \, e^{-\left(\frac{h\nu}{kT}\right)} \quad (1)$$

(e.g., Lang 1986), where $g(\nu,T)$ is the usual Gaunt factor, i.e., a logarithmic term which takes into account the finite range of possible impact parameters during an electron-atom collision.

The X-ray luminosity of an emitting volume V in a frequency bandwidth $[\nu_1, \nu_2]$ is

$$L_X(\nu_1, \nu_2) = 4\pi \int_\nu \int_V \eta_\nu \, e^{-\tau_\nu} \, d\nu \, dV, \quad (2)$$

where τ_ν is the opacity coefficient. If the absorption along the line of sight is negligible, the total luminosity can then be numerically expressed as

$$L_{X,tot} \simeq 1.4 \times 10^{-27} \, T^{1/2} g \, Z^2 \int_V n_i n_e \, dV \quad (erg \, s^{-1}). \quad (3)$$

For a totally ionized plasma (which is the case at X-ray temperatures), $n_i = n_e$, and the emission measure is:

$$EM = \int_V n_e^2 \, dV, \qquad (4)$$

so that the X-ray luminosity is simply proportional to EM, i.e. to the emitting volume if the electron density is uniform.

For the imaging instruments aboard *Einstein*, v_1 and v_2 correspond to the limits of ~ 0.4 keV to ~ 4 keV mentioned above, but the actual numbers are a function of the gain of the detectors and may vary from one observation to the next. For $T \sim 1 - 2$ keV, $L_X(v_1, v_2) \simeq L_{X,\text{tot}}$. For smaller values of T, however, one must correct for the increasingly large contribution of line emission by heavy elements (e.g., Kato 1976), which, in view of the poor energy resolution of the imaging instruments, dominate the continuous spectrum for $T \lesssim 0.2$ keV.

2.4. DERIVED MAGNETIC STRUCTURES

X-ray flares are observed to last for up to several hours, implying that the emitting plasma must be somehow confined. This can be done only by magnetic loops, as strongly suggested by the *Skylab* results on the Sun (e.g., Moore et al. 1980) in view of the close analogy between PMS flares and solar flares. In this case, one usually approximates the confining volume by $V \approx (4\pi/3)\, l^3$, where l is characteristic height of the plasma above the stellar surface. The temperature T (and the absorption column density N_H) is determined from a χ^2 fit to the observed X-ray spectrum; g and Z can then be calculated, knowing the (standard) elemental abundances. The electron density n_e and the characteristic size l are the remaining parameters. In the absence of any further information, they cannot be determined separately. However, it is possible to calculate them independently, in the case of X-ray flares, if the light curve is known well enough. For single events, a rapid heating phase is followed by cooling by electron conduction and/or by radiation, with e-folding times :

$$t_{c,\,\text{rad}} = 3kT/[n_e\, P(T)], \qquad (5)$$
$$t_{c,\,\text{cond}} = 3k\beta\, n_e\, l^2 / \kappa(T), \qquad (6)$$

where $P(T)$ is the radiative cooling function (e.g. Gaetz and Salpeter 1983), $\approx 2 \times 10^{-23}$ erg cm^{-3} s^{-1} for $T \sim 10^7$ K, $\kappa(T)$ is the thermal conductivity, $\simeq 10^{-6}\, T^{5/2}$ erg s^{-1} K^{-1} cm^{-1}, and β is a geometrical factor accounting for the transport of electrons out of the loop, usually taken equal to 5 - 10 for the solar case. From the observed cooling phase of the light curve, it is possible to compute n_e, or $n_e l^2$ (and l via EM, eq. 4), depending on which of the above cases applies. The value of the magnetic field B (or, rather, its average over the volume V in the simple model above) is then obtained by equating the magnetic pressure ($B^2/8\pi$) and the gas pressure ($\simeq 2n_e kT$).

A few individual rapid X-ray flares that fortuitously occurred during

Einstein and EXOSAT satellite exposures have been analyzed in this context (Feigelson and DeCampli 1981, Montmerle *et al.* 1983, Walter and Kuhi 1984, Tagliaferri *et al.* 1988). The largest X-ray flares are characterized by rise times of minutes to hours, fall times of several hours, and peak luminosities up to 5×10^{31} erg.s^{-1} (see Fig. 1). Application of the simple loop models described above suggest that the temperatures ($T \sim 10^7$ K), plasma densities ($n_e \sim 10^{10}$ cm^{-3}) and magnetic field strengths ($B \sim 10^3$ G at the base of the loop) are not remarkably different from solar flares. But the main difference with solar flares is that the loop heights l_X inferred from the high X-ray luminosities are up to 10^3 times larger ($l_X \lesssim 10^{11}$ cm). The size of the X-ray emitting region is thus in general comparable to the stellar radius R_* ($\sim 2 R_\odot$, typically), and sometimes as large as $2 R_*$.

3. Evidence for magnetic activity in PMS stars : radio (cm range)

3.1. BASIC PHYSICAL MECHANISMS

The emission mechanisms in the radio (cm, or GHz) range have a lot in common with those in X-rays. In particular, there are thermal and nonthermal mechanisms. However, for radio stars, both kinds have to be taken into account, and the role of magnetic fields is of fundamental importance. (For details, see, e.g., Dulk 1985, and André 1987.)

3.1.1. Bremsstrahlung (free-free) emission. The main thermal mechanism, as in X-rays, is bremsstrahlung, i.e., inelastic collisions between electrons and ions ($E_e \gg E_{ion}$). While the plasma densities may be comparable ($n_e \sim 10^{10}$ cm^{-3}), emission in the GHz range requires the temperature to be much lower ($T \sim 10^4$ K).

The absorption coefficient is then given approximately by :

$$\alpha_\nu = 3.6 \times 10^{-17} [n_e/(10^5 \text{ cm}^{-3})]^2 [\nu / 5 \text{ GHz}]^{-2.1} [T/(10^4 \text{ K})]^{-1.35} \text{ cm}^{-1}. \tag{7}$$

Since at GHz frequencies, $h\nu/kT \ll 1$, we are in the Rayleigh-Jeans regime, and the source function is

$$B_\nu(T) = 2 kT \nu^2/c^2, \tag{8}$$

and the emission coefficient is

$$\eta_\nu = B_\nu(T) \alpha_\nu \propto n_e^2 \nu^{-0.1} T^{-0.35}. \tag{9}$$

By radiative transfer along a line of sight, one has $I_\nu = \int B_\nu(T) e^{-\tau_\nu} d\tau_\nu$, or in terms of the brightness temperature

$$T_b = \int T_{\text{eff}}(\nu) e^{-\tau_\nu} d\tau_\nu, \qquad (10)$$

where T_{eff} is the effective temperature of the source. Thus, in the optically thin case ($\tau_\nu \ll 1$), one has $T_b = T_{\text{eff}} \tau_\nu$. In the optically thick case ($\tau_\nu \gg 1$), $T_b = T_{\text{eff}}$, i.e., the emission becomes that of a black body.

A particularly useful application is that of the radio emission by an ionized, spherically symmetric sphere, like an HII region or a stellar wind, which one finds by integrating over the lines of sight crossing the sphere (e.g., Panagia and Felli 1975). In most cases, the emitting sphere has an optically thick core of radius R_c, and the effective emitting region is a thin superficial layer such that $0 < \tau < 1$. The core radius R_c depends on the opacity, and is therefore frequency dependent.

For an electron density law of the form $n_e(r) \propto r^{-\beta}$, the flux density is

$$S_\nu \propto \nu^\alpha, \text{ with } \alpha = 2 - 4.2/(2\beta - 1). \qquad (11)$$

For a "standard", fully ionized wind, one has $\beta = 2$, so that $\alpha = 0.6$. Conversely, given this density law, one finds the (mass-loss) rate:

$$\dot{M}_w/10^{-8} M_\odot \text{ yr}^{-1}$$
$$= 2.6 \, [S_\nu/\text{mJy}]^{0.75} \, [v_w/100 \text{ km.s}^{-1}] \, [\nu/5 \text{ GHz}]^{-0.45} \, [T_e/10^4 \text{ K}]^{-0.08}$$
$$\times [d/150 \text{ pc}]^{-1.5}. \qquad (12)$$

Hence the minimum $\dot{M}$ detectable by the VLA is $\lesssim 10^{-8} M_\odot \text{ yr}^{-1}$.

In the case of an accretion flow, $\beta = 1.5$, and $\alpha = -0.1$, so that the (accretion) rate is given by:

$$\dot{M}_{acc}/10^{-8} M_\odot \text{ yr}^{-1}$$
$$= 0.77 \, [S_\nu/\text{mJy}]^{-0.5} \, [\nu/5 \text{ GHz}]^{-0.05} \, [T_e/10^4 \text{ K}]^{0.17} \, [M/M_\odot]^{0.5}$$
$$\times [d/150 \text{ pc}] \, \{1+2\ln[r_{\max}/r_1(\nu)]\}^{-0.5}. \qquad (13)$$

$r_{\max}$ and $r_1(\nu)$ are integration limits necessary to get a finite flux (for details, see Panagia and Felli 1975, and Panagia, this volume).

3.1.2. "Non-thermal" mechanisms. This corresponds to the radio emission of a population of electrons spiralling in a magnetic field. This is again free-free emission, but this time turning around magnetic field lines, instead of around charged atoms.

Generally, the electrons have a power-law distribution of their energy, of the form $N_e(E) \propto E^{-\delta}$ for $E_{\min} < E < E_{\max}$. In the Sun, one has $\delta \sim 3$, and E is typically in the MeV-GeV range. They are generally accelerated to these energies by various shock processes, for instance chaotic winds, accretion on a compact companion, flares, etc. These electrons spiral in a magnetic field B, and are immersed in a plasma of density n_e and temperature T, which may act as an absorber (see below). If the electrons of energy $E = \gamma m_e c^2$ are mildly relativistic (Lorentz factor $\gamma \sim 1$), the emission mechanism is called gyrosynchrotron, and synchrotron if $\gamma \gg 1$.

The fundamental frequency corresponds one cycle around a magnetic field line, and is :

$$\nu_0 = (1/\gamma)\,\nu_B \quad (\nu_B = \text{cyclotron, or Larmor, frequency} = eB/[4\pi m_e c]). \tag{14}$$

At low energies ($\gamma \ll 1$), the "cyclotron" emission consists in a line spectrum of harmonics of ν_0 ; at high frequencies, corresponding to energetic electrons ($\gamma \gg 1$), the spectrum is a continuum peaking at $\nu_s = (3/2)\,\gamma^2\,\nu_B$. If the magnetic field is not uniform in space (like a dipolar loop), the resulting spectrum is sometimes called "inhomogeneous" (e.g., Klein and Chiuderi-Drago 1987). In this case, depending on the electron energies and the value of the magnetic field, the maximum emission comes from different regions (see André 1987), and there is a correspondence between frequency and space, which can be used to infer the morphology of distant magnetic fields.

Typically, for stars, GHz frequencies are produced by MeV electrons in a (local) ~ 1 G field. The emission at a given frequency ν comes primarily from electrons having a Lorentz factor :

$$\gamma_m = 3.5\,[\nu/5\text{ GHz}]^{0.5}\,[B/100\text{ G}]^{-0.5}. \tag{15}$$

The calculation of the emission and absorption coefficients depends on the details of the physical processes, and is much more complicated than in the thermal case. However, it is always possible to define an "brightness temperature" T_b, i.e., that of a blackbody radiating the same power at a given frequency within the same volume. This is one way of realizing how much greater the radiation efficiency of non-termal mechanisms is compared to that of thermal mechanisms : as we shall see in the following sections, for instance, gyrosynchrotron is typically 10^6 times more efficient than thermal bremstrahlung. Translated in terms of T_b, one finds very high values, on the order of 10^8-10^9 K (in fact much higher than the temperature of the X-rays themselves). Conversely, finding observational evidence for such high values of T_b is an immediate indication of a non-thermal mechanism (which then remains to be identified).

As a final note, it is important to realize that the so-called "non-thermal" nature of the emission is linked only with the presence of a magnetic field. For instance, another process is the "thermal gyro-synchrotron" mechanism, which occurs whenever a thermal plasma is

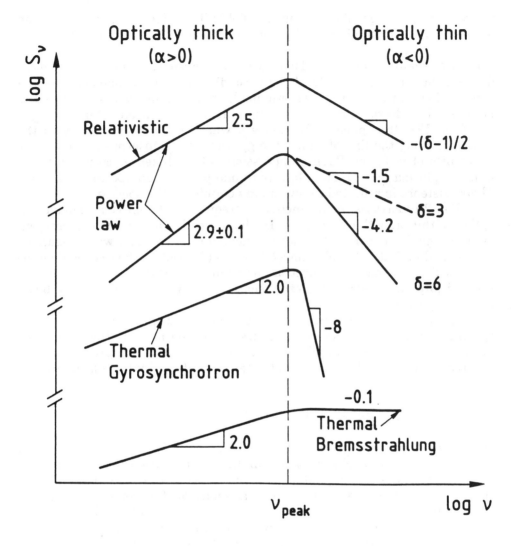

Fig. 2. *Summary of the various radio emission processes, in the optically thick (resp. optically thin) regimes. The three upper curves refer to gyrosynchrotron emission by electrons having a power-law distribution with index δ. The radio spectra are also power laws, of index α ($S_\nu \propto \nu^\alpha$). (From André 1987, after Dulk 1985.)*

permeated by a large enough magnetic field : it is a form of gyrosynchrotron mechanism, hence fundamentally non-thermal in spite of its adjective.

3.1.3. Absorption processes. There are many absorption processes, depending on the various possible physical conditions (plasma density, electron distribution, temperature, magnetic field, etc.), and are too complicated to be described here. One can quote plasma absorption (for $\nu < \nu_p = [n_e e^2/\pi m_e]^{1/2}$, the "plasma frequency"), which can take place within the source or along the line of sight ; the gyroresonance absorption (for ν = low harmonics of ν_B) ; the Razin suppression (which takes place if refraction index of plasma is $\neq 1$), etc. More than one process can operate at a time. Their existence is reflected in low-energy cut-offs in the spectra.

Fig. 2 illustrates, as a function of frequency, the spectral behavior of optically thin and thick regimes. In the optically thin regime, corresponding to high frequencies, the spectral slope α ($S_\nu \propto \nu^\alpha$) is always negative : it is related to that of the electrons δ by $\alpha = -(\delta - 1)/2$, in the case of a power-law distribution and, if the electron distribution is Maxwellian, $\alpha = -0.1$. In the optically thick regime, the exact slope depends on the mechanism, but α is always positive.

An important consequence of this situation is that the slope of radio spectra reflects the opacity of the emission, not necessarily the underlying mechanism. The only unambiguous case is that of a steeply descending spectrum, which will always indicate optically thin (gyro)synchrotron emission.

3.2. BASIC OBSERVATIONAL RESULTS

Given the magnetic nature of the X-ray emission, it is particularly interesting to make observations in the radio (cm) range to look for nonthermal emission associated with flares (as is the case of the Sun at these wavelengths), as well as thermal emission for instance associated with ionized winds. This was done by a number of workers, taking advantage of the increased sensitivity and imaging capability in the cm range offered by the Very Large Array (VLA) interferometer in New Mexico, by following two approaches.

In the first approach, known young [pre-main sequence] stellar objects ("YSOs"), selected according to a variety of criteria, are the targets of pointed observations : optically visible TTS or WTTS found in early catalogs (e.g. Bieging, Cohen, and Schwartz 1984; O'Neal et al 1990), sources of molecular outflows (Snell and Bally 1986 ; Rodríguez *et al.* 1989), or Herbig-Haro objects (e.g., Curiel *et al.* 1989). This approach gives evidence that ~ 10% of stars now known as CTTS or WTTS, as well as the majority of outflow exciting sources, are radio emitters at the level of a flux density sensitivity ~ 0.5 mJy, while almost all the classical, low-luminosity Herbig-

Haro objects remain undetected at this level (a major exception is the HH 1-2 system, see § 3.3).

The second approach makes no assumption about the nature of the putative radio emitters, and consists in surveying large areas (several square degrees) overlapping nearby dark clouds. Such surveys have revealed the existence of hitherto unknown objects, in ρ Oph (André, Montmerle, and Feigelson 1987), CrA (Brown 1987), Orion (Trapezium region : Garay, Moran, and Reid 1987 ; Churchwell et al. 1987 ; and L1641 : Morgan, Snell, and Strom 1990). Several stellar radio sources had been previously found in X-rays, but many were discovered to be so deeply embedded in the clouds that even the X-rays, if present, would be absorbed. Most of these embedded sources are YSOs associated with optically invisible near-IR sources, while a few of them are "radio-discovered" YSOs, so far undetected even in the IR (like the outflow source VLA 1623 in ρ Oph, André et al. 1990a). The number of radio-emitting YSOs found down to a flux density of ~ 1 mJy is small, on the order of 10 % of known near-IR sources (K-magnitude limit typically 12). As the sensitivity increases, however, the number of radio-emitting objects also increases significantly, to literally form embedded clusters of radio stars. The fraction of identifications with near-IR sources reaches ~ 50 % when the minimum detectable flux density reaches ~ 0.1 mJy (Leous et al. 1991). Remarkably, the sources with radio flux densities above ~ 1 mJy do not show evidence for a near-IR excess, and therefore appear as "embedded WTTS".

Long-term monitoring of the sources has proved necessary to understand the radio emission mechanism. As in X-rays, strong variability has been found by way of long or repeated observations : the discovery of the first radio flare in a PMS star came as early as 1983 (on the ρ Oph star DoAr 21, Feigelson and Montmerle 1985), and was followed by other similar findings (e.g., Stine et al. 1988). However, as can be seen in Fig. 3, the frequency of occurrence and/or duty cycle of PMS radio flares seems much lower than in X-rays (Stine et al. 1988, Bieging and Cohen 1989). For lack of adequate coverage, the exact timescale for variability is unknown, but variations in flux of factors of at least 2 have been found in a few hours, and up to 10 between observations separated by a few months (e.g., Cohen and Bieging 1986).

3.3. INTERPRETATION

Understanding the nature of the emission from sources detected through unbiased surveys is difficult, precisely because it turns out that a wide range of objects are represented.

In the case of CTTS and outflow sources, the radio emission is most likely of thermal origin, as shown in particular by the fact that some of

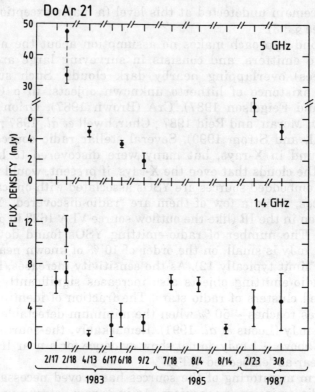

Fig. 3. *Monitoring with the VLA of the 5 GHz and 1.4 GHz flux densities of the star DoAr 21 in ρ Oph, from 1983 to 1987, over timescales of hours to months. A radio flare occurred on Feb. 18, 1983, and was visible at both frequencies for a few hours.* (From Stine et al. 1988.)

these sources are resolved by the VLA on a scale of ~ 0.1 - 10" (i.e., ~ 10^{14} - 10^{16} cm, or ~ 10^3 - $10^5\,R_*$, at the distance of the nearest molecular clouds). This confirms the existence of dense ionized stellar winds around CTTS, with mass-loss rates $\dot{M} \lesssim 10^{-8}\,M_\odot.\mathrm{yr}^{-1}$, already suspected to be present on the basis of some Hα and NaD (e.g., Mundt 1984) or [OIII] (Edwards et al. 1987) emission line profiles. However, there are clear examples, like the exciting source of the HH 1–2 system (Rodríguez et al. 1990), where the simple, spherical model generally used in the wind interpretation does not fit the observed radio spectrum and source shape, suggesting that the wind is rather a bipolar, confined jet (e.g., Reynolds 1986).

On the other hand, as mentioned above, the majority of the stellar radio sources emitting above ~ 1 mJy are unresolved and do not exhibit an IR excess. Since the timescales for their variability are relatively short (hours to weeks), their radio emission must be nonthermal, because the interpretation in terms of an ionized wind implies large, resolvable emission sizes, $\gtrsim 1000\,R_*$, and thus longer variability timescales (months to years). (Sources detected in the radio with a lower flux have in most cases been observed at only one wavelength, and their nature is more uncertain at present; see discussion in Leous et al. 1991.) Strictly speaking, only such a variability can be considered as evidence for non-thermal, "solar-like" activity in the form of flares. However, since it turns out that most of the time the stellar radio flux is not strongly variable, but more or less quiescent, one could then think of generally considering the spectral index α ($S_\nu \propto \nu^\alpha$) to find the emission mechanism, but, as discussed above, this procedure is often unreliable. It is then necessary to do polarization and mapping studies (e.g., VLBI), as discussed in detail in the following section, to demonstrate that the radio emission of several quiescent sources is also non-thermal. In this case, the term "solar activity" should be understood in a much broader, looser sense, only because modelling the quiescent radio emission still relies mostly on loop models devised for the Sun.

3.4. DERIVED AND OBSERVED MAGNETIC STRUCTURES

3.4.1. Radio polarization data. The most direct signature of the presence of magnetic fields that radio observations can provide is the detection of circular and/or linear polarization. Circularly polarized radio emission generally traces nonrelativistic or mildly relativistic electrons radiating in magnetic fields of ~ 1 - 1000 G through the gyroresonance, gyrosynchrotron, or even free-free processes, while linearly polarized emission generally traces synchrotron-emitting, ultrarelativistic electrons in weaker magnetic fields, << 0.1 G (e.g., Dulk 1985). The main difficulty of this approach is that detecting radio polarization at a level of a few percents on sources with a flux density at most a few mJy is often a challenge. However, the power of this technique has recently been illustrated in two different stellar cases.

First, André et al. (1988) detected weak (~ 7 %) *circularly* polarized radio emission in the young, embedded B3 star S1 belonging to the ρ Ophiuchi cloud core. The modelling of the emission led to the suggestion that S1 was surrounded by an axisymmetric (possibly dipolar) extended ($2l_{radio} \lesssim 8 - 10\,R_*$) magnetosphere, oriented essentially pole-on, and with a large-scale surface magnetic field similar to what prevails on the well-known magnetic B stars ($B_* \sim 10$ kG). This interpretation has later been confirmed by VLBI measurements (see below), and the level of polarization has repeatedly been detected in subsequent observations spread over several years.

Second, Yusef-Zadeh et al. (1990) discovered weak (6-8 %) *linearly* polarized emission at 6 cm near the core of an HH-like object, known as the

Orion "streamers", in the L1641 cloud. This detection implies the presence of a weak organized magnetic field of at least ~ 5 x 10^{-4} G on a scale of 2500 AU. The authors suggest that this magnetic field is generated by a rapidly rotating, deeply embedded pre-main sequence star featuring a magnetosphere rather similar to that proposed for S1. In both sources, it is important to stress that the radio emission is essentially non-variable, suggesting a permanent or at least long-lived magnetic field configuration.

3.4.2. VLBI measurements. We have seen above (§ 2.3) that the X-rays were likely confined in large ($l_X \lesssim 2 R_*$) magnetic loops, as can be indirectly shown from the X-ray emission measures and variability timescales, and that modelling the radio emission led to even larger sizes of magnetic structures ($2l_{radio} \lesssim 10 R_*$).

But direct measurements of these sizes can be obtained in the radio (cm) range, by means of Very Long Baseline Interferometry (VLBI) observations (see, e.g., Lestrade 1988). VLBI techniques makes use of radio telescopes separated by hundreds or thousands of kilometers, to resolve emission regions on milliarcsecond angular scales, corresponding to linear scales around 10^{12} cm (i.e., a few R_*) in nearby star-forming regions. By combining VLBI with simultaneous measurements on arcsecond scales using km-sized interferometers like the VLA, one can determine whether the radio emission is also produced in regions several hundred times larger, such as in ionized winds.

The already mentioned star S1 in ρ Oph was the first YSO detected by this technique (André *et al.* 1991). The overall size (~ 13 R_*) and brightness temperature (~ 2 x 10^8 K) which were measured at 6 cm are consistent with the main lines of the magnetospheric model proposed on the basis of the circular polarization detection. The combination of polarization and size measurements for this source provides good and almost direct constraints on the value of the magnetic field : assuming a dipolar field, a value of $B_* \sim$ 2 kG is derived at the stellar surface.

Extensive VLBI studies of the "bright" ($S_{6cm} > 1$ mJy) stellar radio sources of Taurus and Ophiuchus (mostly WTTS or embedded objects without IR excess) are now in progress, and already prove extremely successful. The first set of observations has demonstrated that most or all of WTTS radio emission is nonthermal, with brightness temperatures between 4 x 10^8 to > 2 x 10^9 K (Lawrence *et al.* 1990). Two radio-bright WTTS, DoAr 21 and HD283447, were mapped in detail and were spatially resolved, with the radio emission coming from regions ~ 6 - 30 R_* large. In these cases, the size of the emitting region (but not necessarily that of the magnetic structure) also varies on timescales of days or hours. As in the X-rays, the radio luminosities are very large, roughly 10^5 times that of the strongest contemporary solar flares.

In any case, the main conclusion which can be drawn from these recent VLBI results is that they confirm inferences based on earlier nonimaging variability characteristics of WTTS radio emission, which, as discussed in § 3.2, suggested that the extent of the magnetic fields surrounding nonthermal young radio stars is much larger than on the Sun and reaches several stellar radii. In particular, this excludes an alternative model, initially proposed for RS CVn radio flares, and which involves magnetic fields erupting throughout the stellar surface rather than in more localized large magnetic structures (Mullan 1985). By contrast, the derived strength of the surface magnetic field (between a few 100 G and a few kG), which agrees well with the first direct measurement recently obtained by Basri and Marcy (1990) on a WTTS ($<B_*>$ = 1700 ± 500 G), again appears to be not much higher than the peak values measured on the Sun. This means that the "filling factor" for strong magnetic fields is much larger than on the Sun, a result consistent with that derived from X-ray flares, and optical and UV starspot observations (see next section). In other words, the magnetic field of young radio stars is probably much more organized than the tangled fields observed in solar active regions.

4. Evidence for magnetic activity in PMS stars : optical and UV

4.1. FLARES

T Tauri stars have long been known to be variable from near-UV to near-IR wavelengths, on timescales ranging from a few minutes to a few decades. In the early fifties, they have been put in the category of "flare stars" by such pioneers as Haro and Ambartsumian. The flare activity is best observed in the near-UV band, since the contrast with the essentially red photosphere is enhanced. Analysis of the time structure of these flares (Worden *et al.* 1981) has shown that they are distributed according to a power law, smaller flares being more frequent than larger ones. The optical flares are solar-like, but enhanced several thousand times with respect to solar flares. According to a recent study, there is some difference in the time distribution of UV flares between CTTS and WTTS (Gahm 1990).

4.2. STARSPOTS

It has been known for more than 25 years that, in addition to flaring activity, some TTS exhibit quasi-cyclic light variations on a timescale of a few days. Yet, it is only recently that systematic photometric studies have been undertaken, leading to the detection of periodic light-curves for a number of TTS (CTTS and WTTS alike). The amplitude of the photometric wave, usually of the order of a few tenths of a magnitude, decreases from near-UV to near-IR wavelengths and can often be modelled by a single large starspot (or an assembly of smaller starspots covering typically 10% of the stellar surface), cooler than the stellar photosphere by a few 100 K (see, e.g., Bouvier, Bertout, and Bouchet 1986 ; Vrba *et al.* 1986 ; Vrba *et al.* 1989 ; Bouvier and Bertout 1990). These starspot properties are very similar to those found on other types of magnetically active stars, such as RS CVn systems and BY Draconis stars.

By analogy with the Sun, chromospheric plages are expected to be associated with cool starspots. Spectroscopic evidence for such chromospheric plages has been reported by Herbig and Soderblom (1980), and the coexistence of cool and hot spots at the surface of TTS is sometimes required to model their periodic light-curves (Vrba *et al.* 1986). In some cases, only hot spots can account for the large amplitude (up to several magnitudes) of the stellar modulation observed at near-UV wavelengths. Then, the luminosity of these spots usually amounts to a significant fraction of the luminosity of the star, and they have been interpreted as accretion shock regions near the stellar surface, where the material, presumably accreted from a circumstellar disk, is channelled along mid-latitude magnetic field lines (Bertout, Basri, and Bouvier 1988). That WTTS only have cool spots while hot spots have only been found on CTTS so far supports this idea.

The lifetime of TTS spots greatly varies from one star to another. Long-term photometric monitoring of the WTTS V410 Tau (Vrba, Herbst, and Booth 1988) has shown that two large, cool spots survived for several years at its surface. In other stars, however, the photometric wave has been found to disappear from one season to another. Short spot lifetimes (less than a few days) may explain why rotational modulation was detected in only about one third of the TTS in which it was searched for. In fact, both periodic and non-periodic variability on a timescale of days usually have comparable amplitudes and similar wavelength dependence. This suggests that modulation of the stellar luminosity by spots analogous to sunspots is indeed the main mechanism responsible for day-to-day variability in these stars, whether periodic or not (there are some exceptions, however, see, e.g., Gahm *et al.* 1989).

5. Origin of magnetism

5.1. THE DYNAMO EFFECT

The X-ray, radio, UV and optical evidence, essentially based on variability studies but also on more detailed modelling, therefore largely supports the existence, at the surface of low-mass PMS stars, of a magnetic activity very similar to the solar activity.

On the Sun, the origin of the magnetism is widely thought to originate in the outer convective zone, via the so-called "dynamo mechanism". It is therefore not surprising that TTS, dMe, RS CVn or BY Dra stars, which share this structural property with the Sun because they are all late-type, cool stars, should, to varying degrees, have magnetic properties similar to those of the Sun. The idea underlying the dynamo mechanism is that seed magnetic fields in an electrically charged medium are amplified by the combined action of convection and differential rotation (the so-called "$\alpha\omega$" mechanism ; see, e.g., Gilman 1983).

In essence, and just as for a bicycle generator, the starting point is Ohm's law, $\boldsymbol{j} = \sigma\,(\boldsymbol{E} + \boldsymbol{v}\times\boldsymbol{B})$, σ being the electrical conductivity. Using Maxwell's equations (without radiation), one can write the variation of the magnetic field $\boldsymbol{B}$ with time due to moving charges as :

$$\partial \boldsymbol{B}/\partial t + \boldsymbol{v}\nabla \boldsymbol{B} = \boldsymbol{B}\nabla \boldsymbol{v} + \lambda\Delta \boldsymbol{B}\,, \tag{16}$$

where λ is the magnetic diffusivity (expressing the tendency of the magnetic field to leak out of the medium). If $\boldsymbol{B}\nabla \boldsymbol{v} > \lambda\Delta \boldsymbol{B}$, there is magnetic energy generation, at the expense of kinetic energy.

Unfortunately, the dynamo theory is not developed well enough in stars to lead to any quantitative prediction. One simply expects some correlation between the level of magnetic activity and rotation (itself correlated somehow with differential rotation), both being defined in a broad sense. Since, in the case of embedded YSOs, there is currently no observational access to rotation (as could be given, for instance, by high resolution near-IR lines), the existing evidence for a dynamo mechanism must be drawn from observations of optically visible PMS stars.

5.2. LINKS WITH ROTATION

There are various definitions of the stellar "rotation" parameter : it can be the projected rotational velocity ($v\sin i$, where the projection factor is generally unknown), the rotational period (which can be directly measured from rotational modulation associated with starspots), or even the so-called "Rossby number" (see, e.g., Hartmann and Noyes 1987), which is a mixture of rotation and convection (and equal to the ratio of the observed rotation period and of the computed convective turnover time). As for the "magnetic activity" parameter, it is not better defined : it can be the X-ray luminosity, or chromospheric activity tracers such as the MgII H + K or CaII h + k line fluxes (e.g., Rutten and Schrijver 1987, Bouvier 1990a).

It turns out, however, that all combinations tend to yield the same rather reassuring qualitative trend, namely that the "magnetic activity" decreases with "rotation" (velocity), as expected from dynamo theory (for a more extensive discussion, see e.g., Montmerle 1987). As an example, Fig. 4 shows a nearly identical correlation between the X-ray luminosity and rotational period for a variety of late-type stars, including the more evolved, magnetically active RS CVn binary systems. The existence of such a correlation reinforces the conclusion that surface magnetic fields in active regions of TTS, CTTS and WTTS alike, are of the same order, i.e., a few kG.

Other activity diagnostics, such as the MgII and CaII lines, are more ambiguous. In WTTS, these lines are weak and narrow, like that of Hα (for which EW < 5 - 10 Å), and their fluxes are consistent with a purely chromospheric origin. In CTTS, however, the broad profiles of emission lines such as Hα cannot be reproduced by chromospheric models, and the measured line fluxes are far too large to be accounted for by a solar-type dynamo mechanism (Calvet and Albarrán 1984, Bouvier 1990a). We will return to this question below (§ 6.1).

5.3. DEPENDENCE WITH AGE

In the course of their PMS evolution, the low-mass stars progressively spin down (e.g., Stauffer 1988, Soderblom and Stauffer 1990, Bouvier 1990b). Consequently, according to the dynamo mechanism, we also expect their activity to decrease with time. As a matter of fact, the average PMS star emits 10^3 times more in X-rays than old disk stars, and 10 times more than young main sequence stars such as the Pleiades. However, the exact form of the relation between stellar age and X-ray luminosity L_X is still debated. Walter et al. (1988) choose selected main sequence stars and Taurus-Auriga X-ray detected PMS stars to derive a relation of the form $L_X \propto \exp(-t/t_0)$ with $t_0 = 4 \times 10^8$ yr. Feigelson and Kriss (1989) use unbiased X-ray luminosity functions of optically-selected samples of main sequence stars and Chamaeleon PMS stars, treating X-ray nondetections as well as detections, to derive the relation $L_X \propto t^{-0.6}$ for the entire range $10^6 \leq t \leq 10^{10}$ yr, which is close to the so-called "Skumanitch law" for the stellar spindown (angular velocity $\propto t^{-0.5}$ for $t \gtrsim 10^7$ yrs, Skumanich 1972). It is therefore unclear at present whether the X-ray activity decays with age throughout the PMS phase, or whether it is relatively constant in the early part of this phase in which the star descends the Hayashi track. The problem is complicated by the fact that the star not only spins down, but also undergoes gravitational contraction, so that the available area for X-ray emission effectively decreases (see Fig. 5). At any rate, this emission appears to be present as soon as convection begins when stars emerge from their accreting envelopes at the "birthline" (Stahler 1988 ; also Stahler and Walter 1991), and is seen to decay with time, which does not contradict the expectation from the dynamo mechanism.

5.4. TOWARD HIGHER MASSES

Since the dynamo mechanism operates only in moving, charged media,

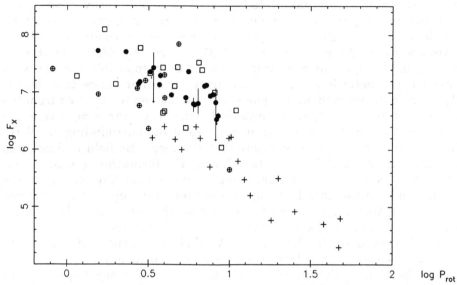

Fig. 4. *The X-ray surface flux* F_X *is plotted against the rotation period* P_{rot}, *for various kinds of late-type stars. Key:* ● T Tauri stars ; ⊕ dKe and dMe emission-line stars ; + late-type dwarfs ; ☐ RS CVn close binary systems. *The data points span more than 2 orders of magnitude. A correlation is seen, almost independently of the stellar type, which gives a qualitative support to the dynamo mechanism as generating the magnetic fields associated with the X-ray activity.* (From Bouvier 1990a.)

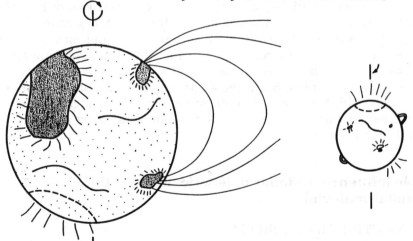

Fig. 5. *Sketch of the compared sizes and activity regions of the young Sun (at a late, diskless stage, ~ a few* 10^6 *yrs old, presumably after the formation of the solar system), and of the present-day Sun.* (After Feigelson, Giampapa, and Vrba 1991.)

hence is applicable to low-mass, late-type stars with outer convective layers, one should not expect signs of magnetic activity in higher-mass stars, which have outer *radiative* layers. Yet we have already quoted the extensive work on the radio emission of S1, a ~ 20 $M_\odot$ ZAMS B3 star. In such a case, the dynamo mechanism cannot be responsible for the well-established magnetic field. It does not have to, either, since this field is likely a stable dipole, and such a global field is not really expected from the dynamo mechanism. Rather, and especially since this star is very young (a few thousand years, as deduced from the size of the surrounding small HII region, André et al. 1988), it is likely that the magnetic field is fossil, i.e., has remained trapped in the star during its formation phase. Older "magnetic" B stars are common, which means that the decay of the magnetic field (for instance by Ohmic losses) may take up to a few 10^6 yrs. Note, however, that a dynamo contribution from the deep convective interior is not excluded (see, e.g., Borra, Landstreet, and Mestel 1982).

Another problem is that of the A stars, or, rather, of their PMS counterparts known as the Herbig Ae-Be stars, which have a mass of ~ 2 $M_\odot$ and above (e.g., Catala 1989). These are complex objects, which, as shown by the example of AB Aur, may have strong, fast rotating winds. In the few known cases, the radio emission is thermal. But the large sample of X-ray-discovered young A stars found by Strom *et al.* (1990) in L1641, which share many X-ray properties of the lower-mass stars, may seem puzzling : according to the standard models, these stars are already entirely radiative in their PMS phase. One possibility is that all have invisible, X-ray emitting low-mass companions, as is often found for main-sequence A stars. A perhaps more exciting possibility is suggested by very recent work by Palla and Stahler (1991) on PMS evolution of stars in the ~ 1 - 8 $M_\odot$ range. For a mass $\gtrsim$ 2.5 $M_\odot$, deuterium burns in a subsurface shell, and may drive an outer convective zone. It is therefore conceivable that the dynamo mechanism may operate in such conditions and give rise to the observed magnetic activity traced by the X-rays. Since the deuterium burning phase found by Palla and Stahler (1991) has a limited duration, it would be a crucial test of the theory to see whether the age of the X-ray luminous young A stars does correspond to such a phase.

6. Possible influence of stellar magnetic fields on the circumstellar material

6.1. AN EVOLUTIONARY PROBLEM

In the current "standard" view, the CTTS spectral features from the UV to the near-IR, and in particular the Hα line, arise as a consequence of the presence of an accretion disk (reponsible for the near-IR excess) and of a boundary layer (responsible for the UV excess and emission lines) between the slowly rotating star and the fast rotating disk. Now WTTS lack such features. Then the question arises : is the absence of CTTS features in WTTS spectra really due to the absence of an accretion disk or of some form of circumstellar material ? Or are WTTS simply ex-CTTS, in other words do they represent diskless, evolved PMS stars ?

First, many PMS stars display no IR excess : according to the IR classification (Lada 1987 ; Wilking, Lada, and Young 1989), they are "Class III" (i.e., pure blackbody) objects, in other words they resemble embedded WTTS. Yet they must be very young, because of they are located just at the edge of dense cores in dark clouds : this is the case of the radio-emitting stars found by André et al. (1987) in the ρ Oph cloud, and by O'Neal et al. (1990) in the Taurus clouds. As an example, Fig. 6 shows the spatial distribution of the radio-emitting sources (> 1 mJy) in the ρ Oph cloud core region (André et al. 1987) ; a similar distribution is observed in the L1495E and L1495W clouds (André, Loren, and Wootten 1991). Strictly speaking, of course, no age can be attributed to sources not seen in the optical, because they cannot be placed on an H-R diagram. But when these radio-emitting stars are optically visible, they are indeed found to be among the youngest TTS ($\sim$ 5 x 10^5 to $\sim$ 1.5 x 10^6 yrs old, on average ; see O'Neal et al. 1990).

Second, a related problem exists more generally when considering optically visible CTTS and WTTS as a whole. Indeed, when put on an H-R diagram, these two classes of PMS stars appear mixed (see, e.g., Strom et al. 1989), whereas, according to "standard" evolutionary models (e.g., Adams, Lada, and Shu 1987, Shu, Adams, and Lizano 1987), CTTS, being surrounded by accretion disks, should be younger, and WTTS, deprived of such disks, should represent a more advanced evolutionary stage.

So we are faced with a problem : how is it that very young stars, still embedded in the vicinity of cloud cores, apparently do not show evidence for circumstellar material ? Has this anything to do with the presence of extended magnetic structures ?

A first answer lies in the fact that the "standard" scenario of early stellar evolution is essentially based on near-IR data, which, in current disk models, only traces warm material comparatively close to the star ($\lesssim$ 1 AU). The outer (> 10 AU), cold (<< 100 K) material of these disks is visible only in the mm range, either in the form of gas (molecular lines) or in the form of dust grains (continuum), and therefore the conclusions drawn from near-IR data alone may not apply to these regions.

6.2. COLD DUST AROUND YOUNG STELLAR OBJECTS

Standard (i.e., viscous keplerian) accretion disk models assume a continuous temperature distribution from the star out, generally taken as a

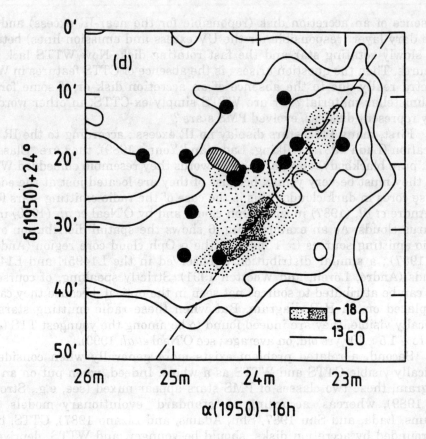

Fig. 6. *Spatial distribution of the ρ Oph stellar radio sources (open circles; with $S_{6cm} \gtrsim 1$ mJy), compared with the gas distribution within the core of the cloud. Note how the sources are concentrated in the central region. No other source is known farther out. (From André, Montmerle, and Feigelson 1987.)*

power law $T(r) \propto r^{-q}$ ($q < 1$), as well as a continuous power-law surface density distribution $\Sigma(r) \propto r^{-p}$ ($p < 1$). Since the near-IR emission is optically thick, it is a probe of the temperature of the inner regions; on the other hand, the mm range is optically thin, and therefore is a measure of the mass of the emitting material. One thus expects, via the temperature and density distributions, a tight correlation between the "warm" inner near-IR regions ($T \lesssim 1000$ K) and the "cold" outer mm-emitting regions ($T \lesssim 50$ K), if the same disk structure applies to all sources.

Following the pioneering work of Beckwith et al. (1986) and Sargent and Beckwith (1987) on the extreme TTS HL Tau, sensitive continuum mm studies of several tens of T Tauri stars and embedded YSOs have recently been reported and have allowed to probe their cold dust content. With the James Clerk Maxwell telescope in Hawaii, operating in atmospheric windows from 0.35 to 1.1 mm, Weintraub, Sandell, and Duncan (1989) and Adams, Emerson, and Fuller (1990) selected small samples comprising almost exclusively CTTS, and found a very high detection rate. Beckwith et al. (1990) used the IRAM 30-m radio telescope near Granada, Spain, equipped with the MPIfR bolometer operating at 1.3 mm, to study a much larger sample of 86 CTTS and WTTS stars in Taurus-Auriga : 53 % of the CTTS, and 29% of the WTTS were detected. André et al. (1990b) have independently undertaken a study similar to that of Beckwith et al. (1990) in the ρ Oph core using the IRAM 30-m telescope, but selecting a sample of embedded sources spanning a broader range of near-IR properties than CTTS and WTTS, from Class III to Class I (displaying a large IR excess above 2 μm, see Lada 1987), with a very good sensitivity (down to an r.m.s. of ~ 1 mJy). Out of 18 such objects, 14 were detected, belonging to all classes and spanning a large range in fluxes, from ~ 30 to ~ 400 mJy. In addition, the sample included two exceptional objects exciting molecular outflows and believed to be extremely young (a few 10^3 yrs), IRAS16293 (Walker et al. 1988) and VLA1623 (André et al. 1990a) ; both are very bright at mm wavelengths (resp. 5 and 0.9 Jy).

While the high detection rate of Weintraub et al. (1989) and Adams et al. (1990) was not unexpected, because there was already from IR data strong independent evidence for circumstellar material around their program stars, the results of Beckwith et al. (1990) are more surprising. Indeed, many WTTS turn out to be surrounded by cold material, whereas there are near-IR indications that no or little warm material is present. On the other hand, André et al. (1990b) find a lack of obvious correlation between the mm emission and near-IR excess of the objects in the sample. In particular, as shown on Fig. 7, heavily obscured (Class I) objects do not appear brighter at 1.3 mm than CTTS (Class II) in general. On the other hand, there is one case of an object without near-IR excess (Class III) showing a mm emission.

Although the samples of Beckwith et al. (1990) and André et al. (1990b) are qualitatively different, the two results are quite consistent with each other. In the framework of the disk interpretation of the emission of YSOs at long wavelengths, it is found that, in many cases, the near-IR/mm correlation expected for "standard" accretion disks does not hold. More precisely, many WTTS show substantial mm emission (~ 30 % of the objects). Since these stars have no UV excess nor significant Hα emission, hence no boundary layer, implying that there is no physical contact between the star and the disk, they must however be surrounded by hollow, "45 r.p.m." cold disks. The absence of near-IR emission, i.e., of warm material close to the star confirms this picture, and gives sizes ≈ 1 AU for the radius of the inner hole (see Skrutskie et al. 1989). Conversely, CTTS with weak or absent mm emission (~ 50 %) would be surrounded by more or less "compact" disks, at most a few 10 AU in size.

Whether "hollow" or "compact", the overall size of such disks must be < 1000 AU, because none of the detected sources is resolved within the ~ 12" resolution of the 30-m observations. We note that the derived disk masses (gas + dust) are ~ 0.005 to 0.1 $M_\odot$, and thus encompass that of the primitive solar nebula (for a discussion on the mass determinations, and other consequences, see, e.g., André et al. 1990b, Beckwith et al. 1990, and Montmerle et al. 1991).

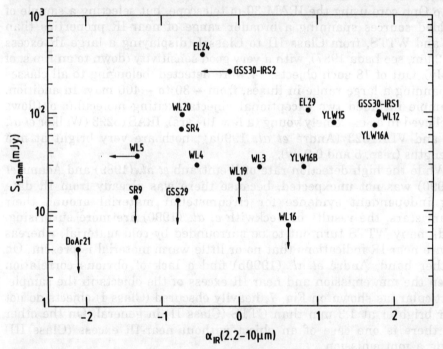

Fig. 7. *Scatter diagram of the 1.3 mm flux density vs. the near-IR spectral index α_{IR} (slope of the λF_λ spectrum between 2.2 µm and 10 µm, Lada 1987) for a selected sample of sources in the ρ Oph cloud. Class I, Class II, and Class III IR sources lie to the right, center, and left parts of this diagram, respectively. There is no obvious correlation between the mm flux (which traces optically thin, cold dust, at several AU from the central star) and the near-IR spectral energy distribution (which traces the optically thick, warm material within 1 AU), suggesting the existence of different structural features in the accretion disks. (From André et al. 1990b)*

6.3. MAGNETIC FIELDS AND STABILITY OF ACCRETION DISKS

From the combined IR and mm data, the existence of circumstellar disks exhibiting central cavities appears well established and frequent (~ 1/3 of the WTTS population). It is therefore important to inquire how such disks may form, if only to have a better understanding of PMS evolution with accretion disks, and, possibly, to establish a link with planet formation (see the discussions in, e.g., Montmerle and André 1989, Montmerle et al. 1991, Skrutskie et al. 1989).

The central idea is that accretion disks around single stars are intrinsically unstable, and some promising theoretical developments have recently appeared. Adams, Ruden, and Shu (1989) and Tagger et al. (1990) have shown that spatially thin disks (scaleheight << size) are subject to efficient dynamical instabilities ("swing amplification" phenomenon). Supplementing the initial work of Papaloizou and Pringle (1987) by including magnetic fields (here assumed to be locally perpendicular to the disk) in addition to gravitation, Tagger et al. (1990) have shown that, depending on the respective strength of the magnetic field and the local gravity, several instabilities are possible. (The magnetic field may be of stellar origin, or produced locally by some kind of dynamo resulting from differential rotation.)

If the magnetic field dominates gravity, which may be the case in the inner regions if it is stellar and increases more rapidly toward the star than the matter density (as does a dipolar field, for instance), these instabilities may result in the formation of a cavity in the vicinity of the star and thus explain the existence of hollow disks. Support for this situation may be found in the existence of generally large magnetic structures around the non-thermal radio sources, and of strong surface magnetic fields indicated by the widespread X-ray activity of WTTS. Farther out, other instabilities are possible : if the magnetic field is negligible and the self-gravity small, the hydrodynamical instabilities of Papaloizou and Pringle (1987) occur, and if self-gravity is large enough, the spiral instabilities suggested to be present in galactic disks take over (Pellat, Tagger, and Sygnet 1990). Work is in progress to determine the timescales over which such instabilities develop in circumstellar disks, but obviously such considerations leave ample room for many possible, time-dependent configurations (including spiral arms in the outer regions), and have important implications on the evolution of YSOs.

7. Conclusions

In contrast to the difficult measurements of stellar magnetic fields in the optical (by Zeeman line splitting, for instance), especially if the stars are as far away as T Tauri stars, the observations in X-rays and in the radio (cm) range have allowed unprecedented access to their strengths and structure. The basic reason lies in the extremely high luminosities found at these wavelengths (typically 10^3 - 10^6 times the solar values), which more than compensate for distance. The solar nature of the X-ray and radio emission seems to be well established by the existing data, and supports a dynamo origin for the magnetic field, although it is fair to say that there is a large body of evidence for *analogies* with the Sun, and still very few direct *proofs*. The recent VLBI measurements of several radio-emitting PMS stars constitute a very important first step towards directly establishing the existence of the magnetic fields, although no high-resolution image is yet available at radio wavelengths (let alone in X-rays) which could help us to "see" these magnetic fields. Also, the quantitative conclusions on field strengths, sizes, and configurations, while essentially confirmed by VLBI, are still largely model-dependent.

In spite of all these provisos, a fairly clear unified picture emerges. The X-ray evidence shows that large magnetic fields (size $\lesssim 2 R_*$) exist in a majority, if not all, of low-mass PMS stars, actually in many more than were known before the launch of the *Einstein* satellite. In a more limited number of cases, this size can be so large (up to $\approx 10 R_*$) that the gyro-synchrotron emission of a non-thermal population of electrons becomes visible in the radio (cm) range. In spite of their large sizes, however, these fields do not appear stronger than, say, in solar active regions. Fig. 8 puts together the plausible magnetic field configurations deduced from the X-rays and the radio.

The non-thermal radio stars make up a relatively small, but very interesting stellar population. All the present evidence indicates that they are very young objects (in particular their apparently systematic location in the vicinity of molecular cloud cores). Yet they do not possess the classical signposts of youth, i.e., the near-IR excess testifying of the presence of warm, dusty circumstellar material close to the star. Rather, as radio observations (this times in the mm range) have shown, at least some of them are indeed surrounded by circumstellar material, but cold and more than a few AU away from the star. This is also the case for ~ 1/3 of the so-called "weak-line" T Tauri stars (most of them X-ray discovered), whereas it was initially thought that such stars did not have any circumstellar material at all. It is tempting to suggest that the large-scale magnetic fields seen in X-rays, and sometimes in the radio (cm) range, play a role in "excavating" the accretion disk present in "classical" T Tauri stars, and in so doing turn them into "weak-line" T Tauri stars.

The ongoing theoretical work offers exciting possibilities in terms of disk stability analysis, and should help clarify the role that the resulting evolution of accretion disks, particularly taking into account magnetic fields, may play in the pre-main sequence evolution of the stars themselves.

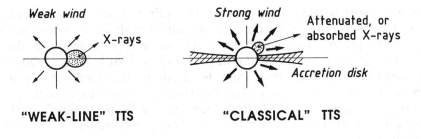
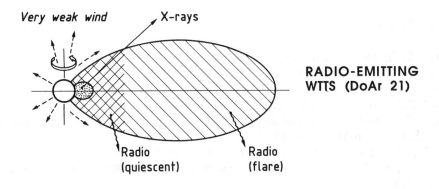
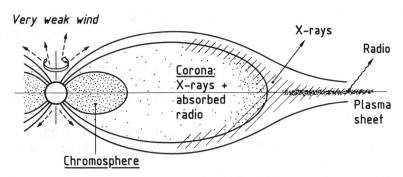

Fig. 8. *Compared magnetic structures (to scale) of young stars: the X-ray emitting "weak-line" and "classical" T Tauri stars, a late-type star (DoAr 21, see Fig. 3), and an early-type star (S1 in ρ Oph), both emitting in the radio (cm) range by the gyrosynchrotron mechanism. The magnetic structures of these two stars have been resolved by VLBI at a milli-arcsecond scale. (From Montmerle and André 1988.)*

Acknowledgements. I am grateful to Philippe André, Jérôme Bouvier, and Eric Feigelson, who recently contributed to reviews on which several sections of the present paper are based, and to Isabelle Joncour for discussions on the manuscript. It is also a pleasure to thank Nick Kylafis and Charlie Lada for their kind invitation, for a stimulating meeting in a warm and friendly atmosphere, and also for their patience in expecting this manuscript. I learned as much from "juniors" as from "seniors", and this is, after all, everything that a school is about, particularly in the Greek sense of the term.

REFERENCES

Adams, F.C, Emerson, J., and Fuller, G. 1990, *Astrophys. J.,* **357**, 606.
Adams, F.C., Lada, C.J., and Shu F.H. 1987, *Astrophys. J.* **312**, 788.
Adams, F.C., Ruden, S.P., and Shu, F.H. 1989, *Astrophys. J.* **347**, 959.
Agrawal, P.C. , Rao, A.R., and Riegler, G.R. 1986, *Mon. Not. Roy. Astr. Soc.* **219**, 777.
André, P. 1987, in *Protostars and Molecular Clouds,* eds. T. Montmerle and C. Bertout (C.E.A. : Gif-sur-Yvette), p. 143.
André, P., Loren, R.B., and Wootten, H.A. 1991, in preparation.
André, P., Martín-Pintado, J., Despois, D., and Montmerle, T. 1990a, *Astron. Astrophys.* **236**, 180.
André, P., Montmerle, T., Feigelson, E.D., and Steppe, H. 1990b, *Astron. Astrophys.,* **240**, 321.
André, P., Montmerle, T., and Feigelson, E.D. 1987, *Astron. J.* **93**, 1182.
André, P., Montmerle, T., Feigelson, E.D., Stine, P.C., and Klein, K.L. 1988, *Astrophys. J.* **335**, 940.
André, P., Phillips, R.B., Lestrade, J.F., and Klein, K.L. 1991, *Astrophys. J.,* in press.
Bai, T., and Sturrock, P.A. 1989, *Ann. Rev. Astr. Astrophys.* **27**, 421.
Basri, G., and Marcy, G. 1990, in *The Sun and Cool Stars : Activity, Magnetism, Dynamos,* Proc. of Helsinki Conf., preprint.
Beckwith, S.V.W., Sargent, A.I., Chini, R.S., and Güsten, R. 1990, *Astron. J.,* **99**, 924.
Beckwith, S., Sargent, A.I., Scoville, N.Z., Masson, C.R., Zuckerman, B., and Phillips, T.G. 1986, *Astrophys. J.* **309**, 755.
Bertout, C. 1989, *Ann. Rev. Astr. Astrophys.* **27**, 351.
Bertout, C., Basri, G., and Bouvier, J. 1988, *Astrophys. J.* **330**, 350.
Bieging, J. H. and Cohen, M. 1989, *Astron. J.* **98**, 1686.
Bieging, J.H., Cohen, M., and Schwartz, P.R. 1984, *Astrophys. J.* **282**, 699.
Borra, E.F., Landstreet, J.D., and Mestel L. 1982, *Ann. Rev. Astr. Astrophys.* **20**, 191.
Bouvier, J. 1990a, *Astron. J.* **99**, 946.
Bouvier, J. 1990b, in *Frontiers in Stellar Structure Theory,* Proc. Les Houches Meeting, *Lect. Notes in Phys.,* in press.
Bouvier, J., and Bertout, C. 1989, *Astron. Astrophys.* **211**, 99. (Erratum in *Astron. Astrophys.* **218**, 337.)

Bouvier, J., Bertout, C., and Bouchet, P. 1986, *Astron. Astrophys.* **158**, 149.
Brown, A. 1987, *Astrophys. J. (Letters)* **322**, L31.
Caillault, J.P., and Zoonematkermani, S. 1989, *Astrophys. J. (Letters)* **338**, L57.
Calvet, N., and Albarrán, J. 1984, *Rev. Mexicana Astron. Astrof.* **9**, 35.
Catala, C. 1989, in *Low Mass Star Formation and Pre-Main Sequence Objects,* Proc. ESO Workshop, ed. B. Reipurth (Garching : ESO), p. 471.
Churchwell, E., Felli, M., Wood, D.O.S., and Massi, M. 1987, *Astrophys. J.* **321**, 516.
Cohen, M., and Bieging, J.H. 1986, *Astron. J.* **92**, 1396.
Cohen, M., and Kuhi, L. V. 1979, *Astrophys. J. Suppl.* **41**, 743.
Curiel, S., Rodríguez, L.F., Canto´, J., and Torrelles, J.M. 1989, *Rev. Mexicana Astron. Astrof.* **17**, 137.
de Korte, P.A.J., *et al.* 1981, *Sp. Sci. Rev.* **30**, 495.
Dulk, G.A. 1985, *Ann. Rev. Astron. Astrophys.* **23**, 169.
Edwards, S., Cabrit, S., Strom, S.E., Heyer, I., Strom, K.M., and Anderson, E. 1987, *Astrophys. J.* **321**, 473.
Evans, N., Levreault, R., Beckwith, S.V.W., and Skrutskie, M. 1987, *Astrophys. J.* **320**, 364.
Feigelson, E. D. and DeCampli, W. M. 1981, *Astrophys. J. (Letters)* **243**, L89.
Feigelson, E. D., Giampapa, M. S. and Vrba, F. J. 1991, in *The Sun in Time,* eds. C. Sonett and M. Giampapa, (Tucson : U. Arizona Press), in press.
Feigelson, E.D., Jackson, J.M., Mathieu, R.D., Myers, P.C., and Walter, F.M. 1988, *Astron. J.* **94**, 1251.
Feigelson, E. D. and Kriss, G. A. 1989, *Astrophys. J.* **338**, 262.
Feigelson, E. D. and Montmerle, T. 1985, *Astrophys. J. (Letters)* **289**, L19.
Gaetz, T.J., and Salpeter, E.E. 1983, *Astrophys. J. Suppl.* **52**, 155.
Gahm, G.F. 1981, *Astrophys. J. (Letters)* **242**, L163.
Gahm, G. F. 1988, in *Formation and Evolution of Low Mass Stars,* eds. A. K. Dupree and M. T. Lago (Dordrecht : Kluwer), p. 295.
Gahm, G.F. 1990, in *Flare Stars in Star Clusters, Associations, and the Solar Vicinity,* IAU Symp. 137 (Dordrecht : Kluwer), p. 193.
Gahm, G.F., Fischerstrom , C., Liseau, R., and Lindroos, K.P. 1989, *Astron. Astrophys.* **211**, 115.
Garay, G., Moran, J.M., and Ried, M.J. 1987, *Astrophys. J.* **314**, 535.
Giacconi, R., *et al.,* 1981, in *Telescopes for the 1980s* (Palo Alto : Annual Reviews Inc.)
Gilman, D.A. 1983, in *Solar and Stellar Magnetic Fields : Origin and Coronal Effects,* ed. J.O. Stenflo (Dordrecht : Reidel), p. 247.
Hartmann, L., and Noyes R.W. 1987, *Ann. Rev. Astron. Astrophys.* **25**, 271.
Herbig, G. H. and Bell, K. R. 1988. Third Catalog of Emission-Line Stars of the Orion Population. *Lick Obs. Bull.* **1111**, pp. 1-90.
Herbig, G.H., and Soderblom, D.R. 1980, *Astrophys. J.* **242**, 628.
Kato, T. 1976, *Astrophys. J. Suppl.* **30**, 397.
Klein, K.L., and Chiuderi-Drago, F. 1987, *Astron. Astrophys.* **175**, 179.
Ku, W. H.-M., Righini-Cohen, G., and Simon, M. 1982, *Science,* **215**, 61.
Lada, C.J. 1987, in *Star Forming Regions,* IAU Symp. 115, ed. M. Peimbert and J. Jugaku (Dordrecht : Reidel), p. 1.

Lang, K.R. 1986, *Astrophysical Formulae* (Heidelberg : Springer).
Lawrence, S. L., Shaughnessey, P. J., Phillips, R. B., Lonsdale, C. J. and Feigelson, E. D. 1990, in preparation.
Leous, J. A., Feigelson, E. D., André, P. and Montmerle, T. 1991, *Astrophys. J.,* in press.
Lestrade, J.-F. 1988, in *The Impact of VLBI on Astrophysics and Geophysics,* IAU Symp. 129, eds. J. Moran and M. Reid (Dordrecht : Reidel), p. 265.
Montmerle, T. 1987, in *Solar and Stellar Physics,* Proc. 5th Eur. Solar Meeting, ed. E.H. Schröter and M. Schüssler (Berlin : Springer), *Lect. Notes in Phys.* **292**, 117.
Montmerle, T. and André, P. 1988, in *Formation and Evolution of Low-Mass Stars,* eds. A. Dupree and T. Lago (Dordrecht : Reidel), p. 225.
Montmerle, T., and André, P. 1989, in *Low Mass Star Formation and Pre-Main Sequence Objects,* Proc. ESO Workshop, ed. B. Reipurth (Garching : ESO), p. 407.
Montmerle, T., Feigelson, E.D., Bouvier, J., and André, P. 1991, in *Protostars and Planets III,* eds. E. Levy and M. Mathews (Tucson : U. Arizona Press), in press.
Montmerle, T., Koch-Miramond, L., Falgarone, E., and Grindlay, J. E. 1983, *Astrophys. J.* **269**, 182.
Montmerle, T., Koch-Miramond, L., Falgarone, E., and Grindlay, J. E. 1984, *Physica Scripta* **T7**, 59.
Moore, R., *et al.* 1980, in *Solar Flares,* ed. P.A. Sturrock (Boulder : Colorado Associated U. Press), p. 341.
Morgan, J.A., Snell, R.L., and Strom, K.M. 1990, *Astrophys. J.* **362**, 274.
Mullan, D. J. 1985, *Astrophys. J.* **295**, 628.
Mundt, R. 1984, *Astrophys. J.* **280**, 749.
O'Neal, D., Feigelson, E. D., Mathieu, R. D., and Myers, P. C. 1990, *Astron. J.* **100**, 1610.
Palla, F., and Stahler S.W. 1991, *Astrophys. J.,* in press.
Panagia, N., and Felli, M. 1975, *Astron. Astrophys.* **39**, 1.
Papaloizou, J.C.B., and Pringle, J.E. 1987, *Mon. Not. Roy. Astr. Soc.* **225**, 267.
Pellat, R., Tagger, M., and Sygnet, J.F. 1990, *Astron. Astrophys.* **231**, 347.
Reynolds, S.P. 1986, *Astrophys. J.* **304**, 713.
Rodríguez, L.F. 1988, in *Galactic and Extragalactic Star Formation,* ed. M. Pudritz and M. Fich, (Dordrecht : Kluwer), p. 97.
Rodríguez, L.F., Ho, P.T.P., Torrelles, J.M., Curiel, S., and Canto´, J. 1990, *Astrophys. J.* **352**, 645.
Rodríguez, L.F., Myers, P.C., Cruz-Gonzalez, I., and Terebey, S. 1989, *Astrophys. J.* **347**, 461.
Rutten, R.G.M., and Schrijver, C.J. 1987, *Astron. Astrophys.* **177**, 155.
Rybicki, G.B., and Lightman, A.P. 1979, *Radiation Processes in Astrophysics* (New York : Wiley Interscience).
Ryter, C., Cesarsky, C.J., and Audouze, J. 1975, *Astrophys. J.* **198**, 103.
Sargent, A.I., and Beckwith, S. 1987, *Astrophys. J.* **323**, 294.
Shu, F.H., Adams, F.C., and Lizano, S. 1987, *Ann. Rev. Astron. Astrophys.* **25**, 23.
Skrutskie, M.F., Dutkevitch, D., Strom, S.E., Edwards, S., Strom, K.M., and Shure, M.A. 1989, *Astron. J.* **99**, 1187.

Skumanich, A. 1972, *Astrophys. J.* **171**, 565.
Snell, R.L., and Bally, J.B. 1986, *Astrophys. J.* **303**, 683.
Soderblom, D.R., and Stauffer, J.R. 1990, in *Inside the Sun*, eds. G. Berthomieu and M. Cribier (Dordrecht : Kluwer), p. 403.
Stahler, S. W. 1988, *Astrophys. J.* **332**, 804.
Stahler, S.W., and Walter, F.M. 1991, in *Protostars and Planets III*, eds. E. Levy and M. Mathews (Tucson : U. Arizona Press), in press.
Stauffer, J.R. 1988, in *Formation and Evolution of Low-Mass Stars*, eds. A. Dupree and T. Lago (Dordrecht : Reidel), p. 331.
Stine, P.C., Feigelson, E.D., André, P., Montmerle, T. 1988, *Astron. J.* **96**, 1394.
Strom, K. M., Strom, S. E., Edwards, S., Cabrit, S., Skrutskie, M. F. 1989, *Astron. J.* **97**, 1451.
Strom, K.M., et al. 1990, *Astrophys. J.* **362**, 168.
Tagger, M., Henriksen, R.N., Sygnet, J.F., and Pellat, R. 1990, *Astrophys. J.* **353**, 654.
Tagliaferri, G., Giommi, P., Angelini, L., Osborne, J. P. and Pallavicini, R. 1988, *Astrophys. J. (Letters)* **331**, L113.
Trümper, J. 1984, *Physica Scripta* **T7**, 209.
Vrba, F.J., Herbst, W., and Booth, J.F. 1988, *Astron. J.* **96**, 1032.
Vrba, F.J., Rydgren, A.E., Chugainov, P.F., Shakovskaya, N.I., and Zak, D.S. 1986, *Astrophys. J.* **306**, 199.
Vrba, F.J., Rydgren, A.E., Chugainov, P.F., Shakovskaya, N.I., and Weaver, W.B. 1989, *Astron. J.* **97**, 483.
Walker, C.K., Lada, C.J., Young, E.T., and Margulis, M. 1988, *Astrophys. J.* **332**, 335.
Walker, M.F. 1987, *Pub. Astr. Soc. Pacific* **99**, 392.
Walter, F.M. 1988, *Pub. Astr. Soc. Pacific* **99**, 31.
Walter, F. M., Brown, A., Mathieu, R. D., Myers, P. C. and Vrba, F. J. 1988, *Astron. J.* **96**, 297.
Walter, F. M. and Kuhi, L. V. 1981, *Astrophys. J.* **250**, 254.
Walter, F. M. and Kuhi, L. V. 1984, *Astrophys. J.* **284**, 194.
Weaver, W.B. 1989, *Astron. J.* **97**, 483.
Weintraub, D.A., Sandell, G., and Duncan, W.D. 1989, *Astrophys. J. (Letters)* **340**, L69.
Wilking, B.A., Lada, C.J., Young, E.T. 1989, *Astrophys. J.* **340**, 823.
Worden, S.P., Schneeberger, T.J., Kuhn, J.R., and Africano, J.L. 1981, *Astrophys. J.* **244**, 520.
Yusef-Zadeh, F., Cornwell, T.J., Reipurth, B., and Roth, M. 1990, *Astrophys. J. (Letters)* **348**, L61.

Thierry Montmerle and Steve Ruden

POLARIZATION OF LIGHT AND MODELS OF THE CIRCUMSTELLAR ENVIRONMENT OF YOUNG STELLAR OBJECTS

P. BASTIEN
Département de physique and Observatoire du mont Mégantic
Université de Montréal
B. P. 6128, Succursale A
Montréal, Québec H3C 3J7
Canada

"Most of observational work in astronomy and astrophysics is concerned with only two aspects of radiation received from the observed objects: its *direction* and its *intensity*, considered as a function of wavelength and time. A third aspect of this radiation is its *polarization*, also considered as a function of wavelength and time. The importance of information revealed by studies of this third aspect of radiation from astronomical objects has recently begun to be appreciated." (K. Serkowski, 1974).

ABSTRACT. This paper shows that polarization is the key to understanding important information about Young Stellar Objects (YSOs) and their circumstellar disks. First, the properties of polarized of light are reviewed with particular concern about astronomical observations. Then two polarization mechanisms of interest for the study of star forming regions are discussed in some detail: dichroic extinction and scattering of light. Finally, these mechanisms are discussed in the context of models of YSOs. Parameters of disks around YSOs can now be deduced from linear polarization maps. The distribution of the circumstellar material around YSOs is discussed and the various theoretical density distributions which have been proposed are reviewed.

1. Polarization of Light

The above quotation still applies today. However, the situation has improved as more and more astronomers are taking advantage of polarimetry. In this paper, we will see many examples of information that can be deduced by making use of the polarization of light. This subject is not covered appropriately in most textbooks on optics. However, one of them treats polarization quite well and I recommend it for additional information: Hecht (1987). We start with some general properties of polarized light (1.1) and its representation (1.2).

1.1 WHAT POLARIZED LIGHT IS

Light is a transverse electromagnetic wave with perpendicular electric and magnetic field vectors. Usually, we can treat the electric field vector only, since this is essentially what we measure with our instruments.

One can add two orthogonal disturbances, both traveling in the same direction, with the same frequency:

$$\mathbf{E}_x(z, t) = \hat{\mathbf{i}} E_{0x} \cos(kz - \omega t)$$

$$\mathbf{E}_y(z, t) = \hat{\mathbf{j}} E_{0y} \cos(kz - \omega t + \varepsilon)$$

Depending on the ratio of the 2 amplitudes E_{0x} and E_{0y}, and the value of the phase difference ε, linear, circular and in general elliptical polarization states can be produced:

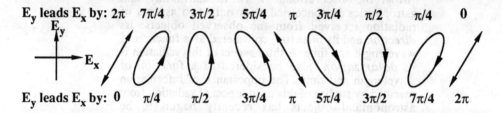

Figure 1. Addition of two linearly polarized beams with various phase differences between them to produce various forms of elliptically polarized light. The amplitude E_{0y} was taken to be twice that of E_{0x}.

Natural light is a superposition of a very large number of individually polarized short wavetrains of the same frequency at a very rapid rate to render the resultant polarization state indiscernible.

In general, light is made up of a part of natural (or unpolarized) light and a part of elliptically polarized light.

At the microscopic level, all individual **photons** have a spin, $+\hbar$ or $-\hbar$. Right or left circularly polarized light is made up of photons with $-\hbar$ or $+\hbar$ respectively.

Classically, linearly polarized light can be represented by the sum of 2 right and left circularly polarized beams. However, according to Quantum Mechanics, one can not say that a beam is made up of equal amounts of right- and left-handed photons. Rather, each individual photon exists in either state with equal probability. If the probability between the 2 states differs, then elliptically polarized light will result.

If, when measuring the intensity of light through a polarizer at different orientations, the intensity varies from I_{max} to I_{min}, then light is polarized linearly by an amount given by:

$$P = \frac{I_{max} - I_{min}}{I_{max} + I_{min}}$$

The total intensity of this beam is:

$$I = I_{max} + I_{min}.$$

The orientation of the polarizer corresponding to I_{max} is the position angle of the electric vector of polarized light, θ, measured with respect to a known, fixed, direction.

1.2 REPRESENTATION OF POLARIZED LIGHT

In Astronomy, polarization is usually measured in %. Magnitudes were used in the past but they are no longer used. Units of 0.001 (i.e., 0.1%) are sometimes used for solar system objects, but this practice is becoming rare. The convention is to measure position angles eastward from North.

To make progress, one needs a quantitative representation of polarized light. Four parameters are necessary and sufficient for describing all states of polarization and therefore give a complete description of (monochromatic) light. There are many different representations available (Chandrasekhar 1950), but one is much more useful and more widely used than the others: the **Stokes parameters**.

The Stokes parameters are: the total intensity I, which includes both natural and polarized light, Q and U for describing linearly polarized light, and V for the circular polarization. They satisfy this simple relation:

$$I \geq \left(Q^2 + U^2 + V^2\right)^{1/2}$$

where equality holds when light is 100% elliptically polarized. The Stokes parameters are additive, i.e., when combining two beams, the parameters of the resulting beam is simply the sum of the parameters of each individual beam. This property of the Stokes parameters explains why they are so useful for treating problems dealing with polarized light.

The transformation between (P, θ), and (Q, U) is:

$$P = \frac{\sqrt{Q^2 + U^2}}{I} \qquad Q = P \cos 2\theta$$

$$\theta = \tan^{-1}\left(\frac{U}{Q}\right) \qquad U = P \sin 2\theta$$

It is very useful to represent linearly polarized light in the so-called (Q, U) plane because the polarization vectors add vectorially in this plane (see Figure 2b).

The astronomical convention is to define U so as to be compatible with the convention of position angles in Astronomy (i.e., increasing counterclockwise). Therefore, Q positive is for polarization in the North-South direction (0°), Q is negative when the polarization is in the East-West direction (90°), while U is positive for a position angle of 45°, and when U is negative, the position angle is 135°.

But, in many physics books, (e.g., Chandrasekhar 1950, van de Hulst 1957), U is defined to give a position angle which increases clockwise, contrary to the astronomical convention. One must therefore make sure which convention is being followed in a given context. This is particularly important when using the matrices described below. Authors should always make clear which convention they follow.

On the other hand, the definition of V is ambiguous: the two different signs have been used in Astronomy. The predominant use (e.g., Serkowski 1962, Lang 1974) is to have right circularly polarized radiation, i.e., rotating clockwise when seen by an observer looking at the beam coming towards him (her), positive. Obviously then, left circular polarization is negative and counterclockwise when looking at the beam from the same vantage point. This is illustrated in Figure 3. An example of a paper which followed the opposite convention is that by Shafter and Jura (1980).

The various conventions and practices for both linear and circular polarization have been discussed in details by Gehrels (1974) and Clarke (1974). Recommendations for which ones to follow and for the nomenclature to use are also given there.

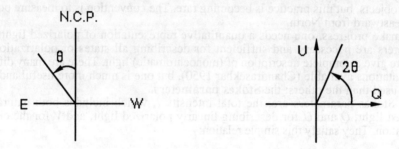

Figure 2. Representation of polarized light in the plane of the sky (a) and in the (Q, U)-plane. The length of the polarization vector is the same in both representations. The position angle θ is measured East from North in the plane of the sky and varies from 0° to 180°. In the (Q, U)-plane, the angle measured from the Q-axis is 2θ, and this quantity varies from 0° to 360° so that the whole plane is covered.

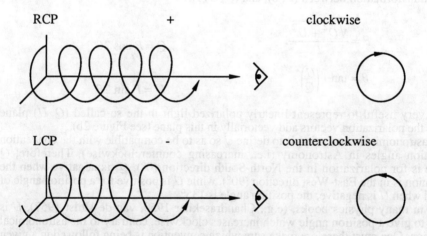

Figure 3. Representation of right and left circularly polarized light. If one imagines a plane perpendicular to the direction of propagation, light will cross it in a clockwise (counterclockwise) manner for RCP (LCP).

2. Polarization Mechanisms

There are many polarization mechanisms of interest in Astronomy. Lists have been given by, e. g., Serkowski (1974), and Angel (1974). Here, we will be concerned with only two of them. But even with two mechanisms, it is not obvious, and in fact there are confusions, as to which one is the appropriate one in different situations!

2.1 DICHROIC EXTINCTION

Measurements of linear polarization in the visible have been used to map the morphology of the magnetic field in the plane of the sky. Such maps have been obtained for the general interstellar medium (e. g. Serkowski, Mathewson, and Ford 1975), and also for dark molecular clouds. In this later case, stars background to the clouds are observed.

Interstellar (IS) linear polarization was discovered accidentally in 1949 by Hall and Hiltner. Light from stars close to the galactic plane tends in general to be polarized parallel to the galactic plane. The wavelength dependence of IS linear polarization is smooth and follows an empirical law which is sometimes called the Serkowski (1971, 1973) law:

$$P_{IS} = P_{max} \exp\left[-K \ln^2\left(\frac{\lambda_{max}}{\lambda}\right)\right],$$

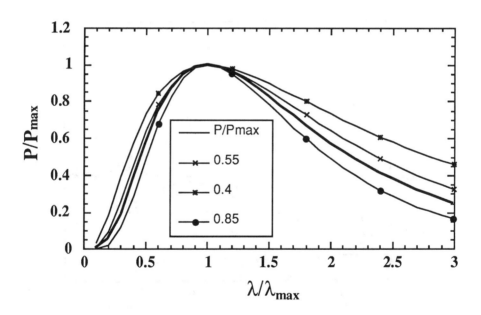

Figure 4. Wavelength dependence of interstellar polarization normalized to the maximum polarization and the wavelength at which this maximum polarization occurs. The heavy curve is for $K = 1.15$, and the others are for the expression below with the values of λ_{max} as indicated in microns.

where λ is the effective wavelength of the bandpass, λ_{max} is the wavelength of the maximum polarization, typically in the range 0.45 - 0.80 µm. The parameter λ_{max} is a characteristic of IS grains (Coyne, Gehrels, and Serkowski 1974). P_{max} is the maximum linear polarization and the constant $K = 1.15$ (Coyne, Gehrels, and Serkowski 1974, Serkowski, Mathewson, and Ford 1975). This wavelength dependence is illustrated in Figure 4. More recently, deviations from this empirical law have been found in the near infrared, leading to a different value of K (Wilking et al. 1980; Wilking, Lebofsky and Rieke 1982):

$$K = -0.10 + 1.86 \lambda_{max}$$

The main effect of the dependence of K on λ_{max} is to produce broader polarization curves for stars with smaller λ_{max}, as can be seen in Figure 4. The maximum level of IS polarization observed so far is $\approx 10\%$ (for VI Cyg No. 12), but values of $P_{max} > 5\%$ are rather rare.

The position angle is usually assumed to be independent of wavelength, but in fact rotations of the polarization vector with wavelength have been observed in $\approx 13\%$ of the stars for which the polarization is predominantly IS (Coyne 1974). These rotations of position angle have been interpreted as being due to changing grain alignment along the

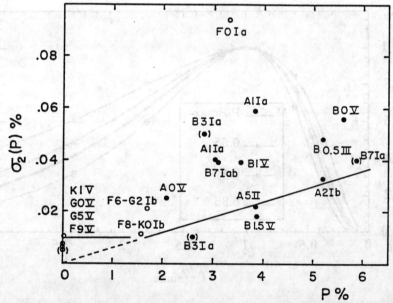

Figure 5. Observed scatter versus mean polarization for polarized stars with spectral types indicated. B-A stars ares shown by dark circles, F-G-K stars by light circles. The straight line envelope corresponds to $\sigma_2 (P) / P = 0.006$; it intersects the constant instrumental value $\sigma_2 (P) \approx 0.01\%$, below $P = 2\%$ (from Bastien et al. 1988).

line of sight to the star, as would be the case if many IS clouds were present in that line of sight.

Small variations in both P ($\approx 0.01 - 0.03\%$) and θ ($\approx 1° - 5°$) have been observed in many polarized standard stars (stars used for calibrating the origin of position angles) on time scales of the order of 5 to 10 days (e.g., Bastien et al. 1988, and references therein). In that paper, the possibility that part of these variations might be of interstellar origin was raised (see Figure 5). However, the data currently available do not allow any firm conclusion on this issue yet.

Linear polarization measurements of stars background to dark clouds give information about the structure of the magnetic field in the clouds. Measurements in the optical region of the spectrum are restricted to the periphery of the clouds because of the large optical depths through these dark clouds. However, measurements in the near-infrared allow us to penetrate deeper to the cores of the molecular clouds.

Many clouds have been mapped so far, yielding various results. In some cases the component of the magnetic field perpendicular to our line of sight is perpendicular to the cloud elongation, in other cases it is parallel, and there are other clouds where neither of these cases is found. These results are summarized in Table 1.

Table 1. Polarization observations of cloud regions.

		P (or $B_\perp$) and the cloud elongation are:			
Perpendicular		Parallel		Neither perpendicular nor parallel	
B216/217 (in Tau)	1	B42 (in Oph)	2	B18 (L1529, in Tau)	1
Lupus 1	3	R CrA cloud (Eastern part)	4	L1506 (in Tau)	5
L204	6	GF 7 (in Cyg):	7	Perseus[a]	5
		L1641 (in Ori)	8		
		L1755 (in Oph)	5		

Notes: [a]There are two different groups of polarization vectors; the group with the largest polarization makes an angle of 75° with respect to the general cloud elongation.
References: (1) Heyer et al. 1987; (2) Vrba, Strom and Strom 1976; (3) Strom, Strom and Edwards 1988; (4) Vrba, Coyne and Tapia 1981; (5) Goodman et al. 1990; (6) McCutcheon et al. 1986; (7) McDavid 1984; (8) Vrba, Strom and Strom 1988.

The dispersion in θ for entire cloud complexes is found to be smaller than the variations in position angle of the filamentary cloud elongations (see e. g., Figure 6). This has led Goodman et al. (1990) to suggest that the magnetic field does not dominate cloud structure on scales ≥ 1 pc. The dispersion in θ can be combined with Zeeman measurements in the radio (which give the magnitude of the longitudinal component of the field) to yield estimates of the magnitude of the magnetic field and its inclination to the line of sight (Myers and Goodman 1990).

The linear polarization observed in both the general IS medium and near the periphery of dark molecular clouds is due to aligned nonspherical grains. Light suffers its maximum extinction in the (projected) direction of the grains' longest axis; this is the dichroic property of the matter in the region containing nonspherical dust grains.

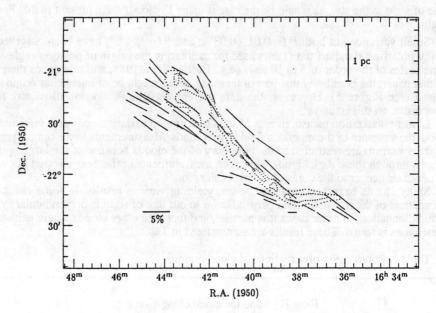

Figure 6. Polarization map of stars background to L1755 in Ophiuchus superposed on $^{13}CO(J = 1-0)$ contours of the cloud. Notice that the bend in the SW part of the cloud does not affect the alignment of the polarization vectors (from Goodman *et al.* 1990).

Grain alignment mechanisms are still not well understood. In general, grains tend to align themselves with their shortest axis parallel to the magnetic field (Purcell 1979). As a consequence of dichroic extinction, which is greatest along the longest axis of the grains, the polarization is parallel to the magnetic field projected on the plane of the sky. Various alignment mechanisms have been proposed: paramagnetic relaxation of thermally rotating grains (Davis and Greenstein 1951), or paramagnetic relaxation of grains in suprathermal rotation (Purcell 1979), or relaxation of superparamagnetic grains (Jones and Spitzer 1967; Mathis 1986). Other mechanisms are even less efficient than these. Hildebrand (1988) has reviewed recently grain alignment mechanisms.

2.2 SCATTERING OF LIGHT

Another mechanism which is very important in the environments of young stars is the scattering of light by particles. These particles can be electrons, molecules and grains. A photon may be scattered once, or many times, before escaping. In the latter case one is dealing with multiple scattering. Of course, the outcome of single scattering is easier to understand. Once the basic ideas are understood, one may say that single scattering may become "intuitive". However, this is far from being the case with multiple scattering which must be computed to find out what happens in a given situation.

2.2.1 *Single scattering.* The scattering process is best described when referred to the scattering plane, i.e., the plane which contains the incident (I_0) and scattered (I) beams.

The scattering angle, χ is the angle between the incident and the outgoing beams.

A light beam is usually polarized perpendicularly to the scattering plane. This is explained by Figure 8 - we only need to remember that light is a transverse electromagnetic wave.

There are some exceptions to this rule. For small dielectric grains, i.e. grains with a small imaginary part in their index of refraction, e.g. silicates, the polarization is parallel to the scattering plane. The constraints are fairly severe for this phenomenon to occur: the range in size is rather narrow, in addition to the constraint on the composition.

The intensity of scattered light is given by:

$$I = I_0 \frac{F(\chi, \phi)}{k^2 r^2}, \text{ where } k = \frac{2\pi}{\lambda}.$$

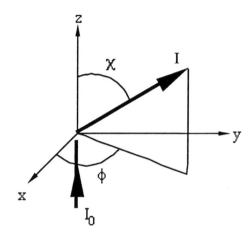

Figure 7. Representation of the scattering geometry.

The function $F(\chi, \phi)$ is called the **scattering function**. In many instances, one sees also the phase function:

$$\frac{F(\chi, \phi)}{k^2 C_{sca}} \equiv \text{phase function}.$$

A **scattering diagram** is a plot of $F(\chi, \phi)$ or I as a function of χ for a given value of ϕ. Many examples are given in Bohren and Huffman (1983). We give one in Figure 9.

The **scattering cross section** is defined as follows:

$$C_{sca} = \frac{1}{k^2} \int F(\chi, \phi) \, d\Omega.$$

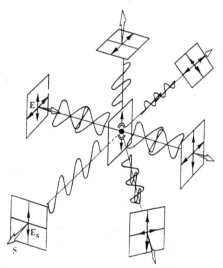

Figure 8. Polarization of light when scattered by a particle. The polarization is perpendicular to the scattering plane.

The absorption and extinction cross sections are defined similarly. In general, these cross sections are functions of the orientation of the particle, and the polarization state of incident light, in addition to depend on the optical properties, the size and shape of the particle.

By conservation of energy, this relation between the cross sections holds:

$$C_{ext} = C_{sca} + C_{abs}.$$

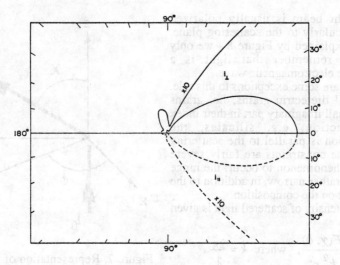

Figure 9. Example of a scattering diagram for a sphere with $x = ka = 3$, where a is the radius. The index of refraction is $m = 1.33 + i10^{-8}$. From Bohren and Huffman (1983).

The scattering process is described in all its details by the scattering matrix which is given by:

$$\begin{bmatrix} I \\ Q \\ U \\ V \end{bmatrix} = \begin{bmatrix} a_1 & b_1 & b_3 & b_5 \\ c_1 & a_2 & b_2 & b_4 \\ c_3 & c_2 & a_3 & b_6 \\ c_5 & c_4 & c_6 & a_4 \end{bmatrix} \begin{bmatrix} I_0 \\ Q_0 \\ U_0 \\ V_0 \end{bmatrix}$$

If the incident beam is made of natural (i.e., unpolarized) light, in general the result of the scattering process is polarized light, i.e., $Q, U, V \neq 0$. On the other hand, if the incident light is polarized (i.e., $Q_0, U_0, V_0 \neq 0$), there is in general a contribution to I coming from each one of Q_0, U_0, and V_0. A consequence of this fact is that the intensity map for a given object will be wrong if polarization is neglected!

All the physics involved in the process is hidden in the 16 elements of the scattering matrix. These elements are not independent as there are 9 relations between them (van de Hulst 1957). The elements depend on the optical properties of the material (index of refraction or dielectric constant), the shape, the size and the orientation of the scatterers. Clearly, it is impossible to know all this information for all the grains in the circumstellar environment of a YSO, and we want to examine possible ways to simplify the problem.

A possible simplification is to consider spherical particles of arbitrary size, for which the scattering matrix is much simpler:

$$\begin{bmatrix} a_1 & b_1 & 0 & 0 \\ b_1 & a_1 & 0 & 0 \\ 0 & 0 & a_3 & -b_6 \\ 0 & 0 & b_6 & a_3 \end{bmatrix}.$$

When the particles are small compared to the wavelength of light, another simplification can be used. This simplification is applicable to Thomson and Rayleigh scattering. If

$$|m|x = |m|ka \ll 1,$$

where m is the complex index of refraction, is satisfied, then the scattering matrix is

$$F = \frac{\sigma_0}{r^2} \begin{bmatrix} 1 + \cos^2\chi & 1 - \cos^2\chi & 0 & 0 \\ 1 - \cos^2\chi & 1 + \cos^2\chi & 0 & 0 \\ 0 & 0 & 2\cos\chi\cos\delta & -2\cos\chi\sin\delta \\ 0 & 0 & 2\cos\chi\sin\delta & 2\cos\chi\cos\delta \end{bmatrix},$$

where χ is the scattering angle, and δ and is the phase. For Thomson scattering (i.e., scattering by electrons), the cross section σ_0 is given by:

$$\sigma_0 = \frac{3}{16\pi} \sigma_T, \text{ and } \delta = 0,$$

where, σ_T is the usual Thomson cross section. A consequence of the phase being zero for electron scattering is that no circular polarization can be produced, as can seen by examining the matrix above. This fact can be used for identifying the type of particles responsible for the scattering in doubtful cases. For example, if in some objects both linear and circular polarization are observed, then particles other than electrons, most likely grains, must be present.

Mie developed in 1908 a theory for treating the scattering of light by spherical particles; it is now called the Mie scattering theory. This is the problem of the scattering of a plane wave by a homogeneous sphere of arbitrary size. To solve it, one must solve Maxwell's equations and apply the proper boundary conditions at the surface of the sphere. The solution is not simple, but it is analytical and can be handled easily with modern computers.

The solution of the problem yields the two complex amplitude functions $S_1(\chi)$ and $S_2(\chi)$. These are related to the van de Hulst intensities i_1 and i_2 by:

$$i_1 = |S_1(\chi)|^2, \quad i_2 = |S_2(\chi)|^2.$$

From these intensities, it is easy to compute the intensity and the linear polarization of the outgoing beam:

$$I = I_0 \frac{(i_1 + i_2)}{2k^2 r^2},$$

and

$$P = \frac{i_1 - i_2}{i_1 + i_2}.$$

The elements of the scattering matrix can be computed from the amplitude functions or the van de Hulst intensities. The details about this and other scattering matrices can be found in the books by Bohren and Huffman (1983) and by van de Hulst (1957). Mie scattering theory has been used extensively in Astronomy for computing the results of scattering by dust grains. The hypothesis that the grains are spherical is usually made.

However, in most situations of astrophysical interest, the dust grains are not spherical. For example, the Brownlee particles, interplanetary dust particles captured in the upper atmosphere, are far from spherical as can be seen in Figure 10. Similarly, the polarization of comets, of zodiacal light, and of IS dust all imply that dust grains are usually **not** spherical. Therefore, improvements on the Mie scattering theory is needed for treating scattering by arbitrary particles. A practical approach has been developed recently (Bastien and Levasseur-Regourd 1991). It is based on photometric and polarimetric observations of comets and on theoretical relations between the elements of the scattering matrix which are not all independent as mentioned above. The advantage of cometary grains in the present context is that these grains are probably the grains which resemble most closely the grains present in a protostellar nebula since cometary nuclei have been formed at the same time than the protosolar nebula and have remained essentially unaltered at large distances from the Sun for most of their lifetime.

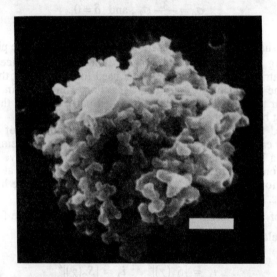

Figure 10. Photograph of a chondritic (i.e. of solar composition) interplanetary dust particle. The scale bar is 1 μm. From Brownlee (1978).

2.2.2 *Multiple scattering*. When the optical depth is large, the photons are scattered many times before they can escape. When $\tau \geq 0.1$, second and higher order scatterings are no longer negligible (see 2.2.3 below). A good criterion for determining if multiple scattering is important is the following. If, when the density of matter is doubled, the intensity of scattered light is also doubled, then single scattering only is appropriate. Otherwise, multiple scattering should be used.

Contrary to what is reported in many places in the astronomical literature, when light is multiply scattered, the polarization is decreased, but it **is not** zero (this is sometimes called depolarization). It can still be substantial (≈ 20%). The fact that significant linear polarization was observed in some sources was used, wrongly, by some authors to rule out multiple scattering simply because they claimed that the observed polarization was too high. This is wrong and the case of multiple scattering in a circumstellar disk (see below in 3.5) is a good case for demonstrating this fact.

Another consequence of multiple scattering is that it may lead to a case where "what you see is not what you think you see" (or when WYSIWYG goes wrong). In the problem of multiple scattering in a circumstellar disk, the photocenter is displaced differently according to the wavelength of observation, as a result of multiple scattering. This is a result of the optical depth being greater at shorter wavelengths, leading to greater displacements of the photocenter from the "real position" for shorter wavelengths.

Some nice examples of this effect have been given recently: GL 490 (Campbell, Persson, and McGregor 1986), L1551 IRS5 (Campbell *et al.* 1988), and NGC 7538 (Campbell and Persson 1988). The case of L1551 IRS 5 is shown in Figure 11. The offset of the photocenter in various bandpasses in right ascension and declination is displayed. The radio position is the most likely position of the star. This phenomenon is probably occurring also in R Mon because of the very large linear polarization which is measured right at the photocenter. In that case, what has previously been thought to be the star is probably just a reflection of light from the star by the thick circumstellar disk. HL Tau is another similar case. All optically thick disks seen edge-on or close to edge-on are expected to present a similar effect.

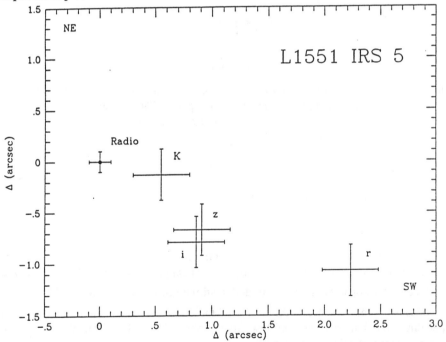

Figure 11. Position offsets of the photocenter in different bandpasses for the source L1551 IRS 5. The "true" position of the star is given very closely by the radio position. Note that shorter wavelengths are further away from this position, as expected. From Campbell *et al.* (1988).

Another consequence of multiple scattering is circular polarization. When linearly polarized light is scattered by grains (either spherical or of arbitrary shape), circular polarization is produced. This can be seen by examining the scattering matrix for spherical particles above. Notice that for grains, even spherical, the phase δ which appears in the matrix is not zero in general.

Circular polarization has been detected recently in T Tauri stars (TTS) (Nadeau and Bastien 1986, Bastien, Robert and Nadeau 1989) and in young Ae/Be stars (Ménard, Bastien and Robert 1988, Gravel and Bastien 1991). The interpretation is in terms of multiple scattering by grains in a disk geometry. Dichroic extinction by aligned grains in the circumstellar disk can be ruled out by these observations of circular polarization. In order to get a large linear polarization, the grains would need to be aligned all in the same direction, but on the other hand, the maximum circular polarization is obtained when there is a rotation of 45° in the alignment of the grains, in which case the linear polarization is very severely reduced. See Ménard, Bastien and Robert (1988) for observations of R Mon and for a full discussion of this problem.

2.2.3 *How Is the Intensity Affected?* To get a better understanding of how the intensity is affected in the presence of multiple scattering, it is necessary to consider some equations and to distinguish between absorption and extinction. The absorption optical depth is defined as:

$$\tau_{abs} = \int N C_{abs} \, ds ,$$

where

$$C_{abs} = Q_{abs} A .$$

The geometrical cross sectional area is A, C_{abs} is the absorption cross section as seen above and Q_{abs} is the absorption efficiency of the grains. The scattering optical depth, and the emission and scattering cross sections are defined similarly.

The intensity observed is

$$I = I_0 \, e^{-(\tau_{abs} + \tau_{sca})} + I_{sca} .$$

However, the effective opacity is given by:

$$I = I_0 \, e^{-\tau_{eff}} ,$$

where I_0 is the intensity emitted by the star. The effective opacity is what can be determined by an observer, by comparing what is expected, I_0, with what is observed, I. But the effective opacity is clearly an underestimate of the total opacity which is $\tau_{abs} + \tau_{sca}$, because of the presence of the additional radiation I_{sca}, which was emitted initially in other directions but was scattered towards the observer. In the IS medium, I_{sca} is zero, but not in a circumstellar environment.

In calculations with a Monte Carlo code which treats multiple scattering properly, Lefèvre & Daniel (1988) found for a spherical shell model with silicate grains of radii $a = 0.1$ μm at $\lambda = 5000$Å, $\tau_{ext} = 1.1$ which corresponds to $\tau_{eff} = 0.15$ only! The statistics of this specific model are as follows: direct photons received from the star 33%, photons absorbed in the shell 14%, scattered photons 53%. Of the scattered photons, the number of scatterings is as follows: scattered once 45%, scattered twice 28%, and scattered 3 times 14%. The main conclusion then is that singly scattered photons are not predominant even with τ_{eff} as low as 0.15. Observers would say in this case that $A_V = 0.16$. [Recall: $A_V = 2.5 \, (\log_{10} e) \tau_{eff, V}$]. Therefore, this means that the use of single scattering only is appropriate only in the less extincted cases.

How should one use A_V properly then? Using A_V is fine for the IS medium, but **not** in the circumstellar environment, because there is light leaving the star in other directions (than toward the observer) and which is scattered towards the observer. There are at least two important consequences of this fact. Because of the additional light scattered toward the observer, some stars may have a negative A_V, i.e. we receive more light, because of the presence of the circumstellar material, than we would receive if there was no circumstellar matter. There are about 20 such stars in the catalog by Cohen and Kuhi (1979); they have been given $A_V = 0$.

Another consequence of misusing A_V is that in many star+disk models (e. g., Adams, Lada, and Shu 1987; Bertout, Basri, and Bouvier 1988; Kenyon and Hartmann 1987), the luminosity of the disk is an important fraction of the total luminosity. However in many of these models, the parameters derived depend critically on the adopted value of A_V. This value, as demonstrated here, has in fact little physical meaning.

Effects of circumstellar material on derived extinction curves have been discussed for some "classical" Be stars from the point of view of a photometrist by Krelowski, Papaj, and Wegner (1990). This is again the same phenomenon as discussed here, but presented differently.

To summarize the points given above, the proper interpretation of A_V values is:

> A_V as determined from observations of stars with circumstellar material should be used only as a lower limit to the true A_V.

3. Models of the Circumstellar Environment

We first review the observations (3.1) which provide information about the circumstellar environment, examine possible density distributions (3.2), and consider the models with single (3.3 and 3.4) and multiple (3.5) scattering.

Scattering by dust particles was identified as the mechanism responsible for linear polarization in T Tauri stars (Bastien and Landstreet 1979), and this is also true for the more massive YSOs. There are not enough electrons around T Tauri stars to produce significant linear polarization, i.e., > 0.1 %. Therefore we will consider only dust in the following.

3.1 OBSERVATIONS RELATING TO THE CIRCUMSTELLAR ENVIRONMENT

We will be concerned here mostly with polarization data, as other types of data related to the circumstellar environment of YSOs are treated by other authors in this book. It is convenient to divide the data into two parts: spatially unresolved (3.1.1) and spatially resolved (3.1.2).

3.1.1 *Observations in the visible and near-infrared through a diaphragm.* For a detailed review of these observations, see Bastien (1988a). The main properties are summarized here, and an update is given.

Polarization variability is a very common phenomenon in YSOs: ≥ 85% in TTS (Bastien 1982, 1985, 1988a; Drissen, Bastien, and St.-Louis 1989; Ménard and Bastien 1990), and ≥ 73% in HES (Vrba, Schmidt, and Hintzen 1979; Bastien 1988a)

$P(\lambda)$ curves have been observed for more than 50 TTS and young Ae/Be stars. Clearly, there is no standard wavelength dependence for the polarization. In some stars, P increases toward the ultraviolet and in others it increases toward the red. In general, one has: $P = P(\lambda, t)$, and $\theta_*(\lambda, t)$.

The linear polarization can be quite large: the TTS HL Tau has 12% (Vrba, Strom, and Strom 1976, Bastien 1982), while V376 Cas, a young Ae/Be star, has 21% (Bastien *et al.* 1989; Asselin, Bastien, and Ménard 1991). Such large values of linear polarization can not be explained by models with single scattering only.

There is a good correlation between polarization in the visual and the infrared color excess (Bastien 1982, 1985). This is not surprising since dust is responsible for the infrared excess (Cohen and Kuhi 1979) and for the linear polarization (Bastien and Landstreet 1979).

The linear polarization vectors are found to be generally perpendicular to the bipolar outflow (CO and/or optical). This correlation was studied for many outflow sources (Bastien 1987). The most recent evaluation (Appenzeller and Mundt 1990) gives 75% of 22 outflow sources with an outflow perpendicular to the linear polarization to within 30°. This property is a natural consequence of the model discussed below (3.5).

More recent results on the polarization of YSOs are the following (see Bastien and Ménard 1990b for a brief account).

The polarization distribution for a sample of classical TTS (CTTS) is significantly different than that of the weak-line TTS (WTTS) (Ménard, Bertout, and Bastien 1991). From a polarization survey for WTTS and CTTS, the polarization of CTTS is quite significant (average ≈ 2 %), whereas the WTTS show a strong peak at zero polarization (see the histogram in Bastien and Ménard 1990b). This implies that in most (but not all) WTTS there is very little circumstellar material close to the central star. This agrees with the fact that WTTS have weaker infrared excesses (Strom *et al.* 1989).

The polarization surveys of YSOs carried to the 13th magnitude previously (Bastien 1982, Bastien 1985) for both hemispheres are currently being extended to fainter magnitudes (Ménard and Bastien 1990, and unpublished data). This has allowed the discovery of more stars with a large polarization (≥ 4%). Direct images of these more highly polarized sources have been obtained and they are in almost all cases associated with extended nebulosity and in some cases with jets and/or outflows.

A polarization survey in the K band (2.2 µm) has been published for TTS in the Taurus-Auriga region by Tamura and Sato (1989). They found a good correlation between the optical and the infrared polarization. The position angles coincide between the visual and the infrared for all but only three stars. Similar conclusions are therefore derived from the infrared and the visual polarization.

3.1.2 *Maps: Intensity, color and polarization maps.*

With the advent of sensitive two-dimensional detectors, it has been possible to obtain spatial information about the distribution of polarization in the nebulosities associated with YSOs. This is in addition to the numerous direct images of these sources which have been obtained, in many cases in relation with the search for optical jets. As will be clear in the model discussion, all the intensity maps, color maps, and polarization maps are very useful in constraining the models. A polarization map of HL Tau is presented in Figure 12. A nice polarization map of the Mon R2 region in the K band was presented at this conference by Aspin and Walther (1990).

Here are the basic characteristics of observed linear polarization maps:

(1) An extended centrosymmetric pattern, reaching up to 70%. By taking perpendiculars to the polarization vectors, one finds the position of the central source, a consequence of the fact that polarization is perpendicular to the scattering plane (cf. 2.2.1). Alterations to the pattern near the edges of the bipolar region have been observed in some sources. Good examples of such alterations are R Mon (Scarrott, Draper, and Warren-Smith 1989), and PV Cep (Gledhill, Warren-Smith, and Scarrott 1987).

(2) A pattern of aligned vectors close to the central source. The vectors can be either well aligned or, as in many cases, form an elliptical rather than centrosymmetric pattern. In this region of the map, the polarization is typically 10 to 15%, and up to 30% in one case (V645 Cyg; Lenzen 1987).

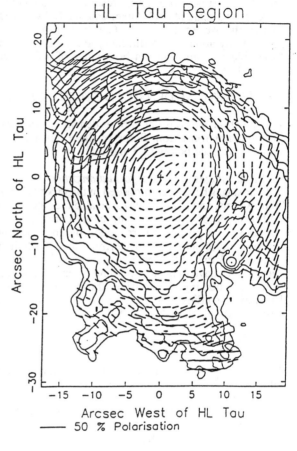

Figure 12. Linear polarization map of HL Tau. The deformation of the circular pattern near the central position marked by a cross is clearly noticeable. From Gledhill and Scarrott (1989).

(3) Polarization null points. Two points, usually on either side of the central source, where the polarization goes to zero. They are located between the region with aligned vectors and the centrosymmetric pattern, going out in the projected plane of the disk.

A compilation of polarization maps for 56 galactic sources, all but a few being YSOs, has been made by Bastien and Ménard (1990a). In this list, 11 (20%) have a centrosymmetric pattern only, 34 (60%) have a pattern of aligned vectors, and 11 (20%) are peculiar. In some of the centrosymmetric maps where the spatial resolution used is rather low, it is possible that a pattern of aligned vectors may be found with a higher spatial resolution. We caution that many of these maps merely show that there is interesting structure (i.e., a disk) to be studied, but the spatial resolution is not adequate for a detailed comparison with model calculations. For these studies, it is always best to get the highest possible spatial resolution. As an example, images obtained at the CFH telescope at a scale of 0.11"/pixel with the RCA2 CCD at the Cassegrain focus allowed to resolve the Ae/Be star V376 Cas (which has $P \approx 21\%$) into two individual peaks separated by about 1". Preliminary analyses suggests that we are seeing light from the same object coming out on both sides of a disk seen edge-on (Asselin, Bastien and Ménard 1991).

3.2 THEORETICAL DENSITY DISTRIBUTIONS IN DISKS AND ENVELOPES

In this section, we consider various density distributions of the material at different stages during stellar formation. Some of these density distributions will be required for the models discussed below (3.3 and 3.5).

3.2.1 *Main stages in the star formation process.*

Four stages have been proposed in the process leading from a molecular cloud to a star (Shu, Adams, and Lizano 1987; see also the figure given by Shu in this book). These are: (1) formation of slowly rotating cloud cores, (2) "inside-out" dynamical collapse and formation of a protostar surrounded by an accretion disk, (3) mass loss by a bipolar outflow and/or jets, and (4) termination of infall, revealing the central star and its circumstellar disk. Most objects that we will be concerned with are in stages (3) and (4). These stages have to be kept in mind when considering the possible density distributions.

In the following, we consider the case of spherical symmetry (3.2.2) and the disks. Theoretical studies usually distinguish between geometrically thin disks (3.2.3) and geometrically thick disks (3.2.4).

3.2.2 *Spherical symmetry.*

Results of spherically symmetric collapses have been obtained from asymptotic similarity solutions (Larson 1969, Penston 1979, Shu 1977, Hunter 1977, Tohline 1982). They have been confirmed by numerous hydrodynamic calculations. It is possible to distinguish at least three different stages in the evolution. For each of them, we give the density and velocity distributions.

1. The collapse of a uniform density cloud to form a centrally condensed object

$$\rho(r,t) = C_0 |t|^{-2}, \qquad v(r,t) = -C_1 |t|^{-1} r^1.$$

2. The formation of a core at the center. The density and velocity in the envelope are given by:

$$\rho(r,t) = C_2 c_s^2 r^{-2}, \qquad v(r,t) = -C_3 c_s.$$

3. The accretion flow of the isothermal envelope on the new stellar core is described by:

$$\rho(r,t) = C_4 c_s^{3/2} t^{-3/2} r^{-3/2}, \qquad v(r,t) = -C_5 c_s^{-1/2} t^{1/2} r^{-1/2}.$$

In the above, C_0-C_5 are constants and r, t, ρ, v and c_s are the radial distance, the time, the density, the radial velocity, and the sound speed respectively. In summary, accretion flows are characterized by a $r^{-3/2}$ density distribution, and pure collapse flows by r^{-2}.

3.2.3. Thin disks.

The theory of thin accretion disks has been developed over many years in contexts other than star formation. Reviews can be found in Frank, King, and Raine (1985), King (1989). See also the proceedings of conferences on accretion disks by Meyer et al. (1989) and Bertout et al. (1991).

The assumptions which are made are:

(1) Azimuthal symmetry. A direct consequence of this assumption is that no gravitational torques are possible for transporting angular momentum.

(2) The disk is geometrically thin, i.e. the scale height $H \ll r$, or $v_\phi \gg c_s$, where c_s is the sound speed and v_ϕ the azimuthal velocity component. As a result, pressure forces are neglected and the vertical component of the velocity, v_z, satisfies: $v_z \ll v_r$

(3) The optical depth perpendicularly to the disk is very large, i.e.,

$$\tau = \rho H \kappa_R(\rho, T_c, \Sigma, \alpha, ...) = \Sigma \kappa_R \gg 1,$$

where κ_R is the Rosseland opacity, and Σ is the surface density.

When time derivatives can be neglected, a stationary solution is obtained:

$$\nu \Sigma = \frac{\dot{M}}{3\pi} \left[1 - \left(\frac{R_*}{r}\right)^{1/2} \right],$$

where

$$\nu = \nu(\rho, T_c, \Sigma, \alpha, ...),$$

and

$$\dot{M} = 2\pi r \Sigma(-v_r)$$

are the viscosity coefficient and the accretion rate respectively.

The vertical density distribution obeys hydrostatic equilibrium, the disk mass being negligible:

$$\rho(r, z) = \rho_c(r) \exp\left(-\frac{z^2}{2H^2}\right), \quad \rho_c = \frac{\Sigma}{H}.$$

A better treatment of the vertical distribution has been given recently by Hubeny (1990).

To make further progress with the stationary solution, a prescription for handling the viscosity is needed. One of the most widely used formulation, the α prescription, which is due to Shakura et Sunyaev (1973),

$$\nu = \alpha c_s H,$$

gives

$$\Sigma \propto r^{-3/4} f^{14/5},$$

and

$$H \propto r^{9/8} f^{3/5},$$

where

$$\rho = \frac{\Sigma}{H} \propto r^{-15/8} f^{11/5},$$

and

$$f = 1 - \left(\frac{R_*}{r}\right)^{1/2}.$$

When $r \gg R_*$, then $f \to 1$. Hence at large distances from the star, the surface density decreases as $r^{-3/4}$. We also notice that the disk scale height increases as $r^{9/8}$ at large distances. An increase of the disk thickness with radius is the property of flaring disks which have been used as a means of increasing the total disk luminosity and producing flatter infrared spectra in models of reprocessing disks. However, the scatterings going on in these models should also be included.

3.2.4. *Thick Disks.* Thick, differentially rotating, disks with constant specific angular momentum have been shown to be strongly unstable to non-axisymmetric modes (Papaloizou and Pringle 1984). See Narayan and Goodman (1989), and Narayan (1991) for recent reviews. It is not clear at the present time which types of thick disks may be of astrophysical relevance. This problem requires 3-D hydrodynamic calculations for a solution. While numerical calculations should give a more accurate representation of the density distribution and velocity field in the circumstellar environment, it would be useful in most cases to have a fit to the results with simple analytic formulae for further study in radiative transfer models.

3.3 SINGLE SCATTERING MODELS

Models with single scattering should be restricted to small optical depths (at least $\tau < 0.1$), as discussed above in Section 2.2. This makes their applicability to YSOs rather limited.

Because a net linear polarization is observed from YSOs, the density distribution can not be spherically symmetric. The asymmetry needed to produce the polarization must come from the distribution of the scatterers. Various distributions have been used, from flattened ellipsoids to prolate clouds.

Many dust scattering models in circumstellar shells have been produced to compute the wavelength dependence of the polarization, $P(\lambda)$, for a variety of stars: Zellner (1971) for novae, Shawl(1975) for red giant stars, Bastien (1981) and Ménard and Bastien (1987) for T Tauri stars. In all these models, the central star is assumed to be a point source. As an example, the polarization of T Tau on JD 2443078 (see Fig. 2 in Bastien and Landstreet 1979) can be fit by an ellipsoidal envelope with iron grains with a diameter of 0.3-0.4 μm or with graphite grains of 0.3 to >0.6 μm diameter. For RY Tau on JD 2443817, grains with a diameter > 0.3 μm are required while on JD 2441739 for RY Tau no good fit could be obtained.

A general result of these calculations is that the wavelength dependence of the polarization is relatively independent of the distribution of the scatterers as long as there are as many scatterers "in front" as "behind" the plane of the sky (Shawl 1975).

A model was proposed by Elsässer and Staude (1978) to account for the large polarization of the Becklin-Neugebauer object in Orion. The geometry in this model is that of a bipolar nebula seen edge-on. The light leaves from the poles and is scattered by the material on either side of the disk. The disk is used only to block direct light from the star in this model.

Using the bipolar geometry suggested by Elsässer and Staude (1978), additional models were computed by Heckert and Zeilick (1985) to reproduce infrared polarization observations of YSOs. A small optically thick torus was used to block the direct star light to the observer, but the star could illuminate the bipolar lobes where single scattering was assumed to take place. They showed that their data are generally consistent with models of scattering off large dust grains.

However, there are many problems with these models. First, they are not unique since many models may give the same result. Also, polarization values ≥ 2 % can not be obtained with single scattering only, unless obscuration in front of the star is used. This implies that multiple scattering must be occurring in that region. Finally, multiple scattering is required for at least those cases where circular polarization has been detected, and probably for most sources as indicated by, e. g., tables of A_V values.

3.4 POLARIZATION AND BINARY STARS

Many models with single scattering have been developed for the circumstellar environment of binary stars. The orbital motion in these systems allows us to see the distribution of scatterers from different points of view, which provides us with very important information. The polarization of binary stars describes a double loop in the (Q, U) plane for one orbital period. These temporal variations of the linear polarization of binary stars have been used to find the inclination of the orbital plane with respect to the line of sight, the orientation of the line of nodes, and moments of the density distribution of the scatterers. Knowledge of the inclination, combined with appropriate spectroscopic observations can yield the mass of the stellar components. This has been done so far only for stars with electron scattering, most of them Wolf-Rayet stars. The basic method has been developed by Brown, McLean, and Emslie (1978), with some additions, corrections and clarifications given by Drissen et al. 1986) and by Bastien (1988b). Simmons (1983) extended the method to scattering mechanisms other than Thomson scattering, as long as the scatterers are spherical; therefore, the method should be applicable to Mie scattering, which is more appropriate to YSOs than Thomson scattering.

There are very few young spectroscopic binaries known so far (Zinnecker 1989). The polarization of V826 Tau (= FK1, P2) has been found (Ménard and Bastien 1990) to show very small polarization variations, if at all, ($\leq 0.1\%$), so that this method would be very difficult to apply. The precision of the inclination angle derived by this method depends, among other things, on the amplitude of the observed polarization curve. The star RY Lup (Bastien et al. unpublished observations) has shown large amplitude polarization (> 1%) and photometric (> 1 mag) variations which can be interpreted by a binary model.

3.5 MODELS WITH MULTIPLE SCATTERING

Previous models in which multiple scattering has been taken into account are those by Sandford (1973), Sandford and Pauls (1973), and Daniel (1978). However, these models are not spatially resolved: a value for the polarization integrated over the star plus the whole nebula is derived. These models have been used to compute the wavelength dependence of linear polarization and also the so-called polarization reversal which is a rotation of the position angle by 90° as a function of wavelength. In order to explain the polarization maps, models with multiple scattering and the spatial resolution are needed. More recently, Lefèvre and Daniel (1988) produced spatially resolved calculations (see 2.2.3), but these had a spherical shell around the star.

Until 1988, the only explanation proposed for the patterns of aligned vectors observed in polarization maps was that of dichroic extinction by aligned nonspherical grains in a circumstellar disk. Various alignment mechanisms are possible, but alignment by a toroidal magnetic field is usually preferred. No model calculations have been performed to confirm this explanation; only qualitative arguments comparing with interstellar polarization have been invoked.

A model with two scatterings only has been presented by Bastien and Ménard (1988) to show that a pattern of aligned vectors can be obtained without having to resort to aligned grains, i.e., dichroic extinction is not necessary to get a pattern of aligned vectors. A comparison of this model with an observed polarization map is shown in Figure 13.

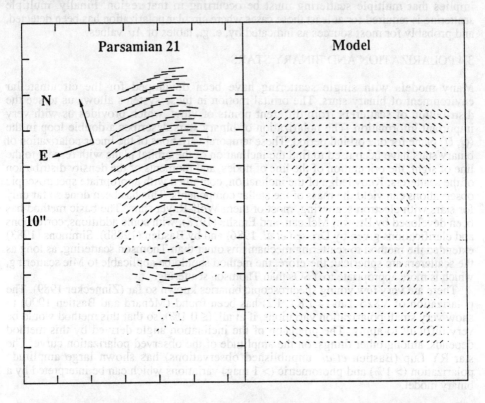

Figure 13. Comparison between the observed polarization map of Parsamyan 21 and results from the two-scattering model for a disk seen edge-on. (From Bastien and Ménard 1990a).

A detailed discussion of the question of the mechanism responsible for producing the polarization maps has been given by Bastien and Ménard (1990a). Arguments in favor of multiple scattering and against aligned grains are given. The main arguments against aligned grains is that the polarization is too high to be explained this way since the alignment efficiency is not sufficiently high. The wavelength dependence of polarization observed in YSOs can not be explained by aligned grains. The circular polarization observed in the region of aligned polarization vectors near R Mon can not be explained with aligned grains. On the other hand, multiple scattering can explain all the observed polarization properties of these stars.

The two-scattering model has been used to derive inclination angles for 29 disks associated with sources which have a pattern of aligned vectors (Bastien and Ménard 1990a). The other objects either have too poor a spatial resolution to allow such a determination of the inclination, or they have a centrosymmetric pattern which is obtained for all inclinations ≤ 45°. For two objects only, L1551 IRS 5 and HL Tau, had the inclination been estimated by other methods. Adams, Lada, and Shu (1987) determined an inclination 70° for L1551 IRS 5 from the radial and tangential motions of the associated Herbig-Haro objects. The values from the two methods are compatible to within the errors. Monin *et al.* (1989) found an inclination for HL Tau from an infrared color map. However, they shifted the images so that the peak in the isophotes in the two filters coincide. As demonstrated above, the photocenters should be displaced in most cases, so that this method gives only a lower limit to the inclination. The histogram of all inclination angles determined by Bastien and Ménard (1990a) with their method agrees with the expected theoretical sin i distribution.

The size of the region where aligned vectors are observed can also be determined from the observed maps. In this simplistic model, this size would correspond to the size of the disk where the optical depth is ≥ 1. It remains to be verified if this is still true in the more rigorous model. In particular, one should make sure that a particular density distribution outside the disk region does not keep a pattern of aligned vectors because of the polarized photons it receives.

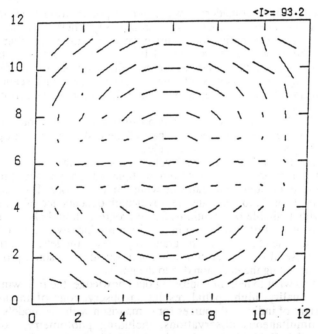

Figure 14. Polarization map computed with the Monte Carlo code. The region shown corresponds to 10 AU on the side. Compare this with Figure 13 which shows an observed map and the two-scattering model. From Bastien and Ménard (1990b).

The results from the two-scattering model have now been confirmed by rigorous calculations taking multiple scattering properly into account (Ménard 1989, Ménard 1991,

Ménard and Bastien 1991). The density distributions which have been calculated are those of a disk with a surface density varying as r^{-n}; various values of n from 2 to 0.1 have been tested. In the vertical direction, the density decreases exponentially. Two different compositions have been computed: silicate grains and graphite grains of radii 0.2 µm. In Figure 14, a polarization map for an average inclination angle of 93.2°, i.e. almost edge-on, is given. Graphite grains with a r^{-1} density law were used. The optical depth in the plane of the disk is 36.

The pattern of aligned vectors is easily obtained when the optical depth in the disk is sufficiently large. If the optical depth is too small, then a centrosymmetric pattern is obtained, even when looking at the disk edge-on.

This Monte Carlo model allows the computation of polarization, intensity and color maps for a given density distribution. Therefore, comparison with high spatial resolution observations of all this information for a particular object will put stringent constraints on the numerous parameters involved, and yield information about the geometry of the circumstellar environment.

4. Conclusions and Prospects

In this review, we have considered two different polarization mechanisms and also different astrophysical environments. We can now say where each mechanism can be found "at work". Dichroic extinction occurs in general in the interstellar medium and in dark clouds, where the grains are aligned by a magnetic field. On the other hand, scattering is the dominant process in the circumstellar environment of YSOs. There, one finds that both single and multiple scattering are occurring. Multiple scattering is essential in order to give an accurate representation of intensity, color and polarization maps of the disks around YSOs.

From the models currently available, it is possible to get the inclination of the disks, an idea about the size of the optically thick part of the disks, and also in general a better understanding of geometry for individual objects.

One can expect for the near future developments along the following lines. Detailed polarization, intensity and color maps from multiple scattering models; these will be at a high spatial resolution, and with grains obeying the Mie theory. The multiple scattering models will also be used to compute the wavelength dependence of the polarization for comparison with available observations. Some progress should also be made at getting information about density distribution in the circumstellar environment. We should also see the start of the use of more realistic grain properties in models; the Mie theory should be viewed as a useful first order approximation. Eventually, some progress should be made towards including the emission due to the grains.

For those who want to help in improving our knowledge, the following is particularly needed. High quality, high spatial resolution observations of intensity, color and polarization maps of individual sources for comparison with the models currently being developed. Simultaneous observations, including photometry, polarimetry, and spectroscopy of individual stars (single stars with spots, or binaries), to get the inclination of the orbit for binaries, and information about the density distribution. One needs also a better understanding of the physics involved in disks, including the density distribution for modelling purposes. Finally, there should also be more work on grain properties in order to improve the models which strongly depend on the grain properties used.

5. References

Adams, F. C., Lada, C. J., and Shu, F. H. 1987, *Ap. J.*, **312**, 788.
Angel, J. R. P. 1974, in *Planets, Stars, and Nebulae Studied With Photopolarimetry.*, T. Gehrels (ed.), Univ. of Arizona Press, Tucson, 54.
Appenzeller, I., and Mundt, R. 1990, *Astr. Astrophys. Rev.*, **1**, 291.
Aspin, C. and Walther, D. 1990, poster presented at this conference.
Asselin, L., Bastien, P., and Ménard, F. 1991, in preparation.
Bastien, P. 1981, *Astr. Ap. .*, **94**, 294.
Bastien, P. 1982, *Astr. Ap. Suppl.*, **48**, 153, and 513.
Bastien, P. 1985, *Ap. J. Suppl.*, **59**, 277.
Bastien, P. 1987, *Ap. J.*, **317**, 231.
Bastien, P. 1988a, in *Polarized Radiation of Circumstellar Origin*, Proceedings of a Vatican Observatory Conference, G. V. Coyne, et al.(eds.), Vatican Press, Vatican City, 541.
Bastien, P. 1988b, in *Polarized Radiation of Circumstellar Origin*, Proceedings of a Vatican Observatory Conference, G. V. Coyne, et al. (eds.), Vatican Press, Vatican City, 595.
Bastien, P., Drissen, L., Ménard, F., Moffat, A. F. J., Robert, C., and St-Louis, N. 1988, *A. J.*, **95**, 900.
Bastien, P., and Landstreet, J. D., 1979, *Ap. J.(Letters)*, **229**, L137.
Bastien, P., and Levasseur-Regourd, A.-C. 1991, in preparation.
Bastien, P., and Ménard, F. 1988, *Ap. J.*, **326**, 334.
Bastien, P., and Ménard, F. 1990a, *Ap. J.*, **364**, 232.
Bastien, P., and Ménard, F. 1990b, in *IAU Symposium No. 137, Flare Stars in Star Clusters, Associations and the Solar Vicinity*, L. V. Mirzoyan, B. R. Pettersen, and Tsvetkov, M. K. (eds.), Kluwer Academic Publishers, Dordrecht, 179.
Bastien, P., Ménard, F., Asselin, L., and Turbide, L., 1989, in *Modeling the Circumstellar Environment: How and Why*, Proc. of a Colloquium in honor of Jean-Claude Pecker, P. Delache, S. Laloë, C. Magnan and J. Tran Thanh Van (eds.), Éditions Frontières, Gif-sur-Yvette, 185.
Bastien, P., Robert, C., and Nadeau, R. 1989, *Ap. J.*, **339**, 1089.
Bertout, C., Basri, G., and Bouvier, J. 1988, *Ap. J.*, **330**, 350.
Bertout, C., Collin-Souffrin, S., Lasota, J. P., and Tran Than Van, J. (eds.) 1991, *Structure and Emission Properties of Accretion Disks*, Proc. of 6th I. A. P. Colloquium/I. A. U. Colloquium No. 129, Éditions Frontières, Gif-sur-Yvette, in press.
Bohren, C. F. et Huffman, D. R. 1983, *Absorption and Scattering of Light by Small Particles*, John Wiley and Sons, Inc., New York, 530 p.
Brown, J. C., McLean, I. S., and Emslie, A. G. 1978, *Astr. Ap.*, **68**, 415.
Brownlee, D. E. 1978, in *Protostars and Planets*, T. Gehrels (ed.), Univ. of Arizona Press, Tucson, 134.
Campbell, B., and Persson, S. E. 1988, *A. J.*, **95**, 1185.
Campbell, B., Persson, S. E., and McGregor, P. J. 1986, *Ap. J.*, **305**, 336.
Campbell, B., Persson, S. E., Strom, S. E., and Grasdalen, G. L. 1988, *A. J.*, **95**, 1173.
Chandrasekhar, S. 1950, *Radiative Transfer*, Oxford University Press, Oxford. Dover Publications Inc., New York, 1960, 393 p.
Clarke, D. 1974, in *Planets, Stars, and Nebulae Studied With Photopolarimetry*, T. Gehrels (ed.), Univ. of Arizona Press, Tucson, 45.

Cohen, M., and Kuhi, L. V. 1979, *Ap. J. Suppl.*, **41**, 743.
Coyne, G. V. 1974, *A. J.*, **79**, 565.
Coyne, G. V., Gehrels, T., and Serkowski, K. 1974, *A. J.*, **79**, 581.
Daniel, J.-Y. 1978, *Astr. Ap.*, **67**, 345.
Davis, L., and Greenstein, J. L. 1951, *Ap. J.*, **114**, 206.
Drissen, L., Bastien, P., and St.-Louis, N. 1989, *A. J.*, **97**, 814.
Drissen, L., Lamontagne, R., Moffat, A. F. J., Bastien, P., and Séguin, M., 1986, *Astrophys. J.*, **304**, 188.
Elsässer, H., and Staude, H. J. 1978, *Astr. Ap.*, **70**, L3.
Frank, J., King, A. R., and Raine, D. J. 1985, *Accretion Power in Astrophysics*, Cambridge University Press, Cambridge, 273 p.
Gehrels, T. 1974, in *Planets, Stars, and Nebulae Studied With Photopolarimetry*, T. Gehrels (ed.), Univ. of Arizona Press, Tucson, 3.
Gledhill, T. M., and Scarrott, S. M. 1989, *M. N. R. A. S.*, **236**, 139.
Gledhill, T. M., Warren-Smith, R. F. and Scarrott, S. M. 1987, *M. N. R. A. S.*, **229**, 643.
Goodman, A. A., Bastien, P., Myers, P. C., and Ménard, F. 1990, *Ap. J.*, **359**, 363.
Gravel, P., and Bastien, P. 1990, poster presented at this conference, and in preparation..
Hecht, E. 1987, *Optics*, 2nd ed., Addison-Wesley, Reading, MA, 676 p.
Heckert, P. A., and Zeilik, M. 1985, *A. J.*, **90**, 2291.
Heyer, M. H., Vrba, F. J., Snell, R. L., Schloerb, F. P., Strom, S. E., Goldsmith, P. F., and Strom, K. M. 1987, *Ap. J.*, **321**, 855.
Hildebrand, R. H. 1988, *Q. Jl. R. Astr. Soc.*, **29**, 327.
Hubeny, I. 1990, *Ap. J.*, **351**, 632.
Hunter, C. 1977, *Ap. J.*, **218**, 834.
Jones, R. V., and Spitzer, L. 1967, *Ap. J.*, **147**, 943.
Kenyon, S. J., and Hartmann, L. 1987, *Ap. J.*, **323**, 714.
King, A. R. 1989, in *Accretion Disks and Magnetic Fields in Astrophysics*, G. Belvedere (ed.), Kluwer Academic Publishers, Dordrecht, 43.
Krelowski, J., Papaj, J., and Wegner, W. 1990, in *IAU Symposium No. 137, Flare Stars in Star Clusters, Associations and the Solar Vicinity*, L. V. Mirzoyan, B. R. Pettersen, and Tsvetkov, M. K. (eds.), Kluwer Academic Publishers, Dordrecht, 293.
Lang, K. R. 1974, *Astrophysical Formulae*, Springer-Verlag, Berlin, p. 11.
Larson, R. B. 1969, *M. N. R. A. S.*, **145**, 271.
Lefèvre, J. and Daniel, J.-Y. 1988, in *Polarized Radiation of Circumstellar Origin*, Proceedings of a Vatican Observatory Conference, G. V. Coyne, *et al.* (eds.), Vatican Press, Vatican City, 523.
Lenzen, R. 1987, *Astr. Ap.*, **173**, 124.
Mathis, M. S. 1986, *Ap. J.*, **308**, 281.
McCutcheon, W. H., Vrba, F. J., Dickman, R. L., and Clemens, D. P. 1986, *Ap. J.*, **309**, 619.
McDavid, D. 1984, *Ap. J.*, **284**, 141.
Ménard, F. 1989, *Ph. D. Thesis*, Université de Montréal.
Ménard, F. 1991, in preparation.
Ménard, F., and Bastien, P. 1987, in *I.A.U. Symposium 122, Circumstellar Matter*, I. Appenzeller and C. Jordan (eds.), Reidel, Dordrecht, 133.
Ménard, F., and Bastien, P. 1990, *A. J.*, submitted.
Ménard, F., and Bastien, P. 1991, in preparation.
Ménard, F., Bastien, P., and Robert, C. 1988, *Ap. J.*, **335**, 290.

Ménard, F., Bertout, C., and Bastien, P., 1991, in preparation.
Meyer, F., Duschl, W. J., Frank, J., and Meyer-Hofmeister, E. (eds.) 1989, in *Theory of Accretion Disks*, Kluwer Academic Publishers, Dordrecht, 477 p.
Monin, J. L., Pudritz, R. E., Rouan, D., and Lacombe, F. 1989, *Astr. Ap.*, **215**, L1.
Myers, P. C., and Goodman, A. A. 1990, preprint.
Nadeau, R., and Bastien, P. 1986, *Ap. J. (Letters)*, **307**, L5.
Narayan, R. and Goodman, J. 1989, in *Theory of Accretion Disks*, F. Meyer et al. (eds.), Kluwer Academic Publishers, Dordrecht, 231.
Narayan, R. 1991, in *Structure and Emission Properties of Accretion Disks*, Proc. of 6th I. A. P. Colloquium/I. A. U. Colloquium No. 129, Bertout, C., Collin-Souffrin, S., Lasota, J. P., and Tran Than Van, J. (eds.), Éditions Frontières, Gif-sur-Yvette, in press.
Papaloizou, J. C. B., and Pringle, J. E. 1984, *M. N. R. A. S.*, **208**, 721.
Penston, M. V. 1969, *M. N. R. A. S.*, **144**, 425.
Purcell, E. M. 1979, *Ap. J.*, **231**, 404.
Sandford, M. T. II 1973, *Ap. J.*, **183**, 555.
Sandford, M. T. II, and Pauls, T. A. 1973, *Ap. J.*, **179**, 875.
Scarrott, S. M., Draper, P. W., and Warren-Smith, R. F. 1989, *M. N. R. A. S.*, **237**, 621.
Serkowski, K. 1962, *Adv. Astr. Ap.*, **1**, 289.
Serkowski, K., 1971, in *I. A. U. Coll. No. 15: New Directions and Frontiers in Variable Star Research*, Veröff. Remeis-Sternwarte Bamberg 9, No. 100, 11.
Serkowski, K. 1973, in *I. A. U. Symp. No. 52: Interstellar Dust and Related Topics*, J. M. Greenberg and H. C. van de Hulst (eds.), Reidel, Dordrecht, 145.
Serkowski, K., 1974, in *Methods of Experimental Physics, vol. 12 Astrophysics*, Academic Press, New York, 361.
Serkowski, K., Mathewson, D. S., and Ford, V. L. 1975, *Ap. J.*, **196**, 261.
Shafter, A., and Jura, M. 1980, *A. J.*, **85**, 1513.
Shakura, N. I., and Sunyaev, R. A. 1973, *Astr. Ap.*, **24**, 337.
Shawl, S. J. 1975, *A. J.*, **80**, 595.
Shu, F. H. 1977, *Ap. J.*, **214**, 488.
Shu, F. H., Adams, F. C., and Lizano, S. 1987, *Ann. Rev. Astr. Ap.*, **25**, 23.
Simmons, J. F. L. 1983, *M. N. R. A. S.*, **205**, 153.
Strom, S. E., Strom, K. M., and Edwards, S. 1988, in *Galactic and Extragalactic Star Formation*, R. Pudritz and M. Fich (eds.), Kluwer Academic Publishers, Dordrecht, p. 53.
Strom, S. E., Strom, K. M., Edwards, S., Cabrit, S., and Skrutskie, M. F. 1989, *A. J.*, **97**, 1451.
Tamura, M., and Sato, S. 1989, *A. J.*, **98**, 1368.
Tohline, J. E. 1982, *Fund. Cosmic Phys.*, **8**, 1.
van de Hulst, H. C. 1957, *Light Scattering by Small Particles*, J. Wiley and Sons, New York, 470 p.
Vrba, F. J., Coyne, G. V., and Tapia, S. 1981, *Ap. J.*, **243**, 189.
Vrba, F. J., Schmidt, G. D., and Hintzen, P. M. 1979, *Ap. J.*, **227**, 185.
Vrba, F. J., Strom, S. E., and Strom, K. M. 1976, *A. J.*, **81**, 958.
Vrba, F. J., Strom, S. E., and Strom, K. M. 1988, *A. J.*, **96**, 680.
Wilking, B. A., Lebofsky, M. J., Martin, P. G., Rieke, G. H., and Kemp, J. C. 1980, *Ap. J.*, **235**, 905.
Wilking, B. A., Lebofsky, M. J., and Rieke, G. H. 1982, *A. J.*, **87**, 695.
Zellner, B. 1971, *A. J.*, **76**, 651.

Zinnecker, H. 1989, in *Low Mass Star Formation and Pre-Main Sequence Objects*, B. Reipurth, (ed.), European Southern Observatory, Garching bei München, 447.

Rolf Güsten and Alyssa Goodman

Nick Kylafis and Charlie Lada

Nick Kyhgkjs and Charlie Lada

Index

Accretion, 223, 379, 415, 460-461, 577, 628
   time dependent, 638
Accretion Disks (see Circumstellar Disks)
Accretion Flows, 414
   structure equations, 421
   turbulent viscosity, 419
   (see also cloud collapse)
Accretion Luminosity, 223, 353, 657, 660
Accretion Shock, 380, 415
Accretion Timescale, 416
Alfvén Radius, 452, 545
Alfvén Surface, 389, 545, 558
Alfvén Velocity, 545, 617
Alfvén Waves, 49, 452, 554-555, 617
   damping, 452
   torsional, 452
   trapping, 465-466
Alfvénic Mach Number, 545
Ambipolar Diffusion, 63, 69-73, 369, 449-451, 454, 460-461
   heating of clouds by, 208
   timescale, 451, 457, 463
Angular Momentum, 454
   classical problem, 87-88, 450, 452
   conservation, 465
   in GMC's, 12-15
   loss of, 87, 88, 388, 400, 555
   extraction from disks, 555
   redistribution by magnetic breaking, 87, 372, 388
   (see also magnetic breaking)
Associations, 125, 337
   ages, 140
   CAS-TAU, 133, 135
   Ophiuchus-Scorpius-Centaurus (OSCA), 127-133
   Lac OBI, 15
   Ori OBI, 136-137
   internal stellar motions, 126
   locations in galaxy, 138
   run-away OB stars, 148
   stellar content, 140
   subgroups, 127, 129, 135, 140
   T-associations, 337

B-ρ Relation, 449, 454, 455-457
   derivation, 455
   plot, 459, 461
   relation to fragmentation, 457-459
Binary Systems
   binary star periods, 450, 464, 465
   formation, 88, 93, 402, 437-446, 464-465
   formation by capture, 441
   formation by fission, 438
   formation by fragmentation, 444
   in associations, 145
   independent condensations, 442
Bipolar Outflows:
   Observations, 369, 471
   around ultracompact HII regions, 262, 474
   basic properties, 472-474
   driving sources, 360, 472

energetics, 474, 484
extreme high velocity (EHV) flows, 480
line widths, 473
linewidth-luminosity relation, 485
masses, 473
mechanical luminosity, 484-485, 560
mechanical-stellar luminosity relation, 487, 540, 589
models, 490-493
momentum, 474,484
opacity, 473
relation to HH objects, 488, 528
relation to H_2 emission, 489
relation to star formation, 359, 366, 369-371
structure, 371, 472, 490-493
velocity structure, 490, 492
thrust-luminosity relation, 486, 540, 589
timescales, 473
(see also Herbig-Haro objects)
Bipolar Outflows: Theory
centrifugally driven, 386, 543
Ekman pumping, 393-394
hydromagnetic disk wind, 372-373, 539-562
hydromagnetic stellar wind, 386-392
magnetic collimation, 551, 561
mass loss rate, 360, 388, 559, 597
mechanical luminosity, 560
origin of, 385-392, 542-545
thrust, 560
x-celerator, 386-391
Birthline, 377,380
Bimodal Opposition to Gravity, 455
Bok Globules, 314
Boundary Layer, 637, 658-660
Bow Shocks
around ultracompact HII regions, 251-257
H-H objects, 502, 513

Cacoon Stars, 415
Centrifugal Force, 465, 501-503
Circumstellar Disks
accretion disks, 349, 383, 396-402., 529, 628, 656-671, 727
accretion luminosity of, 660
accretion viscosity, 384, 639
angular momentum transport in, 400, 629, 656-657
differential rotation of, 631
binary star formation, 402
flared, 350, 398
formation of, 354, 381, 464
gravitational instability, 384, 402, 641
hollow, 350-351, 699-700
line profiles, 631
masses, 350, 541, 700
optical depth, 727
planet formation in, 402
protostellar, 354, 383-385, 460-461

reprocessing, 349, 397-398, 628
scale height, 398-399, 636
sizes, 350
spectral signatures, 348, 397, 628-629
(see also SED's, Class II)
spiral density waves (m=1) in, 400-402
surface density, 636
temperature distribution, 348, 398, 541, 629, 658, 698
vertical density distribution, 636
Cloud Collapse
Alfvénic, 452, 464, 465
axisymmetric, 431-432, 460-461
dynamical, 74-76, 414, 452, 465
free-fall timescale, 223, 329, 416, 453
gravitational instability, 43, 453
inside-out, 63, 376, 379, 413-415
isothermal sphere, 353, 376
mass infall rate, 353, 376, 379, 683
quasi-static, 73, 450, 457, 460, 463
self-initiated, 61, 63, 76, 449, 460-461, 464
stages of, 414-415
triggers of, 450
(see also accretion flows)
(see also numerical models of cloud collapse)
Clusters, Embedded, 338-342, 454
Clusters, Origin, 454
Cometary Globules, 315-317

Cosmic Rays, 79-81
heating clouds by, 206
Cyclotron Emission, 684

Dark Clouds, 287
see also Molecular Clouds
Dense Cores see Molecular Cloud Substructure
Disk winds, 372, 529, 539-562
Dust
circumstellar, 343, 697
cocoon, 247
dust photosphere, 351
opacity, 179, 350, 351
scattering-- see Scattering
Dynamo, 386, 692

Effective Temperature, 456
Extinction, 346, 713

Fast Magnetosonic Mach Number, 545
Fragmentation (see molecular cloud substructure)
Fuors (FU Orionis Objects), 531, 623
accretion-rate, 636
mass loss rate, 642
light curves, 625
outburst mechanisms, 638
spectra, 632, 633, 643
spectral energy distribution, 627

Gamma Rays, 183
Giant Molecular Clouds (GMC's), 3-31, 35, 61, 144
basic properties, 4-5, 8-10, 287, 453
formation mechanisms, 40-52, 81-87, 462
lifetimes, 8, 52-55
magnetic fields, 163

masses, 4, 180
mass spectrum, 10, 46, 47, 48
relation to atomic gas, 19-20
self-gravitating, 39
structure, 21-27, 159
see also Molecular Clouds
Gould's Belt, 150
Gravitational Instability, 43, 81, 453, 641

Herbig-Haro Objects, 488, 497
driving sources, 530
HH 1-2, 321, 500, 506
HH 7-11, 319, 488
HH 34, 319
HH 46/47, 520
HH 111, 524
images of, 508-513
infrared spectra, 504
optical spectra, 501-504
radio observations, 506
relation to bipolar flows, 488, 528
relation to Fuors, 531, 645
role of accretion disks, 529
ultraviolet spectra, 505
shock waves in, 502, 503
see also Jets
H-R Diagram, 346, 376, 377
class I sources, 357
class III sources, 346
Hayashi track, 377
Heyney track, 377
T-Tauri stars, 377
HII Regions
see Ultracompact HII Regions

Initial Luminosity Function (ILF)
see Initial Mass Function

Initial Mass Function (IMF), 330-337, 452-455, 463
for embedded clusters, 335-337
for field stars, 333
infrared, 336
Initial Mass Spectrum
see Initial Mass Function

Jeans Instability, 81, 413, 460
Jeans Mass, 413
Jets, 513-515, 524
Collimation, 519, 561
Kinematics, 515
magnetic fields in, 516
models, 516
origin and evolution, 560
working surface, 518

Kelvin-Helmholtz contraction, 223, 330, 378, 415, 416
Keplerian Disk, 542, 631, 657

Larmor Frequency, 684
L1551, 358, 369-372, 531, 721

Magnetic Breaking, 87, 89-104, 388, 450, 458, 462, 465
effect of ambipolar diffusion on, 104
Magnetic Detachment, 466
Magnetic Fields, 49-51, 61, 163, 309, 375, 450
ambipolar diffusion of (see ambipolar diffusion)
in accretion disks, 473, 539, 701
critical mass to flux ratio, 74, 368, 453, 460
flux-freezing, 67-69, 450

origin in stellar dynamo, 386, 693
protostellar, 386, 387, 461
relation to stellar winds, 372, 385-396
stellar, 691, 695, 696, 703
Magnetic Length Scale, 449, 452
Magnetic Support, 450
Magnetic Tension, 450
Magnetohydrodynamics, 78-79, 546
Masers, 269
   absorption coefficient, 271
   amplification, 271
   apparent size, 273
   beaming, 272
   H_2O, 280
   OH, 282
   polarization, 276
   proper motions, 275
   pumping, 279
   saturation, 271
   spectral profiles, 275
   theory, 278
   thermalization, 272
   variability, 275
Molecular Clouds
   column densities, 10, 180
   heating and cooling, 184-207, 292
   linewidth-size relation, 9, 306
   magnetic fields, 309
   masses, 180-184
   mass spectra, 10, 163
   rotation of, 88
   temperatures, 184, 292
   volume density, 184, 295-300
   volume density profile, 300
   see also Giant Molecular Clouds

Molecular Cloud Substructure, 21-27, 159, 305
   clump mass spectrum, 23-24, 161-162, 332, 341
   clump stability, 27-29
   dense cores, 288, 310-313, 374
   effect on B-ρ relation, 457-459
   efficiency of fragmentation, 457
   formation of dense cores, 375, 449, 450, 457-459
   fragmentation processes, 309-310, 457-459
   hierarchical clumping, 307-308, 450, 451, 458
   magnetized cores, 375
   structure of a dense core, 375
   rotating dense cores, 374
   see also Bok Globules
Molecules
   carbon-chain, 303
   density probes, 184, 295-300
   excitation, collisional, 171
   excitation, radiative, 175
   mass tracers, 181-183, 304
   temperature probes, 292-295

Numerical Models of Cloud Collapse, 411-436, 460-461
   adaptive grid, 427
   artificial viscosity, 426
   axially symmetric (2-D), 431, 460-461
   finite differences, 423
   spherically symmetric (1-D), 428
   three dimensional (3-D), 432
   see also cloud collapse

Ophiuchi Cloud Complex, 130, 342
ORION Molecular Cloud, 156-164
ORION OB Association
see Associations

Parker Instability, 82, 451
Photodissociation Region (PDR), 188-194
Polarization, 709
  circumstellar, 724-725
  interstellar, 713
  in binary stars, 729
  in circumstellar disks, 728-731
  in YSO's, 724
  Stokes parameters, 711
  radio, 689
Pre-Main Sequence Stars, 330, 346, 376, 415, 565, 623, 675
  (see also T-Tauri Stars)
  (see also Fuors)
Pressure
  gravitational, 455
  magnetic, 455
  thermal, 455
  wave, 456
Protostars
  accretion luminosity, 223, 353
  evolution, 357, 383-386, 449-468
  equation of state, 417
  formation, 376
  high mass, 211
  low mass, 310, 353
  masses, 452, 454
  opacity, 419
  outflows from, 313, 359-361
  models, 353

rotation, 353, 381
spectra (see also SED's: Class I), 351-353, 358, 380-381
structure: infalling envelope, 351-353
structure: accreting core, 378-380, 416

Radiative Transfer, 165, 224, 279, 351
  for dust, 160, 179, 351
  for HII regions, 224
  for ionized winds, 566-568
  line trapping, 177
  Sobolev approximation (LVG), 572, 599
Radio Stars, 682
Retrograde Rotation, 100, 451

Scattering
  diagram, 717
  cross section, 717
  function, 717
  matrix, 718
  multiple, 729
  single, 717
Shock Waves
  C-shocks, 202, 204
  J-shocks, 199, 204
Spectral Energy Distributions (SED's), 343-345
  class I, 351, 584-585
  class II, 347, 675-676
  class IID, 345, 374
  class III, 346, 584-585
  class IIID, 345, 347, 350, 699
  extreme class I, 354
  infrared spectral index, 343, 348
  spectral evolution, 357

spectral sequence, 343, 585, 699
Spectra
  continuous emission from dust, 178
  energy distributions (see Source Energy Distribution)
  fine-structure lines, 170
  free-free emission, 224, 569, 682
  HH objects, 501-506
  infrared-line, 167
  masers, 275-277
  molecular line, 168
  radio continuum, 685
  radio-line, 167
  line profiles, 178
Star Formation
  bimodal, 337, 368
  four stages of, 357-359, 366-368
  high mass stars, 155, 269
  in dense gas, 161, 310, 339-340
  in embedded clusters, 338
  in magnetized clouds, 61, 374
  low mass stars, 310, 329, 365
  relation to bipolar outflow, 312-314, 359-361, 366-367, 369-376
  sequential, 128
  theory, 62, 353, 365, 411, 437, 449
Star Formation Efficiency (SFE), 340, 342, 450
Stellar Rotational Velocities, 452
Stellar Winds
  adiabatic expansion cooling of, 616

bow shocks, 254-257, 502, 513
centrifugally driven, 543
continuous spectrum, 569
excitation, 605
formation (see Bipolar Outflows: Theory)
frictional heating of, 616
fully ionized, 565
heat equation for, 612
heating by Alfvén wave dissipation, 617
heating by H_2 formation, 618
ionization, 575, 598-604
ionization luminosity, 586
mass loss rate, 360, 388, 559, 576, 596, 683, 688
mass loss rate-- luminosity relation, 589
MHD winds, 643-648
neutral winds, 595
non-thermal emission, 582
radiative cooling, 613
radiative heating, 613
radiative luminosity, 586
radio observations of, 582-584
spectral line emission, 572, 605-607
temperature, 612
thrust--luminosity relation, 589
time variability, 576
Weber-Davis solution, 506
see also Bipolar Flows
Supernova, Shock Wave, 462
Synchronous Galactocentric Orbit, 451
S106, 581

T-Tauri Stars, 649
  classical (CTTS), 628, 679
  emission lines, 650, 651
  evolution, 357, 694-697

flares, 691
infrared excess, 344, 347, 393, 653
lithium in, 670
line profiles, 651
location on H-R diagram, 377
mass-loss rates in, 626, 662
modeling spectra, 665
naked, 345, 628
P-Cygni profiles, 653
post, 345
rotation of, 393, 659, 693
spectral signature, 345, 347, 350, 655
spectral veiling, 666
starspots, 691
UV excess, 344, 393, 654
variability, 652, 691
weak-line (WTTS) (see also SED's: Class III), 628, 679
X-ray emission from, 654, 675
(see also SED's: Class II)
(see also SED's: Class III)
(see also Circumstellar Disks)
Taurus Molecular Complex, 289
HCL2, 289
HCL2C, 289, 290
mass, 305
structure, 305
TMC1, 303, 313
see also Associations: Cas-Tau
Thermal Instability, 40, 82
Turbulence, 49, 456
hierarchical, 308

Ultracompact HII Regions, 221

champagne flows, 258
clustering, 232
continuous spectrum, 224-228
dust envelopes, 242
exciting stars, 229
galactic distribution, 235
structure, 231

Veiling (see T-Tauri Stars: Spectral Veiling)
Virial Theorem, 115-118, 378, 416
VLBI, 690

Weak-Line T-Tauri Stars (WTTS) (see T-Tauri Stars)

X-Rays from PMS stars, 654, 675
energies, 677
flares, 691
thermal bremsstrahlung, 680
variability, 654, 677-678

Young Stellar Object (YSO), 334, 343, 565, 649, 675
(see also Protostars, Fuors, T-Tauri Stars, YY Orionis stars)
YY Orionis Stars, 652, 668

Zeeman Splitting, 675